Data Analysis and Probability

Instructional programs from prekindergarten through grade 12 should enable all students to—

- formulate questons that can be addressed with data and collect, organize, and display relevant data to answer them;
- select and use appropriate statistical methods to analyze data;
- develop and evaluate inferences and predictions that are based on data;
- understand and apply basic concepts of probability.

Problem Solving

Instructional programs from prekindergarten through grade 12 should enable all students to—

- build new mathematical knowledge through problem solving;
- solve problems that arise in mathematics and in other contexts;
- apply and adapt a variety of appropriate strategies to solve problems;
- monitor and reflect on the process of mathematical problem solving.

Reasoning and Proof

Instructional programs from prekindergarten through grade 12 should enable all students to—

- recognize reasoning and proof as fundamental aspects of mathematics;
- make and investigate mathematical conjectures;
- develop and evaluate mathematical arguments and proofs;
- select and use various types of reasoning and methods of proof.

Communication

Instructional programs from prekindergarten through grade 12 should enable all students to—

- organize and consolidate their mathematical thinking through communication;
- communicate their mathematical thinking coherently and clearly to peers, teachers, and others;
- analyze and evaluate the mathematical thinking and strategies of others;
- use the language of mathematics to express mathematical ideas precisely.

Connections

Instructional programs from prekindergarten through grade 12 should enable all students to—

- recognize and use connections among mathematical ideas;
- understand how mathematical ideas interconnect and build on one another to produce a coherent whole;
- recognize and apply mathematics in contexts outside of mathematics.

Representation

Instructional programs from prekindergarten through grade 12 should enable all students to—

- create and use representations to organize, record, and communicate mathematical ideas;
- select, apply, and translate among mathematical representations to solve problems;
- use representations to model and interpret physical, social, and mathematical phenomena.

A Problem Solving Approach to
MATHEMATICS

Seventh Edition

A Problem Solving Approach to
MATHEMATICS

Seventh Edition

Rick Billstein
University of Montana

Shlomo Libeskind
University of Oregon

Johnny W. Lott
University of Montana

Boston San Francisco New York
London Toronto Sydney Tokyo Singapore Madrid
Mexico City Munich Paris Cape Town Hong Kong Montreal

Acquisitions Editor • *Bill Poole*
Editorial Project Manager • *Rachel Reeve*
Managing Editor • *Karen M. Guardino*
Editorial Production Services • *Jennifer Bagdigian*
Text Designer • *Karen Rappaport*
Marketing Manager • *Carter Fenton*
Manufacturing Buyer • *Evelyn Beaton*
Senior Prepress Buyer • *Caroline Fell*
Cover Designer • *Barbara Atkinson*
Composition/Prepress Services • *Typo-Graphics, Inc.*
Text Illustrations • *Typo-Graphics, Inc.*
Chapter Opener Illustrator • *Darwen Hennings*

Library of Congress Cataloging-in-Publication Data

Billstein, Rick.
    A problem solving approach to mathematics for elementary school teachers / Rick Billstein, Shlomo Libeskind, Johnny W. Lott.—7th ed.
      p. cm.
    Includes bibliographical references and index.
    ISBN 0-201-34730-X—ISBN 0-201-38408-6
    1. Mathematics   Study and teaching (Elementary)   2. Problem solving.   I. Libeskind, Shlomo.   II. Lott, Johnny W., 1944–   III. Title.

QA135.5.B49   2000
372.7—dc21   99-054354

Copyright © 2001 by Addison Wesley Longman. All rights reserved.

No part of this work may be reproduced or transmitted in any form or by any means, electronic or mechanical, including photocopying and recording, or by any information storage or retrieval system without the prior written permission of Addison Wesley Longman unless such copying is expressly permitted by federal copyright law. Address inquiries to Addison Wesley Longman, One Jacob Way, Reading, MA 01867.

Printed in the U.S.A.

23456789-QUV-03 02 01 00

Preface

The seventh edition of *A Problem Solving Approach to Mathematics* continues to strive to meet the expectations for the mathematics education of elementary teachers in the new century.

Standards of the NCTM

In the seventh edition we focus on the 2000 National Council of Teachers of Mathematics (NCTM) publication, *Principles and Standards for School Mathematics* (hereafter referred to as *Principles and Standards*). We also continue to emphasize the need for the teaching of mathematics to include

- logic and mathematical evidence as verification;
- mathematical reasoning;
- conjecturing, inventing, and problem solving; and
- connecting mathematics, its ideas, and its applications.

The seventh edition

- allows instructors a variety of approaches to teaching,
- encourages discussion and collaboration among students and with their instructors,
- allows for the integration of projects into the curriculum, and
- promotes discovery and active learning.

Goals

In the seventh edition our goals are

- to present appropriate mathematics in an intellectually honest and mathematically correct manner.
- to use problem solving as an integral part of mathematics.
- to approach mathematics in a sequence that instills confidence, then challenges students.
- to provide opportunities for alternative forms of teaching and learning.
- to offer communication problems to develop writing skills and allow students to practice explanation.
- to encourage integration of technology tools.
- to provide core mathematics for prospective elementary school teachers that challenges them to determine why mathematics is done the way it is.
- to provide core mathematics that allows instructors to use methods integrated with content.

Problem Solving in the Seventh Edition

We showcase problem-solving skills by

- devoting Chapter 1 to problem-solving skills, and emphasizing deductive versus inductive reasoning.
- using a four-step problem-solving process to solve problems throughout the text.
- beginning each chapter with a preliminary problem that poses a question students can answer with the skills mastered from that chapter.

We encourage teachers to point out and discuss the preliminary problem at the beginning of each chapter to show how the techniques therein are necessary to solve the problem.

Features Retained in this Edition

We continue to incorporate various study aids and features that facilitate learning:

- **Historical Notes** add context and humanize the mathematics.
- **Brainteasers** provide a different avenue for problem solving. They are solved in the instructor's guide, and may be assigned or used by the teacher to challenge students.
- **Laboratory Activities** are integrated throughout the book to provide hands-on learning exercises. A separate activities book is also available.
- **Cartoons** teach or emphasize important material, and add levity.
- **Key terms** are presented in the margins for quick review.
- **Definitions** are either set off in text or presented as key terms in the margin.
- Optional sections as well as problems based on these sections are marked with an asterisk (*). **More difficult problems** are marked with a star (★). **Problems numbered in color** have answers at the back of the book.
- **Questions from the Classroom** allow instructors to use questions posed by students when building a course syllabus. Instructors may require students to write two answers to the questions—one mathematical and one pedagogical—using student texts and professional journals for research.
- **Chapter Outlines** at the end of each chapter help students review the chapter.
- **Chapter Reviews** at the end of each chapter allow students to test themselves.
- **Selected Bibliographies** have been updated and revised, and appear at the end of each chapter.
- **Problem-solving strategies** are often highlighted in italics, and indicated by ▞.
- **Relevant quotes** from the *Principles and Standards* are incorporated throughout the text, and are marked by the standard icon ◆.
- **Student pages** are included to show how the mathematics is actually introduced to the K–8 student.
- **Communication Problems** require students to explain or justify their answers.
- **Full color** has been used for pedagogical reasons and to help students visualize concepts better. Figures are more modern, attractive, and easy to follow. All of the pages taken from elementary mathematics texts are presented in full color.
- **Problem sets** contain six types of problems: (1) ongoing assessment, (2) communication, (3) open-ended, (4) cooperative learning, (5) technology, and (6) review. **Communication, Open-Ended,** and **Cooperative Learning** sections are included to conform with the major *Standards'* processes.
- **Relevant and realistic problems** are more accessible and appealing to students of diverse backgrounds.
- **Technology Corners** include use of Logo, spreadsheets, both graphing and scientific calculators, Geometer's Sketchpad, and computer activities.
- **Now Try This** activities appear throughout each chapter, and are intended to help students become more involved in their learning, to facilitate the development and improvement of their critical thinking and problem-solving skills, and to stimulate both in-class and out-of-class discussion.

Content

The streamlined seventh edition retains the core of the sixth edition and includes some new content.

Chapter 1 An Introduction to Problem Solving

Chapter 1 has more on inductive versus deductive reasoning, and has a new section on algebraic thinking.

Chapter 2 Sets, Whole Numbers, and Functions

In Section 2-3 we introduce whole numbers and basic operations on whole numbers using set concepts.

We have rewritten Section 2-5 on functions from a more concrete and application-oriented perspective. Relations are introduced in the problem set as generalizations of functions.

Chapter 3 Whole-Number Computation

Chapter 3 introduces numeration systems and other number bases. We explore algorithms for addition, subtraction, multiplication, and addition, along with mental mathematics and estimation for whole-number computation.

Chapter 4 Integers and Number Theory

In Chapter 4 we introduce the system of integers and develop an understanding of basic number theory. In several historical notes we tell about recent developments in number theory.

Chapter 5 Rational Numbers as Fractions

We reorganized Chapter 5. The topic of comparing rational numbers is now placed at the beginning of the chapter. In addition, this chapter contains more pictures to help describe the concepts and to assist students with visualization. There is a new section on proportional reasoning.

Chapter 6 Decimals, Percents, and Real Numbers

Chapter 6 now provides a full treatment of decimals, percents, and real numbers. Computing interest is covered as an application of percents and decimals.

Chapter 7 Probability

Chapter 7 introduces elementary probability and methods of counting.

Chapter 8 Statistics: An Introduction

Chapter 8 includes an introduction to different types of graphs. There is a section on measures of central tendency and variation, as well as a section on abuses of statistics.

Chapter 9 Introductory Geometry

This revised chapter has less emphasis on definitions.

Chapter 10 Constructions, Congruence, and Similarity

Chapter 10 includes a reorganized discussion of constructions and congruence. Some material on coordinate geometry has been moved into a section on similarity to show how slope and equations of lines is developed.

Chapter 11 Concepts of Measurement

Chapter 11 now covers linear measure along with areas, surface areas, the Pythagorean Theorem, volume, mass, and temperature.

Chapter 12 Motion Geometry and Tessellations

This chapter has been revised slightly from the sixth edition.

Appendices

Appendices in this edition include

- graphing calculators,
- Geometry Utility (based on Geometer's Sketchpad),
- spreadsheets (based on Microsoft Excel), and
- Logo Turtle Graphics.

Use of Calculators

As the *Principles and Standards* state, coverage of calculators is necessary and timely. The use of the graphing calculator is presented, where relevant, in the Technology Corners. In addition, problems involving the use of both scientific/fraction and graphing calculators appear in the problem sets.

Supplements for the Student

Student's Solutions Manual, ISBN 0-201-61141-4, by Louis Levy, contains detailed solutions to all odd-numbered exercises.

Activities Manual—Mathematics Activities for Elementary School Teachers: A Problem Solving Approach, Fourth Edition, ISBN 0-201-61321-2, by Daniel Dolan, Jim Williamson, and Mari Muri. This revised edition features activities that can be used to develop, reinforce, and apply mathematical concepts. The activities for each concept are ordered by developmental level in each chapter.

NEW: *A Problem Solving Approach to Mathematics for Elementary School Teachers* **Videotapes**   A complete set of videotapes for use by students is available to departments that adopt the seventh edition. All of the basic concepts from the text are reinforced in these videotapes. An instructor experienced in mathematics for elementary school teachers works through detailed examples taken from the text. Videos are available to departments through your Addison Wesley Longman representative.

NEW: InterAct Math CD Tutorial Software   ISBN 0-201-61319-0 Available in Windows and Macintosh versions, InterAct Math Tutorial Software includes exercises that are linked with every objective in the textbook and require the same computational and problem-solving skills as their companion exercises in the text. Each exercise has an example and an interactive guided solution that are designed to involve students in the solution process and to help them identify precisely where they are having trouble. In addition, the software recognizes common student errors and provides students appropriate customized feedback. With its sophisticated answer-recognition capabilities, InterAct Math Tutorial Software recognizes appropriate forms of the same answer for any kind of input.

It also tracks student activity and scores. Contact your Addison Wesley Longman representative.

Supplements for the Instructor

NEW: Instructor's Edition includes sequential answers to all text exercises in a special section at the back of the book.

Instructor's Solutions Manual, ISBN 0-201-61142-2, by Louis Levy, contains detailed solutions to all exercises.

Instructor's Resource Guide, ISBN 0-201-61143-0, includes two forms of chapter assessments with answers for each chapter, suggested answers to Questions from the Classroom, Solutions to the Brainteasers, and suggested answers to the Now Try This activities.

Instructor's Guide to *Mathematics Activities for Elementary School Teachers: A Problem Solving Approach, Fourth Edition,* ISBN 0-201-61322-0, by Daniel Dolan, Jim Williamson, and Mari Muri, contains answers for all activities, as well as additional teaching suggestions for some activities.

NEW: TestGen-EQ CD with QuizMaster-EQ ISBN 0-201-61317-4   TestGen-EQ is a computerized test generator with algorithmically defined problems organized specifically for this textbook. Its user-friendly graphical interface enables instructors to select, view, edit, and add test items, then print tests in a variety of fonts and forms. Seven types of questions are available, and search and sort features let the instructor quickly locate questions and arrange them in a preferred order. A built-in question editor gives the user the power to create graphs, import graphics, insert mathematical symbols and templates, and insert variable numbers or text. An "Export to HTML" feature lets instructors create practice tests that can be posted to a Web site. Tests created with TestGen-EQ can be used with QuizMaster-EQ, which enables students to take exams on a computer network. QuizMaster-EQ automatically grades the exams, stores results on disk, and allows the instructor to view or print a variety of reports for individual students, classes, or courses. This program is available in Windows and Macintosh formats. Contact your Addison Wesley Longman representative.

NEW: Web site http://www.awl.com/Billstein
The Web site contains additional resources for instructors and students.

Acknowledgments

Reviewers of This and Previous Editions

The authors wish to thank the following for their helpful comments and suggestions for this and previous editions of the text.

Leon J. Ablon
G. L. Alexanderson
Haldon Anderson
Bernadette Antkoviak
Sue H. Baker
Jane Barnard
Joann Becker
Cindy Bernlohr
James Bierden
Jim Boone
Sue Boren
Barbara Britton
Beverly R. Broomell
Jane Buerger
Maurice Burke
David Bush
Laura Cameron
Louis J. Chatterley
Phyllis Chinn
Donald J. Dessart
Ronald Dettmers
Jackie Dewar
Amy Edwards
Margaret Ehringer
Albert Filano
Marjorie Fitting
Michael Flom
Martha Gady
Sandy Geiger
Glenadine Gibb
Don Gilmore
Elizabeth Gray
Jerrold Grossman
Alice Guckin
Boyd Henry
Alan Hoffer
E. John Hornsby, Jr.
Judith E. Jacobs
Donald James
Jerry Johnson
Wilburn C. Jones
Robert Kalin
Herbert E. Kasube
Sarah Kennedy
Steven D. Kerr
Leland Knauf
Margret F. Kothmann
Kathryn E. Lenz
Hester Lewellen
Ralph A. Liguori
Don Loftsgaarden
Sharon Louvier
Stanley Lukawecki
Barbara Moses
Charles Nelson
Glenn Nelson
Kathy Nickell
Dale Oliver
Linda Padilla
Dennis Parker
Clyde Paul
Keith Peck
Barbara Pence
Glenn L. Pfeifer
Jack Porter
Edward Rathnell
Jennifer Rutherford
Helen R. Santiz
Jane Schielack
Barbara Shabell
M. Geralda Shaefer
Nancy Shell
Wade H. Sherard
Gwen Shufelt
Ron Smit
Joe K. Smith
William Sparks
Virginia Strawderman
Mary M. Sullivan
Viji Sundar
Sharon Taylor
C. Ralph Verno
Hubert Voltz
John Wagner
Virginia Warfield
Lettie Watford
Mark F. Weiner
Grayson Wheatley
Ken Yoder
Jerry L. Young

BRIEF CONTENTS

Chapter 1   An Introduction to Problem Solving   1
Chapter 2   Sets, Whole Numbers, and Functions   58
Chapter 3   Whole-Number Computation   123
Chapter 4   Integers and Number Theory   169
Chapter 5   Rational Numbers as Fractions   245
Chapter 6   Decimals, Percents, and Real Numbers   293
Chapter 7   Probability   348
Chapter 8   Statistics: An Introduction   408
Chapter 9   Introductory Geometry   461
Chapter 10   Constructions, Congruence, and Similarity   518
Chapter 11   Concepts of Measurement   591
Chapter 12   Motion Geometry and Tessellations   668
Appendix I   Logo Turtle Graphics   724
Appendix II   Graphing Calculators   743
Appendix III   Using a Geometry Drawing Utility   751
Appendix IV   Using a Spreadsheet   759
    Answers to Selected Problems   767
    Index   813

Solution to the cover puzzle follows the index.

CONTENTS

An Introduction to Problem Solving 1

Preliminary Problem 1
1-1 Explorations with Patterns 3
1-2 Mathematics and Problem Solving 18
1-3 Algebraic Thinking 35
*__1-4__ Logic: An Introduction 45
Hint for Solving the Preliminary Problem 54
Questions from the Classroom 54
Chapter Outline 55
Chapter Review 55
Selected Bibliography 57

Sets, Whole Numbers, and Functions 58

Preliminary Problem 58
2-1 Describing Sets 59
2-2 Other Set Operations and Their Properties 72
2-3 Addition and Subtraction of Whole Numbers 82
2-4 Multiplication and Division of Whole Numbers 92
2-5 Functions 102
Hint for Solving the Preliminary Problem 119
Questions from the Classroom 119
Chapter Outline 120
Chapter Review 121
Selected Bibliography 122

Whole-Number Computation 123

Preliminary Problem 123
3-1 Numeration Systems 124
3-2 Algorithms for Whole-Number Addition and Subtraction 135
3-3 Algorithms for Whole-Number Multiplication and Division of Whole Numbers 146
3-4 Mental Mathematics and Estimation for Whole-Number Operations 157

Hint for Solving the Preliminary Problem   165
Questions from the Classroom   165
Chapter Outline   166
Chapter Review   166
Selected Bibliography   167

Integers and Number Theory   169

Preliminary Problem   169
- **4-1** Integers and the Operations of Addition and Subtraction   171
- **4-2** Multiplication and Division of Integers   183
- **4-3** Divisibility   194
- **4-4** Prime and Composite Numbers   206
- **4-5** Greatest Common Divisor and Least Common Multiple   220
- ***4-6** Clock and Modular Arithmetic   234

Hint for Solving the Preliminary Problem   240
Questions from the Classroom   241
Chapter Outline   241
Chapter Review   243
Selected Bibliography   244

Rational Numbers and Fractions   245

Preliminary Problem   245
- **5-1** The Set of Rational Numbers   246
- **5-2** Addition and Subtraction of Rational Numbers   256
- **5-3** Multiplication and Division of Rational Numbers   268
- **5-4** Proportional Reasoning   281

Hint for Solving the Preliminary Problem   288
Questions from the Classroom   289
Chapter Outline   289
Chapter Review   290
Selected Bibliography   292

Decimals, Percents, and Real Numbers   293

Preliminary Problem   293
- **6-1** Introduction to Decimals   295

- **6-2** Operations on Decimals   302
- **6-3** Nonterminating Decimals   314
- **6-4** Percents   321
- ***6-5** Computing Interest   331
- **6-6** Real Numbers   337

Hint for Solving the Preliminary Problem   345
Questions from the Classroom   345
Chapter Review   346
Chapter Outline   347
Selected Bibliography   347

7 Probability   348

Preliminary Problem   348
- **7-1** How Probabilities Are Determined   349
- **7-2** Multistage Experiments with Tree Diagrams and Geometric Probabilities   362
- **7-3** Using Simulations in Probability   378
- **7-4** Odds and Expected Value   385
- **7-5** Methods of Counting   392

Hint for Solving the Preliminary Problem   403
Questions from the Classroom   404
Chapter Outline   404
Chapter Review   405
Selected Bibliography   407

8 Statistics: An Introduction   408

Preliminary Problem   408
- **8-1** Statistical Graphs   410
- **8-2** Measures of Central Tendency and Variation   427
- **8-3** Abuses of Statistics   447

Hint for Solving the Preliminary Problem   455
Questions from the Classroom   456
Chapter Outline   456
Chapter Review   457
Selected Bibliography   459

9 Introductory Geometry   461

Preliminary Problem   461
- **9-1** Basic Notions   462

9-2 Polygons 476
9-3 More about Angles 484
9-4 Geometry in Three Dimensions 496
*__9-5__ Networks 506
Hint for Solving the Preliminary Problem 514
Questions from the Classroom 514
Chapter Outline 515
Chapter Review 516
Selected Bibliography 517

10 Constructions, Congruence, and Similarity 518

Preliminary Problem 518
10-1 Congruence Through Constructions 519
10-2 Other Congruence Properties 533
10-3 Other Constructions 540
10-4 Similar Triangles and Similar Figures 551
10-5 Lines in a Cartesian Coordinate System 567
Hint for Solving the Preliminary Problem 585
Questions from the Classroom 585
Chapter Outline 586
Chapter Review 587
Selected Bibliography 589

11 Concepts of Measurement 591

Preliminary Problem 591
11-1 Linear Measure 592
11-2 Areas of Polygons and Circles 604
11-3 The Pythagorean Theorem 621
11-4 Surface Areas 635
11-5 Volume, Mass, and Temperature 644
Hint for Solving the Preliminary Problem 662
Questions from the Classroom 663
Chapter Outline 663
Chapter Review 665
Selected Bibliography 667

12 Motion Geometry and Tessellations 668

Preliminary Problem 668
12-1 Translations and Rotations 669

12-2 Reflections and Glide Reflections 684
12-3 Size Transformations 696
12-4 Symmetries 704
***12-5** Tessellation of the Plane 713
Hint for Solving the Preliminary Problem 719
Questions from the Classroom 719
Chapter Outline 720
Chapter Review 720
Selected Bibliography 723

APPENDIX I 724

Logo Turtle Graphics

APPENDIX II 743

Graphing Calculators

APPENDIX III 751

Using a Geometry Drawing Tool

APPENDIX IV 759

Using a Spreadsheet

Answers to Selected Problems 767

Index 813

Solution to the cover puzzle follows the index.

*indicates optional section

1
An Introduction to Problem Solving

Preliminary Problem
In a fourth-grade class election, there were five candidates for class president. To try to guarantee a fair election, Ms. Pendergast, the teacher, decided to have a series of head-to-head elections in which each candidate ran against each other candidate. The class president would be the person who won the most head-to-head elections. How many elections did there have to be?

problem solving has long been recognized as one of the hallmarks of mathematics. One of the greatest goals of mathematics education is to have students become good problem solvers. We do not mean doing exercises that are routine practice for skill building. What does *problem solving* mean?

George Polya (1887–1985), one of the great mathematicians and teachers of the twentieth century, pointed out that "Solving a problem means finding a way out of difficulty, a way around an obstacle, attaining an aim which was not immediately attainable." (Polya 1981, p. ix) In *Principles and Standards for School Mathematics* (hereafter referred to as the *Principles and Standards*), we find the following:

▲ *Problem solving is the cornerstone of school mathematics. Without the ability to solve problems, the usefulness and power of mathematical ideas, knowledge, and skills are severely limited. … The goal of school mathematics should be for all students to become increasingly able and willing to engage with and solve problems* (p. 182).

Exercises or practice problems serve a purpose in learning mathematics, but problem solving must be a focus of school mathematics. Your mathematical experience often determines whether situations are *problems* or *exercises*. In the "Shoe" cartoon, the math test contains what might be a problem for some, but is an exercise for others. Would you expect to find this test item in a middle-school mathematics text?

In this text, you will have many opportunities to solve problems. Each chapter opens with a problem that can be solved by using the concepts developed in the chapter.

A hint for the solution to the problem is given at the end of each chapter. Throughout the text, there are numerous other problems solved using a four-step process and others solved using other formats.

Working with other students to solve problems can enhance your problem-solving ability and communication skills. In this text, we encourage *cooperative learning* and encourage students to work in groups whenever possible. To encourage group work and help identify when cooperative learning might be useful, we identify activities that might involve tasks where it would be helpful to have several people gathering data, or the problems might be such that group discussions might lead to strategies for solving the problem.

Students in a group or individually may use strategies explained in the remaining sections of this chapter. One problem-solving strategy, *looking for a pattern,* is used so often that it is the focus of the next section.

Section 1-1 — Explorations with Patterns

Mathematics has been described as the study of patterns. We see patterns everywhere—in wallpaper, tiles, traffic, and even television schedules. Police investigators study case files to find the modus operandi, or pattern of operation, when a series of crimes is committed. Scientists look for patterns in order to isolate variables so that they can reach valid conclusions in their research. In the *Principles and Standards,* we find the following,

> … *students should investigate numerical and geometric patterns and express them mathematically in words or in symbols. They should analyze the structure of the pattern and how it grows or changes, organize this information systematically, and use their analysis to develop generalizations about the mathematical relationships in the pattern* (p. 159).

Patterns can be surprising and aesthetically pleasing. Consider Example 1-1.

Example 1-1

a. Describe the following pattern:

$$1 + 0 \cdot 9 = 1$$
$$2 + 1 \cdot 9 = 11$$
$$3 + 12 \cdot 9 = 111$$
$$4 + 123 \cdot 9 = 1111$$
$$5 + 1234 \cdot 9 = 11111$$

b. Does the above pattern continue? Why or why not?

Solution a. There are several patterns in the example. The following descriptions are intuitive at this stage. The numbers on the far left are **natural numbers;** that is, numbers from the set 1, 2, 3, 4, 5, … . The pattern starts with 1 and continues to the next greater natural number in each successive line. The numbers "in the middle" are products of two numbers, the second of which is 9. The first number in the first product is 0; after that the first number is formed using natural numbers and including one more in each successive line. The resulting numbers on the right are formed using 1s and including an additional 1 in each successive line.

b. The pattern appears to continue for a number of cases, but it does not continue in general. For example,

$$13 + 123456789101112 \cdot 9 = 1{,}111{,}111{,}101{,}910{,}021$$

The pattern breaks down when the pattern of digits in the number being multiplied by 9 contains previously used digits.

As seen in Example 1-1, determining a pattern on the basis of a few cases is not reliable.

Patterns do not always have to be numerical, as shown in Now Try This 1-1.

NOW TRY THIS 1-1

- **a.** Find the next three terms to complete a pattern:
  o, Δ, Δ, o, Δ, Δ, o, ____ , ____ , ____
- **b.** Describe the pattern in words.

Inductive and Deductive Reasoning

inductive reasoning

Scientists make observations and propose general laws based on these observations and patterns seen. Statisticians use patterns when they form conclusions based on collected data. This process, **inductive reasoning,** is the method of making generalizations based on observations and patterns. While inductive reasoning may lead to new discoveries, its weakness is that conclusions are drawn only from the collected evidence. If not all cases have been checked, the possibility exists that another case will prove the conclusion false.

conjecture

Inductive reasoning may lead to a **conjecture,** a statement thought to be true but not yet proved true or false. For example, considering only that $0^2 = 0$ and that $1^2 = 1$, a conjecture might be that *every number squared is equal to itself*. When we find an example that

counterexample

contradicts the conjecture, we have provided a **counterexample.** To show that the above conjecture is not true, it is enough to exhibit only one counterexample, for example, $2^2 = 4$. Sometimes finding a counterexample is difficult, but the lack of a counterexample does not automatically make a conjecture true.

An example of inductive and deductive reasoning is seen on the student page from *Scott-Foresman Addison Wesley Middle School Math Course 2*, 1999.

The following discussion illustrates the danger of making a conjecture based on a few cases. In Figure 1-1, we choose points on a circle and connect them to form distinct, nonoverlapping regions. In this figure, 2 points determine 2 regions, 3 points determine 4 regions, and 4 points determine 8 regions. What is the maximum number of regions that would be determined by 10 points?

Figure 1-1

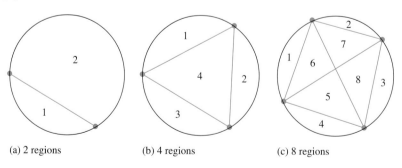

(a) 2 regions   (b) 4 regions   (c) 8 regions

The data from Figure 1-1 are recorded in Table 1-1. It appears that each time we increase the number of points by 1, we double the number of regions. If this were true, then for 5 points we would have 16, or 2^4, regions; for 6 points we would have 32, or 2^5, regions;

Extend Key Ideas ● Geometry

Inductive and Deductive Reasoning

Most people do not believe everything they hear. Inductive and deductive reasoning are different ways to convince people that a statement is true.

When you test an idea many times and look for a pattern in the results, you use *inductive reasoning*. When you use logic to show that an idea is true, you use *deductive reasoning*. Mathematicians use deductive reasoning to prove that theorems, like the Pythagorean Theorem, must be true.

Here's how you might use the two types of reasoning to show that *the acute angles of a right triangle are complementary*.

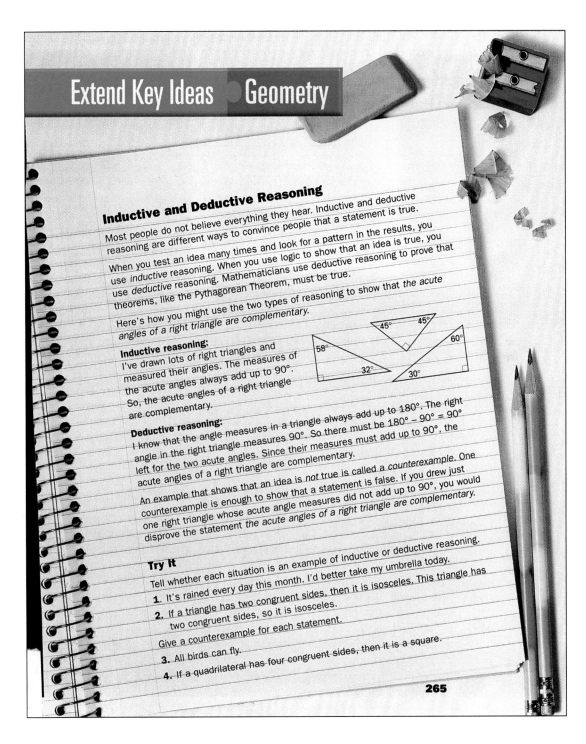

Inductive reasoning:
I've drawn lots of right triangles and measured their angles. The measures of the acute angles always add up to 90°. So, the acute angles of a right triangle are complementary.

Deductive reasoning:
I know that the angle measures in a triangle always add up to 180°. The right angle in the right triangle measures 90°. So there must be 180° − 90° = 90° left for the two acute angles. Since their measures must add up to 90°, the acute angles of a right triangle are complementary.

An example that shows that an idea is *not* true is called a *counterexample*. One counterexample is enough to show that a statement is false. If you drew just one right triangle whose acute angle measures did not add up to 90°, you would disprove the statement *the acute angles of a right triangle are complementary*.

Try It
Tell whether each situation is an example of inductive or deductive reasoning.
1. It's rained every day this month. I'd better take my umbrella today.
2. If a triangle has two congruent sides, then it is isosceles. This triangle has two congruent sides, so it is isosceles.

Give a counterexample for each statement.
3. All birds can fly.
4. If a quadrilateral has four congruent sides, then it is a square.

and so on. If we base our solution on this pattern, we could conjecture that for 10 points, we would have 512, or 2^9, regions.

Table 1-1

Number of Points	2	3	4	5	6	...	10
Maximum Number of Regions	$2 = 2^1$	$4 = 2^2$	$8 = 2^3$				?

Figure 1-2

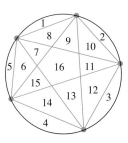

Before reading on, check to see whether we obtain 16 regions for 5 points. We obtain a figure similar to that in Figure 1-2, and our guess of 16 regions is verified. For 6 points, the pattern predicts that the number of regions will be 32. Choose the points so that they are not equally spaced and count the regions carefully. You should obtain 31 regions and not 32 regions as predicted. No matter how the points are located on the circle, the guess of 32 regions is not correct. Therefore the pattern breaks down, and the conjecture of 512 regions for 10 points that is based on this pattern is probably incorrect.

A natural-looking pattern 2, 4, 8, 16, ... is suggested in this example, but the pattern does not continue, as shown when actual pictures are drawn. If we look at only the first four terms of the sequence 2, 4, 8, 16 without context, the doubling pattern is logical. In the context of looking for regions of a circle, however, the pattern is incorrect. An important message here is, use all the pertinent information when solving problems.

Arithmetic Sequences

sequence   A **sequence** is an ordered arrangement of numbers, figures, or objects. A sequence has items or terms identified as *1st, 2nd, 3rd*, and so on. Often, sequences can be classified by their properties. For example, what property do the following first three sequences have that the fourth does not?

a. 1, 2, 3, 4, 5, 6, ...
b. 0, 5, 10, 15, 20, 25, ...
c. 2, 6, 10, 14, 18, 22, ...
d. 1, 11, 111, 1111, 11111, 111111, ...

In each of the first three sequences, each term—starting from the second—is obtained from the preceding one by adding a fixed number. In other words, the difference between consecutive numbers in the sequences is always the same. In part (d), the difference between the first two terms is $11 - 1$, or 10; the second difference is $111 - 11$, or 100; the third difference is $1111 - 111$, or 1000; and so on. Sequences such as the first three

arithmetic sequence   are arithmetic sequences. An **arithmetic sequence** is one in which each successive term is obtained from the previous term by the addition or subtraction of a fixed number, the

difference   **difference.** The difference in part (c) is 4 because 4 is the fixed number that is added each time to obtain the next number.

Numerical sequences can be generated from objects, as shown in Example 1-2.

Example 1-2   Find a pattern in the number of matchsticks required to continue the pattern shown in Figure 1-3.

Figure 1-3

Solution Assume the matchsticks are arranged so that each figure has one more square than the preceding figure. Note that the addition of a square to an arrangement requires the addition of three matchsticks each time. Thus, the numerical pattern obtained is 4, 7, 10, 13, 16, 19, ... This is an arithmetic sequence with a difference of 3.

REMARK An informal description of an arithmetic sequence is one that can be described as an "add d" pattern where d is the common difference.

In the language of children, the pattern in Example 1-2 is "add 3," meaning that 3 must be added to any term after the first to find the next. This is an example of a **recursive pattern**. In a recursive pattern, after one or more consecutive terms are given to start, each successive term of the sequence is obtained from the previous term(s). For example, 3, 6, 9, ... is another "add 3" sequence starting with 3, while 1, 2, 3, 5, 8, 13, ... is a recursive pattern in which the next term is obtained from adding the two previous terms.

A recursive pattern is typically used in a spreadsheet, as seen in Figure 1-4 where column A is the *index column* that tracks the order of the terms; the headers for the columns are A, B, etc. The first entry in the B column (in the B1 cell) is 4; and to find the term in the B2 cell, we use the number in the B1 cell and add 3. Once the B2 cell entry is found, the pattern is continued using the *Fill Down* command. In spreadsheet language, the formula is =B1+3 to find any term after the first by adding 3 to the previous term. The formula is based on a recursive pattern and is a **recursive formula**. [For more explicit directions on using a spreadsheet, see Appendix IV.]

Figure 1-4

A	B
1	4
2	7
3	10
4	13
5	16
6	19
7	22
8	25
9	28

If you want to find the number of matchsticks in the 100th figure in Example 1-2 then the spreadsheet may be used or we can find a different type of general rule for finding the number of matchsticks when given the number of the term. The problem-solving strategy of *making a table* is again helpful here.

The spreadsheet in Figure 1-4 gives an easy way to form a table. The index column (column A) gives the numbers of the terms and column B gives the terms of the sequence. If one is building such a table without a spreadsheet, it might look like Table 1-2. An *ellipsis,* denoted by three dots, indicates that the sequence continues in the same manner.

Table 1-2

Number of Term	Term
1	4
2	$7 = 4 + 3 = 4 + 1 \cdot 3$
3	$10 = (4 + 3) + 3 = 4 + 2 \cdot 3$
4	$13 = (4 + 2 \cdot 3) + 3 = 4 + 3 \cdot 3$
⋮	⋮
n	$4 + (n - 1)3$

*n*th term

REMARK The **nth term** of the sequence is $4 + (n - 1)3$, or $3n + 1$.

We have already seen that the common difference of the sequence in Table 1-2 is 3. A natural sequence of numbers with this common difference is found in the elementary grades by "skip-counting." For example, if a child skip-counts by 3, we would hear the child saying, "3, 6, 9, 12, 15, ..." Suppose we append the sequence of skip-count terms listed in a third column of Table 1-2 as seen in Table 1-3.

Table 1-3

Number of Term	Term	Skip-Count Term
1	4	3
2	7	6
3	10	9
4	13	12
5	16	15
6	19	18
7	22	21
8	25	24
9	28	27
⋮	⋮	⋮
n	$3n + 1$	$3n$

To find the *n*th term of the original sequence, the skip-count terms are helpful. The skip-count terms are just the multiples of 3 with the first being $1 \cdot 3$, the second being $2 \cdot 3$, the third being $3 \cdot 3$, and so on. Thus the *n*th term in the skip-count column is $3n$. Note that the terms in the middle column are 1 more than the corresponding terms in the skip-count column. Thus, a general pattern for the numbers in the middle column is $3n + 1$. With this pattern, the 100th term is $3 \cdot 100 + 1$ or 301.

A different approach to counting the matchsticks in the 100th term of Figure 1-4 might be as follows: If the matchstick figure has 100 squares, we could find the total number of matchsticks by adding together the number of horizontal and vertical sticks. There are $2 \cdot 100$ placed horizontally. Why? Notice that in the first figure, there are 2 matchsticks placed vertically; in the second, 3; and in the third, 4. In the one hundredth figure, there should be $100 + 1$ vertical matchsticks. Altogether there will be $2 \cdot 100 + (100 + 1)$, or 301 matchsticks in the one hundredth figure. Similarly, in the nth figure, there would be $2n$ horizontal and $(n + 1)$ vertical matchsticks for a total of $3n + 1$. This discussion is summarized in Table 1-4.

Are the nth terms generated by using both techniques equivalent? That is, does $4 + (n - 1) \cdot 3 = 3n + 1$?

If we were given the value of the term, we could use the formula for the nth term in Table 1-4 to work backward to find the number of the term. For example, given the term 1798, we know that $3n + 1 = 1798$. Therefore $3n = 1797$ and $n = 599$. Consequently, the 599th term is 1798. We could obtain the same answer by solving $4 + (n - 1) \cdot 3 = 1798$.

Table 1-4

Number of Term	Number of Matchsticks Horizontally	Number of Matchsticks Vertically	Total
1	2	2	4
2	4	3	7
3	6	4	10
4	8	5	13
.	.	.	.
.	.	.	.
.	.	.	.
100	200	101	301
.	.	.	.
.	.	.	.
.	.	.	.
n	$2n$	$n + 1$	$2n + (n + 1) = 3n + 1$

In the matchstick problem, we found the nth term of a sequence. If the nth term of a sequence is given, we can find any term of the sequence, as shown in Example 1-3.

Example 1-3

Find the first four terms of a sequence whose nth term is given by the following and determine whether the sequence is arithmetic.

a. $4n + 3$ **b.** $n^2 - 1$

Solution **a.** To find the first term, substitute $n = 1$ in the formula $4n + 3$ to obtain $4 \cdot 1 + 3$, or 7. Similarly, substituting $n = 2, 3, 4$, we obtain, respectively, $4 \cdot 2 + 3$, or 11; $4 \cdot 3 + 3$, or 15; and $4 \cdot 4 + 3$, or 19. Hence, the first four terms of the sequence are 7, 11, 15, and 19, and this sequence is arithmetic with a difference of 4.

b. Substituting $n = 1, 2, 3, 4$ in the formula $n^2 - 1$, we obtain, respectively, $1^2 - 1$, or 0; $2^2 - 1$, or 3; $3^2 - 1$, or 8; and $4^2 - 1$, or 15. Thus the first four terms of the sequence are 0, 3, 8, and 15. This sequence is not arithmetic, having no common difference.

NOW TRY THIS 1-2

● If you have an arithmetic sequence with the 2nd term 11 and the 5th term 23, find the 100th term. ●

To generalize our work with arithmetic sequences, suppose the first term in an arithmetic sequence is a and the difference is d. The strategy of *making a table* can be used to investigate the general term for the sequence $a, a + d, a + 2d, a + 3d, \ldots$, as shown in Table 1-5. *The nth term of any sequence with first term a and difference d is given by $a + (n - 1)d$.* For example, in the arithmetic sequence $5, 9, 13, 17, 21, 25, \ldots$, the first term is 5 and the difference is 4. Thus the nth term is given by $a + (n - 1)d = 5 + (n - 1)4$. Simplifying algebraically, we obtain $5 + (n - 1)4 = 5 + 4n - 4 = 4n + 1$.

Table 1-5

Number of Term	Term
1	a
2	$a + d$
3	$a + 2d$
4	$a + 3d$
5	$a + 4d$
.	.
.	.
.	.
n	$a + (n - 1)d$

Example 1-4 The diagrams in Figure 1-5 show the molecular structure of alkanes, a class of hydrocarbons. C represents a carbon atom and H a hydrogen atom. A connecting segment shows a chemical bond.

Figure 1-5

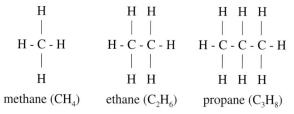

methane (CH$_4$)   ethane (C$_2$H$_6$)   propane (C$_3$H$_8$)

a. Hectane is an alkane with 100 carbon atoms. How many hydrogen atoms does it have?
b. Write a general rule for alkanes C$_n$H$_m$ showing the relationship between m and n.

Solution a. To determine the relationship between the number of carbon and hydrogen atoms, we study the drawing of the alkanes and disregard the extreme left and right hydrogen atoms in each. We can see that for every carbon atom, there are two hydrogen atoms. Therefore there are twice as many hydrogen atoms as carbon atoms plus the two hydrogen atoms at the extremes. For example, when there are 3 carbon atoms, there are (2 · 3) + 2, or 8, hydrogen atoms. This notion is summarized in Table 1-6. If we extend the table for 4 carbon atoms, we get (2 · 4) + 2, or 10, hydrogen atoms. For 100 carbon atoms, there are (2 · 100) + 2, or 202, hydrogen atoms.

Table 1-6

No. of Carbon Atoms	No. of Hydrogen Atoms
1	4
2	6
3	8
.	.
.	.
100	?
.	.
.	.
.	.
n	m

b. In general, for n carbon atoms there would be n hydrogen atoms attached above, n attached below, and 2 attached on the sides. Hence, the total number of hydrogen atoms would be $2n + 2$. Because the number of hydrogen atoms was designated by m, it follows that $m = 2n + 2$.

Example 1-5

A theater is set up in such a way that there are 20 seats in the first row and 4 additional seats in each consecutive row. The last row has 144 seats. How many rows are there in the theater?

Solution Because there are 4 additional seats in each consecutive row, the number of seats in the rows form an arithmetic sequence. The first term of the sequence is 20 and the difference, d, is 4. The last term in the sequence is 144. A computerized spreadsheet could easily be used to count the number of terms in the sequence, 20, 24, 28, … , 144. However, without technology, we count the terms as follows: In an arithmetic sequence, the nth term is $a + (n - 1)d$, where a is the first term, d is the difference, and n is the number of the term. In this case, $a = 20$ and $d = 4$. Therefore

$$a + (n - 1)d = 20 + (n - 1)4.$$

We now want to find the number of the term, n, when the nth term, $20 + (n - 1)4$, is equal to 144. Therefore

$$20 + (n - 1)4 = 144$$
$$(n - 1)4 = 124$$
$$n - 1 = 31$$
$$n = 32.$$

This tells us that when $n = 32$, the value of the term is 144. This implies that there are 32 rows in the theater. Instead of using a formula, we could have made a table and looked for a pattern.

Geometric Sequences

geometric sequence

ratio

A child has 2 biological parents, 4 grandparents, 8 great grandparents, 16 great-great grandparents, and so on. The number of ancestors form the **geometric sequence** 2, 4, 8, 16, 32, ... Each successive term of a geometric sequence is obtained from its predecessor by multiplying by a fixed number, the **ratio.** In this example, both the 1st term and the ratio are 2. (The ratio is 2 because each person has two parents.) To find the *n*th term, examine the pattern in Table 1-7.

Table 1-7

Number of Term	Term
1	$2 = 2^1$
2	$4 = 2 \cdot 2 = 2^2$
3	$8 = (2 \cdot 2) \cdot 2 = 2^3$
4	$16 = (2 \cdot 2 \cdot 2) \cdot 2 = 2^4$
5	$32 = (2 \cdot 2 \cdot 2 \cdot 2) \cdot 2 = 2^5$
⋮	⋮

Definition

If n is a natural number, then $a^n = \overbrace{a \cdot a \cdot a \cdot \ldots \cdot a}^{n \text{ terms}}$

When the given term is written as a power of 2, the number of the term is the exponent. Following this pattern, the 10th term is 2^{10}, or 1024, the 100th term is 2^{100}, and the *n*th term is 2^n. Thus the number of ancestors in the *n*th previous generation is 2^n.

Finding the *n*th Term for a Geometric Sequence

It is possible to find the *n*th term of any geometric sequence when given the first term and the ratio. If the first term is a and the ratio is r, then the terms are as listed in Table 1-8.

Table 1-8

Number of Term	Term
1	a
2	ar
3	ar^2
4	ar^3
5	ar^4
⋮	⋮
n	ar^{n-1}

Notice that the second term is ar, the third term is ar^2, the fourth term is ar^3. The power of r in each term is 1 less than the number of the term. This pattern continues since we multiply by r to get the next term. Thus, the nth term is ar^{n-1}. For $n = 1$, we have $ar^{1-1} = ar^0$. Because the 1st term is a, then $ar^0 = a$. This implies that $r^0 = 1$. This is true for all numbers $r \neq 0$, as discussed in Chapter 6. Thus when $n = 1$ and $r \neq 0$, we have $ar^0 = a(1) = a$. For the geometric sequence 3, 12, 48, 192, ..., the first term is 3 and the ratio is 4, and so the nth term is given by $ar^{n-1} = 3 \cdot 4^{n-1}$.

NOW TRY THIS 1-3

- **a.** Two bacteria are in a dish. The number of bacteria triples every hour. Following this pattern, find the number of bacteria in the dish after 10 hours and after n hours.
- **b.** Suppose that instead of increasing geometrically as in part (a), the number of bacteria increases arithmetically by 3 each hour. Compare the growth after 10 hours and after n hours. Comment on the difference in growth of a geometric sequence versus an arithmetic sequence.

Other Sequences

figurate numbers

Figurate numbers provide examples of sequences that are neither arithmetic nor geometric. Such numbers can be represented by dots arranged in the shape of certain geometric figures. The number 1 is the beginning of most patterns involving figurate numbers. The array in Figure 1-6 represents the first four terms of the sequence of **triangular numbers.**

triangular numbers

Figure 1-6

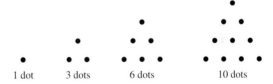

1 dot   3 dots   6 dots   10 dots

The triangular numbers can be written numerically as 1, 3, 6, 10, 15, ... The sequence 1, 3, 6, 10, 15, ... is not an arithmetic sequence because there is no common difference, as Figure 1-7 shows.

Figure 1-7

$$\begin{array}{ccccccccc} 1 & & 3 & & 6 & & 10 & & 15 \\ & \searrow\swarrow & & \searrow\swarrow & & \searrow\swarrow & & \searrow\swarrow & \\ \text{(First difference)} & 2 & & 3 & & 4 & & 5 & \end{array}$$

However, the sequence of differences 2, 3, 4, 5, ... is an arithmetic sequence with difference 1, as Figure 1-8 shows.

Figure 1-8

$$\begin{array}{ccccccccccccc} 1 & & 3 & & 6 & & 10 & & 15 & & 21 & & 28 \\ & \searrow\swarrow & & \searrow\swarrow & & \searrow\swarrow & & \searrow\swarrow & & \searrow\swarrow & & \searrow\swarrow & \\ \text{(First difference)} & 2 & & 3 & & 4 & & 5 & & 6 & & 7 \\ & & \searrow\swarrow & & \searrow\swarrow & & \searrow\swarrow & & \searrow\swarrow & & \searrow\swarrow & & \\ \text{(Second difference)} & & 1 & & 1 & & 1 & & 1 & & 1 & & \end{array}$$

Successive terms for the original sequence are shown in color.

Table 1-9 suggests a pattern for finding the next terms and the nth term for the triangular numbers. The second term is obtained from the first term by adding 2; the third term is obtained from the second term by adding 3; and so on. In general, because the nth triangular number has n dots in the nth row, it is equal to the sum of the dots in the previous triangular number (the $(n - 1)$st one) plus the n dots in the nth row. Following this pattern, the 10th term is $1 + 2 + 3 + 4 + 5 + 6 + 7 + 8 + 9 + 10$, or 55, and the nth term is $1 + 2 + 3 + 4 + 5 + \ldots + (n - 1) + n$.

Table 1-9

Number of Term	Term
1	1
2	$3 = 1 + 2$
3	$6 = 1 + 2 + 3$
4	$10 = 1 + 2 + 3 + 4$
5	$15 = 1 + 2 + 3 + 4 + 5$
.	.
.	.
.	.
10	$55 = 1 + 2 + 3 + 4 + 5 + 6 + 7 + 8 + 9 + 10$

Next consider the first four *square numbers* in Figure 1-9. These square numbers, 1, 4, 9, 16, ..., can be written as $1^2, 2^2, 3^2, 4^2$, and so on. The number of dots in the 10th array is 10^2; the number of dots in the 100th array is 100^2; and the number of dots in the nth array is n^2. Notice that the sequence of square numbers is neither arithmetic nor geometric.

Figure 1-9

1 dot   4 dots   9 dots   16 dots

Example 1-6 Assuming that the pattern you discover continues, find the seventh term in each of the following sequences:

a. 5, 6, 14, 29, 51, 80, ...   **b.** 2, 3, 9, 23, 48, 87, ...

Solution **a.** Following is the sequence of first differences:

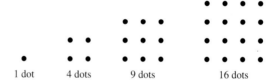

To discover a pattern for the original sequence, we try to find a pattern for the sequence of differences 1, 8, 15, 22, 29, ... This sequence is an arithmetic sequence with fixed difference 7:

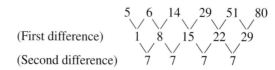

Thus the 6th term in the first difference row is 29 + 7, or 36, and the 7th term in the original sequence is 80 + 36, or 116. What number follows 116?

b. Because the second difference is not a fixed number, we go on to the third difference:

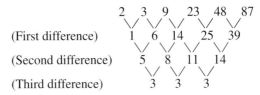

The third difference is a fixed number; therefore the second difference is an arithmetic sequence. The 5th term in the second-difference sequence is 14 + 3, or 17; the 6th term in the first-difference sequence is 39 + 17, or 56; and the 7th term in the original sequence is 87 + 56, or 143.

When asked to find a pattern for a given sequence, first look for some easily recognizable pattern and determine whether the sequence is arithmetic or geometric. If a pattern is unclear, taking successive differences may help. *It is possible that none of the methods described reveal a pattern.*

BRAIN TEASER Female bees are born from fertilized eggs, and male bees are born from unfertilized eggs. This means that a male bee has only a mother, while a female bee has a mother and a father. If the ancestry of a male bee is traced 10 generations back, how many bees are there in all 10 generations?

ONGOING ASSESSMENT 1-1

1. For each of the following sequences of figures, determine a possible pattern and draw the next figure according to that pattern:

   a.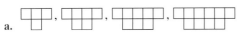

   b.

   c.

   d.

   e.

   f.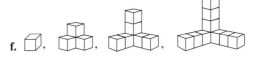

2. In each of the following, list terms that continue a possible pattern. Which of the sequences are arithmetic, which are geometric, and which are neither?
   a. 1, 3, 5, 7, 9
   b. 0, 50, 100, 150, 200
   c. 3, 6, 12, 24, 48
   d. 10, 100, 1000, 10,000, 100,000
   e. 9, 13, 17, 21, 25, 29
   f. 1, 8, 27, 64, 125
3. Find the 100th term and the nth term for each of the sequences in problem 2 above.
4. Use a traditional clock face to determine the next three terms in the following sequence:

   1, 6, 11, 4, 9, …

5. Observe the following pattern:
   $1 + 3 = 2^2$.
   $1 + 3 + 5 = 3^2$.
   $1 + 3 + 5 + 7 = 4^2$.

   a. Use inductive reasoning to state a generalization based on this pattern.
   b. Based on your generalization in (a), find

   $1 + 3 + 5 + 7 + … + 35$.

6. Following is a pattern of circular shapes:

   If you make this pattern until you have 10 black circles in a row, how many open circles and how many black circles will you need?
7. The following geometric arrays suggest a sequence of numbers:

   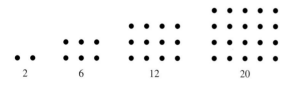

   a. Find the next three terms.
   b. Find the 100th term.
   c. Find the nth term.
8. In the following pattern, one hexagon takes 6 toothpicks to build, two hexagons take 11 toothpicks to build, and so on. How many toothpicks would it take to build (a) 10 hexagons? (b) n hexagons?

   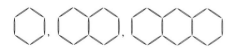

9. The first windmill takes 5 matchstick squares to build, the second takes 9 to build, and the third takes 13 to build, as shown. How many matchstick squares will it take to build (a) the 10th windmill? (b) the nth windmill?

   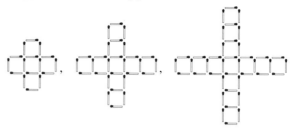

10. Each of the following figures is made of small triangles like the first one in the sequence. (The second figure is made of 4 triangles.) Make a conjecture concerning the number of small triangles needed to make (a) the 100th figure and (b) the nth figure.

11. In the following sequence, the figures are made of cubes that are glued together. If the exposed surface needs to be painted, how many squares will be painted in (a) the 10th figure? (b) the nth figure?

12. The school population for a certain school was predicted to increase by 50 students per year for the next 10 years. If the current enrollment is 700 students, what will the enrollment be after 10 years?
13. A tank contains 15,360 L of water. At the end of each day, half of the water is removed and not replaced. How much water is left in the tank after 10 days?
14. Joe's annual income has been increasing each year by the same amount. The first year his income was $24,000, and the ninth year his income was $31,680. In which year was his income $45,120?
15. The first difference of a sequence is 2, 4, 6, 8, …. Find the first six terms of the original sequence in each of the following cases:
    a. The first term of the original sequence is 3.
    b. The sum of the first two terms of the original sequence is 10.
    c. The fifth term of the original sequence is 35.
16. List the next three terms to continue a pattern in each of the following. (Finding differences may be helpful.)
    a. 5, 6, 14, 32, 64, 115, 191
    b. 0, 2, 6, 12, 20, 30, 42
    ★c. 10, 8, 3, 0, 4, 20, 53
17. How many terms are there in each of the following sequences?
    a. 51, 52, 53, 54, … , 151   b. 1, 2, 2^2, 2^3, … , 2^{60}
    c. 10, 20, 30, 40, … , 2000   d. 9, 13, 17, 21, 25, … , 353
    e. 1, 2, 4, 8, 16, 32, … , 1024

18. Find the first five terms of the sequence whose nth term is as follows:
    a. $n^2 + 2$
    b. $5n - 1$
    c. $10^n - 1$
    d. $3n + 2$
19. The number of petals on many flowers generates the following sequence:

    $$1, 1, 2, 3, 5, 8, 13, 21, \ldots,$$

    which is a *Fibonacci sequence*. This sequence is named after the great Italian mathematician Leonardo Fibonacci, who lived in the twelfth and thirteenth centuries.
    a. Write the first 12 terms of the sequence.
    b. Notice that the sum of the first three terms in the sequence is one less than the fifth term of the sequence. Does a similar relationship hold for the sum of the first four terms, five terms, and six terms?
    c. Guess the sum of the first 10 terms of the sequence.
    ★d. Make a conjecture concerning the sum of the first n terms of the sequence.
20. The Fibonacci sequence defined in problem 19 can start with arbitrary first and second terms. If the 1st term is 2 and the 2nd is 4, then the sequence is 2, 4, 6, 10, 16, 26, 42, ... Answer the questions in problem 19(a) and (c) for this sequence.
21. Consider the following sequences:
    a. 300, 500, 700, 900, 1100, 1300, ...
    b. 2, 4, 8, 16, 32, 64, ...

    Find the number of the term in which the geometric sequence becomes greater than the arithmetic sequence.
22. Cut a piece of paper into 5 pieces. Take any one of the pieces and cut it into 5 pieces, and so on.
    a. What number of pieces can be obtained in this way?
    b. What is the number of pieces obtained in the nth cut?
23. The sequence 32, a, b, c, 512 is a geometric sequence. Find a, b, c.
24. Each box in the following row had a number written in it such that the sum of any three numbers in succession in the row was 15. Many of the numbers were erased. Determine the missing numbers for each square.

    | 6 | | | | | 4 | | |

Communication

25. Explain how to show that each of these statements is false:
    a. All rectangles have diagonals that are perpendicular.
    b. The sum of two even numbers is always divisible by 4.
26. Elsa met 3 university students from Montana, and they all wore cowboy boots. Elsa generalized that all Montana university students wear cowboy boots. How could you show that her generalization is false?
27. Give two examples of how inductive reasoning might be used in your everyday life. Is a conclusion based on inductive reasoning certain?
28. a. If a fixed number is added to each term of an arithmetic sequence, is the resulting sequence an arithmetic sequence? Justify your answer.
    b. If each term of an arithmetic sequence is multiplied by a fixed number, will the resulting sequence always be an arithmetic sequence? Justify your answer.
    c. If you add the corresponding terms of two arithmetic sequences, is the resulting sequence arithmetic?

Open-Ended

29. Patterns can be used to count the number of dots on the Chinese checkerboard; two patterns are shown below. Determine several other patterns to count the dots.

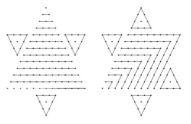

$1 + 2 + 3 + \ldots + 13 + 3(10)$ $1 + 3 + 5 + 7 + \ldots + 17 + 4(10)$

30. Make up a pattern involving figurate numbers and find a formula for the 100th term. Describe your pattern and how you found the 100th term.
31. List some operations on each term of an arithmetic sequence that do not result in an arithmetic sequence.

Cooperative Learning

32. Each person in the group follows this set of directions:
    Pick a number.
    Multiply by 3.
    Add 6.
    Divide by 3.
    Subtract your original number.
    a. Compare your answers and make a generalization based on your results.
    b. What type of reasoning did you use in making your generalization?
    c. In your group, make up a number puzzle where the answer is always 17 and write an explanation of why your puzzle works.
    d. Write another number puzzle; try it with another group.
33. If the pattern below continued indefinitely, the resulting figure would be called the *Sierpinski triangle*, or *Sierpinski gasket*.

In a group, determine each of the following. Discuss different counting strategies.
a. How many black triangles would be in the fifth figure?
b. How many white triangles would be in the fifth figure?
c. If the pattern is continued for n figures, how many black triangles will there be?
d. If the pattern is continued for n figures, how many white triangles will there be?

TECHNOLOGY CORNER

In the spreadsheet of Figure 1-10, entries in the A and B columns were entered somewhat randomly. The numbers in successive columns were determined by performing some operation(s) on the numbers in the A and B columns. For example, column C is the result of adding 3 to the number in the A column. The first entry in the C column is 4 because $1 + 3 = 4$. The value 4 is said to be located in cell C1. The value of cell C2 is 5 because $2 + 3 = 5$. The rules for other columns involve entries from both A and B. Determine rules for finding the entries in columns D through I.

Figure 1-10

	A	B	C	D	E	F	G	H	I
1	1	2	4	3	1	2	8	4	3
2	2	3	5	5	1	6	10	7	7
3	3	4	6	7	1	12	12	10	13
4	5	5	8	10	0	25	16	15	26
5	8	0	11	8	−8	0	22	16	1
6	0	−2	3	−2	−2	0	6	−2	1
7	−2	5	1	3	7	−10	2	1	−9

Section 1-2 — Mathematics and Problem Solving

George Polya (1887–1985) described the experience of problem solving in his book, *How to Solve It* (p. v):

A great discovery solves a great problem, but there is a grain of discovery in the solution of any problem. Your problem may be modest; but if it challenges your curiosity and brings into play your inventive facilities, and if you solve it by your own means, you may experience the tension and enjoy the triumph of discovery.

To solve a problem, we must first understand both the task and the given information. Next, it is helpful to determine a strategy to accomplish the task. Once we arrive at a solution, we should determine whether the solution makes sense and is reasonable. This process of problem solving can be described using a four-step process similar to the one developed by George Polya.

Four-Step Problem-Solving Process

1. **Understanding the problem**
   a. Can you state the problem in your own words?
   b. What are you trying to find or do?
   c. What are the unknowns?
   d. What information do you obtain from the problem?
   e. What information, if any, is missing or not needed?
2. **Devising a plan**
   The following list of strategies, although not exhaustive, is very useful:
   a. Look for a pattern.
   b. Examine related problems and determine if the same technique applied to them can be applied here.
   c. Examine a simpler or special case of the problem to gain insight into the solution of the original problem.
   d. Make a table.
   e. Make a diagram.
   f. Write an equation.
   g. Use guess and check.
   h. Work backward.
   i. Identify a subgoal.
   j. Use indirect reasoning.
   k. Use direct reasoning.
3. **Carrying out the plan**
   a. Implement the strategy or strategies in step 2 and perform any necessary actions or computations.
   b. Check each step of the plan as you proceed. This may be intuitive checking or a formal proof of each step.
   c. Keep an accurate record of your work.
4. **Looking back**
   a. Check the results in the original problem. (In some cases, this will require a proof.)
   b. Interpret the solution in terms of the original problem. Does your answer make sense? Is it reasonable? Does it answer the question that was asked?
   c. Determine whether there is another method of finding the solution.
   d. If possible, determine other related or more general problems for which the techniques will work.

Avoiding Mind Sets

If you approach problems in only one way, you risk forming a mind set. For example, consider the following. Spell the word "spot" three times out loud. "S-P-O-T! S-P-O-T! S-P-O-T!" Now answer the question, "What do you do when you come to a green light?" Write your answer. If you answered "Stop," you may be guilty of having formed a mind set. You do not stop at a *green* light.

Consider the following problem: "A shepherd had 36 sheep. All but 10 died. How many lived?"

Did you answer "10"? If you did, you are catching on and are ready to try some problems. If you did not answer "10," then you did not understand the question. *Understanding the problem* is the first step in the four-step problem-solving process. Using the four-step process does not guarantee a solution to a problem, but it does provide a systematic means of attacking problems.

Strategies for Problem Solving

Strategies are tools that might be used to discover or construct the means to achieve a goal. For each strategy described next, an example is given that can be solved with that strategy. Often, problems can be solved in more than one way. You may devise a different strategy to solve the example problems. There is no one best strategy to use.

Strategy: Look for a Pattern

The strategy of looking for a pattern was examined in the previous section, where we concentrated on sequences of numbers. We continue that investigation here.

HISTORICAL NOTE

Carl Gauss (1777–1855) is regarded as the greatest mathematician of the nineteenth century and one of the greatest mathematicians of all time. Born to humble parents in Brunswick, Germany, he was an infant prodigy who, it is said, at age 3 corrected an arithmetic error in his father's bookkeeping. Gauss claimed that he could figure before he could talk.

Gauss made contributions in the areas of astronomy, geodesy, and electricity. After Gauss's death, the King of Hanover ordered a commemorative medal prepared in his honor. On the medal was an inscription referring to Gauss as the "Prince of Mathematics," a title that has stayed with his name.

Problem Solving **Gauss's Problem**

When the German mathematician Carl Gauss was a child, his teacher required the students to find the sum of the first 100 natural numbers. The teacher expected this problem to keep the class occupied for some time. Gauss gave the answer almost immediately. Can you?

- **Understanding the Problem** The natural numbers are $1, 2, 3, 4, \ldots$. Thus the problem is to find the sum $1 + 2 + 3 + 4 + \ldots + 100$.

- **Devising a Plan** One strategy is to *look for a pattern*. If we consider $1 + 100, 2 + 99, 3 + 98, \ldots, 50 + 51$, it is evident that there are 50 pairs of numbers, each with a sum of 101, as shown in Figure 1-11.

Figure 1-11

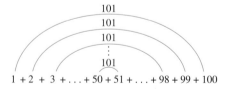

- **Carrying Out the Plan** There are 50 pairs, each with the sum 101. Thus the total can be found by multiplying $50 \cdot 101$, or 5050.

- **Looking Back** The method is mathematically correct because addition can be performed in any order, and multiplication is repeated addition. A more general problem is to find the sum of the first n natural numbers, $1 + 2 + 3 + 4 + 5 + 6 + \ldots + n$, where n is any natural number. We use the same plan as before and notice the relationship in Figure 1-12. If n is an even natural number, there are $n/2$ pairs of numbers. The sum of each pair is $n + 1$. Therefore the sum $1 + 2 + 3 + \ldots + n$ is given by $(n/2)(n + 1)$.

Figure 1-12

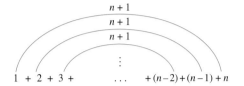

A different strategy for finding the sum $1 + 2 + 3 + \ldots + n$ involves *making a diagram* and thinking of the sum geometrically as a stack of blocks. To find the sum, consider the stack in Figure 1-13 (a) and a stack of the same size placed differently, as in Figure 1-13 (b). The total number of blocks in the stack in Figure 1-13 (b) is $n(n + 1)$, which is twice the desired sum. Thus the desired sum is $n(n + 1)/2$.

Figure 1-13

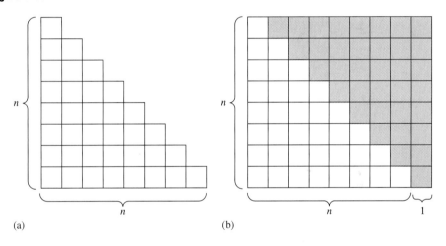

Notice that this problem is connected to the problems in the previous section that involved triangular numbers.

Strategy: Make a Table

A table can be used to summarize data or to help us see a pattern. It can also help us to consider all possible cases in a given problem.

Problem Solving **Sums of Even Natural Numbers**

Find the sum of the even natural numbers less than or equal to 100. Devise a strategy for finding that sum and generalize the result.

• **Understanding the Problem** We begin to find the sum of the even natural numbers $2 + 4 + 6 + 8 + \ldots + 100$ by searching for a pattern that is used to form the number in this sum. The numbers are an arithmetic sequence with a difference of 2, as seen in Table 1-10.

Table 1-10

Number of Term	Term
1	$2 = 2 \cdot 1$
2	$4 = 2 \cdot 2$
3	$6 = 2 \cdot 3$
4	$8 = 2 \cdot 4$
.	.
.	.
.	.
50	$100 = 2 \cdot 50$

Thus we want to find the sum of

$$2 \cdot 1 + 2 \cdot 2 + 2 \cdot 3 + 2 \cdot 4 + \ldots + 2 \cdot 50$$

- **Devising a Plan** This problem is much like Gauss's original problem if we write the sum as follows:

$$2 \cdot 1 + 2 \cdot 2 + 2 \cdot 3 + 2 \cdot 4 + \ldots + 2 \cdot 50 = 2(1 + 2 + 3 + 4 + \ldots + 50)$$

Seeing this, we can find the sum of the expression in parentheses by *looking at a related problem,* using Gauss's technique and doubling the result.

- **Carrying Out the Plan** The sum $1 + 2 + 3 + 4 + \ldots + 50 = \dfrac{50(50 + 1)}{2}$, or 1275.

Then we have $2 \cdot 1275$, or 2550, for the sum.

- **Looking Back** A different way to consider this problem is to realize that there are 25 pairs of even numbers, each of whose sum is seen in Figure 1-14 as 102.

Figure 1-14

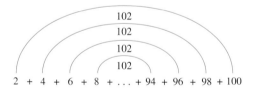

Thus the sum is $25 \cdot 102$, or 2550.

To generalize the result for finding the sum of the first n consecutive even natural numbers, $2 \cdot 1 + 2 \cdot 2 + \ldots + 2n$, we can treat this as twice the sum of the first $n/2$ natural numbers and use the technique of Gauss. What other generalizations can we make?

> **NOW TRY THIS 1-4**
>
> ● Jean wants to build a rectangular picture frame that has an area of 120 cm² (square centimeters). She wants the length and width to be natural numbers, {1, 2, 3, 4, …}. What dimensions give the least perimeter? ●

Strategy: Examine a Simpler Case

One strategy for solving a complex problem is to *examine a simpler case* of the problem and then consider other parts of the complex problem.

Problem Solving   **Ms. Danielson's Problem**

In *A Collection of Performance Tasks and Rubrics: Middle School Mathematics* (1997), Charlotte Danielson poses the following problem:

> *In an old folk tale, a poor peasant is offered as much land as he can walk around from sunup to sundown. If you were given that offer, how much land could you claim? What shape would it be?* (p. 102).

Answer Ms. Danielson's problem.

- Understanding the Problem   This is a complex problem with many answers depending on assumptions that are made. Among the things that we must consider are (1) how fast the person can walk, (2) where the person is walking, (3) how long the day is from sunup to sundown which in turn depends on the time of year the walk is to be made, (4) what shape of pattern the person walks, and (5) how to find the area of different shapes walked.

- Devising a Plan   One strategy for answering the problem is *to examine a simpler problem* and once that is understood to add more assumptions until a satisfactory solution is reached. (In doing this, it should be noted that a satisfactory solution depends on the person solving the problem.)

We first assume that the person can walk at about 3 miles per hour on a street, the streets are laid out in square blocks of 1/12 miles on each side, the person can walk about 14 hours a day in the middle of summer in daylight (16 hours of daylight with 2 hours of rest), and to make a closed shape the person walks a square. With these assumptions, we can arrive at a solution.

- Carrying Out the Plan   If a person can walk 3 miles per hour for 14 hours, the person walks 3 · 14, or 42, miles. If there are 1/12 miles to each side of a block, then the person can walk 42/(1/12), or about 504, blocks. The 504 blocks form the perimeter of the square that we assume the person walked. A square has four sides the same length, so the person has walked four times the length of one side of the square to walk the perimeter, or 504/4 = 126 blocks for the length of one side of the square shape. If the shape walked is a square with one side 126 blocks long, then the area of the shape is the area of the square that is 126 blocks by 126 blocks long. Thus the area is 126 · 126, or a total of 15,876, square blocks for the area.

- Looking Back   If the area of the shape is 15,786 square blocks and each block has area (1/12 mi)(1/12 mi), or 1/144 square miles, then the area of the shape is approximately 110 square miles.

The simplifications of the problem are arbitrary in order to allow us to find a solution. The problem should not stop here. If we now change the problem to say that the person is walking in a flat field at the same rate, the same number of hours but walks in a circle, we have increased the difficulty of the problem of finding the area of the shape. The next step would be to try other shapes and to try to find the areas of irregular shapes. These more difficult problems are left to the reader. Because different assumptions lead to different answers, there is no absolute answer to the original problem.

NOW TRY THIS 1-5

- Nikki is setting up tables for a noon luncheon in the gym. She has 25 small square tables that hold one person to a side. She plans to put 25 of these tables in a row to make one long rectangular table that is only one table wide. If 60 people will be attending the lunch, will she have enough space to seat them all using this plan? If she placed all 25 tables in the form of a big square, how many people could she seat?

Strategy: Identify a Subgoal

As we attempt to devise a plan for solving some problems, it may become apparent that the problem could be solved if the solution to a somewhat easier or more familiar related problem could be found. In such a case, finding the solution to the easier problem may become a subgoal of the primary goal of solving the original problem. The following Magic Square problem shows an example of this.

Problem Solving — **A Magic Square**

Arrange the numbers 1 through 9 into a square subdivided into nine smaller squares like the one shown in Figure 1-15 so that the sum of every row, column, and main diagonal is the same. (The result is called a *magic square*.)

Figure 1-15

- **Understanding the Problem** We need to put each of the nine numbers $1, 2, 3, \ldots, 9$ in the small squares, a different number in each square, so that the sum of the numbers in each row, in each column, and in each of the two diagonals is the same.

- **Devising a Plan** If we knew the fixed sum of the numbers in each row, column, and diagonal, we would have a better idea of which numbers can appear together in a single row, column, or diagonal. Thus our *subgoal* is to find that fixed sum. The sum of the nine numbers, $1 + 2 + 3 + \ldots + 9$, equals 3 times the sum in one row. Consequently, the fixed sum is obtained by dividing $1 + 2 + 3 + \ldots + 9$ by 3. Using the process developed by Gauss, we have $(1 + 2 + 3 + \ldots + 9) \div 3 = \dfrac{9 \cdot 10}{2} \div 3$, or $45 \div 3 = 15$, so the sum in each row, column, and diagonal must be 15. Next, we need to decide what numbers could occupy the various squares. The number in the center space will appear in 4 sums, each adding to 15 (two diagonals, the second row, and the second column). Each number in the corners will appear in three sums of 15. (Do you see why?) If we write 15 as a sum of three different numbers 1 through 9 in all possible ways, we could then count how many sums contain each of the numbers 1 through 9. The numbers that appear in at least four sums are candidates for placement in the center square, whereas the numbers that appear in at least three sums are candidates for the corner squares. Thus our new *subgoal* is to write 15 in as many ways as possible as a sum of three different numbers from the set $\{1, 2, 3, \ldots, 9\}$.

- **Carrying Out the Plan**   The sums of 15 can be written systematically as follows:

$$9 + 5 + 1$$
$$9 + 4 + 2$$
$$8 + 6 + 1$$
$$8 + 5 + 2$$
$$8 + 4 + 3$$
$$7 + 6 + 2$$
$$7 + 5 + 3$$
$$6 + 5 + 4$$

Notice that the order in each sum is not important. (Do you see why?) Hence, $1 + 5 + 9$ and $5 + 1 + 9$, for example, are counted as the same. Notice that 1 appears in only two sums, 2 in three sums, 3 in two sums, and so on. Table 1-11 summarizes this pattern.

Table 1-11

Number	1	2	3	4	5	6	7	8	9
Number of Sums Containing the Number	2	3	2	3	4	3	2	3	2

The only number that appears in four sums is 5; hence, 5 must be in the center of the square. (Do you see why?) Because 2, 4, 6, and 8 appear three times each, they must go in the corners. Suppose we choose 2 for the upper left corner. Then 8 must be in the lower right corner. (Why?) This is shown in Figure 1-16(a). Now we could place 6 in the lower left corner or upper right corner. If we choose the upper right corner, we obtain the result in Figure 1-16(b). The magic square can now be completed, as shown in Figure 1-16(c).

Figure 1-16

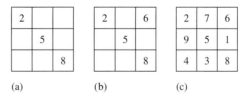

(a)   (b)   (c)

- **Looking Back**   We have seen that 5 was the only number among the given numbers that could appear in the center. However, we had various choices for a corner, and hence it seems that the magic square we found is not the only one possible. Can you find all the others?

Another way to see that 5 must be in the center square is to consider the sums $1 + 9$, $2 + 8, 3 + 7, 4 + 6$, as shown in Figure 1-17. We could add 5 to each to obtain 15.

Figure 1-17

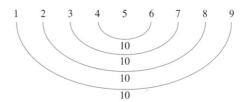

NOW TRY THIS 1-6

• Five friends decided to give a party and split the costs equally. Al spent $4.75 on invitations, Betty spent $12 for drinks and $5.25 on vegetables, Carl spent $24 for pizza, Dani spent $6 on paper plates and napkins, and Ellen spent $13 on decorations. Determine who owes money to whom and how the money can be paid. •

Strategy: Examine a Related Problem

Sometimes a problem is similar to a previously encountered problem. If so, it is often possible to apply a similar approach to solve the new problem. Such is the case in the next problem.

Problem Solving   Matchstick Patterns

Ryan is building matchstick square sequences, as shown in Figure 1-18. He used 67 matchsticks to form the last figure in his sequence. How many matchsticks will he use for the entire project?

Figure 1-18

• **Understanding the Problem**   From experience with patterns, we can recognize the sequence generated by the matchsticks as 4, 7, 10, 13, ..., 67. The last number is 67 because Ryan used 67 matchsticks to form the last figure in the sequence. This is an arithmetic sequence with difference 3. Find the sum of the numbers in this sequence.

• **Devising a Plan**   A *related problem* is Gauss's problem of finding the sum $1 + 2 + 3 + 4 + \ldots + 100$. In that problem, we paired 1 with 100, 2 with 99, 3 with 98, and so on, and observed that there were 50 pairs of numbers, each with a sum of 101. A similar approach in the present problem yields a sum of 71 for each pair. To find the total, we need to know the number of pairs in Figure 1-19.

Figure 1-19

To find the number of pairs, we need the number of terms in the sequence. Thus we have identified a *subgoal,* which is to find the number of terms in the sequence. From the discussion after Example 1-2, we see the nth term of this sequence is $4 + (n - 1)3$. To find the number of the term corresponding to 67, we solve the equation $4 + (n - 1)3 = 67$ and obtain $n = 22$. Thus there are 22 terms in the given sequence.

• **Carrying Out the Plan**   Because the number of terms is 22, we have 11 pairs of matchstick figures, each of whose sum is 71 matchsticks. Therefore the total is $11 \cdot 71$, or 781 matchsticks.

- **Looking Back** Using the outlined procedure, we should be able to find the sum of any arithmetic sequence in which we know the first two terms and the last term.

Strategy: Write an Equation

A problem-solving strategy commonly used in algebraic thinking consists of *writing an equation.* An example that combines using a table and writing an equation is seen on the student page from *Scott Foresman-Addison Wesley Middle School Math Course 2,* 1999.

Strategy: Draw a Diagram

It has often been said that a picture is worth a thousand words. This is particularly true in problem solving. In geometry, drawing a picture often provides the insight necessary to solve a problem. The following problem is a nongeometric problem that can be solved by drawing a diagram.

Problem Solving **Paper-Measuring Problem**

Miguel has a single sheet of 8 1/2-by-11-in. paper. He needs to measure exactly 6 in. Can he do it using the sheet of paper?

- **Understanding the Problem** The problem is to use only an 8 1/2-by-11-in. paper to measure something of exactly 6 in. There are the two edge lengths with which to work.

- **Devising a Plan** A natural thing to ask is, what combinations of lengths can be made from the two given lengths? We could fold to halve the lengths, but this would lead to such fractional lengths as 4 1/4 in., 2 1/8 in., 5 1/2 in., and so on, that appear unhelpful. Another idea is to consider other folds that could be used to combine 8 1/2 in. and 11 in. to get 6 in. For example, consider how to obtain 6 from 8 1/2 and 11.

$$11 - 8\ 1/2 = 2\ 1/2$$
$$8\ 1/2 - 2\ 1/2 = 6$$

If we could fold the paper to obtain 2 1/2 in. from the two given lengths and then "subtract" the 2 1/2 from 8 1/2, then we could obtain the desired result of 6. One strategy to investigate how to obtain 2 1/2 in. from the 11 in. and 8 1/2 in. lengths is to *draw a diagram.*

- **Carrying Out the Plan** By folding the paper as shown by the arrows in Figure 1-20, we can obtain a length of 6 in.

Figure 1-20

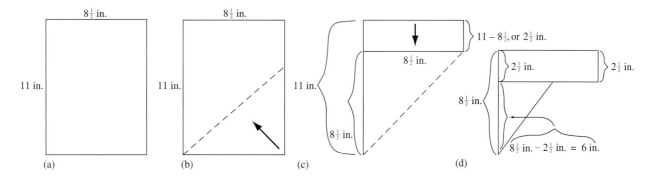

28 CHAPTER 1 *An Introduction to Problem Solving*

With the fold in Figure 1-20(b), we obtain a length of $11 - 8\ 1/2$, or $2\ 1/2$ in. With the fold in Figure 1-20(c), we obtain $8\ 1/2 - 2\ 1/2$, or 6 in.

• **Looking Back** Though there are other ways to solve this problem, drawing a diagram showing the folds of the paper combines notions of geometry and gives a way to fold a square from a rectangle. An entirely different way to solve the problem is possible if we have more than one sheet of paper. That problem is left to the reader.

Understanding and Writing Equations — 10-4

▶ **Lesson Link** You've written expressions to describe sequences. Now you'll see how to write equations to describe the relationship between two quantities.

You'll Learn …
■ to write an equation from a table of values

… How It's Used
Equations allow spreadsheets to calculate values automatically.

Explore Showing Relationships with Tables

Faster than a Speeding Beetle?

Some of the fastest-moving insects are beetles. The table shows how fast one type of beetle can crawl.

Time (sec)	1	2	3	4	5	…	t
Distance (in.)	4	8	12	16	20	…	?

1. Describe any patterns you see in the table. Write an expression using *t* for the final box in the distance row.

2. Use words to describe the relationship between the distance traveled and the time.

3. If this beetle crawled for 60 seconds, how far would it go?

4. How long would it take this beetle to crawl 36 inches?

5. Explain how you found your answers to Steps 3 and 4. What is different about the questions asked in these steps?

Learn Understanding and Writing Equations

When two expressions name the same quantity, we can link them with an equation. Writing an equation is an important way to show a relationship between variables.

The geometric formulas you've worked with are equations. For instance, since the perimeter of a square is four times the length of a side, we write $p = 4s$. The equal sign shows that p and $4s$ always have the same value.

$$p = 4s$$

10-4 • *Understanding and Writing Equations* **495**

> **NOW TRY THIS 1-7**
>
> ● An elevator stopped at the middle floor of a building. It then moved up 4 floors and stopped. It then moved down 6 floors, and then moved up 10 floors and stopped. The elevator was now three floors from the top floor. How many floors does the building have? ●

Strategy: Guess and Check

In the strategy of *guess and check,* we first guess at a solution using as reasonable a guess as possible. Then we check to see whether the guess is correct. If not, the next step is to learn as much as possible about the solution based on the guess before making the next guess. This strategy can be regarded as a form of trial and error, where the information about the error helps us choose what trial to make next.

The guess-and-check strategy is often used when a student does not know how to solve the problem more efficiently or if the student does not yet have the tools to solve the problem in a faster way. The guess-and-check strategy is demonstrated in the Postcards and Letters problem.

Problem Solving   Postcards and Letters

Marques mailed 32 postcards and letters. The bill at the post office was $8.35. The postcards cost $.20 each and letters cost $.33 each. How many of each kind did he mail?

- **Understanding the Problem**   The total bill for postage for 32 postcards and letters was $8.35. Postcards cost $.20 to mail and letters cost $.33. We are to determine how many postcards and how many letters were mailed.

- **Devising a Plan**   When we have a problem with a limited number of possible answers, *guess and check* is a possible strategy. We can make a guess and then see how close our guess is to the correct answer. Then this information can be used to make a better guess the next time. Suppose we guess that the number of postcards is 10. This implies that the number of letters is $32 - 10$, or 22. If this were true, then the total bill would be

$$(10 \cdot \$.20) + (22 \cdot \$.33) = \$9.26.$$

This answer tells us that 10 postcards and 22 letters would cost more than Marques spent and that the next guess should involve more postcards and fewer letters. Suppose the next guess is 20 postcards and therefore 12 letters. Then the amount spent is

$$(20 \cdot \$.20) + (12 \cdot \$.33) = \$7.96.$$

This amount is less than the required $8.35, but we are getting closer. At this point, we know that the correct number of postcards is between 10 and 20. We also know that the number of postcards is closer to 20 than to 10. Why? We can continue guessing and checking to find the correct answer.

- **Carrying Out the Plan** If we continue guessing and checking in this manner, we will find that

$$(17 \cdot \$0.20) + (15 \cdot \$0.33) = \$8.35.$$

Therefore the number of postcards is 17 and the number of letters is 15.

- **Looking Back** Remember to check the answer in terms of the original conditions in order to see if the solution is correct. Is the answer found the only answer possible?

A different method of finding the answer to this problem involves using a graphing calculator and drawing a set of graphs. (Information on use of a graphing calculator is found in Appendix II.) Thinking about the problem algebraically, we use the strategy of *writing an equation*. We do not know how many postcards were used, but we know that the total number of postcards and letters is 32. Thus if x represented the number of postcards, then $32 - x$ would be the number of letters. The cost of mailing is $0.20 for a postcard and $0.33 for a letter. Thus, the total cost, y, of mailing both the postcards and the letters is $0.20x + \$0.33(32 - x)$. Since we know that this total cost is $8.35, we could solve the equation below:

$$\$0.20x + \$0.33(32 - x) = \$8.35$$

Using the graphing calculator, we could graph both sides of the equation as lines using the following:

$$y1 = 0.20x + 0.33(32 - x)$$
$$y2 = 8.35$$

The graph is in Figure 1-21.

Figure 1-21

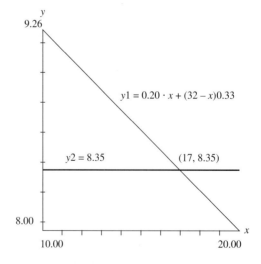

By graphing both of these and using the trace function, we can estimate that the number of postcards is 17 and then the number of letters is $32 - 17$, or 15.

Yet another way to solve the problem is to realize that 32 items cost at least $.20, for a total of $6.40. That leaves $1.95 extra to come from the $.13 per letter. Because $1.95/.13 = 15$, there are 15 letters.

SECTION 1-2 *Mathematics and Problem Solving* **31**

Strategy: Work Backward

In some problems, it is easier to start with the result and to work backward. This is demonstrated on the following student page from the *Scott Foresman-Addison Wesley Problem Solving Handbook,* 1999.

Problem Solving STRATEGIES
- Look for a Pattern
- Make an Organized List
- Make a Table
- Guess and Check
- Work Backward
- Use Logical Reasoning
- Draw a Diagram
- Solve a Simpler Problem

Work Backward

A problem may give you the result of a series of steps and ask you to find the initial value. To solve the problem, you can work your way backward, step by step, to the beginning. ◄

Example

The Astro calculator was introduced in 1993. In 1994, the price was raised $8. In 1995, the price was lowered $14 because of lower demand. In 1996, the price was halved to $18 because of competition from a new calculator. Find the original price.

The problem describes three steps occurring in order (price raised, price lowered, price halved). It also tells you the end result (the final price was $18). To solve the problem, work backward to the beginning.

Step	What Happened	Conclusion
3	The price was halved to $18.	Before this step, the price was *twice* $18, or $36.
2	The price was lowered $14 to $36.	Before this step, the price was $14 *greater than* $36, or $50.
1	The price was raised $8 to $50.	Before this step, the price was $8 *less than* $50, or $42.

The original price of the calculator was $42.

Try It

a. At a sale, T-shirts were marked down $4. Pei bought 3 T-shirts. A sales tax of $2 was added to her bill, bringing the total cost to $26. Find the price of T-shirts before the sale.

b. Mount Whitney, Harney Peak, Mount Davis, and Woodall Mountain are the highest points in California, South Dakota, Pennsylvania, and Mississippi, respectively. Mount Whitney is about twice as tall as Harney Peak. Mount Davis is about 4020 feet less tall than Harney Peak and 4 times as tall as 806 ft Woodall Mountain. About how tall is Mount Whitney?

Strategy: Use Indirect Reasoning

To show that a statement is true, it is sometimes easier to show that it is impossible for the statement to be false. This can be done by showing that if the statement were false, something contradictory or impossible would follow. This approach is useful when it is difficult to start a direct argument and when negating the given statement gives us something tangible with which to work. An example follows.

Problem Solving   **Checkerboard Problem**

In Figure 1-22, you are given a checkerboard with the two squares on opposite corners removed and a set of dominoes such that each domino can cover 2 adjacent squares on the board. Can the dominoes be arranged in such a way that all the remaining squares on the board can be covered with no dominoes hanging off the board? If not, why not?

Figure 1-22

- **Understanding the Problem**   Two red spaces on opposite corners were removed from the checkerboard in Figure 1-22. We are asked whether it is possible to cover the remaining 62 squares with dominoes the size of 2 squares.

- **Devising a Plan**   If we try to cover the board in Figure 1-22 with dominoes, we will find that the dominoes do not fit and some squares will remain uncovered. To show that there is no way to cover the board with dominoes, we use *indirect reasoning*. If the remaining 62 squares could be covered with dominoes, it would take 31 dominoes to accomplish the task. We want to show that this implies something impossible.

- **Carrying Out the Plan**   Each domino must cover 1 black and 1 red square. Hence, 31 dominoes would cover 31 red and 31 black squares. This is impossible, however, because the board in Figure 1-22 has 30 red and 32 black squares. Consequently, our assumption that the board in Figure 1-22 can be covered with dominoes is wrong.

- **Looking Back**   The counting of black and red squares implies that if we remove any number of squares from a checkerboard so that the number of remaining red squares differs from the number of remaining black squares, the board cannot be covered with dominoes. (Do you see why?) We could also investigate what happens when two squares of the same color are removed from an 8-by-7 board and other sized boards. Also, is it always possible to cover the remaining board if two squares of opposite colors are removed?

Strategy: Use Direct Reasoning

Problem Solving — **Checker Games**

Each of two people won three games of checkers. It is possible that only five games were played?

• **Solution** To understand this problem, we know that each person won 3 games. *Reasoning directly* we see that if the people each won three games and they played each other, then there would have had to be six games played. Thus they could not each play each other in all games and have three wins each. Could they in fact each win three games while playing only five total games and not have played each other? The answer is no and the situation is impossible.

ONGOING ASSESSMENT 1-2

1. Use the approach in Gauss's Problem on page 20 to find the following sums (do not use formulas):
   a. $1 + 2 + 3 + 4 + \ldots + 99$
   b. $1 + 2 + 3 + 4 + \ldots + n$ where n is odd
   c. $1 + 3 + 5 + 7 + \ldots + 1001$
2. How many cuts does it take to divide a log into
   a. 5 equal-sized pieces?
   b. 6 equal-sized pieces?
   c. n equal-sized pieces?
3. How many ways can you make change for a $50 bill using $5, $10, and $20 bills?
4. Cookies are sold singly or in packages of 2 or 6. How many ways can you buy a dozen cookies?
5. The sign says you are leaving Missoula, Butte is 120 miles away, and Bozeman is 200 miles away. There is a rest stop halfway between Butte and Bozeman. How far is the rest stop from Missoula?
6. Alababa, Bubba, Cory, and Dandy are in a horse race. Bubba is the slowest, Cory is faster than Alababa but slower than Dandy. Name the finishing order of the horses.
7. Frankie and Johnny began reading a novel on the same day. Frankie reads 8 pages a day and Johnny reads 5 pages a day. If Frankie is on page 72, what page is Johnny on?
8. How many squares are in the following figure?

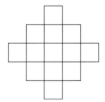

9. What is the largest sum of money—all in coins and no silver dollars—that you could have in your pocket without being able to give change for a dollar, a half-dollar, a quarter, a dime, or a nickel?

10. If 0.2 oz of catsup is used on each of 22 billion hamburgers, how many 16-oz bottles of catsup are needed?
11. a. Place the digits 1, 2, 4, 5, and 7 in the following boxes so that in (i) the greatest product is obtained and in (ii) the greatest quotient is obtained:

   b. Use the same digits as in (a) to obtain (i) the least product and (ii) the least quotient.
12. How many terms are there in the following sequence? 1, 8, 15, 22, ... , 113
13. Suppose you could spend $10 every minute, night and day. How much could you spend in a year? (Assume there are 365 days in a year.)
14. Refer to the following pattern and answer questions (a) through (c):

$$1 = 2^1 - 1$$
$$1 + 2 = 2^2 - 1$$
$$1 + 2 + 2^2 = 2^3 - 1$$
$$1 + 2 + 2^2 + 2^3 = 2^4 - 1$$
$$1 + 2 + 2^2 + 2^3 + 2^4 = 2^5 - 1$$

   a. Write a simpler expression for $1 + 2 + 2^2 + 2^3 + 2^4 + 2^5$. Justify your answer.
   b. Write a simpler expression for the sum in the nth row in the above pattern.
   c. Use a spreadsheet or a calculator to check your answer in (b) for $n = 15$.
15. How many four-digit numbers have the same digits as 1993?
16. A compass and a ruler together cost $4. The compass costs 90¢ more than the ruler. How much does the compass cost?
17. Two houses on the same street are separated by a large, empty field. The first house is numbered 29, and the other is numbered 211. An architect is designing 13 new houses to be built between the two existing houses.

a. What should the numbers of the new houses be if along with the existing houses, the numbers need to form an arithmetic sequence?
b. What is the difference of this sequence?

18. Same-sized cubes are glued together to form a staircaselike sequence of solids as shown:

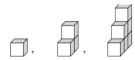

All of the faces of the cubes not glued together need to be painted. How many squares will need to be painted in
(a) the 100th solid? (b) the nth solid?

19. Marc goes to the store with exactly $1.00 in change. He has at least one of each coin less than a half-dollar coin, but he does not have a half-dollar coin.
    a. What is the least number of coins he could have?
    b. What is the greatest number of coins he could have?

20. A farmer needs to fence a rectangular piece of land. She wants the length of the field to be 80 ft longer than the width. If she has 1080 ft of fencing material, what should the length and the width of the field be?

21. Find a 3-by-3 magic square using the numbers 3, 5, 7, 9, 11, 13, 15, 17, and 19.

22. Find the following sums:
    a. $1 + 6 + 11 + 16 + 21 + \ldots + 1001$
    b. $3 + 7 + 11 + 15 + 19 + \ldots + 403$

★23. **a.** Using the existing lines on the checkerboard shown below, how many squares are there?

b. If the number of rows and columns of the checkerboard is doubled, is the number of squares doubled? Justify your answer.

24. Eight marbles look alike, but one is slightly heavier than the others. Using a balance scale, explain how you can determine the heavier one in exactly
    a. 3 weighings   **b.** 2 weighings

25. **a.** Find the sum of all the numbers in the following array.

1	2	3	4	5	6	…	100
2	4	6	8	10	12	…	200
3	6	9	12	15	18	…	300
			⋮				
100	200	300	400	500	600	…	100 · 100

    b. Generalize part (a) to a similar array in which each row has n numbers and there are n rows.

26. Use the strategy of *indirect reasoning* to justify these:
    a. If the product of two positive numbers is greater than 82, then at least one of the numbers is greater than 9.
    b. If the product of two positive numbers is greater than 81, then at least one of the numbers is greater than 9.

Communication

27. A different version of the story of Gauss's computing $1 + 2 + 3 + \ldots + 100$ reports that he simply listed the numbers in the following way to discover the sum:

    $$\begin{array}{r}1 + 2 + 3 + 4 + 5 + \ldots + 98 + 99 + 100 \\ 100 + 99 + 98 + 97 + 96 + \ldots + 3 + 2 + 1 \\ \hline 101 + 101 + 101 + 101 + 101 + \ldots + 101 + 101 + 101\end{array}$$

    Does this method give the same answer? Discuss the advantages of this method over the one described in the text.

28. Explain why it is impossible to have a 3-by-3 magic square with numbers 1, 3, 4, 5, 6, 7, 8, 9, and 10.

Open-Ended

29. How many breaths do you take in a year?
30. Choose a problem-solving strategy and make up a problem that would use this strategy. Write the solution using Polya's four-step approach.

Cooperative Learning

31. The distance around the world is approximately 40,000 km. Approximately how many people of average size in your group would it take to stretch around the world if they were holding hands?

32. Work in pairs on the following version of a game called NIM. A calculator is needed for each pair.
    a. Player 1 presses $\boxed{1}$ and $\boxed{+}$ or $\boxed{2}$ and $\boxed{+}$. Player 2 does the same. The players take turns until the target number of 21 is reached. The first player to make the display read 21 is the winner. Determine a strategy for deciding who always wins.
    b. Try a game of NIM using the digits 1, 2, 3, and 4, with a target number of 104. The first player to reach 104 wins. What is the winning strategy?
    c. Try a game of NIM using the digits 3, 5, and 7, with a target number of 73. The first player to exceed 73 loses. What is the winning strategy?
    d. Now play Reverse NIM with the keys $\boxed{1}$ and $\boxed{2}$. Instead of $\boxed{+}$, use $\boxed{-}$. Put 21 on the display. Let the target number be 0. Determine a strategy for winning Reverse NIM.
    e. Try Reverse NIM using the digits 1, 2, and 3 and starting with 24 on the display. The target number is 0. What is the winning strategy?
    f. Try Reverse NIM using the digits 3, 5, and 7 and starting with 73 on the display. The first player to display a negative number loses. What is the winning strategy?

Review Problems

33. List three terms to complete possible patterns:
    a. 3, 6, 9, 12, 15, 18, _____, _____, _____
    b. 1, 2, 3, 2, 9, 2, 27, 2, 81, 2, _____, _____, _____
34. Find the *n*th term for the sequence 22, 32, 42, 52, ...
35. How many terms are in the sequence 3, 7, 11, 15, 19, ... , 83?
36. Find the sums of the terms in the sequence in problem 35.

37. Examine the following:
    $2 \cdot 9 = 18.$ $5 \cdot 9 = 45.$
    $3 \cdot 9 = 27.$ $6 \cdot 9 = 54.$
    $4 \cdot 9 = 36.$

    a. What patterns do you see? Check to see if your patterns work for $7 \cdot 9, 8 \cdot 9, 9 \cdot 9$, and $10 \cdot 9$.
    b. Explain how the patterns you noticed might be helpful in remembering the basic multiplication table for 9s.

BRAIN TEASER

Ten women are fishing all in a row in a boat. One seat in the center of the boat is empty. The 5 women in the front of the boat want to change seats with the 5 women in the back of the boat. A person can move from her seat to the next empty seat or she can step over one person without capsizing the boat. What is the minimum number of moves needed for the 5 women in front to change places with the 5 in back?

LABORATORY ACTIVITY

Place a half-dollar, quarter, and nickel in position *A* as shown in Figure 1-23. Try to move these coins, one at a time, to position *C*. At no time may a larger coin be placed on a smaller coin. Coins may be placed in position *B*. How many moves does this take? Now add a penny to the pile and see how many moves this takes. This is a simple case of the famous Tower of Hanoi problem, in which ancient Brahman priests were required to move a pile of 64 disks of decreasing size, after which the world would end. How long will this take at a rate of one move per second?

Figure 1-23

Section 1-3 — Algebraic Thinking

One of the strategies of the previous section is *write an equation*. Writing an equation is normally a part of algebraic thinking, as are generalizing a pattern, writing an expression for a function rule, defining subtraction in terms of addition, graphing equations, and using recursive patterns and formulas. Because algebraic thinking is so important in mathematics at all levels, from the early grades on, we chose to include a separate section on it. (This section is not intended to be a review of a high school algebra course.)

For PreK–2 the *Principles and Standards* note,

 The recognition, comparison, and analysis of patterns are important components of a student's intellectual development. When students notice that operations seem to have particular properties, they are beginning to think algebraically (p. 91).

At the 3–5 level the *Principles and Standards* say,

 ... algebraic ideas should emerge and be investigated as students identify or build numerical and geometric patterns ... (p. 159).

At the 6–8 level,

Students in the middle grades should learn algebra both as a set of concepts and competencies tied to the representation of quantitative relationships and as a style of mathematical thinking for formalizing patterns. In the middle grades, students should work more frequently with algebraic symbols than in lower grades. It is essential that they become comfortable in relating symbolic expressions containing variables to verbal, tabular, and graphical representations of numerical and quantitative relationships (p. 223).

Developing Algebraic Thinking Skills

To apply algebra in solving problems, you frequently need to translate given information into a symbolic expression involving variables designated by letters or words. In all such examples, you may name the variables as you choose. Frequently, letters are chosen over words to make it easy to avoid confusion if letters are repeated in words or to make expressions shorter.

Example 1-7 In each of the following, translate the given information into a symbolic expression involving quantities designated by letters:

a. One weekend, a music store sold twice as many CDs as cassettes and 25 fewer records than CDs. If the music store sold c cassettes, how many records and CDs did it sell?
b. French fries have about 12 calories apiece. A hamburger has about 600 calories. Akiva is on a diet of 2000 calories per day. If he ate f french fries and one hamburger, how many more calories can he consume that day?
c. First class postage in 1999 is 33¢ for the first ounce and 22¢ for each additional ounce. At this rate, what is the cost of mailing a letter that weighs z oz?

Solution a. Because c cassettes were sold, twice as many CDs as cassettes implies $2c$ CDs. Thus 25 fewer records than CDs implies $2c - 25$ records.
b. First, find how many calories Akiva consumed eating f french fries and one hamburger. Then, to find how many more calories he can consume, subtract this expression from 2000.

| 1 french fry | 12 calories |
| f french fries | $12f$ calories |

Therefore the number of calories in f french fries and one hamburger is

$$600 + 12f.$$

The number of calories left for the day is $2000 - (600 + 12f)$, or $2000 - 600 - 12f$, or $1400 - 12f$.

c. If a letter weighs z ounces, the first ounce costs 33¢ and the remaining $(z - 1)$ oz cost 22¢ each. The cost of the $(z - 1)$ oz therefore is $22(z - 1)$ cents. We have the following:

Cost of mailing a z-ounce letter = cost of first ounce + cost of next $(z - 1)$ oz

$$= 33 + 22(z - 1) \text{ cents}$$
$$= 11 + 22z \text{ cents}$$

HISTORICAL NOTE

The word *algebra* comes from the Arabic book *Al-jabr wa'l muqabalah* written by Mohammed al-Khowârizmî (ca. 825). Algebra was introduced in Europe in the thirteenth and fourteenth centuries by Leonardo of Pisa (also called Fibonacci). Algebra was occasionally referred to as *Ars Magna*, or "the great art." Both Diophantus (ca. A.D. 250) and Francois Viète (1540–1603) have been called "fathers of algebra." Little is known about Diophantus, a Greek, except that he is supposed to have lived to be 84 years old and that he wrote *Arithmetica*, a treatise originally in thirteen books. Viète was a French lawyer who devoted his leisure time to mathematics. Not liking the word *algebra*, he referred to the subject as "the analytic art."

Algebraic thinking is frequently needed to understand how simple number tricks work, as seen in Example 1-8.

Example 1-8 A teacher instructed her class as follows:

Take any number and add 15 to it. Now multiply that sum by 4. Next subtract 8 and divide the difference by 4. Now subtract 12 from the quotient and tell me the answer, I will tell the original number.

Analyze the instructions to see how the teacher was able to determine the original number.

Solution Translate the information into an algebraic form.

Instructions	Discussion	Symbols
Take any number.	Since any number is used, we need a variable to represent the number. Let n be that variable.	n
Add 15 to it.	We are told to add 15 to "it." "It" refers to the variable n.	$n + 15$
Multiply that sum by 4.	We are told to multiply "that sum" by 4. "That sum" is $n + 15$.	$4(n + 15)$
Subtract 8.	We are told to subtract 8 from the product.	$4(n + 15) - 8$
Divide the difference by 4.	The difference is $4(n + 15) - 8$. Divide it by 4.	$\dfrac{4(n + 15) - 8}{4}$
Subtract 12 from the quotient and tell me the answer.	We are told to subtract 12 from the quotient.	$\dfrac{4(n + 15) - 8}{4} - 12$

Having translated what the teacher told the class to do results in the algebraic expression $\dfrac{4(n + 15) - 8}{4} - 12$. We are also told that we have to tell the teacher the answer obtained and then she produces the original number. Let's use the strategy of working backward to see if we can tell what happens. Suppose we tell the teacher that our final result is r. Think about how r was obtained. Just before we told the teacher "r," we had subtracted 12. To reverse that operation, we could add 12 to obtain $r + 12$. Prior to that we had divided by 4. To reverse that, we could multiply by 4 to obtain $4r + 48$. To get that result we had subtracted 8, so that now we add 8 to obtain $4r + 56$. Just previous to that we had multiplied by 4 so that now we divide $4r + 56$ by 4 to obtain $r + 14$. The first operation had been to add 15 so now we subtract 15 from $r + 14$ to get $r - 1$. Thus the teacher knows when we tell her that our final result is r, it is 1 less than the number with which we started, or the number with which we started, n, is $1 +$ the result.

This can be shown as follows:

$$\frac{4(n + 15) - 8}{4} - 12 = \frac{4n + 60 - 8 - 4 \cdot 12}{4}.$$
$$= \frac{4n + 4}{4}.$$
$$= \frac{4(n + 1)}{4}.$$
$$= n + 1.$$

The solutions of many problems involve the strategy of *writing an equation*. Thus we first need a basic knowledge of solving equations.

Properties of Equations

To solve equations, we need several properties of equality. Children discover many of these by using a balance scale. For example, consider two weights of amounts a and b on the balances, as in Figure 1-24(a). If the balance is level, then $a = b$. When we add an equal amount of weight c to both sides, the balance is still level, as in Figure 1-24(b).

Figure 1-24

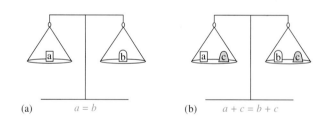

(a) $a = b$ (b) $a + c = b + c$

This demonstrates that if $a = b$, then $a + c = b + c$, which is the addition property of equality.

Similarly, if the scale is balanced with amounts a and b, as in Figure 1-25(a), and we put additional a's on one side and an equal number of b's on the other side, the scale remains level, as in Figure 1-25(b).

Figure 1-25

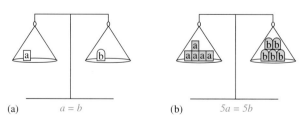

(a) $a = b$ (b) $5a = 5b$

Figure 1-25 suggests that if c is any real number and $a = b$, then $ac = bc$, which is the multiplication property of equality. These properties are summarized next.

Properties

The addition property of equality For any numbers a, b, and c, if $a = b$, then $a + c = b + c$.
The multiplication property of equality For any numbers a, b, and c, if $a = b$, then $ac = bc$.

The properties imply that we may add the same number to both sides of an equation or multiply both sides of the equation by the same number without affecting the equality. A new statement results from reversing the order of the *if* and *then* parts of the addition property of equality. The new statement is the *converse* of the original statement. In the case of the addition property, the converse is a true statement. The converse of the multiplication property of equality is also true when $c \neq 0$. These properties are summarized next.

Cancellation Properties of Equality

1. For any numbers a, b, and c, if $a + c = b + c$, then $a = b$.
2. For any numbers a, b, and c, with $c \neq 0$ if $ac = bc$, then $a = b$.

substitution property

Equality is not affected if we substitute a number for its equal. This property is referred to as the **substitution property.** Examples of substitution follow:

1. If $a + b = c + d$ and $d = 5$, then $a + b = c + 5$.
2. If $a + b = c + d$, if $b = e$, and if $d = f$, then $a + e = c + f$.

Solving Equations

Part of algebraic thinking involves operations on numbers and other elements represented by symbols. Finding solutions to equations is one part of algebra.

To solve equations, we may use the properties of equality developed earlier. A balance scale can be used to demonstrate solving equations. Consider $3x - 14 = 1$. Put the equal expressions on the opposite pans of the balance scale. Since the expressions are equal, the pans should be level, as in Figure 1-26.

Figure 1-26

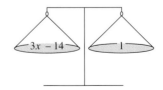

To solve for x, we use the properties of equality to manipulate the expressions on the scale so that after each step, the scale remains level and, at the final step, only an x remains on one side of the scale. The number on the other side of the scale represents the solution to the original equation. To find x in the equation of Figure 1-26, consider the scales pictured in successive steps in Figure 1-27. In Figure 1-27, each successive scale represents an equation that is equivalent to the original equation; that is, each has the same solution as the original. The last scale shows $x = 5$. To check that 5 is the correct solution, we substitute 5 for x in the original equation. Because $3 \cdot 5 - 14 = 1$ is a true statement, 5 is the solution to the original equation.

Figure 1-27

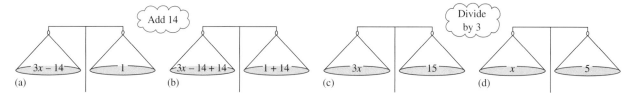

Example 1-9

Solve each of the following for x:

a. $x + 4 = -6$.
b. $-x - 5 = 8$.

Solution The solutions that follow show all the steps in the process:

a.
$$x + 4 = -6.$$
$$(x + 4) + {}^-4 = -6 + -4.$$
$$x + (4 + -4) = -6 + -4.$$
$$x + 0 = -10.$$
$$x = -10.$$

b.
$$-x - 5 = 8.$$
$$(-x + -5) + 5 = 8 + 5.$$
$$-x + (-5 + 5) = 13.$$
$$-x + 0 = 13.$$
$$-x = 13.$$
$$(-x)(-1) = 13(-1).$$
$$x = -13.$$

HISTORICAL NOTE

Mary Fairfax Somerville (1780–1872) was born in Scotland of upper-class parents. Her introduction to algebra came at about age 13 while reading a ladies' fashion magazine that contained some puzzles. Although not allowed to study mathematics formally, at age 27, widowed, and with two children, she bought and studied a set of mathematics books. In her autobiography, she wrote, "I was sometimes annoyed when in the midst of a difficult problem someone would enter and say, 'I have come to spend a few hours with you.'" Shortly before her death, she wrote, "I am now in my ninety-second year, ... , I am extremely deaf, and my memory of ordinary events, and especially of the names of people, is failing, but not for mathematical and scientific subjects. I am still able to read books on the higher algebra for four or five hours in the morning and even to solve the problems. Sometimes I find them difficult, but my old obstinacy remains, for if I do not succeed today, I attack them again tomorrow."

Application Problems

The following simple model demonstrates a method for solving application problems. Formulate the problem as a mathematical problem, solve the mathematical problem, and then interpret the solution in terms of the original problem.

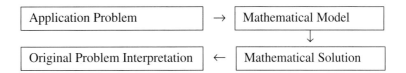

At the third-grade level, an example of this model appears as follows:

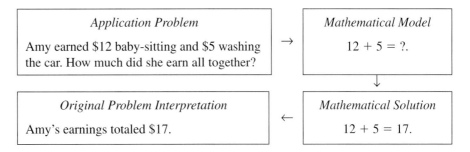

We can apply Polya's four-step problem-solving process to solving word problems in which the use of algebraic thinking is appropriate.

In Understanding the Problem, we identify what is given and what is to be found. In Devising a Plan, we assign letters to the unknown quantities and translate the information in the problem into a model involving equations. In Carrying Out the Plan, we solve the equations or inequalities. In Looking Back, we interpret and check the solution in terms of the original problem.

In the following problems, we demonstrate Polya's four-step problem-solving process.

Problem Solving **Overdue Books**

Bruno has five books overdue at the library. The fine for overdue books is 10¢ a day per book. He remembers that he checked out an astronomy book a week earlier than four novels. If his total fine was $8.70, how long was each book overdue?

- **Understanding the Problem** Bruno has five books overdue. He checked out an astronomy book seven days earlier than the four novels, so the astronomy book is overdue seven days more than the novels. The fine per day for each book is 10¢, and the total fine was $8.70. We need to find out how many days each book is overdue.

- **Devising a Plan** Let d be the number of days that each of the four novels is overdue. The astronomy book is overdue seven days longer, that is, $d + 7$ days. To *write an equation* for d, we express the total fine in two ways. The total fine is $8.70. This fine in cents equals the fine for the astronomy book plus the fine for the four novels.

Fine for each of the novels = $\underbrace{\text{fine per day}}_{10} \cdot \underbrace{\text{times the number of overdue days}}_{d}$.

Fine for the four novels = $\underbrace{\text{1 day's fine for 4 novels}}_{4 \cdot 10}$ times $\underbrace{\text{number of overdue days}}_{d}$.

$= (4 \cdot 10)d.$
$= 40d.$

Fine for the astronomy book = $\underbrace{\text{fine per day}}_{10}$ times $\underbrace{\text{the number of overdue days}}_{(d + 7)}$.

$= 10 \cdot (d + 7)$ (in cents).

Because each of the above expressions is in cents, we need to write the total fine of $8.70 as 870¢ to produce the following:

Fine for the 4 novels + fine for the astronomy book = total fine.
$40d \quad\quad + \quad\quad 10(d + 7) \quad\quad = 870.$

- **Carrying Out the Plan** Solve the equation for d.

$$40d + 10(d + 7) = 870.$$
$$40d + 10d + 70 = 870.$$
$$50d = 870 - 70.$$
$$50d = 800.$$
$$d = 16.$$

Thus each of the 4 novels was 16 days overdue, and the astronomy book was overdue $d + 7$, or 23, days.

- **Looking Back** To check the answer, follow the original information. Each of the four novels was 16 days overdue, and the astronomy book was 23 days overdue. Because the fine was 10¢ per day per book, the fine for each of the novels was $16 \cdot 10$¢, or 160¢. Hence, the fine for all four novels was $4 \cdot 160$¢, or 640¢. The fine for the astronomy book was $23 \cdot 10$¢, or 230¢. Consequently, the total fine was 640¢ + 230¢, or 870¢, which agrees with the given information of $8.70 as the total fine.

Problem Solving **Newspaper Delivery**

In a small town, three children deliver all the newspapers. Abby delivers three times as many papers as Bob, and Connie delivers 13 more than Abby. If the three children delivered a total of 496 papers, how many papers does each deliver?

- **Understanding the Problem** The problem asks for the number of papers that each child delivers. It gives information that compares the number of papers that each child delivers as well as the total number of papers delivered in the town.

- **Devising a Plan** Let a, b, and c be the number of papers delivered by Abby, Bob, and Connie, respectively. We translate the given information into *equations* as follows:

Abby delivers three times as many papers as Bob: $a = 3b$.
Connie delivers 13 more papers than Abby: $c = a + 13$.
Total delivery is 496: $a + b + c = 496$.

To reduce the number of variables, substitute $3b$ for a in the second and third equations:

$$c = a + 13 \text{ becomes } c = 3b + 13.$$
$$a + b + c = 496 \text{ becomes } 3b + b + c = 496.$$

Next, make an equation in one variable, b, by substituting $3b + 13$ for c in the equation $3b + b + c = 496$, solve for b, and then find a and c.

- **Carrying Out the Plan**

$$3b + b + 3b + 13 = 496.$$
$$7b + 13 = 496.$$
$$7b = 483.$$
$$b = 69.$$

Thus $a = 3b = 3 \cdot 69 = 207$. Also, $c = a + 13 = 207 + 13 = 220$. So, Abby delivers 207 papers, Bob delivers 69 papers, and Connie delivers 220 papers.

- **Looking Back**   To check the answers, follow the original information, using $a = 207$, $b = 69$, and $c = 220$. The information in the first sentence, "Abby delivers three times as many papers as Bob" checks, since $207 = 3 \cdot 69$. The second sentence, "Connie delivers 13 more papers than Abby" is true because $220 = 207 + 13$. The information on the total delivery checks, since $207 + 69 + 220 = 496$.

ONGOING ASSESSMENT 1-3

1. In the following, write an expression in terms of the given variable that represents the indicated quantity (that is, write the quantity as a function of the given variable).
   a. The distance traveled at a constant speed of 60 mph during t hours
   b. The cost of having a plumber spend h hours at your house if the plumber charges $20 for coming to the house and $25 per hour for labor
   c. The amount of money in cents in a jar containing d dimes and some nickels and quarters, if there are 3 times as many nickels as dimes and twice as many quarters as nickels
   d. The sum of three consecutive integers if the least integer is x
   e. The amount of bacteria after n minutes if the initial amount of bacteria is q and the amount of bacteria doubles every minute. (*Hint:* The answer should contain q as well as n.)
   f. The temperature after t hours if the initial temperature is 40°F and each hour it drops by 3°F
   g. Pawel's total salary after 3 yr if the first year his salary was s dollars, the second year it was $5000 higher, and the third year it was twice as much as the second year
   h. The sum of three consecutive odd natural numbers if the least is x
   i. The sum of three consecutive natural numbers if the middle is m

2. Pluto, once thought to be the farthest planet from the sun in our solar system, was also the smallest. The following table gives the weight P on Pluto for a given weight E on Earth measured in the same units.

Weight on Earth (E)	1	2	3	4	5	10	100
Weight on Pluto (P)	0.04	0.08	0.12	0.16	0.2	0.4	4

   Based on the information given in the table, answer each of the following:
   a. Find an equation for P in terms of E.
   b. Express E in terms of P.
   c. What is Debbie's weight on Pluto if her weight on Earth is 135 lb?
   d. Find the weight on Earth of an object that weighs 100 lb on Pluto.

3. If 1 U.S. dollar is worth 1.35 Swiss francs and 1 Swiss franc is worth 1.22 Dutch gilders, find a formula that will give a U.S. tourist the amount of Dutch gilders g for U.S. dollars d.

4. Joel is designing squares made of matchsticks. The following are 1 × 1, 2 × 2, and 3 × 3 such squares. How many matchsticks will he need for
   a. a 10 × 10 square?
   b. an $n \times n$ square?

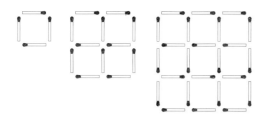

5. The formula for converting degrees Celsius (C) to degrees Fahrenheit (F) is $F = \left(\frac{9}{5}\right) \cdot C + 32$. Samantha reads that the temperature is 32°C in Spain. What is the Fahrenheit temperature?
6. To convert a temperature on the Celsius (C) scale to Kelvin (K) scale, you need to add 273.15 to the Celsius temperature.
   a. Write a formula for K in terms of C.
   b. Write a formula for converting from the Kelvin scale to the Celsius scale.
7. Write an equation relating the variables described in each of the following situations:
   a. The pay P for t hours if you are paid $8 an hour
   b. The pay P for t hours if you are paid $15 for the first hour and $10 for each additional hour
   c. The total pay P for a visit and t hours of gardening if you are paid $20 for the visit and $10 for each hour of gardening
   d. The total cost C of membership in a health club that charges a $300 initiation fee and $4 for each of n days attended
   e. The cost C of renting a midsized car for 1 day of driving m miles if the rent is $30 per day plus 35¢ per mile.
8. Solve each of the following:
   a. $3 - x = -15$
   b. $-x - 3 = 15$
9. A teacher instructed her class as follows: *Take any number, multiply it by 3, add 49, and divide the result by 7. Subtract 7 from the quotient, divide the new result by 3, and tell me your answer. I will tell you the original number.* To determine each student's original number, the teacher multiplied each answer by 7. Explain how the teacher was able to tell each student's original number.
10. David has three times as much money as Rick. Together, they have $400. How much does each have?
11. For a certain event, 812 tickets were sold for a total of $1912. If students paid $2 per ticket and nonstudents paid $3 per ticket, how many student tickets were sold?
12. A man left an estate of $64,000 to three children. The eldest child received three times as much as the youngest. The middle child received $14,000 more than the youngest. How much did each child receive?

Communication

13. Students were asked to write an algebraic expression for the sum of three consecutive natural numbers. One student wrote $x + (x + 1) + (x + 2) = 3x + 3$. Another wrote $(x - 1) + x + (x + 1) = 3x$. Explain who is correct and why.
14. A rod of length l was cut into two pieces of equal length. Then one of the pieces was halved again (and so on.) Find the length of the smallest piece after n cuts in terms of l and n. Explain your reasoning.
15. a. Find the sum of the following five expressions:

    $$x - 2, x - 1, x, x + 1, x + 2$$

    b. Based on your answer to part (**a**), state in words the property that the answer demonstrates.

Open-Ended

16. In Example 1-8, a teacher instructed her class to take any number and perform a series of computations using that number. The teacher was able to tell each student's original number by subtracting 1 from the student's answer. Create similar instructions for students so that the teacher needs only to do the following to obtain the student's original number:
    a. Add 1 to the answer.
    b. Multiply the answer by 2.
    c. Multiply the answer times itself.

Cooperative Learning

17. Examine several elementary school textbooks for grades 1 through 4 and report on which algebraic concepts are introduced in each and how they are introduced.

Review Problems

18. List the terms to continue a possible pattern in the following sequences:
    a. 7, 14, 21, 28, …
    b. 4, 1, 8, 1, 12, …
19. Find the nth term for the following arithmetic sequence: 12, 32, 52, 72, …
20. Find how many terms are in the following arithmetic sequence: 6, 10, 14, 18, … , 86
21. In how many ways can you make change for $.21?

*Section 1-4 Logic: An Introduction

statement Logic is a tool used in mathematical thinking and problem solving. It is essential for reasoning. In logic, a **statement** *is a sentence that is either true or false, but not both.* The following expressions are not statements because their truth values cannot be determined without more information.

1. She has blue eyes.
2. $x + 7 = 18$.
3. $2y + 7 > 1$.
4. $2 + 3$

Expressions (**1**), (**2**), and (**3**) become statements if, for (**1**), "she" is identified, and for (**2**) and (**3**), values are assigned to x and y, respectively. However, an expression involving he or she or x or y may already be a statement. For example, "If he is over 210 cm tall, then he is over 2 m tall," and "$2(x + y) = 2x + 2y$" are both statements because they are true no matter who he is or what the numerical values of x and y are.

negation From a given statement, it is possible to create a new statement by forming a **negation.** The negation of a statement is a statement with the opposite truth value of the given statement. If a statement is true, its negation is false, and if a statement is false, its negation is true. Consider the statement "It is snowing." The negation of this statement may be stated simply as "It is not snowing."

Example 1-10 Negate each of the following statements:

a. $2 + 3 = 5$
b. A hexagon has 6 sides.

Solution a. $2 + 3 \neq 5$
b. A hexagon does not have 6 sides.

Sentences like "The shirt is blue" and "The shirt is green" are statements if put in context. However, they are not negations of each other. A statement and its negation must have opposite truth values. If the shirt is actually red, then both of the above statements are false and, hence, cannot be negations of each other. However, the statements "The shirt is blue" and "The shirt is not blue" are negations of each other because they have opposite truth values no matter what color the shirt really is.

quantifiers Some statements involve **quantifiers** and are more complicated to negate. Quantifiers include words such as *all, some, every,* and *there exists.*

- The quantifiers *all, every,* and *no* refer to each and every element in a set.
- The quantifiers *some* and *there exists at least one* refer to one or more, or possibly all, of the elements in a set.
- *All, every,* and *each* have the same mathematical meaning as do *some* and *there exists at least one.*

Consider the following statement involving the existential quantifier *some* and known to be true. "Some professors at Paxson University have blue eyes." This means that at least one professor at Paxson University has blue eyes. It does not rule out the possibilities that all the Paxson professors have blue eyes or that some of the Paxson professors do not have blue eyes. Because the negation of a true statement is false, neither "Some professors at Paxson University do not have blue eyes" nor "All professors at Paxson have blue eyes" are negations of the original statement. One possible negation of the original statement is "No professors at Paxson University have blue eyes."

To discover if one statement is a negation of another, we use arguments similar to the preceding one to determine if they have opposite truth values in all possible cases.

General forms of qualified statements with their negations follow:

Statement	Negation
Some a are b.	No a is b.
Some a are not b.	All a are b.
All a are b.	Some a are not b.
No a is b.	Some a are b.

Example 1-11 Negate each of the following regardless of their truth value.

a. All students like hamburgers.
b. Some people like mathematics.
c. There exists a counting number x such that $3x = 6$.
d. For all counting numbers, $3x = 3x$.

Solution
a. Some students do not like hamburgers.
b. No people like mathematics.
c. For all counting numbers x, $3x \neq 6$.
d. There exists a counting number x such that $3x \neq 3x$.

Table 1-12

Statement p	Negation $\sim p$
T	F
F	T

truth table

There is a symbolic system defined to help in the study of logic. If p represents a statement, the negation of the statement p is denoted by $\sim p$ is read as "not p." **Truth tables** are often used to show all possible true-false patterns for statements. Table 1-12 summarizes the truth tables for p and $\sim p$.

Compound Statements

compound statement

Table 1-13

p	q	Conjunction $p \wedge q$
T	T	T
T	F	F
F	T	F
F	F	F

conjunction

From two given statements, it is possible to create a new, **compound statement** by using a connective such as *and*. For example, "It is snowing" and "The ski run is open" together with *and* give "It is snowing and the ski run is open." Other compound statements can be obtained by using the connective *or*. For example, "It is snowing or the ski run is open."

The symbols $\wedge$ and $\vee$ are used to represent the connectives *and* and *or*, respectively. For example, if p represents "It is snowing" and q represents "The ski run is open," then "It is snowing and the ski run is open" is denoted by $p \wedge q$. Similarly, "It is snowing or the ski run is open" is denoted by $p \vee q$.

The truth value of any compound statement, such as $p \wedge q$, is defined using the truth table of each of the simple statements. Because each of the statements p and q may be either true or false, there are four distinct possibilities for the truth of $p \wedge q$, as shown in Table 1-13. The compound statement $p \wedge q$ is the **conjunction** of p and q and is defined to be true if, and only if, both p and q are true. Otherwise, it is false.

disjunction   The compound statement $p \lor q$—that is, p or q—is a **disjunction.** In everyday language, *or* is not always interpreted in the same way. In logic, we use an *inclusive or*. The statement "I will go to a movie or I will read a book" means I will either go to a movie, or read a book, or do both. Hence, in logic, p or q, symbolized as $p \lor q$, is defined to be false if both p and q are false and true in all other cases. This is summarized in Table 1-14.

Example 1-12 Classify each of the following as true or false.

$$p: 2 + 3 = 5. \qquad q: 2 \cdot 3 = 6. \qquad r: 5 + 3 = 9.$$

a. $p \land q$
b. $q \lor r$

Solution **a.** p is true and q is true, so $p \land q$ is true.
b. q is true and r is false, so $q \lor r$ is true.

logically equivalent   Truth tables are used not only to summarize the truth values of compound statements; they also are used to determine if two statements are logically equivalent. Two statements are **logically equivalent** if, and only if, they have the same truth values. If p and q are logically equivalent, we write $p \equiv q$.

Conditionals and Biconditionals

conditionals • implications   Statements expressed in the form "if p, then q" are called **conditionals,** or **implications,** and are denoted by $p \rightarrow q$. Such statements also can be read "p implies q." The "if" part of a conditional is called the **hypothesis** of the implication and the "then" part is called the **conclusion.**

hypothesis
conclusion

Table 1-14

Many types of statements can be put in "if-then" form. An example follows:

Statement: All equilateral triangles have acute angles.

If-then form: If a triangle is equilateral, then it has acute angles.

An implication may also be thought of as a promise. Suppose Betty makes the promise, "If I get a raise, then I will take you to dinner." If Betty keeps her promise, the implication is true; if Betty breaks her promise, the implication is false. Consider the following four possibilities:

	p	q	
(1)	T	T	Betty gets the raise; she takes you to dinner.
(2)	T	F	Betty gets the raise; she does not take you to dinner.
(3)	F	T	Betty does not get the raise; she takes you to dinner.
(4)	F	F	Betty does not get the raise; she does not take you to dinner.

The only case in which Betty breaks her promise is when she gets her raise and fails to take you to dinner, case (2). If she does not get the raise, she can either take you to dinner or not without breaking her promise. The definition of implication is summarized in Table 1-15. Observe that the only case for which the implication is false is when p is true and q is false.

Table 1-14

p	q	Disjunction $p \lor q$
T	T	T
T	F	T
F	T	T
F	F	F

Table 1-15

p	q	Implication $p \rightarrow q$
T	T	T
T	F	F
F	T	T
F	F	T

An implication can be worded in several equivalent ways, as follows:

1. If the sun shines, then the swimming pool is open. (If p, then q.)
2. If the sun shines, the swimming pool is open. (If p, q.)
3. The swimming pool is open if the sun shines. (q if p.)
4. The sun is shining implies the swimming pool is open. (p implies q.)
5. The sun is shining only if the pool is open. (p only if q.)
6. The sun's shining is a sufficient condition for the swimming pool to be open. (p is a sufficient condition for q.)
7. The swimming pool's being open is a necessary condition for the sun to be shining. (q is a necessary condition for p.)

Any implication $p \rightarrow q$ has three related implication statements, as follows:

Statement:	If p, then q.	$p \rightarrow q$
Converse:	If q, then p.	$q \rightarrow p$
Inverse:	If not p, then not q.	$\sim p \rightarrow \sim q$
Contrapositive:	If not q, then not p.	$\sim q \rightarrow \sim p$

Example 1-13 Write the converse, the inverse, and the contrapositive for the following statement:

If I am in San Francisco, then I am in California.

Solution *Converse:* If I am in California, then I am in San Francisco.
Inverse: If I am not in San Francisco, then I am not in California.
Contrapositive: If I am not in California, then I am not in San Francisco.

Example 1-13 can be used to show that if an implication is true, its converse and inverse are not necessarily true. However, the contrapositive is true. Let's check these observations on the following: *If a number is of the form 2^n, where n is a natural number, the number is even.* We check the truth of the converse, inverse, and contrapositive.

Inverse: *If a number is not of the form 2^n where n is a natural number, then it is not even.* This is false, since 6 is not of the form 2^n, but it is even.

Converse: *If a number is even, then it is of the form 2^n, where n is a natural number.* This is false, since 6 is even but not a power of 2.

Contrapositive: *If a number is not even, then it is not of the form 2^n, where n is a natural number.* This is true for the following reason: A natural number that is not even must be odd. An odd natural number cannot be written as a product of 2 and any number and hence cannot be of the form 2^n.

The contrapositive of the last statement is the original statement. Hence the above discussion suggests that if $p \rightarrow q$ is true, its contrapositive $\sim q \rightarrow \sim p$ is also true and if the contrapositive is true, the original statement must be true. It follows that a statement and its contrapositive cannot have opposite truth values. We summarize this in the following property.

> **Property**
>
> **Equivalence of a statement and its contrapositive** The implication $p \rightarrow q$ and its contrapositive $\sim q \rightarrow \sim p$ are logically equivalent.

REMARK The fact that a statement and its contrapositive are logically equivalent can also be established by showing that they have the same truth tables.

---

if and only if
biconditional

Connecting a statement and its converse with the connective *and* gives $(p \rightarrow q) \land (q \rightarrow p)$. This compound statement can be written as $p \leftrightarrow q$ and usually is read **"p if and only if q."** The statement "p if and only if q" is **biconditional.** When is a biconditional statement true?

Valid Reasoning

valid reasoning

In problem solving, the reasoning is said to be **valid** if the conclusion follows unavoidably from true hypotheses. Consider the following example:

Hypotheses: All roses are red.
                    This flower is a rose.

Conclusion: Therefore this flower is red.

Euler diagram

The statement "All roses are red" can be written as the implication "If a flower is a rose, then it is red" and pictured with the **Euler diagram** in Figure 1-28.

Figure 1-28

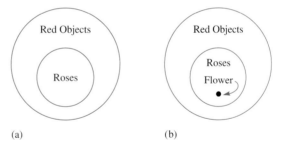

(a)   (b)

The information "This flower is a rose" implies that this flower must belong to the circle containing roses, as pictured in Figure 1-28(b). This flower also must belong to the circle containing red objects. Thus the reasoning is valid because it is impossible to draw a picture that satisfies the hypotheses and contradicts the conclusion.

Consider the following argument with true hypotheses:

Hypotheses: All elementary school teachers are mathematically literate.
                    Some mathematically literate people are not children.

Conclusion: Therefore no elementary school teacher is a child.

Let E be the set of elementary school teachers, M be the set of mathematically literate people, and C be the set of children. Then the statement "All elementary school teachers are mathematically literate" can be pictured as in Figure 1-29(a). The statement "Some mathematically literate people are not children" can be pictured in several ways. Three of these are illustrated in Figure 1-29(b) through (d).

50 CHAPTER 1 *An Introduction to Problem Solving*

According to Figure 1-29(d), it is possible that some elementary school teachers are children, and yet the given statements are satisfied. Therefore the conclusion that "No elementary school teacher is a child" does not follow from the given hypotheses. Hence, the reasoning is not valid.

Figure 1-29

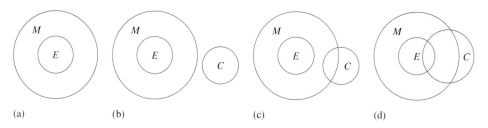

(a)   (b)   (c)   (d)

If only one picture can be drawn to satisfy the hypotheses of an argument and contradict the conclusion, the argument is not valid. However, to show that an argument is valid, *all* possible pictures must show that there are no contradictions. There must be no way to satisfy the hypotheses and contradict the conclusion if the argument is valid.

Example 1-14 Determine if the following argument is valid:

Hypotheses: In Washington, D.C., all lobbyists wear suits.
No one in Washington, D.C., over 6 ft tall wears a suit.

Conclusion: Persons over 6 ft tall are not lobbyists in Washington, D.C.

Solution If L represents the lobbyists in Washington, D.C., and S the people who wear suits, the first hypothesis is pictured as shown in Figure 1-30(a). If W represents the people in Washington, D.C., over 6 ft tall, the second hypothesis is pictured in Figure 1-30(b). Because people over 6 ft tall are outside the circle representing suit wearers and lobbyists are in the circle S, the conclusion is valid and no person over 6 ft tall is a lobbyist in Washington, D.C.

Figure 1-30

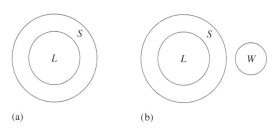

(a)   (b)

direct reasoning
law of detachment/
modus ponens

A different method for determining whether an argument is valid uses **direct reasoning** and a form of argument called the **law of detachment** (or ***modus ponens***). For example, consider the following true statements:

If the sun is shining, then we shall take a trip.
The sun is shining.

Using these two statements, we can conclude that we shall take a trip. In general, the law of detachment is stated as follows:

If the statement "if p, then q" is true, and p is true, then q must be true.

Example 1-15 Determine whether the following argument is valid when the hypotheses are true:

Hypotheses: If you eat spinach, then you will be strong.
You eat spinach.

Conclusion: Therefore you will be strong.

Solution Using the law of detachment, we see that the conclusion is valid.

indirect reasoning
modus tollens

A different type of reasoning, **indirect reasoning,** uses a form of argument called *modus tollens.* For example, consider the following true statements:

If a figure is a square, then it is a rectangle.

The figure is not a rectangle.

The conclusion is that the figure cannot be a square.
Modus tollens can be interpreted as follows:

If we have a conditional accepted as true, and we know the conclusion is false, then the hypothesis must be false.

Example 1-16 Determine conclusions for each of the following true statements:

a. If a person lives in Boston, then the person lives in Massachusetts. Jessica does not live in Massachusetts.
b. If $x = 3$, then $2x \neq 7$. We know that $2x = 7$.

Solution **a.** Jessica does not live in Boston.
   b. $x \neq 3$.

chain rule

The final reasoning argument to be considered here involves the **chain rule.** Consider the following statements:

If I save, I will retire early.

If I retire early, I will become lazy.

What is the conclusion? The conclusion is that if I save, I will become lazy. In general, the chain rule can be stated as follows:

If "if p, then q" and "if q, then r" are true, then "if p, then r" is true.

People often make invalid conclusions based on advertising or other information. Consider, for example, the true statement "Healthy people eat Super-Bran cereal." Are the following conclusions valid?

If a person eats Super-Bran cereal, then the person is healthy.

If a person is not healthy, the person does not eat Super-Bran cereal.

If the original statement is denoted by $p \rightarrow q$, where p is "a person is healthy" and q is "a person eats Super-Bran cereal," then the first conclusion is the converse of $p \rightarrow q$, that is, $q \rightarrow p$, and the second conclusion is the inverse if $p \rightarrow q$, that is, $\sim p \rightarrow \sim q$. Hence, neither is valid.

Example 1-17 Determine conclusions for the following true statements:

a. If a triangle is equilateral, then it is isosceles. If a triangle is isosceles, it has two congruent sides.

b. If natural numbers are whole numbers, then natural numbers are integers. If numbers are integers, then the numbers are rational numbers. If numbers are rational numbers, then the numbers are real numbers.

Solution a. If a triangle is equilateral, then it has two congruent sides.

b. If a number is a natural number, then it is a real number.

ONGOING ASSESSMENT 1-4

1. Determine which of the following are statements and then classify each statement as true or false:
   a. $2 + 4 = 8$.
   b. Shut the window.
   c. Los Angeles is a state.
   d. He is in town.
   e. What time is it?
   f. $5x = 15$.
   g. $3 \cdot 2 = 6$.
   h. $2x^2 > x$.
   i. This statement is false.
   j. Stay put!

2. Use quantifiers to make each of the following true, where x is a natural number:
   a. $x + 8 = 11$.
   b. $x + 0 = x$.
   c. $x^2 = 4$.
   d. $x + 1 = x + 2$.
   e. $x + 3 = 3 + x$.
   f. $3(x + 2) = 12$.
   g. $5x + 4x = 9x$.

3. Use quantifiers to make each equation in problem 2 false.

4. Write the negation for each of the following statements:
   a. This book has 500 pages.
   b. Six is less than 8.
   c. $3 \cdot 5 = 15$.
   d. Some people have blond hair.
   e. All dogs have 4 legs.
   f. Some cats do not have 9 lives.
   g. All squares are rectangles.
   h. Not all rectangles are squares.

5. Complete each of the following truth tables:

   a.

p	$\sim p$	$\sim(\sim p)$
T		
F		

   b.

p	$\sim p$	$p \vee \sim p$	$p \wedge \sim p$
T			
F			

   c. Based on part (a), is p logically equivalent to $\sim(\sim p)$?

   d. Based on part (b), is $p \vee \sim p$ logically equivalent to $p \wedge \sim p$?

6. If q stands for "This course is easy" and r stands for "Lazy students do not study," write each of the following in symbolic form:
   a. This course is easy, and lazy students do not study.
   b. Lazy students do not study, or this course is not easy.
   c. It is false that both this course is easy and lazy students do not study.
   d. This course is not easy.

7. If p is false and q is true, find the truth values for each of the following:
   a. $p \wedge q$
   b. $p \vee q$
   c. $\sim p$
   d. $\sim q$
   e. $\sim(\sim p)$
   f. $\sim p \vee q$
   g. $p \wedge \sim q$
   h. $\sim(p \vee q)$
   i. $\sim(\sim p \wedge q)$
   j. $\sim q \wedge \sim p$

8. Find the truth value for each statement in problem 7 if p is false and q is false.

9. For each of the following, is the pair of statements logically equivalent?
   a. $\sim(p \vee q)$ and $\sim p \vee \sim q$ b. $\sim(p \vee q)$ and $\sim p \wedge \sim q$
   c. $\sim(p \wedge q)$ and $\sim p \vee \sim q$ d. $\sim(p \wedge q)$ and $\sim p \vee \sim q$

10. Complete the following truth table:

p	q	$\sim p$	$\sim q$	$\sim p \vee \sim q$
T	T			
T	F			
F	T			
F	F			

11. Restate the following in a logically equivalent form:
    a. It is not true that today is Wednesday and the month is June.
    b. It is not true that yesterday I both ate breakfast and watched television.
    c. It is not raining, or it is not July.

12. Write each of the following in symbolic form if p is the statement "It is raining" and q is the statement "The grass is wet."
    a. If it is raining, then the grass is wet.
    b. If it is not raining, then the grass is wet.
    c. If it is raining, then the grass is not wet.
    d. The grass is wet if it is raining.
    e. The grass is not wet implies that it is not raining.
    f. The grass is wet if, and only if, it is raining.

13. For each of the following implications, state the converse, inverse, and contrapositive:
    a. If $x = 5$, then $2x = 10$.
    b. If you do not like this book, then you do not like mathematics.
    c. If you do not use Ultra Brush toothpaste, then you have cavities.
    d. If you are good at logic, then your grades are high.

14. Iris makes the true statement, "If it rains, then I am going to the movies." Does it follow logically that if it does not rain, then Iris does not go to the movies?

15. Consider the statement "If every digit of a number is 6, then the number is divisible by 3." Which of the following is logically equivalent to the statement?
    a. If every digit of a number is not 6, then the number is not divisible by 3.
    b. If a number is not divisible by 3, then some digit of the number is not 6.
    c. If a number is divisible by 3, then every digit of the number is 6.

16. Write a statement logically equivalent to the statement "If a number is a multiple of 8, then it is a multiple of 4."

17. Investigate the validity of each of the following arguments:
    a. All women are mortal.
       Hypatia was a woman.
       Therefore Hypatia was mortal.
    b. All squares are quadrilaterals.
       All quadrilaterals are polygons.
       Therefore all squares are polygons.
    c. All teachers are intelligent.
       Some teachers are rich.
       Therefore some intelligent people are rich.
    d. If a student is a freshman, then the student takes mathematics.
       Jane is a sophomore.
       Therefore Jane does not take mathematics.

18. For each of the following, form a conclusion that follows logically from the given statements:
    a. All college students are poor.
       Helen is a college student.
    b. Some freshmen like mathematics.
       All people who like mathematics are intelligent.
    c. If I study for the final, then I will pass the final.
       If I pass the final, then I will pass the course.
       If I pass the course, then I will look for a teaching job.
    d. Every equilateral triangle is isosceles.
       There exist triangles that are isosceles.

19. Write the following in if-then form:
    a. Every figure that is a square is a rectangle.
    b. All integers are rational numbers.
    c. Figures with exactly 3 sides may be triangles.
    d. It rains only if it is cloudy.

20. Write two logical equivalences discovered in problem 9(a) through (d). These equivalences are called DeMorgan's laws for *and* and *or*.

Communication

21. Translate each of the following statements into symbolic form. Give the meanings of the symbols that you use.
    a. If Mary's little lamb follows her to school, then its appearance there will break the rules and Mary will be sent home.
    b. If it is not the case that Jack is nimble and quick, then Jack will not make it over the candlestick.
    c. If the apple had not hit Isaac Newton on the head, then the laws of gravity would not have been discovered.

22. Determine the validity of the following conclusions. Explain your reasoning.
    a. If you study hard, you will get at least a B in this course.
       If you get at least a B in this course, you will graduate.
       Therefore if you study hard, you will graduate.
    b. All teachers are college graduates. Therefore if a person did not graduate from college, the person is not a teacher.
    c. All ducks have feathers. No mammals are ducks. Therefore no mammals have feathers.

23. Write a valid conclusion based on the following statements. Explain why the conclusion is valid.
    a. We go shopping if and only if I get a bonus. I got a bonus.
    b. All rectangles are parallelograms. This figure is not a parallelogram.

c. If the day is sunny, we go hiking. If it is freezing, we don't go hiking.

Open-Ended

24. Give two examples from mathematics for each of the following:
    a. A statement and its converse are true.
    b. A statement is true, but its converse is false.
    c. An "if and only if" true statement.
    d. An "if and only if" false statement.

Cooperative Learning

25. Each person in a group makes 5 statements similar to the ones in Examples 1-14 through 1-17 but concerning mathematical objects, each with a valid or invalid conclusion. The statements should be as varied as possible. Each group member exchanges his or her statements with another person—not revealing which are valid and which are not—and determines which of the other person's statements are valid and which are not. The two group members compare their answers and discuss any discrepancies.

26. Discuss the paradox arising from the following:
    a. This textbook is 1000 pages long.
    b. The author of this textbook is Dante.
    c. The statements (a), (b), and (c) are all false.

HINT FOR SOLVING THE PRELIMINARY PROBLEM

If there are 5 candidates in the election, consider how many candidates each candidate has to run against. Duplicates should be eliminated, and making a list may be an appropriate strategy.

QUESTIONS FROM THE CLASSROOM

1. A student claims she checked that $n^{50} > 2^n$ (the symbol $>$ designates "greater than") for $n = 1, 2, 3, \ldots, 50$. Hence, she claims that $n^{50} > 2^n$ should be true for all values of n. How do you respond?

2. A student says she read that Thomas Robert Malthus (1766–1834), a renowned British economist and demographer, claimed that the increase of population will take place, if unchecked, in a geometric sequence, while the supply of food will increase in only an arithmetic sequence. This theory implies that population increases faster than food production. The student is wondering why. How do you respond?

3. A student claims that the sequence 6, 6, 6, 6, 6, … never changes, so it is neither arithmetic nor geometric. How do you respond?

4. A student claims that two terms are enough to determine any sequence. For example, 3, 6, … means the sequence would be 3, 6, 9, 12, 15, …. What is your response?

5. A student claims that algebra cannot be taught to students before grade 8. How do you respond?

CHAPTER OUTLINE

I. Mathematical patterns
   A. Patterns are an important part of problem solving.
   B. Patterns are used in **inductive reasoning** to form conjectures. Inductive reasoning is the method of making generalizations based on observations and patterns. A **conjecture** is a statement that is thought to be true but that has not yet been proved to be true or false.
   C. A **sequence** is a group of terms in a definite order.
      1. **Arithmetic sequence:** Each successive term is obtained from the previous one by the addition of a fixed number called the **difference.** The nth term is given by $a + (n - 1)d$, where a is the first term and d is the difference.
      2. **Geometric sequence:** Each successive term is obtained from its predecessor by multiplying it by a fixed number called the **ratio.** The nth term is given by ar^{n-1}, where a is the first term and r is the ratio.
      3. $a^n = \underbrace{a \cdot a \cdot a \cdot a \cdot \ldots \cdot a}_{n \text{ terms}}$

4. $a^0 = 1$, where a is a natural number.
5. Finding differences for a sequence is one technique for finding the next terms.

II. Problem solving
   A. Problem solving can be guided by the following four-step process:
      1. Understanding the problem
      2. Devising a plan
      3. Carrying out the plan
      4. Looking back
   B. Important problem-solving strategies include the following:
      1. Look for a pattern.
      2. Make a table.
      3. Examine a simpler or special case of the problem to gain insight into the solution of the original problem.
      4. Identify a subgoal.
      5. Examine related problems and determine if the same technique can be applied.
      6. Work backward.
      7. Write an equation.
      8. Draw a diagram.
      9. Guess and check.
      10. Use indirect reasoning.
      11. Use direct reasoning.
   C. Beware of mind sets!

III. Properties of equality
   A. Addition property: For numbers a, b, and c, if $a = b$, then $a + c = b + c$.
   B. Multiplication property: For numbers a, b, and c, if $a = b$, then $ac = bc$.
   C. Cancellation properties: For numbers a, b, and c
      1. if $a + c = b + c$, then $a = b$.
      2. if $c \neq 0$, and $ac = bc$, then $a = b$.
   D. Equality is not affected if we substitute a number for its equal.

*IV. Logic
   A. A **statement** is a sentence that is either true or false but not both.
   B. The **negation** of a statement is a statement with the opposite truth value of the given statement. The negation of p is denoted by $\sim p$.
   C. The **compound statement** $p \wedge q$ is called the **conjunction** of p and q and is defined to be true if, and only if, both p and q are true.
   D. The compound statement $p \vee q$ is called the **disjunction** of p and q and is true if either p or q or both are true.
   E. Statements of the form "if p, then q" are called **conditionals** or **implications** and are false only if p is true and q is false.
   F. Given the conditional $p \rightarrow q$, the following can be found:
      1. Converse: $q \rightarrow p$
      2. Inverse: $\sim p \rightarrow \sim q$
      3. Contrapositive: $\sim q \rightarrow \sim p$
   G. If $p \rightarrow q$ is true, the converse and the inverse are not necessarily true, but the contrapositive is true.
   H. Two statements are **logically equivalent** if, and only if, they have the same truth value. An implication and its contrapositive are logically equivalent.
   I. The statement "$p \rightarrow q$ and $q \rightarrow p$" is written $p \leftrightarrow q$, a **biconditional,** and referred to as "p if and only if q."
   J. Laws to determine the validity of arguments include the **law of detachment,** *modus tollens,* and the **chain rule.**

CHAPTER REVIEW

1. List three more terms that complete a pattern in each of the following:
   a. 0, 1, 3, 6, 10, ___, ___, ___
   b. 52, 47, 42, 37, ___, ___, ___
   c. 6400, 3200, 1600, 800, ___, ___, ___
   d. 1, 2, 3, 5, 8, 13, ___, ___, ___
   e. 2, 5, 8, 11, 14, ___, ___, ___
   f. 1, 4, 16, 64, ___, ___, ___
   g. 0, 4, 8, 12, ___, ___, ___
   h. 1, 8, 27, 64, ___, ___, ___
2. Classify each sequence in problem 1 as arithmetic, geometric, or neither.
3. Find a possible nth term in each of the following:
   a. 5, 8, 11, 14, …
   b. 1, 8, 27, 64, …
   c. 3, 9, 27, 81, 243, …
4. Find the first five terms of the sequences whose nth term is given as follows:
   a. $3n + 2$ b. $n^2 + n$ c. $4n - 1$
5. Find the following sums:
   a. $2 + 4 + 6 + 8 + 10 + \ldots + 200$
   b. $51 + 52 + 53 + 54 + \ldots + 151$
6. a. Determine a possible pattern in the sequence 1, 12, 123, 1234, 12345, … .
   b. If the tenth term of the sequence in part (a) is supposed to have 10 digits, what is the tenth term? Explain your reasoning.

7. Complete the following magic square; that is, complete the square so that the sum in each row, column, and diagonal is the same.

16	3	2	13
	10		
9		7	12
4		14	

8. How many people can be seated at 12 square tables lined up end to end if each table individually holds four persons?
9. A shirt and a tie sell for $9.50. The shirt costs $5.50 more than the tie. What is the cost of the tie?
10. If fence posts are to be placed in a row 5 m apart, how many posts are needed for 100 m of fence?
11. A total of 129 players entered a single-elimination handball tournament. In the first round of play, the top-seeded player received a bye and the remaining 128 players played in 64 matches. Thus 65 players entered the second round of play. How many matches must be played to determine the tournament champion?
12. Given the six numbers 3, 5, 7, 9, 11, and 13, pick five of them that when multiplied together give 19,305.
13. If a complete turn of a car tire moves a car forward 6 ft, how many turns of the tire occur before the tire goes off its 50,000-mi warranty?
14. The members of Mrs. Grant's class are standing in a circle; they are evenly spaced and are numbered in order. The student with number 7 is standing directly across from the student with number 17. How many students are in the class?
15. A carpenter has three large boxes. Inside each large box are two medium-sized boxes. Inside each medium-sized box are five small boxes. How many boxes are there altogether?
16. How many triangles are there in the following figure? Explain your reasoning.

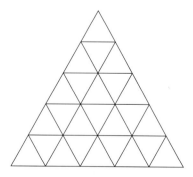

17. Mary left her home and averaged 16 km/hr riding her bicycle on an uphill trip to Larry's house. On the return trip over the same route, she averaged 20 km/hr. If it took 4 hr to make the return trip, how much cycling time did the entire trip take?

18. The perimeter of a rectangle is 68 ft. The length of the rectangle is 4 ft more than twice the width. Find the length and width of the rectangle. Explain your reasoning.
19. An ant farm can hold a total of 100,000 ants. If the farm held 1500 ants on the first day, 3000 ants on the second day, 6000 ants on the third day, and so on in this manner, in how many days will the farm be full?
20. Toma's team entered a mathematics contest where teams of students compete by answering questions that are worth either 3 points or 5 points. No partial credit was given. Toma's team scored 44 points on 12 questions. How many 5-point questions did the team answer correctly?
21. There are three baskets sitting next to each other on a high shelf so that you cannot see the contents of any basket. Under the first basket is a sign that says APPLES. Under the second basket is a sign that says ORANGES. And under the third basket is a sign that says APPLES AND ORANGES. Each basket is incorrectly labeled. One basket contains all apples, one all oranges, and one a combination of apples and oranges. Is it possible to reach up on the shelf and without looking into any of the baskets select one piece of fruit and on the basis of knowing what that piece of fruit is correctly label all three baskets? Explain your reasoning.
22. Solve the following for x if possible:
    a. $3x + 5 = 17$
    b. $4 - 2x = 0$
    c. $3(x + 1) = 3x$
*23. Which of the following are statements?
    a. The moon is inhabited.
    b. $3 + 5 = 8$
    c. $x + 7 = 15$
    d. Some women have Ph.Ds in mathematics.
*24. Negate each of the following:
    a. Some women smoke.
    b. $3 + 5 = 8$
    c. All heavy-metal rock is loud.
    d. Beethoven wrote only classical music.
*25. Write the converse, inverse, and contrapositive of the following: If we have a rock concert, someone will faint.
*26. Find valid conclusions for the following arguments:
    a. All Americans love Mom and apple pie.
       Joe Czernyu is an American.
    b. Steel eventually rusts.
       The Statue of Liberty has a steel structure.
    c. Albertina will pass Math 100 or be a dropout.
       Albertina is not a dropout.
*27. Write the following argument symbolically and then determine its validity:
    If you are fair-skinned, you will sunburn.
    If you sunburn, you will not go to the dance.
    If you do not go to the dance, your parents will want to know why you didn't go to the dance.
    Your parents do not want to know why you didn't go to the dance.
    Therefore you are not fair-skinned.

SELECTED BIBLIOGRAPHY

Bledsoe, G. "Hook Your Students on Problem Solving." *Arithmetic Teacher* 37 (December 1989): 16–20.

Bradley, E. "Is Algebra in the Cards?" *Mathematics Teaching in the Middle School* 2 (May 1997): 398–403.

Brown, S., and M. Walter. *The Art of Problem Posing.* Philadelphia: Franklin Institute Press, 1991.

Curcio, F. "Exploring Patterns in Nonroutine Problems." *Mathematics Teaching in the Middle School* 2 (February 1997): 262–269.

Danielson, C. *A Collection of Performance Tasks and Rubrics: Middle School Mathematics.* Larchmont, NY: Eye on Education, 1997.

Day, R., and G. Jones. "Building Bridges to Algebraic Thinking." *Mathematics Teaching in the Middle School* 2 (February 1997): 208–212.

English, L. "Promoting a Problem-Posing Classroom." *Teaching Children Mathematics* 4 (November 1997): 172–179.

Ferrini-Mundy, J., G. Lappan, and E. Phillips. "Experiences with Patterning." *Teaching Children Mathematics* 3 (February 1997): 262–268.

Fouche, K. "Algebra for Everyone: Start Early." *Mathematics Teaching in the Middle School* 2 (February 1997): 226–229.

Gill, A. "Multiple Strategies: Product of Reasoning and Communication." *Arithmetic Teacher* 40 (March 1993): 380–386.

Kenney, P. "Probing the Foundations of Algebra: Grade-4 Pattern Items in NAEP." *Teaching Children Mathematics* 3 (February 1997): 268–274.

Kersch, M., and J. McDonald. "How Do I Solve Thee? Let Me Count the Ways." *Arithmetic Teacher* 39 (October 1991): 38–41.

Krulik, S., and J. Rudnick. "For Better Problem Solving and Reasoning." *Teaching Children Mathematics* 1 (February 1994): 334–338.

Lambkin-Kroll, D., J. Masingila, and S. Mau. "Grading Cooperative Problem Solving." *Mathematics Teacher* 85 (November 1992): 619–627.

Malloy, C. "Mathematics Projects Promote Students' Algebraic Thinking." *Mathematics Teaching in the Middle School* 2 (February 1997): 282–288.

Mikusa, M. G. "Problem Solving Is More Than Solving Problems." *Mathematics Teaching in the Middle School* 4 (November 1998): 20–25.

Moody, W. "A Program in Middle School Problem Solving." *Arithmetic Teacher* 38 (December 1990): 6–11.

Norman, F. "Figurate Numbers in the Classroom." *Arithmetic Teacher* 38 (March 1991): 42–45.

Ploger, D. "Spreadsheets, Patterns, and Algebraic Thinking." *Teaching Children Mathematics* 3 (February 1997): 330–334.

Polya, G. *How to Solve It.* Princeton, N.J.: Princeton University Press, 1957.

———. *Mathematical Discovery, Combined Edition.* New York: John Wiley & Sons, Inc., 1981.

Scheibelhut, C. "I Do and I Understand, I Reflect and I Improve." *Teaching Children Mathematics* 1 (December 1994): 242–246.

Talton, C. "Let's Solve the Problem Before We Find the Answer." *Arithmetic Teacher* 36 (September 1988): 40–45.

Thompson, A. "On Patterns, Conjectures, and Proof: Developing Students." *Arithmetic Teacher* 33 (September 1985): 20–23.

Usiskin, Z. "Doing Algebra in Grades K–4." *Teaching Children Mathematics* 3 (February 1997): 346–356.

van Reeuwijk, M., and M. Wijers. "Students' Construction of Formulas in Context." *Mathematics Teaching in the Middle School* 2 (February 1997): 230–236.

Vance, J. "Number Operations from an Algebraic Perspective." *Teaching Children Mathematics* 4 (January 1998): 282–285.

Verzoni, K. "Turning Students into Problem Solvers." *Mathematics Teaching in the Middle School* 3 (October 1997): 102–107.

Zoest, L., and A. Enyart. "Discourse, of Course: Encouraging Genuine Mathematical Conversation." *Mathematics Teaching in the Middle School* 4 (November–December 1998): 150–157.

2

Sets, Whole Numbers, and Functions

Preliminary Problem

The World's Largest Pizza Parlour offers a pizza that is 3 feet in diameter and comes with any combination of 40 different toppings. In its advertisements the establishment boasts that its mushroom and cheese toppings are so good that every customer will order at least those two toppings. The ad also says that even if every person in the world ordered a different combination of toppings, assuming cheese and mushroom were always chosen, there would still be many unordered combinations of toppings. Is the ad truthful?

Georg Cantor, in the years from 1871 through 1884, created *set theory*, a new area of mathematics. His theories had a profound effect on research and mathematics teaching.

HISTORICAL NOTE

Georg Cantor, 1845–1918, a German mathematician, was born of Danish parents in St. Petersburg, Russia. His family moved to Frankfurt when he was 11. Against his father's advice, Cantor pursued a career in mathematics and obtained his doctorate in Berlin at age 22. Most of his academic work was spent at the University of Halle. His hope of becoming a professor at the University of Berlin did not materialize, because his work gained little recognition during his lifetime.

Cantor's work was praised as an "astonishing product of mathematical thought, one of the most beautiful realizations of human activity. . . ."

The language of set theory and basic set operations clarifies and unifies many mathematical concepts and is useful for teachers in understanding the mathematics covered in elementary school. However, if not well understood, set theory can be confusing. The following "Peanuts" cartoon demonstrates how a lack of understanding can lead to frustration.

Sets and relations between sets are used in elementary school textbooks to teach children the *concept* of whole numbers and the concept of "less than" as well as addition, subtraction, and multiplication of whole numbers. After introducing set notation, relations between sets, set operations and their properties, we will show how some of these concepts can be used to understand the concept of a whole number, operations on whole numbers, and the properties of these operations. We also use the concept of a set to define relations and functions.

Section 2-1 — Describing Sets

set
elements • members

Set is undefined, but it is understood to be any collection of objects. Individual objects in a set are **elements,** or **members,** of the set. For example, each letter is an element of the set of letters in the English language.

We use braces to enclose the elements of a set and label the set with a capital letter for easy reference. The set of letters of the English alphabet can be written as

$A = \{a, b, c, d, e, f, g, h, i, j, k, l, m, n, o, p, q, r, s, t, u, v, w, x, y, z\}$.

The order in which the elements are written makes no difference, and *each element is listed only once.* For example, the set of letters in the word *book* could be written as {b, o, k}, {o, b, k}, or {k, o, b}.

We symbolize an element belonging to a set by using the symbol ∈. For example, $b \in A$. The fact that A does not contain the Greek letter α (alpha) is written as $\alpha \notin A$.

well defined

For a given set to be useful in mathematics, it must be **well defined.** If we are given a set and some particular object, the object does or does not belong to the set. For example, the set of all citizens of Pasadena, California, who ate rice on January 1, 2000, is well defined. We may not know if a particular resident of Pasadena ate rice or not, but that resident either belongs or does not belong to the set. On the other hand, the set of all tall people is not well defined because we do not know which particular people qualify as "tall" people.

natural numbers

We may use sets to define mathematical terms. For example, the set N of **natural numbers** is defined by the following:

$$N = \{1, 2, 3, 4, \ldots\}.$$

set-builder notation

Sometimes the individual elements of a set are not known or they are too numerous to list. In these cases, the elements may be indicated by using **set-builder notation.** For example, the set of decimals between 0 and 1 can be written as

$$D = \{x \mid x \text{ is a decimal between 0 and 1}\}.$$

This is read "D is the set of all elements x such that x is a decimal between 0 and 1." The vertical line is read "such that." It would be impossible to list all the elements of D. Hence the set-builder notation is indispensable here.

Example 2-1 Write the following sets using set-builder notation:

  a. {51, 52, 53, 54, … , 498, 499}
  b. {2, 4, 6, 8, 10, …}
  c. {1, 3, 5, 7, …}
  d. {$1^2, 2^2, 3^2, 4^2, \ldots$}

Solution
  a. $\{x \mid x \text{ is a natural number greater than 50 and less than 500}\}$, or $\{x \mid 50 < x < 500, x \in N\}$.
  b. $\{x \mid x \text{ is an even natural number}\}$. Or because every even natural number can be written as 2 times some natural number, this set can be written as $\{x \mid x = 2n, n \in N\}$ or, in a somewhat simpler form, as $\{2n \mid n \in N\}$.
  c. $\{x \mid x \text{ is an odd natural number}\}$. Or because every odd natural number can be written as some even number minus one, this set can be written as $\{x \mid x = 2n - 1, n \in N\}$ or $\{2n - 1 \mid n \in N\}$.
  d. $\{x \mid x \text{ is a square of a natural number}\}$ or $\{n^2 \mid n \in N\}$.

Example 2-2 Each of the following sets is described in set-builder notation. Write each of the sets by listing its elements.

  a. $A = \{2k + 1 \mid k \in N, 2 < k < 6\}$.
  b. $B = \{a^2 + b^2 \mid a, b \in N, 1 < a < 4 \text{ and } 1 < b < 5\}$.

Table 2-1

k	$n = 2k + 1$
3	$2 \cdot 3 + 1 = 7$
4	$2 \cdot 4 + 1 = 9$
5	$2 \cdot 5 + 1 = 11$

Solution a. Because k is a natural number and $2 < k < 6$, we substitute $k = 3, 4, 5$ in $2k + 1$ and obtain the corresponding values of n shown in Table 2-1. Thus $A = \{7, 9, 11\}$.

b. Here $a = 2$ or 3 and $b = 2, 3$, or 4. Table 2-2 shows all possible combinations of a and b and the corresponding values of $a^2 + b^2$. Thus $B = \{8, 13, 20, 18, 25\}$. Notice that 13 appears twice in the table but only once in the set. Why?

Table 2-2

a \ b	2	3	4
2	$2^2 + 2^2 = 8$	$2^2 + 3^2 = 13$	$2^2 + 4^2 = 20$
3	$3^2 + 2^2 = 13$	$3^2 + 3^2 = 18$	$3^2 + 4^2 = 25$

Sets can have other sets as their members. For example, consider the set E of all countries in the European Economic Community. Denmark is an element of E, whereas a citizen of Denmark is not an element of E.

equal sets Two sets are **equal** if, and only if, they contain exactly the same elements. The order in which the elements are listed does not matter. If A and B are equal, written $A = B$, then every element of A is an element of B, and every element of B is an element of A. If A does not equal B, we write $A \neq B$. Consider sets $D = \{1, 2, 3\}$, $E = \{2, 5, 1\}$, and $F = \{1, 2, 5\}$. Sets D and E are not equal; sets E and F are equal.

One-to-One Correspondence

Consider the set of people $P = \{\text{Tomas, Dick, Mari}\}$ and the set of swimming lanes $S = \{1, 2, 3\}$. Suppose each person in P is to swim in a lane numbered 1, 2, or 3 so that no **one-to-one correspondence** two people swim in the same lane. Such a person-lane pairing is a **one-to-one correspondence**. One way to exhibit this one-to-one correspondence is Tomas $\leftrightarrow$ 1, Dick $\leftrightarrow$ 2, and Mari $\leftrightarrow$ 3, as shown in Figure 2-1.

Figure 2-1

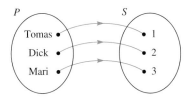

Other possible one-to-one correspondences exist between the sets P and S. There are several schemes for exhibiting them. For example, all six possible one-to-one correspondences between sets P and S can be listed as follows:

1. Tomas $\leftrightarrow$ 1   **2.** Tomas $\leftrightarrow$ 1   **3.** Tomas $\leftrightarrow$ 2
   Dick $\leftrightarrow$ 2   Dick $\leftrightarrow$ 3   Dick $\leftrightarrow$ 1
   Mari $\leftrightarrow$ 3   Mari $\leftrightarrow$ 2   Mari $\leftrightarrow$ 3

4. Tomas ↔ 2		**5.** Tomas ↔ 3		**6.** Tomas ↔ 3	
Dick ↔ 3		Dick ↔ 1		Dick ↔ 2	
Mari ↔ 1		Mari ↔ 2		Mari ↔ 1	

Notice that the diagram in (**1**) as well as Figure 2-1 represent a single one-to-one correspondence between the sets P and S. The correspondence Tomas ↔ 1 can also be a one-to-one correspondence but between two different sets, namely the sets {Tomas} and {1}.

Definition of One-to-One Correspondence

If the elements of sets P and S can be paired so that for each element of P there is exactly one element of S and for each element of S there is exactly one element of P, then the two sets P and S are said to be in **one-to-one correspondence**.

NOW TRY THIS 2-1

- Consider a set of four people and a set of four swimming lanes.
    a. Exhibit all the one-to-one correspondences between the two sets.
    b. How many such one-to-one correspondences are there?
    c. Conjecture the number of one-to-one correspondences between two sets with five elements each and explain how you arrived at your conjecture.

Another method of demonstrating a one-to-one correspondence is to use a table, such as Table 2-3, where the lane numbers are listed across the top of the table and the possible pairings of swimmers to lanes are listed in the table.

Table 2-3

1	2	3
Tomas	Dick	Mari
Tomas	Mari	Dick
Dick	Tomas	Mari
Dick	Mari	Tomas
Mari	Tomas	Dick
Mari	Dick	Tomas

We can also use a tree diagram to list the possible one-to-one correspondences, as Figure 2-2 shows. To read the tree diagram and see the one-to-one correspondence, we follow each branch. The person occupying a specific lane in a correspondence is listed below the lane number. For example, the top branch gives the pairing (Tomas, 1), (Dick, 2), and (Mari, 3).

Figure 2-2

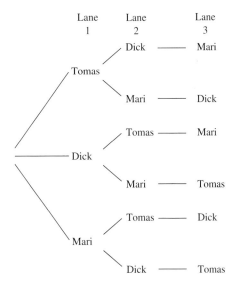

Observe in Figure 2-2 that, in assigning a swimmer to lane 1, we have a choice of three people: Tomas, Dick, or Mari. If we put Tomas in lane 1, then he cannot be in lane 2, and hence the second lane must be occupied by either Dick or Mari. In the same way, we see that if Dick is in lane 1, then there are two choices for lane 2: Tomas or Mari. Similarly if Mari is in lane 1, then again there are two choices for the second lane: Tomas or Dick. Thus for each of the three ways we can fill the first lane there are two subsequent ways to fill the second lane, and hence there are $2 + 2 + 2$, or $3 \cdot 2$, or 6 ways to arrange the swimmers in the first two lanes.

Notice that for each of these arrangements of the swimmers in the first two lanes, there remains only one possible swimmer to fill the third lane. That is, if Mari fills the first lane and Dick fills the second, then Tomas must be in the third. Thus the total number of arrangements for the three swimmers is equal to $3 \cdot 2$, or 6.

Similar reasoning can be used to find how many ice cream arrangements are possible on a two-scoop cone if ten flavors are offered. If we count chocolate and vanilla (chocolate on bottom and vanilla on top) different from vanilla and chocolate (vanilla on bottom and chocolate on top) and allow two scoops to be of the same flavor, we can proceed as follows. There are 10 choices for the first scoop and for each of these 10 choices there are 10 subsequent choices for the second scoop. Thus the total number of arrangements is $10 \cdot 10$, or 100.

The counting argument used to find the number of possible one-to-one correspondences between the set of swimmers and the set of lanes and the previous problem about ice cream scoop arrangements are examples of the fundamental counting principle.

Property

Fundamental Counting Principle If event M can occur in m ways and, after it has occurred, event N can occur in n ways, then event M followed by event N can occur in $m \cdot n$ ways.

NOW TRY THIS 2-2

- Extend the Fundamental Counting Principle to *n* events.

Equivalent Sets

Suppose a room contains 20 chairs and one student is sitting in each chair with no one standing. There is a one-to-one correspondence between the set of chairs and the set of students in the room. In this case, the set of chairs and the set of students are **equivalent sets**.

equivalent sets

Definition of Equivalent Sets

Two sets A and B are **equivalent**, written $A \sim B$, if and only if there exists a one-to-one correspondence between the sets.

The term *equivalent* should not be confused with *equal*. The difference should be made clear by Example 2-3.

Example 2-3 Let

$$A = \{p, q, r, s\}, \quad B = \{a, b, c\}, \quad C = \{x, y, z\}, \quad \text{and} \quad D = \{b, a, c\}.$$

Compare the sets, using the terms *equal* and *equivalent*.

Solution Each set is both equivalent to and equal to itself.
 Sets A and B are not equivalent ($A \nsim B$) and not equal ($A \neq B$).
 Sets A and C are not equivalent ($A \nsim C$) and not equal ($A \neq C$).
 Sets A and D are not equivalent ($A \nsim D$) and not equal ($A \neq D$).
 Sets B and C are equivalent ($B \sim C$) but not equal ($B \neq C$).
 Sets B and D are equivalent ($B \sim D$) and equal ($B = D$).
 Sets C and D are equivalent ($C \sim D$) but not equal ($C \neq D$).

NOW TRY THIS 2-3

- If two sets are equivalent, are they necessarily equal? Explain why or why not.

Cardinal Numbers

The concept of one-to-one correspondence can be used to introduce the notion of two sets having the same number of elements. Suppose a child knows how to count only to three. The child might still tell that there are as many fingers on the left hand as on the right hand by matching the fingers on one hand with the fingers on the other hand. Naturally placing the fingers so that the left thumb touches the right thumb, the left index finger touches the right index finger, and so on, exhibits a one-to-one correspondence between the fingers of the two hands. Similarly, without counting, children realize that if every student in a class sits in a chair and no chairs are empty, there are as many chairs as students.

One-to-one correspondence between sets is often used to introduce the concept of a number as follows. (In elementary school, the approach is similar but without the abstract notation.) The five sets $\{a, b\}$, $\{p, q\}$, $\{x, y\}$, $\{b, a\}$, and $\{*, \#\}$ are equivalent to one another and share the property of "twoness." These sets have the same cardinal number, namely, 2.

cardinal number The **cardinal number** of a set X, denoted by $n(X)$, indicates the number of elements in the set X. If $D = \{a, b\}$, the cardinal number of D is 2, and we write $n(D) = 2$. If A is equivalent to B, then A and B have the same cardinal number; that is, $n(A) = n(B)$.

finite set A set is a **finite set** if the cardinal number of the set is zero or a natural number. For example, the set of letters in the English alphabet is a finite set because it contains exactly 26 elements. Another way to think of this is that the set of letters in the English alphabet can be put into a one-to-one correspondence with the set $\{1, 2, 3, \ldots, 26\}$. The set of nat-

infinite set ural numbers N is an example of an **infinite set,** a set that is not finite.

More About Sets

empty set · null set A set that contains no elements has cardinal number 0 and is an **empty,** or **null, set.** The empty set is designated by the symbol $\varnothing$ or $\{\ \}$. Two examples of sets with no elements are the following:

$$C = \{x \mid x \text{ was a state of the United States before A.D. 1200}\}.$$

$$D = \{x \mid x \text{ is a natural number less than } 1\}.$$

REMARK The empty set is often incorrectly recorded as $\{\varnothing\}$. This set is not empty but contains one element. Likewise, $\{0\}$ does not represent the empty set. Why?

---

universal set · universe The **universal set,** or the **universe,** denoted by U, is the set that contains all elements being considered in a given discussion. For this reason, you should know what the universal set is in any given problem. Suppose $U = \{x \mid x \text{ is a person living in California}\}$ and $F = \{x \mid x \text{ is a female living in California}\}$. The universal set and set F can be represented by a diagram, as in Figure 2-3(a). The universal set is usually indicated by a large rectangle, and particular sets are indicated by geometric figures inside the rectangle, as shown in

Venn diagram Figure 2-3(a). This figure is an example of a **Venn diagram,** named after the Englishman John Venn (1834–1923) who used such diagrams to illustrate ideas in logic. The set of elements in the universe that are not in F is the set of males living in California and is the **complement** of F. It is represented by the shaded region in Figure 2-3(b).

Figure 2-3

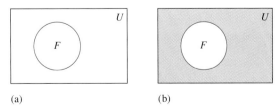

(a) (b)

Definition of Set Complement

The **complement** of a set F, written $\overline{F}$, is the set of all elements in the universal set U that are not in F; that is $\overline{F} = \{x \mid x \in U \text{ and } x \notin F\}$.

Example 2-4

a. If $U = \{a, b, c, d\}$ and $B = \{c, d\}$, find (i) $\overline{B}$; (ii) $\overline{U}$; (iii) $\overline{\varnothing}$.
b. If $U = \{x \mid x \text{ is an animal in the zoo}\}$ and $S = \{x \mid x \text{ is a snake in the zoo}\}$, describe $\overline{S}$.
c. If $U = N$, $E = \{2, 4, 6, 8, \ldots\}$, and $O = \{1, 3, 5, 7, \ldots\}$, find (i) $\overline{E}$; (ii) $\overline{O}$.

Solution

a. (i) $\overline{B} = \{a, b\}$; (ii) $\overline{U} = \varnothing$; (iii) $\overline{\varnothing} = U$.
b. Because the individual animals in the zoo are not known, $\overline{S}$ must be described using set-builder notation:

$\overline{S} = \{x \mid x \text{ is a zoo animal that is not a snake}\}$.

c. (i) $\overline{E} = O$; (ii) $\overline{O} = E$.

Subsets

subset

Consider the sets $A = \{1, 2, 3, 4, 5, 6\}$ and $B = \{2, 4, 6\}$. All the elements of B are contained in A and B is a **subset** of A. We write $B \subseteq A$. In general, we have the following definition.

Definition of Subset

B is a **subset** of A, written $B \subseteq A$, if and only if every element of B is an element of A.

This definition allows B to be equal to A. The definition is written with the phrase "if and only if," which means "if B is a subset of A, then every element of B is an element of A, and if every element of B is an element of A, then B is a subset of A." If both $A \subseteq B$ and $B \subseteq A$, then $A = B$.

proper subset

If B is a subset of A and B is not equal to A, then B is a **proper subset** of A, written $B \subset A$. This means that every element of B is contained in A and there is at least one element of A that is not in B.

To indicate this, sometimes a Venn diagram like the one shown in Figure 2-4 is used.

Figure 2-4

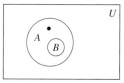

Example 2-5 Given $A = \{1, 2, 3, 4, 5\}, B = \{1, 3\}, P = \{x \mid x = 2^n - 1, n \in N\}$:

a. Which sets are subsets of each other?
b. Which sets are proper subsets of each other?

Solution **a.** Because $2^1 - 1 = 1, 2^2 - 1 = 3, 2^3 - 1 = 7, 2^4 - 1 = 15,$ and $2^5 - 1 = 31,$ $P = \{1, 3, 7, 15, 31, \ldots\}$. Thus $B \subseteq P$. Also $B \subseteq A, A \subseteq A, B \subseteq B,$ and $P \subseteq P$.
b. $B \subset A$ and $B \subset P$.

NOW TRY THIS 2-4

a. Suppose $A \subset B$. Can we always conclude that $A \subseteq B$?

b. If $A \subseteq B$, does it follow that $A \subset B$?

When a set A is not a subset of another set B, we write $A \not\subseteq B$. To show that $A \not\subseteq B$, we must find at least one element of A that is not in B. If $A = \{1, 3, 5\}$ and $B = \{1, 2, 3\}$, then A is not a subset of B because 5 is an element of A but not of B. Likewise, $B \not\subseteq A$ because 2 belongs to B but not to A.

Figures 2-4 and 2-5 show three relationships between two sets. In Figure 2-4 $B \subset A$. In Figure 2-5(a), A and B have no elements in common, in (b) A and B have at least one element in common.

Figure 2-5

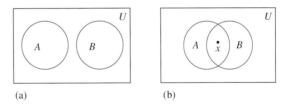

(a) (b)

It is not obvious how the empty set fits the definition of a subset because no elements in the empty set are elements of another set. To investigate this problem, we use the strategies of *indirect reasoning* and *looking at a special case*.

For the set $\{1, 2\}$, either $\emptyset \subseteq \{1, 2\}$ or $\emptyset \not\subseteq \{1, 2\}$. Suppose $\emptyset \not\subseteq \{1, 2\}$. Then there must be some element in $\emptyset$ that is not in $\{1, 2\}$. Because the empty set has no elements, there cannot be an element in the empty set that is not in $\{1, 2\}$. Consequently, $\emptyset \not\subseteq \{1, 2\}$ is false. Therefore the only other possibility, $\emptyset \subseteq \{1, 2\}$, is true.

The same reasoning can be applied in the case of the empty set and any other set. In particular, note that the empty set is a subset of itself and a proper subset of any set other than itself.

Subsets and elements of sets are often confused. We say that $2 \in \{1, 2, 3\}$. But because 2 is not a set, we cannot substitute the symbol $\subseteq$ for $\in$. However, $\{2\} \subseteq \{1, 2, 3\}$ and $\{2\} \subset \{1, 2, 3\}$. Conversely, the symbol $\in$ should not be used between $\{2\}$ and $\{1, 2, 3\}$.

Inequalities: An Application of Set Concepts

The notion of a proper subset and the concept of one-to-one correspondence can be used to define the concept of "less than" among natural numbers. The set $\{a, b, c\}$ has fewer elements than the set $\{w, x, y, z\}$ because when we try to pair the elements of the two sets, as in

$$\{a, b, c\}$$
$$| \ | \ |$$
$$\{x, y, z, w\},$$

we see that there is an element of the second set that is not paired with an element of the first set. The set $\{a, b, c\}$ is equivalent to a proper subset of the set $\{x, y, z, w\}$.

less than
greater than

In general, if A and B are finite sets, A has fewer elements than B and $n(A)$ is **less than** B, written $n(A) < n(B)$, if A is equivalent to a proper subset of B. We say that a is **greater than** b, written $a > b$, if and only if $b < a$. Defining the concept of "less than or equal to" in a similar way is explored in the Ongoing Assessment.

We have just seen that if A and B are finite sets and $A \subset B$, then A has fewer elements than B and it is not possible to find a one-to-one correspondence between the sets. Consequently, A and B are not equivalent. However, when both sets are infinite and $A \subset B$, the sets could be equivalent. For example, consider the set N of counting numbers and the set E of even counting numbers. We have $E \subset N$, but it is still possible to find a one-to-one correspondence between the sets. To do so, we correspond each number in set N to a number in set E that is twice as large. That is, $n \in N$ corresponds to $2n \in E$ as shown below.

$$N = \{1, 2, 3, 4, \ 5, \ldots n \ldots\}.$$
$$\updownarrow \updownarrow \updownarrow \updownarrow \ \updownarrow$$
$$E = \{2, 4, 6, 8, 10, \ldots 2n \ldots\}.$$

Notice that in the above correspondence, every element of N corresponds to a unique element in E and, conversely, every element of E corresponds to a unique element in N. For example, 11 in N corresponds to $2 \cdot 11$, or 22, in E. And 100 in E corresponds to $100 \div 2$, or 50, in N. Thus $N \approx E$; that is, N and E are equivalent.

Problem Solving   Passing a Senate Measure

A committee of senators consists of Abel, Baro, Carni, and Davis. Suppose each member of the committee has one vote and a simple majority is needed either to pass or reject any measure. A measure that is neither passed nor rejected is considered to be blocked and will be voted on again. Determine the number of ways a measure could be passed or rejected and the number of ways a measure could be blocked.

- **Understanding the Problem**   We are asked to determine how many ways the committee of four could pass or reject a proposal and how many ways the committee of four could block a proposal. To pass or reject a proposal requires a winning coalition, that is, a group of senators that can pass or reject the proposal, regardless of what the others do. To block a proposal, there must be a blocking coalition, that is, a group that can prevent any proposal from passing but that cannot reject the measure.

- **Devising a Plan**   To solve the problem, we can *make a list* of subsets of the set of senators. Any subset of the set of senators with three or four members will form a winning coalition. Any subset of the set of senators with exactly two members will form a blocking coalition.

- **Carrying Out the Plan** We list all subsets of the set $S = \{$Abel, Baro, Carni, Davis$\}$ that have at least three elements and all subsets that have exactly two elements. For ease, we identify the members as follows: A—Abel, B—Baro, C—Carni, D—Davis. All the subsets are given next:

$\emptyset$	$\{A, B\}$	$\{A, B, C\}$	$\{A, B, C, D\}$
$\{A\}$	$\{A, C\}$	$\{A, B, D\}$	
$\{B\}$	$\{A, D\}$	$\{A, C, D\}$	
$\{C\}$	$\{B, C\}$	$\{B, C, D\}$	
$\{D\}$	$\{B, D\}$		
	$\{C, D\}$		

There are five subsets with at least three members that can form a winning coalition and pass or reject a measure and six subsets with exactly two members that can block a measure.

- **Looking Back** Other questions that might be considered include the following:

  1. How many minimal winning coalitions are there? In other words, how many subsets are there of which no proper subset could pass a measure?
  2. Devise a method to solve this problem without listing all subsets.
  3. In "Carrying Out the Plan," 16 subsets of $\{A, B, C, D\}$ are listed. Use that result to systematically list all the subsets of a committee of five senators. Can you find the number of subsets of the five-member committee without actually counting the subsets?

NOW TRY THIS 2-5

- Suppose a committee of U.S. senators consists of five members.
  a. How many winning coalitions are there now? (See the Passing a Senate Measure problem on page 68 for a definition of *winning coalition*.)
  b. Compare the number of winning coalitions having exactly four members with the number of senators on the committee. What is the reason for the result?
  c. Compare the number of winning coalitions having exactly three members with the number of subsets of the committee having exactly two members. What is the reason for the result?

Number of Subsets of a Set

How many subsets can be made from a set containing n elements? To obtain a general formula, we use the strategy of *trying simpler cases* first.

1. If $P = \{a\}$, then P has two subsets, $\emptyset$ and $\{a\}$.
2. If $Q = \{a, b\}$, then Q has four subsets, $\emptyset$, $\{a\}$, $\{b\}$, and $\{a, b\}$.
3. If $R = \{a, b, c\}$, then R has eight subsets, $\emptyset$, $\{a\}$, $\{b\}$, $\{c\}$, $\{a, b\}$, $\{a, c\}$, $\{b, c\}$, and $\{a, b, c\}$.

Using the information from these cases, we *make a table and search for a pattern*, as in Table 2-4.

Table 2-4

Number of Elements	Number of Subsets
1	2, or 2^1
2	4, or 2^2
3	8, or 2^3
.	.
.	.
.	.

Table 2-4 suggests that for 4 elements, there are 2^4, or 16, subsets. Is this guess correct? If $E = \{a, b, c, d\}$, then all the subsets of $D = \{a, b, c\}$ are also subsets of E. Eight new subsets are also formed by adjoining the element d to each of the eight subsets of D. The eight new subsets are $\{d\}$, $\{a, d\}$, $\{b, d\}$, $\{c, d\}$, $\{a, b, d\}$, $\{a, c, d\}$, $\{b, c, d\}$, and $\{a, b, c, d\}$. Thus there are twice as many subsets of set E (with four elements) as there are of set D (with three elements). Consequently, there are $2 \cdot 8$, or 2^4, subsets of a set with four elements. In a similar way, there are $2 \cdot 2^4$, or 2^5, subsets of a set with five elements. Because including one more element in a finite set doubles the number of possible subsets of the new set, a set with 6 elements will have $2 \cdot 2^5$, or 2^6, subsets and so on. In each case, the number of elements and the power of 2 used to obtain the number of subsets are equal. *Therefore, if there are n elements in a set, 2^n subsets can be formed.* If we apply the above result to the empty set—that is when $n = 0$—then we have $2^0 = 1$ because the empty set has only one subset—itself.

BRAIN TEASER Bertrand Russell's antinomy, "Is the set of all sets that are not members of themselves a member of itself?" has become popularized in the following paradox: The town barber shaves all those males, and only those males, who do not shave themselves. Assuming the barber is a male who shaves, who shaves the barber?

ONGOING ASSESSMENT 2-1

1. Write the following sets by listing their members or using set-builder notation:
   a. The set of letters in the word *mathematics*
   b. The set of states in the continental United States
   c. The set of natural numbers greater than 20
   d. The set of states in the United States that border the Pacific Ocean
2. Rewrite the following statements using mathematical symbols:
   a. P is equal to the set whose elements are a, b, c, and d.
   b. The set consisting of the elements 1 and 2 is a proper subset of $\{1, 2, 3, 4\}$.
   c. The set consisting of the elements 0 and 1 is not a subset of $\{1, 2, 3, 4\}$.
   d. 0 is not an element of the empty set.
   e. The set whose only element is 0 is not equal to the empty set.
3. Which of the following pairs of sets can be placed in one-to-one correspondence?
   a. $\{1, 2, 3, 4, 5\}$ and $\{m, n, o, p, q\}$
   b. $\{m, a, t, h\}$ and $\{f, u, n\}$
   c. $\{a, b, c, d, e, f, \ldots, m\}$ and $\{1, 2, 3, 4, 5, 6, \ldots, 13\}$
   d. $\{x \mid x$ is a letter in the word *mathematics*$\}$ and $\{1, 2, 3, 4, \ldots, 11\}$
   e. $\{\bigcirc, \triangle\}$ and $\{2\}$

4. How many one-to-one correspondences are there between two sets with
   a. 5 elements each?
   b. 6 elements each?
   c. n elements each?
5. How many one-to-one correspondences are there between the sets $\{x, y, z, u, v\}$ and $\{1, 2, 3, 4, 5\}$ if in each correspondence
   a. x must correspond to 5?
   b. x must correspond to 5 and y to 1?
   c. $x, y,$ and z must correspond to odd numbers?
6. Which of the following represent equal sets?
   $A = \{a, b, c, d\}$.
   $B = \{x, y, z, w\}$.
   $C = \{c, d, a, b\}$.
   $D = \{x \mid 1 \leq x \leq 4, x \in N\}$.
   $E = \varnothing$.
   $F = \{\varnothing\}$.
   $G = \{0\}$.
   $H = \{\ \}$.
   $I = \{x \mid x = 2n + 1, n \in W\}$, where $W = \{0, 1, 2, 3, \ldots\}$.
   $J = \{x \mid x = 2n - 1, n \in N\}$.
7. Find the cardinal number of each of the following sets:
   a. $\{101, 102, 103, \ldots, 1100\}$
   b. $\{1, 3, 5, \ldots, 1001\}$
   c. $\{1, 2, 4, 8, 16, \ldots, 1024\}$
   d. $\{x \mid x = k^2, k = 1, 2, 3, \ldots, 100\}$
   e. $\{\{1, 2\}, \{3, 4\}, \{5, 6\}\}$
   f. $\{i + j \mid i \in \{1, 2, 3\} \text{ and } j \in \{1, 2, 3\}\}$
8. If U is the set of all college students and A is the set of all college students with a straight A average, describe $\overline{A}$.
9. Suppose B is a proper subset of C.
   a. If $n(C) = 8$, what is the maximum number of elements in B?
   b. What is the least possible number of elements in B?
10. Suppose C is a subset of D and D is a subset of C.
    a. If $n(C) = 5$, find $n(D)$.
    b. What other relationship exists between sets C and D?
11. Indicate which symbol, $\in$ or $\notin$, makes each of the following statements true:
    a. 3 ____ $\{1, 2, 3\}$.
    b. 0 ____ $\varnothing$.
    c. $\{1\}$ ____ $\{1, 2\}$.
    d. $\varnothing$ ____ $\varnothing$.
    e. $\{1, 2\}$ ____ $\{1, 2\}$.
12. Indicate which symbol, $\subseteq$ or $\nsubseteq$, makes each part of problem 11 true.
13. Answer each of the following. If your answer is *no*, tell why.
    a. If $A = B$, can we always conclude that $A \subseteq B$?
    b. If $A \subseteq B$, can we always conclude that $A \subset B$?
    c. If $A \subset B$, can we always conclude that $A \subseteq B$?
    d. If $A \subseteq B$, can we always conclude that $A = B$?
14. Use the definition of *less than* to show each of the following:
    a. $2 < 4$.   b. $3 < 100$.   c. $0 < 3$.
15. On a certain senate committee there are seven senators: Abel, Brooke, Cox, Dean, Eggers, Funk, and Gage. Three of these members are to be appointed to a subcommittee. How many possible subcommittees are there?

Communication

16. Which of the following sets are well defined? Explain your answers.
    a. The set of wealthy school teachers
    b. The set of great books
    c. The set of natural numbers greater than 100
    d. The set of subsets of $\{1, 2, 3, 4, 5, 6\}$
    e. The set $\{x \mid x \neq x \text{ and } x \in N\}$
17. Is $\varnothing$ a proper subset of every set? Explain!
18. Define *less than or equal to* in a way similar to the definition of *less than*.
19. a. Give three examples of sets A and B and a universal set U such that $A \subset B$, find $\overline{A}$ and $\overline{B}$.
    b. Based on your observations, conjecture a relationship between $\overline{B}$ and $\overline{A}$.
    c. Justify your conjecture in (b) using a Venn diagram.

Open-Ended

20. a. Describe three sets of which you are a member.
    b. Describe three sets that have no members.
21. Find an infinite set A such that
    a. $\overline{A}$ is finite.
    b. $\overline{A}$ is infinite.
22. Describe two sets from real-life situations such that it is clear from using one-to-one correspondence, and not from counting, that one set has fewer elements than the other.

Cooperative Learning

23. a. Use a calculator if necessary to estimate the time it would take a computer to list all the subsets of $\{1, 2, 3, \ldots, 64\}$. Assume the fastest computer can list one subset in approximately 1 microsecond (one millionth of a second).
    b. Estimate the time it would take the computer to exhibit all the one-to-one correspondences between the sets $\{1, 2, 3, \ldots, 64\}$ and $\{65, 66, 67, \ldots, 128\}$.

Section 2-2 — Other Set Operations and Their Properties

Finding the complement of a set is an operation that acts on only one set at a time. In this section, we consider operations that act on two sets at a time.

Set Intersection

Suppose that during the fall quarter, a college wants to mail a survey to all its students who are enrolled in both art and biology classes. To do this, the school officials must identify those students who are taking both classes. If A and B are the sets of students taking art courses and the set of students taking biology courses during the fall quarter, respectively, then the desired set of students includes those common to A and B, or the **intersection** of A and B. The intersection of sets A and B is the shaded region in Figure 2-6.

intersection

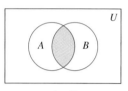

$A \cap B$

Figure 2-6

> **Definition of Set Intersection**
>
> The **intersection** of two sets A and B, written $A \cap B$, is the set of all elements common to both A and B. $A \cap B = \{x \mid x \in A \text{ and } x \in B\}$.

The key word in the definition of *intersection* is *and*. In everyday language, as in mathematics, *and* implies that both conditions must be met. In the above example, the desired set is the set of those students enrolled in both art and biology.

disjoint sets

If sets such as A and B have no elements in common, we call them **disjoint sets.** In other words, two sets A and B are disjoint if, and only if, $A \cap B = \emptyset$. For example, the set of males taking biology and the set of females taking biology are disjoint. The Venn diagram in Figure 2-7 implies that sets A and B are disjoint.

If A represents all students enrolled in art classes and B all students enrolled in biology classes, we may use a Venn diagram, taking into account that some students are enrolled in both subjects. If we know that 100 students are enrolled in art and 200 in biology and that 20 of these students are enrolled in both art and biology, we can record this information as in Figure 2-8. Notice that the total number of students in set A is 100 and the total in set B is 200.

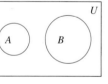

Figure 2-7

Example 2-6 Find $A \cap B$ in each of the following:

a. $A = \{1, 2, 3, 4\}, B = \{3, 4, 5, 6\}$
b. $A = \{0, 2, 4, 6, \ldots\}, B = \{1, 3, 5, 7, \ldots\}$
c. $A = \{2, 4, 6, 8, \ldots\}, B = \{1, 2, 3, 4, \ldots\}$

Solution
a. $A \cap B = \{3, 4\}$.
b. $A \cap B = \emptyset$; therefore A and B are disjoint.
c. $A \cap B = A$ because all the elements of A are also in B.

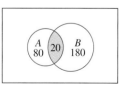

Figure 2-8

Set Union

If A is the set of students taking art courses during the fall quarter and B is the set of students taking biology courses during the fall quarter, then the set of students taking art or biology or both is the **union** of sets A and B. The union of sets A and B is pictured in Figure 2-9.

union

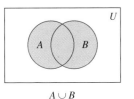

$A \cup B$

Figure 2-9

Definition of Set Union

The **union** of two sets A and B, written $A \cup B$, is the set of all elements in A or in B, $A \cup B = \{x \mid x \in A \text{ or } x \in B\}$.

The key word in the definition of *union* is *or*. In mathematics, *or* usually means "one or the other or both." This is known as the *inclusive or*.

Example 2-7 Find $A \cup B$ for each of the following:

a. $A = \{1, 2, 3, 4\}, B = \{3, 4, 5, 6\}$ b. $A = \{0, 2, 4, 6, \ldots\}, B = \{1, 3, 5, 7, \ldots\}$
c. $A = \{2, 4, 6, 8, \ldots\}, B = \{1, 2, 3, 4, \ldots\}$

Solution a. $A \cup B = \{1, 2, 3, 4, 5, 6\}$
b. $A \cup B = \{0, 1, 2, 3, 4, \ldots\}$
c. Because every element of A is already in B we have $A \cup B = B$.

NOW TRY THIS 2-6

- Notice that in Figure 2-8, $n(A \cup B) = 80 + 20 + 180 = 280$, but $n(A) + n(B) = 100 + 200 = 300$; hence in general $n(A \cup B) \neq n(A) + n(B)$. Use the concept of intersection of sets to write a formula for $n(A \cup B)$.

Set Difference

If A is the set of students taking art classes during the fall quarter and B is the set of students taking biology classes, then the set of all students taking biology but not art is called the **complement of A relative to B**, or the **set difference** of B and A.

complement of A relative to B · set difference

Definition of Relative Complement

The **complement of A relative to B**, written $B - A$, is the set of all elements in B that are not in A, $B - A = \{x \mid x \in B \text{ and } x \notin A\}$.

A Venn diagram representing $B - A$ is shown in Figure 2-10 (a). The shaded region represents all the elements that are in B but not in A. A Venn diagram for $B \cap \overline{A}$ is given in Figure 2-10 (b). The shaded region represents all the elements that are in B and in $\overline{A}$.

Notice that $B \cap \overline{A} = B - A$ because $B \cap \overline{A}$ is, by definition of intersection and complement, the set of all elements in B and not in A.

Figure 2-10

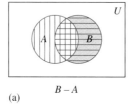
(a) $B - A$

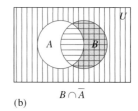
(b) $B \cap \overline{A}$

Example 2-8 If $U = \{a, b, c, d, e, f, g\}$, $A = \{d, e, f\}$, $B = \{a, b, c, d, e, f\}$, and $C = \{a, b, c,\}$, find each of the following:

a. $A - B$
b. $B - A$
c. $B - C$
d. $C - B$

Solution **a.** $A - B = \emptyset$
b. $B - A = \{a, b, c\}$
c. $B - C = \{d, e, f\}$
d. $C - B = \emptyset$

Properties of Set Operations

commutative property of set union

commutative property of set intersection

Because the order of elements in a set is not important, $A \cup B$ is equal to $B \cup A$. This is the **commutative property of set union.** It does not matter in which order we write the sets when the union of two sets is involved. Similarly, $A \cap B = B \cap A$. This is the **commutative property of set intersection.**

NOW TRY THIS 2-7

- Use Venn diagrams and other means to find whether grouping is important when the same operation is involved. For example, is it always true that $A \cap (B \cap C) = (A \cap B) \cap C$? Similar questions should be investigated involving union and set difference.

Example 2-9 Is grouping important when two set operations are involved? For example, is it true that $A \cap (B \cup C) = (A \cap B) \cup C$?

Solution To investigate this, we let $A = \{a, b, c, d\}$, $B = \{c, d, e\}$, and $C = \{d, e, f, g\}$. Then

$$A \cap (B \cup C) = \{a, b, c, d\} \cap (\{c, d, e\} \cup \{d, e, f, g\})$$
$$= \{a, b, c, d\} \cap \{c, d, e, f, g\}$$
$$= \{c, d\}$$
$$(A \cap B) \cup C = (\{a, b, c, d\} \cap \{c, d, e\}) \cup \{d, e, f, g\}$$
$$= \{c, d\} \cup \{d, e, f, g\}$$
$$= \{c, d, e, f, g\}.$$

In this case, $A \cap (B \cup C) \neq (A \cap B) \cup C$. So we have found a counterexample, that is, an example illustrating that the general statement is not always true.

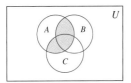

Figure 2-11

To discover an expression that is equal to $A \cap (B \cup C)$, consider the Venn diagram for $A \cap (B \cup C)$ shown by the shaded region in Figure 2-11. In the figure, $A \cap C$ and $A \cap B$ are subsets of the shaded region. The union of $A \cap C$ and $A \cap B$ is the entire shaded region. Thus $A \cap (B \cup C) = (A \cap B) \cup (A \cap C)$.

Property

Distributive property of set intersection over union. For all sets A, B, and C,
$$A \cap (B \cup C) = (A \cap B) \cup (A \cap C).$$

Example 2-10 Use set notation to describe the shaded portions of the Venn diagrams in Figure 2-12.

Figure 2-12

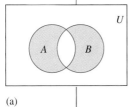 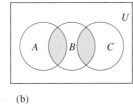 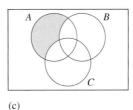 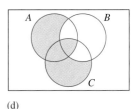

(a)   (b)   (c)   (d)

Solution The solutions can be described in many different, but equivalent, forms. The following are possible answers:

a. $(A \cup B) - (A \cap B)$, or $(A \cup B) \cap \overline{(A \cap B)}$
b. $(A \cap B) \cup (B \cap C)$, or $B \cap (A \cup C)$
c. $(A - B) - C$, or $A - (B \cup C)$, or $(A - (A \cap B)) - (A \cap C)$
d. $((A \cup C) - B)) \cup (A \cap B \cap C)$, or $(A - (B \cup C)) \cup (C - (A \cup B)) \cup (A \cap C)$

NOW TRY THIS 2-8

- If, on both sides of the equation in the Distributive Property of Set Intersection over Union, the symbol ∩ is replaced by ∪ and the symbol ∪ is replaced by ∩, is the new property true? Explain why or why not.

Using Venn Diagrams as a Problem-Solving Tool

Venn diagrams can be used as a problem-solving tool for modeling information, as shown in the following examples.

Example 2-11 Suppose M is the set of all students taking mathematics and E is the set of all students taking English. Identify the students described by each region in Figure 2-13.

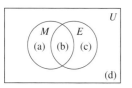

Figure 2-13

Solution

Region (a) contains all students taking mathematics but not English.
Region (b) contains all students taking both mathematics and English.
Region (c) contains all students taking English but not mathematics.
Region (d) contains all students taking neither mathematics nor English.

The following student page from *McDougal Littell Middle Grades MATH Thematics, Book 1*, 1999 is another example of the use of Venn diagrams in modeling information. Answer questions 48 through 56 on the student page.

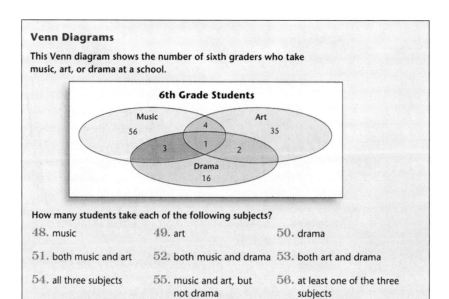

Example 2-12

In a survey of 110 college freshmen investigating their high school backgrounds, the following information was gathered:

25 took physics
45 took biology
48 took mathematics
10 took physics and mathematics
8 took biology and mathematics
6 took physics and biology
5 took all 3 subjects

a. How many students took biology but neither physics nor mathematics?
b. How many took physics, biology, or mathematics?
c. How many did not take any of the 3 subjects?

Solution To solve this problem, we *build a model* using sets. Because there are three distinct subjects, we should use three circles. The maximum number of regions of a Venn diagram determined by three circles is eight. In Figure 2-14, P is the set of students taking physics, B is the set taking biology, and M is the set taking mathematics. The shaded region represents the five students who took all three subjects. The lined region represents the students who took physics and mathematics, but who did not take biology.

In part (a) we are asked for the number of students in the subset of B that has no elements in common with either P or M. That is, $B - (P \cup M)$. In part (b) we are asked for the number of elements in $P \cup B \cup M$. Finally, in part (c) we are asked for the number of students in $\overline{P \cup B \cup M}$, or $U - (P \cup B \cup M)$. Our strategy is to find the number of students in each of the eight non-overlapping regions.

One mind set to beware of in this problem is thinking that the 25 who took physics, for example, took only physics. That is not necessarily the case. If those students had been taking only physics, then we should have been told so.

a. Because a total of 10 students took physics and mathematics and 5 of those also took biology, $10 - 5$, or 5, students took physics and math but not biology. Similarly, because 8 students took biology and mathematics and 5 took all 3 subjects, $8 - 5$, or 3, took biology and mathematics but not physics. Also $6 - 5$, or 1, student took physics and biology but not mathematics. To find the number of students who took biology but neither physics nor mathematics, we subtract from 45 (the total number that took biology) the number of those that are in the distinct regions that include biology and other subjects, that is, $1 + 5 + 3$, or 9. Because $45 - 9 = 36$, we know that 36 students took biology but neither physics nor mathematics.

b. To find the number of students in all the distinct regions in P, M, or B, we proceed as follows. The number of students who took physics but neither mathematics nor biology is $25 - (1 + 5 + 5)$, or 14. The number of students who took mathematics but neither physics nor biology is $48 - (5 + 5 + 3)$, or 35. Hence the number of students who took mathematics, physics, or biology is $35 + 14 + 36 + 3 + 5 + 5 + 1$, or 99.

c. Because the total number of students is 110, the number that did not take any of the three subjects is $110 - 99$, or 11.

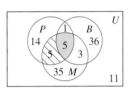

Figure 2-14

Cartesian Products

Cartesian product Another way to produce a set from two given sets is by forming the **Cartesian product.** This formation pairs the elements of one set with the elements of another set in a specific way. Suppose a person has three pairs of pants, $P = \{\text{blue, white, green}\}$, and two shirts, $S = \{\text{blue, red}\}$. According to the Fundamental Counting Principle, there are $3 \cdot 2$, or 6, possible different pant-and-shirt pairs as shown in Figure 2-15.

Figure 2-15

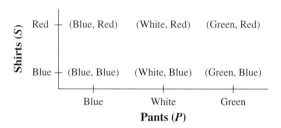

The pairs of pants and shirts form a set of all possible pairs in which the first member of the pair is an element of set P and the second member is an element of set S. The set of all possible pairs is given in Figure 2-15. Because the first component in each pair represents pants and the second component in each pair represents shirts, the order in which the components are written is important. Thus (green, blue) represents green pants and a blue shirt, whereas (blue, green) represents blue pants and a green shirt. Therefore the two pairs represent different outfits. Because the order in each pair is important, the pairs are called **ordered pairs.** The positions that the ordered pairs occupy within the set of outfits is immaterial. Only the order of the **components** within each pair is significant.

ordered pairs
components

The pant-and-shirt pairs suggest the following definition of equality for ordered pairs: $(x, y) = (m, n)$ *if, and only if, the first components are equal and the second components are equal.* A set consisting of ordered pairs such as the ones in the pants-and-shirt example is the Cartesian product of the set of pants and the set of shirts.

> ### Definition of Cartesian Product
>
> For any sets A and B, the **Cartesian product** of A and B, written $A \times B$ is the set of all ordered pairs such that the first component of each pair is an element of A and the second component of each pair is an element of B.
>
> $$A \times B = \{(x, y) \mid x \in A \text{ and } y \in B\}$$

REMARK $A \times B$ is commonly read as "A cross B."

Example 2-13 If $A = \{a, b, c\}$ and $B = \{1, 2, 3\}$, find each of the following:

  a. $A \times B$ **b.** $B \times A$ **c.** $A \times A$

Solution **a.** $A \times B = \{(a, 1), (a, 2), (a, 3), (b, 1), (b, 2), (b, 3), (c, 1), (c, 2), (c, 3)\}$.
         b. $B \times A = \{(1, a), (1, b), (1, c), (2, a), (2, b), (2, c), (3, a), (3, b), (3, c)\}$.
         c. $A \times A = \{(a, a), (a, b), (a, c), (b, a), (b, b), (b, c), (c, a), (c, b), (c, c)\}$.

ONGOING ASSESSMENT 2-2

1. Use a Venn diagram or any other approach to decide whether the following pairs of sets are equal.
   a. $A \cap B$ and $B \cap A$
   b. $A \cup B$ and $B \cup A$
   c. $A \cup (B \cup C)$ and $(A \cup B) \cup C$
   d. $A \cup \emptyset$ and A
   e. $A \cup A$ and $A \cup \emptyset$
   f. $(A \cap A)$ and $(A \cap \emptyset)$

2. Tell whether each of the following is true or false for all sets A, B, or C. If false, give a counterexample.
   a. $A \cup \emptyset = A$.
   b. $A - B = B - A$.
   c. $A \cup A = A$.
   d. $\overline{A \cap B} = \overline{A} \cap \overline{B}$.
   e. $(A \cup B) \cup C = A \cup (B \cup C)$.
   f. $(A \cup B) - A = B$.
   g. $(A - B) \cup A = (A - B) \cup (B - A)$.

3. If $B \subseteq A$, find a simpler expression for each of the following:
   a. $A \cap B$ b. $A \cup B$

4. For each of the following, shade the portion of the Venn diagram that illustrates the set:
   a. $A \cup B$
   b. $A \cap \overline{B}$
   c. $\overline{A \cap B}$
   d. $(A \cap B) \cup (A \cap C)$
   e. $\overline{A} \cap B$
   f. $(A \cup B) \cap \overline{C}$
   g. $(A \cap B) \cup C$
   h. $(\overline{A} \cap B) \cup C$

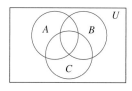

5. If S is a subset of universe U, find each of the following:
   a. $S \cup \overline{S}$ b. $\emptyset \cup S$ c. $\overline{U}$
   d. $\overline{\emptyset}$ e. $S \cap \overline{S}$ f. $\emptyset \cap S$

6. For each of the following conditions, find $A - B$:
   a. $A \cap B = \emptyset$.   b. $B = U$.
   c. $A = B$.   d. $A \subseteq B$.

7. a. Give two examples of sets A and B for which $A - B = \emptyset$. Show that in each example $A \subseteq B$.
   b. If for sets A and B we know that $A - B = \emptyset$, is it necessarily true that $A \subseteq B$? Justify your answer.

8. Use set notation to identify each of the following shaded regions:

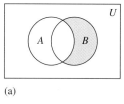

(a)

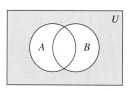

(b)

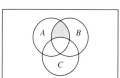

(c)

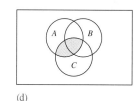

(d)

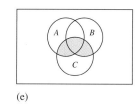

(e)

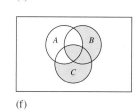

(f)

9. In the following, shade the portion of the diagram that represents the given sets:

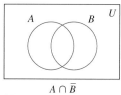

(a) $A \cap \overline{B}$

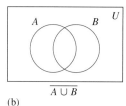

(b) $\overline{A \cup B}$

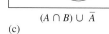

(c) $(A \cap B) \cup \overline{A}$

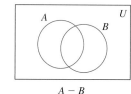

(d) $A - B$

10. Use Venn diagrams to determine if each of the following is true:
    a. $A \cup (B \cap C) = (A \cup B) \cap C$.
    b. $A \cap (B \cup C) = (A \cap B) \cup C$.
    c. $A - (B - C) = (A - B) - C$.
11. For each of the following pairs of sets, explain which is a subset of the other. If neither is a subset of the other, explain why.
    a. $A \cap B$ and $A \cap B \cap C$
    b. $A \cup B$ and $A \cup B \cup C$
    c. $A \cup B$ and $(A \cup B) \cap C$
    d. $A - B$ and $B - A$
12. a. If A has three elements and B has two elements, what is the greatest number of elements possible in (i) $A \cup B$? (ii) $A \cap B$? (iii) $B - A$? (iv) $A - B$?
    b. If A has n elements and B has m elements, what is the greatest number of elements possible in (i) $A \cup B$? (ii) $A \cap B$? (iii) $B - A$? (iv) $A - B$?
13. If $n(A) = 4$, $n(B) = 5$, and $n(C) = 6$, what is the greatest and least number of elements in
    a. $A \cup B \cup C$?
    b. $A \cap B \cap C$?
14. The equation $\overline{A \cup B} = \overline{A} \cap \overline{B}$ and a similar equation for $\overline{A \cap B}$ are referred to as *DeMorgan's laws* in honor of the famous British mathematician who first discovered them.
    a. Use Venn diagrams to show that $\overline{A \cup B} = \overline{A} \cap \overline{B}$.
    b. Discover an equation similar to the one in part (a) involving $\overline{A \cap B}$, $\overline{A}$, and $\overline{B}$. Use Venn diagrams to show that the equation holds.
    c. Verify the equations in (a) and (b) for specific sets.
15. Given that the universe is the set of all humans, $B = \{x \mid x$ is a college basketball player$\}$, and $S = \{x \mid x$ is a college student more than 200 cm tall$\}$, describe each of the following in words:
    a. $B \cap S$
    b. $\overline{S}$
    c. $B \cup S$
    d. $\overline{B \cup S}$
    e. $\overline{B} \cap S$
    f. $B \cap \overline{S}$
16. Suppose P is the set of all eighth-grade students at the Paxson School, with B the set of all students in the band and C the set of all students in the choir. Identify in words the students described by each region of the following figure:

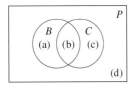

17. Of the eighth graders at the Paxson School, 7 played basketball, 9 played volleyball, 10 played soccer, 1 played basketball and volleyball only, 1 played basketball and soccer only, 2 played volleyball and soccer only, and 2 played volleyball, basketball, and soccer. How many played one or more of the three sports?
18. In a fraternity with 30 members, 18 take mathematics, 5 take both mathematics and biology, and 8 take neither mathematics nor biology. How many take biology but not mathematics?
19. Write the letters in the appropriate sections of the following Venn diagram using the following information:

    Set A contains the letters in the word *Iowa*.
    Set B contains the letters in the word *Hawaii*.
    Set C contains the letters in the word *Ohio*.

    The universal set contains the letters in the word *Washington*.

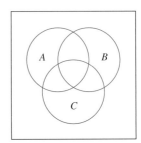

20. In Paul's bicycle shop, 40 bicycles are inspected. If 20 needed new tires and 30 needed gear repairs, answer the following:
    a. What is the greatest number of bikes that could have needed both?
    b. What is the least number of bikes that could have needed both?
    c. What is the greatest number of bikes that could have needed neither?
21. A pollster interviewed 500 university seniors who owned credit cards. She reported that 240 owned Goldcard, 290 had Supercard, and 270 had Thriftcard. Of those seniors, the report said that 80 owned a Goldcard and a Supercard, 70 owned a Goldcard and a Thriftcard, 60 owned a Supercard and a Thriftcard, and 50 owned all 3 cards. When the report was submitted for publication in the local campus newspaper, the editor refused to publish it, claiming the poll was not accurate. Was the editor right? Why or why not?
22. The Red Cross looks for three types of antigens in blood tests: A, B, and Rh. When the antigen A or B is present, it is listed, but if both these antigens are absent, the blood is type O. If the Rh antigen is present, the blood is positive, otherwise, it is negative. If a laboratory technician reports the following results after testing the blood samples of 100 people, how many were classified as O negative? Explain your reasoning.

Number of Samples	Antigen in Blood
40	A
18	B
82	Rh
5	A and B
31	A and Rh
11	B and Rh
4	A, B, and Rh

23. Classify the following as true or false. If false, give a counterexample. Assume that A and B are finite sets.
    a. If $n(A) = n(B)$, then $A = B$.
    b. If $A \sim B$, then $A \cup B$ is not equivalent to B.
    c. If $A - B = \emptyset$, then $A = B$.
    d. If $B - A = \emptyset$, then $B \subseteq A$.
    e. If $A \subset B$, then $n(A) < n(B)$.
    f. If $n(A) < n(B)$, then $A \subset B$.

24. Three announcers each tried to predict the winners of Sunday's professional football games. The only team not picked that is playing Sunday was the Giants. The choices for each person were as follows:

    Phyllis: Cowboys, Steelers, Vikings, Bills
    Paula: Steelers, Packers, Cowboys, Redskins
    Rashid: Redskins, Vikings, Jets, Cowboys

    If the only teams playing Sunday are those just mentioned, which teams will play which other teams?

25. Let $A = \{x, y\}$, $B = \{a, b, c\}$, and $C = \{0\}$. Find each of the following:
    a. $A \times B$
    b. $B \times A$
    c. $B \times \emptyset$
    d. $(A \cup B) \times C$
    e. $A \cup (B \times C)$

26. For each of the following, the Cartesian product $C \times D$ is given by the following sets. Find C and D.
    a. $\{(a, b), (a, c), (a, d), (a, e)\}$
    b. $\{(1, 1), (1, 2), (1, 3), (2, 1), (2, 2), (2, 3)\}$
    c. $\{(0, 1), (0, 0), (1, 1), (1, 0)\}$

27. Answer each of the following:
    a. If A has five elements and B has four elements, how many elements are in $A \times B$?
    b. If A has m elements and B has n elements, how many elements are in $A \times B$?
    c. If A has m elements, B has n elements, and C has p elements, how many elements are in $(A \times B) \times C$?

28. If there are six teams in the Alpha league and five teams in the Beta league and if each team from one league plays each team from the other league exactly once, how many games are played?

29. José has four pairs of slacks, five shirts, and three sweaters. From how many combinations can he choose if he chooses a pair of slacks, a shirt, and a sweater each day?

Communication

30. Answer each of the following and justify your answer:
    a. If $a \in A \cap B$, is it true that $a \in A \cup B$?
    b. If $a \in A \cup B$, is it true that $a \in A \cap B$?

31. The primary colors are red, blue, and yellow. If each is considered a set, write an explanation of what we would expect to get with the intersection of each of these sets from the regions pictured in the following figure:

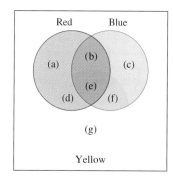

32. a. Is the operation of forming Cartesian products commutative? Explain why or why not.
    b. Is the operation of forming Cartesian products associative? Explain why or why not.

Open-Ended

33. Make up and solve a story problem concerning specific sets A, B, and C for which $n(A \cup B \cup C)$ is known and it is required to find $n(A)$, $n(B)$, and $n(C)$.

34. Describe a real-life situation that can be represented by each of the following:
    a. $A \cap \overline{B}$
    b. $A \cap B \cap C$
    c. $A - (B \cup C)$

Cooperative Learning

35. List several properties of set operations. They might include properties mentioned in the text or problems as well as properties that your group discovers. Compare your group's properties with those listed by others. Be prepared to show why your properties are true.

36. Use set operations like union, intersection, complement, and set difference to describe the shaded region in the following figure in as many ways as possible. Compare your expressions with those of other groups to see which has the most. What is the total number of different expressions found by all the groups? Which expressions appeared in all the groups?

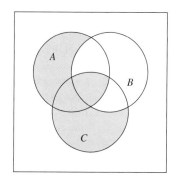

Review Problems

37. Find the number of elements in the following sets:
    a. $\{x \mid x \text{ is a letter in } commonsense\}$
    b. the set of letters appearing in the word *committee*
38. If $A = \{1, 2, 3, 4\}$ and $B = \{1, 2, 3, 4, 5\}$, answer the following questions:
    a. How many subsets of A do not contain the element 1?
    b. How many subsets of A contain the element 1?
    c. How many subsets of A contain either the element 1 or 2?
    d. How many subsets of A contain neither the element 1 nor 2?
    e. How many subsets of B contain the element 5 and how many do not?
    f. If all the subsets of A are known, how can all the subsets of B be listed systematically? How many subsets of B are there?
39. a. Which of the following sets are equal?
    b. Which sets are proper subsets of the other sets?

    $A = \{2, 4, 6, 8, 10, \ldots\}$.
    $B = \{x \mid x = 2n + 2, n = 0, 1, 2, 3, 4, \ldots\}$.
    $C = \{x \mid x = 4n, n \in N\}$.

40. Give examples from real life for each of the following:
    a. A one-to-one correspondence between two sets
    b. A correspondence between two sets that is not one-to-one

LABORATORY ACTIVITY

A set of attribute blocks consists of 32 blocks. Each block is identified by its own shape, size, and color. The 4 shapes in a set are square, triangle, rhombus, and circle; the 4 colors are red, yellow, blue, and green; the 2 sizes are large and small. In addition to the blocks, each set contains a group of 20 cards. Ten of the cards specify one of the attributes of the blocks (for example, red, large, square). The other 10 cards are negation cards and specify the lack of an attribute (for example, not green, not circle). Many set-type problems can be studied with these blocks. For example, let A be the set of all green blocks and B be the set of all large blocks. Using the set of all blocks as the universal set, describe elements in each set listed below to determine which are equal:

1. $A \cup B; B \cup A$
2. $\overline{A \cap B}; \overline{A} \cap \overline{B}$
3. $\overline{A \cap B}; \overline{A} \cup \overline{B}$
4. $A - B; A \cap \overline{B}$

Section 2-3 — Addition and Subtraction of Whole Numbers

In Section 2-1 we saw that the concept of one-to-one correspondence between sets can be used to introduce children to the concept of a number. The *Principles and Standards* say that in grades pre-K–2, students should:

 … associate number words with small collections of objects and gradually learn to count and keep track of objects in larger groups. They should learn that counting objects in a different order does not alter their result, and they may notice that the next whole number in the counting sequence is one more than the number just named. Children should learn that the last number named represents the last object as well as the total number of objects in the collection (p. 79).

In the following "Peanuts" cartoon, it seems that Linus's little brother has not yet learned to associate number words with a collection of objects.

whole numbers

When zero is joined with the set of natural numbers, $N = \{1, 2, 3, 4, 5, \ldots\}$, we have the set of numbers called **whole numbers,** denoted by $W = \{0, 1, 2, 3, 4, 5, \ldots\}$. In this section, we provide a variety of models for teaching computational skills and allow students to revisit mathematics at a level they must know in order to be competent teachers.

HISTORICAL NOTE

Historians think that the word *zero* originated from the Hindu word *sūnya,* which means "void." Then *sūnya* was translated into the Arabic *sifr,* which when translated to Latin became *zephirum,* from which the word *zero* was derived.

Addition of Whole Numbers

Children encounter addition in preschool years by combining objects and wanting to know how many objects there are in the combined set. A set model is one way to represent addition of whole numbers.

Set Model

Suppose Jane has 4 boxes in one pile and 3 in another. If she combines the two groups of boxes, how many boxes are there in the combined group? Figure 2-16 shows the solution as it might appear in an elementary school text. The combined set of boxes is the union of

Figure 2-16

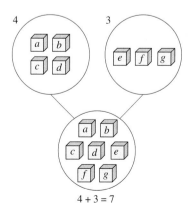

$4 + 3 = 7$

the disjoint sets of 4 boxes and 3 boxes. After the sets have been combined, children count the boxes to determine that there are 7 boxes in all. Note the importance of the sets having no elements in common. If the sets have common elements, then an incorrect conclusion can be drawn.

Using these ideas we can use set terminology to define addition formally.

> **Definition of Addition of Whole Numbers**
>
> Let A and B be two disjoint finite sets. If $n(A) = a$ and $n(B) = b$, then $a + b = n(A \cup B)$.

addends · sum   The numbers a and b in $a + b$ are the **addends** and $(a + b)$ is the **sum.**

> **NOW TRY THIS 2-9**
>
> ● If the sets in the preceding definition of addition of whole numbers are not disjoint, explain why the definition is incorrect. ●

HISTORICAL NOTE

The symbol "+" first appeared in a 1417 manuscript and was a short way of writing the Latin word *et*, which means "and." However, Johann Widmann wrote a book in 1498 that made use of the + and − symbols for addition and subtraction. The word *minus* means "less" in Latin. First written as an *m*, it was later shortened to a horizontal bar.

Number-Line Model

For some problems, the set model for addition that we have developed is not the appropriate model. For example, consider the following questions:

1. Josh has 4 feet of red ribbon and 3 feet of white ribbon. How many feet of ribbon does he have altogether?
2. One day, Gail drank 4 ounces of orange juice in the morning and 3 ounces at lunch time. If she drank no other orange juice that day, how many ounces of orange juice did she drink for the entire day?

A number line can be used to model whole-number addition. Any line marked with two fundamental points, one representing 0 and the other representing 1, can be turned into a number line. The points representing 0 and 1 mark the ends of a *unit segment*. Other points can be marked and labeled, as shown in Figure 2-17. Any two consecutive points in Figure 2-17 mark the ends of a segment that has the same length as the unit segment.

Figure 2-17

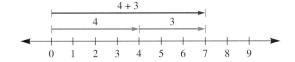

Addition problems can be modeled using directed arrows on the number line. For example, the sum of 4 + 3 is shown in Figure 2-17. Arrows representing the addends, 4 and 3, are combined into one arrow representing the sum. Figure 2-17 poses an inherent problem for students. If an arrow starting at 0 and ending at 3 represents 3, why should an arrow starting at 4 and ending at 7 represent 3?

In Section 2-1 we used the concept of a set and the concept of a one-to-one correspondence to define *greater-than* relations. A number line can also be used to describe **greater-than** and **less-than** relations on the set of whole numbers. For example, in Figure 2-17, notice that 4 is to the left of 7 on the number line. We say, "four is less than seven," and we write $4 < 7$. We can also say "seven is greater than four" and write $7 > 4$. Since 4 is to the left of 7, there is a natural number that can be added to 4 to get 7, namely, 3. Thus $4 < 7$ because $4 + 3 = 7$. We can generalize this discussion to form the following definition of *less than*.

greater than · less than

Definition of Less Than

For any whole numbers a and b, a is **less than** b, written $a < b$, if and only if there exists a natural number k such that $a + k = b$.

greater than or equal to · less than or equal to

Sometimes equality is combined with the inequalities, greater than and less than, to give the relations **greater than or equal to** and **less than or equal to,** denoted by $\geq$ and $\leq$. Thus, $a \leq b$ means $a < b$ or $a = b$. The emphasis with respect to these symbols is on the *or*. Observe that "$3 < 5$ or $3 = 5$" is a true statement, so $3 \leq 5$ is true. Both $5 \geq 3$ and $3 \geq 3$ are true statements.

Whole-Number Addition Properties

Any time two whole numbers are added, we are guaranteed that a unique whole number will be obtained. This property is sometimes referred to as the *closure property of addition of whole numbers*. We say that "the set of whole numbers is closed under addition." Figure 2-18(a) shows two additions. Pictured above the number line is $3 + 5$ and below the number line is $5 + 3$. The sums are exactly the same. Figure 2-18(b) shows the same sums obtained with colored rods with the result being the same. Both illustrations in Figure 2-18 demonstrate the idea that two whole numbers can be added in either order. This property is true in general and is the *commutative property of addition of whole numbers*. We say that "addition of whole numbers is commutative."

Figure 2-18

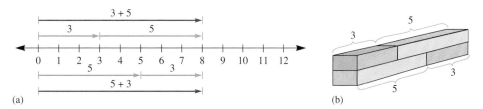

(a) (b)

> **Property**
>
> **Commutative property of addition of whole numbers** If a and b are any whole numbers, then $a + b = b + a$.

The commutative property of addition of whole numbers is not obvious to many young children. They may be able to find the sum $9 + 2$ and not be able to find the sum $2 + 9$. This is because one of the techniques used to teach addition is *counting on*. Using this technique, $9 + 2$ can be computed by starting at 9 and then counting on two more as "ten" and "eleven." To compute $2 + 9$, the *counting on* is more involved. Students need to understand that $2 + 9$ is another name for $9 + 2$.

Another property of addition is demonstrated when we select the order in which to add three or more numbers. For example, we could compute $24 + 8 + 2$ by grouping the 24 and the 8 together: $(24 + 8) + 2 = 32 + 2 = 34$. (The parentheses indicate that the first two numbers are grouped together.) We might also recognize that it is easy to add any number to 10 and compute it as $24 + (8 + 2) = 24 + 10 = 34$. This example illustrates the *associative property of addition of whole numbers*.

> **Property**
>
> **Associative property of addition of whole numbers** If a, b, and c are any whole numbers, then $(a + b) + c = a + (b + c)$.

When several numbers are being added, the parentheses are usually omitted, since the grouping does not alter the result.

Another property of addition of whole numbers operates when one addend is 0. In Figure 2-19; set A has 5 blocks and set B has 0 blocks. The union of sets A and B has only 5 blocks.

Figure 2-19

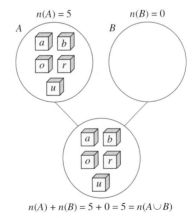

$n(A) + n(B) = 5 + 0 = 5 = n(A \cup B)$

This example illustrates the following property of whole numbers.

Property

Identity property of addition of whole numbers There is a unique whole number 0, the **additive identity,** such that for any whole number a, $a + 0 = a = 0 + a$.

Example 2-14 Which properties justify each of the following?

a. $5 + 7 = 7 + 5$.
b. $1001 + 733$ is a whole number.
c. $(3 + 5) + 7 = (5 + 3) + 7$.
d. $(8 + 5) + 2 = 2 + (8 + 5) = (2 + 8) + 5$.
e. $(10 + 5) + (10 + 3) = (10 + 10) + (5 + 3)$.

Solution
a. Commutative property of addition
b. Closure property of addition
c. Commutative property of addition
d. Commutative and associative properties of addition
e. Commutative and associative properties of addition combined

Mastering Basic Addition Facts

Certain mathematical facts are *basic addition facts*. Basic addition facts are those involving a single digit plus a single digit. As the *Principles and Standards* point out:

Knowing basic number combinations—the single-digit addition and multiplication pairs and their counterparts for subtraction and division—is essential (p. 32).

One method of learning the basic facts is to organize them according to different strategies, listed as follows:

1. *Counting On.* The strategy of counting on from the greater of the addends can be used any time we need to add whole numbers, but is inefficient. It is usually used when the other addend is 1, 2, or 3. For example, $5 + 3$ can be computed by starting at 5 and then counting on 6, 7, and 8. Likewise, $2 + 8$ would be computed by starting at 8 and then counting on 9 and 10.
2. *Doubles.* The next strategy considered involves the use of *doubles*. Doubles such as $4 + 4$ and $6 + 6$ receive special attention with students. After doubles are mastered, *doubles + 1* and *doubles + 2* can be learned easily. For example, if a student knows $6 + 6 = 12$, then $6 + 7$ is $(6 + 6) + 1$, or one more than the double of 6, or 13. Likewise, $7 + 9$ is $(7 + 7) + 2$ or two more than the double of 7, or 16.
3. *Making 10.* Another strategy is that of *making 10* and then adding any leftover. For example, we could think of $8 + 5$ as shown in Figure 2-20. Notice that we are really using the associative property of addition.

Figure 2-20

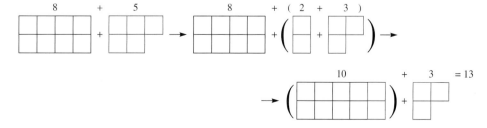

Many basic facts might be classified under more than one strategy. For example, we could find 9 + 8 using *making 10* as 9 + (1 + 7) = (9 + 1) + 7 = 10 + 7 = 17 or using a *double plus 1* as (8 + 8) + 1.

Subtraction of Whole Numbers

Subtraction of whole numbers can be modeled in several ways: the *set (take-away)* model, the *missing-addend* model, the *comparison* model, and the *number-line* model.

Take-Away Model

One way to think about subtraction is this: Instead of imagining a second set of objects as being joined to a first set (as in addition), consider the second set as being *taken away* from a first set. For example, suppose we have 8 blocks and take away 3 of them, as shown in Figure 2-21. We record this process as 8 − 3 = 5.

Figure 2-21

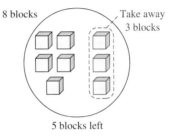

Missing-Addend Model

A second model for subtraction, the *missing-addend* model relates subtraction and addition. Recall that in Figure 2-21, 8 − 3 is pictured as 8 blocks "take away" 3 blocks. The number of blocks left is the number 8 − 3, or 5. This can also be thought of as the number of blocks that could be added to 3 blocks in order to get 8 blocks, that is,

$$\boxed{8 - 3} + 3 = 8.$$

missing addend The number 8 − 3, or 5, is the **missing addend** in the equation

$$\square + 3 = 8.$$

The missing-addend model gives elementary school students an opportunity to begin algebraic thinking. An unknown is a major part of the problem of trying to decide the difference of 8 minus 3.

Cashiers often use the missing-addend model. For example, if the bill for a movie is $8 and you pay $10, the cashier might calculate the change by saying "8 and 2 is 10." This idea can be generalized: for any whole numbers a and b, such that $a \geq b$, $a - b$ is the unique whole number such that $(a - b) + b = a$. That is, $a - b$ is the unique solution of the equation $\square + b = a$. The definition can be written more formally as follows:

Definition of Subtraction of Whole Numbers

For any whole numbers a and b, such that $a \geq b$, $a - b$ is the unique whole number c such that $a = b + c$.

Comparison Model

A third way to consider subtraction is by using a *comparison* model. Suppose we have 8 blocks and 3 balls and we want to know how many more blocks we have than balls. We can pair the blocks and balls, as shown in Figure 2-22, and determine that there are 5 more blocks than balls. We also write this as $8 - 3 = 5$.

Figure 2-22

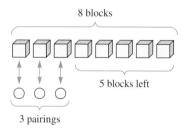

Number-Line Model

Subtraction can also be modeled on a number line, as suggested in Figure 2-23, where it is shown that $5 - 3 = 2$.

Figure 2-23

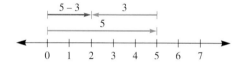

Properties of Subtraction

In an attempt to find $3 - 5$, we use the definition of subtraction: $3 - 5 = c$ if there is a solution to $c + 5 = 3$. Since there is no whole number c that satisfies the equation, $3 - 5$ is not meaningful in the set of whole numbers. In general, it can be shown that if $a < b$, then $a - b$ is not meaningful in the set of whole numbers.

ONGOING ASSESSMENT 2-3

1. Give an example to show why, in the definition of addition, sets A and B must be disjoint.
2. Explain whether the following given sets are closed under addition:
   a. $B = \{0\}$.
   b. $T = \{0, 3, 6, 9, 12, \ldots\}$.
   c. $N = \{1, 2, 3, 4, 5, \ldots\}$.
   d. $V = \{3, 5, 7\}$.
   e. $\{x \mid x \in W, x > 10\}$.
3. Draw a picture of colored rods illustrating the commutative property of addition.
4. Rewrite each of the following subtraction problems as an equivalent addition problem:
   a. $21 - 7 = x$.
   b. $x - 119 = 213$.
   c. $213 - x = 119$.
   d. $213 - 119 = x$.
5. Each of the following is an example of one of the properties for addition of whole numbers. Identify the property illustrated.
   a. $6 + 3 = 3 + 6$.
   b. $(6 + 3) + 5 = 6 + (3 + 5)$.
   c. $(6 + 3) + 5 = (3 + 6) + 5$.

6. In the definition of *less than,* can the natural number *k* be replaced by the whole number *k*? Why or why not?
7. **a.** Recall how we have defined *less-than* and *greater-than* relations and give an equivalent definition using the concept of subtraction for each of the following:
   (i) $a < b$. (ii) $a > b$.
   b. Use subtraction to define $a \geq b$.
8. Find the next three terms in each of the following arithmetic sequences:
   a. 8, 13, 18, 23, 28, _____, _____, _____
   b. 98, 91, 84, 77, 70, 63, _____, _____, _____
9. If *A*, *B*, and *C* each stand for a different single digit from 1 to 9, answer the following if
   $$A + B = C.$$
   a. What is the greatest digit that *C* could be? Why?
   b. What is the greatest digit that *A* could be? Why?
   c. What is the smallest digit that *C* could be? Why?
   d. If *A*, *B*, and *C* are even, what number(s) could *C* be? Why?
   e. If *C* is 5 more than *A*, what number(s) could *B* be? Why?
   f. If *A* is three times as great as *B*, then what number(s) could *C* be? Why?
   g. If *A* is odd and *A* is 5 more than *B*, what number(s) could *C* be? Why?
10. If *A*, *B*, *C*, and *D* each stand for a different single digit from 1 to 9, answer each of the following if

    $$\begin{array}{r} A \\ + B \\ \hline CD \end{array}$$

    a. What is the value of *C*? Why?
    b. Can *D* be 1? Why?
    c. If *D* is 7, what values can *A* be?
    d. If *A* is 6 greater than *B*, then what is the value of *D*?
11. Find the total of the terms in the 50th row in the following figure:

    ```
    1                      1st row
    1 − 1                  2nd row
    1 − 1 + 1              3rd row
    1 − 1 + 1 − 1          4th row
    1 − 1 + 1 − 1 + 1      5th row
    ```

12. Make each of the following a magic square:

    a.
	1	6
	5	7
4		2

    b.
17	10	
	14	
13	18	

13. **a.** Place whole numbers in the following four squares so that each pair has the sum shown. Note that one diagonal sum must be 20 and the other diagonal sum must be 7.

    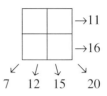

    b. Will examples similar to the one in part (a) always be possible with any numbers written for the sums as shown? Explain your answer.
14. **a.** Place the numbers 1, 2, 3, 4, 5, and 6 in the following boxes so that no square has a number greater than the one directly below it or directly to the right of it:

    b. Can you find more than one way to place the numbers?
15. **a.** A domino set contains all number pairs from double-zero to double-six, with each number pair occurring only once; that is, the following domino counts as two-four and four-two. How many dominoes are in the set?

    b. When considering the sum of all dots on a single domino in an ordinary set of dominoes, explain how the commutative property might be important.
16. Make a calculator count to 100 as follows (use a constant operation if available):
    a. By ones **b.** By twos **c.** By fives
17. Make a calculator count backward from 27 to 0 as follows (use a constant operation if possible):
    a. By ones **b.** By threes **c.** By nines
18. If a calculator is made to count by twos starting at 2, what is the thirteenth number in the sequence?
19. **a.** At a volleyball game, the players stood in a row ordered by height. If Kent is shorter than Mischa, Sally is taller than Mischa, and Vera is taller than Sally, who is the tallest and who is the shortest?
    b. Write possible heights for the players in part (a).

Communication

20. In Figure 2-17, arrows were used to represent numbers in completing an addition. Explain whether you think an arrow starting at 0 and ending at 3 represents the same number as an arrow starting at 4 and ending at 7. How would you explain this to students?

21. When subtraction and addition appear in an expression without parentheses, it is agreed that the operations are performed in order of their appearance from left to right. Taking this into account, answer the following:
    a. Use an appropriate model for subtraction to explain why
    $$a - b - c = a - c - b,$$
    assuming that all expressions are meaningful.
    b. Use an appropriate model for subtraction to explain why
    $$a - b - c = a - (b + c).$$

22. Explain whether it is important for elementary students to learn various properties of addition of whole numbers.

23. Explain whether it is important for elementary students to learn more than one model for performing the operations of addition and subtraction.

24. Do elementary students still have to learn their basic facts when the calculator is a part of the curriculum? Why or why not?

25. Explain how the following model can be used to illustrate each of the following addition and subtraction facts:
    a. $9 + 4 = 13$. b. $4 + 9 = 13$.
    c. $4 = 13 - 9$. d. $9 = 13 - 4$.

Open-Ended

26. Describe any model not in this text that you might use to teach addition to students.

27. Investigate whether your calculator has constant addition and subtraction features. Write an example to show how each feature works.

28. Suppose $A \subseteq B$. If $n(A) = a$ and $n(B) = b$, then $b - a$ could be defined as $n(B - A)$. Choose two sets A and B and illustrate this definition.

Cooperative Learning

29. Use groups of people to illustrate properties of addition.

30. Work with a group of other students in base five to determine what the addition facts might be. Use the addition models of this section to illustrate how addition facts in base five might be found.

Review Problems

31. Determine whether the following sets are equivalent. Justify your answers.
    a. $\{2, 4, 6, 8, \ldots, 1000\}$ and $\{3, 6, 9, 12, 15, \ldots, 1500\}$
    b. $\{1, 4, 7, 10, 13, \ldots, 2998\}$ and $\{1, 2, 3, 4, \ldots, 1000\}$
    c. $\{0, 1, 2, 3, 4, \ldots\}$ and $\{1, 2, 3, 4, 5, \ldots\}$
    d. $S = \{x \mid x = 2n, n \in N\}$ and $W = \{x \mid x = 4n, n \in N\}$

32. Decide which of the following are always true. If true, justify the statement and if false, provide a counterexample.
    a. $A - (B \cup C) = (A - B) \cap (A - C)$.
    b. If $A \cup B = B$, then $A \subseteq B$.
    c. If $A \cap B = \emptyset$, then $\overline{A} \cup \overline{B} = U$.
    d. If $A \not\subseteq B$ and $B \not\subseteq C$, then $A \not\subseteq C$.

33. a. How many one-to-one correspondences are possible between $A = \{a, b, c\}$ and $B = \{1, 2, 3\}$?
    b. How many elements are there in $A \times B$?

34. Classify each of the following as true or false. If true, justify the statement, and if false, provide a counterexample.
    a. If $A \cup B = A \cup C$, then $B = C$.
    b. If $A \cap B = A \cap C$, then $B = C$.
    c. If $A \cap B = \emptyset$ and $B \cap C = \emptyset$, then $A \cap C = \emptyset$.

35. A committee of senators has 6 members. A two-thirds vote is needed to carry any proposal. How many winning coalitions are there?

36. Oakridge has a population of 4800 and only one movie theater. One week the movie *The Lion King* was shown, and 3100 people went to see it. Next week, the movie *Apollo 13* was shown, and 2200 residents went to see it.
    a. What is the greatest number of townspeople that could have seen both movies? Justify your answer.
    b. What is the least number of people that could have seen both movies? Justify your answer.

37. If $U = \{a, b, c, d\}$, $A = \{a, b, c\}$, $B = \{b, c\}$, and $C = \{d\}$, find each of the following:
    a. $A \cup \overline{B}$
    b. $\overline{A \cap B}$
    c. $A \cap \emptyset$
    d. $B \cap C$
    e. $B - A$

BRAIN TEASER Use Figure 2-24 to design an *unmagic square*. That is, use each of the digits 1, 2, 3, 4, 5, 6, 7, 8, and 9 exactly once so that every column, row, and diagonal adds to a different sum.

Figure 2-24

Section 2-4 — Multiplication and Division of Whole Numbers

Multiplication of Whole Numbers

In this section, we use three models to discuss multiplication: the *repeated-addition* model, the *array* model, and the *Cartesian-product* model.

Repeated-Addition Model

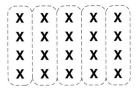

Figure 2-25

Suppose we have a classroom with 5 columns of 4 chairs each, as shown in Figure 2-25. How many chairs are there altogether? We can think of this as combining 5 sets of 4 objects into a single set.

The 5 columns of 4 suggest the following addition:

$$\underbrace{4 + 4 + 4 + 4 + 4}_{\text{five 4s}} = 20$$

We write $4 + 4 + 4 + 4 + 4$ as $5 \cdot 4$ and say "5 times 4" or "5 multiplied by 4." The advantage of the multiplication notation is evident when the number of addends is large. Thus if we have 25 columns of 4 chairs each, we can find the total number of chairs by adding 25 fours, or $25 \cdot 4$.

HISTORICAL NOTE

William Oughtred (1575–1660), an English mathematician, placed emphasis on mathematical symbols. He first introduced the use of "St. Andrew's cross" ($\times$) as the symbol for multiplication. This symbol was not readily adopted because, as Gottfried Wilhelm von Leibnitz (1646–1716) objected, it was too easily confused with the letter *x*. Leibnitz adopted the use of the dot ($\cdot$) for multiplication, which then became commonly used.

The *repeated-addition* model can be illustrated in several ways, including the use of a number line and the use of arrays. For example, using colored rods of length 4, we could show that the combined length of five 4-rods has a combined length that can be found by joining the rods end-to-end, as in Figure 2-26(a). Figure 2-26(b) shows the process using arrows on a number line.

Figure 2-26

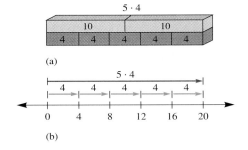

The Array Model

Another representation of repeated addition that is useful in exploring multiplication of whole numbers is an *array*. In Figure 2-27(a), we cross sticks to create intersection points thus forming an array of points. The number of points on a single vertical stick is 4 and there are 5 sticks, forming a total of 5 · 4 points in the array. In Figure 2-27(b), the array is shown as a 4-by-5 grid. The number of squares required to fill in the grid is 20. These models motivate the following definition of multiplication of whole numbers.

Figure 2-27

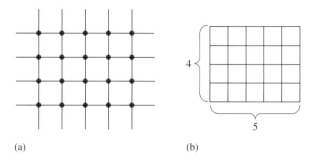

(a)   (b)

Definition of Multiplication of Whole Numbers

For any whole numbers a and $n \neq 0$,

$$n \cdot a = \underbrace{a + a + a + \cdots + a}_{n \text{ terms}}.$$

If $n = 0$, then $0 \cdot a = 0$.

Cartesian-Product Model

The *Cartesian-product* model offers another way to discuss multiplication. Suppose you can order a soyburger on light or dark bread with one condiment: mustard, mayonnaise, or horseradish. To show the number of different soyburger orders that a waiter could write for the cook, we use a *tree diagram*. The ways of writing the order are listed in Figure 2-28 where the bread is chosen from the set $B = \{\text{light, dark}\}$ and the condiment is chosen from the set $C = \{\text{mustard, mayonnaise, horseradish}\}$.

Figure 2-28

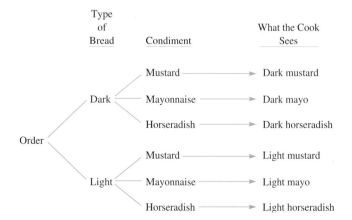

Each order can be written as an ordered pair, for example, (dark, mustard). The set of ordered pairs forms the Cartesian product $B \times C$. The fundamental counting principle tells us that the number of ordered pairs in $B \times C$ is $2 \cdot 3$.

The preceding discussion demonstrates how multiplication can be defined in terms of Cartesian products. An alternative definition of multiplication of whole numbers follows.

> **Alternative Definition of Multiplication of Whole Numbers**
>
> For finite sets A and B, if $n(A) = a$ and $n(B) = b$, then $a \cdot b = n(A \times B)$.

product · factors

> **REMARK** In this definition, sets A and B do not have to be disjoint. The expression $a \cdot b$ is the **product** of a and b, and a and b are **factors**. Also, note that $A \times B$ indicates the Cartesian product, not multiplication. We multiply numbers, not sets.

> **NOW TRY THIS 2-10**
>
> ● How would you use the repeated-addition definition of multiplication to explain to a child unfamiliar with both the Fundamental Counting Principle and the concept of cross product, that the number of possible outfits consisting of a shirt and pants combination—given 6 shirts and 5 pairs of pants—is $6 \cdot 5$? ●

Properties of Whole-Number Multiplication

As with addition, multiplication on the set of whole numbers is a binary function. Because of this, each ordered pair of whole numbers is paired with a unique whole number. As a result, the set of whole numbers is *closed* under multiplication. That is, if we multiply any two whole numbers, the result is a whole number. This property is referred to as the closure property of multiplication of whole numbers. In addition, multiplication on the set of whole numbers, like addition, has the commutative, associative, and identity properties.

> **Properties of Multiplication of Whole Numbers**
>
> **Closure property of multiplication of whole numbers** For any whole numbers a and b, $a \cdot b$ is a unique whole number.
> **Commutative property of multiplication of whole numbers** For any whole numbers a and b, $a \cdot b = b \cdot a$.
> **Associative property of multiplication of whole numbers** For any whole numbers a, b, and c, $(a \cdot b) \cdot c = a \cdot (b \cdot c)$.
> **Identity property of multiplication of whole numbers** There is a unique whole number 1 such that for any whole number a, $a \cdot 1 = a = 1 \cdot a$.
> **Zero multiplication property of whole numbers** For any whole number a, $a \cdot 0 = 0 = 0 \cdot a$.

The commutative property of multiplication of whole numbers is illustrated easily by building a 3 × 5 grid and then turning it sideways, as shown in Figure 2-29. We see that the number of 1 × 1 squares present in either case is 15, that is, $3 \cdot 5 = 15 = 5 \cdot 3$. The commutative property can be verified by recalling that $n(A \times B) = n(B \times A)$.

The associative property of multiplication of whole numbers can be illustrated as follows. Suppose $a = 3$, $b = 5$, and $c = 4$. In Figure 2-30(a), we see a picture of $3 \cdot (5 \cdot 4)$ blocks. In Figure 2-30(b), we see the same blocks, this time arranged as $(3 \cdot 5) \cdot 4$. Because both sets of blocks in Figure 2-30(a) and (b) compress to the set shown in Figure 2-30(c), we see that $3 \cdot (5 \cdot 4) = (3 \cdot 5) \cdot 4$. The associative property is useful in computations such as the following:

$$3 \cdot 40 = 3 \cdot (4 \cdot 10) = (3 \cdot 4) \cdot 10 = 12 \cdot 10 = 120.$$

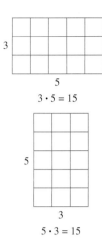

Figure 2-29

Figure 2-30

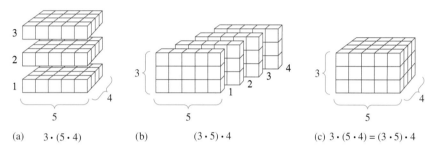

The multiplicative identity for whole numbers is 1. For example, $3 \cdot 1 = 1 + 1 + 1 = 3$. In general, for any whole number a,

$$a \cdot 1 = \underbrace{1 + 1 + 1 + \cdots + 1}_{a \text{ terms}} = a.$$

Thus $a \cdot 1 = a$, which, along with the commutative property for multiplication, implies that $a \cdot 1 = a = 1 \cdot a$. Cartesian products can also be used to show that $a \cdot 1 = a = 1 \cdot a$.

Next, consider multiplication involving 0. For example $6 \cdot 0 = 0 + 0 + 0 + 0 + 0 + 0 = 0$. Thus we see that multiplying 0 by 6 yields a product of 0 and, by commutativity, $0 \cdot 6 = 0$. This is an example of the zero multiplication property. This property can also be verified by using the definition of multiplication in terms of Cartesian products.

The Distributive Property of Multiplication over Addition

The next property we investigate is the basis for understanding multiplication algorithms. The area of the large rectangle in Figure 2-31 equals the sum of the areas of the two smaller rectangles and hence $5 \cdot (3 + 4) = 5 \cdot 3 + 5 \cdot 4$.

Figure 2-31

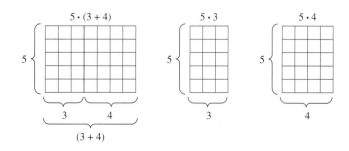

The properties of addition and multiplication also can be used to justify this result:

$$5 \cdot (3 + 4) = \underbrace{(3 + 4) + (3 + 4) + (3 + 4) + (3 + 4) + (3 + 4)}_{5 \text{ terms}}$$

Definition of multiplication

$$= (3 + 3 + 3 + 3 + 3) + (4 + 4 + 4 + 4 + 4)$$

Commutative and associative properties of addition

$$= 5 \cdot 3 + 5 \cdot 4 \quad \text{Definition of multiplication}$$

This example illustrates the *distributive property of multiplication over addition* for whole numbers. Because in algebra it is customary to write $a \cdot b$ as ab, we state the *distributive property of multiplication over addition* as follows:

Property

Distributive property of multiplication over addition for whole numbers For any whole numbers a, b, and c,

$$a(b + c) = ab + ac.$$

REMARK Because the commutative property of multiplication of whole numbers holds, the distributive property of multiplication over addition can be rewritten as $(b + c)a = ba + ca$.

The distributive property can be generalized to any finite number of terms. For example, $a(b + c + d) = ab + ac + ad$.

Students find the distributive property of multiplication over addition useful when doing mental mathematics. For example, $13 \cdot 7 = (10 + 3) \cdot 7 = 10 \cdot 7 + 3 \cdot 7 = 70 + 21 = 91$. This property is used to combine like terms when we work with variables; for example, $3ab + 2ab = (3 + 2)ab = 5ab$.

Example 2-15
a. Use an area model to show that $(x + y)(z + w) = xz + xw + yz + yw$.
b. Use the distributive property of multiplication over addition to justify the result in part (a).

Solution a. Consider the rectangle in Figure 2-32 whose width is $x + y$ and whose length is $z + w$. The area of the entire rectangle is $(x + y)(z + w)$. If we divide the rectangle into smaller rectangles as shown, we notice that the sum of the areas of the four smaller rectangles is $xz + xw + yz + yw$. Because the area of the original rectangle equals the sum of the areas of the smaller rectangles, the result follows.
b. To apply the distributive property of multiplication over addition, we think about $x + y$ as one number and proceed as follows:

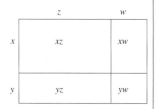

Figure 2-32

$$(x + y)(z + w) = (x + y)z + (x + y)w$$
$$= xz + yz + xw + yw \quad \text{The distributive property of multiplication over addition}$$
$$= xz + xw + yz + yw \quad \text{The commutative property of addition}$$

Order of Operations

Difficulties involving the order of arithmetic operations sometimes arise. For example, many students will treat $2 + 3 \cdot 6$ as $(2 + 3)6$, while others will treat it as $2 + (3 \cdot 6)$. In the first case, the value is 30; in the second case, the value is 20. To avoid confusion, mathematicians agree that when no parentheses are present, multiplications are performed *before* additions. Thus, $2 + 3 \cdot 6 = 2 + 18 = 20$. This order of operations is not built into calculators that display an incorrect answer of 30.

Division of Whole Numbers

We discuss division using three models: the *set (partition)* model, the *missing-factor* model, and the *repeated-subtraction* model.

Set (Partition) Model

Suppose we have 18 cookies and want to give an equal number of cookies to each of 3 friends: Bob, Dean, and Charlie. How many should each person receive? If we draw a picture, we can see that we can divide (or partition) the 18 cookies into 3 *sets*, with an equal number of cookies in each set. Figure 2-33 shows that each friend received 6 cookies.

Figure 2-33

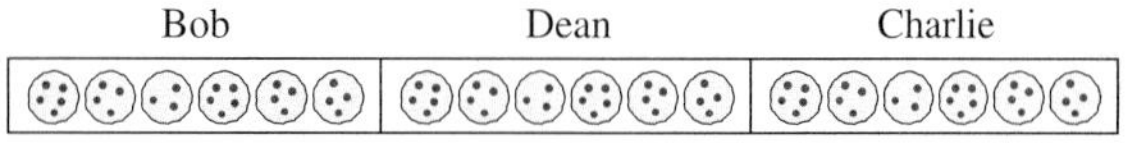

Bob Dean Charlie

The answer may be symbolized as $18 \div 3 = 6$. Thus $18 \div 3$ is the number of cookies in each of 3 disjoint sets whose union has 18 cookies. In this approach to division, we partition a set into a number of equivalent subsets.

Missing-Factor Model

Another strategy for dividing 18 cookies among 3 friends is to use the *missing-factor* model. If each friend receives c cookies, then the 3 friends receive $3c$, or 18, cookies. Hence, $3c = 18$. Since $3 \cdot 6 = 18$, then $c = 6$. We have answered the division computation by using multiplication. This leads us to the following definition of division of whole numbers.

Definition of Division of Whole Numbers

For any whole numbers a and b, with $b \neq 0$, $a \div b = c$ if, and only if, c is the unique whole number such that $b \cdot c = a$.

dividend · divisor · quotient

> REMARK The number a is the **dividend,** b is the **divisor,** and c is the **quotient.** Note that $a \div b$ can also be written as $\frac{a}{b}$ or $b\overline{)a}$.

Repeated-Subtraction Model

Suppose we have 18 cookies and want to package them in cookie boxes that hold 6 cookies each. How many boxes are needed? We could reason that if one box is filled, then we would have $18 - 6$ (or 12) cookies left. If one more box is filled, then there are $12 - 6$ (or 6) cookies left. Finally, we could place the last 6 cookies in a third box. This discussion can be summarized by writing $18 - 6 - 6 - 6 = 0$. We have found by repeated subtraction that $18 \div 6 = 3$.

Calculators can be used to show that division of whole numbers can be thought of as repeated subtraction. For example, consider $135 \div 15$. If the calculator has a constant key, press $\boxed{1}\,\boxed{5}\,\boxed{-}\,\boxed{K}\,\boxed{1}\,\boxed{3}\,\boxed{5}\,\boxed{=}$ ... and then count how many times you must press the $\boxed{=}$ key in order to make the display read 0. Calculators with a different constant feature may require a different sequence of entries. For example, if the calculator has an automatic constant, we can press $\boxed{1}\,\boxed{3}\,\boxed{5}\,\boxed{-}\,\boxed{1}\,\boxed{5}\,\boxed{=}$ and then count the number of times we press the $\boxed{=}$ key to make the display read 0. Compare your answer with the one achieved by pressing this sequence of keys:

$$\boxed{1}\,\boxed{3}\,\boxed{5}\,\boxed{\div}\,\boxed{1}\,\boxed{5}\,\boxed{=}$$

The Division Algorithm

Just as subtraction of whole numbers is not always meaningful, division of whole numbers is not always meaningful. For example, to find $383 \div 57$ we look for a whole number c such that $57c = 383$.

Table 2-5 shows several products of whole numbers times 57. Since 383 is between 342 and 399, there is no whole number c such that $57c = 383$. Because no whole number c satisfies this equation, we see that $383 \div 57$ has no meaning in the set of whole numbers, and the set of whole numbers is not closed under division.

Table 2-5

$57 \cdot 1$	$57 \cdot 2$	$57 \cdot 3$	$57 \cdot 4$	$57 \cdot 5$	$57 \cdot 6$	$57 \cdot 7$
57	114	171	228	285	342	399

remainder

division algorithm

However, if 383 apples were to be divided among 57 students, each student would receive 6 apples and 41 apples would remain. The number 41 is the **remainder.** Thus 383 contains six 57s with a remainder of 41. Observe that the remainder is a whole number less than 57. The concept illustrated is the **division algorithm.**

> **Division Algorithm**
>
> Given any whole numbers a and b with $b \neq 0$, there exist unique whole numbers q (quotient) and r (remainder) such that
>
> $$a = bq + r \quad \text{with} \quad 0 \leq r < b.$$

When a is "divided" by b and the remainder is 0 we say that a is *divisible* by b or that b is a *divisor* of a or that b divides a. By the *division algorithm* a is divisible by b if $a = bq$ for a unique whole number q. Thus 63 is divisible by 9 because $63 = 9 \cdot 7$. Notice that 63 is also divisible by 7.

Example 2-16 If 123 is divided by a number and the remainder is 13, what are the possible divisors?

Solution If 123 is divided by b, then from the division algorithm we have

$$123 = bq + 13 \quad \text{and} \quad b > 13.$$

Table 2-6

1	110
2	55
5	22
10	11

Using the definition of subtraction, we have $bq = 123 - 13$, and hence $110 = bq$. Now we are looking for two numbers whose product is 110, where one number is greater than 13. Table 2-6 shows the pairs of divisors of 110.

We see that 110, 55, and 22 are possible divisors because each is greater than 13. The numbers 1, 2, 5, 10, and 11 cannot be divisors.

An alternative method of solving Example 2-16 is to use the integer division key on your calculator.

Division by 0 and 1

The whole numbers 0 and 1 deserve special attention with respect to division of whole numbers. Before reading on, try to find the values of the following three expressions:

1. $3 \div 0$ **2.** $0 \div 3$ **3.** $0 \div 0$

Consider the following explanations:

1. By definition, $3 \div 0 = c$ if there is a unique number c such that $0 \cdot c = 3$. Since the zero property of multiplication states that $0 \cdot c = 0$ for any whole number c, there is no whole number c such that $0 \cdot c = 3$. Thus $3 \div 0$ is undefined because there is no answer to the equivalent multiplication problem.
2. By definition, $0 \div 3 = c$ if there exists a unique number such that $3 \cdot c = 0$. Because any number times 0 is 0, and in particular $3 \cdot 0 = 0$, then $c = 0$ and $0 \div 3 = 0$. Note that $c = 0$ is the only number that satisfies $3 \cdot c = 0$.
3. By definition, $0 \div 0 = c$ if there is a unique whole number c such that $0 \cdot c = 0$. Notice that for *any* c, $0 \cdot c = 0$. According to the definition of division, c must be unique. Since there is no *unique* number c such that $0 \cdot c = 0$, it follows that $0 \div 0$ is indeterminate, or undefined.

Division involving 0 may be summarized as follows. Let n be any natural number. Then

1. $n \div 0$ is undefined;
2. $0 \div n = 0$;
3. $0 \div 0$ is indeterminate, or undefined.

Recall that $n \cdot 1 = n$ for any whole number n. Thus, by the definition of division, $n \div 1 = n$. For example, $3 \div 1 = 3$, $1 \div 1 = 1$, and $0 \div 1 = 0$.

ONGOING ASSESSMENT 2-4

1. For each of the following, find, if possible, the whole numbers that make the equations true:
   a. $3 \cdot \square = 15$.
   b. $18 = 6 + 3 \cdot \square$.
   c. $\square \cdot (5 + 6) = \square \cdot 5 + \square \cdot 6$.

2. In terms of set theory, the product na could be thought of as the number of elements in the union of n sets with a elements in each. If this were the case, what must be true about the sets?

3. Determine if the following sets are closed under multiplication:
   a. $\{0, 1\}$
   b. $\{0\}$
   c. $\{2, 4, 6, 8, 10, \ldots\}$
   d. $\{1, 3, 5, 7, 9, \ldots\}$
   e. $\{1, 4, 7, 10, 13, \ldots\}$
   f. $\{0, 1, 2\}$

4. a. If 5 is removed from the set of whole numbers, is the set closed with respect to addition?
   b. If 5 is removed from the set of whole numbers, is the set closed with respect to multiplication?
   c. Answer the same questions as (a) and (b) if 6 is removed from the set of whole numbers.

5. Rename each of the following using the distributive property for multiplication over addition so that there are no parentheses in the final answer:
   a. $(a + b)(c + d)$
   b. $3(x + y + 5)$
   c. $\square(\triangle + \bigcirc)$
   d. $(x + y)(x + y + z)$

6. Place parentheses, if needed, to make each of the following equations true:
   a. $4 + 3 \cdot 2 = 14$.
   b. $9 \div 3 + 1 = 4$.
   c. $5 + 4 + 9 \div 3 = 6$.
   d. $3 + 6 - 2 \div 1 = 7$.

7. The generalized distributive property for three terms states that for any whole numbers a, b, c, and d, $a(b + c + d) = ab + ac + ad$. Justify this property using the distributive property for two terms.

8. For each of the following, find whole numbers to make the statement true, if possible:
   a. $18 \div 3 = \square$.
   b. $\square \div 76 = 0$.
   c. $28 \div \square = 7$.

9. Rewrite each of the following division problems as a multiplication problem:
   a. $40 \div 8 = 5$.
   b. $326 \div 2 = x$.
   c. $48 \div x = 16$.
   d. $x \div 5 = 17$.

10. Think of a number. Multiply it by 2. Add 2. Divide by 2. Subtract 1. How does the result compare with your original number? Will this work all the time? Explain your answer.

11. Show that, in general, each of the following is false if a, b, and c are whole numbers:
    a. $(a \div b) \div c = a \div (b \div c)$.
    b. $a \div (b + c) = (a \div b) + (a \div c)$.

12. Show that $(a + b)^2 = a^2 + 2ab + b^2$ using
    a. the distributive property of multiplication over addition.
    b. an area model.

13. Show that if $b > c$ then $a(b - c) = ab - ac$ using
    a. an area model suggested by the given figure (express the shaded area in two different ways).

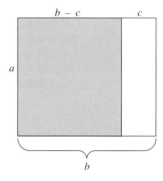

   b. the definition of subtraction in terms of addition and the distributive property of multiplication over addition.
   c. Name the property you have justified in parts (a) and (b).

14. Suppose c is a divisor of a and of b. Show that $(a + b) \div c = (a \div c) + (b \div c)$ using
    a. a model.
    b. the definition of division in terms of multiplication and the distributive property of multiplication over addition.

15. Millie and Samantha began saving money at the same time. Millie plans to save $3 a month, and Samantha plans to save $5 a month. After how many months will Samantha have exactly $10 more than Millie?

16. String art is formed by connecting evenly spaced nails on the vertical and horizontal axes by line segments. Connect the nail farthest from the origin on the vertical axis with the nail closest to the origin on the horizontal axis. Continue until all nails are connected, as shown in the figure that follows. How many intersection points are created with 10 nails on each axis?

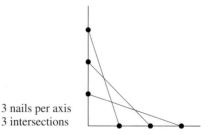

3 nails per axis
3 intersections

17. There were 17 sandwiches for 7 people on a picnic. How many whole sandwiches were there for each person if they were divided equally? How many were left over?

18. **a.** Find all pairs of whole numbers whose product is 36.
    b. Plot the points found in (a) on a grid.
    c. Compare the pattern shape formed by the points to the pattern shape that could be found by adding all pairs of whole numbers whose sum is 36.

19. A new model of car is available in 4 exterior colors and 3 interior colors. Use a tree diagram and specific colors to show how many color schemes are possible for the car.

20. Students were divided into 8 teams with 9 on each team. Later the same students were divided into teams with 6 on each team. How many teams were there then?

21. To find $7 \div 5$ on the calculator, press $\boxed{7}\ \boxed{\div}\ \boxed{5}\ \boxed{=}$, which yields 1.4. To find the whole-number remainder, ignore the decimal portion of 1.4, multiple $5 \cdot 1$, and subtract this product from 7. The result is the remainder. Use a calculator to find the whole-number remainder for each of the following divisions:
    a. $28 \div 5$
    b. $32 \div 10$
    c. $29 \div 3$
    d. $41 \div 7$
    e. $49,382 \div 14$

22. In the following problems, use only the designated number keys on the calculator. You may use any function keys.
    a. Use the keys $\boxed{1}$, $\boxed{9}$, and $\boxed{7}$ exactly once each in any order and use any operations available to write as many of the whole numbers as possible from 1 to 20. For example, $9 - 7 - 1 = 1$ and $1 \cdot 9 - 7 = 2$.
    b. Use the $\boxed{4}$ key as many times as desired with any operations to display 13.
    c. Use the $\boxed{2}$ key three times with any operations to display 24.
    d. Use the $\boxed{1}$ key five times with any operations to display 100.

23. In each of the following, tell what computation must be done last:
    a. $5(16 - 7) - 18$
    b. $54/(10 - 5 + 4)$
    c. $(14 - 3) + (24 \cdot 2)$
    d. $21,045/345 + 8$

24. Is x/x always equal to 1? Explain your answer.

25. Is $x \cdot x$ ever equal to x? Explain your answer.

26. Describe all pairs of whole numbers whose sum and product are the same.

27. Write an algebraic expression for each of the following:
    a. Width of a rectangle whose area is A and length is l
    b. Feet, f, in yards
    c. Hours, h, in minutes
    d. Days, d, in weeks

28. Find infinitely many whole numbers that leave remainder 3 upon division by 5.

★29. The binary operation $\odot$ is defined on the set $S = \{a, b, c\}$, as shown in the following table. For example, $a \odot b = b$ and $b \odot a = b$.

$\odot$	a	b	c
a	a	b	c
b	b	a	c
c	c	c	c

   a. Is S closed with respect to $\odot$?
   b. Is $\odot$ commutative on S?
   c. Is there an identity for $\odot$ on S? If yes, what is it?
   d. Is $\odot$ associative on S?

Communication

30. Suppose a student argued that $0/0 = 0$ because "nothing divided by nothing" is "nothing." How would you help that person?

31. Sue claims the following is true by the distributive law, where a and b are whole numbers:
$$3(ab) = (3a)(3b).$$
How might you help her?

32. Can 0 be the identity for multiplication? Explain why or why not.

33. Why do you think the use of a dot for multiplication is more popular in mathematics than the symbol $\times$?

Open-Ended

34. Investigate whether your calculator has constant addition and subtraction features. Write an example to show how each feature works.

35. Describe a real-life situation that could be represented by the expression $5 + 8 \cdot 6$.

36. How would you explain to a child that an even number has the form $2q$ and an odd number has the form $2q + 1$ where q is a whole number?

37. Suppose $A \subseteq B$. If $n(A) = a$ and $n(B) = b$, then $b - a$ could be defined as $n(B - A)$. Choose two sets A and B and illustrate this definition.

Cooperative Learning

38. Multiplication facts that most children have memorized can be stated in a table that is partially filled:

	0	1	2	3	4	5	6	7	8	9
1										
2										
3										
4					16					
5										
6										
7										
8										72
9										81

Fill out the table and find as many patterns as you can. List all the patterns that your group discovered and explain to the rest of the group why some of those patterns occur in the table.

Review Problems

39. Give a set that is not closed under addition.

40. Are the whole numbers commutative under subtraction? If not, give a counterexample.

41. Illustrate $11 - 3$ using a number-line model.

42. Explain why if a, b, and c are whole numbers, and $c \neq 0$, then $a < b$ implies $ac < bc$.

LABORATORY ACTIVITY

Enter a number less than 20 on the calculator. If the number is even, divide it by 2; if it is odd, multiply it by 3 and add 1. Next, use the number on the display. Follow the given directions. Repeat the process.

1. Will the display eventually reach 1?
2. Which number less than 20 takes the most steps before reaching 1?
3. Do even or odd numbers reach 1 more quickly?
4. Investigate what happens with numbers greater than 20.

Section 2-5 — **Functions**

The *Principles and Standards* for grades pre-K–2 point out that:

… systematic experience with patterns can build up to an understanding of the idea of function, and experience with number and their properties lays a foundation for later work with symbols and algebraic expressions (p. 37).

For grades 6–8, the *Principles and Standards* say that all students should:

… represent and analyze patterns and functions, using words, tables, and graphs (p. 158).

The following is an example of a game called "guess my rule," often used to introduce the concept of a function.

When Tom said 2, Noah said 5. When Dick said 4, Noah said 7. When Mary said 10, Noah said 13. When Liz said 6, what did Noah say? What is Noah's rule?

The answer to the first question may be 9, and the rule could be, "Take the original number and add 3"; that is, for any number n, Noah's answer is $n + 3$.

Example 2-17 Guess the teacher's rule for the following responses:

(a)
You	Teacher
1	3
0	0
4	12
10	30

(b)
You	Teacher
2	5
3	7
5	11
10	21

(c)
You	Teacher
2	0
4	0
7	1
21	1

Solution
a. The teacher's rule could be, "Multiply the given number n by 3," that is, $3n$.
b. The teacher's rule could be, "Double the original number n and add 1," that is, $2n + 1$.
c. The teacher's rule could be, "If the number n is even, answer 0; if the number is odd, answer 1."

HISTORICAL NOTE

The Babylonians (ca. 2000 B.C.) probably had a working idea of what a function was. To them, it was a table or a correspondence. René Descartes (1637), Gottfried Wilhelm von Leibnitz (1692), Johann Bernoulli (1718), Leonhard Euler (1750), Joseph Louis Lagrange (1800), and Jean Joseph Fourier (1822) were among the mathematicians contributing to the notion of a function. Leonhard Euler in 1734 first used the notation $f(x)$. In the late 1800s, Georg Cantor and others began to use the modern definition.

Another way to prepare students for the concept of a function is by using a "function machine." The following student page from *Scott Foresman Addison-Wesley Middle School MATH Course 2,* 1999, shows an example of a function machine. What goes in the machine is referred to as input and what comes out as output. Thus, on the student page, if the input to the first function machine is 8, the output is 16. In later grades, a special notation for the output is used. For any input element x, the output is denoted by $f(x)$, read "f of x." For the function machine pictured on the student page, when the input is 17, the output would be written as $f(17)$. Because the output is 34, we have $f(17) = 34$. Because the machine works according to the rule "double it," $f(x) = 2x$ and $f(17) = 2 \cdot 17 = 34$.

Do problems 1 through 6 on the following student page.

Extend Key Ideas — Algebra

Functions

A function is a relationship between numbers. You can think of the function as taking a number and transforming it into another number.

Function machines can be a useful way to think about functions. This machine seems to be using the rule "double it" to decide which number it puts out.

The equation that represents the "double it" function can be written as $y = 2x$. The input number is x and the output number is y.

When you know the equation for the function, you can substitute an input (x) value to find the output (y) value that goes with it.

If $y = 3x + 5$, what's the y-value for an x-value of 2?
$y = 3(2) + 5$ Substitute 2 for x.
$y = 6 + 5$ Multiply.
$y = 11$ Add.

Try It

Evaluate each function for the given values.

1. $y = 5x$ for $x = 1, 2,$ and 3
2. $y = x + 2$ for $x = 6, 8,$ and 10
3. $y = 2x - 1$ for $x = 5, 7,$ and 9
4. $y = 4x + 2$ for $x = 2, 3,$ and 4

Think of each table as a function machine. Copy and complete each one. Then write an equation for the table.

5.
x	1	2	3	4	5
y	5	10	15		

6.
x	1	2	3	4	5
y	3	5	7		

Example 2-18 Consider the function machine in Figure 2-34. What will happen if the numbers 0, 1, 3, and 6 are entered?

Solution If the numbers output are denoted by $f(x)$, the corresponding values can be described using Table 2-7.

Figure 2-34

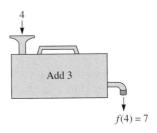

$f(4) = 7$

Table 2-7

x	$f(x)$
0	3
1	4
3	6
6	9

We can write an equation to depict the rule in Example 2-18 as follows. If the input is x, the output is $x + 3$; that is, $f(x) = x + 3$. The output values can be obtained by substituting the values 0, 1, 3, 4, and 6 for x in $f(x) = x + 3$, as shown:

$$f(0) = 0 + 3 = 3.$$
$$f(1) = 1 + 3 = 4.$$
$$f(3) = 3 + 3 = 6.$$
$$f(4) = 4 + 3 = 7.$$
$$f(6) = 6 + 3 = 9.$$

In many applications, both the inputs and the outputs of a function machine are numbers. However, inputs and outputs can be any objects. For example, consider a particular candy machine that accepts only 25¢, 50¢, or 75¢ and outputs one of three types of candy. A function machine associates *exactly one output with each input*. If you enter some element x as input and obtain $f(x)$ as output, then every time you enter the same x as input, you will obtain the same $f(x)$ as output. The idea of a function machine associating exactly one output with each input according to some rule leads to the following definition.

Definition of Function

A **function** from set A to set B is a correspondence from A to B in which each element of A is paired with one, and only one, element of B.

domain

range

The set A in the previous definition is the set of all allowable inputs and is the **domain** of the function. The set B is any set that includes all the possible outputs. The set of all outputs is the **range** of the function. Set B in the definition is any set that includes the range and can be the range itself. The distinction is made for convenience sake, since sometimes the range is not easy to find. For example, consider corresponding to each student at a university the student's I.D. number. This is a function from the set of all students to the set S

106 CHAPTER 2 Sets, Whole Numbers, and Functions

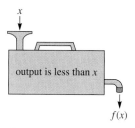

Figure 2-35

of natural numbers. The range in this case is all the I.D. numbers of students who are enrolled at the university. The range is a proper subset of the set S. Normally if no domain is given to describe a function, then the domain is assumed to have the most elements for which the rule is meaningful.

A calculator is a function machine. Suppose a student enters $\boxed{9}\ \boxed{\times}\ \boxed{K}$ on the calculator that has a constant key, $\boxed{K}$. The student then presses $\boxed{0}$ and hands the calculator to another student. The other student is to determine the rule by entering various numbers followed by the $\boxed{=}$ key. Machines with an automatic constant feature can also be used.

Other buttons on a calculator are function buttons. For example, the $\boxed{\pi}$ button always displays an approximation for π, such as 3.1415927; the $\boxed{+/-}$ button either displays a negative sign in front of a number or removes an existing negative sign; and the $\boxed{x^2}$ and $\boxed{\sqrt{\ }}$ buttons square numbers and take the square root of numbers, respectively.

Are all input-output machines function machines? Consider the machine in Figure 2-35. For any natural-number input x, the machine outputs a number that is less than x. If, for example, you input the number 10, the machine may output 9, since 9 is less than 10. If you input 10 again, the machine may output 3, since 3 is less than 10. Such a machine is not a function machine because the same input may give different outputs.

Example 2-19 Which, if any, of the parts of Figure 2-36 exhibits a function from A to B? If a correspondence is a function from A to B, find the range of the function.

Figure 2-36

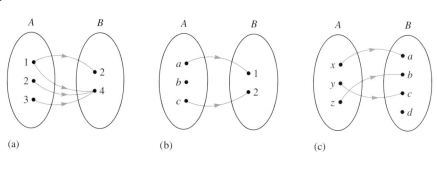

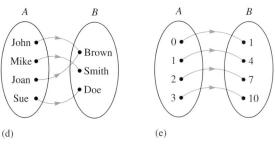

Solution **a.** Figure 2-36(a) does not define a function from A to B, since the element 1 is paired with both 2 and 4.
b. Figure 2-36(b) does not define a function from A to B, since the element b is not paired with any element of B. (It is a function from a subset of A to B.)

c. Figure 2-36(c) does define a function from *A* to *B*, since there is one and only one arrow leaving each element of *A*. The fact that *d*, an element of *B*, is not paired with any element in the domain does not violate the definition. The range is {*a*, *b*, *c*} and does not include *d* because *d* is not an output of this function, as no element of *A* is paired with *d*.
  d. Figure 2-36(d) illustrates a function, since there is only one arrow leaving each element in *A*. It does not matter that an element of set *B*, Brown, has two arrows pointing to it. The range is {Brown, Smith, Doe}.
  e. Figure 2-36(e) illustrates a function whose range is {1, 4, 7, 10}.

Figure 2-36(e) also illustrates a one-to-one correspondence between *A* and *B*. In fact, any one-to-one correspondence between *A* and *B* defines a function from *A* to *B* as well as a function from *B* to *A*.

NOW TRY THIS 2-11

- Determine which of the following are functions from the set of natural numbers to {0, 1}. Justify your reasoning.
  a. For every natural-number input, *x* the output is 0.
  b. For every natural-number input, the output is 0 if the input is an even number and the output is 1 if the input is an odd number.

A function can be represented in a variety of ways. A useful way to describe a function is in a table. Consider the information in Table 2-8 relating the amount spent on advertising and the resulting sales in a given month for a small business. If $A = \{0, 1, 2, 3, 4\}$ and $S = \{1, 3, 6, 8, 10\}$, the table describes a function from *A* to *S*.

The information in Table 2-8 can also be given using ordered pairs. When 0 is the input and 1 is the output, that is recorded as the ordered pair (0, 1). Similarly, the information in the second row is recorded as (1, 3) and the rest of the information as (2, 6), (3, 8), and (4, 10). The first component in the ordered pair is always an element in the domain and the second is the corresponding output.

Table 2-8

Amount of Advertising (In $1000s)	Amount of Sales (In $1000s)
0	1
1	3
2	6
3	8
4	10

Example 2-20 Which of the following sets of ordered pairs represent functions? If a set represents a function, give its domain and range. If it does not, explain why.

a. {(1, 2), (1, 3), (2, 3), (3, 4)} **b.** {(1, 2), (2, 3), (3, 4), (4, 5)}
c. {(1, 0), (2, 0), (3, 0), (4, 4)} **d.** $\{(a, b) \mid a \in N, b = 2a\}$

Solution
a. This is not a function because the input 1 has two different outputs.
b. This is a function with domain {1, 2, 3, 4} and range {2, 3, 4, 5}.
c. This is a function with domain {1, 2, 3, 4} and range {0, 4}. The output 0 appears more than once, but this does not contradict the definition of a function in that each input corresponds to only one output.
d. This is a function with domain N and range E, the set of all even natural numbers.

Perhaps one of the most widely recognized representations of a function is a graph. Graphs are visual representations of functions and appear in newspapers and books and on television. To graph the function in Table 2-8, consider the set of ordered pairs {(0, 1), (1, 3), (2, 6), (4, 10)} and correspond to each ordered pair a point on the grid in Figure 2-37. We use the horizontal scale for the inputs and the vertical scale for the outputs and mark the point corresponding to (0, 1) by starting at 0 on the horizontal scale and going up 1 unit on the vertical scale. To mark the point that corresponds to (1, 3), we start at 0 and move 1 unit horizontally and then 3 units vertically. Marking the point that corresponds to an ordered pair is referred to as **graphing** the ordered pair. The set of all points that correspond to all the ordered pairs is the graph of the function.

Figure 2-37

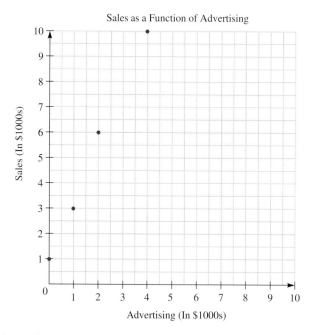

In applications of mathematics it is often convenient to have different scales for vertical and horizontal axes. For example, in the following student page from McDougal Littel *Middle Grade MATH Thematics Book 3*, 1999, on the horizontal axis, one unit represents two weeks (each unit is one side of one small square), while on the vertical axis each unit represents $25. The points on the graphs on the student page have been connected. This implies that we could find how much Lynda and Maria owe after any number of weeks. For example, from the graphs we see that for the first of four weeks, Maria owes more than Lynda. (Why?)

Modeling Linear Change (pp. 483–485)

When a quantity changes by the same amount at regular intervals, the quantity shows linear change. You can use a linear equation to model linear change. You can also use a table or a graph.

Example Lynda borrowed $175 from her parents. She pays them back $10 a week. Her sister Maria borrowed $200 and pays back $15 a week. Who will finish paying off her loan first?

Write an equation for each person. Let y = the amount the person owes after x weeks.

Lynda
$y = 175 - 10x$

Maria
$y = 200 - 15x$

Number of weeks	Amount Lynda owes	Amount Maria owes
0	175	200
1	165	185
2	155	170
3	145	155
4	135	140
5	125	125
6	115	110
…	…	…
13	45	5
14	35	0

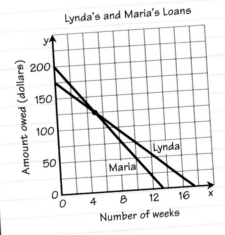

Maria only has to pay $5 in the 14th week.

After five weeks, they owe the same amount of money. After that, Maria is paying more per week than Lynda, so she will pay off her loan first.

Example 2-21 Explain why a telephone company would not set rates for telephone calls as depicted on the graph in Figure 2-38.

Figure 2-38

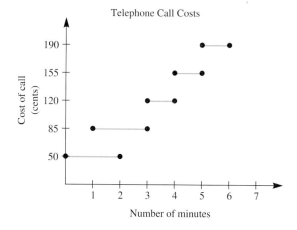

Solution The graph does not depict a function. For example, a customer could be charged either $.50 or $.85 for a 2-minute call.

More on Applications of Functions

Functions have many real-life applications. For example, on direct-dial long-distance calls you pay only for the minutes you talk. Suppose the weekday rate for a long-distance phone call from Madison, Wisconsin, to Urbana, Illinois, is 27¢ per minute. We have seen that one way to describe a function is by writing an equation. Based on the information in Table 2-9, the equation relating time to cost is $C = t \cdot 27$, where t is a natural number. This could also be written as $f(t) = 27t$, where $f(t)$ is the cost of the call in cents. If we restrict the time

Table 2-9

Number of Minutes Talked	Total Cost in Cents
1	$1 \cdot 27 = 27$
2	$2 \cdot 27 = 54$
3	$3 \cdot 27 = 81$
4	$4 \cdot 27 = 108$
5	$5 \cdot 27 = 135$
.	.
.	.
.	.
t	$t \cdot 27$

in minutes to the first 5 natural numbers, the function can be described as the set of ordered pairs {(1, 27), (2, 54), (3, 81), (4, 108), (5, 135)}. Figure 2-39(a) shows the graph of the function. The graph consists of 5 points that are not connected, since the phone company charges for any part of a minute t as if it were a full minute. The break in the vertical axis shown by ⌇ indicates that part of the vertical axis between 0 and 27 has been compressed and that the scale is not accurate between 0 and 27.

In graphing the function in Figure 2-39(a), we assumed the domain to be the set of natural numbers. Because the phone company charges for any part of a minute as if it were a full minute, the charge for any call that is 1 min long or less is 27¢. The charge for any call less than or equal to 2 min but longer than 1 min is 54¢. The cost is calculated similarly for calls between two consecutive minutes. The graph reflecting this information is shown in Figure 2-39(b). The filled dot at the right endpoint of each segment in the graph indicates that the point belongs to the graph, while the empty dot at the left endpoint of each segment indicates that the point does not belong to the graph. The domain of the function graphed in Figure 2-39(b) is the set of all positive numbers.

Figure 2-39

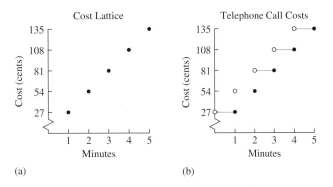

NOW TRY THIS 2-12

- Can the cost of 10.5 min of phone conversation from Madison, Wisconsin, to Urbana, Illinois, be found by substituting 10.5 into the formula $C = t \cdot 27$ developed above? Why or why not?

Functions and their graphs are used frequently in the business world, as the following example illustrates.

Example 2-22 In Figure 2-40 the blue graph shows the cost C in dollars of producing a given number of tee shirts. The red graph shows the revenue R in dollars from selling any number of tee shirts.

Figure 2-40

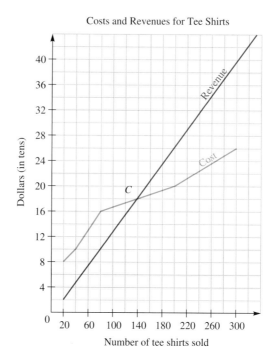

The graphs technically should consist only of disconnected points because a factory produces only a whole number of tee shirts. However, for visual effect, it is customary to connect the points to produce continuous segments. Based on the information in the graphs, find the following:

a. The cost of producing the first 300 tee shirts
b. The revenue from the sale of the first 80 tee shirts
c. The profit or loss if the first 80 tee shirts are produced and sold
d. The break-even point, that is, the number of items that must be produced and sold in order for the net profit to be $0.

Solution **a.** From the blue graph in Figure 2-40 we see that the cost corresponding to 300 tee shirts is $260.
b. From the red graph, we see that the revenue from the sale of the first 80 tee shirts is $100.
c. The cost of producing 80 tee shirts is $160. Because the profit is the difference between the cost and the revenue, the loss in this case is $160.00 − $100, or $60.00.
d. The break-even point is at point C, where the graphs intersect. At that point, the cost and the revenue are the same. The number of tee shirts corresponding to point C is 140.

Operations on Functions

Consider the function machines in Figure 2-41. If 2 is entered in the top machine, then $f(2) = 2 + 4 = 6$. Six is then entered in the second machine and $g(6) = 2 \cdot 6 = 12$. The functions in Figure 2-41 illustrate the **composition of two functions.** In the composi-

composition of two functions

tion of two functions, the range of the first function becomes the domain of the second function.

Figure 2-41

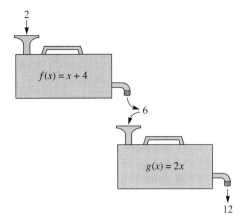

If the first function f is followed by a second function g, as in Figure 2-41, symbolize the composition of the functions as $g \circ f$. If we input 3 in the function machines of Figure 2-41 then the output is symbolized by $(g \circ f)(3)$. Because f acts first on 3, to compute $(g \circ f)(3)$ we find $f(3) = 3 + 4 = 7$ and then $g(7) = 2 \cdot 7 = 14$. Hence, $(g \circ f)(3) = 14$ and $(g \circ f)(3) = g(f(3))$. Also note that $(g \circ f)(x) = g(f(x)) = 2 \cdot f(x) = 2(x + 4)$ and hence $g(f(3)) = 2(3 + 4) = 14$.

Example 2-23   If $f(x) = 2x + 3$ and $g(x) = x - 3$, find the following:

a. $(f \circ g)(3)$ b. $(g \circ f)(3)$ c. $(f \circ g)(x)$ d. $(g \circ f)(x)$

Solution
a. $(f \circ g)(3) = f(0) = 2 \cdot 0 + 3 = 3$.
b. $(g \circ f)(3) = g(9) = 9 - 3 = 6$.
c. $(f \circ g)(x) = f(g(x)) = 2 \cdot g(x) + 3 = 2(x - 3) + 3 = 2x - 6 + 3 = 2x - 3$.
d. $(g \circ f)(x) = g(f(x)) = f(x) - 3 = (2x + 3) - 3 = 2x$.

REMARK   Example 2-23 shows that composition of functions is not commutative, since $(f \circ g)(3) \neq (g \circ f)(3)$.

We have seen that a function can be represented in a variety of ways: a table, a function machine, pictures of sets and arrows, a set of ordered pairs, and a graph. Pictures of sets with arrows and function machines are used mostly as pedagogical devices in learning the concept of a function. The most common representations are a table, an equation, and a graph. Depending on the situation, one representation may be more useful than another. For example, if the domain of a function is a large set, a table is not a convenient representation. An equation is a compact way to represent a function, but if one is given a graph of a function, it is not always possible to find an equation that represents the function. In later chapters, we learn how to graph certain kinds of equations. Graphing calculators are capable of graphing most functions given by equations with specified domains.

TECHNOLOGY CORNER

A graphing calculator sketch of the function $y = 2x + 1$ for x between 0 and 5 is shown in Figure 2-42. Use a graphing calculator to sketch the graphs of $y = 2x + b$ for three choices of b. What do the graphs seem to have in common? Why?

Figure 2-42

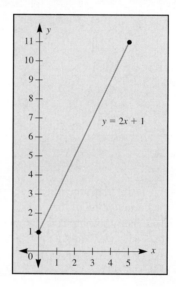

ONGOING ASSESSMENT 2-5

1. The following sets of ordered pairs are functions. Give a rule that could describe each function.
   a. $\{(2, 4), (3, 6), (9, 18), (12, 24)\}$
   b. $\{(5, 3), (7, 5), (11, 9), (14, 12)\}$
   c. $\{(2, 8), (5, 11), (7, 13), (4, 10)\}$
   d. $\{(2, 5), (3, 10), (4, 17), (5, 26)\}$
2. Which of the following are functions from the set $\{1, 2, 3\}$ to the set $\{a, b, c, d\}$? If the set of ordered pairs is not a function, explain why not.
   a. $\{(1, a), (2, b), (3, c), (1, d)\}$ b. $\{(1, c), (3, d)\}$
   c. $\{(1, a), (2, b), (3, a)\}$ d. $\{(1, a), (1, b), (1, c)\}$
3. a. Draw a diagram of a function with domain $\{1, 2, 3, 4, 5\}$ and range $\{a, b\}$.
   b. How many possible functions are there in part (a)?
4. Suppose $f(x) = 2x + 1$ and the domain is $\{0, 1, 2, 3, 4\}$. Describe the function in the following ways:
   a. Draw an arrow diagram involving two sets.
   b. Use ordered pairs.   c. Make a table.
   d. Draw a graph to depict the function.
5. Determine which of the following are functions from $W = \{0, 1, 2, 3, \ldots\}$ to W. If your answer is that it is not a function, explain why not.
   a. $f(x) = 2$ for all $x \in W$.
   b. $f(x) = 0$ if $x \in \{0, 1, 2, 3\}$, and $f(x) = 3$ if $x \notin \{0, 1, 2, 3\}$.
   c. $f(x) = x$.
   d. $f(x) = 0$ for all $x \in W$ and $f(x) = 1$ if $x \in \{3, 4, 5, 6, \ldots\}$.
   e. $f(x)$ is the sum of the digits in x and $x \in W$.
6. a. Write a rule for computing the cost of mailing a first-class letter based on its weight.
   b. Find the cost of mailing a 3-oz letter.
   c. Show that the rule you found in part (a) is a function whose inputs are weights in ounces. Find the range of this function.
   d. Graph the function in part (b).
7. The dosage of a certain drug is related to the weight of a child as follows: 50 mg of the drug and an additional 15 mg for each 2 lb or fraction of 2 lb of body weight above 30 lb. Sketch the graph of the dosage as a function of the weight of a child for children who weigh between 20 and 40 lb.
8. According to wildlife experts, the rate at which crickets chirp is a function of the temperature; that is, $C = T - 40$, where C is the number of chirps every 15 sec and T is the temperature in degrees Fahrenheit.
   a. How many chirps does the cricket make per second if the temperature is 70°F?
   b. What is the temperature if the cricket chirps 40 times in 1 min?
9. If taxi fares are $3.50 for the first half mile and $0.75 for each additional quarter mile, answer the following:

a. What is the fare for a 2-mi trip?
b. Write a rule for computing the fare for an *n*-mile trip by taxi.

10. For each of the following, guess what might be Latifah's rule. In each case, if *n* is your input and $L(n)$ is Latifah's answer, express $L(n)$ in terms of *n*.

a.

You	Latifah
3	8
4	11
5	14
10	29

b.

You	Latifah
0	1
3	10
5	26
8	65

c.

You	Latifah
6	42
0	0
8	72
2	6

11. The following graph shows arithmetic achievement-test scores for students of a sixth-grade class. From the graph, estimate the following:
    a. The frequency of the score made most often
    b. The highest score obtained
    c. The number of boys who would have had to score 54 on the test in order to match the number of girls who scored 54

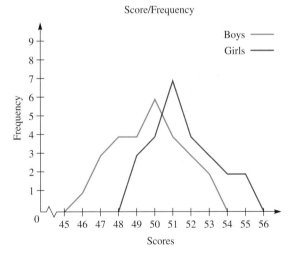

Score/Frequency

12. For each of the following sequences, find a possible function whose domain is the set of natural numbers and whose outputs are the terms of the sequence.
    a. 3, 8, 13, 18, 23, …
    b. 3, 9, 27, 81, 243, …
    c. 2, 4, 6, 8, 10, …

13. Consider two function machines that are placed as shown. Find the final output for each of the following inputs:
    a. 5   **b.** 3   **c.** 10

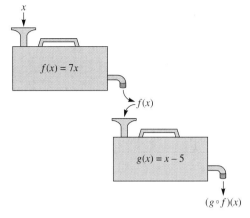

14. Let $t(n)$ represent the *n*th term of a sequence for $n \in N$. Answer the following:
    a. If $t(n) = 4n - 3$, which of the following are values of the function whose domain is *N*?
    (i) 1   (ii) 385   (iii) 389   (iv) 392
    b. If $t(n) = n^2$, which of the following are values of the function whose domain is *N*?
    (i) 1   (ii) 4   (iii) 9   (iv) 10   (v) 900
    c. If $t(n) = n(n + 1)$, which are in the range of the function whose domain is *N*?
    (i) 2   (ii) 12   (iii) 2550   (iv) 2600

15. Consider a function machine that accepts inputs as ordered pairs. Suppose the components of the ordered pairs are natural numbers and the first component is the length of the rectangle and the second is its width. The following machine computes the perimeter (the distance around a figure) of the rectangle. Thus for a rectangle whose length, *l*, is 3 and whose width, *w*, is 2, the input is (3, 2) and the output is $2 \cdot 3 + 2 \cdot 2$, or 10. Answer each of the following:
    a. For each of the following inputs, find the corresponding output: (1, 7), (2, 6), (6, 2), (5, 5).
    b. Find the set of all the inputs for which the output is 20.
    c. What is the domain and the range of the function?

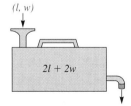

16. The following graph shows Mandy's height in centimeters measured every month starting 1 month after her fourteenth birthday. Answer the following:
    a. What was the change in Mandy's height from the first to the second month after her birthday?
    b. During which consecutive months was the change in Mandy's height the greatest?

c. What missing information would enable you to find Mandy's change in height 1 month after her birthday?

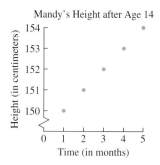

17. The following graph shows the relationship between the number of cars on a certain road at different times between 5:00 A.M. and 9:00 A.M.
    a. What was the increase in the number of cars on the road between 6:30 A.M. and 7:00 A.M.?
    b. During which half hour was the increase in the number of cars the greatest?
    c. What was the increase in the number of cars between 8:00 A.M. and 8:30 A.M.?
    d. During which half hour(s) did the number of cars decrease? By how much?

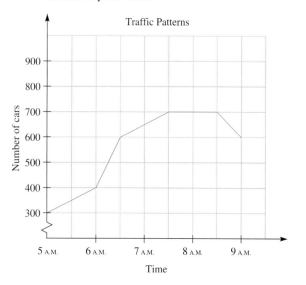

18. A health club charges a one-time initiation fee of $100 plus a membership fee of $40 per month.
    a. Write an expression for the cost function $C(x)$ that gives the total cost for membership at the health club for x months.
    b. Draw the graph of the function in (a).
    c. The health club decided to give its members an option of a higher initiation fee but a lower monthly membership charge. If the initiation fee is $300 and the monthly membership fee is $30, use a different color and draw on the same set of axes the cost graph under this plan.
    d. Determine from the graphs after how many months the second plan is less expensive for the member.

19. A particle is thrown straight up. We know its height H in feet after t seconds is given by the function $H(t) = 128t - 16t^2$.
    a. Find $H(2)$, $H(6)$, $H(3)$, and $H(5)$. Why are some of the outputs equal?
    b. Graph the function and from the graph find at what instant the ball is at its highest point. What is its height at that instant?
    c. How long will it take the particle to hit the ground?
    d. What is the domain of H?
    e. What is the range of H?

20. A rectangular plot is to be bound on one side by a straight river and on the opposite side by a fence. Suppose 900 yd of fence are available and the length of the side of the rectangle parallel to the river is denoted by x.
    a. Find an expression for the area $A(x)$ in terms of x.
    b. Graph $A(x)$.
    c. Use the graph in (b) or your calculator to estimate the length and width of the rectangle for which the area will be the largest.

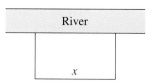

21. For each of the following sequences of matchstick figures, let $S(n)$ be the function giving the total number of matchsticks in the nth figure.
    a. For each of the following, find the total number of matchsticks in the 4th figure.
    b. For each of the following, find as simple a formula as possible for $S(n)$ in terms of n.

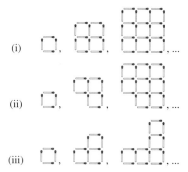

22. A function can be represented as a set of ordered pairs where the set of all the first components is the domain and the set of all the second components is the range. Is the converse also true? That is, is every set of ordered pairs a function whose domain is the set of first components and whose range is the second components? Justify your answer.

23. Suppose each point in the figure represents a child on a playground, the letters represent their names, and an arrow going from I to J means that I "is the sister of" J.

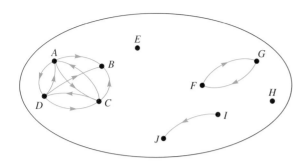

a. Based on the information in the figure, who are definitely girls and who are definitely boys?
b. Suppose we write "A is the sister of B" as an ordered pair (A, B). Based on the information in the diagram, write the set of all such ordered pairs.
c. Is the set of all ordered pairs in (b) a function with domain equal to the set of all first components of the ordered pairs and with the range equal to the set of all second components?

24. A generalization of the concept of a function is the concept of a *relation*. Given two sets A and B, a relation from A to B is any set of ordered pairs in which the first components are from A and the second are from B. Which of the following are functions and which are relations but not functions from the set of first components of the ordered pairs to the set of second components?
   a. {(Montana, Helena), (Oregon, Salem), (Illinois, Springfield), (Arkansas, Little Rock)}
   b. {(Pennsylvania, Philadelphia), (New York, New York), (New York, Niagara Falls), (Florida, Ft. Lauderdale)}
   c. $\{(x, y) \mid x$ resides in Birmingham, Alabama, and x is the mother of y, where y is a U.S. resident$\}$
   d. $\{(1, 1), (2, 4), (3, 9), (4, 16)\}$
   e. $\{(x, y) \mid$ where x and y are natural numbers and $x + y$ is an even number$\}$

25. a. Is the rule "has as mother" a function whose domain is the set of all people?
   b. Is the relation "has as brother" a function from the set of all boys to the set of all boys?

The following definitions are needed to answer problems 26 and 27.
A *relation on a set* X. This is a relation from X to X. A relation on X may have one or more of the following properties:
The reflexive property. A relation on a set X is reflexive if, and only if, for all a in X, a is related to a, that is (a, a) is in the set of ordered pairs.
The symmetric property. A relation on a set X is symmetric if, and only if, for all elements a and b in X, whenever a is related to b, then b is also related to a; that is, if (a, b) is in the set of ordered pairs, so is (b, a).
The transitive property. A relation on a set X is transitive if, and only if, for all elements $a, b,$ and c in X, whenever a is related to b and b is related to c, then a is related to c; that is, if (a, b) and (b, c) are in the set of ordered pairs, then (a, c) is also in the same set ($a, b,$ and c do not have to be different).

An **equivalence relation** on a set X is any relation on X that satisfies the reflexive, symmetric, and transitive properties.

26. Tell whether each of the following is reflexive, symmetric, or transitive on the set of all people. Which are equivalence relations?
   a. "Is a parent of"
   b. "Is the same age as"
   c. "Has the same last name as"
   d. "Is the same height as"
   e. "Is married to"
   f. "Lives within 10 mi of"
   g. "Is older than"

27. Tell whether each of the following is reflexive, symmetric, or transitive on the set of subsets of a nonempty set. Which are equivalence relations?
   a. "Is equal to"
   b. "Is a proper subset of"
   c. "Is not equal to"
   d. "Has the same cardinal number as"

Communication

28. Does the diagram define a function from A to B? Why or why not?

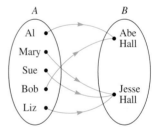

29. Is a one-to-one correspondence a function? Explain your answer and give an example.

30. Which of the following are functions from A to B? If your answer is "not a function," explain why not.
   a. A is the set of mathematics faculty at the university. B is the set of all mathematics classes. To each mathematics faculty member, we associate the class that that person is teaching during a given term.
   b. A is the set of mathematics classes at the university. B is the set of mathematics faculty. To each mathematics class, we associate the teacher who is teaching the class.
   c. A is the set of all U.S. senators and B is the set of all Senate committees. We associate each senator to the committee of which the senator is chairperson.

31. When a boat is put in the water, its hull is partially above and partially below the water. The part below the water is the "draft" and the part above the water is the "freeboard." The following table shows the relationship between the draft and freeboard for a 50-cm-deep boat:

Draft	Freeboard
5	45
10	40
15	35
20	30
30	20

a. Graph the freeboard as a function of the draft. Is it meaningful to connect the points graphed? Explain.
b. If d stands for the draft in centimeters and $f(d)$ for the freeboard, write an equation expressing $f(d)$ in terms of d.

Open-Ended

32. Examine several newspapers and magazines and describe at least three examples of functions that appear. What is the domain and range of each function?
33. Give at least three examples of functions from A to B where neither A nor B are sets of numbers.
34. Draw a sequence of figures of matchsticks and describe the pattern in words. Find as simple an expression as possible for $S(n)$, the total number of matchsticks in the nth figure.
35. A function whose output is always the same regardless of the input is a constant function. Give several examples of constant functions from real life.
36. A function whose output is the same as its input is an identity function. Give several concrete examples of identity functions.

Cooperative Learning

37. Each person in a group picks a natural number and uses it as an input in the following function machine:

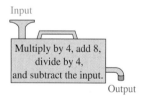

a. Compare your answers. Based on the answers, make a conjecture about the range of the function.
b. Based on your answer in (a), graph the function.
c. Write the function in the simplest possible way using $f(x)$ notation.
d. Justify your conjecture in (a).
e. Make up similar function machines and try different inputs in your group.
f. Devise a function machine in which the machine performs several operations, but the output is always the same as the input. Exchange your answer with someone in the group and check that the other person's function machine performs as required.
g. Write the functions the group came up with in the simplest way using $f(x)$ notation and graph them.

38. In the following two-person (or two-group) activity, one person presents the graphs of several arithmetic and several geometric sequences. The second person is required to find what kind of sequence corresponds to each graph and the simplest possible expression for $T(n)$, the nth term of the sequence. Compare your expressions for $T(n)$ and make a conjecture on how to tell from the graph if a sequence is arithmetic.

Review Problems

39. Rewrite each of the following in equivalent form using only multiplication or addition operations.
   a. $109 - 11 = 98$. b. $x - y = z$.
   c. $60 \div 4 = 15$. d. $x \div 3 = y$.
   e. $(x - 3) \div 5 = 10$.
40. Is the number of elements in $A \cup B$ always equal to the sum of the number of elements in A and the number of elements in B? Why or why not?
41. Shade the portion of the Venn diagram that illustrates the set $(B - C) \cup (A \cap B)$.

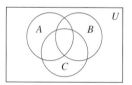

BRAIN TEASER Only 10 rooms were vacant in the Village Hotel. Eleven men went into the hotel at the same time, each wanting a separate room. The clerk, settling the argument, said, "I'll tell you what I'll do. I'll put two men in Room 1 with the understanding that I will come back and get one of them a few minutes later." The men agreed to this. The clerk continued, "I will put the rest of you men in rooms as follows: the third man in Room 2, the fourth man in Room 3, the fifth man in Room 4, the sixth man in Room 5, the seventh man in Room 6, the eighth man in Room 7, the ninth man in Room 8, and the tenth man in Room 9." Then the clerk went back and got the extra man he had left in Room 1 and put him in Room 10. Everybody was happy. What is wrong with this plan?

HINT FOR SOLVING THE PRELIMINARY PROBLEM

There are 40 toppings, but since everyone orders mushroom and cheese, there are 38 other toppings to choose from. Find the number of subsets of the 38 toppings. Each of these subsets will exclude the mushroom and cheese toppings. Adding these two toppings to each of the subsets of the 38 toppings will give us all the toppings, including mushroom and cheese. Compare the number of the subsets of the set of 40 toppings that include cheese and mushrooms to the population of the world, which is about 6 billion—that is, 6,000,000,000.

QUESTIONS FROM THE CLASSROOM

1. A student argues that $\{\varnothing\}$ is the proper notation for the empty set. What is your response?
2. A student asks, "If $A = \{a, b, c\}$ and $B = \{b, c, d\}$, why isn't it true that $A \cup B = \{a, b, c, b, c, d\}$?" What is your response?
3. A student says that she can show that if $A \cap B = A \cap C$, then it is not necessarily true that $B = C$; but she thinks that whenever $A \cap B = A \cap C$ and $A \cup B = A \cup C$, then $B = C$. What is your response?
4. A student claims that a finite set of numbers is any set that has a greatest element. Do you agree?
5. A student claims that the complement bar can be broken over the operation of intersections; that is, $\overline{A \cap B} = \overline{A} \cap \overline{B}$. What is your response?
6. A student claims that $\overline{A} \cap \overline{B}$ includes all elements that are not in A. What is your response?
7. A student asks whether a formula and a function are the same. What is your response?
8. A student states that either $A \subseteq B$ or $B \subseteq A$. Is the student correct?
9. A student is asked to find all one-to-one correspondences between 2 given sets. He finds the Cartesian product of the sets and claims that his answer is correct because it includes all possible pairings between the elements of the sets. How do you respond?
10. A student argues that adding 2 sets A and B, or $A + B$, and taking the union of 2 sets, $A \cup B$, is the same thing. How do you respond?
11. A student asks "Does $2(3 \cdot 4)$ equal $(2 \cdot 3)(2 \cdot 4)$?" Is there a distributive property of multiplication over multiplication?
12. Since $39 + 41 = 40 + 40$, is it true that $39 \cdot 41 = 40 \cdot 40$?
13. The division algorithm, $a = bq + r$, holds for $a \geq b$; $a, b, q, r \in W$, $r < b$, and $b \neq 0$. Does the division algorithm hold when $a < b$?
14. Can we define $0 \div 0$ as 1? Why or why not?
15. a. A student claims that for all whole numbers $(a \cdot b) \div b = a$. How do you respond?
    b. The student in part (a) claims that $0 \div 0 = 0$. The student's reasoning is, "if $a = 0$ and $b = 0$ are substituted in the equation in part (a), the result is $0 \cdot 0 \div 0 = 0$. But because $0 \cdot 0 = 0$, it follows that $0 \div 0 = 0$. How do you respond?
16. A student asks if division on the set of whole numbers is distributive over subtraction. How do you respond?
17. A student says that 0 is the identity for subtraction. How do you respond?
18. A student claims that on the following number line, the arrow doesn't really represent 3 because the end of the arrow does not start at 0. How do you respond?

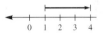

19. A student asks whether a function from A to B is related to the cross product $A \times B$. How do you respond?
20. A student claims that the following machine does not represent a function machine because it accepts two inputs at once rather than a single input. How do you respond?

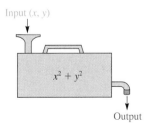

CHAPTER OUTLINE

I. Set definitions and notation
  A. A **set** can be described as any collection of objects.
  B. Sets should be **well defined** so that an object either does or does not belong to the set.
  C. An **element** is any **member** of a set.
  D. Sets can be specified by either listing all the elements or using **set-builder notation.**
  E. The **empty set,** written $\emptyset$, contains no elements.
  F. The **universal set** contains all the elements being discussed.

II. Relationships and operations on sets
  A. Two sets are **equal** if, and only if, they have exactly the same elements.
  B. Two sets A and B are in **one-to-one correspondence** if, and only if, each element of A can be paired with exactly one element of B and each element of B can be paired with exactly 1 element of A.
  C. Two sets A and B are **equivalent** if, and only if, their elements can be placed into one-to-one correspondence (written $A \sim B$).
  D. Set A is a **subset** of set B if, and only if, every element of A is an element of B (written $A \subseteq B$).
  E. Set A is a **proper subset** of set B if, and only if, every element of A is an element of B and there is at least one element of B that is not in A (written $A \subset B$).
  F. A set containing n elements has 2^n subsets.
  G. The **union** of two sets A and B is the set of all elements in A, in B, or in both A and B (written $A \cup B$).
  H. The **intersection** of two sets A and B is the set of all elements belonging to both A and B (written $A \cap B$).
  I. The **cardinal number** of a finite set S, $n(S)$ indicates the number of elements in the set.
  J. A set is **finite** if the number of elements in the set is zero or a natural number. Otherwise, the set is **infinite**.
  K. Two sets A and B are **disjoint** if they have no elements in common.
  L. The **complement** of a set A is the set consisting of the elements of the universal set not in A (written $\overline{A}$).
  M. The **complement of set A relative to set B** (set difference) is the set of all elements in B that are not in A (written $B - A$).
  N. The **Cartesian product** of sets A and B, written $A \times B$, is the set of all ordered pairs such that the first element in each pair is from A and the second element of each pair is from B.
  O. Properties of set operations.
    1. Commutative property of set union.
    2. Commutative property of set intersection.
    3. Distributive property of set intersection over union.

III. Whole numbers
  A. The set of **whole numbers** W is $\{0, 1, 2, 3, \ldots\}$.
  B. The basic operations for whole numbers are addition, subtraction, multiplication, and division.
    1. Addition: If $n(A) = a$ and $n(B) = b$, where $A \cap B = \emptyset$, then $a + b = n(A \cup B)$. The numbers a and b are **addends** and $a + b$ is the **sum.**
    2. Subtraction: If a and b are any whole numbers, then $a - b$ is the unique whole number c such that $a = b + c$.
    3. Multiplication: If a and b are any whole numbers, then
    $$ab = \underbrace{b + b + b + \ldots + b}_{a \text{ terms}},$$
    where a and b are **factors** and ab is the **product.**
    4. Multiplication: If A and B are sets such that $n(A) = a$ and $n(B) = b$, then $ab = n(A \times B)$.
    5. Division: If a and b are any whole numbers with $b \neq 0$, $a \div b$ is the unique whole number c such that $bc = a$. The number a is the **dividend,** b is the **divisor,** and c is the **quotient.**
    6. **Division algorithm:** Given any whole numbers a and b, with $b \neq 0$, there exist unique whole numbers q and r such that $a = bq + r$, with $0 \leq r < b$.
  C. Properties of addition and multiplication of whole numbers
    1. Closure: If $a, b \in W$, then $a + b \in W$ and $ab \in W$.
    2. Commutative: If $a, b \in W$, then $a + b = b + a$ and $ab = ba$.
    3. Associative: If $a, b, c \in W$, then $(a + b) + c = a + (b + c)$ and $a(bc) = (ab)c$.
    4. Identity: 0 is the unique identity element for addition of whole numbers; 1 is the unique identity element for multiplication.
    5. Distributive property of multiplication over addition: If $a, b, c \in W$, then $a(b + c) = ab + ac$.
    6. Zero multiplication property: For any whole number a, $a \cdot 0 = 0 = 0 \cdot a$.
  D. Relations on whole numbers
    1. $a < b$ if, and only if, there is a natural number c such that $a + c = b$.
    2. $a > b$ if, and only if, $b < a$.

III. Functions
  A. A **function** from set A to B is a correspondence in which each element $a \in A$ is paired with one, and only one, element $b \in B$. If the function is denoted by f, we write $f(a) = b$. The element $a \in A$ is the input, and $f(a)$ is the output. A is the **domain** of the function. B is any set containing all the outputs. The set of all the outputs is the **range** of the function.
  B. A function can be represented by a table, an equation, a function machine, a set of ordered pairs, or a graph.

CHAPTER REVIEW

1. List all the subsets of $\{m, a, t, h\}$.
2. Let
   $U = \{u, n, i, v, e, r, s, a, l\}$,
   $A = \{r, a, v, e\}$, $C = \{l, i, n, e\}$, and
   $B = \{a, r, e\}$, $D = \{s, a, l, e\}$.
   Find each of the following:
   a. $A \cup B$ b. $C \cap D$
   c. $\overline{D}$ d. $A \cap \overline{D}$
   e. $\overline{B \cup C}$ f. $(B \cup C) \cap D$
   g. $(\overline{A \cup B}) \cap (C \cap \overline{D})$ h. $(C \cap D) \cap A$
   i. $n(\overline{C})$ j. $n(C \times D)$
3. Indicate the following sets by shading the figure:

   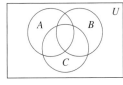 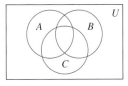
   (a) $A \cap (B \cup C)$ (b) $(\overline{A \cup B}) \cap C$

4. Suppose you are playing a word game with seven letters. How many seven-letter words can there be?
5. a. Show one possible one-to-one correspondence between sets D and E if $D = \{t, h, e\}$ and $E = \{e, n, d\}$.
   b. How many one-to-one correspondences between sets D and E are possible?
6. Use a Venn diagram to determine whether $A \cap (B \cup C) = (A \cap B) \cup C$ for all sets A, B, and C.
7. According to a student survey, 16 students liked history, 19 liked English, 18 liked mathematics, 8 liked mathematics and English, 5 liked history and English, 7 liked history and mathematics, 3 liked all three subjects, and every student liked at least one of the subjects. Draw a Venn diagram describing this information and answer the following questions:
   a. How many students were in the survey?
   b. How many students liked only mathematics?
   c. How many students liked English and mathematics but not history?
8. Describe, using symbols, the shaded portion in each of the following figures:

   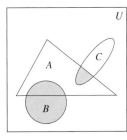 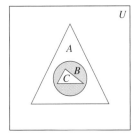
   (a)   (b)

9. Classify each of the following as true or false. If false, tell why.
   a. For all sets A and B, either $A \subseteq B$ or $B \subseteq A$.
   b. The empty set is a proper subset of every set.
   c. For all sets A and B, if $A \sim B$, then $A = B$.
   d. The set $\{5, 10, 15, 20, \ldots\}$ is a finite set.
   e. No set is equivalent to a proper subset of itself.
   f. If A is an infinite set and $B \subseteq A$, then B also is an infinite set.
   g. For all finite sets A and B, if $A \cap B \neq \emptyset$, then $n(A \cup B) \neq n(A) + n(B)$.
   h. If A and B are sets such that $A \cap B = \emptyset$, then $A = \emptyset$ or $B = \emptyset$.
10. Suppose P and Q are equivalent sets and $n(P) = 17$.
    a. What is the minimum number of elements in $P \cup Q$?
    b. What is the maximum number of elements in $P \cup Q$?
    c. What is the minimum number of elements in $P \cap Q$?
    d. What is the maximum number of elements in $P \cap Q$?
11. Case Eastern Junior College awarded 26 varsity letters in crew, 15 in swimming, and 16 in soccer. If awards went to 46 students and only 2 lettered in all sports, how many students lettered in two of the 3 sports?
12. Consider the set of northwestern states or provinces $\{$Montana, Washington, Idaho, Oregon, Alaska, British Columbia, Alberta$\}$. If a person chooses one element, show that in three yes or no questions, we can determine the element.
13. For each of the following, identify the properties of the operation(s) for whole numbers illustrated:
    a. $3 \cdot (a + b) = 3 \cdot a + 3 \cdot b$.
    b. $2 + a = a + 2$.
    c. $16 \cdot 1 = 1 \cdot 16 = 16$.
    d. $6 \cdot (12 + 3) = 6 \cdot 12 + 6 \cdot 3$.
    e. $3 \cdot (a \cdot 2) = 3 \cdot (2 \cdot a)$.
    f. $3 \cdot (2 \cdot a) = (3 \cdot 2) \cdot a$.
14. Using the definitions of less than or greater than, prove that each of the following inequalities is true:
    a. $3 < 13$. b. $12 > 9$.
15. For each of the following, find all possible replacements to make the following statements true for whole numbers:
    a. $4 \cdot \square - 37 < 27$.
    b. $398 = \square \cdot 37 + 28$.
    c. $\square \cdot (3 + 4) = \square \cdot 3 + \square \cdot 4$.
    d. $42 - \square \geq 16$.
16. Use the distributive property of multiplication and addition facts, if possible, to rename each of the following:
    a. $3a + 7a + 5a$
    b. $3x^2 + 7x^2 - 5x^2$
    c. $x(a + b + y)$
    d. $(x + 5)3 + (x + 5)y$
17. How many 12-oz cans of juice would it take to give 60 people one 8-oz serving each?

18. Heidi has a brown and a gray pair of slacks; a brown, a yellow, and a white blouse; and a blue and a white sweater. How many different outfits does she have if each outfit she wears consists of slacks, a blouse, and a sweater?
19. I am thinking of a whole number. If I divide it by 13, then multiply the answer by 12, then subtract 20, and then add 89, I end up with 93. What was my original number?
20. A ski resort offers a weekend ski package for $80 per person or $6000 for a group of 80 people. Which would be the cheaper option for a group of 80?
21. Josi has a job in which she works 30 hr/wk and gets paid $5/hr. If she works more than 30 hr in a week, she receives $8/hr for each hour over 30 hr. If she worked 38 hr this week, how much did she earn?
22. In a television game show, there are five questions to answer. Each question is worth twice as much as the previous question. If the last question was worth $6400, what was the first question worth?
23. Which of the following sets of ordered pairs are functions from the set of first components to the set of second components?
    a. $\{(a, b), (c, d), (e, a), (f, g)\}$
    b. $\{(a, b), (a, c), (b, b), (b, c)\}$
    c. $\{(a, b), (b, a)\}$
24. Given the following function rules and the domains, find the associated ranges:
    a. $f(x) = x + 3$; domain = $\{0, 1, 2, 3\}$.
    b. $f(x) = 3x - 1$; domain = $\{5, 10, 15, 20\}$.
    c. $f(x) = x^2$; domain = $\{0, 1, 2, 3, 4\}$.
    d. $f(x) = x^2 + 3x + 5$; domain = $\{0, 1, 2\}$.
25. Which of the following correspondences from A to B describe a function? If a correspondence is a function, find its range. Justify your answers.
    a. A is the set of college students, and B is the set of majors. To each college student corresponds his or her major.
    b. A is the set of books in the library, and B is the set N of natural numbers. To each book corresponds the number of pages in the book.
    c. $A = \{(a, b) \mid a \in N \text{ and } b \in N\}$, and $B = N$. To each element of A corresponds the number $4a + 2b$.
    d. $A = N$ and $B = N$. If x is even, then $f(x) = 0$ and if x is odd, then $f(x) = 1$.
    e. $A = N$ and $B = N$. To each natural number corresponds the sum of its digits.
26. A health club charges an initiation fee of $200, 1 month of free membership and then $55 per month.
    a. If $C(x)$ is the total cost of membership in the club for x months, express $C(x)$ in terms of x.
    b. Graph $C(x)$ for the first 12 mo.
    c. Use the graph in (b) to find when the total cost of membership in the club will exceed $600.
    d. When will the total cost of membership exceed $6000?

SELECTED BIBLIOGRAPHY

Cantlon, D. "Kids + Conjecture = Mathematics Power" *Teaching Children Mathematics* (October 1998): 108–112.

Crouse, R., and A. Alison. "Tips for Beginners: The Human Coordinate System." *Mathematics Teacher* 84 (February 1991): 108–109.

Greenwood, J. "Name That Graph." *Mathematics Teacher* 88 (January 1995): 8–11.

Huinker, D. "Multiplication and Division Word Problems: Improving Students' Understanding." *Arithmetic Teacher* 37 (October 1989): 8–12.

Johnston, A. "Introducing Function and Its Notation." *Mathematics Teacher* 80 (October 1987): 558–560.

Kline, K. "Kindergarten Is More Than Counting." *Teaching Children Mathematics* (October 1998): 84–87.

McGinty, R., and J. Van Beynen. "Deductive and Analytical Thinking." *Mathematics Teacher* 78 (March 1985): 188–194.

Mercer, J. "Teaching Graphing Concepts with Graphing Calculators." *Mathematics Teacher* 88 (April 1995): 268–273.

O'Daffer, P. "Inductive and Deductive Reasoning." *Mathematics Teacher* 83 (May 1990): 378–384.

O'Regan, P. "Intuition and Logic." *Mathematics Teacher* 81 (November 1988): 664–668.

Payne, J., and D. Huinker. "Early Numbers and Numeration." In *Research Ideas for the Classroom: Early Childhood Mathematics,* edited by R. Jensen. New York, Macmillan, 1993.

Rubenstein, R. "The Function Game." In *Mathematics Teaching in the Middle School* (November–December 1996): 74–78.

Sand, M. "A Function Is a Mail Carrier." In *Mathematics Teacher* (September 1996): 468–469.

Sanders, W., and R. Antes. "Teaching Logic with Logic Boxes." *Mathematics Teacher* 81 (November 1988): 643–647.

Schloemer, C. "Tips for Teaching Cartesian Graphing: Linking Concepts and Procedures." *Teaching Children Mathematics* (September 1994): 20–23.

Schulman, L., and R. Eston. "A Problem Worth Revisiting." *Teaching Children Mathematics* (October 1998): 72–77.

Shealy, B. "Becoming Flexible with Functions: Investigating United States Population Growth." *Mathematics Teacher* (May 1996): 414–418.

Spence, L. "How Many Elements Are in a Union of Sets?" *Mathematics Teacher* 80 (November 1987): 666–670, 681.

Sundar, V. "Thou Shalt Not Divide by Zero." *Arithmetic Teacher* 37 (March 1991): 50–51.

Thornton, C., and P. Smith. "Action Research: Strategies for Learning Subtraction Facts." *Arithmetic Teacher* 35 (April 1988): 8–12.

Van de Walle, J. "Concepts of Number." *Mathematics for the Young Child,* edited by Joseph Payne. Reston, Va., National Council of Teachers of Mathematics, 1990.

3

Whole-Number Computation

Preliminary Problem
The Washington School PTA set up a phone tree in order to reach all of its members. Each person's responsibility, after receiving a call, is to call two other assigned members until all members have been called. If we assume that everyone is at home and answers the phone, and that each phone call takes 30 seconds, what is the least amount of time necessary to reach all 85 members in the group?

The *Principles and Standards for School Mathematics* defines number sense as *the ability to decompose numbers naturally, use particular numbers like 100 or 1/2 as referents, use the relationships among arithmetic operations to solve problems, understand the base-ten number system, estimate, make sense of numbers, and recognize the relative and absolute magnitude of numbers* (p. 32).

Also in the *Principles and Standards* we find,

Knowing basic number combinations—the single-digit addition and multiplication pairs and their counterparts for subtraction and division—is essential. Equally essential is computational fluency—having and using efficient and accurate methods for computing (p. 32).

In the first section of this chapter, we introduce various number systems and compare them to the Hindu-Arabic system of numbers that we use today. By comparing the Hindu-Arabic system with ancient systems that used other bases, we may develop a clearer picture of how we do whole number computation. The Hindu-Arabic system relies on ten digits: 0, 1, 2, 3, 4, 5, 6, 7, 8, and 9 that represent the cardinal numbers of sets equivalent to those shown in Table 3-1.

Table 3-1

Digits for Cardinal Number of Set	Set
0	$\emptyset$
1	{a}
2	{a, b}
3	{a, b, c}
4	{a, b, c, d}
5	{a, b, c, d, e}
6	{a, b, c, d, e, f}
7	{a, b, c, d, e, f, g}
8	{a, b, c, d, e, f, g, h}
9	{a, b, c, d, e, f, g, h, i}

Section 3-1 — Numeration Systems

numeration system

Written symbols such as 2 or 5 are *numerals*. Using numerals to represent numbers greater than 9 requires a numeration system. A **numeration system** is a collection of properties and symbols agreed upon to represent numbers systematically. The Hindu-Arabic system relies on the following properties:

1. All numerals are constructed from the ten digits.
2. Place value is based on powers of 10, the number base of the system.

place value

face value

expanded form

factor

Because the Hindu-Arabic system is based on powers of 10, the system is sometimes called a base-ten, or a decimal, system. **Place value** assigns a value of a digit depending on its placement in a numeral. To find the value of a digit in a whole number, we multiply the place value of the digit by its **face value,** where the face value is a digit. For example, in the numeral 5984, the 5 has place value "thousands," the 9 has place value "hundreds," the 8 has place value "tens," and the 4 has place value "units," as seen in Figure 3-1. We could write 5984 in **expanded form** as $5 \cdot 10^3 + 9 \cdot 10^2 + 8 \cdot 10 + 4 \cdot 1$. In the expanded form of 5984, exponents have been used. For example, 1000, or $10 \cdot 10 \cdot 10$, is written as 10^3. In this case, 10 is a **factor** of the product. In general, we have the following definition.

SECTION 3-1 *Numeration Systems* 125

Figure 3-1

> **Definition of a^n**
>
> If a is any number and n is any natural number, then
> $$a^n = \underbrace{a \cdot a \cdot a \cdot \ldots \cdot a}_{n \text{ factors}}.$$

Numbers have been recorded in many ways over the ages. The Babylonians used wedge-shaped marks pressed in wet clay. The Egyptians used papyrus and ink-filled brushes and based their system on tally marks to represent objects being counted. The Mayans introduced a symbol for zero. Perhaps the prisoner in the following cartoon felt that the college grad's new symbols were very sophisticated for that era. Table 3-2 shows some other ways that numbers have been recorded.

Table 3-2

	1	2	3	4	5	6	7	8	9	10
Babylonian	▼	▼▼	▼▼▼	▼▼▼▼	▼▼▼ ▼▼	▼▼▼ ▼▼▼	▼▼▼▼ ▼▼▼	▼▼▼▼ ▼▼▼▼	▼▼▼▼▼ ▼▼▼▼	<
Egyptian	l	ll	lll	llll	lll ll	lll lll	llll lll	llll llll	lll lll lll	∩
Mayan	〇	•	••	•••	••••	—	•̇	••̇	•••̇	••••̇	═
Greek	α	β	γ	δ	ε	ϕ	ζ	η	θ	ι
Roman	I	II	III	IV	V	VI	VII	VIII	IX	X
Hindu	0	1	2	3	8	4	6	7	8	9
Arabic	·	١	٢	٣	٤	٥	٦	٧	٨	٩
Hindu-Arabic	0	1	2	3	4	5	6	7	8	9	10

Egyptian Numeration System

The Egyptian numeration system, which dates back to about 3400 B.C., used *tally marks*. In a *tally numeration system,* there is a one-to-one correspondence between the marks and the items being counted. The first nine numerals in the Egyptian system in Table 3-2 show the use of tally marks. The Egyptians improved on the system based only on tally marks by developing a *grouping system* to represent certain sets of numbers. This makes the numbers easier to record. For example, the Egyptians used a heel bone symbol, ∩, to stand for a grouping of ten tally marks.

$$||||||||| \rightarrow \cap$$

Table 3-3 shows other numerals that the Egyptians used in their system.

Table 3-3

Egyptian Numeral	Description	Hindu-Arabic Equivalent
		Vertical staff	1
∩	Heel bone	10
9	Scroll	100
↑	Lotus flower	1,000
∕	Pointing finger	10,000
⌒	Polliwog or burbot	100,000
⚱	Astonished man	1,000,000

additive property   The Egyptian system involved an **additive property;** that is, the value of a number was the sum of the face values of the numerals. The Egyptians customarily wrote the numerals in decreasing order from left to right as in ⌒999∩∩||. The number can be converted to base ten as shown below.

⌒	represents	100,000	
999	represents	300	(100 + 100 + 100)
∩∩	represents	20	(10 + 10)
||	represents	2	(1 + 1)
⌒999∩∩||	represents	100,322	

NOW TRY THIS 3-1

• **a.** Use the Egyptian system to represent 1,312,322.
b. Use the Hindu-Arabic system to represent ⌒⌒↑↑↑∩∩||||.

c. What disadvantages do you see of the Egyptian system when compared to the Hindu-Arabic system? •

Babylonian Numeration System

The Babylonian numeration system was developed at about the same time as the Egyptian system. The symbols in Table 3-4 were made using a stylus either vertically or horizontally on clay tablets. The Babylonian system has been well preserved because clay tablets were used rather than papyrus.

Table 3-4

Babylonian Numeral	Hindu-Arabic Equivalent
▼	1
＜	10

The Babylonian numerals 1 through 59 were similar to the Egyptian numerals, but the vertical staff and the heel bone were replaced by the symbols shown in Table 3-4. For example, ＜＜ ▼▼ represented 22.

The Babylonian numeration system used a place value system. Numbers greater than 59 were represented by repeated groupings of 60, much as we use groupings of 10 today. For example ▼▼ ＜＜ might represent $2 \cdot 60 + 20$, or 140. The space indicates that ▼▼ represents $2 \cdot 60$ rather than 2. It is thought that the Babylonians chose to work with 60 because it can be evenly divided by many numbers. This simplifies division and operations with fractions.

The initial Babylonian system was inadequate by today's standards. For example, the symbol ▼▼ could have represented 2 or $2 \cdot 60$ because the Babylonian system lacked a symbol for zero until after 300 B.C.

Numerals immediately to the left of a second space have a value $60 \cdot 60$ times their face value, and so on.

＜＜ ▼	represents	$20 \cdot 60 + 1$, or 1201
＜▼ ＜▼ ▼	represents	$11 \cdot 60 \cdot 60 + 11 \cdot 60 + 1$, or $11 \cdot 60^2 + 11 \cdot 60 + 1$, or 40,261
▼ ＜▼ ＜▼ ▼	represents	$1 \cdot 60 \cdot 60 \cdot 60 + 11 \cdot 60 \cdot 60 + 11 \cdot 60 + 1$, or $1 \cdot 60^3 + 11 \cdot 60^2 + 11 \cdot 60 + 1$, or 256,261

NOW TRY THIS 3-2

- **a.** Use the Babylonian system to represent 12,321.
- **b.** Use the Hindu-Arabic system to represent ▼▼ ＜▼ ▼.
- **c.** What advantages does the Hindu-Arabic system have over the Babylonian system?

Mayan Numeration System

In the early development of numeration systems, people frequently used parts of their bodies to count. Fingers could be matched to objects to stand for one, two, three, four, or five objects. Two hands could then stand for a set of ten objects. In warmer climates where people went barefoot, people may have used their toes as well as their fingers for counting. The Mayans introduced an attribute that was not present in the Egyptian or early Babylonian systems, namely, a symbol for zero. The Mayan system used only three symbols, which Table 3-5 shows.

Table 3-5

Mayan Numeral	Hindu-Arabic Equivalent
•	1
—	5
⊖	0

The symbols for the first ten numerals in the Mayan system are shown in Table 3-2. Notice the groupings of five, where each horizontal bar represents a group of five. Thus the symbol for 19 was ≝, or 3 fives and 4 ones. The symbol for 20 was ⊖, which represents 1 group of twenty plus 0 ones. The Mayans wrote numbers vertically with the greatest place value on top. In Figure 3-2(a), we have $2 \cdot 5 + 3 \cdot 1$, or 13 groups of twenty plus $2 \cdot 5 + 1 \cdot 1$, or 11 ones, for a total of 271. In Figure 3-2(b), we have $3 \cdot 5 + 1 \cdot 1$, or 16, groups of twenty and 0 ones, for a total of 320.

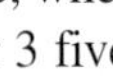

Figure 3-2

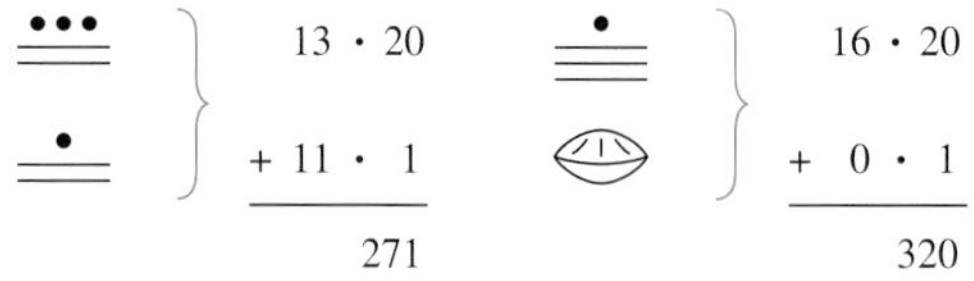

$$\begin{array}{cc} 13 \cdot 20 \\ + 11 \cdot 1 \\ \hline 271 \end{array} \qquad \begin{array}{cc} 16 \cdot 20 \\ + 0 \cdot 1 \\ \hline 320 \end{array}$$

(a) \qquad (b)

In a true base-twenty system, the place value of the symbols in the third position vertically from the bottom should be 20^2, or 400. However, the Mayans used $20 \cdot 18$, or 360, instead of 400. (The number 360 is an approximation of the length of a calendar year, which consisted of 18 months of 20 days each, plus 5 "unlucky" days.) Thus, instead of place values of 1, 20, 20^2, 20^3, 20^4, and so on, the Mayans used 1, 20, $20 \cdot 18$, $20^2 \cdot 18$, $20^3 \cdot 18$, and so on. For example, in Figure 3-3(a), we have $5 + 1$ (or 6) groups of 360, plus $5 + 5 + 2$ (or 12) groups of 20, plus $5 + 4$ (or 9) groups of 1, for a total of 2409. In Figure 3-3(b), we have $2 \cdot 5$ (or 10) groups of 360, plus 0 groups of 20, plus 2 ones, for a total of 3602. Spacing is important in the Mayan system. For example, if two horizontal bars are placed close together, as in ≡, the symbols represent $5 + 5 = 10$. If the bars are spaced apart, as in =, then the value is $5 \cdot 20 + 5 \cdot 1 = 105$.

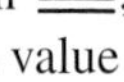

Figure 3-3

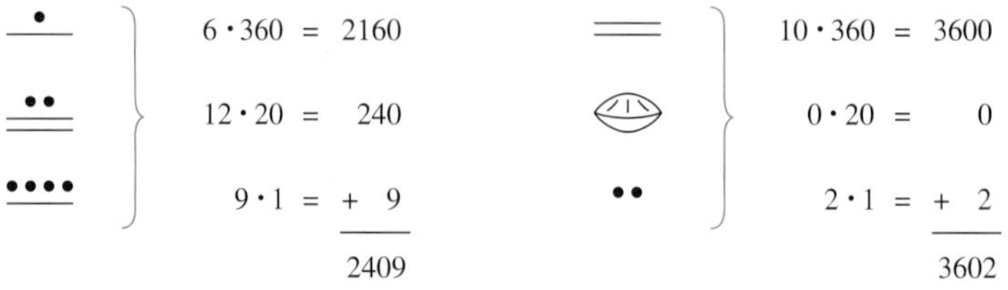

$$\begin{array}{rl} 6 \cdot 360 = & 2160 \\ 12 \cdot 20 = & 240 \\ 9 \cdot 1 = & + 9 \\ \hline & 2409 \end{array} \qquad \begin{array}{rl} 10 \cdot 360 = & 3600 \\ 0 \cdot 20 = & 0 \\ 2 \cdot 1 = & + 2 \\ \hline & 3602 \end{array}$$

(a) \qquad (b)

Roman Numeration System

The Roman numeration system was used in Europe in its early form from the third century B.C. It remains in use today, as seen on cornerstones, on the opening pages of books, and on faces of some clocks. The Roman system used only a few symbols, as shown in Table 3-6.

Table 3-6

Roman Numeral	Hindu-Arabic Equivalent
I	1
V	5
X	10
L	50
C	100
D	500
M	1000

Roman numerals can be combined by using an additive property. For example, MDCLXVI represents $1000 + 500 + 100 + 50 + 10 + 5 + 1 = 1666$, CCCXXVIII represents 328, and VI represents 6.

subtractive property   To avoid repeating a symbol more than three times, as in IIII, a **subtractive property** was introduced in the Middle Ages. For example, I is less than V, so if it is to the left of V, it is subtracted. Thus IV has a value of $5 - 1$, or 4, and XC represents $100 - 10$, or 90.

Some extensions of the subtractive property could lead to ambiguous results. For example, IXC could be 91 or 89. By custom, 91 is written XCI and 89 is written LXXXIX. In general, only one smaller number symbol can be to the left of a larger number symbol, and the pair must be one of those listed in Table 3-7.

Table 3-7

Roman Numeral	Hindu-Arabic Equivalent
IV	$5 - 1$, or 4
IX	$10 - 1$, or 9
XL	$50 - 10$, or 40
XC	$100 - 10$, or 90
CD	$500 - 100$, or 400
CM	$1000 - 100$, or 900

multiplicative property   In the Middle Ages, a bar was placed over a Roman number to multiply it by 1000. The use of bars is based on a **multiplicative property**. For example, $\overline{V}$ represents $5 \cdot 1000$, or 5000, and $\overline{CDX}$ represents $410 \cdot 1000$, or 410,000. To indicate even greater numbers, more bars appear. For example, $\overline{\overline{V}}$ represents $(5 \cdot 1000) \cdot 1000$, or 5,000,000; $\overline{\overline{\overline{CXI}}}$ represents $111 \cdot 1000^3$, or 111,000,000,000; and $\overline{CXI}$ represents $110 \cdot 1000 + 1$, or 110,001.

Several properties might be used to represent some numbers, for example:

$$\overline{DCLIX} = \underbrace{(500 \cdot 1000)}_{\text{Multiplicative}} + \underbrace{(100 + 50)}_{\text{Additive}} + \underbrace{(10 - 1)}_{\text{Subtractive}} = 500{,}159.$$

Other Number Bases

The Luo peoples of Kenya used a *quinary*, or base five, system. A system of this type can be modeled by counting with only one hand. The digits available for counting are 0, 1, 2, 3, and 4. In the "one-hand system," or base-five system, you count 1, 2, 3, 4, 10, where 10 represents one hand and no fingers. Counting in base five proceeds as shown in Figure 3-4. We write the small "five" below the numeral as a reminder that the number is written in base five. If no base is written, a number is assumed to be in base ten.

Figure 3-4

Base-Five Symbol	Base-Five Grouping	One-Hand System
0_{five}		0 fingers
1_{five}	x	1 finger
2_{five}	xx	2 fingers
3_{five}	xxx	3 fingers
4_{five}	xxxx	4 fingers
10_{five}	(xxxxx)	1 hand and 0 fingers
11_{five}	(xxxxx) x	1 hand and 1 finger
12_{five}	(xxxxx) xx	1 hand and 2 fingers
13_{five}	(xxxxx) xxx	1 hand and 3 fingers
14_{five}	(xxxxx) xxxx	1 hand and 4 fingers
20_{five}	(xxxxx)(xxxxx)	2 hands and 0 fingers
21_{five}	(xxxxx)(xxxxx) x	2 hands and 1 finger

What number follows 44_{five}? There are no more two-digit numbers in the system after 44_{five}. In base ten, the same situation occurs at 99. We use 100 to represent 10 tens, or 1 hundred. In the base-five system, we need a symbol to represent 5 fives. To continue the analogy with base ten, we use 100_{five} to represent 1 group of 5 fives, 0 groups of five, and 0 units. To distinguish from "one hundred" in base ten, the name for 100_{five} is read "one-zero-zero base five." The number 100 means $1 \cdot 10^2 + 0 \cdot 10^1 + 0$, whereas the number 100_{five} means $(1 \cdot 10^2 + 0 \cdot 10^1 + 0)_{\text{five}}$, or $(1 \cdot 5^2 + 0 \cdot 5^1 + 0)_{\text{ten}}$, or 25.

Example 3-1 Convert 11244_{five} to base ten.

Solution
$$11244_{\text{five}} = 1 \cdot 5^4 + 1 \cdot 5^3 + 2 \cdot 5^2 + 4 \cdot 5^1 + 4 \cdot 1$$
$$= 1 \cdot 625 + 1 \cdot 125 + 2 \cdot 25 + 4 \cdot 5 + 4 \cdot 1$$
$$= 625 + 125 + 50 + 20 + 4$$
$$= 824$$

Example 3-1 suggests a method for changing a base-ten number to a base-five number using powers of five. To convert 824 to base five, we divide by successive powers of five. A shorthand method for illustrating this conversion is in the following:

$$
\begin{array}{r|r|l}
625 & 824 & 1 \\
& -625 & \\
\end{array}
\quad \text{How many groups of 625 in 824?}
$$

$$
\begin{array}{r|r|l}
125 & 199 & 1 \\
& -125 & \\
\end{array}
\quad \text{How many groups of 125 in 199?}
$$

$$
\begin{array}{r|r|l}
25 & 74 & 2 \\
& -50 & \\
\end{array}
\quad \text{How many groups of 25 in 74?}
$$

$$
\begin{array}{r|r|l}
5 & 24 & 4 \\
& -20 & \\
\end{array}
\quad \text{How many groups of 5 in 24?}
$$

$$
\begin{array}{r|r|l}
1 & 4 & 4 \\
& -4 & \\
& 0 & \\
\end{array}
\quad \text{How many 1s in 4?}
$$

Thus $824 = 11244_{\text{five}}$.

NOW TRY THIS 3-3

- A different method of converting 824 to base five is shown using successive divisions by 5. The quotient in each case is placed below the dividend and the remainder is placed on the right, on the same line with the quotient. The answer is read from bottom to top, that is, as 11244_{five}. Why does it work?

$$
\begin{array}{r|r|l}
5 & 824 & \\
5 & 164 & 4 \\
5 & 32 & 4 \\
5 & 6 & 2 \\
& 1 & 1 \\
\end{array}
$$

Calculators with the integer division feature—$\boxed{\text{INT}\div}$ on a Texas Instrument calculator or $\boxed{\div R}$ on a Casio—can be used to change base-ten numbers to different number bases. For example, to convert 8 to base five, we enter $8\ \boxed{\text{INT}\div}\ 5\ \boxed{=}$ and obtain $\underset{Q}{1}\ \underset{R}{3}$. This implies that $8 = 13_{\text{five}}$. Will this technique work to convert 34 to base five? Why or why not?

Historians tell of early tribes that used base two. Some Australian tribes still count "one, two, two and one, two twos, two twos and one, ..." Because base two has only two digits, it is called the **binary system.** Base two is especially important because of its use in computers. One of the two digits is represented by the presence of an electrical signal and the other by the absence of an electrical signal. Although base two works well for computers, it is inefficient for everyday use because multidigit numbers are reached very rapidly in counting in this system, as seen in the following cartoon where Peter is reading Jason's counting to five.

binary system

FOXTROT © 1996 Bill Amend. Reprinted with permission of UNIVERSAL PRESS SYNDICATE. All rights reserved.

Conversions from base two to base ten, and vice versa, can be accomplished in a manner similar to that used for base-five conversions.

Example 3-2

a. Convert 10111_{two} to base ten.
b. Convert 27 to base two.

Solution

a. $10111_{two} = 1 \cdot 2^4 + 0 \cdot 2^3 + 1 \cdot 2^2 + 1 \cdot 2^1 + 1$
$= 16 + 0 + 4 + 2 + 1$
$= 23.$

Alternative Solution:

b.

16	27	1	How many groups of 16 in 27?
	−16		
8	11	1	How many groups of 8 in 11?
	−8		
4	3	0	How many groups of 4 in 3?
	−0		
2	3	1	How many groups of 2 in 3?
	−2		
1	1	1	How many 1s in 1?
	−1		
	0		

Alternative:
2 | 27
2 | 13 1
2 | 6 1
2 | 3 0
   | 1 1

Thus 27 is equivalent to 11011_{two}.

Another commonly used number base system is the base-twelve, or the duodecimal ("dozens") system. Eggs are bought by the dozen, and pencils are bought by the *gross* (a dozen dozens). In base twelve, there are twelve digits, just as there are ten digits in base ten, five digits in base five, and two digits in base two. In base twelve, new symbols are needed to represent the following groups of x's:

$$\underbrace{x\,x\,x\,x\,x\,x\,x\,x\,x\,x}_{10\ x\text{'s}} \quad \text{and} \quad \underbrace{x\,x\,x\,x\,x\,x\,x\,x\,x\,x\,x}_{11\ x\text{'s}}$$

The new symbols chosen are T and E, respectively, so that the base-twelve digits are 0, 1, 2, 3, 4, 5, 6, 7, 8, 9, T, and E. Thus in base twelve we count "1, 2, 3, 4, 5, 6, 7, 8, 9, T, E, 10, 11, 12, ... , 17, 18, 19, $1T$, $1E$, 20, 21, 22, ... , 28, 29, $2T$, $2E$, 30, ..."

Example 3-3 a. Convert $E2T_{twelve}$ to base ten. b. Convert 1277 to base twelve.

Solution

a. $E2T_{twelve} = 11 \cdot 12^2 + 2 \cdot 12^1 + 10 \cdot 1$
$= 11 \cdot 144 + 24 + 10$
$= 1584 + 24 + 10$
$= 1618$

b.
```
  144 | 1277  | 8    How many groups of 144 in 1277?
       - 1152
       12 | 125  | T    How many groups of 12 in 125
            - 120
            1 | 5 | 5   How many 1s in 5?
               - 5
                 0
```

Thus $1277 = 8T5_{twelve}$.

Example 3-4 Rob used base twelve to write the following:

$$g36_{twelve} = 1050_{ten}$$

What is the value of g?

Solution Using expanded form we could write the following equations.

$$g \cdot 12^2 + 3 \cdot 12 + 6 \cdot 1 = 1050$$
$$144g + 36 + 6 = 1050$$
$$144g + 42 = 1050$$
$$144g = 1008$$
$$g = 7$$

ONGOING ASSESSMENT 3-1

1. For each of the following, tell which numeral represents the greater number and why:
   a. $\overline{MCDXXIV}$ and $\overline{\overline{MCDXXIV}}$
   b. 4632 and 46,032
   c. <▼▼ and < ▼▼
   d. 999∩∩|| and 𓆼∩|
   e. ⋮⋮⋮ and 👁

2. For each of the following, name both the succeeding and preceding numbers (one more and one less):
   a. MCMXLIX   b. MI   c. CMXCIX
   d. << <▼   e. 𓆼99   f. ⋮⋮⋮

3. If the cornerstone represents when a building was built and it reads MCMXXII, when was this building built?

4. Write each of the following in Roman symbols:
   a. 121   b. 42
   c. 89   d. 5282

5. Write each of the following in Egyptian symbols:
   a. 52   b. 103
   c. 100,003   d. 38

6. Complete the following table, which compares symbols for numbers in different numeration systems:

	Hindu-Arabic	Babylonian	Egyptian	Roman	Mayan
a.	72				
b.		< ▼▼			
c.			𓆼99∩∩|||		

7. For each of the following decimal numerals, give the place value of the underlined numeral:
   a. 827,3̲67
   b. 8̲,421,000
   c. 97,9̲98
   d. 810,4̲85
8. Rewrite each of the following as a base-ten numeral:
   a. $3 \cdot 10^6 + 4 \cdot 10^3 + 5$
   b. $2 \cdot 10^4 + 1$
   c. $3 \cdot 10^3 + 5 \cdot 10^2 + 6 \cdot 10$
   d. $9 \cdot 10^6 + 9 \cdot 10 + 9$
9. Study the following counting frame. In the frame, the value of each dot is represented by the number in the box below the dot. For example, the following figure represents the number 154:

••	•••	••
64	8	1

   What numbers are represented in the frames in (a) and (b)?

   a.
•••	••	•
25	5	1

   b.
•		••	•
8	4	2	1

10. A certain 3-digit whole number has the following properties: The hundreds digit is greater than 7; the tens digit is an odd number; and the sum of the digits is 10. What could the number be?
11. Write the first 15 counting numbers for each of the following bases:
    a. Base two
    b. Base three
    c. Base four
    d. Base eight
12. How many different digits are needed for base twenty?
13. Write 2032_{four} in expanded base-four notation.
14. Determine the greatest three-digit number in each of the following bases:
    a. Base two
    b. Base six
    c. Base ten
    d. Base twelve
15. Find the numbers preceding and succeeding each of the following:
    a. $EE0_{twelve}$
    b. 100000_{two}
    c. 555_{six}
    d. 100_{seven}
    e. 1000_{five}
    f. 110_{two}
16. What, if anything, is wrong with the following numerals:
    a. 204_{four}
    b. 607_{five}
    c. $T12_{three}$
17. Convert each of the following base-ten numbers to numbers in the indicated bases:
    a. 432 to base five
    b. 1963 to base twelve
    c. 404 to base four
    d. 37 to base two
18. Change 42_{eight} to base two.
19. Write each of the following numbers in base ten:
    a. 432_{five}
    b. 101101_{two}
    c. $92E_{twelve}$
    d. $T0E_{twelve}$
    e. 111_{twelve}
    f. 346_{seven}
20. Suppose you have two quarters, four nickels, and two pennies. What is the value of your money in cents? Write a base-five representation of this amount.
21. You are asked to distribute $900 in prize money. The dollar amounts for the prizes are $625, $125, $25, $5, and $1. How should this $900 be distributed in order to give the fewest number of prizes?
22. What is the minimum number of quarters, nickels, and pennies necessary to make 97¢?
23. Convert each of the following:
    a. 58 days to weeks and days
    b. 54 months to years and months
    c. 29 hours to days and hours
    d. 68 inches to feet and inches
24. A bookstore ordered 11 gross, 6 dozen, and 6 pencils. Express the number of pencils in base twelve and in base ten.
25. For each of the following, find b if possible. If not possible, tell why.
    a. $b2_{seven} = 44_{ten}$
    b. $5b2_{twelve} = 734_{ten}$
    c. $23_{ten} = 25_b$
    d. $b2_{seven} = 2b_{ten}$
26. Write $12^5 + 25 \cdot 12^4 + 23$ in base-twelve notation without multiplying out 12^5 and 12^4.
27. The Chinese abacus, depicted as follows, shows the number 5857:

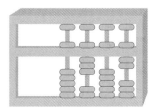

    Discuss how the number 5857 is depicted and show how the number 4869 could be depicted.
28. On a calculator, use only the keys 1, 2, 3, 4, 5, 6, 7, 8, and 9, and fill the calculator's display to show each of the following:
    a. The greatest number possible if each key may be used only once
    b. The least number possible if each key may be used only once
    c. The greatest number possible if a key may be used more than once
    d. The least number possible if a key may be used more than once
29. In a game called WIPEOUT, we are to "wipe out" digits from a calculator's display without changing any of the other digits. "Wipeout" in this case means to replace the chosen digit(s) with a 0. For example, if the initial number is 54,321 and we are to wipe out the 4, we could subtract 4000 to obtain 50,321. Complete the following two problems and then try other numbers or challenge another person to wipe out a number from the number you have placed on the screen:
    a. Wipe out the 2s from 32,420.
    b. Wipe out the 5 from 67,357.

Communication

30. Ben claims that zero is the same as nothing. Explain how you as a teacher would respond to Ben's statement.

31. What are the major drawbacks to each of the following systems?
    a. Egyptian **b.** Babylonian **c.** Roman
32. Why are large numbers written with commas separating groups of three digits?

Open-Ended

33. An inspector of weights and measures uses a special set of weights to check the accuracy of scales. Various weights are placed on a scale to check accuracy of any amount from 1 oz through 15 oz. What is the least number of weights the inspector needs? What weights are needed to check the accuracy of scales from 1 oz through 15 oz? From 1 oz through 31 oz?

Cooperative Learning

34. **a.** Create a numeration system with unique symbols and write a paragraph explaining the properties of the system.
    b. Complete the following table using the system:

Hindu-Arabic Numeral	Your System Numeral
1	
5	
10	
50	
100	
5,000	
10,000	
115,280	

BRAIN TEASER There are 3 nickels and 3 dimes concealed inside 3 boxes. Two coins are placed in each of the boxes, which are labeled 10¢, 15¢, and 20¢. The coins are placed in such a way that no box contains the amount of money shown on its label; for example, the box labeled 10¢ does not really have a total of 10¢ in it. What is the minimum number of coins that you would have to remove from a box, and from which box or boxes, to determine which coins are in which boxes?

Section 3-2 Algorithms for Whole-Number Addition and Subtraction

According to the *Principles and Standards,*

Students exhibit computational fluency when they demonstrate flexibility in the computational methods they choose, understand and can explain these methods, and produce accurate answers efficiently. The computational methods that a student uses should be based on mathematical ideas that the student understands well, including the structure of the base-ten number system, properties of multiplication and division, and number relationships (p. 152).

algorithm

The previous chapter introduced definitions of addition and subtraction. These definitions, along with knowledge of basic facts and properties, are necessary to perform more involved additions and subtractions. Computations involving greater numbers are commonly done by applying various algorithms. An **algorithm** (named for the ninth-century Arabian mathematician Mohammed al Khowarizmi) is a systematic procedure used to accomplish an operation. This section focuses on developing and understanding some algorithms involved in addition and subtraction of whole numbers.

Addition Algorithms

Paper-and-pencil algorithms need to be taught developmentally; that is, they must proceed from the concrete stage to the abstract stage at appropriate times. The use of concrete teaching aids—such as chips, bean sticks, an abacus, or base-ten blocks—helps provide insight into the creation of algorithms for addition. A set of base-ten blocks, shown in Figure 3-5, consists of *units, longs, flats,* and *blocks,* representing 1, 10, 100, and 1000, respectively.

Figure 3-5

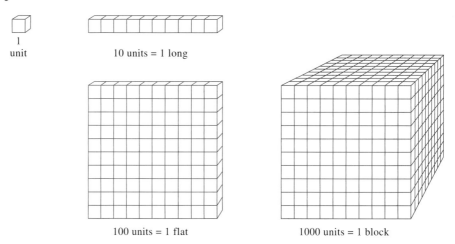

Students trade blocks by regrouping. That is, they take a set of base-ten blocks representing a number and trade them until they have the fewest possible pieces representing the same number. For example, suppose you have 58 units and want to trade them. The units can be grouped into tens to form longs. Five sets of 10 units each can be traded for 5 longs. Thus, 58 units can be traded for 5 longs and 8 units. In terms of numbers, this is analogous to rewriting 58 as $5 \cdot 10 + 8$.

Example 3-5   What is the fewest number of pieces you can receive in a fair exchange for 11 flats, 17 longs, and 16 units?

Solution   The 16 units can be traded for 1 long and 6 units.

11 flats	17 longs	~~16 units~~		(16 units = 1 long and 6 units)
	1 long	6 units		(Trade)
11 flats	18 longs	6 units		(After the first trade)

11 flats	~~18 longs~~	6 units		(18 longs = 1 flat and 8 longs)
1 flat	8 longs			(Trade)
12 flats	8 longs	6 units		(After the second trade)

	~~12 flats~~	8 longs	6 units	(12 flats = 1 block and 2 flats)
1 block	2 flats			(Trade)
1 block	2 flats	8 longs	6 units	(After the third trade)

Therefore, the fewest number of pieces is $1 + 2 + 8 + 6 = 17$. This trading is analogous to rewriting $11 \cdot 10^2 + 17 \cdot 10 + 16$ as $1 \cdot 10^3 + 2 \cdot 10^2 + 8 \cdot 10 + 6$, which implies that there are 1286 units.

We now use base-ten blocks to help develop an algorithm for whole-number addition. Suppose we wish to add $14 + 23$. We show this computation with a concrete model in

Figure 3-6(a), with the expanded algorithm in Figure 3-6(b) and the standard algorithm in Figure 3-6(c).

Figure 3-6

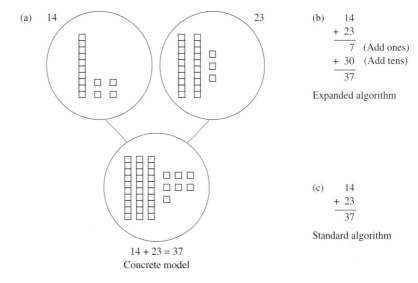

14 + 23 = 37
Concrete model

(b)
```
   14
 + 23
 ----
    7   (Add ones)
 + 30   (Add tens)
 ----
   37
```
Expanded algorithm

(c)
```
   14
 + 23
 ----
   37
```
Standard algorithm

A more formal justification for this addition not usually presented at the elementary level is the following:

$14 + 23 = (1 \cdot 10 + 4) + (2 \cdot 10 + 3)$ Expanded form

$= (1 \cdot 10 + 2 \cdot 10) + (4 + 3)$ Commutative and associative properties of addition

$= (1 + 2) \cdot 10 + (4 + 3)$ Distributive property of multiplication over addition

$= 3 \cdot 10 + 7$ Single-digit addition facts

$= 37$ Place value

An example using base-ten blocks with regrouping is shown on the student page that follows from *Scott Foresman-Addison Wesley Math,* Grade 3, 1999. Answer the "Talk About It" questions on the bottom of the page.

After using concrete models with regrouping as on the student page, children should be ready to use the expanded and standard algorithms. Figure 3-7 shows the computation 37 + 28 using both algorithms. Notice in Figure 3-7(b) that when there were more than ten ones we regrouped ten ones as a ten and then added the tens. This process could be repeated. Notice that the word *regroup* was used rather than *carry.* The words *regroup* or *trade* are used in the elementary school to describe the actions that take place.

Figure 3-7

(a)
```
    37
  +28
  ---
    15   (Add ones)
  +50   (Add tens)
  ---
    65
```
Expanded algorithm

(b)
```
    1
    37
  +28
  ---
    65   (Add the ones, regroup and add the tens)
```
Standard algorithm

138 CHAPTER 3 *Whole-Number Computation*

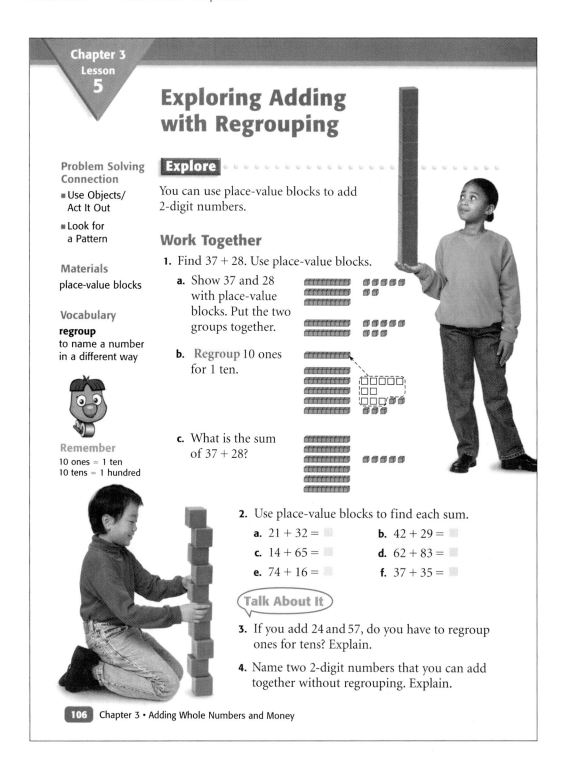

Scratch Addition

scratch addition

The **scratch addition** algorithm allows students to perform complicated additions by doing a series of additions that involve only two single digits. An example follows:

1. 87
   $6\cancel{5}_2$
   $+49$

   Add the numbers in the units place starting at the top. When the sum is 10 or more, record this sum by scratching a line through the last digit added and writing the number of units next to the scratched digit. For example, since $7 + 5 = 12$, the "scratch" represents 10 and the 2 represents the units.

2. 87
   $6\cancel{5}_2$
   $+4\cancel{9}_1$

   Continue adding the units, including any new digits written down. When the addition again results in a sum of 10 or more, as with $2 + 9 = 11$, repeat the process described in (1).

3. ${}^{2}87$
   $6\cancel{5}_2$
   $+4\cancel{9}_1$
   $\overline{1}$

   When the first column of additions is completed, write the number of units, 1, below the addition line. Count the number of scratches, 2, and add this number to the second column.

4. ${}^{2}\cancel{8}_0 7$
   $6\,\cancel{5}_2$
   $\cancel{4}_0\cancel{9}_1$
   $\overline{201}$

   Repeat the procedure for each successive column.

Subtraction Algorithms

As with addition, base-ten blocks can provide a concrete model for subtraction. Consider how the base-ten blocks are used to perform the subtraction in the student page on p. 140 from *Scott Foresman-Addison Wesley Math*, Grade 3, 1999. Answer questions 3 and 4 from the student page.

After students have worked with base-ten blocks for subtraction they are ready to proceed to the standard algorithm as shown below on the partial student page from *Scott Foresman-Addison Wesley Math*, Grade 3, 1999.

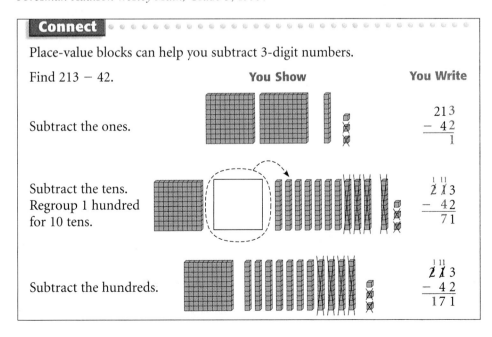

140 CHAPTER 3 *Whole-Number Computation*

Notice that in the standard algorithm, shown on the right in the partial student page on p. 139, when there are not enough tens available to subtract you regroup one hundred as 10 tens and then subtract the tens. Regrouping is used as needed to perform subtractions using the standard algorithm.

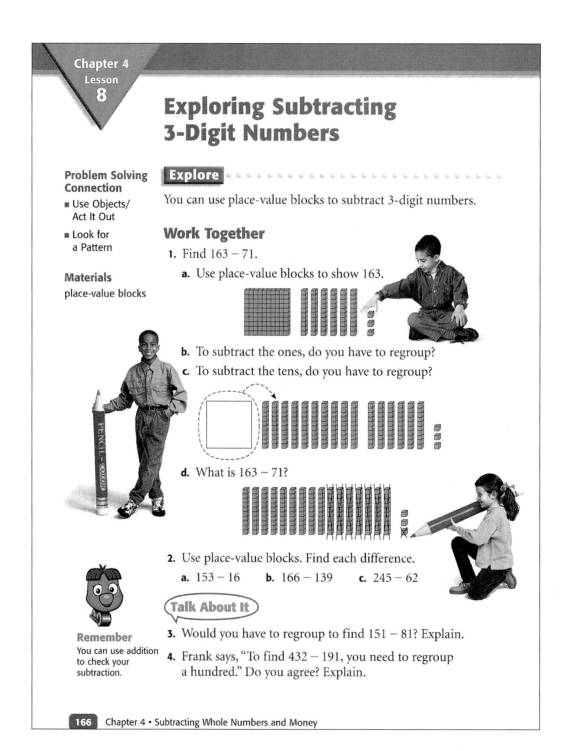

Chapter 4 Lesson 8

Exploring Subtracting 3-Digit Numbers

Problem Solving Connection
- Use Objects/Act It Out
- Look for a Pattern

Materials
place-value blocks

Explore

You can use place-value blocks to subtract 3-digit numbers.

Work Together

1. Find 163 − 71.
   a. Use place-value blocks to show 163.
   b. To subtract the ones, do you have to regroup?
   c. To subtract the tens, do you have to regroup?
   d. What is 163 − 71?

2. Use place-value blocks. Find each difference.
   a. 153 − 16   b. 166 − 139   c. 245 − 62

Talk About It

3. Would you have to regroup to find 151 − 81? Explain.

4. Frank says, "To find 432 − 191, you need to regroup a hundred." Do you agree? Explain.

Remember
You can use addition to check your subtraction.

166 Chapter 4 • Subtracting Whole Numbers and Money

Addition and Subtraction in Bases Other Than Ten

A look at computation in other bases may provide insight into computation in base ten. Use of multibase blocks may be helpful in building an addition table for different bases and is highly recommended. Table 3-8 is a base-five addition table.

Table 3-8 Base-Five Addition Table

+	0	1	2	3	4
0	0	1	2	3	4
1	1	2	3	4	10
2	2	3	4	10	11
3	3	4	10	11	12
4	4	10	11	12	13

NOW TRY THIS 3-4

- Determine how a number line might be used to model addition in base five.

Using the addition facts in Table 3-8, we develop algorithms for base-five addition similar to those for base-ten addition. Suppose we wish to add $12_{\text{five}} + 31_{\text{five}}$. We show the computation using a concrete model in Figure 3-8(a), an expanded algorithm in Figure 3-8(b), and the standard algorithm in Figure 3-8(c). Additions in other number bases can be handled similarly.

Figure 3-8

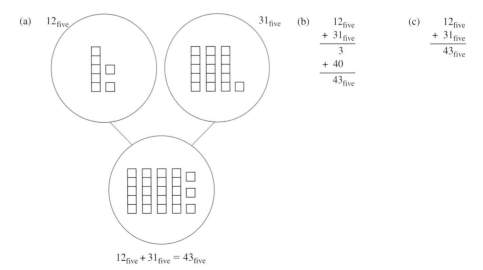

The subtraction facts for base five can also be derived from the addition-facts table by using the definition of subtraction. For example, to find $12_{\text{five}} - 4_{\text{five}}$, recall that $12_{\text{five}} - 4_{\text{five}} = c_{\text{five}}$ if, and only if, $c_{\text{five}} + 4_{\text{five}} = 12_{\text{five}}$. From Table 3-8, we see that $c = 3_{\text{five}}$. An example of subtraction involving regrouping, $32_{\text{five}} - 14_{\text{five}}$, is developed in Figure 3-9.

Figure 3-9

NOW TRY THIS 3-5

- **a.** Build an addition table for base two.
- **b.** Perform the subtraction $1101_{\text{two}} - 111_{\text{two}}$.

BRAIN TEASER The number on a license plate consists of five digits. When the license plate is looked at upside down, you can still read it, but the value of the upside-down number is 78,633 greater than the real license number. What is the license number?

ONGOING ASSESSMENT 3-2

1. Find the missing numbers in each of the following:

   a.
   ```
       _ _ 1
   +   4 2 _
   _____
     _ 4 0 2
   ```

   b.
   ```
       _ 0 2 5
       1 1 _ 6
   +   3 1 4 8
   _____
       6 _ 6 _
   ```

2. Find the missing numbers in each of the following:

   a.
   ```
       3 _ _
   -   1 5 9
   _____
       _ 2 4
   ```

   b.
   ```
       1 _ _ _ 6
   -     8 3 0 9
   _____
         4 9 8 7
   ```

3. Place the digits 7, 6, 8, 3, 5, and 2 in the boxes to obtain
   a. the greatest sum.   b. the least sum.

4. Place the digits 7, 6, 8, 3, 5, and 2 in the boxes to obtain
   a. the greatest difference.   b. the least difference.

5. Tom's diet allows only 1500 calories per day. For breakfast, Tom had skim milk (90 calories), a waffle with no syrup (120 calories), and a banana (119 calories). For lunch, he had $\frac{1}{2}$ cup of salad (185 calories) with mayonnaise (110 calories), and tea (0 calories). Then he had pecan pie (570 calories). Can he have dinner consisting of steak (250 calories), a salad with no mayonnaise, and tea?

6. Wally kept track of last week's money transactions. His salary was $150 plus $54 in overtime and $260 in tips. His transportation expenses were $22, his food expenses were $60, his laundry costs were $15, his entertainment expenditures were $58, and his rent was $185. Did he save any money last week? If so, how much?

7. In the following problem, the sum is correct but the order of the numbers in each addend has been scrambled. Correct the addends to obtain the correct sum.

$$\begin{array}{r} 2\ 8\ 3\ 4 \\ +\ 6\ 3\ 1\ 5 \\ \hline 9\ 0\ 5\ 9 \end{array} \qquad \begin{array}{r} \square\ \square\ \square\ \square \\ +\ \square\ \square\ \square\ \square \\ \hline 9\ 0\ 5\ 9 \end{array}$$

8. If 1 mo is approximately 4 wk and 1 yr is approximately 365 days or 52 wk, answer the following:
   a. Lewis and Clark spent approximately 2 yr, 4 mo, and 9 days exploring the territory in the Northwest. What is this time in weeks?
   b. It took Magellan 1126 days to circle the world. How many years is this?
   c. How many seconds old are you?
   d. Approximately how many times does your heart beat in 1 yr?

9. Janet worked her addition problems by placing the partial sums as shown here:

$$\begin{array}{r} 569 \\ +\ 645 \\ \hline 14 \\ 10 \\ 11 \\ \hline 1214 \end{array}$$

   a. Use this method to work the following:
      (i) 687   (ii) 359
         + 549   + 673
   b. Explain why this algorithm works.

10. Analyze the following computations. Explain what is wrong in each case.
    a. 135   b.   87   c.   57   d.   56
      + 47   + 25   − 38   − 18
      ─────   ─────   ─────   ─────
       172   1012   21   48

11. George is cooking an elaborate meal for Thanksgiving. He can cook only one thing at a time in his microwave oven. His turkey takes 75 min; the pumpkin pie takes 18 min; rolls take 45 sec; and a cup of coffee takes 30 sec to heat. How much time does he need to cook the meal?

12. Perform each of the following operations using the bases shown:
    a. $43_{five} + 23_{five}$ b. $43_{five} - 23_{five}$
    c. $432_{five} + 23_{five}$ d. $42_{five} - 23_{five}$
    e. $110_{two} + 11_{two}$ f. $10001_{two} - 111_{two}$

13. Construct an addition table for base eight.

14. Perform each of the following operations:
    a.   3 hr 36 min 58 sec
       + 5 hr 56 min 27 sec
    b.   5 hr 36 min 38 sec
       − 3 hr 56 min 58 sec

15. Perform each of the following operations (2 c = 1 pt, 2 pt = 1 qt, 4 qt = 1 gal):
    a.   1 qt 1 pt 1 c   b.   1 qt 1 c
       +   1 pt 1 c   −   1 pt 1 c
    c.   1 gal 3 qt 1 c
       −   4 qt 2 c

16. Mari is going to invite 20 friends to a party. She would like to have at least 2 c of cider for each guest. If cider is sold only by the gallon, how many gallons should she buy?

17. A palindrome is any number that reads the same backward as forward, for example, 121 and 2332. Try the following. Begin with any number. Is it a palindrome? If not, reverse the digits and add this reversed number to the original number. Is the result a palindrome? If not, repeat the above procedure until a palindrome is obtained. For example, start with 78. Because 78 is not a palindrome, we add: 78 + 87 = 165. Because 165 is not a palindrome, we add: 165 + 561 = 726. Again, 726 is not a palindrome, so we add 726 + 627 to obtain 1353. Finally, 1353 + 3531 yields 4884, which is a palindrome.
    a. Try this method with the following numbers:
       (i) 93   (ii) 588   (iii) 2003
    b. Find a number for which the procedure described takes more than five steps to form a palindrome.

18. a. Place the numbers 24 through 32 in the following circles so that the sums are the same in each direction:

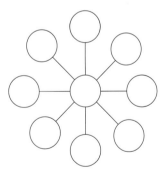

    b. How many different numbers can be placed in the middle to obtain a solution?

19. Andrew's calculator was not functioning properly, When he pressed $\boxed{8}\boxed{+}\boxed{6}\boxed{=}$, the numeral 20 appeared on on the display. When he pressed $\boxed{5}\boxed{+}\boxed{4}\boxed{=}$, 13 was displayed. When he pressed $\boxed{1}\boxed{5}\boxed{-}\boxed{3}\boxed{=}$, 9 was displayed. What do you think Andrew's calculator was doing?

20. The following is a supermagic square taken from an engraving called *Melancholia* by Dürer that includes the year (1514) it was constructed:

16	3	2	13
5	10	11	8
9	6	7	12
4	15	14	1

   a. Find the sum of each row, the sum of each column, and the sum of each diagonal.
   b. Find the sum of the four numbers in the center.
   c. Find the sum of the four numbers in each corner.
   d. Add 11 to each number in the square. Is the square still a magic square? Explain your answer.
   e. Subtract 11 from each number in the square. Is the square still a magic square?

21. Use scratch addition to perform the following:

   a. $\quad 432_{ten}$
   $\quad\quad 976_{ten}$
   $\quad +1418_{ten}$

   b. $\quad 32_{five}$
   $\quad\quad 13_{five}$
   $\quad\quad 22_{five}$
   $\quad\quad 43_{five}$
   $\quad\quad 23_{five}$
   $\quad + 12_{five}$

22. Perform each of the following operations:

   a. $\quad$ 4 gross 4 dz 6 ones
   $\quad -\quad\quad\quad$ 5 dz 9 ones

   b. $\quad$ 2 gross 9 dz 7 ones
   $\quad +$ 3 gross 5 dz 9 ones

23. In a small rural community, the elementary school has no refrigerators. Through a federally financed program, the school provides 1 c of milk per day for each student. Milk for the day is purchased at the local store each morning, and the school buys the exact amount necessary. The milk is available in gallons, half-gallons, quarts, pints, or cups, and the larger containers are better buys.
   a. If 1 gal, 1 qt, and 1 pt of milk were purchased on Tuesday, how many students were at school that day?
   b. On Wednesday, 31 students were at school. How much milk was purchased that day to make the best buy?

24. A *score* is equal to 20. Indicate each of the following as a base-ten number:
   a. Three score and ten
   b. Four score and seven

25. Determine what is wrong with the following:
   $\quad\quad 22_{five}$
   $\quad + 33_{five}$
   $\quad\quad\overline{\;55_{five}\;}$

26. Fill in the missing numbers in each of the following:

   a. $\quad 2\_\;\_\_{five}$
   $\quad -\;2\;2_{five}$
   $\quad\overline{\;\_\;0\;3_{five}\;}$

   b. $\quad 2\;0\;0\;1\;0_{three}$
   $\quad -\;\;\;2\_\;2\_\_{three}$
   $\quad\overline{\;1\_\;2\_\;1_{three}\;}$

27. The Hawks played the Elks in a basketball game. Based on the information below, complete the scoreboard showing the number of points scored by each team during each quarter and the final score of the game.

TEAMS	QUARTERS				FINAL SCORE
	1	2	3	4	
Hawks					
Elks					

   a. The Hawks scored 15 points in the first quarter.
   b. The Hawks were behind by 5 points at the end of the first quarter.
   c. The Elks scored 5 more points in the second quarter than they did in the first quarter.
   d. The Hawks scored 7 more points than the Elks in the second quarter.
   e. The Elks outscored the Hawks by 6 points in the fourth quarter.
   f. The Hawks scored a total of 120 points in the game.
   g. The Hawks scored twice as many points in the third quarter as the Elks did in the first quarter.
   h. The Elks scored as many points in the third quarter as the Hawks did in the first two quarters combined.

Communication

28. Cathy found her own algorithm for subtraction. She subtracted as follows:
   $\quad\quad 97$
   $\quad - 28$
   $\quad\overline{\;- 1\;}$
   $\quad + 70$
   $\quad\overline{\;\;69\;}$

   How would you respond if you were her teacher?

29. Discuss why the words *regroup* and *trade* are used rather than *carry* and *borrow* for whole-number addition and subtraction algorithms.

30. Discuss whether children should be encouraged to develop and use their own algorithms for whole-number addition and subtraction or whether they should be taught only one algorithm per operation and all students should use only the one algorithm.

31. Explain why the scratch addition algorithm works.

32. The *equal addends* algorithm is sometimes taught in elementary school. The following shows how this algorithm works for 1464 − 687.

$$\begin{array}{r}1\,4\,6\,{}^{1}4\\ -\ 6\,{}^{9}8\,7\\ \hline 7\end{array}$$
(Add 10 to the 4 ones in the top number to get 14 ones.)
(Add 1 ten to the 8 tens in the bottom number to get 9 tens.)
(Subtract the ones.)

Now we move to the next column.

$$\begin{array}{r}1\,4\,{}^{1}6\,{}^{1}4\\ -\ {}^{7}6\,{}^{9}8\,7\\ \hline 7\,7\,7\end{array}$$
(Add 10 tens to the 6 tens in the top number to get 16 tens.)
(Add 1 hundred to the 6 hundreds in the bottom number to get 7 hundreds.)
(Subtract the 9 tens from the 16 tens and then the 7 hundreds from the 14 hundreds.)

a. Try the technique on three more subtractions.
b. Explain why the *equal addends* algorithm works.

Open-Ended

33. Search for or develop an algorithm for whole-number addition or subtraction and write a description of your algorithm so that others can understand and use it.

Cooperative Learning

34. Investigate the money system of some country and decide whether the system is a base-ten or other base system. Decide whether values in the money system you chose can be added and subtracted easily.

Review Problems

35. Investigate the measuring of lengths in the metric system. Develop a plan for using place value with lengths to convert among different metric units.
36. Write 5280 in expanded form.
37. What is the value of MCDX in Hindu-Arabic numerals?
38. Convert each of the following to base ten.
  a. EOT_{twelve}
  b. 1011_{two}
  c. 43_{five}

LABORATORY ACTIVITY

1. One type of Japanese abacus, *soroban,* is shown in Figure 3-10(a). In this abacus, a bar separates two sets of bead counters. Each counter above the bar represents five times the counter below the bar. Numbers are illustrated by moving the counter toward the bar. The number 7632 is pictured. Practice demonstrating and adding numbers on this abacus.

Figure 3-10

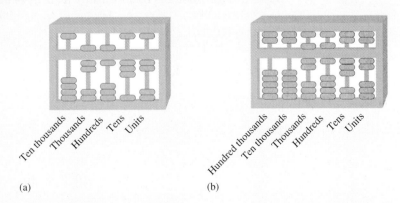

(a)   (b)

2. The Chinese abacus, *suan pan* (see Figure 3-10(b)), is still in use today. This abacus is similar to the Japanese abacus but has two counters above the bar. The number 7632 is also pictured on it. Practice demonstrating and adding numbers on this abacus. Compare the ease of using the two versions.

Section 3-3 — Algorithms for Whole-Number Multiplication and Division

Multiplication Algorithms

To develop algorithms for multiplying multidigit whole numbers, we use the strategy of *examining simpler computations first*. Consider $4 \cdot 12$. This computation could be pictured as in Figure 3-11(a) with 4 rows of 12 blocks, or 48 blocks. The blocks in Figure 3-11(a) can also be partitioned to show that $4 \cdot 12 = 4 \cdot (10 + 2) = 4 \cdot 10 + 4 \cdot 2$. The numbers $4 \cdot 10$ and $4 \cdot 2$ are *partial products*.

Figure 3-11

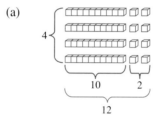

Figure 3-11(a) illustrates the distributive property of multiplication over addition on the set of whole numbers. The process leading to an algorithm for multiplying $4 \cdot 12$ is seen in Figure 3-11(b).

Multiplication by 10^n

To compute products such as $3 \cdot 200$, we proceed as follows:

$$3 \cdot 200 = 3 \cdot (2 \cdot 10^2)$$
$$= (3 \cdot 2) \cdot 10^2$$
$$= 6 \cdot 10^2$$
$$= 6 \cdot 10^2 + 0 \cdot 10^1 + 0 \cdot 1$$
$$= 600$$

We see that multiplying 6 by 10^2 results in annexing two zeros to 6. This idea can be generalized to the statement that *multiplication of any natural number by 10^n, where n is a natural number, results in annexing n zeros to the number.*

When multiplying powers of 10, an extension of the definition of exponents is used. For example, $10^2 \cdot 10^1 = (10 \cdot 10) \cdot 10 = 10^3$, or 10^{2+1}. In general, where a is a natural number and m and n are whole numbers, $a^m \cdot a^n$ is given by the following:

$$a^m \cdot a^n = \underbrace{(a \cdot a \cdot a \cdot \ldots \cdot a)}_{m \text{ factors}} \cdot \underbrace{(a \cdot a \cdot a \cdot \ldots \cdot a)}_{n \text{ factors}}$$

$$= \underbrace{a \cdot a \cdot a \cdot \ldots \cdot a}_{m + n \text{ factors}} = a^{m+n}.$$

Consequently, $a^m \cdot a^n = a^{m+n}$.

NOW TRY THIS 3-6

- Does $a^n + a^m$ ever equal a^{m+n}? If so, when?

Computations with Two-Digit Factors

Consider $14 \cdot 23$. We first model this computation using base-ten blocks as shown in Figure 3-12(a), and then showing all the partial products and adding, as shown in Figure 3-12(b).

Figure 3-12

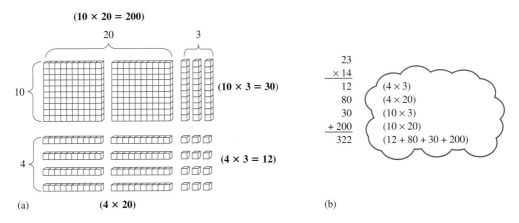

Another approach is to write 14 as $10 + 4$ and use the distributive property of multiplication over addition, as follows:

$$14 \cdot 23 = (10 + 4) \cdot 23$$
$$= 10 \cdot 23 + 4 \cdot 23$$
$$= 230 + 92$$
$$= 322$$

This last approach leads to an algorithm for multiplication:

$$
\begin{array}{r}
23 \\
\times 14 \\
\hline
92 \\
230 \\
\hline
322
\end{array}
\quad
\begin{array}{l}
10 + 4 \\
(4 \cdot 23) \\
(10 \cdot 23)
\end{array}
\quad \text{or} \quad
\begin{array}{r}
23 \\
\times 14 \\
\hline
92 \\
23 \\
\hline
322
\end{array}
$$

We are accustomed to seeing the partial product 230 written without the zero, as 23. The placement of 23 with 3 in the tens column obviates having to write the 0 in the units column. When children first learn multiplication algorithms, they should be encouraged to include the zero in order to avoid errors and promote better understanding. Children should also be encouraged to *estimate* whether their answers are reasonable. In this exercise, we know that the answer must be between $10 \cdot 20 = 200$ and $20 \cdot 30 = 600$ because $10 < 14 < 20$ and $20 < 23 < 30$. Because 322 is between 200 and 600, the answer is reasonable.

Lattice Multiplication

lattice multiplication

The **lattice multiplication** algorithm for multiplying 14 and 23 is shown in Figure 3-13. (Determining the reasons why lattice multiplication works is left as an exercise.)

Figure 3-13

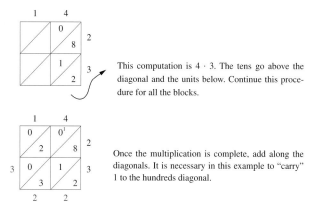

This computation is $4 \cdot 3$. The tens go above the diagonal and the units below. Continue this procedure for all the blocks.

Once the multiplication is complete, add along the diagonals. It is necessary in this example to "carry" 1 to the hundreds diagonal.

Division Algorithms

Algorithms for division of whole numbers can be developed by using repeated subtraction. Consider the following:

A shopkeeper is packaging juice in cartons that hold 6 bottles each. She has 726 bottles. How many cartons does she need?

We might reason that if 1 carton holds 6 bottles, then 10 cartons hold 60 bottles and 100 cartons hold 600 bottles. If 100 cartons are filled, there are $726 - 100 \cdot 6$, or 126, bottles remaining. If 10 more cartons are filled, then $126 - 10 \cdot 6$, or 66, bottles remain. Similarly, if 10 more cartons are filled, $66 - 10 \cdot 6$, or 6, bottles remain. Finally, 1 carton will hold the remaining 6 bottles. The total number of cartons necessary is $100 + 10 + 10 + 1$, or 121. This procedure is summarized in Figure 3-14(a). A more efficient method is shown in Figure 3-14(b).

Figure 3-14

```
(a) 6)726            (b) 6)726
     -600   100 sixes     -600   100 sixes
      126                  126
      -60    10 sixes      -120    20 sixes
       66                    6
      -60    10 sixes       -6     1 six
        6                    0   121 sixes
       -6     1 six
        0   121 sixes
```

Divisions such as the one in Figure 3-14 are usually shown in elementary school texts in the most efficient form, as in Figure 3-15(b), in which the numbers in color in Figure 3-15(a) are omitted. The technique used in Figure 3-15(a) is often called "scaffolding" and may be used as a preliminary step to achieving Figure 3-15(b).

Figure 3-15

$$
\text{(a)} \quad \begin{array}{r} 121 \\ \hline 1 \\ 20 \\ 100 \\ 6 \overline{)726} \\ -600 \\ \hline 126 \\ -120 \\ \hline 6 \\ -6 \\ \hline 0 \end{array} \qquad \text{(b)} \quad \begin{array}{r} 121 \\ 6 \overline{)726} \\ -6 \\ \hline 12 \\ -12 \\ \hline 6 \\ -6 \\ \hline 0 \end{array}
$$

Short Division

The process used in Figure 3-15(b) is usually referred to as "long" division. Another technique, called "short" division, can be used when the divisor is a one-digit number and most of the work is done mentally. An example of short division is given next:

Decide where to start.	Divide the hundreds. Write the remainder by the tens.	Divide the tens. Write the remainder by the ones.	Divide the ones.

$$
\begin{array}{r} 5 \\ 5\overline{)2\ 8\ 8\ 0} \end{array} \qquad \begin{array}{r} 5 \\ 5\overline{)2\ 8^3 8\ 0} \end{array} \qquad \begin{array}{r} 5\ 7 \\ 5\overline{)2\ 8^3 8^3 0} \end{array} \qquad \begin{array}{r} 5\ 7\ 6 \\ 5\overline{)2\ 8^3 8^3 0} \end{array}
$$

$\{$ Not enough thousands, so divide the hundreds. $\}$ $\quad$ $\{28 \div 5 = 5\ R3\}$ $\quad$ $\{38 \div 5 = 7\ R3\}$ $\quad$ $\{30 \div 5 = 6\ R0\}$

Division by a Two-Digit Divisor

An example of division by a divisor of more than one digit is given next. Consider $32\overline{)2618}$.

1. Estimate the quotient in $32\overline{)2618}$. Because $1 \cdot 32 = 32$, $10 \cdot 32 = 320$, $100 \cdot 32 = 3200$, we see that the quotient is between 10 and 100.
2. Find the number of tens in the quotient. Because $26 \div 3$ is approximately 8, then 26 hundreds divided by 3 tens is approximately 8 tens. We then write the 8 in the tens place, as shown:

$$
\begin{array}{r} 80 \\ 32\overline{)2618} \\ -2560 \quad (32 \cdot 80) \\ \hline 58 \end{array}
$$

3. Find the number of units in the quotient. Because $5 \div 3$ is approximately 1, then 5 tens divided by 3 tens is approximately 1. This is shown below on the left, with the more efficient form shown on the right.

$$
\begin{array}{r}
\frac{81}{1} \\
80 \\
32\overline{)2618} \\
-2560 \\
\hline
58 \\
-32 \quad (32 \cdot 1) \\
\hline
26
\end{array}
\qquad
\begin{array}{r}
81 \text{ R}26 \\
32\overline{)2618} \\
-256 \\
\hline
58 \\
-32 \\
\hline
26
\end{array}
$$

4. Check: $32 \cdot 81 + 26 = 2618$.

Normally in grade-school books, we see the format shown on the right, which places the remainder beside the quotient.

Using the division algorithm introduced in Chapter 2, we can write the above division as

$$2618 = 32 \cdot 81 + 26.$$

Use the ideas of division and the division algorithm to work through Now Try This 3-7.

NOW TRY THIS 3-7

● Suppose $a = 173 \cdot 34 + 40$. Without multiplying $173 \cdot 34$, find the remainder when

    a. a is divided by 173.
    b. a is divided by 34.
    c. a is divided by 17.
    d. a is divided by 2. ●

Division in many elementary texts is taught using a four-step algorithm: *estimate, multiply, subtract,* and *compare*. This is demonstrated in the student page that follows from *Scott Foresman-Addison Wesley Math*, Grade 4, 1999. Study the student page and answer the questions at the bottom of the page.

Multiplication and Division in Different Bases

As with addition and subtraction, we need to identify the basic facts of multiplication before we can use algorithms. The multiplication facts for base five are given in Table 3-9. These facts can be derived by using repeated addition.

Table 3-9 Base-Five Multiplication Table

×	0	1	2	3	4
0	0	0	0	0	0
1	0	1	2	3	4
2	0	2	4	11	13
3	0	3	11	14	22
4	0	4	13	22	31

Chapter 7 Lesson 7

2- or 3-Digit Quotients

You Will Learn

how to find 2- or 3-digit quotients

Math Tip

Always estimate the quotient to see if you can start dividing a 3-digit dividend in the hundreds place.

Learn

What does a lime fruit drink have in common with a coral reef in Florida? How about a mango drink and a rain forest? Lots of cents!

The fruit drink company that Catherine Page works for gives part of its profits to environmental groups. The money will help save endangered sites.

Catherine Page works for a natural fruit drink company in Charlotte, North Carolina.

Example

Catherine's company stacks 114 cases onto pallets. Cases are stacked 6 layers high. How many cases are in each layer?

Find 114 ÷ 6.

Step 1	Step 2	Step 3
Estimate to decide where to start dividing. **Think:** 114 is close to 120. $$20\\6\overline{)120}$$ Start by dividing tens. Your quotient will have 2 digits.	Divide the tens. $$1\\6\overline{)114}\\\underline{-6}\\5$$ Multiply. 1 × 6 = 6 Subtract. 11 − 6 = 5 Compare. 5 < 6	Bring down the ones and divide. $$19\\6\overline{)114}\\\underline{-6}\\54\\\underline{-54}\\0$$ Multiply. 9 × 6 = 54 Subtract. 54 − 54 = 0 Compare. 0 < 6

There are 19 cases in each layer on the pallet.

Talk About It

1. How did estimating help you make sure you divided in the right place?
2. How can you tell that 375 ÷ 7 will have a 2-digit quotient?

310 Chapter 7 • Dividing by 1-Digit Divisors

There are various ways to do the multiplication $21_{five} \cdot 3_{five}$.

Five	Ones
2	1
×	3

$\rightarrow$

$$(20 + 1)_{five} \qquad 21_{five} \qquad 21_{five}$$
$$\times \quad 3_{five} \rightarrow \times \quad 3_{five} \rightarrow \times \quad 3_{five}$$
$$(110 + 3)_{five} \quad \rightarrow 3 \quad \quad \quad 113_{five}$$
$$\rightarrow 110$$
$$113_{five}$$

The multiplication of a two-digit number by a two-digit number is developed next:

$$\begin{array}{r} 23_{five} \\ \times\ 14_{five} \\ \hline 22 \\ 130 \\ 30 \\ 200 \\ \hline 432_{five} \end{array} \qquad \begin{array}{l} (10 + 4)_{five} \\ (4 \cdot 3)_{five} \\ (4 \cdot 20)_{five} \\ (10 \cdot 3)_{five} \\ (10 \cdot 20)_{five} \end{array} \qquad \begin{array}{r} 23_{five} \\ \times\ 14_{five} \\ \hline 202 \\ 230 \\ \hline 432_{five} \end{array}$$

Lattice multiplication can also be used to multiply numbers in various number bases. This is explored in Ongoing Assessment 3-3.

Division in different bases can be performed using the multiplication facts and the definition of division. For example, $22_{five} \div 3_{five} = c$ if, and only if, $c \cdot 3_{five} = 22_{five}$. From Table 3-9, we see that $c = 4_{five}$. As in base ten, computing multidigit divisions efficiently in different bases requires practice. The ideas behind the algorithms for division can be developed by using repeated subtraction. For example, $3241_{five} \div 43_{five}$ is computed by means of the repeated-subtraction technique in Figure 3-16(a) and by means of the conventional algorithm in Figure 3-16(b). Thus $3241_{five} \div 43_{five} = 34_{five}$ with remainder 14_{five}.

Figure 3-16

(a)
$$\begin{array}{r} 43_{five} \overline{\smash{)}3241_{five}} \\ -\ 430 \\ \hline 2311 \\ -\ 430 \\ \hline 1331 \\ -\ 430 \\ \hline 401 \\ -\ 141 \\ -\ 210 \\ -\ 141 \\ \hline 14 \end{array} \qquad \begin{array}{l} (10 \cdot 43)_{five} \\ \\ (10 \cdot 43)_{five} \\ \\ (10 \cdot 43)_{five} \\ \\ (2 \cdot 43)_{five} \\ (2 \cdot 43)_{five} \\ (34 \cdot 43)_{five} \end{array}$$

(b)
$$\begin{array}{r} 34_{five} \\ 43_{five} \overline{\smash{)}3241_{five}} \\ -\ 234 \\ \hline 401 \\ -\ 332 \\ \hline 14 \end{array}$$

Computations involving base two are demonstrated in Example 3-6.

Example 3-6

a. Multiply:

$$\begin{array}{r} 101_{two} \\ \times\ 11_{two} \end{array}$$

b. Divide:

$$101_{two} \overline{\smash{)}110110_{two}}$$

Solution a. $\begin{array}{r} 101_{two} \\ \times\ 11_{two} \\ \hline 101 \\ 101 \\ \hline 1111_{two} \end{array}$ b. $\begin{array}{r} 1010_{two} \\ 101_{two}\overline{)110110_{two}} \\ -\ 101 \\ \hline 111 \\ -\ 101 \\ \hline 100 \end{array}$

ONGOING ASSESSMENT 3-3

1. Fill in the missing numbers in each of the following:
   a. $\begin{array}{r} 4\_6 \\ \times\ 783 \\ \hline 1\_78 \\ 3408 \\ \underline{982} \\ 3335\_8 \end{array}$
   b. $\begin{array}{r} 327 \\ \times\ 9\_1 \\ \hline 327 \\ 1\_08 \\ \underline{9\_3} \\ 30\_\_07 \end{array}$

2. Perform the following multiplications using the lattice multiplication algorithm:
   a. $\begin{array}{r} 728 \\ \times\ 94 \\ \hline \end{array}$
   b. $\begin{array}{r} 306 \\ \times\ 24 \\ \hline \end{array}$

3. Explain why the lattice multiplication algorithm works.

4. The following chart gives average water usage for 1 person for one day:

Use	Average Amount
Taking bath	110 L (liters)
Taking shower	75 L
Flushing toilet	22 L
Washing hands, face	7 L
Getting a drink	1 L
Brushing teeth	1 L
Doing dishes (one meal)	30 L
Cooking (one meal)	18 L

   a. Use the chart to calculate how much water you use each day.
   b. The average American uses approximately 200 L of water per day. Are you average?
   c. If there are 266,000,000 people in the United States, on average approximately how much water is used in the United States per day?

5. Simplify each of the following using properties of exponents. Leave answers as powers.
   a. $5^7 \cdot 5^{12}$
   b. $6^{10} \cdot 6^2 \cdot 6^3$
   c. $10^{296} \cdot 10^{17}$
   d. $2^7 \cdot 10^5 \cdot 5^7$

6. a. Which is greater, $2^{80} + 2^{80}$ or 2^{100}? Why?
   b. Which is greatest, 2^{101}, $3 \cdot 2^{100}$, or 2^{102}? Why?

7. The following model illustrates $14 \cdot 23$:

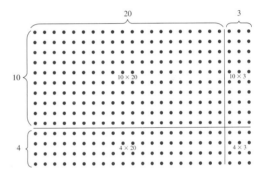

   Draw similar models illustrating each of the following:
   a. $6 \cdot 23$
   b. $18 \cdot 25$

8. Consider the following:

   $\begin{array}{r} 476 \\ \times\ 293 \\ \hline 952 \\ 4284 \\ \underline{1428} \\ 139468 \end{array}$ $\begin{array}{l} (2 \cdot 476) \\ (9 \cdot 476) \\ (3 \cdot 476) \end{array}$

   a. Use the conventional algorithm to show that the answer is correct.
   b. Explain why the algorithm works.
   c. Try the method to multiply 84×363.

9. The Russian peasant algorithm for multiplying 27×68 follows. (Disregard remainders when halving.)

	Halves		Doubles	
→	27	×	68	
Halve 27	→	13	136	Double 68.
Halve 13		6	272	Double 136.
Halve 6	→	3	544	Double 272.
Halve 3	→	1	1088	Double 544.

In the "Halves" column, choose the odd numbers. In the "Doubles" column, circle the numbers paired with the odds from the "Halves" column. Add the circled numbers.

68
136
544
<u>1088</u>
1836   This is the product of 27 · 68.

Try this algorithm for 17 · 63 and other numbers.

10. Complete the following table:

a	b	a · b	a + b
	56	3752	
32			110
		270	33

11. Answer the following questions based on the activity chart given next:

Activity	Calories Burned per Hour
Playing tennis	462
Snowshoeing	708
Cross-country skiing	444
Playing volleyball	198

  a. How many calories are burned during 3 hr of cross-country skiing?
  b. Jane played tennis for 2 hr while Carolyn played volleyball for 3 hr. Who burned more calories, and how many more?
  c. Lyle went snowshoeing for 3 hr and Maurice went cross-country skiing for 5 hr. Who burned more calories, and how many more?

12. On a 14-day vacation, Glenn increased his caloric intake by 1500 calories per day. He also worked out more than usual by swimming 2 hr a day. Swimming burns 666 calories per hour, and a net gain of 3500 calories adds 1 lb of weight. Did Glenn gain at least 1 lb during his vacation?

13. Sue purchased a $30,000 life-insurance policy at the price of $24 for each $1000 of coverage. If she pays the premium in 12 monthly installments, how much is each installment?

14. Perform each of the following divisions using both the repeated-subtraction and standard algorithms:
  a. 8)623   b. 36)298   c. 391)4001

15. Place the digits 7, 6, 8, and 3 in the boxes □)□□□ to obtain
  a. the greatest quotient.   b. the least quotient.

16. A 1K computer memory chip can store 1024 bits of information. How many bits of information can be stored in a 64K chip?

17. Using a calculator, Ralph multiplied by 10 when he should have divided by 10. The display read 300. What should the correct answer be?

18. Twenty members of the band plan to attend a festival. The band members washed 245 cars at $2 per car to help cover expenses. The school will match every dollar the band raises with a dollar from the school budget. The cost of renting the bus to take the band is 72¢ per mile and the round trip is 350 mi. The band members can stay in the dorm for 2 nights at $5 per person per night. Meals for the trip will cost $28 per person. Has the band raised enough money yet? If not, how many more cars do they have to wash?

19. The following figure shows four function machines. The output from one machine becomes the input for the one below it. Complete the accompanying chart.

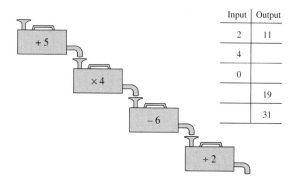

20. Choose three different digits.
  a. Form six different two-digit numbers from the numbers you chose. Each number can be used only once.
  b. Add the six numbers.
  c. Add the three digits you chose.
  d. Divide the answer in (b) by the answer in (c).
  e. Repeat (a) through (d) with three different numbers.
  f. Is the final result always the same? Why?

21. Consider the following multiplications. Notice that when the digits in the factors are reversed, the products are the same.

$$\begin{array}{r} 36 \\ \times\ 42 \\ \hline 1512 \end{array} \qquad \begin{array}{r} 63 \\ \times\ 24 \\ \hline 1512 \end{array}$$

  a. Find other multiplications where this procedure works.
  b. Find a pattern for the numbers that work in this way.

22. Molly read 160 pages in her book in 4 hr. Her sister Karly took 4 hr to read 100 pages in the same book. If the book is 200 pages long and if the two girls continued to read at these rates, how much longer would it take Karly to read the book than Molly?

23. Discuss possible error patterns in each of the following:

  a. $\begin{array}{r} 34 \\ \times\ 8 \\ \hline 2432 \end{array}$ b. $\begin{array}{r} 35 \\ \times\ 26 \\ \hline 90 \end{array}$ c. $\begin{array}{r} 34 \\ \times\ 6 \\ \hline 114 \end{array}$ d. $\begin{array}{r} 5\ 3 \\ 5\overline{)2515} \\ -\ 25 \\ \hline 15 \\ -\ 15 \\ \hline 0 \end{array}$

24. To transport the complete student body of 1672 students to a talk given by the governor, the school plans to rent buses that can hold 29 students each. How many buses are needed? Will all the buses be full?
25. **a.** If $a = 131 \cdot 4789 + 200$ without multiplying, mentally find the quotient and remainder when a is divided by 131.
    b. Write the answer for part (a) in the form $a = bq + r$ where $0 \leq r < 131$.
26. What happens when you multiply any two-digit number by 101? Discuss why this happens.
27. Perform each of the following operations using the bases shown:
    a. $32_{five} \cdot 4_{five}$
    b. $32_{five} \div 4_{five}$
    c. $43_{five} \cdot 23_{five}$
    d. $143_{five} \div 3_{five}$
    e. $10010_{two} \div 11_{two}$
    f. $10110_{two} \cdot 101_{two}$
28. For what possible bases are each of the following computations correct?

    a. $\begin{array}{r} 213 \\ + 308 \\ \hline 522 \end{array}$ **b.** $\begin{array}{r} 322 \\ - 233 \\ \hline 23 \end{array}$ **c.** $\begin{array}{r} 213 \\ \times 32 \\ \hline 430 \\ 1043 \\ \hline 11300 \end{array}$ **d.** $\begin{array}{r} 101 \\ 11\overline{)1111} \\ -11 \\ \hline 11 \\ -11 \\ \hline 0 \end{array}$

29. **a.** Use lattice multiplication to compute $(323_{five}) \cdot (42_{five})$.
    b. Find the smallest values of a and b so that $32_a = 23_b$.
30. Place the digits 7, 6, 8, and 3 in the boxes □□□ × □ to obtain
    a. the greatest product. **b.** the least product.
31. Place the digits 7, 6, 8, 3, and 2 in the boxes □□□ × □□ to obtain
    a. the greatest product. **b.** the least product.
32. Find the missing numbers in the following:

    a. $\begin{array}{r} 37 \\ \times\ 43 \\ \hline \_\_\_ \\ \_\_\_\_ \\ \hline \_591 \end{array}$ **b.** $\begin{array}{r} \_\_ \\ \times\ 36 \\ \hline 558 \\ 2790 \\ \hline \_\_\_\_ \end{array}$ **c.** $\begin{array}{r} \_\overline{)123} \\ -\ 9 \\ \hline 33 \\ -27 \\ \hline 6 \end{array}$

33. Find the products of the following and describe the pattern that emerges:
    a. 1×1
    11×11
    111×111
    1111×1111
    b. 99×99
    999×999
    9999×9999
    c. Test the patterns discovered for 30 terms using a spreadsheet. If the patterns do not continue as expected, determine when the patterns stop.
34. When $12 \times 483 = 5796$ is multiplied, every digit 1 through 9 is used either in one of the factors or in the product.
    a. Show that this also happens in the following:
    (i) 27×198
    (ii) 48×159
    (iii) 39×186
    b. Find other examples where all digits are used that have the following as factors:
    (i) 1963 (ii) 483 (iii) 297
    c. Which whole numbers 1 through 9, if any, cannot be the units digit of a factor when using every digit in multiplication as described above. Why?

Communication

35. Explain any connections you see between operations in base five and base ten.
36. Choose what you consider the "best" algorithm studied in this section. Explain the reasoning behind your choice.
37. Long division has been recommended for reduced attention in elementary classrooms. Do you agree or disagree? Defend your answer.
38. Pick a number. Double it. Multiply the result by 3. Add 24. Divide by 6. Subtract your original number. Is the result always the same? Write a convincing argument for your answer.

Open-Ended

39. If a student presented a new "algorithm" for computing with whole numbers, describe the process you would recommend to the student to determine if the algorithm would always work.

Cooperative Learning

40. The traditional sequence for teaching operations in the elementary school is first addition, then subtraction, followed by multiplication, and finally division. Some educators advocate teaching addition followed by multiplication, then subtraction followed by division. Within your group prepare arguments for teaching the operations in either order listed above.

Review Problems

41. Write the number succeeding 673 in Egyptian numerals.
42. Write $3 \cdot 10^5 + 2 \cdot 10^2 + 6 \cdot 10$ as a Hindu-Arabic numeral.
43. Illustrate the identity property of addition for whole numbers.
44. Rename each of the following using the distributive property of multiplication over addition:
    a. $ax + bx + 2x$
    b. $3(a + b) + x(a + b)$
45. At the beginning of a trip, the odometer registered 52,281. At the end of the trip, the odometer registered 59,260. How many miles were traveled on this trip?

BRAIN TEASER

For each of the following, replace the letters with digits in such a way that the computation is correct. Each letter may represent only one digit.

a. LYNDON
    × B
    ─────────
    JOHNSON

b. MA
    MA
    + MA
    ─────
    EEL

LABORATORY ACTIVITY

1. Messages can be coded on paper tape in base two. A hole in the tape represents 1, whereas a space represents 0. The value of each hole depends on its position; from left to right, 16, 8, 4, 2, 1 (all powers of 2). Letters of the alphabet may be coded in base two according to their position in the alphabet. For example, G is the seventh letter. Since $7 = 1 \cdot 4 + 1 \cdot 2 + 1$, the holes appear as they do in Figure 3-17:

Figure 3-17

a. Decode the following message:

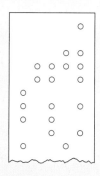

b. Write your name on a tape using base two.

2. The following number game uses base-two arithmetic:

Card E		Card D		Card C		Card B		Card A	
16	24	8	24	4	20	2	18	1	17
17	25	9	25	5	21	3	19	3	19
18	26	10	26	6	22	6	22	5	21
19	27	11	27	7	23	7	23	7	23
20	28	12	28	12	28	10	26	9	25
21	29	13	29	13	29	11	27	11	27
22	30	14	30	14	30	14	30	13	29
23	31	15	31	15	31	15	31	15	31

a. Suppose a person's age appears on cards E, C, and B, and the person is 22. Can you discover how this works and why?

b. Design card F so that the numbers 1 through 63 can be used in the game. Note that cards A through E must also be changed.

Section 3-4 Mental Mathematics and Estimation for Whole-Number Operations

The *Principles and Standards* point out that,

When students leave grade 5, they should be able to solve problems involving whole-number computation and should recognize that each operation will help them solve many different types of problems. They should be able to solve many problems mentally, to estimate a reasonable result for a problem, to efficiently recall or derive the basic number combinations for each operation, and to compute fluently with multidigit whole numbers (p. 149).

mental mathematics

computational estimation

Mental mathematics and estimation strategies are valuable tools for students in elementary school. *Mental mathematics* is the process of producing an answer to a computation without using external computational aids. **Computational estimation** is the process of forming an approximate answer to a numerical problem. Estimation strategies can be used to tell whether answers are reasonable or not. In the cartoon below we see that Calvin was not able to tell whether his result was reasonable.

Mental mathematics makes use of a variety of strategies and properties. For example, consider 8×26. Students may think of this computation in a variety of ways, as shown below:

- 8×20 is 160 and 8×6 is 48, so 8×26 is $160 + 48$, or 208.
- 8×25 is 200, then 8×1 is 8 more, so 8×26 is $200 + 8$, or 208.
- 8×30 is 240, then take off $8 \times 4 = 32$, so 8×26 is $240 - 32 = 208$.

Next we consider several of the most commonly used strategies for performing operations mentally on whole numbers.

Mental Mathematics: Addition

1. *Adding from the left*

   a. 67
      + 36

   $60 + 30 = 90$ (Add the tens.)
   $7 + 6 = 13$ (Add the units.)
   $90 + 13 = 103$ (Add the two sums.)

   b. 36
      + 36

   $30 + 30 = 60$ (Double 30.)
   $6 + 6 = 12$ (Double 6.)
   $60 + 12 = 72$ (Add the doubles.)

2. *Breaking up and bridging*

$$\begin{array}{r} 67 \\ +\ 36 \\ \hline \end{array}$$
$67 + 30 = 97$ (Add the first number to the tens in the second number.)
$97 + 6 = 103$ (Add this sum to the units in the second number.)

3. *Trading off*

   a. $$\begin{array}{r} 67 \\ +\ 36 \\ \hline \end{array}$$
   $67 + 3 = 70$ (Add 3 to make a multiple of 10.)
   $36 - 3 = 33$ (Subtract 3 to compensate for the 3 that was added.)
   $70 + 33 = 103$ (Add the two numbers.)

   b. $$\begin{array}{r} 67 \\ +\ 29 \\ \hline \end{array}$$
   $67 + 30 = 97$ (Add 30 (next multiple of 10 greater than 29.))
   $97 - 1 = 96$ (Subtract 1 to compensate for the extra 1 that was added.)

4. *Using compatible numbers*

   Compatible numbers are numbers whose sums are easy to calculate mentally.

   130, 50, 70, 20, + 50 → 200, 100

   $130 + 70 = 200$
   $50 + 50 = 100$
   $100 + 200 = 300$
   $300 + 20 = 320$

5. *Making compatible numbers*

   $$\begin{array}{r} 25 \\ +\ 79 \\ \hline \end{array}$$
   $25 + 75 = 100$ (25 + 75 adds to 100.)
   $100 + 4 = 104$ (Add 4 more units.)

Mental Mathematics: Subtraction

1. *Breaking up and bridging*

   $$\begin{array}{r} 67 \\ -\ 36 \\ \hline \end{array}$$
   $67 - 30 = 37$ (Subtract the tens in the second number from the first number.)
   $37 - 6 = 31$ (Subtract the units in the second number from the difference.)

2. *Trading off*

   $$\begin{array}{r} 71 \\ -\ 39 \\ \hline \end{array}$$
   $(71 + 1) = 72; (39 + 1) = 40$ (Add 1 to both numbers. Perform the subtraction, which is easier than the original problem.)
   $(72 - 40) = 32$

   Notice that adding 1 to both numbers does not change the answer. Why?

3. *Drop the zeros*

   $$\begin{array}{r} 8700 \\ -\ 500 \\ \hline \end{array}$$
   $87 - 5 = 82$ (Notice that there are two zeros in each number. Drop these zeros and perform the computation. Then replace the two zeros to obtain proper place value.)
   $82 \rightarrow 8200$

   Another mental-mathematics technique for subtraction is called "adding up." This method is based on the *missing addend* approach and is sometimes referred to as the "cashier's algorithm." An example of the cashier's algorithm follows.

Example 3-7

Noah owed $11 for his groceries. He used a $50 check to pay the bill. While handing Noah the change, the cashier said, "$11, $12, $13, $14, $15, $20, $30, $50." How much change did Noah receive?

Solution Table 3-10 shows what the cashier said and how much money Noah received each time. Since $11 plus $1 is $12, Noah must have received $1 when the cashier said $12. The same reasoning follows for $13, $14, and so on. Thus the total amount of change that Noah received is given by $1 + $1 + $1 + $1 + $5 + $10 + $20 = $39. In other words, $50 − $11 = $39 because $39 + $11 = $50.

Table 3-10

What the Cashier Said	$11	$12	$13	$14	$15	$20	$30	$50
Amount of Money Noah Received Each Time	0	$1	$1	$1	$1	$5	$10	$20

NOW TRY THIS 3-8

- Perform each of the following computations mentally and explain what technique you used to find the answer.

  a. $40 + 160 + 29 + 31$
  b. $3679 - 474$
  c. $75 + 28$
  d. $2500 - 700$

Mental Mathematics: Multiplication

As with addition and subtraction, mental mathematics is useful for multiplication. Several examples are given next:

1. *Front-end multiplying*

   $$\begin{array}{r} 64 \\ \times\ 5 \\ \hline \end{array}$$

   $60 \times 5 = 300$ (Multiply the number of tens in the first number by 5.)
   $4 \times 5 = 20$ (Multiply the number of units in the first number by 5.)
   $300 + 20 = 320$ (Add the two products.)

2. *Using compatible numbers*

   $2 \times 9 \times 5 \times 20 \times 5$ Rearrange as $9 \times (2 \times 5) \times (20 \times 5) = 9 \times 10 \times 100 = 9000$

3. *Thinking money*

   a. $$\begin{array}{r} 64 \\ \times\ 5 \\ \hline \end{array}$$

   Think of the product as 64 nickels, which can be thought of as 32 dimes, which is $32 \times 10 = 320$ cents.

b. 64
×50

Think of the product as 64 half-dollars, which is 32 dollars, or 3200 cents.

c. 64
×25

Think of the product as 64 quarters, which is 32 half-dollars, or 16 dollars. Thus we have 1600 cents.

Mental Mathematics: Division

1. *Breaking up the dividend*

   $7\overline{)4256}$ $7\overline{)42|56}$ (Break up the dividend into parts.)

   $\dfrac{600 + 8}{7\overline{)4200 + 56}}$ (Divide both parts by 7.)

   $600 + 8 = 608$ (Add the answers together.)

2. *Using compatible numbers*

   a. $3\overline{)105}$ $105 = 90 + 15$ (Look for numbers that you recognize as divisible by 3 and having a sum of 105.)

   $\dfrac{30 + 5 = 35}{3\overline{)90 + 15}}$ (Divide both parts and add the answers.)

   b. $8\overline{)232}$ $232 = 240 - 8$ (Look for numbers that are easily divisible by 8 and whose difference is 232.)

   $\dfrac{30 - 1 = 29}{8\overline{)240 - 8}}$ (Divide both parts and take the difference.)

NOW TRY THIS 3-9

- Perform each of the following computations mentally and explain what technique you used to find the answer.

  a. $25 \cdot 32 \cdot 4$ **b.** $123 \cdot 3$
  c. $25 \cdot 35$ **d.** $5075 \div 25$

Computational Estimation

Computational estimation may help determine whether an answer is reasonable or not. This is especially useful when the computation is done on a calculator. Some of the common estimation strategies for addition are given next.

1. *Front-end*

   Front-end estimation begins by focusing on the lead, or front, digits of the addition. These front, or lead, digits are added and assigned an appropriate place value. At this point we may have an underestimate that needs to be adjusted. The adjustment is made by focusing on the next group of digits. The following example shows how front-end estimation works.

SECTION 3-4   *Mental Mathematics and Estimation for Whole-Number Operations*   161

Steps:
1. **Add front-end digits**
   $4 + 3 + 5 = 12$.
2. **Place value** = 1200.
3. **Adjust** $61 + 38 \approx 100$ and $20 + 100$ is 120.
4. **Adjusted estimate** is $1200 + 120 = 1320$.

The student page from *Scott Foresman-Addison Wesley Math,* Grade 4, 1999 shows how front-end estimation can be used to check whether computations are reasonable.

2. *Grouping to nice numbers*

   The strategy used to obtain the adjustment in the preceding example is the *grouping to nice numbers* strategy, which means that numbers that "nicely" fit together are grouped. Another example is given here.

   About 100 ⟨ 23, 39 ⟩ 32 ⟨ 64, + 49 ⟩ About 100   Therefore the sum is about $100 + 100$, or 200.

3. *Clustering*

   Clustering is used when a group of numbers cluster around a common value. This strategy is limited to certain kinds of computations. In the next example, the numbers seem to cluster around 6000.

   ```
   6200
   5842
   6512
   5521
  +6319
   ```

   1. Estimate the "average"—about 6000.
   2. Multiply the "average" by the number of values to obtain $5 \cdot 6000 = 30{,}000$.

4. *Rounding*

   Rounding is a way of cleaning up numbers so that they are easier to handle. Rounding enables us to find approximate answers to calculations, as follows:

   ```
    4724      5000      Round 4724 to 5000.
   +3192     +3000      Round 3192 to 3000.
             ─────
              8000      Add the rounded numbers.

    1267      1300      Round 1267 to 1300.
   - 510     - 500      Round 510 to 500.
             ─────
               800      Subtract the rounded numbers.
   ```

   Performing estimations requires a knowledge of place value and rounding techniques. We illustrate a rounding procedure that can be generalized to all rounding situations. For example, suppose we wish to round 4724 to the nearest thousand. We may proceed in four steps (see also Figure 3-18).

   a. Determine between which two consecutive thousands the number lies.
   b. Determine the midpoint between the thousands.

Figure 3-18

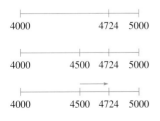

Chapter 3 Lesson 6

Column Addition

You Will Learn
how to add 3 or 4 addends

Vocabulary
addends
numbers that are added together to make a sum

front-end estimation
a way to estimate by first looking at the leading digits

Learn

"Don't get in over your head," reads Jonathan's ad. His ad reaches many people who don't know how to use computers. Jonathan prints 442 flyers, 339 brochures, and 628 bookmarks. How many items does he print in all?

Jonathan runs a desktop publishing business in Farmersville, Illinois.

You can add three **addends** to find the total number of items. Use **front-end estimation** to check.

Example
Find 442 + 339 + 628.

Step 1	Step 2	Step 3
Add the ones. Regroup as needed.	Add the tens. Regroup as needed.	Add the hundreds.
$^{1}$ 442 339 + 628 ───── 9	$^{1\,1}$ 442 339 + 628 ───── 09	$^{1\,1}$ 442 339 + 628 ───── 1,409

Round to the front-end, or leading, digits.

```
442  →   400
339  →   300
+ 628  → + 600
         -----
         1,300
```

Adjust your estimate.

```
442  →    40
339  →    40
+ 628  → + 30
         -----
          110
```

1,300 + 110 = 1,410
Since 1,409 is close to 1,410, the answer is reasonable.
Jonathan needs to print 1,409 items.

Talk About It

Why do you line up the ones, tens, and hundreds before you add?

108 Chapter 3 • Adding and Subtracting Whole Numbers and Money

---

    c. Determine which thousand the number is closer to by observing whether it is greater than or less than the midpoint. (Not all texts use the same rule for rounding when a number falls at a midpoint.)

    d. If the number to be rounded is greater than or equal to the midpoint, round the given number to the greater thousand; otherwise, round to the lesser thousand. In this case, we round 4724 to 5000.

5. *Using the range*

It is often useful to know into what *range* an answer falls. The range is determined by finding a low estimate and a high estimate and reporting that the answer falls in this interval. An example follows:

Problem	Low Estimate	High Estimate
378	300	400
+ 524	+ 500	+ 600
	800	1000

Thus a range for this problem is from 800 to 1000.

Estimation: Multiplication and Division

Examples of estimation strategies for multiplication and division are given next.

1. *Front-end*

   524
   × 8

   $500 \times 8 = 4000$ (Start multiplying at the front to obtain a first estimate.)
   $20 \times 8 = 160$ (Multiply the next important digit 8 times.)
   $4000 + 160 = 4160$ (Adjust the first estimate by adding the two numbers.)

2. *Compatible numbers*

   5)4163

   5)4000

   800
   5)4000

   (Change 4163 to a number close to it that you know is divisible by 5.)
   (Carry out the division and obtain the first estimate of 800. Various techniques can be used to adjust the first estimate.)

NOW TRY THIS 3-10

● Estimate each of the following mentally and explain what technique you used to find the answer.

a. A sold-out concert was held in a theater with a capacity of 4,525 people. Tickets were sold for $9 each. How much money was collected?

b. Fliers are to be delivered to 3,625 houses and there are 42 people who will be doing the distribution. How many houses will each person visit? ●

ONGOING ASSESSMENT 3-4

1. Use compatible numbers to compute each of the following mentally.
   a. $2 \cdot 9 \cdot 5 \cdot 6$
   b. $8 \cdot 25 \cdot 7 \cdot 4$
   c. $5 \cdot 11 \cdot 3 \cdot 20$
   d. $82 + 37 + 18 + 13$

2. Supply reasons for each of the first four steps given below.

   $(525 + 37) + 75 = 525 + (37 + 75)$
   $= 525 + (75 + 37)$
   $= (525 + 75) + 37$
   $= 600 + 37$
   $= 637$

3. Use breaking and bridging or front-end multiplying to compute each of the following mentally.
   a. 567 + 38   b. 321 · 3
   c. 997 − 32   d. 56 · 30
4. Use trading off to compute each of the following mentally.
   a. 85 − 49   b. 87 + 33
   c. 19 · 6   d. 58 + 39
5. A car trip took 8 hours of driving at an average of 62 mph. Mentally compute the total number of miles traveled. Describe your method.
6. Mentally compute each of the following using the cashier's algorithm. Describe your method.
   a. 53 − 28   b. 63 − 47
7. Round each number to the place value indicated by the digit in bold.
   a. 5**2**80   b. **1**15,234
   c. 1**1**5,234 d. 2,3**2**5
8. Estimate each answer by rounding.
   a. 878 ÷ 29
   b. 25,201 − 19,987
   c. 32 · 28
   d. 2215 + 3023 + 5967 + 975
9. Use front-end estimation with adjustment to estimate each of the following:
   a. 2215 + 3023 + 5987 + 975
   b. 234 + 478 + 987 + 319 + 469
10. a. Would the clustering strategy of estimation be a good one to use in each of the following cases? Why or why not?

   (i)   474   (ii)   483
         1467   475
           64   530
        + 2445   503
                       + 528

   b. Estimate each part of (a) using the following stategies:
      (i) Front-end
      (ii) Grouping to nice numbers
      (iii) Rounding
11. Use the range strategy to estimate each of the following. Explain how you arrived at your estimates.
    a. 22 · 38
    b. 145 + 678
    c. 278 + 36
12. Tom estimated 31 · 179 in the three ways shown below.
    (i) 30 · 200 = 6000
    (ii) 30 · 180 = 5400
    (iii) 31 · 200 = 6200
    Without finding the actual product, which estimate do you think is closer to the actual product? Why?
13. About 3,540 calories must be burned to lose one pound of body weight. Estimate how many calories must be burned to lose 6 pounds.
14. Suppose you had a balance of $3287 in your checking account and you wrote checks for $85, $297, $403, and $523. Estimate your balance and tell what you did and tell whether you think your estimate is too high or too low.
15. A theater has 38 rows with 23 seats in each row. Estimate the number of seats in the theater and tell how you arrived at your estimate.
16. Without computing tell which of the following have the same answer. Describe your reasoning.
    a. 44 · 22 and 22 · 11
    b. 22 · 32 and 11 · 64
    c. 13 · 33 and 39 · 11
17. The following is a list of the areas in square miles of Europe's largest countries. Mentally use this information to decide if each of the given statements is true.

    | France | 211,207 |
    | Spain | 194,896 |
    | Sweden | 173,731 |
    | Finland | 130,119 |
    | Norway | 125,181 |

    a. Sweden is less than 40,000 mi^2 larger than Finland.
    b. France is more than twice the size of Norway.
    c. France is more than 100,000 mi^2 larger than Norway.
    d. Spain is about 21,000 mi^2 larger than Sweden.
18. The attendance at a World's Fair for one week is given below.

    | Monday | 72,250 |
    | Tuesday | 63,891 |
    | Wednesday | 67,490 |
    | Thursday | 73,180 |
    | Friday | 74,918 |
    | Saturday | 68,480 |

    Estimate the week's attendance and tell what strategy you used and why you used it.
19. Use your calculator to calculate 25^2, 35^2, 45^2, and 55^2 and then see if you can find a pattern that will let you find 65^2 and 75^2 mentally.
20. Use your calculator to multiply several two-digit numbers times 99. Then see if you can find a pattern that will let you find the product of any two-digit number and 99 mentally.

Communication

21. What is the difference between mental mathematics and computational estimation?
22. Is the front-end estimate for addition before adjustment always less than the exact sum? Explain why or why not.
23. In the new textbooks, there is an emphasis on mental mathematics and estimation. Do you think these topics are important for today's students? Why?
24. Suppose x and y are positive whole numbers. If x is greater than y and you estimate $x − y$ by rounding x up and y down, will your estimate always be too high or too low or could it be either? Explain.

Open-Ended

25. Give several examples from real-world situations where an estimate, rather than an exact answer, is close enough.
26. Give a numerical example of when front-end estimation and rounding can produce the same estimate. Give an example of when they can produce a different estimate.

Cooperative Learning

27. Have each person in your group choose a different grade (3–6) textbook and make a list of mental math or estimation strategies done for each grade level. How do the lists compare?
28. As a group, without multiplying decide whether $19{,}876 \cdot 43$ or $19{,}875 \cdot 44$ is greater. Prepare a group response to present to the rest of the class.

HINT FOR SOLVING THE PRELIMINARY PROBLEM

It is very helpful to *draw a model* of what is happening when the phone calls are made. A *table* is also useful to record the results of your model. A key to solving the problem is to realize that when a caller calls one of his two people, the first person being called does not wait for the second person to be called before this first person can begin calling his two people. This problem can be extended and generalized.

QUESTIONS FROM THE CLASSROOM

1. **a.** A student asks if $39 + 41 = 40 + 40$, is it true that $39 \cdot 41 = 40 \cdot 40$. How do you reply?
   b. Another student says that he knows that $39 \cdot 41 \neq 40 \cdot 40$ but he found that $39 \cdot 41 = 40 \cdot 40 - 1$. He also found that $49 \cdot 51 = 50 \cdot 50 - 1$. He wants to know if this pattern continues. How would you respond?
2. A student asks why he has to learn about any estimation strategy other than rounding. What is your response?
3. While studying different number bases a student asks if it is possible to have a negative number for a base. What do you tell this student?
4. A student divides as follows. How would you help?

$$\begin{array}{r} 4\ 5 \\ 3\overline{)1215} \\ -\ 12 \\ \hline 15 \\ -\ 15 \\ \hline 0 \end{array}$$

5. A student claims that the Roman system is a base-ten system since it has symbols for 10, 100, and 1000. How do you respond?
6. A student divides as follows. How do you help?

$$\begin{array}{r} 15 \\ 6\overline{)36} \\ \underline{6} \\ 30 \\ \underline{30} \end{array}$$

7. When using Roman numerals, a student asks whether it is correct to write $\overline{II}$, as well as MI, for 1001. How do you respond?
8. A parent complains about the use of manipulatives in the classroom and likens their use to the use of fingers to count. How do you respond?
9. A student asks how you can find the quotient and the remainder in a division problem like $592 \div 36$ using a calculator without an integer division button?

CHAPTER OUTLINE

I. Numeration systems
   A. Properties of numeration systems give basic structure to the systems
      1. Additive property
      2. Place-value property
      3. Subtractive property
      4. Multiplicative property
II. Exponents
   A. For any whole number a and any natural number n,
   $$a^n = \underbrace{a \cdot a \cdot a \cdot \ldots \cdot a}_{n \text{ factors}},$$
   where a is the **base** and n is the **exponent**.
   B. $a^0 = 1, a \in N$
   C. For any natural number a, with whole numbers m and n, $a^m \cdot a^n = a^{m+n}$.
III. Algorithms for whole-number operations
   A. Addition and subtraction algorithms
      1. Concrete models
      2. Expanded algorithms
      3. Standard algorithms
      4. Addition and subtraction with regrouping
      5. Scratch addition
      6. Addition and subtraction in different number bases
   B. Multiplication and division algorithms
      1. Concrete models
      2. Expanded algorithms
      3. Standard algorithms
      4. Lattice multiplication
      5. Scaffolding with division
      6. Short division
      7. Multiplication and division in different number bases
IV. Mental mathematics and computational estimation strategies
   A. Mental mathematics
      1. Adding from the left
      2. Breaking up and bridging
      3. Trading off
      4. Using compatible numbers
      5. Making compatible numbers
      6. Dropping the zeroes
      7. Using the cashier's algorithm
      8. Front-end multiplying
      9. Thinking money
      10. Breaking up the dividend
   B. Computational estimation strategies
      1. Front-end
      2. Grouping to nice numbers
      3. Clustering
      4. Rounding
      5. Range
      6. Compatible numbers

CHAPTER REVIEW

1. Convert each of the following to base ten:
   a. $\overline{\text{CDXLIV}}$ b. 432_{five} c. $ET0_{\text{twelve}}$
   d. 1011_{two} e. 4136_{seven}
2. Convert each of the following numbers to numbers in the indicated system:
   a. 999 to Roman
   b. 86 to Egyptian
   c. 123 to Mayan
   d. 346_{ten} to base five
   e. 27_{ten} to base two
3. Simplify each of the following, if possible. Write your answers in exponential form, a^b.
   a. $3^4 \cdot 3^7 \cdot 3^6$ b. $2^{10} \cdot 2^{11}$ c. $3^4 + 2 \cdot 3^4$
   d. $2^{80} + 3 \cdot 2^{80}$ e. $2^{100} + 2^{100}$
4. Use both the scratch and the traditional algorithms to perform the following:

   $$\begin{array}{r} 316 \\ 712 \\ +\ 91 \\ \hline \end{array}$$

5. Use both the traditional and the lattice multiplication algorithms to perform each of the following:

   $$\begin{array}{r} 613 \\ \times\ 98 \\ \hline \end{array}$$

6. Use both the repeated-subtraction and the conventional algorithms to perform the following:
   a. $912\overline{)4803}$ b. $11\overline{)1011}$
   c. $23_{\text{five}}\overline{)3312_{\text{five}}}$ d. $11_{\text{two}}\overline{)1011_{\text{two}}}$
7. Use the division algorithm to check your answers in problem 6.
8. For each of the following base-ten numbers, tell the place value for each of the circled digits:
   a. 4③2   b. ③432   c. 19③24
9. You had a balance in your checking account of $720 before writing checks for $162, $158, and $33 and making a deposit of $28. What is your new balance?
10. Jim was paid $320 a month for 6 mo and $410 a month for 6 mo. What were his total earnings for the year?

11. A soft drink manufacturer produces 15,600 cans of his product each hour. Cans are packed 24 to a case. How many cases could be filled with the cans produced in 4 hr?
12. A limited partnership of 120 investors sold a piece of land for $461,040. If divided equally, how much did each investor receive?
13. How many 12-oz cans of juice would it take to give 60 people one 8-oz serving each?
14. I am thinking of a whole number. If I divide it by 13, then multiply the answer by 12, then subtract 20, and then add 89, I end up with 93. What was my original number?
15. Apples normally sell for 32¢ each. They go on sale for 3 for 69¢. How much money is saved if you purchase 2 dz apples while they are on sale?
16. A ski resort offers a weekend ski package for $80 per person or $6000 for a group of 80 people. Which would be the cheaper option for a group of 80?
17. The owner of a bicycle shop reported his inventory of bicycles and tricycles in an unusual way. He said he counted 126 wheels and 108 pedals. How many bikes and how many trikes did he have?
18. Josi has a job in which she works 30 hr/wk and gets paid $5/hr. If she works more than 30 hr in a week, she receives $8/hr for each hour over 30 hr. If she worked 38 hr at her job this week, how much did she earn?
19. In a television game show, there are five questions to answer. Each question is worth twice as much as the previous question. If the last question was worth $6400, what was the first question worth?
20. Write an example of a base other than ten used in a real-life situation. How is it used?
21. Describe the important characteristics of each of the following systems:
    a. Egyptian
    b. Babylonian
    c. Roman
    d. Hindu-Arabic
22. Write 128 in each of the following bases:
    a. five   b. two   c. twelve
23. Perform each of the following computations:
    a. $123_{\text{five}} + 34_{\text{five}}$
    b. $1010_{\text{two}} - 101_{\text{two}}$
    c. $23_{\text{five}} \times 34_{\text{five}}$
    d. $1001_{\text{two}} \times 101_{\text{two}}$
24. Tell how to use compatible numbers mentally to perform each of the following:
    a. $26 + 37 + 24 - 7$ b. $4 \cdot 7 \cdot 9 \cdot 25$
25. Compute each of the following mentally. Name the strategy you used to perform your mental math.
    a. $63 \cdot 7$
    b. $85 - 49$
    c. $(18 \cdot 5) \cdot 2$
    d. $2436 \div 6$
26. Estimate the following addition using (a) front-end estimation with adjustment and (b) rounding. Compare your answers and tell which you think is closer to the exact answer.

    $$\begin{array}{r} 543 \\ 398 \\ 255 \\ 408 \\ + 998 \end{array}$$

27. Using clustering, estimate the sum $2345 + 2854 + 2234 + 2203$.
28. Explain how the standard division algorithm works for the division given below.

    $$\begin{array}{r} 23 \\ 14\overline{)322} \\ -28 \\ \hline 42 \\ -42 \\ \hline 0 \end{array}$$

SELECTED BIBLIOGRAPHY

Bates, T., and L. Rousseau. "Will the Real Division Algorithm Please Stand Up?" *Arithmetic Teacher* 33 (March 1987): 42–46.

Bobis, J. "Using a Calculator to Develop Number Sense." *Arithmetic Teacher* 38 (January 1991): 42–45.

Bohan, H., and S. Bohan. "Extending the Regular Curriculum Through Creative Problem Solving." *Arithmetic Teacher* 41 (October 1993): 83–87.

Broadent, F. "Lattice Multiplication and Division." *Arithmetic Teacher* 34 (January 1987): 28–31.

Burns, M. "Introducing Division Through Problem-Solving Experiences." *Arithmetic Teacher* 38 (April 1991): 14–18.

Curcio, F., and S. Schwartz. "There Are No Algorithms for Teaching Algorithms." *Teaching Children Mathematics* 5 (September 1998): 26–30.

Englert, G., and R. Sinicrope. "Making Connections with Two-Digit Multiplication." *Arithmetic Teacher* 41 (April 1994): 446–448.

Gluck, D. "Helping Students Understand Place Value." *Arithmetic Teacher* 38 (March 1991): 10–13.

Graeber, A. "Misconceptions about Multiplication and Division." *Arithmetic Teacher* 40 (March 1993): 408–411.

Greenes, C., L. Schulman, and R. Spungin. "Developing Sense About Numbers." *Arithmetic Teacher* 40 (January 1993): 279–284.

Guershon, H., and M. Behr. "Ed's Strategy for Solving Division Problems." *Arithmetic Teacher* 39 (November 1991): 38–40.

Hope, J. "Promoting Number Sense in School." *Arithmetic Teacher* 36 (February 1989): 12–16.

Hope, J., B. Reys, and R. Reys. *Mental Math in the Middle Grades.* Palo Alto: Dale Seymour Publishing, 1987.

Huinker, D. "Multiplication and Division Word Problems: Improving Students' Understanding." *Arithmetic Teacher* 37 (October 1989): 8–12.

Kami, C., and L. Joseph. "Teaching Place Value and Double-Column Addition." *Arithmetic Teacher* 35 (February 1998): 48–52.

Kami, C., and B. Lewis. "The Harmful Effects of Algorithms." *Teaching K–8* 23 (January 1993): 36–38.

Kami, C., B. Lewis, and S. Livingston. "Primary Arithmetic: Children Inventing Their Own Procedures." *Arithmetic Teacher* 41 (December 1993): 200–203.

Kouba, V., and K. Franklin. "Multiplication and Division: Sense Making and Meaning." *Teaching Children Mathematics* 1 (May 1995): 574–577.

Moore, T. "More on Mental Computation." *Mathematics Teacher* 79 (March 1987): 168–169.

Nagel, N., and C. Swingen. "Students' Explanations of Place Value in Addition and Subtraction." *Teaching Children Mathematics* 5 (November 1998): 164–170.

Reys, R. *Computational Estimation (Grades 6, 7, and 8).* Palo Alto: Dale Seymour Publishing, 1987.

Reys, R. "Computation Versus Number Sense." *Mathematics Teaching in the Middle School* 4 (October 1998): 110–112.

Reys, B., and R. Reys. "Computation in the Elementary Curriculum: Shifting the Emphasis." *Teaching Children Mathematics* 5 (December 1998): 236–241.

Reys, R., and D. Yang. "Relationships Between Computational Performance and Number Sense Among Sixth- and Eighth-Grade Students in Taiwan." *Journal for Research in Mathematics Education* 29 (March 1998): 225–237.

Sowder, J. "Mental Computation and Number Sense." *Arithmetic Teacher* 37 (March 1990): 18–20.

Stanic, G., and W. McKillip. "Developmental Algorithms Have a Place in Elementary School Mathematics." *Arithmetic Teacher* 36 (January 1989): 14–16.

Thornton, C., G. Jones, and J. Neal. "The 100's Chart: A Stepping Stone to Mental Mathematics." *Teaching Children Mathematics* 1 (April 1995): 480–483.

Van de Walle, J. "Redefining Computation." *Arithmetic Teacher* 38 (January 1991): 46–51.

Wearne, D., and J. Hiebert. "Place Value and Addition and Subtraction." *Arithmetic Teacher* 41 (January 1994): 272–274.

4
Integers and Number Theory

Preliminary Problem
Two friends, Tira and Ben, met at a high school reunion after not having seen each other for many years. As they talked, Tira asked, "How many children do you have and what are their ages?"
"I have three children; the product of their ages is 36 and the sum of their ages is your street address," answered Ben. Tira thought for a moment and then said, "I need more information to solve the problem."
"Oh yes," replied Ben. "My oldest child is a girl." With this additional information Tira immediately found the ages of the children. How did Tira figure out the ages of the children, and what were their ages?

The *Principles and Standards* point out that:

Negative numbers should be introduced at this level (grades 3–5) through the use of familiar models such as temperature or owing money. The number line is also an appropriate and helpful model, and students should recognize that points to the left of 0 on a horizontal line can be represented by numbers less than 0 (p. 151).

In the middle grades, students should extend these initial understandings of integers. Positive and negative integers should be seen as useful for noting relative changes or values (p. 218).

In this chapter, we examine the system of integers and develop an understanding of number theory.

Negative numbers are useful in everyday life. For example, Mount Everest (the highest point on Earth) is 29,028 ft above sea level, while the Dead Sea (the lowest point on Earth) is 1293 ft below sea level. We may symbolize these elevations as 29,028 and $^{-}1293$. In mathematics, the need for integers, including negative whole numbers, arises because subtractions cannot always be performed in only the set of whole numbers. To compute $4 - 6$ using the definition of subtraction for whole numbers, we must find a whole number n such that $6 + n = 4$. Because there is no such whole number n, the subtraction cannot be completed in the set of whole numbers. To perform the computation, we must invent a new number. This new number is a *negative integer*. If we attempt to calculate $4 - 6$ on a number line, then we must draw intervals to the left of 0. In Figure 4-1, $4 - 6$ is pictured as an arrow that starts at 0 and ends 2 units to the left of 0. The new number that corresponds to a point 2 units to the left of 0 is *negative two*, symbolized by $^{-}2$. Other numbers to the left of 0 are created similarly. The new set of numbers $\{^{-}1, ^{-}2, ^{-}3, ^{-}4, \ldots\}$ is the set of

negative integers   **negative integers.**

Figure 4-1

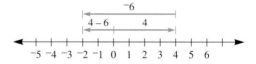

integers   The union of the set of negative integers and the set of whole numbers is the set of **integers.** The set of integers is denoted by *I*:

$$I = \{\ldots, ^{-}4, ^{-}3, ^{-}2, ^{-}1, 0, 1, 2, 3, 4, \ldots\}$$

The *Principles and Standards* emphasize the importance of studying concepts from number theory in the middle grades:

Students can also work with whole numbers in their study of number theory. Tasks, such as the following, involving factors, multiples, prime numbers, and divisibility, can afford opportunities for problem solving and reasoning.

  1. *Explain why the sum of the digits on any multiple of 3 is itself divisible by 3* (p. 217).

The study of integers, their properties, and relations among them is *number theory*. As a field of study, number theory started to flourish in the seventeenth century with the work of Pierre de Fermat (1605–1665). Topics in number theory that occur in the elementary school curriculum include factors, multiples, divisibility tests, prime numbers, prime factorizations, greatest common divisors, and least common multiples. The topic of congruences, introduced by Karl Gauss (1777–1855), is also incorporated into the elementary curriculum through clock arithmetic and modular arithmetic. This topic of congruences gives students a look at another mathematical system.

HISTORICAL NOTE

The Hindu mathematician Bhramagupta (ca. 598 A.D.–665 A.D.) provided the first systematic treatment of negative numbers and of zero. The European mathematicians, not being aware of Bhramagupta's work for almost 1000 years after his death, shunned negative numbers. The Italian mathematician Gerolamo Cardano (1501–1576) was the first among European mathematicians to consider negative solutions to certain equations, but, still being uncomfortable with the concept of negative numbers, he called them "fictitious" numbers.

Section 4-1 — Integers and the Operations of Addition and Subtraction

Representations of Integers

In the set of integers, the symbol "$-$" unfortunately is used to indicate both a subtraction and a negative sign. To reduce confusion between the uses of this symbol in this text, a raised "$^-$" sign is used for negative numbers, as in $^-2$, in contrast to the lower sign for subtraction. To emphasize that an integer is positive, some people use a raised plus sign, as in $^+3$. In this text, we use the plus sign for addition only and write $^+3$ simply as 3.

opposites   The negative integers are **opposites** of the positive integers. For example, the opposite of 5 is $^-5$. Similarly, the positive integers are the opposites of the negative integers. Because the opposite of 4 is denoted by $^-4$, the opposite of $^-4$ can be denoted by $^-(^-4)$, or 4. The opposite of 0 is 0.

In the set of integers I, every element has an opposite that is also in I.

Example 4-1   For each of the following, find the opposite of x:

a. $x = 3$.   b. $x = ^-5$.   c. $x = 0$.

Solution   a. $^-x = ^-3$.   b. $^-x = ^-(^-5) = 5$.   c. $^-x = ^-0 = 0$.

The value of ^-x in Example 4-1(b) is 5. Thus, the *term ^-x does not necessarily represent a negative integer.* In other words, x is a variable that can be replaced by some number either positive, zero, or negative.

HISTORICAL NOTE

The dash has not always been used for both the subtraction operation and the negative sign. Other notations were developed but never adopted. One such notation was used by Mohammed al-Khowârizmî (ca. 825), who indicated a negative number by placing a small circle over it. For example, $^-4$ was recorded as $\overset{\circ}{4}$. The Hindus denoted a negative number by enclosing it in a circle; for example, $^-4$ was recorded as ④. The symbols + and − first appeared in print in European mathematics in the late fifteenth century. The symbols referred not to addition or subtraction or positive or negative numbers, but to surpluses and deficits in business problems.

Integer Addition

There are many ways to introduce operations on integers. Before formally defining the operations, we consider a more informal approach.

Chip Model

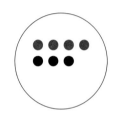

Figure 4-2

In the chip model, positive integers are represented by black chips and negative integers by red chips. One red chip neutralizes one black chip. Hence, the integer $^-1$ can be represented by 1 red chip, or 2 red and 1 black, or 3 red and 2 black, and so on. Similarly, every integer can be represented in many ways using chips. Figure 4-2 shows a chip model for the addition $^-4 + 3$. We put four red chips together with 3 black chips. Because 3 red chips neutralize 3 black ones, Figure 4-2 represents the equivalent of 1 red chip, or $^-1$.

Charged-Field Model

A model similar to the chip model uses positive and negative charges. A field has 0 charge if it has the same number of positive (+) and negative (−) charges. As in the chip model, a given integer can be represented in many ways using the charged-field model. Figure 4-3 uses the model for $3 + {^-5}$. Because 3 positive charges "neutralize" 3 negative charges, the net result is 2 negative ones. Hence, $3 + {^-5} = {^-2}$.

Figure 4-3

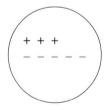

Patterns

Addition of whole numbers was established in Chapter 2. Addition of integers can also be motivated by using patterns of addition of whole numbers. Notice that in the left-hand column, the first four facts are known from whole-number addition. Also notice that the 4 stays fixed and as the numbers added to 4 decrease by 1, the sum decreases by 1. Following this pattern, $4 + {^-1} = 3$ and we can complete the remainder of the first column. Similar reasoning can be used to complete the computations in the right-hand column, where $^-2$ stays fixed and the other numbers decrease by 1 each time.

$$4 + 3 = 7 \qquad {}^-2 + 4 = 2$$
$$4 + 2 = 6 \qquad {}^-2 + 3 = 1$$
$$4 + 1 = 5 \qquad {}^-2 + 2 = 0$$
$$4 + 0 = 4 \qquad {}^-2 + 1 = {}^-1$$
$$4 + {}^-1 = 3 \qquad {}^-2 + 0 = {}^-2$$
$$4 + {}^-2 = 2 \qquad {}^-2 + {}^-1 = {}^-3$$
$$4 + {}^-3 = 1 \qquad {}^-2 + {}^-2 = {}^-4$$
$$4 + {}^-4 = 0 \qquad {}^-2 + {}^-3 = {}^-5$$
$$4 + {}^-5 = {}^-1 \qquad {}^-2 + {}^-4 = {}^-6$$
$$4 + {}^-6 = {}^-2 \qquad {}^-2 + {}^-5 = {}^-7$$

TECHNOLOGY CORNER

On a spreadsheet, in column A enter 4 and fill down 20 rows. In column B, enter 3 as the first entry and then write a formula to add $^-1$ to 3 for the second entry, add $^-1$ to the second entry to get the third entry, and fill down continuing the pattern. In column C, find the sum of the respective entries in columns A and B. What patterns do you observe? Repeat the problem by changing the entries in column A to $^-4$ and repeating the process.

Number-Line Model

Another model for addition of integers involves a number line used with a moving object. For example, the car in Figure 4-4(a) starts at 0, facing in a positive direction (to the right). To represent a positive integer, the car moves forward, and to represent a negative integer, it moves in reverse. For example, Figure 4-4(a) through (d) illustrates four different additions.

Figure 4-4

5 + 3 is seen as moving the car forward 5 units and then 3 more units forward for a net move of 8 units to the right from 0. Thus 5 + 3 = 8.

(a)
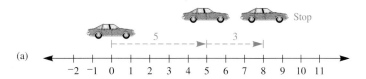

$^-5$ + 3 is seen as moving the car 5 units in reverse and then moving it forward 3 units for a net move of 2 units to the left from 0. Thus $^-5$ + 3 = $^-2$.

(b)
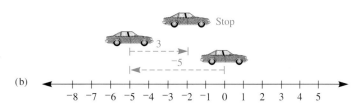

$3 + {}^-5$ is seen as moving the car 3 units forward and then 5 units in reverse for a net move of 2 units to the left from 0. Thus $3 + {}^-5 = {}^-2$.

(c)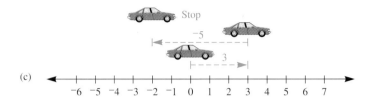

${}^-3 + {}^-5$ is seen as moving the car 3 units in reverse and then 5 more units in reverse for a net move of 8 units to the left from 0. Thus ${}^-3 + {}^-5 = {}^-8$.

(d)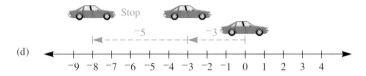

Without the car, ${}^-3 + {}^-5$ can be pictured as in Figure 4-5.

Figure 4-5

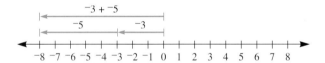

Figure 4-6 similarly depicts integer addition of $3 + {}^-5$.

Figure 4-6

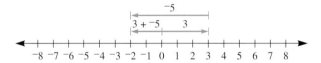

Example 4-2 involves a thermometer with a scale in the form of a vertical number line.

Example 4-2 The temperature was ${}^-4°C$. In an hour, it rose $10°C$. What is the new temperature?

Solution Figure 4-7 shows that the new temperature is $6°C$ and that ${}^-4 + 10 = 6$.

Absolute Value

Because 4 and ${}^-4$ are opposites of each other, they are on opposite sides of 0 on the number line and are the same distance (4 units) from 0, as shown in Figure 4-8.

Figure 4-8

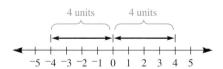

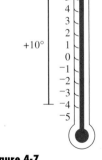

Figure 4-7

absolute value

Distance is always a positive number or zero. The distance between the points corresponding to an integer and 0 is called the **absolute value** of the integer. Thus the absolute value of both 4 and $^-4$ is 4, written as $|4| = 4$ and $|^-4| = 4$, respectively. (A more formal definition of absolute value as a function is given in Ongoing Assessment 4-1.)

Example 4-3

Evaluate each of the following:

a. $|20|$ b. $|^-5|$ c. $|0|$ d. $^-|^-3|$ e. $|2 - 5|$

Solution   a. $|20| = 20$ b. $|^-5| = 5$ c. $|0| = 0$ d. $^-|^-3| = ^-3$
e. $|2 - 5| = |^-3| = 3$

We can describe addition of integers as the process of finding the difference or the sum of the absolute values of these integers and attaching an appropriate sign.

Properties of Integer Addition

Integer addition has all the properties of whole-number addition. These properties are summarized next.

Properties

Given integers a, b, and c:
Closure property of addition of integers $a + b$ is a unique integer.
Commutative property of addition of integers $a + b = b + a$.
Associative property of addition of integers $(a + b) + c = a + (b + c)$.
Identity element of addition of integers   0 is the unique integer such that, for all integers a, $0 + a = a = a + 0$.

---

additive inverse

We have seen that every integer has an opposite. This opposite is also the **additive inverse** of the integer. The fact that each integer has a unique (one and only one) additive inverse is recorded below.

Uniqueness Property of Additive Inverse

For every integer a, there exists a unique integer ^-a, the additive inverse of a, such that $a + {^-a} = 0 = {^-a} + a$.

Observe that the opposite, or the additive inverse, of ^-a can be written as $^-(^-a)$, or a. Because the additive inverse of ^-a must be unique, we have $^-(^-a) = a$. Other properties of addition of integers can be investigated by considering previously developed notions. For example, we saw that $^-2 + {^-4} = {^-6}$, and we know that $^-6$ is the additive inverse of 6, or $2 + 4$. This leads us to the following:

$$^-2 + {^-4} = {^-(2 + 4)}.$$

This relationship is true in general and is stated next.

> **Properties of Additive Inverse**
>
> For any integers, a and b:
>
> 1. $^-(^-a) = a$.
> 2. $^-a + {^-b} = {^-(a+b)}$.

Example 4-4   Find the additive inverse of each of the following:

a. $^-(3 + x)$ b. $(a + {^-4})$ c. $^-3 + ({^-x})$

Solution   a. $3 + x$
b. $^-(a + {^-4})$, which by the above listed property (2), applied from right to left, can be written as $^-(a) + {^-({^-4})}$, or $^-a + 4$.
c. $^-[{^-3} + ({^-x})]$, which can be written as $^-({^-3}) + {^-({^-x})}$, or $3 + x$.

Integer Subtraction

As with integer addition, we explore several models for integer subtraction.

Chip Model

To find $3 - {^-2}$, we want to subtract $^-2$ (or remove 2 red chips) from 3 black chips. We need to represent 3 so that at least 2 red chips are present. In Figure 4-9, 3 is represented using 2 red and 5 black chips. When the 2 red chips are removed, 5 black ones are left and, hence, $3 - {^-2} = 5$.

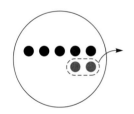

Figure 4-9

Charged-Field Model

Integer subtraction can be modeled with a charged field. For example, consider $^-3 - {^-5}$. To subtract $^-5$ from $^-3$, we must represent $^-3$ so that at least 5 negative charges are present. An example is shown in Figure 4-10(a). To subtract $^-5$, remove the 5 negative charges, leaving 2 positive charges, as in Figure 4-10(b). Hence, $^-3 - {^-5} = 2$.

Figure 4-10

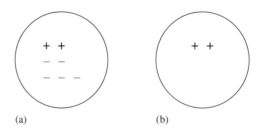

(a)   (b)

Patterns Model

We can find the difference of two integers by considering the following patterns, where we start with subtractions that we already know how to do. Both the following pattern on the left and the pattern on the right start with $3 - 2 = 1$.

$$3 - 2 = 1 \qquad 3 - 2 = 1$$
$$3 - 3 = 0 \qquad 3 - 1 = 2$$
$$3 - 4 = ? \qquad 3 - 0 = 3$$
$$3 - 5 = ? \qquad 3 - {}^-1 = ?$$

In the pattern on the left, the difference decreases by 1. If we continue the pattern, we have $3 - 4 = {}^-1$ and $3 - 5 = {}^-2$. In the pattern on the right, the difference increases by 1. If we continue the pattern, we have $3 - {}^-1 = 4$ and $3 - {}^-2 = 5$.

Number-Line Model

The number-line model used for integer addition can also be used to model integer subtraction. In Figure 4-11, the car starts at 0 and is pointed in a positive direction (to the right). In this model, the operation of subtraction corresponds to facing the car in a negative direction. We subtract a positive integer by moving the car forward and a negative integer by moving the car in reverse. Figure 4-11 shows some examples.

Figure 4-11

5 – 3 first tells you to move the car forward 5 units. The subtraction sign tells you to face the car in the negative direction. Finally, move forward 3 units for a net move of 2 units to the right from 0. Thus 5 – 3 = 2.

(a)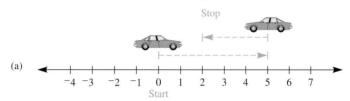

$5 - {}^-3$ first tells you to move the car forward 5 units. Then face it in a negative direction and move it in reverse 3 units. The net move is 8 units to the right of 0. Thus $5 - {}^-3 = 8$.

(b)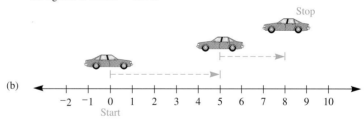

In ${}^-5 - {}^-3$, move the car 5 units in reverse. Then face it in a negative direction and move it in reverse 3 units. The net move is 2 units to the left of 0. Thus ${}^-5 - {}^-3 = {}^-2$.

(c)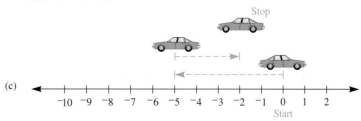

Subtraction as the Inverse of Addition

Subtraction of integers, like subtraction of whole numbers, can be defined in terms of addition. Recall that $5 - 3$ can be computed by finding a whole number n as follows:

$$5 - 3 = n \quad \text{if, and only if,} \quad 5 = 3 + n.$$

Because $3 + 2 = 5$, then $n = 2$.

Similarly, we compute $3 - 5$ as follows:

$$3 - 5 = n \quad \text{if, and only if,} \quad 3 = 5 + n.$$

Because $5 + {}^-2 = 3$, then $n = {}^-2$. In general, for integers a and b, we have the following definition of *subtraction*.

Definition of Subtraction

For integers a and b, $a - b$ is the unique integer n such that $a = b + n$.

From our previous work with addition of integers, we know that $3 - 5 = {}^-2$ and $3 + {}^-5 = {}^-2$. Hence, $3 - 5 = 3 + {}^-5$. In general, the following is true.

Property

For all integers a and b, $a - b = a + ({}^-b)$.

NOW TRY THIS 4-1

- **a.** Is the set of integers closed under subtraction? Why?
- **b.** Do the commutative, associative, or identity properties hold for subtraction of integers? Why or why not?

Example 4-5 Use the definition of *subtraction* to compute the following:

a. $3 - 10$ **b.** ${}^-2 - 10$

Solution
a. Let $3 - 10 = n$. Then $10 + n = 3$, so $n = {}^-7$. Therefore $3 - 10 = {}^-7$.
b. Let ${}^-2 - 10 = n$. Then $10 + n = {}^-2$, so $n = {}^-12$. Therefore ${}^-2 - 10 = {}^-12$.

Many early calculators had a change-of-sign key, either $\boxed{\text{CHS}}$ or $\boxed{+/-}$. Later generations of calculators use $\boxed{(-)}$, a key that allows computation with integers. For example, to compute $8 - ({}^-3)$, we would press $\boxed{8}\boxed{-}\boxed{3}\boxed{+/-}\boxed{=}$. Investigate what happens if you press $\boxed{8}\boxed{-}\boxed{-}\boxed{3}\boxed{=}$.

Example 4-6 Using the fact that $a - b = a + ({}^-b)$, compute each of the following:

a. $2 - 8$ b. $2 - ({}^-8)$ c. ${}^-12 - ({}^-5)$ d. ${}^-12 - 5$

Solution
a. $2 - 8 = 2 + {}^-8 = {}^-6$
b. $2 - ({}^-8) = 2 + {}^-({}^-8) = 2 + 8 = 10$
c. ${}^-12 - ({}^-5) = {}^-12 + {}^-({}^-5) = {}^-12 + 5 = {}^-7$
d. ${}^-12 - 5 = {}^-12 + {}^-5 = {}^-17$

Example 4-7 Use the fact that $a - b = a + ({}^-b)$ and the properties of additive inverse to write expressions equal to each of the following without using parentheses:

a. ${}^-(b - c)$ b. $a - (b + c)$

Solution
a. ${}^-(b - c) = {}^-(b + {}^-c) = {}^-b + {}^-({}^-c) = {}^-b + c$
b. $a - (b + c) = a + {}^-(b + c) = a + ({}^-b + {}^-c) = (a + {}^-b) + {}^-c = a + {}^-b + {}^-c$

REMARK It is possible to simplify the answers in parts (a) and (b) further, as follows: ${}^-b + c = c + {}^-b = c - b$, and $a + {}^-b + {}^-c = (a - b) - c$.

Example 4-8 Simplify each of the following:

a. $2 - (5 - x)$ b. $5 - (x - 3)$ c. ${}^-(x - y) - y$

Solution
a. $2 - (5 - x) = 2 + {}^-(5 + {}^-x)$
$= 2 + {}^-5 + {}^-({}^-x)$
$= 2 + {}^-5 + x$
$= {}^-3 + x$

b. $5 - (x - 3) = 5 + {}^-(x + {}^-3)$
$= 5 + {}^-x + {}^-({}^-3)$
$= 5 + {}^-x + 3$
$= 8 + {}^-x$
$= 8 - x$

c. ${}^-(x - y) - y = {}^-(x + {}^-y) + {}^-y$
$= [{}^-x + {}^-({}^-y)] + {}^-y$
$= ({}^-x + y) + {}^-y$
$= {}^-x + (y + {}^-y)$
$= {}^-x + 0$
$= {}^-x$

Order of Operations

Subtraction on the set of integers is neither commutative nor associative, as illustrated in these counterexamples:

$$5 - 3 \neq 3 - 5 \quad \text{because} \quad 2 \neq {}^-2$$
$$(3 - 15) - 8 \neq 3 - (15 - 8) \quad \text{because} \quad {}^-20 \neq {}^-4$$

Remember, computations within parentheses must be completed before other computations.

An expression such as $3 - 15 - 8$ is ambiguous unless we know in which order to perform the subtractions. Mathematicians agree that $3 - 15 - 8$ means $(3 - 15) - 8$; that is, the subtractions in $3 - 15 - 8$ are performed in order from left to right. Similarly, $3 - 4 + 5$ means $(3 - 4) + 5$ and not $3 - (4 + 5)$. Thus $(a - b) - c$ may be written without parentheses as $a - b - c$.

Example 4-9 Compute each of the following:

a. $2 - 5 - 5$ b. $3 - 7 + 3$ c. $3 - (7 - 3)$

Solution
a. $2 - 5 - 5 = {}^-3 - 5 = {}^-8$
b. $3 - 7 + 3 = {}^-4 + 3 = {}^-1$
c. $3 - (7 - 3) = 3 - 4 = {}^-1$

TECHNOLOGY CORNER

a. On a graphing calculator, graph the function with equation $y = x - {}^-4$.
b. Using the graph in (a), describe what happens as x takes on values that are less than ${}^-4$, equal to ${}^-4$, and greater than ${}^-4$.

ONGOING ASSESSMENT 4-1

1. Find the additive inverse of each of the following integers. Write your answer in the simplest possible form.
   a. 2 b. ${}^-5$ c. m
   d. 0 e. ${}^-m$ f. $a + b$
2. Simplify each of the following:
   a. ${}^-({}^-2)$ b. ${}^-({}^-m)$ c. ${}^-0$
3. Evaluate each of the following:
   a. $|{}^-5|$ b. $|10|$
   c. ${}^-|{}^-5|$ d. ${}^-|5|$
4. Demonstrate each of the following additions using the charged-field or chip model:
   a. $5 + {}^-3$ b. ${}^-2 + 3$ c. ${}^-3 + 2$
   d. ${}^-3 + {}^-2$
5. Compute each of the following:
   a. $3 - {}^-2$ b. ${}^-3 - 2$
   c. ${}^-3 - {}^-2$

6. Demonstrate each of the additions in problem 4 above using a number-line model.
7. Write an addition fact corresponding to each of the following sentences and then answer the question:
   a. A certain stock dropped 17 points and the following day gained 10 points. What was the net change in the stock's worth?
   b. The temperature was ${}^-10°C$ and then it rose by $8°C$. What is the new temperature?
   c. The plane was at 5000 ft and dropped 100 ft. What is the new altitude of the plane?
   d. A visitor in a Las Vegas casino lost $200, won $100, and then lost $50. What is the change in the gambler's net worth?
   e. In four downs, the football team lost 2 yd, gained 7 yd, gained 0 yd, and lost 8 yd. What is the total gain or loss?

8. On January 1, Jane's bank balance was $300. During the month, she wrote checks for $45, $55, $165, $35, and $100 and made deposits of $75, $25, and $400.
   a. If a check is represented by a negative integer and a deposit by a positive integer, express Jane's transactions as a sum of positive and negative integers.
   b. What was the balance in Jane's account at the end of the month?
9. Use a number-line model to find the following:
   a. $^-4 - ^-1$ b. $^-4 - ^-3$
10. Use patterns to show the following:
    a. $^-4 - ^-1 = ^-3$ b. $^-2 - 1 = ^-3$
11. Perform each of the following:
    a. $^-2 + (3 - 10)$ b. $[8 - (^-5)] - 10$
    c. $(^-2 - 7) + 10$ d. $^-2 - (7 + 10)$
    e. $8 - 11 - 10$ f. $^-2 - 7 + 3$
12. In each of the following, write a subtraction problem that corresponds to the question and an addition problem that corresponds to the question and then answer the question:
    a. The temperature is 55°F and is supposed to drop 60°F by midnight. What is the expected midnight temperature?
    b. Moses has overdraft privileges at his bank. If he had $200 in his checking account and he wrote a $220 check, what is his balance?
13. Answer each of the following:
    a. In a game of Triominoes, Jack's scores in five successive turns are 17, $^-8$, $^-9$, 14, and 45. What is his total at the end of five turns?
    b. The largest bubble chamber in the world is 15 ft in diameter and contains 7259 gal of liquid hydrogen at a temperature of $^-247°C$. If the temperature is dropped by 11°C per hour for 2 consecutive hours, what is the new temperature?
    c. The greatest recorded temperature ranges in the world are around the "cold pole" in Siberia. Temperatures in Verkhoyansk have varied from $^-94°F$ to 98°F. What is the difference between the high and low temperatures in Verkhoyansk?
14. Motor oils protect car engines over a range of temperatures. These oils have names like 10W–40 or 5W–30. The following graph shows the temperatures, in degrees Fahrenheit, at which the engine is protected by a particular oil. Using the graph, find which oils can be used for the following temperatures:
    a. Between $^-5°$ and 90°   b. Below $^-20°$
    c. Between $^-10°$ and 50°   d. From $^-20°$ to over 100°
    e. From $^-8°$ to 90°

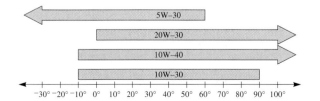

15. Apply the properties discussed in this section to simplify each of the following as much as possible. Show all work, as in Example 4-8.
    a. $3 - (2 - 4x)$ b. $x - (^-x - y)$
    c. $4x - 2 - 3x$ d. $4x - (2 - 3x)$
16. For which integers a, b, c does $a - b - c = a - (b - c)$? Justify your answer.
17. Find all integers x, if there are any, such that the following are true:
    a. ^-x is positive.
    b. ^-x is negative.
    c. $^-x - 1$ is positive.
    d. $|x| = 2$.
    e. $^-|x| = 2$.
    f. $^-|x|$ is negative.
    g. $^-|x|$ is positive.
18. Let W stand for the set of whole numbers, I the set of integers, I^+ the set of positive integers, and I^- the set of negative integers. Find each of the following:
    a. $W \cup I$ b. $W \cap I$ c. $I^+ \cup I^-$
    d. $I^+ \cap I^-$ e. $W - I$ f. $I - W$
    g. $W - I^+$ h. $W - I^-$ i. $I \cap I$
19. Complete the magic square using the following integers: $^-13, ^-10, ^-7, ^-4, 2, 5, 8, 11$.

20. Let $f(x) = ^-x - 1$ with domain I. Find the following:
    a. $f(^-1)$ b. $f(100)$ c. $f(^-2)$
    d. For which values of x will the output be 3?
21. Let $f(x) = |1 - x|$. Find the following:
    a. $f(10)$ b. $f(^-1)$
    c. All the inputs for which the output is 1
    d. The range
22. The following is frequently given as the definition for the absolute value of a function, where the domain is the set of integers:

    If x is a positive integer or 0, then $|x| = x$.
    If x is a negative integer, then $|x| = ^-x$.

    a. What is the range of this function?
    b. Use the definition to evaluate each of the following:
       i. $|5|$ ii. $|^-5|$
       iii. $|0|$ iv. $^-|^-7|$
    c. The absolute value of an integer is never negative. Does this contradict the fact that the absolute value of x could be equal to ^-x? Explain why or why not.
23. Determine how many integers there are between the following given integers (not including the given integers):
    a. 10 and 100   b. $^-30$ and $^-10$
    c. $^-10$ and 10   d. x and y (if $x < y$)

24. Suppose $a = 6$, $b = 5$, $c = 4$, and $d = {}^-3$. Insert parentheses in the expression $a - b - c - d$ to obtain the greatest possible and the least possible values. What are these values?

25. An arithmetic sequence may have a positive or negative difference. In each of the following arithmetic sequences, find the difference and write the next two terms:
    a. $0, {}^-3, {}^-6, {}^-9$
    b. $7, 3, {}^-1, {}^-5$
    c. $x + y, x, x - y$
    d. $1 - 3x, 1 - x, 1 + x$

26. Find the sums of the following arithmetic sequences:
    a. ${}^-20 + {}^-19 + {}^-18 + \ldots + 18 + 19 + 20$
    b. $100 + 99 + 98 + \ldots + {}^-50$
    c. $100 + 98 + 96 + \ldots + {}^-6$

27. In an arithmetic sequence, the eighth term minus the first term equals 21. The sum of the first and the eighth term is ${}^-5$. Find the fifth term of the sequence.

28. Classify each of the following as true or false. If false, give a counterexample.
    a. $|{}^-x| = |x|$
    b. $|x - y| = |y - x|$
    c. $|{}^-x + {}^-y| = |x + y|$
    d. $|x^2| = x^2$
    e. $|x^3| = x^3$
    f. $|x^3| = x^2|x|$

29. Assume gear A has 56 teeth and gear B has 14 teeth. If gear A rotates 7 times per minute, how many times does gear B rotate and in what direction in relation to gear A? Explain your reasoning.

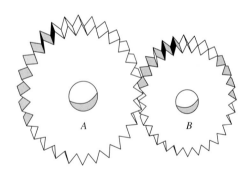

30. Complete each of the following integer arithmetic problems on your calculator, making use of the $\boxed{+/-}$ key (or the $\boxed{(-)}$ key, if you have a later generation calculator). For example, to find ${}^-5 + {}^-4$, press $\boxed{5}\,\boxed{+/-}\,\boxed{+}\,\boxed{4}\,\boxed{+/-}\,\boxed{=}$.
    a. ${}^-12 + {}^-6$
    b. ${}^-7 + {}^-99$
    c. ${}^-12 + 6$
    d. $27 + {}^-5$
    e. $3 + {}^-14$
    f. ${}^-7 - {}^-9$
    g. ${}^-12 - 6$
    h. $16 - {}^-7$

31. Estimate each of the following and then use a calculator to find the actual answer:
    a. $343 + {}^-42 - 402$
    b. ${}^-1992 + 3005 - 497$
    c. $992 - {}^-10003 - 101$
    d. ${}^-301 - {}^-1303 + 4993$

Communication

32. Which model(s) would you use to teach addition and subtraction of integers? Explain why you chose the model you did.

33. A turnpike driver had car trouble. He knew that he had driven 12 mi from milepost 68 before the trouble started. Assuming he is confused and disoriented when he calls on his cellular phone for help, how can he determine his possible location? Explain.

34. a. Show that when $a + b$ is added to ${}^-a + {}^-b$, the result is 0.
    b. Use part (a) and the definition of additive inverse to explain why ${}^-a + {}^-b = {}^-(a + b)$.

35. Dolores claims that the best way to understand that $a - b = a + ({}^-b)$, for all integers a and b, is to show that when you add b to each expression you get the same answer.
    a. Explain why you think Dolores is making this claim.
    b. Do you agree with Dolores that her approach is the "best way"? If not, what is a better approach?

36. Addition of integers with like signs can be described using absolute values as follows:
    To add integers with like signs, add the absolute values of the integers. The sum has the same sign as the integers.
    Describe in a similar way how to add integers with unlike signs.

37. Explain why $b - a$ and $a - b$ are opposites of each other.

Open-Ended

38. Describe a realistic word problem that models ${}^-50 + ({}^-85) - ({}^-30)$.

39. In the library at a well-known university, some floors are below ground level while others are above ground level. If the ground-level floor is designated the zero floor, design a system for using integers to number the floors and then design an operation system for the elevator to model addition and subtraction with integers.

40. Select a current middle-school text that introduces addition and subtraction of integers and discuss which models were used and how effective you think they would be with a group of students.

Cooperative Learning

41. Examine several elementary mathematics textbooks. Report on how addition and subtraction of integers is treated, and on how various properties are justified. Discuss in your group how the treatment of addition and subtraction of integers presented in this section compares to the treatment in elementary textbooks.

BRAIN TEASER

If the digits 1 through 9 are written in order, it is possible to place plus and minus signs between the numbers or to use no operation symbol at all to obtain a total of 100. For example,

$$1 + 2 + 3 + {}^-4 + 5 + 6 + 78 + 9 = 100.$$

Can you obtain a total of 100 using fewer plus or minus signs than in the given example? Note that digits, such as 7 and 8, may be combined.

TECHNOLOGY CORNER

Edit the following Logo program and type it into your computer.

```
TO ABS :X
    IF :X < 0 THEN OUTPUT (-:X)
    OUTPUT :X
END
```

Run this program and input the following values:

a. $^-7$ b. 0 c. 140 d. $^-21$

Section 4-2 — Multiplication and Division of Integers

We approach multiplication of integers through a variety of models: *patterns, charged-field, chip,* and *number-line*.

Patterns Model

First, we approach multiplication of integers by using repeated addition. For example, if a running back lost 2 yd on each of three carries in a football game, then he had a net loss of $^-2 + {}^-2 + {}^-2$, or $^-6$, yards. Since $^-2 + {}^-2 + {}^-2$ can be written as $3 \cdot ({}^-2)$, using repeated addition, we have $3 \cdot ({}^-2) = {}^-6$.

Consider $({}^-2) \cdot 3$. It is meaningless to say that there are $^-2$ threes in a sum.

But if the commutative property of multiplication is to hold for all integers, we must have $({}^-2) \cdot 3 = 3 \cdot ({}^-2) = {}^-6$.

Next, consider $({}^-3) \cdot ({}^-2)$. We can develop the following pattern:

$$3 \cdot ({}^-2) = {}^-6$$
$$2 \cdot ({}^-2) = {}^-4$$
$$1 \cdot ({}^-2) = {}^-2$$
$$0 \cdot ({}^-2) = 0$$
$$({}^-1) \cdot ({}^-2) = ?$$
$$({}^-2) \cdot ({}^-2) = ?$$
$$({}^-3) \cdot ({}^-2) = ?$$

The first four products, ⁻6, ⁻4, ⁻2, and 0, are terms in an arithmetic sequence with fixed difference 2. If the pattern continues, the next 3 terms in the sequence are 2, 4, and 6. Thus it appears that (⁻3) · (⁻2) = 6. Likewise, (⁻2) · (⁻3) = 6.

Charged-Field Model and Chip Model

The *charged-field model* and *chip model* can be used to illustrate multiplication of integers, although an interpretation must be given to the signs. Consider Figure 4-12, where 3 · (⁻2) is pictured using a chip model.

Figure 4-12

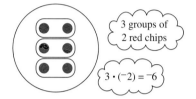

To find (⁻3) · (⁻2) using the charged-field model, we interpret the signs as follows: ⁻3 is taken to mean "*remove 3 groups of*"; ⁻2 is taken to mean "*2 negative charges.*" To do this, we first start with a 0 charged field that includes at least 6 negative charges, as shown in Figure 4-13.

Figure 4-13

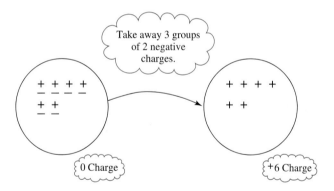

The result is a charge of positive 6, so (⁻3) · (⁻2) = 6.

TECHNOLOGY CORNER

On a spreadsheet, in column A enter 5 as the first entry and then write a formula to add ⁻1 to 5 for the second entry. Then add ⁻1 to the second entry and fill down continuing the pattern. In column B, repeat the process. In column C, find the product of the respective entries in columns A and B. What patterns do you observe?

Number-Line Model

As with addition and subtraction, we demonstrate multiplication by using a car moving along a number line, according to the following rules:

1. Traveling to the left (west) means moving in the negative direction, and traveling to the right (east) means moving in the positive direction.
2. Time in the future is denoted by a positive value, and time in the past is denoted by a negative value.

Consider the number line shown in Figure 4-14. Various cases using this number line are given next.

Figure 4-14

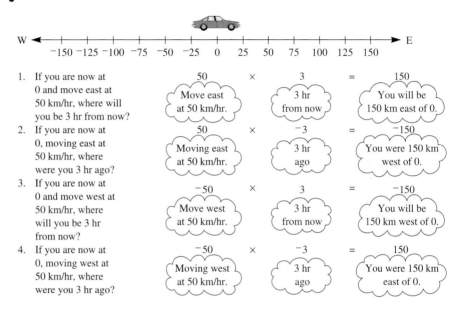

1. If you are now at 0 and move east at 50 km/hr, where will you be 3 hr from now?   50 × 3 = 150   (Move east at 50 km/hr.) (3 hr from now) (You will be 150 km east of 0.)
2. If you are now at 0, moving east at 50 km/hr, where were you 3 hr ago?   50 × ⁻3 = ⁻150   (Moving east at 50 km/hr.) (3 hr ago) (You were 150 km west of 0.)
3. If you are now at 0 and move west at 50 km/hr, where will you be 3 hr from now?   ⁻50 × 3 = ⁻150   (Move west at 50 km/hr.) (3 hr from now) (You will be 150 km west of 0.)
4. If you are now at 0, moving west at 50 km/hr, where were you 3 hr ago?   ⁻50 × ⁻3 = 150   (Moving west at 50 km/hr.) (3 hr ago) (You were 150 km east of 0.)

These models illustrate the following definition of *multiplication of integers*.

Definition of Multiplication of Integers

For any **whole numbers** *a* and *b*, the following holds:

1. $(^-a)(^-b) = ab$.
2. $(^-a)b = b(^-a) = {}^-(ab)$.

REMARK We will show later in this section that the above properties are true for all integers *a* and *b*.

Properties of Integer Multiplication

The set of integers has properties under multiplication analogous to those of the set of whole numbers under multiplication. These properties are summarized next.

Properties of Integer Multiplication

The set of integers I satisfies the following properties of multiplication for all integers a, b, $c \in I$:

Closure property of multiplication of integers ab is a unique integer.
Commutative property of multiplication of integers $ab = ba$.
Associative property of multiplication of integers $(ab)c = a(bc)$.
Multiplicative identity property 1 is the unique integer such that for all integers a, $1 \cdot a = a = a \cdot 1$.
Distributive properties of multiplication over addition for integers $a(b + c) = ab + ac$ and $(b + c)a = ba + ca$.
Zero multiplication property of integers 0 is the unique integer such that for all integers a, $a \cdot 0 = 0 = 0 \cdot a$.

HISTORICAL NOTE

Emmy Noether (1882–1935) is one of the outstanding mathematicians of the twentieth century. She made lasting contributions to the study of *rings,* algebraic systems among which is the set of integers. When she entered the University of Erlanger (Germany) in 1900, Emmy Noether was one of only two women enrolled. After completing her doctorate in 1907 she could not find a suitable job despite her outstanding achievements, because she was a woman. In 1919 she got a university appointment without pay and, only later, a very modest salary. In 1933, along with many other scholars, she was dismissed from the University at Göttingen because she was Jewish. She immigrated to the United States and taught at Bryn Mawr College until her untimely death only 18 months after arriving in the United States.

Another approach to showing that $(^-2) \cdot 3 = {}^-(2 \cdot 3)$ uses the uniqueness property of additive inverses. If we can show that $(^-2) \cdot 3$ and $^-(2 \cdot 3)$ are additive inverses of the same number, then they must be equal. By definition, the additive inverse of $(2 \cdot 3)$ is $^-(2 \cdot 3)$. That $(^-2) \cdot 3$ is also the additive inverse of $2 \cdot 3$ can be proved by showing that $(^-2) \cdot 3 + 2 \cdot 3 = 0$. The proof follows.

$(^-2) \cdot 3 + 2 \cdot 3 = (^-2 + 2) \cdot 3$ Distributive property of multiplication over addition
$\qquad\qquad\qquad = 0 \cdot 3$ Additive inverse
$\qquad\qquad\qquad = 0$ Zero multiplication

Because $(^-2) \cdot 3$ and $^-(2 \cdot 3)$ are both additive inverses of $(2 \cdot 3)$ and because the additive inverse must be unique, $(^-2) \cdot 3 = {}^-(2 \cdot 3)$.

Using this approach, we could show the following property (whose justification is explored in Ongoing Assessment 4-2):

Property

For every integer a, $(^-1)a = {}^-a$.

It is important to keep in mind that $(^-1)a = {}^-a$ is true for all integers a. Thus if we substitute $^-1$ for a we get $(^-1) \cdot (^-1) = {}^-(^-1)$.

Because $^-(^-1) = 1$ we have yet another justification for the fact that $(^-1) \cdot (^-1) = 1$. Using this result, the above property, and the properties of integers listed earlier, we can show that $(^-a)b = {}^-(ab)$ and that $(^-a)(^-b) = ab$ for all integers a and b as follows:

$$(^-a)b = [(^-1)a]b$$
$$= (^-1)(ab)$$
$$= {}^-(ab)$$

Also:

$$(^-a)(^-b) = [(^-1)a] \cdot [(^-1)b]$$
$$= (^-1) \cdot (^-1)(ab)$$
$$= 1(ab)$$
$$= ab$$

We have established the following properties:

Properties

For all integers a and b,

$$(^-a)b = {}^-(ab).$$
$$(^-a)(^-b) = ab.$$

REMARK It is important to note that in these properties, ^-a and ^-b are not necessarily negative and a and b are not necessarily positive.

The distributive property of multiplication over subtraction follows from the distributive property of multiplication over addition:

$$a(b - c) = a(b + {}^-c)$$
$$= ab + a(^-c)$$
$$= ab + {}^-(ac)$$
$$= ab - ac.$$

Consequently, $a(b - c) = ab - ac$. Similarly, we can show that $(b - c)a = ba - ca$.

Property

Distributive property of multiplication over subtraction for integers For any integers a, b, and c,

$$a(b - c) = ab - ac \quad \text{and}$$
$$(b - c)a = ba - ca.$$

Example 4-10 Simplify each of the following so that there are no parentheses in the final answer:

a. $(^-3)(x - 2)$ **b.** $(a + b)(a - b)$

Solution
a. $(^-3)(x - 2) = (^-3)x - (^-3)(2) = {}^-3x - (^-6) = {}^-3x + {}^-(^-6) = {}^-3x + 6$.
b. $(a + b)(a - b) = (a + b)a - (a + b)b$
$= (a^2 + ba) - (ab + b^2)$
$= a^2 + ab - ab - b^2$
$= a^2 - b^2$
Thus $(a + b)(a - b) = a^2 - b^2$.

The result $(a + b)(a - b) = a^2 - b^2$ in Example 4-10(b) is commonly called the **difference-of-squares** formula.

difference-of-squares

Example 4-11 Use the difference-of-squares formula to simplify the following:

a. $(4 + b)(4 - b)$ **b.** $(^-4 + b)(^-4 - b)$

Solution
a. $(4 + b)(4 - b) = 4^2 - b^2 = 16 - b^2$
b. $(^-4 + b)(^-4 - b) = (^-4)^2 - b^2 = 16 - b^2$

NOW TRY THIS 4-2

- Determine how to use the difference-of-squares formula to compute the following mentally: **a.** $22 \cdot 18$ **b.** $24 \cdot 36$ **c.** $998 \cdot 1002$

When the distributive property of multiplication over subtraction is written in reverse order as

$$ab - ac = a(b - c) \quad \text{and}$$
$$ba - ca = (b - c)a,$$

and similarly for addition, the expressions on the right of each equation are in *factored* form. We say that the common factor a, in ab and ac, has been *factored out*.

Both the difference-of-squares formula and the distributive properties of multiplication over addition and subtraction can be used for factoring.

Example 4-12 Factor each of the following completely:

a. $x^2 - 9$ **b.** $(x + y)^2 - z^2$ **c.** $^-3x + 5xy$ **d.** $3x - 6$

Solution
a. $x^2 - 9 = x^2 - 3^2 = (x + 3)(x - 3)$
b. $(x + y)^2 - z^2 = (x + y + z)(x + y - z)$
c. $^-3x + 5xy = x(^-3 + 5y)$
d. $3x - 6 = 3(x - 2)$

Integer Division

In the set of whole numbers, $a \div b$, where $b \neq 0$, is the unique whole number c such that $a = bc$. If such a whole number c does not exist, then $a \div b$ is undefined. Division on the set of integers is defined analogously.

> **Definition of Integer Division**
>
> If a and b are any integers, with $b \neq 0$, then $a \div b$ is the unique integer c, if it exists, such that $a = bc$.

Example 4-13 Use the definition of integer division, if possible, to evaluate each of the following:

a. $12 \div (^-4)$ b. $^-12 \div 4$ c. $^-12 \div (^-4)$ d. $^-12 \div 5$

Solution
a. Let $12 \div (^-4) = c$. Then, $12 = ^-4c$, and consequently, $c = ^-3$. Thus $12 \div (^-4) = ^-3$.
b. Let $^-12 \div 4 = c$. Then, $^-12 = 4c$, and therefore $c = ^-3$. Thus $^-12 \div 4 = ^-3$.
c. Let $^-12 \div (^-4) = c$. Then, $^-12 = ^-4c$, and consequently, $c = 3$. Thus $^-12 \div (^-4) = 3$.
d. Let $^-12 \div 5 = c$. Then, $^-12 = 5c$. Because no integer c exists to satisfy this equation (why?), we say that $^-12 \div 5$ is undefined over the integers.

Example 4-13 suggests that *the quotient of two negative integers, if it exists, is a positive integer and the quotient of a positive and a negative integer, if it exists, or of a negative and a positive integer, if it exists, is negative.*

Order of Operations on Integers

The following rules apply to the order in which arithmetic operations are performed. Recall that when addition and multiplication appear in a problem without parentheses, multiplication is done first.

When addition, subtraction, multiplication, division, and exponentiation appear without parentheses, exponentiation is done first, then multiplications and divisions in the order of their appearance from left to right, and then additions and subtractions in the order of their appearance from left to right. Arithmetic operations that appear inside parentheses must be done first.

Example 4-14 Evaluate each of the following:

a. $2 - 5 \cdot 4 + 1$
b. $(2 - 5) \cdot 4 + 1$
c. $2 - 3 \cdot 4 + 5 \cdot 2 - 1 + 5$
d. $2 + 16 \div 4 \cdot 2 + 8$
e. $(^-3)^4$
f. $^-3^4$

Solution
a. $2 - 5 \cdot 4 + 1 = 2 - 20 + 1 = {}^-18 + 1 = {}^-17$
b. $(2 - 5) \cdot 4 + 1 = {}^-3 \cdot 4 + 1 = {}^-12 + 1 = {}^-11$
c. $2 - 3 \cdot 4 + 5 \cdot 2 - 1 + 5 = 2 - 12 + 10 - 1 + 5 = 4$
d. $2 + 16 \div 4 \cdot 2 + 8 = 2 + 4 \cdot 2 + 8 = 2 + 8 + 8 = 10 + 8 = 18$
e. $({}^-3)^4 = ({}^-3)({}^-3)({}^-3)({}^-3) = 81$
f. ${}^-3^4 = {}^-(3^4) = {}^-(81) = {}^-81$

REMARK Notice that from Example 4-14(e) and (f), we have $({}^-3)^4 \neq {}^-3^4$. By convention, $({}^-3)^4$ means $({}^-3)({}^-3)({}^-3)({}^-3)$ and ${}^-3^4$ means ${}^-(3^4)$, or ${}^-(3 \cdot 3 \cdot 3 \cdot 3)$.

Ordering Integers

As with whole numbers, a number line can be used to describe greater-than and less-than relations for a set of integers. Because ${}^-5$ is to the left of ${}^-3$ on the number line in Figure 4-15, we say that "${}^-5$ is less than ${}^-3$," and we write ${}^-5 < {}^-3$. We can also say that "${}^-3$ is greater than ${}^-5$," and we can write ${}^-3 > {}^-5$.

Figure 4-15

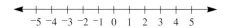

Notice that since ${}^-5$ is to the left of ${}^-3$ there is a positive integer that can be added to ${}^-5$ to get ${}^-3$, namely, 2. Thus ${}^-5 < {}^-3$ because ${}^-5 + 2 = {}^-3$. The definition of *less than* for integers is similar to that for whole numbers.

Definition of *Less Than* for Integers

For any integers a and b, a is less than b, written $a < b$, if and only if there exists a positive integer k such that $a + k = b$.

The last equation implies that $k = b - a$. Thus we have the following property.

Property

$a < b$ (or equivalently, $b > a$) if, and only if, $b - a$ is equal to a positive integer.

For example, using the above property ${}^-5 < {}^-3$ because ${}^-3 - ({}^-5) = {}^-3 + {}^-({}^-5) = {}^-3 + 5 = 2 > 0$. (Note that $a \leq b$ means that $a < b$ or $a = b$. Also $b \geq a$ if, and only if, $a \leq b$)

Example 4-15 Use the above property to justify each of the following for integers x, y, and n.

a. If $x < y$ and n is any integer, then $x + n < y + n$.
b. If $x < y$, then $^-x > {}^-y$.
c. If $x < y$ and $n > 0$, then $n \cdot x < n \cdot y$.
d. If $x < y$ and $n < 0$, then $n \cdot x > n \cdot y$.

Solution
a. Because $x < y$ we have $y - x > 0$. We need to show that $(y + n) - (x + n) > 0$. We have: $y + n - (x + n) = y + n - x - n = y - x$. Because $y - x > 0$, we have $y + n - (x + n) > 0$, and hence $x + n < y + n$.

b. Because $x < y$, $y - x > 0$. We need to show $^-x - ({}^-y) > 0$. We have $^-x - ({}^-y) = {}^-x + {}^-({}^-y) = {}^-x + y = y - x$. Because $y - x > 0$ we have $^-x - ({}^-y) > 0$, and hence $^-x > {}^-y$.

c. Because $x < y$, $y - x > 0$. We need to show that $ny - nx > 0$. We have $ny - nx = n(y - x)$. Because n is a positive integer and $y - x$ is positive, $n \cdot (y - x)$ must also be positive. Because $ny - nx > 0$, we have $nx < ny$.

d. To show that $nx > ny$ we need only show that $nx - ny > 0$. We have $nx - ny = n(x - y)$. Since $y - x > 0$, $x - y < 0$ (why?). Because $n < 0$ and $x - y < 0$, $n(x - y)$ is positive. Thus $ny - nx > 0$ and hence, $nx > ny$.

BRAIN TEASER Express each of the numbers from 1 through 10 using 4 fours and any operations. For example,

$$1 = 44 \div 44, \text{ or}$$
$$1 = (4 \div 4)^{44}, \text{ or}$$
$$1 = {}^-4 + 4 + (4 \div 4).$$

ONGOING ASSESSMENT 4-2

1. Use patterns to show that $(^-1)(^-1) = 1$.
2. Use the charged-field model to show that $(^-4)(^-2) = 8$.
3. Use the number-line model to show that $(^-4)2 = {}^-8$.
4. The number of students eating in the school cafeteria has been decreasing at the rate of 20 per year. Assuming this trend continues, write a multiplication problem that describes the change in the number of students eating in the school cafeteria for each of the following:
   a. The change over the next 4 yr
   b. The situation 4 yr ago
   c. The change over the next n years
   d. The situation n years ago

5. Use the definition of division to find each quotient, if possible. If a quotient is not defined, explain why.
   a. $^-40 \div {}^-8$ b. $143 \div (^-11)$
   c. $^-143 \div 13$ d. $0 \div (^-5)$
   e. $^-5 \div 0$ f. $0 \div 0$

6. Evaluate each of the following, if possible:
   a. $(^-10 \div {}^-2)(^-2)$ b. $(a \div b)b$
   c. $(^-10 \cdot 5) \div 5$ d. $(ab) \div b$
   e. $(^-8 \div {}^-2)(^-8)$ f. $^-8 \div (^-8 + 8)$
   g. $(^-8 + 8) \div 8$ h. $(^-6 + 6) \div (^-2 + 2)$
   i. $(^-36 \div 12) \div 3$ j. $(^-23 - {}^-7) \div 4$
   k. $|^-24| \div (3 - 15)$

7. In a lab, the temperature of various chemical reactions was changing by a fixed number of degrees per minute. Write a multiplication problem that describes each of the following:
   a. The temperature at 8:00 P.M. was 32°C. If it dropped 3°C per minute, what is the temperature at 8:30 P.M.?
   b. The temperature at 8:20 P.M. was 0°C. If it dropped 4°C per minute, what is the temperature at 7:55 P.M.?
   c. The temperature at 8:00 P.M. was $^-20$°C. If it dropped 4°C per minute, what is the temperature at 7:30 P.M.?
   d. The temperature at 8:00 P.M. was 25°C. If it increased every minute by 3°C, what is the temperature at 7:40 P.M.?
   e. The temperature at 8:00 A.M. was $^-5$°C. If it increases by d degrees per minute, what will the temperature be m minutes later?
   f. The temperature at 8:00 P.M. was 0°C. If it dropped d degrees per minute, what was the temperature m minutes before?
   g. The temperature at 8:00 P.M. was 20°C. If it increased every minute by d degrees, what was the temperature m minutes before?

8. a. On each of four consecutive plays in a football game, a team lost 11 yd. If lost yardage is interpreted as a negative integer, write the information as a product of integers and determine the total number of yards lost.
   b. If Jack Jones lost a total of 66 yd in 11 plays, how many yards, on the average, did he lose on each play?

9. In 1989, it was predicted that the farmland acreage lost to family dwellings over the next 9 years would be 12,000 acres per year. If this prediction were true and if this pattern were to continue, how much acreage would be lost to homes by the end of 2000?

10. Show that the distributive property of multiplication over addition, $a(b + c) = ab + ac$ is true for each of the following values of a, b, and c:
    a. $a = {^-1}, b = {^-5}, c = {^-2}$
    b. $a = {^-3}, b = {^-3}, c = 2$
    c. $a = {^-5}, b = 2, c = {^-6}$

11. Compute each of the following:
    a. $({^-2})^3$ b. $({^-2})^4$
    c. $({^-10})^5 \div ({^-10})^2$ d. $({^-3})^5 \div ({^-3})$
    e. $({^-1})^{10}$ f. $({^-1})^{15}$
    g. $({^-1})^{50}$ h. $({^-1})^{151}$

12. Compute each of the following:
    a. $^-2 + 3 \cdot 5 - 1$ b. $10 - 3 \cdot 7 - 4({^-2}) + 3$
    c. $10 - 3 - 12$ d. $10 - (3 - 12)$
    e. $({^-3})^2$ f. $^-3^2$
    g. $^-5^2 + 3({^-2})^2$ h. $^-2^3$
    i. $({^-2})^5$ j. $^-2^4$

13. If x is an integer and $x \neq 0$, which of the following are always positive and which are always negative?
    a. $^-x^2$ b. x^2 c. $({^-x})^2$ d. $^-x^3$
    e. $({^-x})^3$ f. $^-x^4$ g. $({^-x})^4$ h. x^4
    i. x j. ^-x

14. Which of the expressions in problem 13 are equal to each other for all values of x?

15. Identify the property of integers being illustrated in each of the following:
    a. $({^-3}) \cdot (4 + 5) = (4 + 5) \cdot ({^-3})$
    b. $^-4 + {^-7} \in I$
    c. $5 \cdot [4 \cdot ({^-3})] = (5 \cdot 4) \cdot ({^-3})$
    d. $({^-9}) \cdot [5 + ({^-8})] = ({^-9}) \cdot 5 + ({^-9}) \cdot ({^-8})$

16. Simplify each of the following:
    a. $({^-x})({^-y})$ b. $^-2x({^-y})$
    c. $({^-x} + y) + x + y$ d. $^-1 \cdot x$
    e. $x - 2({^-y})$ f. $a - (a - b)$
    g. $y - (y - x)$ h. $^-(x - y) + x$

17. Find all integers x (if possible) that make each of the following true:
    a. $^-3x = 6$ b. $^-3x = {^-6}$
    c. $^-2x = 0$ d. $5x = {^-30}$
    e. $x \div 3 = {^-12}$ f. $x \div ({^-3}) = {^-2}$
    g. $x \div ({^-x}) = {^-1}$ h. $0 \div x = 0$
    i. $x \div 0 = 1$ j. $x^2 = 9$
    k. $x^2 = {^-9}$ l. $^-x \div {^-x} = 1$
    m. $^-x^2$ is negative.   n. $^-(1 - x) = x - 1$
    o. $x - 3x = {^-2x}$

18. Multiply each of the following and combine terms where possible:
    a. $^-2(x - 1)$ b. $^-2(x - y)$
    c. $x(x - y)$ d. $^-x(x - y)$
    e. $^-2(x + y - z)$ f. $^-x(x - y - 3)$
    g. $({^-5} - x)(5 + x)$ h. $(x - y - 1)(x + y + 1)$
    i. $({^-x^2} + 2)(x^2 - 1)$

19. Use the difference-of-squares formula to simplify each of the following, if possible:
    a. $52 \cdot 48$ b. $(5 - 100)(5 + 100)$
    c. $({^-x} - y)({^-x} + y)$ d. $(2 + 3x)(2 - 3x)$
    e. $(x - 1)(1 + x)$ f. $213^2 - 13^2$

20. Factor each of the following expressions completely and then simplify, if possible:
    a. $3x + 5x$ b. $ax + 2x$
    c. $xy + x$ d. $ax - 2x$
    e. $x^2 + xy$ f. $3x - 4x + 7x$
    g. $3xy + 2x - xz$ h. $3x^2 + xy - x$
    i. $abc + ab - a$ j. $(a + b)(c + 1) - (a + b)$
    k. $16 - a^2$ l. $x^2 - 9y^2$
    m. $4x^2 - 25y^2$ n. $(x^2 - y^2) + x + y$

21. a. Develop a formula for $(a - b)^2$.
    b. Use your results from (a) to compute each of the following in your head:
       (i) 98^2 (Hint: Write $98 = 100 - 2$.)
       (ii) 99^2
       (iii) 997^2

22. If x is positive and $y = {^-x}$, then determine which of the following statements is false?
    a. $x^2 y > 0$.   b. $x + y = 0$.
    c. xy is negative.   d. xy^2 is positive.

23. If x and y are integers, classify each of the following as true or false. If true, explain why. If false, give a counterexample.
   a. $|x + y| = |x| + |y|$
   b. $|xy| = |x||y|$
   c. $|x^2| = x^2$
   d. $|x|^2 = x^2$

24. a. Given a calendar for any month of the year, such as the one that follows, pick several 3×3 groups of numbers and find the sum of these numbers. How are the obtained sums related to the middle number?

JULY						
S	M	T	W	T	F	S
		1	2	3	4	5
6	7	8	9	10	11	12
13	14	15	16	17	18	19
20	21	22	23	24	25	26
27	28	29	30	31		

   ★b. Prove that the sum of any 9 digits in any 3×3 set of numbers selected from a monthly calendar will always be equal to 9 times the middle number.

25. In each of the following, find the next two terms. If a sequence is arithmetic or geometric, find its difference or ratio and the nth term.
   a. $^-10, ^-7, ^-4, ^-1, 2, 5, \_, \_$
   b. $10, 7, 4, 1, ^-2, ^-5, \_, \_$
   c. $^-2, ^-4, ^-8, ^-16, ^-32, ^-64, \_, \_$
   d. $^-2, 4, ^-8, 16, ^-32, 64, \_, \_$
   e. $2, ^-2^2, 2^3, ^-2^4, 2^5, ^-2^6, \_, \_$

26. Find the sum of the first 100 terms in parts (a) and (b) of problem 25.

27. Find the first five terms of the sequences whose nth term is:
   a. $n^2 - 10$
   b. $^-5n + 3$
   c. $(^-2)^n - 1$
   d. $(^-2)^n + 2^n$
   e. $n^2(^-1)^n$
   f. $|10 - n^2|$

28. Find the first two terms of an arithmetic sequence in which the fourth term is $^-8$ and the 101st term is $^-493$.

29. Tira noticed that every 30 sec, the temperature of a chemical reaction in her lab was decreasing by the same number of degrees. Initially, the temperature was 28°C and 5 min later, $^-12$°C. In a second experiment, Tira noticed that the temperature of the chemical reaction was initially $^-57$°C and was decreasing by 3°C every minute. If she started the two experiments at the same time, when were the temperatures of the reactions the same? What was that temperature?

30. If x and y are integers, classify each of the following as always true, sometimes true, or never true. Justify your answers.
   a. $xy = ^-|x||y|$
   b. $(^-x)^3 = ^-x^3$
   c. $^-x^2 = x^2$
   d. $|x| > ^-1$
   e. If $x > y$, then $x^2 > y^2$.

Communication

31. Can $(^-x - y)(x + y)$ be multiplied by using the difference-of-squares formula? Explain why or why not.

32. Kahlil said that using the formula $(a + b)^2 = a^2 + 2ab + b^2$, he can find a similar formula for $(a - b)^2$. Examine his argument. If it is correct, supply any missing steps or justifications; if it is incorrect, point out why.

$$(a - b)^2 = [a + (^-b)]^2$$
$$= a^2 + 2a(^-b) + (^-b)^2$$
$$= a^2 - 2ab + b^2$$

33. Seventh-grader Nancy gave the following argument to show that $(^-a)b = ^-(ab)$ for all integers a and b: *I know that* $(^-1)a = ^-a$, *hence:*

$$(^-a)b = [(^-1)a] \cdot b$$
$$= (^-1) \cdot (ab)$$
$$= ^-(ab)$$

If the argument is valid, complete its details; if it is not valid, explain why not.

34. Hosni gave the following argument that $^-(a + b) = ^-a + ^-b$ for all integers a and b. If the argument is correct, supply the missing reasons. If it is incorrect, explain why not.

$$^-(a + b) = (^-1) \cdot (a + b)$$
$$= (^-1)a + (^-1)b$$
$$= ^-a + ^-b$$

35. a. Use the distributive property of multiplication over addition to show that $(^-1) \cdot a + a = 0$. (Hint: write $a = 1 \cdot a$.)
   b. Use part (a) to show that $(^-1)a = ^-a$.

36. The Swiss mathematician Leonhard Euler (1707–1783) argued that $(^-1)(^-1) = 1$ as follows. "The result must be either $^-1$ or 1. If it is $^-1$, then $(^-1)(^-1) = ^-1$. Because $^-1 = (^-1) \cdot 1$, we have $(^-1)(^-1) = (^-1) \cdot 1$. Now dividing both sides of the last equation by $^-1$ we get $^-1 = 1$, which of course cannot be true. Hence $(^-1)(^-1)$ must be equal to 1."
   a. What is your reaction to the above argument? Is it logical? Why or why not?
   b. Can Euler's approach be used to justify other properties of integers? Explain.

37. The following graph shows the development of mathematics in different cultures. Explain the use of positive and negative numbers in the graph.

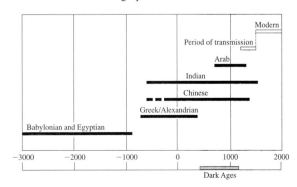

Open-Ended

38. Give examples of situations that cannot be described using only the numbers in the set {1, 2, 3, 4, 5, ...} and explain why.
39. On a national mathematics competition, scoring is accomplished using the formula 4 times the number done correctly minus the number done incorrectly. In this scheme, problems left blank are considered neither correct nor incorrect. Devise a scenario that would allow a student to have a negative score. Use a graph to illustrate different situations in which this could happen.
40. Which model would you use to teach multiplication of integers? Explain why you chose the model.
41. Select a current middle-school text that introduces multiplication and division of integers and discuss any models that were used and how effective you think they would be with a group of students.

Cooperative Learning

42. Devise a scheme for determining a grade-point average for a college student that allows negative quality points for a failing grade.
    a. Use your scheme to determine possible grades for students with positive, zero, and negative grade-point averages.
    b. Compare your scheme with another class group and write a rationale for the best scheme.

43. a. How would you introduce multiplication of integers in a middle-school class and how would you explain that a product of two negative numbers is positive? Write a rationale for your approach.
    b. Present and compare your answers to another class group and together decide about the most appropriate way to introduce the concepts. Write a rationale for your approach.

Review Problems

44. Illustrate $^-8 + {}^-5$ on a number line.
45. Find the additive inverse of each of the following:
    a. $^-5$ b. 7   c. 0
46. Compute each of the following:
    a. $|{}^-14|$ b. $|{}^-14| + 7$
    c. $8 - |{}^-12|$ d. $|11| + |{}^-11|$
47. In the 1400s, European merchants used positive and negative numbers to label barrels of flour. For example, a barrel labeled $^+3$ meant the barrel was 3 lb overweight, whereas a barrel labeled $^-5$ meant the barrel was 5 lb underweight. If the following numbers were found on 100-lb barrels, what was the total weight of the barrels?

BRAIN TEASER If $a, \ldots, z$ are integers, find the product

$$(x - a)(x - b)(x - c) \ldots (x - z).$$

Section 4-3 — Divisibility

The concepts of *even* and *odd* are commonly used. For example, during water shortages in the summer in some parts of the country, houses with even number addresses can water on even-numbered days of the month and houses with odd number addresses can water on odd-numbered days. An even number is a number that has 0 remainder when divided by 2. We say that it is divisible by 2. An odd number is a number that is not divisible by 2. The fact that 12 is divisible by 2 can be stated in the following equivalent statements in the left-hand column:

Example	*General Statement*
12 is divisible by 2.	a is divisible by b.
2 is a divisor of 12.	b is a divisor of a.
12 is a multiple of 2.	a is a multiple of b.
2 is a factor of 12.	b is a factor of a.
2 divides 12.	b divides a.

divides

The statement that "2 divides 12" is written with a vertical segment as in 2|12, where the vertical segment means **divides.** Likewise, "*b* divides *a*" can be written as *b*|*a*. Each statement in the previous right-hand column can be written as *b*|*a*. We write 5∤12 to symbolize that 5 does not divide 12 or that 12 is not divisible by 5. The notation 5∤12 also implies that 12 is not a multiple of 5 and 5 is not a factor of 12.

> **Definition**
>
> If *a* and *b* are any integers, then *b* divides *a*, written *b*|*a*, if, and only if, there is a unique integer *c* such that $a = bc$.

divisor · multiple

If *b*|*a*, then *b* is a factor, or a **divisor,** of *a*, and *a* is a **multiple** of *b*.

Do not confuse *b*|*a* with *b/a*, which is interpreted as $b \div a$. The former, a relation, is either true or false. The latter, an operation, has a numerical value.

> **HISTORICAL NOTE**
>
> Pierre de Fermat (1601–1665) was a lawyer and a magistrate who served in the provincial parliament in Toulouse, France. He devoted his leisure time to mathematics—a subject in which he had no formal training. After his death, his son decided to publish a new edition of Diophantus's *Arithmetica* with Fermat's notes. One of the notes in the margin of Fermat's copy asserted that the equation $x^n + y^n = z^n$ has no positive integer solutions if *n* is an integer greater than 2 and commented, "I have found an admirable proof of this, but the margin is too narrow to contain it." Many great mathematicians spent years trying to prove Fermat's assertion, now called "Fermat's last theorem." In 1995 Andrew Wiles, a Princeton University mathematician, proved Fermat's last theorem. The proof is several hundred pages long.

Example 4-16 Classify each of the following as true or false. Explain your answers.

  a. ⁻3|12   b. 0|2   c. 0 is even.
  d. 8∤2   e. For all integers, *a*, 1|*a*.   f. For all integers *a*, ⁻1|*a*.
  g. 3|6*n* for all integers *n*

Solution a. ⁻3|12 is true because $12 = {}^-4({}^-3)$.
  b. 0|2 is false because there is no integer *c* such that $2 = c \cdot 0$.
  c. 2|0 is true because $0 = 0 \cdot 2$; therefore 0 is even.
  d. 8∤2 is true because there is no integer *c* such that $2 = c \cdot 8$.
  e. 1|*a* is true for all integers *a* because $a = a \cdot 1$.
  f. ⁻1|*a* is true for all integers *a* because $a = ({}^-a)({}^-1)$.
  g. 3|6*n* is true. Because $6n = 3 \cdot 2n$, 6*n* is a multiple of 3 and hence 3|6*n*.

In Example 4-16(g), because 3|6 we were able to show that 3|6*n* for all integers *n*. This can be generalized. If instead of 3|6 we have 3|*a*, then we can conclude that 3 divides any multiple of *a*. Even more generally if *d*|*a* then *d* divides any multiple of *a*. We state this fact in the following theorem.

Theorem 4-1

For any integers a and d, if $d|a$ and n is any integer, then $d|na$.

The above theorem can be stated in an equivalent form:

If d is a factor of a (that is a equals some integer times d), then d is a factor of any multiple of a.

Figure 4-16

We can deduce other notions of divisibility from everyday models. Consider two packages of chewing gum each having five pieces, as in Figure 4-16. We can divide each package of gum evenly among five students. In addition, if we opened both packages and put all of the pieces in a bag, we could still divide the pieces of gum evenly among the five students. To generalize this notion, if we buy gum in larger packages with a pieces in one package and b pieces in a second package with both a and b divisible by 5, we can record the preceding discussion as follows:

If $5|a$ and $5|b$, then $5|(a+b)$.

If the number, a, of pieces of gum in one package is divisible by five, but the number, b, of pieces in the other package is not, then the total, $a + b$, cannot be divided evenly among the five students. This can be recorded as follows:

If $5|a$ and $5\nmid b$, then $5\nmid (a+b)$.

What, if anything, can you conclude if $5\nmid a$ and $5\nmid b$?

Since subtraction can be defined in terms of addition, results similar to addition hold for subtraction. These ideas are listed in Theorem 4-2.

Theorem 4-2

For any integers a, b, and d, the following holds:
a. If $d|a$ and $d|b$, then $d|(a+b)$.
b. If $d|a$ and $d\nmid b$, then $d\nmid (a+b)$.
c. If $d|a$ and $d|b$, then $d|(a-b)$.
d. If $d|a$ and $d\nmid b$, then $d\nmid (a-b)$.

The proofs of most theorems in this section are left as exercises, but the proof of Theorem 4-2(a) is given as an illustration.

Proof. Theorem 4-2(a) is equivalent to the following:

If a is a multiple of d and b is a multiple of d, then $a + b$ is a multiple of d.

Notice that "a is a multiple of d" means $a = m \cdot d$, for some integer m. Similarly "b is a multiple of d" means $b = n \cdot d$ for some integer n. To show that $a + b$ is a multiple of d, we add the above equations as follows:

$$a + b = md + nd.$$

Is $md + nd$ a multiple of d? Notice that $md + nd = (m+n)d$, so $a + b = (m+n)d$. Because $m + n$ is an integer, $a + b$ is a multiple of d and therefore $d|(a+b)$.

Example 4-17 Classify each of the following as true or false, where x, y, and z are integers. If a statement is true, prove it. If a statement is false, provide a counterexample.

a. If $3|x$ and $3|y$, then $3|xy$.
b. If $3|(x+y)$, then $3|x$ and $3|y$.
c. If $9 \nmid a$, then $3 \nmid a$.

Solution
a. True. By Theorem 4-1, if $3|x$, then, for any integer y, $3|yx$ or $3|xy$.
b. False. For example, $3|(7+2)$, but $3 \nmid 7$ and $3 \nmid 2$.
c. False. For example, $9 \nmid 21$, but $3|21$.

NOW TRY THIS 4-3

• In Example 4-17(a), is it true that $3|xy$ regardless of whether $3|y$ or $3 \nmid y$? Why? •

Example 4-18 Five students found a padlocked money box that had a deposit slip attached to it. The deposit slip was water-spotted, so the currency total appeared as shown in Figure 4-17. One student remarked that if the money listed on the deposit slip was in the box, it could easily be divided equally among the 5 students without using coins. How did the student know this?

Figure 4-17

Solution Because the units digit of the amount of the currency is 0, the solution to the problem is to determine whether all natural numbers whose units digit is 0 are divisible by 5. To solve this problem, *look for a pattern*. Natural numbers whose units digit is 0 form a pattern, that is, 10, 20, 30, 40, 50, …. These numbers are multiples of 10. We are to determine whether 5 divides all multiples of 10. Since $5|10$, by Theorem 4-1, 5 divides any multiple of 10. Hence, 5 divides the amount of money in the box, and the student is correct.

Divisibility Rules

As shown in Example 4-18, sometimes it is handy to know if one number is divisible by another just by looking at it or by performing a simple test. We discovered that if a number ends in 0, then the number is divisible by 5. The same argument can be used to show that if a number ends in 5, it is divisible by 5. This is an example of a divisibility rule. Moreover, if the last digit of a number is neither 0 nor 5, then the number is not divisible by 5.

Elementary texts frequently state divisibility rules. However, such rules have limited use except for mental arithmetic. It is possible to determine whether 1734 is divisible by 17, either by using pencil and paper or a calculator. To check divisibility and avoid decimals, we can use a calculator with an integer division button, $\boxed{\text{INT} \div}$. On such a calculator, integer division can be performed using the following sequence of buttons:

$$\boxed{1}\,\boxed{7}\,\boxed{3}\,\boxed{4}\,\boxed{\text{INT} \div}\,\boxed{1}\,\boxed{7}\,\boxed{=}$$

to obtain the display $\underset{Q}{\lfloor 102 \rfloor}$ $\underset{Q}{\lfloor 0 \rfloor}$.

This implies $1734/17 = 102$ with a remainder of 0, which, in turn, implies $17|1734$.

We could have determined this same result mentally by considering the following:

$$1734 = 1700 + 34.$$

Because $17|1700$ and $17|34$, by Theorem 4-2(a), we have $17|(1700 + 34)$, or $17|1734$. Similarly, we could determine mentally that $17 \nmid 1735$.

REMARK Notice that $17|1734$ implies $17|(^{-}1)1734$; that is, $17|{^-}1734$. In general, $d|{^-}a$ if, and only if, $d|a$.

Divisibility Tests for 2, 5, and 10

To determine mentally whether a given integer n is divisible by another integer d, we think of n as the sum or difference of two integers where d divides at least one of these numbers. We try to choose numbers such that one of them is close to n and divisible by d, and the other number is relatively small. As an example, consider the divisibility of 358 by 2:

$$358 = 350 + 8$$
$$= 35(10) + 8.$$

We know that $2|10$, so that $2|35(10)$. We also know that $2|8$, which tells us that $2|(35(10) + 8)$. Because $2|10$, 2 divides any multiple of 10, so to determine the divisibility of any integer by 2, we consider only whether the units digit is divisible by 2. If it is, then by Theorem 4-2(a) the number is divisible by 2. If not, then by Theorem 4-2(b) the number is not divisible by 2.

We can develop a similar test for divisibility by 10. In general, we have the following divisibility rules.

Divisibility Test for 2

An integer is divisible by 2 if, and only if, its units digit is divisible by 2.

Divisibility Test for 5

An integer is divisible by 5 if, and only if, its units digit is divisible by 5, that is, if, and only if, the units digit is 0 or 5.

Divisibility Test for 10

An integer is divisible by 10 if, and only if, its units digit is divisible by 10, that is, if, and only if, the units digit is 0.

Divisibility Test for 4 and 8

When we consider divisibility rules for 4 and 8, we see that $4 \nmid 10$ and $8 \nmid 10$, so it is not a matter of checking the units digit for divisibility by 4 and 8. However, 4 (which is 2^2) divides 10^2, and 8 (which is 2^3) divides 10^3.

We first develop a divisibility rule for 4. Consider any four-digit number n such that $n = a \cdot 10^3 + b \cdot 10^2 + c \cdot 10 + d$. Our *subgoal* is to *write the given number as a sum of two numbers,* one of which is as great as possible and divisible by 4. We know that $4|10^2$ because $10^2 = 4 \cdot 25$ and, consequently, $4|10^3$. Because $4|10^2$, then $4|b \cdot 10^2$ and because

$4|10^3$, then $4|a \cdot 10^3$. Finally, $4|a \cdot 10^3$ and $4|b \cdot 10^2$ imply $4|(a \cdot 10^3 + b \cdot 10^2)$. Now the divisibility of $a \cdot 10^3 + b \cdot 10^2 + c \cdot 10 + d$ by 4 depends on the divisibility of $(c \cdot 10 + d)$ by 4. Notice that $c \cdot 10 + d$ is the number represented by the last two digits in the given number n. We summarize this in the following test.

Divisibility Test for 4

An integer is divisible by 4 if, and only if, the last two digits of the integer represent a number divisible by 4.

To investigate divisibility by 8, we note that the least positive power of 10 divisible by 8 is 10^3 since $10^3 = 8 \cdot 125$. Consequently, all integral powers of 10 greater than 10^3 also are divisible by 8. Hence, the following is a divisibility test for 8.

Divisibility Test for 8

An integer is divisible by 8 if, and only if, the last three digits of the integer represent a number divisible by 8.

Example 4-19

a. Determine whether 97,128 is divisible by 2, 4, and 8.
b. Determine whether 83,026 is divisible by 2, 4, and 8.

Solution a. $2|97{,}128$ because $2|8$.   b. $2|83{,}026$ because $2|6$.
$4|97{,}128$ because $4|28$. $4\nmid 83{,}026$ because $4\nmid 26$.
$8|97{,}128$ because $8|128$. $8\nmid 83{,}026$ because $8\nmid 026$.

REMARK   In Example 4-19(a), it would have been sufficient to check that the given number is divisible by 8 because if $8|a$, then $2|a$ and $4|a$. (Why?) However, if $8\nmid a$, we cannot conclude from this that $4\nmid a$ or $2\nmid a$. (Why?) This relationship is shown in Figure 4-18.

Figure 4-18

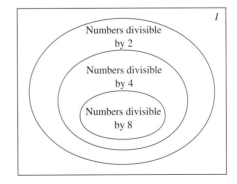

Divisibility Tests for 3 and 9

Next, we consider a divisibility test for 3. No power of 10 is divisible by 3, but the numbers 9, and 99, and 999, and others of this type are close to powers of 10 and are divisible by 3. For example, to determine whether 5721 is divisible by 3, we rewrite the number using 999, 99, and 9, as follows:

$$\begin{aligned} 5721 &= 5 \cdot 10^3 + 7 \cdot 10^2 + 2 \cdot 10 + 1 \\ &= 5(999 + 1) + 7(99 + 1) + 2(9 + 1) + 1 \\ &= 5 \cdot 999 + 5 \cdot 1 + 7 \cdot 99 + 7 \cdot 1 + 2 \cdot 9 + 2 + 1 \\ &= (5 \cdot 999 + 7 \cdot 99 + 2 \cdot 9) + (5 + 7 + 2 + 1) \end{aligned}$$

The sum in the first set of parentheses is divisible by 3, so the divisibility of 5721 by 3 depends on the sum in the second set of parentheses. In this case, $5 + 7 + 2 + 1 = 15$ and $3|15$, so $3|5721$. Hence, to test 5721 for divisibility by 3, we test $5 + 7 + 2 + 1$ for divisibility by 3. Notice that $5 + 7 + 2 + 1$ is the sum of the digits of 5721. The example suggests the following test for divisibility by 3.

> **Divisibility Test for 3**
>
> An integer is divisible by 3 if, and only if, the sum of its digits is divisible by 3.

We can use an argument similar to the one used to demonstrate that $3|5721$ to prove the test for divisibility by 3 on any integer and in particular for any 4-digit number $n = a \cdot 10^3 + b \cdot 10^2 + c \cdot 10 + d$. Even though $a \cdot 10^3 + b \cdot 10^2 + c \cdot 10 + d$ is not necessarily divisible by 3, the number $a \cdot 999 + b \cdot 99 + c \cdot 9$ is close to n and *is* divisible by 3. We have the following:

$$\begin{aligned} a \cdot 10^3 + b \cdot 10^2 + c \cdot 10 + d &= a \cdot 1000 + b \cdot 100 + c \cdot 10 + d \\ &= a(999 + 1) + b(99 + 1) + c(9 + 1) + d \\ &= (a \cdot 999 + b \cdot 99 + c \cdot 9) + (a \cdot 1 + b \cdot 1 + c \cdot 1 + d) \\ &= (a \cdot 999 + b \cdot 99 + c \cdot 9) + (a + b + c + d). \end{aligned}$$

Because $3|9$, $3|99$, and $3|999$, it follows that $3|(a \cdot 999 + b \cdot 99 + c \cdot 9)$. If $3|(a + b + c + d)$, then $3|[(a \cdot 999 + b \cdot 99 + c \cdot 9) + (a + b + c + d)]$; that is, $3|n$. If, on the other hand, $3 \nmid (a + b + c + d)$, it follows from Theorem 4-2(b) that $3 \nmid n$.

Since $9|9$, $9|99$, $9|999$, and so on, a test similar to that for divisibility by 3 applies to divisibility by 9. (Why?)

> **Divisibility Test for 9**
>
> An integer is divisible by 9 if, and only if, the sum of the digits of the integer is divisible by 9.

Example 4-20 Use divisibility tests to determine whether each of the following numbers is divisible by 3 and divisible by 9:

a. 1002   **b.** 14,238

Solution **a.** Because $1 + 0 + 0 + 2 = 3$ and $3|3$, it follows that $3|1002$. Because $9 \nmid 3$, it follows that $9 \nmid 1002$.

b. Because $1 + 4 + 2 + 3 + 8 = 18$ and $3 | 18$, it follows that $3 | 14{,}238$. Because $9 | 18$, it follows that $9 | 14{,}238$.

Example 4-21 The store manager has an invoice for 72 four-function calculators. The first and last digits on the receipt are illegible. The manager can read

$$\$\blacksquare 67.9\blacksquare.$$

What are the missing digits, and what is the cost of each calculator?

Solution Let the missing digits be x and y so that the number is $x67.9y$ dollars, or $x679y$ cents. Because there were 72 calculators sold, the number on the invoice must be divisible by 72. Because the number is divisible by 72, it must be divisible by 8 and 9, which are factors of 72. For the number on the invoice to be divisible by 8, the three-digit number $79y$ must be divisible by 8. Because $79y$ must be divisible by 8, it is an even number. Therefore $79y$ must be either 790, 792, 794, 796, or 798. Only the number 792 is divisible by 8, so we know the last digit, y, on the invoice must be 2.

Because the number on the invoice must be divisible by 9, we know that 9 must divide $x + 6 + 7 + 9 + 2$, or $(x + 24)$. Since 3 is the only single digit that will make $(x + 24)$ divisible by 9, then x must be 3. Therefore the number on the invoice must be $\$367.92$. The calculators must cost $\$367.92/72$, or $\$5.11$, each.

Divisibility Tests for 11 and 6

The divisibility test for 7 is usually harder to use than actually performing the division, so we omit the test. We state the divisibility test for 11 but omit the proof. Interested readers might try to find a proof.

> **Divisibility Test for 11**
>
> An integer is divisible by 11 if, and only if, the sum of the digits in the places that are even powers of 10 minus the sum of the digits in the places that are odd powers of 10 is divisible by 11.

For example, to test whether 8,471,986 is divisible by 11, we check whether 11 divides the difference $(6 + 9 + 7 + 8) - (8 + 1 + 4)$, or 17. Because $11 \nmid 17$, it follows from the divisibility test for 11 that $11 \nmid 8{,}471{,}986$. A number like 2772 is divisible by 11 because $(2 + 7) - (7 + 2) = 9 - 9 = 0$ and 0 is divisible by 11.

The divisibility test for 6 is related to the divisibility tests for 2 and 3. In Section 4-4, we show that if $2 | n$ and $3 | n$, then $2 \cdot 3 | n$. Consequently, the following divisibility test is true.

> **Divisibility Test for 6**
>
> An integer is divisible by 6 if, and only if, the integer is divisible by both 2 and 3.

Divisibility tests for other numbers are explored in the Ongoing Assessment 4-3.

Example 4-22 The number 57,729,364,583 has too many digits for most calculator displays. Determine whether it is divisible by each of the following:

a. 2   b. 3   c. 5   d. 6   e. 8   f. 9   g. 10   h. 11

Solution
a. No, the last digit, 3, is not divisible by 2.
b. No, the sum of the digits is 59, which is not divisible by 3.
c. No, the last digit is neither 0 nor 5.
d. No, because the number is not divisible by 2 and by 3. (Actually, just the fact that the number is not divisible by 2 implies that it is not divisible by 6. Why?)
e. No, because the number formed by the last 3 digits, 583, is not divisible by 8.
f. No, because the sum of the digits is 59, which is not divisible by 9.
g. No, because the units digit is not 0.
h. Yes, because $(3 + 5 + 6 + 9 + 7 + 5) - (8 + 4 + 3 + 2 + 7) = 35 - 24 = 11$ and 11 is divisible by 11.

NOW TRY THIS 4-4

- Fill in the blanks in the following so that the number is divisible by 9. List all possibilities.   12,506,5__ __.

Problem Solving   A Mistake in the Inventory

A class from Washington School visited a neighborhood cannery warehouse. The warehouse manager told the class that there were 11,368 cans of juice in the inventory and that the cans were packed in boxes of 6 or 24, depending on the size of the can. One of the students, Sam, thought for a moment and announced that there was a mistake in the inventory. Is Sam's statement correct? Why or why not?

- **Understanding the Problem** The problem is to determine if the manager's inventory of 11,368 cans was correct. To solve the problem, we must assume there are no partial boxes of cans; that is, a box must contain exactly 6 or exactly 24 cans of juice.

- **Devising a Plan** We know that the boxes contain either 6 cans or 24 cans, but we do not know how many boxes of each type there are. One strategy for solving this problem is to *find an equation* that involves the total number of cans in all the boxes.

  The total number of cans, 11,368, equals the number of cans in all the 6-can boxes plus the number of cans in all the 24-can boxes. If there are n boxes containing 6 cans each, there are $6n$ cans altogether in those boxes. Similarly, if there are m boxes with 24 cans each, these boxes contain a total of $24m$ cans. Because the total was reported to be 11,368 cans, we have the equation $6n + 24m = 11{,}368$. Sam claimed that $6n + 24m \neq 11{,}368$.

  One way to show that $6n + 24m \neq 11{,}368$ is to show that $6n + 24m$ and 11,368 do not have the same divisors. Both $6n$ and $24m$ are divisible by 6. This implies that $6n + 24m$ must be divisible by 6. If 11,368 is not divisible by 6, then Sam is correct.

- **Carrying Out the Plan** The divisibility test for 6 states that a number is divisible by 6 if, and only if, the number is divisible by both 2 and 3. Because 11,368 is an even number, it is divisible by 2. Is it divisible by 3?

The divisibility test for 3 states that a number is divisible by 3 if, and only if, the sum of the digits in the number is divisible by 3. We see that $1 + 1 + 3 + 6 + 8 = 19$, which is not divisible by 3, so 11,368 is not divisible by 3. Hence, Sam is correct.

• **Looking Back** Suppose 11,368 had been divisible by 6. Would that have implied that the manager was correct? The answer is no; it would have implied only that we would have to change our approach to the problem.

As a further Looking Back activity, suppose that, given different data, the manager is correct. Can we determine values for m and n? In fact, this can be done. If a computer is available, a program can be written to determine all possible natural-number values of m and n.

HISTORICAL NOTE

A modern mathematician who worked in the area of number theory was American Julia Robinson (1919–1985). Robinson's work with the Russian mathematician Yuri Matijasevič on Diophantine equations led directly to the solution of the tenth of the famous set of 23 problems the German mathematician David Hilbert posed. Robinson was the first woman mathematician to be elected to the National Academy of Sciences and the first woman president of the American Mathematical Society. She died of leukemia at the age of sixty-five.

BRAIN TEASER

The following is an argument showing that an ant weighs as much as an elephant. What is wrong?

Let e be the weight of the elephant and a the weight of the ant. Let $e - a = d$. Consequently, $e = a + d$. Multiply each side of $e = a + d$ by $e - a$. Then simplify.

$$e(e - a) = (a + d)(e - a)$$
$$e^2 - ea = ae + de - a^2 - da$$
$$e^2 - ea - de = ae - a^2 - da$$
$$e(e - a - d) = a(e - a - d)$$
$$e = a$$

Thus the weight of the elephant equals the weight of the ant.

ONGOING ASSESSMENT 4-3

1. Classify each of the following as true or false. If false, tell why.
   a. 6 is a factor of 30.
   b. 6 is a divisor of 30.
   c. 6|30.
   d. 30 is divisible by 6.
   e. 30 is a multiple of 6.
   f. 6 is a multiple of 30.

2. Using divisibility tests, answer each of the following:
   a. There are 1379 children signed up to play in a baseball league. If exactly 9 players are to be placed on each team, will any team be short of players?
   b. A forester has 43,682 seedlings to be planted. Can these be planted in an equal number of rows with 11 seedlings in each row?
   c. There are 261 students to be assigned to 9 teachers so that each teacher has the same number of students. Is this possible?
   d. Six friends win with a lottery ticket. The payoff is $242,800. Can the money be divided evenly?
   e. Jack owes $7812 on a new car. Can this amount be paid in 12 equal monthly installments?

3. Without using a calculator, test each of the following numbers for divisibility by 2, 3, 4, 5, 6, 8, 9, 10, and 11:
   a. 746,988
   b. 81,342
   c. 15,810

d. 4,201,012
e. 1,001
f. 10,001

4. Determine each of the following without actually performing the division. Explain how you did it in each case.
   a. Is 34,015 divisible by 17?
   b. Is 34,051 divisible by 17?
   c. Is 19,031 divisible by 19?
   d. Is $2 \cdot 3 \cdot 5 \cdot 7$ divisible by 5?
   e. Is $(2 \cdot 3 \cdot 5 \cdot 7) + 1$ divisible by 5?

5. Justify each of the given statements, assuming that a, b, and c are integers. If a statement cannot be justified by one of the theorems in this section, answer "none."
   a. $4 | 20$ implies $4 | 113 \cdot 20$.
   b. $4 | 100$ and $4 \nmid 13$ imply $4 \nmid (100 + 13)$.
   c. $4 | 100$ and $4 \nmid 13$ imply $4 \nmid 1300$.
   d. $3 | (a + b)$ and $3 \nmid c$ imply $3 \nmid (a + b + c)$.
   e. $3 | a$ implies $3 | a^2$.

6. Classify each of the following as true or false:
   a. If every digit of a number is divisible by 3, the number itself is divisible by 3.
   b. If a number is divisible by 3, then every digit of the number is divisible by 3.
   c. A number is divisible by 3 if, and only if, every digit of the number is divisible by 3.
   d. If a number is divisible by 6, then it is divisible by 2 and by 3.
   e. If a number is divisible by 2 and 3, then it is divisible by 6.
   f. If a number is divisible by 2 and 4, then it is divisible by 8.
   g. If a number is divisible by 8, then it is divisible by 2 and 4.

7. Classify each of the statements in problem 6 as sometimes, always, or never true.

8. Devise a test for divisibility by each of the following numbers:
   a. 16
   b. 25

9. When the two missing digits in the following number are replaced, the number is divisible by 99. What is the number?

   $$85\_\_1$$

10. Fill each of the following blanks with the greatest digit that makes the statement true:
    a. $3 | 74\_$
    b. $9 | 83\_45$
    c. $11 | 6\_55$

11. Place a digit in the square, if possible, so that the number

    $$527{,}4\square2$$

    is divisible by
    a. 2   b. 3
    c. 4   d. 9
    e. 11

12. The bookstore marked some notepads down from $2.00 but still kept the price over $1.00. It sold all of them. The total amount of money from the sale of the pads was $31.45. How many notepads were sold?

13. A group of people ordered No-Cal candy bars. The bill was $2.09. If the original price of each was 12¢ but the price has been inflated, how much does each cost?

14. Leap years occur in years that are divisible by 4. However, if the year ends in two zeros, in order for the year to be a leap year, it must be divisible by 400. Determine which of the following are leap years:
    a. 1776   b. 1986   c. 2000   d. 2024

15. In a football game, a touchdown with an extra point is worth 7 points and a field goal is worth 3 points. Suppose that in a game the only scoring done by teams are touchdowns with extra points and field goals.
    a. Which of the scores 1 to 25 are impossible for a team to score?
    b. List all possible ways for a team to score 40 points.
    c. A team scored 57 points with 6 touchdowns and 6 extra points. How many field goals did the team score?

16. Complete the following table where n is the given integer.

	n	Remainder when n is divided by 9	Sum of the digits of n	Remainder when the sum of the digits of n is divided by 9
a.	31			
b.	143			
c.	345			
d.	2987			
e.	7652			

    f. Make a conjecture about the remainder and the sum of the digits in an integer when it is divided by 9.

17. A test for checking computations is called *casting out nines*. Consider the sum $193 + 24 + 786 = 1003$. The remainders when 193, 24, and 786 are divided by 9 are 4, 6, and 3, respectively. The sum of the remainders, 13, has a remainder of 4 when divided by 9, as does 1003. Checking the remainders in this manner provided a quasi-check for the computation. Find the following sums and use casting out nines to check your sums:
    a. $12{,}343 + 4546 + 56$
    b. $987 + 456 + 8765$
    c. $10{,}034 + 3004 + 400 + 20$
    d. Will this check always work for addition? Give an example to illustrate your answer.
    e. Try the check on the subtraction, $1003 - 46$.
    f. Try the check on the multiplication, $345 \cdot 56$.

g. Would it make sense to try the check on division? Why or why not?

18. Classify each of the following as true or false, assuming that $a, b, c,$ and d are integers and $d \neq 0$. If a statement is false, give a counterexample.
  a. If $d|(a + b)$, then $d|a$ and $d|b$.
  b. If $d|(a + b)$, then $d|a$ or $d|b$.
  c. If $d|ab$, then $d|a$ or $d|b$.
  d. If $ab|c$, $a \neq 0$, and $b \neq 0$, then $a|c$ and $b|c$.
  e. If $a|b$ and $b|a$, then $a = b$.
  f. If $d|a$ and $d|b$, then $d|(ax + by)$ for any integers x and y.
  g. If $d \nmid a$ and $d \nmid b$, then $d \nmid (a + b)$.
  h. If $d|a^2$, then $d|a$.
  i. If $d \nmid a$, then $d \nmid a^2$.

★19. Prove Theorem 4-2(b).

20. Prove the test for divisibility by 9 for any five-digit number.

21. a. Choose a two-digit number such that the number in the tens place is one greater than the number in the units place. Reverse the digits in your number, and subtract this number from your original number; for example, $87 - 78 = 9$. Make a conjecture concerning the results of performing these kinds of operations.
  b. Choose any two-digit number such that the number in the tens place is two greater than the number in the units place. Reverse the digits in your number, and subtract this number from your original number; for example, $31 - 13 = 18$. Make a conjecture concerning the results of performing these kinds of operations.
  ★c. Prove that for any two-digit number, if the digits are reversed and the numbers subtracted, the difference is a multiple of 9.
  d. Investigate what happens whenever two-digit numbers with equal digit sums are subtracted; for example, $62 - 35 = 27$.

22. Using only divisibility tests, explain whether 6,868,395 is divisible by 15.

Communication

23. A customer wants to mail a package. The postal clerk determines the cost of the package to be $2.86, but only 6¢ and 15¢ stamps are available. Can the available stamps be used for the exact amount of postage for the package? Why or why not?

24. a. Jim uses his calculator to see if a number n having eight or fewer digits is divisible by a number d. He finds that $n \div d$ has a display of 32. Does $d|n$? Why?
  b. If $n \div d$ gives a display of 16.8, does $d|n$? Why?

25. Which divisibility tests are easiest to use? Why?

26. The numbers x and y are divisible by 5.
  a. Is the sum of x and y divisible by 5? Why?
  b. Is the difference of x and y divisible by 5? Why?
  c. Is the product of x and y divisible by 5? Why?
  d. Is the quotient of x and y divisible by 5? Why?

27. Why is it not always possible to test divisibility on a calculator?

28. Is the area of each of the following rectangles divisible by 4? Explain why or why not.

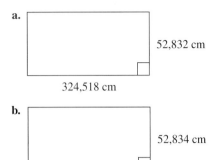

29. a. If 21 divides n, what other natural numbers divide n? Why?
  b. If 16 divides n, what other natural numbers divide n? Why?

30. Can you find three consecutive natural numbers none of which is divisible by 3? Explain your answer.

31. Answer each of the following and justify your answers.
  a. If a number is not divisible by 5, can it be divisible by 10?
  b. If a number is not divisible by 10, can it be divisible by 5?

32. A number in which each digit except 0 appears exactly 3 times is divisible by 3. For example, 777,555,222 and 414,143,313 are divisible by 3. Explain why this statement is true.

33. A palindrome is a number that reads the same forward as backward.
  a. Check the following four-digit palindromes for divisibility by 11:
    i. 4554   ii. 9339   iii. 2002   iv. 2222
  b. Are all four-digit palindromes divisible by 11? Why or why not?
  c. Are all five-digit palindromes divisible by 11? Why or why not?
  d. Are all six-digit palindromes divisible by 11? Why or why not?

34. The numbers 5872 and 2785 are a palindromic pair of numbers because reversing the order of the digits of one number gives the other number. Explain why in a palindromic pair, if one number is divisible by 3, then so is the other.

35. Enter any three-digit number on the calculator; for example, enter 243. Repeat it: 243,243. Divide by 7. Divide by 11. Divide by 13. What is the answer? Try it again with any other three-digit number. Will this always work? Why?

Open-Ended

36. A breakfast food company had a contest in which numbers were placed in breakfast food boxes. A prize of $1000 was awarded to anyone who could collect numbers whose sum was 100. The company had thousands of cards made with the following numbers on them:

3 12 15 18 27 33 45 51 66 75 84 90

a. If the company did not make any more cards, is there a winning combination?

b. If the company is going to add one more number to the list and they want to make sure the contest has at most 1000 winners, suggest a strategy for them to use.

Cooperative Learning

37. In your group, discuss the value of teaching various divisibility tests in middle school. If a teacher decides to discuss the various tests, how should they be introduced?

Review Problems

38. Use the distributive properties of multiplication over addition and over subtraction and other properties of integers to show that $(a + b)(a - b) = a^2 - b^2$.

39. Find all integers x (if possible) that make each of the following true:
   a. $3(^-x) = 6$
   b. $(^-2)|x| = 6$
   c. $(^-x) \div 0 = ^-1$
   d. $^-(x - 1) = 1 - x$
   e. $^-|^-x| = 5$
   f. $^-x < 0$

40. Simplify each of the following.
   a. $3x - (1 - 2x)$
   b. $(^-2x)^2 - 3x^2$
   c. $y - x - 2(y - x)$
   d. $(x - 1)^2 - x^2 + 2x$

TECHNOLOGY CORNER

Edit the following Logo program and type it into your computer. It will determine if a positive integer N is divisible by another integer.

```
TO TESTDIV :N :X
    IF INTEGER (:N/ :X) = :N/ :X PRINT [OKAY] ELSE PRINT
       [NOT DIVISIBLE]
END
```

1. Run this program using various values for :N and :X.
2. How does the program compare to using the $\boxed{\text{INT} \div}$ button on a calculator?

BRAIN TEASER

Dee finds that she has an extraordinary social security number. Its nine digits contain all the numbers from 1 through 9. They also form a number with the following characteristics: when read from left to right, its first two digits form a number divisible by two, its first three digits form a number divisible by 3, its first four digits form a number divisible by 4, and so on, until the complete number is divisible by 9. What is Dee's social security number?

Section 4-4 — Prime and Composite Numbers

When we write $a|b$, we say that a is a divisor of b. One method used in elementary schools to determine the divisors of a number is to use squares of paper and to represent the number as a rectangle. Such a rectangle resembles a candy bar formed with small squares. The dimensions of the rectangle are divisors of the number. For example, Figure 4-19 shows rectangles to represent 12.

Figure 4-19

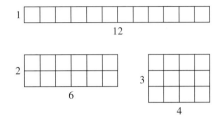

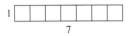

Figure 4-20

As the figure shows, the number 12 has six divisors: 1, 2, 3, 4, 6, and 12. If rectangles were used to find the divisors of 7, then we would find only a 1×7 rectangle, as Figure 4-20 shows. Thus 7 has exactly two divisors: 1 and 7.

To illustrate further the number of divisors of a number, we construct Table 4-1. Below each number listed across the top, we identify numbers less than or equal to 37 that have that number of positive divisors. For example, 12 is in the 6 column because it has six divisors, and 7 is in the 2 column because it has only two divisors.

Table 4-1 Number of Factors

1	2	3	4	5	6	7	8	9
1	2	4	6	16	12		24	36
	3	9	8		18		30	
	5	25	10		20			
	7		14		28			
	11		15		32			
	13		21					
	17		22					
	19		26					
	23		27					
	29		33					
	31		34					
	37		35					

NOW TRY THIS 4-5

- **a.** What patterns do you see forming in Table 4-1?
- **b.** Will there be other entries in the 1 column? Why?
- **c.** What are the next three numbers in the 3 column?
- **d.** Find an entry for the 7 column.
- **e.** What kinds of numbers have an odd number of factors? Why?

prime
composite

The numbers in the 2 column are of particular importance. Notice that they have exactly two divisors, namely, 1 and themselves. Any positive integer with exactly two distinct, positive divisors is a *prime number,* or a **prime.** Any integer greater than 1 that has a positive factor other than 1 and itself is a *composite number,* or a **composite.** For example,

4, 6, and 16 are composites because they have positive factors other than 1 and themselves. The number 1 has only one positive factor, so it is neither prime nor composite. From the 2 column in Table 4-1, we see that the first 12 primes are 2, 3, 5, 7, 11, 13, 17, 19, 23, 29, 31, and 37. Other patterns in the table are explored in the problem set.

Example 4-23

Show that the following numbers are composite:

a. 1564   b. 2781   c. 1001

Solution
a. Since $2|4$, 1564 is divisible by 2 and is composite.
b. Since $3|(2 + 7 + 8 + 1)$, 2781 is divisible by 3 and is composite.
c. Since $11|[(1 + 0) - (0 + 1)]$, 1001 is divisible by 11 and is composite.
d. Because a product of two odd numbers is odd (why?), $3 \cdot 5 \cdot 7 \cdot 11 \cdot 13$ is odd. If we add 1 to an odd number, the sum is even. An even number (other than 2) has a factor of 2 and is therefore composite.

Prime Factorization

Composite numbers can be expressed as products of two or more whole numbers greater than 1. For example, $18 = 2 \cdot 9$, $18 = 3 \cdot 6$, or $18 = 2 \cdot 3 \cdot 3$. Each expression of 18 as a product of factors is a **factorization.**

A factorization containing only prime numbers is a **prime factorization.** To find a prime factorization of a given composite number, first rewrite the number as a product of two smaller numbers greater than 1. Continue the process, factoring the lesser numbers until all factors are primes. For example, consider 260:

$$260 = 26 \cdot 10 = 2 \cdot 13 \cdot 2 \cdot 5 = 2 \cdot 2 \cdot 5 \cdot 13 = 2^2 \cdot 5 \cdot 13.$$

The procedure for finding a prime factorization of a number can be organized using a **factor tree,** as Figure 4-21 (a) demonstrates. The last branches of the tree display the prime factors of 260.

A second way to factor 260 is shown in Figure 4-21(b). The two trees produce the same prime factorization, except for the order in which the primes appear in the products.

Figure 4-21

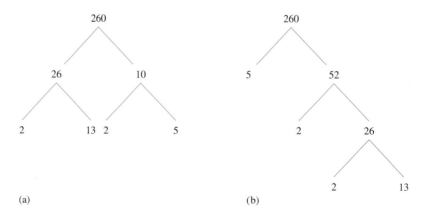

(a)   (b)

The *Fundamental Theorem of Arithmetic,* or the *Unique Factorization Theorem,* states that in general, if order is disregarded, the prime factorization of a number is unique.

Theorem 4-3

Fundamental Theorem of Arithmetic Each composite number can be written as a product of primes in one, and only one, way except for the order of the prime factors in the product.

The Fundamental Theorem of Arithmetic enables us to find the prime factorization of a number. For example, consider 260. We start with the smallest prime, 2, and see whether it divides 260. If not, we try the next greater prime and check for divisibility by this prime. Once we find a prime that divides the number in question, we must find the quotient of the number divided by the prime. This step in the prime factorization of 260 is shown in Figure 4-22 (a). Next we check whether the prime divides the quotient. If so, we repeat the process; if not, we try the next greater prime, 3, and check to see if it divides the quotient. We see that 260 divided by 2 yields 130, as shown in Figure 4-22(b). We continue the procedure, using greater primes, until a quotient of 1 is reached. The original number is the product of all the prime divisors used. The complete procedure for 260 is shown in Figure 4-22(c). An alternative form is shown in Figure 4-22(d).

Figure 4-22

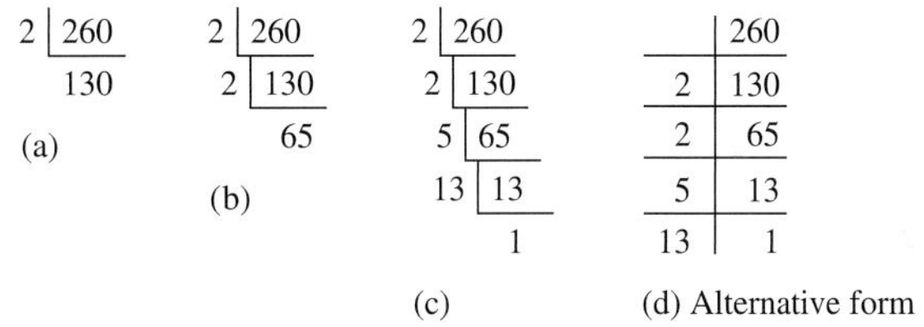

```
 2 | 260      2 | 260      2 | 260          | 260
     130      2 | 130      2 | 130        2 | 130
     (a)          65       5 | 65         2 | 65
                  (b)      13 | 13        5 | 13
                                1        13 | 1
                              (c)       (d) Alternative form
```

The primes in the prime factorization of a number are typically listed in increasing order from left to right and if a prime appears in a product more than once, exponential notation is used. Thus the factorization of 260 is written as $2^2 \cdot 5 \cdot 13$.

Prime factorization is demonstrated in the following student page from *Scott Foresman-Addison Wesley Middle School Math Course 3,* 1999.

Remember

$2^3 = 2 \times 2 \times 2 = 8$
[Page 97]

Prime factorization is the result of writing a number as a product of prime numbers. The prime factorization of a number can be found using a factor tree.

Prime factorization is written in standard form using exponents, with factors in increasing order.

$24 = 2 \times 2 \times 2 \times 3 = 2^3 \times 3$

$2^3 \times 3$ is the prime factorization of 24.

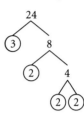

Example 2

You can use your calculator to check for divisibility. For example, to check 138 for divisibility by 8, enter 138 ÷ 8 = into your calculator. 138 is not divisible by 8 because the result is not an integer.

Use a factor tree to find the prime factorization of 140.

	140	
Draw branches for 140.		Factor 140.
Circle 2 because it's prime.	2 70	Factor 70 because it's composite.
Circle 7 because it's prime.	7 10	Factor 10 because it's composite.
Circle 2 because it's prime.	2 5	Circle 5 because it's prime.

$140 = 2 \times 7 \times 2 \times 5$ Write 140 as a product of the prime numbers.

$140 = 2^2 \times 5 \times 7$ Write in standard form.

$2^2 \times 5 \times 7$ is the prime factorization of 140.

Example 3

The message to the right provides the address for a secret party. You were informed to find the prime factorization for the address.

Draw a factor tree for 330.

Write the prime factorization in standard form.

$330 = 2 \times 3 \times 5 \times 11$

The address could be 23511 Oak Street.

Try It

Use a factor tree to find the prime factorization of each number.

a. 54   **b.** 132   **c.** 236   **d.** 354

NOW TRY THIS 4-6

- Colored rods are used in the elementary school classroom to teach many concepts. The rods vary in length from 1 cm to 10 cm. Various lengths have colors associated with them. For example, the 5 rod is yellow. Rods are shown in Figure 4-23 with their appropriate colors.

  A row with all the same color rods is called a one-color train. For example, the following is a one-color train for 18:

a. What other rods can be used to form a one-color train for 18?
b. What one-color trains are possible for 24?
c. How many one-color trains of 2 or more rods are possible for each prime number?
d. If a number can be represented by an all-red train, an all-green train, and an all-yellow train, what is the least number of factors it must have? What are they?

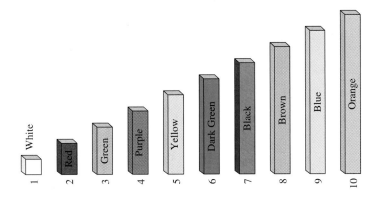

Figure 4-23

Number of Divisors

How many divisors does 24 have? Note that the question asks for the number of divisors, not just prime divisors. To aid in the listing, we group divisors as follows:

$$1, 2, 3, 4, 6, 8, 12, 24$$

The divisors of 24 occur in pairs, where the product of the divisors in each pair is 24. If 3 is a divisor of 24, then 24/3, or 8, is also a divisor of 24. In general, if a natural number k is a divisor of 24, then $24/k$ is also a divisor of 24.

Another way to think of the number of divisors of 24 is to consider the prime factorization $24 = 2^3 \cdot 3$. The divisors of 2^3 are $2^0, 2^1, 2^2,$ and 2^3. The divisors of 3 are 3^0 and 3^1. We know that 2^3 has $(3 + 1)$, or 4, divisors and 3^1 has $(1 + 1)$, or 2, divisors. Because each divisor of 24 is the product of a divisor of 2^3 and a divisor of 3^1, then we use the Fundamental Counting Principle (see Chapter 2) to conclude that 24 has $4 \cdot 2$, or 8, divisors. This is summarized in Table 4-2.

Table 4-2

Divisors of 2^3	$2^0 = 1$	$2^1 = 2$	$2^2 = 4$	$2^3 = 8$
Divisors of 3^1	$3^0 = 1$	$3^1 = 3$		
Divisors of $3^1 \cdot$ Divisors of 2^3 (Divisors of 24)	$3^0 \cdot 2^0 = 1$ $3^1 \cdot 2^0 = 3$	$3^0 \cdot 2^1 = 2$ $3^1 \cdot 2^1 = 6$	$3^0 \cdot 2^2 = 4$ $3^1 \cdot 2^2 = 12$	$3^0 \cdot 2^3 = 8$ $3^1 \cdot 2^3 = 24$

This discussion can be generalized as follows: If p is any prime and n is any natural number, then the divisors of p^n are $p^0, p^1, p^2, p^3, \ldots, p^n$. Therefore there are $(n + 1)$ divisors of p^n. Now, using the Fundamental Counting Principle, we can find the number of divisors of any number whose prime factorization is known. For example, if p and q are different primes then we use the Fundamental Counting Principle to find that $p^n q^m$ will have $(n + 1)(m + 1)$ divisors.

Example 4-24 Find the number of divisors of each of the following:

a. 1,000,000 **b.** 210^{10}

Solution **a.** We first find the prime factorization of 1,000,000.

$$1,000,000 = 1,000 \cdot 1,000 = (2 \cdot 5)(2 \cdot 5)(2 \cdot 5)(2 \cdot 5)(2 \cdot 5)(2 \cdot 5)$$
$$= (2 \cdot 2 \cdot 2 \cdot 2 \cdot 2 \cdot 2) \cdot (5 \cdot 5 \cdot 5 \cdot 5 \cdot 5 \cdot 5)$$
$$= 2^6 \cdot 5^6$$

Because 2^6 has $6 + 1$ divisors and 5^6 has $6 + 1$ divisors, then by the Fundamental Counting Principle $2^6 \cdot 5^6$ has $(6 + 1)(6 + 1)$ or 49 divisors.

b. The prime factorization of 210 is

$$210 = 21 \cdot 10 = 3 \cdot 7 \cdot 2 \cdot 5 = 2 \cdot 3 \cdot 5 \cdot 7,$$
$$210^{10} = (2 \cdot 3 \cdot 5 \cdot 7)^{10} = 2^{10} \cdot 3^{10} \cdot 5^{10} \cdot 7^{10}.$$

By the Fundamental Counting Principle, the number of divisors of 210^{10} is $(10 + 1)(10 + 1)(10 + 1)(10 + 1)$, or 11^4, or 14,641.

NOW TRY THIS 4-7

- Is it necessary to divide 97 by 2, 3, 4, 5, 6, ..., 96 to check if it is prime? Consider the following:

a. If 2 is not a divisor of 97, could any multiple of 2 be a divisor of 97? Why?

b. If 3 is not a divisor of 97, what other numbers could not be divisors of 97? Why?

c. If 5 is not a divisor of 97, what other numbers could not be divisors of 97? Why?

d. If 7 is not a divisor of 97, what other numbers could not be divisors of 97?

e. Conjecture what numbers we have to check for divisibility in order to determine if 97 is prime.

Determining if a Number Is Prime

As depicted in the following cartoon by Sidney Harris, prime numbers have fascinated people of various backgrounds.

In Now Try This 4-7, you might have found that to determine if a number is prime, you must check only divisibility by prime numbers less than the given number. (Why?) However, do we need to check all the primes less than the number? Suppose we want to check if 97 is prime and we find that 2, 3, 5, and 7 do not divide 97. Could a greater prime divide 97? If p is a prime greater than 7, then $p \geq 11$. If $p|97$, then $97/p$ divides 97. However, because $p \geq 11$ then $97/p$ must be less than 10 and hence cannot divide 97. (Why?) So we see that there is no need to check for divisibility by numbers other than 2, 3, 5, and 7. These ideas are generalized in the following theorems.

Theorem 4-4

If d is a divisor of n, then $\dfrac{n}{d}$ is also a divisor of n.

Suppose that p is the *least* divisor of n (greater than 1). Such a divisor must be prime (why?). Then by Theorem 4-4, n/p is also a divisor of n, and because p is the least divisor of n, then $p \leq n/p$. If $p \leq n/p$, then $p^2 \leq n$. This idea is summarized in the following theorem.

> **Theorem 4-5**
>
> If n is composite, then n has a prime factor p such that $p^2 \leq n$.

Theorem 4-5 can be used to help determine whether a given number is prime or composite. For example, consider the number 109. If 109 is composite, it must have a prime divisor p such that $p^2 \leq 109$. The primes whose squares do not exceed 109 are 2, 3, 5, and 7. Mentally, we can see that $2 \nmid 109$, $3 \nmid 109$, $5 \nmid 109$, and $7 \nmid 109$. Hence, 109 is prime. The argument used leads to the following theorem.

> **Theorem 4-6**
>
> If n is an integer greater than 1 and not divisible by any prime p, such that $p^2 \leq n$, then n is prime.

REMARK Because $p^2 \leq n$ implies that $p \leq \sqrt{n}$, Theorem 4-6 implies that to determine if a number n is prime, it is enough to check if any prime less than or equal to $\sqrt{n}$ is a divisor of n.

Example 4-25   **a.** Is 397 composite or prime?   **b.** Is 91 composite or prime?

Solution   **a.** The possible primes p such that $p^2 \leq 397$ are 2, 3, 5, 7, 11, 13, 17, and 19. Because $2 \nmid 397$, $3 \nmid 397$, $5 \nmid 397$, $7 \nmid 397$, $11 \nmid 397$, $13 \nmid 397$, $17 \nmid 397$, $19 \nmid 397$, the number 397 is prime.
   b. The possible primes p such that $p^2 \leq 91$ are 2, 3, 5, and 7. Because 91 is divisible by 7, it is composite.

More About Primes

One way to find all the primes less than a given number is to use the Sieve of Eratosthenes, named after the Greek mathematician Eratosthenes (276–194, or 192 B.C.). If all the natural numbers greater than 1 are considered (or placed in the sieve), the numbers that are not prime are methodically crossed out (or drop through the holes of the sieve). The remaining numbers are prime. The following procedure illustrates this process:

1. In Table 4-3, we cross out 1 because 1 is not prime.
2. Circle 2 because 2 is prime.
3. Cross out other multiples of 2; they are not prime.
4. Circle 3 because 3 is prime.
5. Cross out other multiples of 3.
6. Circle 5 and 7 because they are primes; cross out their multiples.
7. In Table 4-3, we stop after step 6 because 7 is the greatest prime whose square, 49, is less than 100. All the numbers remaining in the list and not crossed out are prime.

Table 4-3

1̶	②	③	4̶	⑤	6̶	⑦	8̶	9̶	1̶0̶
11	1̶2̶	13	1̶4̶	1̶5̶	1̶6̶	17	1̶8̶	19	2̶0̶
2̶1̶	2̶2̶	23	2̶4̶	2̶5̶	2̶6̶	2̶7̶	2̶8̶	29	3̶0̶
31	3̶2̶	3̶3̶	3̶4̶	3̶5̶	3̶6̶	37	3̶8̶	3̶9̶	4̶0̶
41	4̶2̶	43	4̶4̶	4̶5̶	4̶6̶	47	4̶8̶	4̶9̶	5̶0̶
5̶1̶	5̶2̶	53	5̶4̶	5̶5̶	5̶6̶	5̶7̶	5̶8̶	59	6̶0̶
61	6̶2̶	6̶3̶	6̶4̶	6̶5̶	6̶6̶	67	6̶8̶	6̶9̶	7̶0̶
71	7̶2̶	73	7̶4̶	7̶5̶	7̶6̶	7̶7̶	7̶8̶	79	8̶0̶
8̶1̶	8̶2̶	83	8̶4̶	8̶5̶	8̶6̶	87	8̶8̶	89	9̶0̶
9̶1̶	9̶2̶	93	9̶4̶	9̶5̶	9̶6̶	97	9̶8̶	9̶9̶	1̶0̶0̶

Primes have been treasured since antiquity, as the following cartoon depicts.

There are infinitely many whole numbers, infinitely many odd whole numbers, and infinitely many even whole numbers. Are there infinitely many primes? Because prime numbers do not appear in any known pattern, the answer to this question is not obvious. Euclid was the first to prove that there are infinitely many primes.

Mathematicians have long looked for a formula that produces only primes, but no one has found one. One result was the expression $n^2 - n + 41$, where n is a whole number. Substituting 0, 1, 2, 3, ... , 40 for n in the expression always results in a prime number. However, substituting 41 for n gives $41^2 - 41 + 41$, or 41^2, a composite number.

In 1998 Roland Clarkson, a 19-year-old student at California State University, showed that $2^{3021377} - 1$ is prime. The number has 909,526 digits. The full decimal expansion of the number would fill several hundred pages. This is an example of a *Mersenne prime*. A Mersenne prime, named after the French monk Marin Mersenne (1588–1648), is a prime of the form $2^n - 1$, where n is prime. As of July, 1999, there were only 38 known Mersenne primes.

Searching for large primes has led to advances in *distributed computing,* that is, using the Internet to use the unused computing power of great numbers of computers. Searching for Mersenne primes has been used as a test for computer hardware.

Another type of interesting prime is a *Sophie Germain prime,* which is an odd prime p for which $2p + 1$ is also a prime. Notice that $p = 3$ is a Sophie Germain prime, since

$2 \cdot 3 + 1$, or 7, is also a prime. Check that 5, 11, and 23 are also such primes. The primes were named after the French mathematician Sophie Germain. In 1999 Charles Kerchner III discovered the greatest Sophie Germain prime at that time to be $14{,}516{,}877 \cdot (2^{24176} - 1)$.

HISTORICAL NOTE

Sophie Germain (1776–1831) was born in Paris and grew up during the French Revolution. She wanted to study at the prestigious École Polytechnique but women were not allowed as students. Consequently, she studied from lecture notes and Gauss's monograph on number theory. In addition to her work in number theory, she made major contributions to the mathematical theory of elasticity, for which she was awarded the prize of the French Academy of Sciences. Germain's work was highly regarded by Gauss, who recommended her for an honorary degree from the University of Göttingen. She died before the degree could be awarded.

Problem Solving **How Many Bears?**

A large toy store carries one kind of stuffed bear. On Monday the store sold a certain number of the stuffed bears for a total of $1,843 and on Tuesday, without changing the price, the store sold a certain number of the stuffed bears for a total of $1,957. How many toy bears were sold each day if the price of each bear is a whole number and greater than $1?

- **Understanding the Problem**   One day a store sold a number of stuffed bears for $1,843 and on the next day a number of them for a total of $1,957. We need to find the number of bears sold on each day.

- **Devising a Plan**   If x bears were sold the first day and y bears the second day, and if the price of each bear was c dollars, we would have $cx = 1843$ and $cy = 1957$. Thus 1843 and 1957 should have a common factor—the price c. We could factor each number and find the possible factors. If the problem is to have a unique solution, the two numbers should have only one common factor other than 1. Any common factor of 1957 and 1843 will also be a factor of $1957 - 1843 = 114$ and the factors of 114 are easier to find.

- **Carrying Out the Plan**   We have $114 = 2 \cdot 57 = 2 \cdot 3 \cdot 19$. Thus if 1957 and 1843 have a common prime factor, it must be 2, 3, or 19. But neither 2 nor 3 divides the number, hence the only possible common factor is 19. We divide each number by 19 and find:

$$1843 = 19 \cdot 97.$$
$$1957 = 19 \cdot 103.$$

Notice that neither 97 nor 103 is divisible by 2, 3, 5, or 7. Hence 97 and 103 are primes (why?) and therefore the only common factor (greater than 1) of 1843 and 1957 is 19. Consequently, the price of each bear was $19. The first day 97 bears were sold and the next day 103 bears were sold.

- **Looking Back**   Notice that the problem had a unique solution because the only common factor (greater than 1) of the two numbers was 19. We could create similar problems by having the price of the item be a prime number and the numbers of items sold each day also be prime numbers. For example, the total sale on the first day could have been $23 \cdot 101$ or $2,323 and on the second day $23 \cdot 107$ or $2,461 (notice that 23, 101, and 107 are prime numbers).

  To find a common factor of 1957 and 1843, we found all the common factors of $1957 - 1843 = 114 = 2 \cdot 3 \cdot 19$ and checked which of the factors of the difference was a common factor of the original numbers. We have used the following property: If $d|a$ and $d|b$ then $d|(a-b)$. The property assures us that every common factor of a and b will also be a factor of $a-b$.

HISTORICAL NOTE

During World War II, Alan Turing, Peter Hilton, and other British analysts helped crack the codes developed on the German Enigma cipher machine. In the 1970s, determining large prime numbers became extremely useful in coding and decoding secret messages. In all coding and decoding, the letters of an alphabet correspond in some way to nonnegative integers. A "safe" coding system, in which messages are unintelligible to everyone except the intended receiver, was devised by three Massachusetts Institute of Technology scientists (Ronald Rivest, Adi Shamir, and Leonard Adleman) and is referred to as the RSA (their initials) system. The secret deciphering key consists of two large prime numbers chosen by the user. The enciphering key is the product of these two primes. Because it is extremely difficult and time consuming to factor large numbers, it was practically impossible to recover the deciphering key from a known enciphering key. In 1982, new methods for factoring large numbers were invented, which resulted in the use of even greater primes to prevent the breaking of decoding keys.

ONGOING ASSESSMENT 4-4

1. Find the least positive number that is divisible by three primes.
2. **a.** Fill in the missing numbers in the following factor tree:

   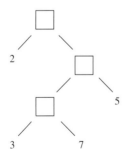

   b. How could you find the top number without finding the other two numbers?
3. Determine which of the following numbers are primes:
   a. 109   **b.** 119   **c.** 33
   d. 101   **e.** 463   **f.** 97
4. What is the greatest prime you must consider to test whether 5669 is prime?
5. Use a factor tree to find the prime factorization for each of the following:
   a. 504   **b.** 2475   **c.** 11,250
6. **a.** When the U.S. flag had 48 stars, the stars formed a 6×8 rectangular array. In what other rectangular arrays could they have been arranged?
   b. How many rectangular arrays of stars could there be if there were only 47 states?
7. If the Spanish Armada had consisted of 177 galleons, could it have sailed in an equal number of small flotillas? If so, how many ships would have been in each?
8. Suppose the 435 members of the House of Representatives are placed on committees consisting of more than 2 members but fewer than 30 members. Each committee is to have an equal number of members and each member is to be on only one committee.
   a. What size committees are possible?
   b. How many committees are there of each size?
9. Mr. Arboreta wants to set out fruit trees in a rectangular array. For each of the following numbers of trees, find all possible numbers of rows if each row is to have the same number of trees:
   a. 36   **b.** 28   **c.** 17   **d.** 144
10. Find the least number divisible by each natural number less than or equal to 12.
11. Some of the divisors of a locker number are 2, 5, and 9. If there are exactly nine additional divisors, what is the locker number?
12. **a.** Use the Fundamental Theorem of Arithmetic to justify that if $2|n$ and $3|n$, then $6|n$.
    b. Is it always true that if $a|n$ and $b|n$, then $ab|n^2$? Either prove the statement or give a counterexample.
13. Find the greatest 4-digit number that has exactly three factors.

14. Extend the Sieve of Eratosthenes to find all primes less than 200.
15. The prime numbers 11 and 13 are called *twin primes* because they differ by 2. Find all the twin primes less than 200. (The existence of infinitely many twin primes has not been proved.)
16. Show that if 1 were considered a prime, every number would have more than one prime factorization.
17. If $42|n$, what other positive integers divide n?
18. Is it possible to find positive integers x, y, and z such that $2^x \cdot 3^y = 5^z$? Why or why not?
19. A prime such as 7331 is a *superprime* because any integers obtained by deleting digits from the right of 7331 are prime; for example, 733, 73, and 7.
    a. For a prime to be a superprime, what digits cannot appear in the number?
    b. Of the digits that can appear in a superprime, what digit cannot be the left-most digit of a superprime?
    c. Find all of the two-digit superprimes.
    d. Find a three-digit superprime.
20. It is not known whether there are infinitely many primes in the infinite sequence consisting only of ones: 1, 11, 111, 1111, ... Find infinitely many composite numbers in the sequence.
21. Show that there are infinitely many composite numbers in the arithmetic sequence 1, 5, 9, 13, 17, ...
22. If $2N = 2^6 \cdot 3^5 \cdot 5^4 \cdot 7^3 \cdot 11^7$, explain why $2 \cdot 3 \cdot 5 \cdot 7 \cdot 11$ is a factor of N.
23. Is $3^2 \cdot 2^4$ a factor of $3^4 \cdot 2^7$? Explain why or why not.
24. Explain why each of the following numbers is composite:
    a. $3 \cdot 5 \cdot 7 \cdot 11 \cdot 13$
    b. $(3 \cdot 4 \cdot 5 \cdot 6 \cdot 7 \cdot 8) + 2$
    c. $(3 \cdot 5 \cdot 7 \cdot 11 \cdot 13) + 5$
    d. $10! + 7$
    e. $10! + k$, where $k = 2, 3, 4, 5, 6, 7, 8, 9,$ or 10
25. Explain why $2^3 \cdot 3^2 \cdot 25^3$ is not a prime factorization and find the prime factorization of the number.

Communication

26. Explain why the product of any three consecutive integers is divisible by 6.
27. Explain why the product of any four consecutive integers is divisible by 24.
28. In order to test for divisibility by 12, one student checked to determine divisibility by 3 and 4, while another checked for divisibility by 2 and 6. Are both students using a correct approach to divisibility by 12? Why or why not?
29. In the Sieve of Eratosthenes in Table 4-3, explain why, after we cross out all the multiples of 2, 3, 5, and 7, the remaining numbers are primes.
30. Let $M = 2 \cdot 3 \cdot 5 \cdot 7 + 11 \cdot 13 \cdot 17 \cdot 19$. Without multiplying, show that none of the primes less than or equal to 19 divides M.
★31. A woman with a basket of eggs finds that if she removes the eggs from the basket 3 or 5 at a time, there is always 1 egg left. However, if she removes the eggs 7 at a time, there are no eggs left. If the basket holds up to 100 eggs, how many eggs does the woman have? Explain your reasoning.
32. Explain why, when a number is composite, its least positive divisor, other than 1, must be prime.
33. a. An eighth grader in Roosevelt Middle School claims that because there are as many even numbers as odd numbers between 1 and 1000, there must be as many numbers that have an even number of divisors as numbers that have an odd number of divisors between 1 and 1000. Is the student correct? Why or why not?
    b. How many numbers between 1 and 1000 have an odd number of divisors?
★34. Euclid proved that, given any list of primes, there exists a prime not in the list. Read the following argument and answer the questions that follow.
    Let $2, 3, 5, 7, \ldots, p$ be a list of all the primes less than or equal to a certain prime p. We will show that there exists a prime not on the list. Consider the product $2 \cdot 3 \cdot 5 \cdot 7 \cdot \ldots \cdot p$. Notice that every prime in our list divides that product. However, if we add 1 to the product; that is, form the number $N = (2 \cdot 3 \cdot 5 \cdot 7 \cdot \ldots \cdot p) + 1$, then none of the primes in the list will divide N. Notice that whether N is prime or composite, some prime q must divide N. Because no prime in our list divides N, q is not one of the primes in our list. Consequently $q > p$. We have shown that there exists a prime greater than p.
    a. Explain why no prime in the list will divide N.
    b. Explain why some prime must divide N.
    c. Someone discovered a prime that has 909,526 digits. How does the above argument assure us that there exists even a larger prime?
    d. Does the above argument show that there are infinitely many primes? Why or why not?
    e. Let $M = 2 \cdot 3 \cdot 5 \cdot 7 \cdot 11 \cdot 13 \cdot 17 \cdot 19 + 1$. Without multiplying, explain why some prime greater than 19 will divide M.

Open-Ended

35. Describe how you could determine the prime factorization of a large number using a calculator. If your calculator has a $\boxed{\text{Simp}}$ button, describe how it might be used to obtain the prime factorization.
36. a. In which of the following intervals do you think there are the most primes? Why? Check to see if you were correct.
    (i) 0–99 (ii) 100–199 (iii) 200–299
    b. What is the longest string of consecutive composite numbers in the intervals?
    c. How many twin primes (see problem 15 above) are there in each interval?
    d. What patterns, if any, do you see for any of the questions above? Predict what might happen in other intervals.
37. It was reported that the greatest prime discovered so far has 65,050 digits. If you wrote this number out, how long a sheet of paper would you need?

38. A number is a *perfect number* if the sum of its factors (other than the number itself) is equal to the number. For example, 6 is a perfect number because its factors sum to 6, that is, 1 + 2 + 3 = 6. An *abundant number* has factors whose sum is greater than the number itself. A *deficient number* is a number with factors whose sum is less than the number itself.
   a. Classify each of the following numbers as perfect, abundant, or deficient:
      (i) 12   (ii) 28   (iii) 35
   b. Find at least one more number that falls in each class.

Cooperative Learning

39. A class of 23 students was using square tiles to build rectangular shapes. Each student had more than 1 tile and each had a different number of tiles. Each student was able to build only one shape of rectangle. All tiles had to be used to build a rectangle and the rectangle could not have holes. For example, a 2 by 6 rectangle uses 12 tiles and is considered the same as a 6 by 2 rectangle but is different from a 3 by 4 rectangle. The class did the activity using the least number of tiles. How many tiles did the class use? Divide the work among the members of your group to explore the various rectangles that could be made.

Review Problems

40. Classify the following as true or false:
    a. 11 is a factor of 189.
    b. 1001 is a multiple of 13.
    c. $7|1001$ and $7 \nmid 12$ imply $7 \nmid (1001 - 12)$.
    d. If a number is divisible by both 7 and 11, then its prime factorization contains 7 and 11.
41. Test each of the following for divisibility by 2, 3, 4, 5, 6, 7, 8, 9, 10, and 11:
    a. 438,162   b. 2,345,678,910
42. Prove that if a number is divisible by 12, then it is divisible by 3.
43. Could $3376 be divided exactly among either 7 or 8 people?

LABORATORY ACTIVITY

In Figure 4-24, a spiral starts with 41 at its center and continues in a counterclockwise direction. Primes are written in and squares that represent composites are shaded. Continue the spiral until you reach the prime 439. Check the primes along the diagonal. Can you find each of the primes from the formula $n^2 + n + 41$ by substituting appropriate values for n?

Figure 4-24

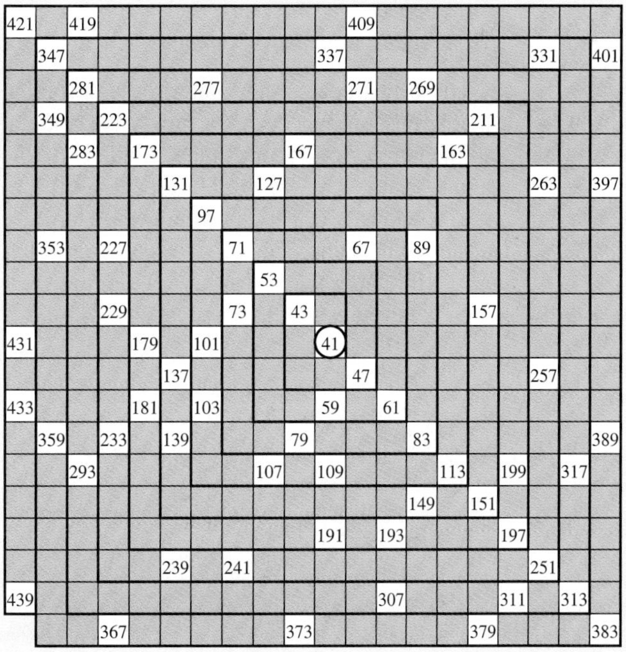

TECHNOLOGY CORNER

Edit the following Logo programs that determine whether a number is prime. Type the programs into your computer and use them to do problem 3 in Ongoing Assessment 4-4. To execute the program, type PRIME with the number you wish to check as input.

```
TO PRIME :N
  CHECK :N INTEGER (SQRT :N)
END

TO CHECK :N :D
  IF :N = 1 PRINT [NEITHER PRIME NOR COMPOSITE] STOP
  IF :D = 1 PRINT "PRIME STOP
  IF REMAINDER :N :D = 0 PRINT "COMPOSITE STOP
  CHECK :N :D - 1
END
```

Section 4-5 — Greatest Common Divisor and Least Common Multiple

Consider the following situation:

Two bands are to be combined to march in a parade. A 24-member band will march behind a 30-member band. The combined bands must have the same number of columns. What is the greatest number of columns in which they can march?

The bands could each march in two columns, and we would have the same number of columns, but this does not satisfy the condition of having the greatest number of columns. The number of columns must divide both 24 and 30. (Why?) Numbers that divide both 24 and 30 are 1, 2, 3, and 6. The greatest of these numbers is 6, so the bands should each march in columns of 6. The first band would have 6 columns with 4 members in each column, and the second band would have 6 columns with 5 members in each column.

In this problem, we have found the greatest number that divides both 24 and 30, that is, the **greatest common divisor (GCD)**.

greatest common divisor

> **Definition**
>
> The **greatest common divisor (GCD)** of two integers a and b is the greatest integer that divides both a and b.

We can find the GCD in many ways. We show several ways next.

Colored Rods Model

We can build a model of two or more integers with colored rods to determine the GCD of two positive integers. For example, consider finding the GCD of 6 and 8 using the 6 rod and the 8 rod, as in Figure 4-25.

Figure 4-25

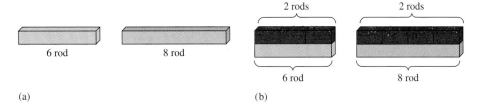

To find the GCD of 6 and 8, we must find the longest rod such that we can use multiples of that rod to build both the 6 rod and the 8 rod. The 1 rods and the 2 rods can be used to build both the 6 and 8 rods, as shown in Figure 4-25(b); the 3 rods can be used to build the 6 rod but not the 8 rod; the 4 rods can be used to build the 8 rod but not the 6 rod; the 5 rods can be used to build neither; and the 6 rods cannot be used to build the 8 rod. Therefore GCD(6, 8) = 2.

> **NOW TRY THIS 4-8**
>
> • Explain how you could use colored rods to solve the marching bands' problem. •

The Intersection-of-Sets Method

In the *intersection-of-sets* method, we list all members of the set of positive divisors of the two positive integers, then find the set of all *common divisors,* and, finally, pick the *greatest* element in that set. For example, to find the GCD of 20 and 32, denote the sets of divisors of 20 and 32 by D_{20} and D_{32}, respectively.

$$D_{20} = \{1, 2, 4, 5, 10, 20\}.$$
$$D_{32} = \{1, 2, 4, 8, 16, 32\}.$$

The set of all common positive divisors of 20 and 32 is

$$D_{20} \cap D_{32} = \{1, 2, 4\}.$$

Because the greatest number in the set of common positive divisors is 4, the GCD of 20 and 32 is 4, written GCD(20, 32) = 4.

The Prime Factorization Method

The intersection-of-sets method is rather time consuming and tedious if the numbers have many divisors. Another, more efficient, method is the prime factorization method. To find GCD(180, 168), first notice that

$$180 = 2 \cdot 2 \cdot 3 \cdot 3 \cdot 5$$
and
$$168 = 2 \cdot 2 \cdot 2 \cdot 3 \cdot 7.$$

We see that 180 and 168 have two factors of 2 and one of 3 in common. These common primes divide both 180 and 168. In fact, the only numbers other than 1 that divide both 180 and 168 must have no more than two 2s and one 3 and no other prime factors in their prime factorizations. The possible common divisors are 1, 2, 2^2, 3, $2 \cdot 3$, and $2^2 \cdot 3$. Hence, the greatest common divisor of 180 and 168 is $2^2 \cdot 3$. The procedure for finding the GCD of two or more numbers by using the prime factorization method is summarized as follows:

To find the GCD of two or more positive integers, first find the prime factorizations of the given numbers and then identify each common prime factor of the given numbers. The GCD is the product of the common factors, each raised to the lowest power of that prime that occurs in either of the prime factorizations.

If we apply the prime factorization technique to finding GCD(4, 9), we see that 4 and 9 have no common prime factors. But that does not mean there is no GCD. We still have 1 as a common divisor, so GCD(4, 9) = 1. Numbers such as 4 and 9, whose GCD is 1, are **relatively prime**

relatively prime. Both the intersection-of-sets method and the prime factorization method are found in *Scott Foresman-Addison Wesley Middle School Math Course 3,* 1999, as the page from the student book on page 223 shows. (GCF in the student page stands for *Greatest Common Factor* which is the same as GCD.)

Example 4-26 Find each of the following:

a. GCD(108, 72)
b. GCD(0, 13)
c. GCD(x, y) if $x = 2^3 \cdot 7^2 \cdot 11 \cdot 13$ and $y = 2 \cdot 7^3 \cdot 13 \cdot 17$
d. GCD(x, y, z) if $z = 2^2 \cdot 7$, using x and y from (b)
e. GCD(x, y) where $x = 5^4 \cdot 13^{10}$ and $y = 3^{10} \cdot 11^{20}$.

Solution
a. Since $108 = 2^2 \cdot 3^3$ and $72 = 2^3 \cdot 3^2$, it follows that GCD(108, 72) = $2^2 \cdot 3^2$ = 36.
b. Because $13|0$ and $13|13$, it follows that GCD(0, 13) = 13.
c. GCD(x, y) = $2 \cdot 7^2 \cdot 13$ = 1274.
d. Because $x = 2^3 \cdot 7^2 \cdot 11 \cdot 13$, $y = 2 \cdot 7^3 \cdot 13 \cdot 17$, and $z = 2^2 \cdot 7$, then GCD(x, y, z) = $2 \cdot 7$ = 14. Notice that GCD(x, y, z) can also be obtained by finding the GCD of z and 1274, the answer from (c).
e. Because x and y have no common prime factors, GCD(x, y) = 1.

Calculator Method

Calculators with a $\boxed{\text{Simp}}$ key can be used to find the GCD of two numbers. For example, to find the GCD(120, 180), use the following sequence of buttons to start: First, press $\boxed{1}\boxed{2}\boxed{0}\boxed{/}\boxed{1}\boxed{8}\boxed{0}\boxed{\text{Simp}}\boxed{=}$ to obtain the display $\boxed{\text{N/D} \rightarrow \text{n/d } 60/90}$. By pressing the $\boxed{x \bigcirc y}$ button, we see $\boxed{2}$ on the display as a common divisor of 120 and 180. By

Example 1

Find the GCF of 36 and 48.

1, 2, 3, 4, 6, 9, 12, 18, 36

1, 2, 3, 4, 6, 8, 12, 16, 24, 48

1, 2, 3, 4, 6, 12

The GCF of 36 and 48 is 12.

Making a list to find the GCF for larger numbers can be impractical.

The GCF can also be found using prime factorization.

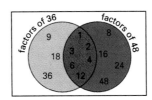

The GCF is found by multiplying together the smallest power of each common prime factor.

Example 2

Use prime factorization to find the GCF of 378 and 180.

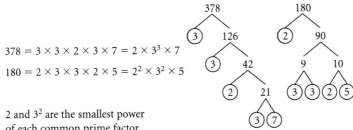

$378 = 3 \times 3 \times 2 \times 3 \times 7 = 2 \times 3^3 \times 7$

$180 = 2 \times 3 \times 3 \times 2 \times 5 = 2^2 \times 3^2 \times 5$

2 and 3^2 are the smallest power of each common prime factor.

$2 \times 3^2 = 18$

The GCF of 378 and 180 is 18.

Try It

Use prime factorization to find the GCF of each pair of numbers.

a. 72 and 138   **b.** 49 and 21   **c.** 276 and 136

You would use the same procedure of multiplying the smallest power of each common prime factor to find the GCF of more than two numbers.

pressing the $\boxed{x \cap y}$ button again and pressing $\boxed{\text{Simp}}$ $\boxed{=}$ $\boxed{x \cap y}$, we see 2 again as a factor. The process is repeated to reveal 3 and 5 as other common factors. The GCD(120, 180) is the product of the common prime factors $2 \cdot 2 \cdot 3 \cdot 5$, or 60.

Euclidean Algorithm Method

Some numbers are hard to factor. For these numbers, another method is more efficient for finding the GCD. For example, suppose we want to find GCD(676, 221). If we could find two smaller numbers whose GCD is the same as GCD(676, 221), our task would be easier. From Theorem 4-2(c), every divisor of 676 and 221 is also a divisor of 676 − 221 and 221. Conversely, every divisor of 676 − 221 and 221 is also a divisor of 676 and 221. Thus the set of all the common divisors of 676 and 221 is the same as the set of all common divisors of 676 − 221 and 221. Consequently, GCD(676, 221) = GCD(676 − 221, 221). This process can be continued to subtract three 221s from 676 so that GCD(676, 221) = GCD(676 − 3 · 221, 221) = GCD(13, 221). To determine how many 221s can be subtracted from 676, we could have divided as follows:

$$\begin{array}{r} 3 \\ 221\overline{)676} \\ 663 \\ \hline 13 \end{array}$$

Continuing, we see that GCD(13, 221) = GCD(0, 13) from the following division:

$$\begin{array}{r} 17 \\ 13\overline{)221} \\ 13 \\ \hline 91 \\ 91 \\ \hline 0 \end{array}$$

Because GCD(0, 13) = 13, the GCD(676, 221) = 13. Based on this illustration, we make the generalization outlined in the following theorem.

Theorem 4-7

If a and b are any whole numbers greater than 0 and $a \geq b$, then GCD (a, b) = GCD(r, b), where r is the remainder when a is divided by b.

Finding the GCD of two numbers by repeatedly using Theorem 4-7 until the remainder 0 is reached is referred to as the **Euclidean algorithm.**

Example 4-27 Use the Euclidean algorithm to find GCD(10,764, 2300).

Solution

$$\begin{array}{r} 4 \\ 2300\overline{)10764} \\ 9200 \\ \hline 1564 \end{array}$$ Thus GCD(10,764, 2300) = GCD(2300, 1564).

$$\begin{array}{r} 1 \\ 1564\overline{)2300} \\ 1564 \\ \hline 736 \end{array}$$ Thus GCD(2300, 1564) = GCD(1564, 736).

SECTION 4-5   *Greatest Common Divisor and Least Common Multiple*

$$\begin{array}{r} 2 \\ 736\overline{)1564} \\ \underline{1472} \\ 92 \end{array}$$ Thus GCD(1564, 736) = GCD(736, 92).

$$\begin{array}{r} 8 \\ 92\overline{)736} \\ \underline{736} \\ 0 \end{array}$$ Thus GCD(736, 92) = GCD(92, 0).

Because GCD(92, 0) = 92, it follows that GCD(10,764, 2300) = 92.

REMARK   The procedure for finding the GCD by using the Euclidean algorithm can be stopped at any step at which the GCD is obvious.

A calculator with the integer division feature can also be used to perform the Euclidean algorithm. This feature yields the quotient and the remainder when doing a division. For example, if the integer division key looks like $\boxed{\text{INT} \div}$, then to find GCD(10,764, 2300) we proceed as follows:

$\boxed{1}\boxed{0}\boxed{7}\boxed{6}\boxed{4}\boxed{\text{INT}\div}\boxed{2}\boxed{3}\boxed{0}\boxed{0}\boxed{=}$ which displays $\underset{Q}{4}$ $\underset{R}{1564}$.

$\boxed{2}\boxed{3}\boxed{0}\boxed{0}\boxed{\text{INT}\div}\boxed{1}\boxed{5}\boxed{6}\boxed{4}\boxed{=}$ which displays $\underset{Q}{1}$ $\underset{R}{736}$.

$\boxed{1}\boxed{5}\boxed{6}\boxed{4}\boxed{\text{INT}\div}\boxed{7}\boxed{3}\boxed{6}\boxed{=}$ which displays $\underset{Q}{2}$ $\underset{R}{92}$.

$\boxed{7}\boxed{3}\boxed{6}\boxed{\text{INT}\div}\boxed{9}\boxed{2}\boxed{=}$ which displays $\underset{Q}{8}$ $\underset{R}{0}$.

The last number we divided by when we obtained a 0 remainder is 92, so GCD (10764, 2300) = 92.

Sometimes shortcuts can be used to find the GCD of two or more numbers, as in the following example.

Example 4-28   Find each of the following:

a. GCD(134791, 6341, 6339)
b. The GCD of any two consecutive integers

Solution   a. Any common divisor of three numbers is also a common divisor of any two of them (why?). Consequently, the GCD of three numbers cannot be greater than the GCD of any two of the numbers. The numbers 6341 and 6339 are close to each other and therefore it is easy to find their GCD:

$$\text{GCD}(6341, 6339) = \text{GCD}(6341 - 6339, 6339)$$
$$= \text{GCD}(2, 6339)$$
$$= 1.$$

Because GCD(134791, 6341, 6339) cannot be greater than 1, it follows that it must equal 1.

b. Notice that GCD(4, 5) = 1, GCD(5, 6) = 1, GCD(6, 7) = 1, and GCD(99, 100) = 1. It seems that the GCD of any two consecutive integers is 1. To justify this conjecture, we need to show that for all integers n, GCD(n, $n + 1$) = 1. We have

$$\text{GCD}(n, n + 1) = \text{GCD}(n + 1, n) = \text{GCD}(n + 1 - n, n)$$
$$= \text{GCD}(1, n)$$
$$= 1.$$

Least Common Multiple

Hot dogs are usually sold 10 to a package, while hot dog buns are usually sold 8 to a package. This mismatch causes troubles when one is trying to match hot dogs and buns. What is the least number of packages of each you could order so that there is an equal number of hot dogs and buns? The numbers of hot dogs that we could have are just the multiples of 10, that is, 10, 20, 30, 40, 50, ... Likewise the possible numbers of buns are 8, 16, 24, 32, 40, 48, ... We can see that the number of hot dogs matches the number of buns whenever 10 and 8 have multiples in common. This occurs at 40, 80, 120, ... In this problem, we are interested in the least of these multiples, 40. Therefore we could obtain the same number of hot dogs and buns in the least amount by buying 4 packages of hot dogs and 5 packages of buns. The answer 40 is the **least common multiple (LCM)** of 8 and 10.

least common multiple (LCM)

Definition

Suppose that a and b are positive integers. Then the least common multiple (LCM) of a and b is the least positive integer that is simultaneously a multiple of a and a multiple of b.

As with GCDs, there are several methods for finding least common multiples.

Colored Rods Method

We can use colored rods to determine the LCM of two numbers. For example, consider the 3 rod and the 4 rod in Figure 4-26(a). We build trains of 3 rods and 4 rods until they are the same length, as shown in Figure 4-26(b). The LCM is the common length of the train.

Figure 4-26

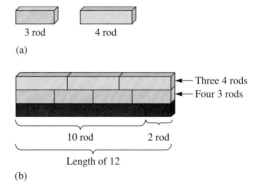

> **NOW TRY THIS 4-9**
>
> - Explain how to use colored rods to solve the hot dog problem, that is, to find LCM(8, 10).

The Intersection-of-Sets Method

In the *intersection-of-sets* method, we first find the set of all positive *multiples* of both the first and second numbers, then find the set of all *common multiples* of both numbers, and finally pick the *least* element in that set. For example, to find the LCM of 8 and 12, denote the sets of positive multiples of 8 and 12 by M_8 and M_{12}, respectively.

$$M_8 = \{8, 16, 24, 32, 40, 48, 56, 64, 72, \ldots\}.$$
$$M_{12} = \{12, 24, 36, 48, 60, 72, 84, 96, 108, \ldots\}.$$

The set of common multiples is

$$M_8 \cap M_{12} = \{24, 48, 72, \ldots\}.$$

Because the least number in $M_8 \cap M_{12}$ is 24, the LCM of 8 and 12 is 24, written LCM(8, 12) = 24.

The Prime Factorization Method

The intersection-of-sets method for finding the LCM is often lengthy, especially when it is used to find the LCM of three or more natural numbers. Another, more efficient method for finding the LCM of several numbers is the *prime factorization method*. For example, to find LCM(40, 12), first find the prime factorizations of 40 and 12, namely, $2^3 \cdot 5$ and $2^2 \cdot 3$, respectively.

If $m = $ LCM(40, 12), then m is a multiple of 40 and must contain both 2^3 and 5 as factors. Also, m is a multiple of 12 and must contain 2^2 and 3 as factors. Since 2^3 is a multiple of 2^2, then $m = 2^3 \cdot 5 \cdot 3 = 120$. In general, we have the following:

To find the LCM of two natural numbers, first find the prime factorization of each number. Then take each of the primes that are factors of either *of the given numbers. The LCM is the product of these primes, each raised to the greatest power of the prime that occurs in either of the prime factorizations.*

Example 4-29 Find the LCM of 2520 and 10,530.

Solution
$$2520 = 2^3 \cdot 3^2 \cdot 5 \cdot 7.$$
$$10{,}530 = 2 \cdot 3^4 \cdot 5 \cdot 13.$$
$$\text{LCM}(2520, 10530) = 2^3 \cdot 3^4 \cdot 5 \cdot 7 \cdot 13.$$

The Euclidean Algorithm Method

To see the connection between the GCD and LCM, consider the GCD and LCM of 6 and 9. Because $6 = 2 \cdot 3$ and $9 = 3^2$, it follows that $GCD(6, 9) = 3$ and $LCM(6, 9) = 18$. Notice that $GCD(6, 9) \cdot LCM(6, 9) = 3 \cdot 18 = 54$, and 54 is the product of the original numbers 6 and 9. In general, for any two natural numbers a and b, the connection between their GCD and LCM is given by Theorem 4-8.

> **Theorem 4-8**
>
> For any two natural numbers a and b,
>
> $$GCD(a, b) \cdot LCM(a, b) = ab.$$

Theorem 4-8 can be justified in several ways. Here is a specific example that suggests how the theorem might be proved.

Suppose $a = 5^{13} \cdot 7^{20} \cdot 11^4$ and
$b = 5^{10} \cdot 7^{25} \cdot 11^6 \cdot 13$.
Then $LCM(a, b) = 5^{13} \cdot 7^{25} \cdot 11^6 \cdot 13$ and
$GCD(a, b) = 5^{10} \cdot 7^{20} \cdot 11^4$.

Now we have

$LCM(a, b) \cdot GCD(a, b) = 5^{13+10} \cdot 7^{25+20} \cdot 11^{6+4} \cdot 13$ and
$ab = 5^{13+10} \cdot 7^{20+25} \cdot 11^{4+6} \cdot 13$.

For the above values of a and b, Theorem 4-8 is true. Notice, however, that in the product $LCM(a, b) \cdot GCD(a, b)$ we have all the powers of the primes appearing in a or in b, because for the LCM we take the greater of the powers of the common primes and for GCD the lesser. Also in ab we have all the powers. Hence Theorem 4-8 is true in general.

Theorem 4-8 is useful for finding the LCM of two numbers a and b when their prime factorizations are not easy to find. $GCD(a, b)$ can be found by the Euclidean algorithm, the product ab can be found by simple multiplication, and $LCM(a, b)$ can be found by division.

Example 4-30 Find $LCM(731, 952)$.

Solution By the Euclidean algorithm, $GCD(731, 952) = 17$. By Theorem 4-8, $17 \cdot LCM(731, 952) = 731 \cdot 952$. Consequently,

$$LCM(731, 952) = \frac{731 \cdot 952}{17} = 40{,}936.$$

Although Theorem 4-8 cannot be used to find the LCM of more than two numbers, it is possible to find the LCM for three or more numbers. For example, to find LCM(12, 108, 120), we can use the prime factorization method.

$$12 = 2^2 \cdot 3$$
$$108 = 2^2 \cdot 3^3$$
$$120 = 2^3 \cdot 3 \cdot 5$$

Then, LCM(12, 108, 120) = $2^3 \cdot 3^3 \cdot 5$ = 1080.

The Division-by-Primes Method

Another procedure for finding the LCM of several natural numbers involves *division by primes.* For example, to find LCM(12, 75, 120), we start with the least prime that divides at least one of the given numbers and divide as follows:

$$\begin{array}{r|rrr} 2 & 12, & 75, & 120 \\ \hline & 6, & 75, & 60 \end{array}$$

Because 2 does not divide 75, simply bring down the 75. To obtain the LCM using this procedure, continue the division process until the row of answers consists of relatively prime numbers.

$$\begin{array}{r|rrr} 2 & 12, & 75, & 120 \\ 2 & 6, & 75, & 60 \\ 2 & 3, & 75, & 30 \\ 3 & 3, & 75, & 15 \\ 5 & 1, & 25, & 5 \\ \hline & 1 & 5, & 1 \end{array} \longrightarrow \text{GCD is 3 (Why?)}$$

Thus LCM(12, 75, 120) = $2 \cdot 2 \cdot 2 \cdot 3 \cdot 5 \cdot 1 \cdot 5 \cdot 1 = 2^3 \cdot 3 \cdot 5^2 = 600$.

Two methods of finding the LCM of two numbers are given on the following student page from *Scott Foresman-Addison Wesley Middle School Math Course 3,* 1999.

Example 1

Paul will water his yard every 3 days and mow the lawn every 7 days. When is the first day he will do both?

Find the least common multiple of 3 and 7.

3, 6, 9, 12, 15, 18, 21, 24, 27, 30, 33, 36, 39, 42, 45, …

7, 14, 21, 28, 35, 42, 49, 56, 63, 70, 77, 84, 91, 98, 105, …

21, 42, …

The first day he will do both is the 21st.

Making a list to find the LCM for larger numbers can be impractical.

The LCM can also be found using prime factorization.
The LCM is found by multiplying the highest power of each prime factor.

Example 2

Use prime factorization to find the LCM of 30 and 45.

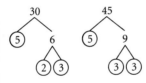

$30 = 2 \times 3 \times 5$

$45 = 3^2 \times 5$

2, 3^2, and 5

$2 \times 3^2 \times 5 = 2 \times 9 \times 5 = 90$

The LCM of 30 and 45 is 90.

Try It

Use prime factorization to find the LCM of the following pairs of numbers.

a. 6 and 5 **b.** 49 and 21 **c.** 276 and 136

You would use the same procedure to find the LCM of more than two numbers.

ONGOING ASSESSMENT 4-5

1. Find the GCD and the LCM for each of the following using the intersection-of-sets method:
   a. 18 and 10
   b. 24 and 36
   c. 8, 24, and 52
   d. 7 and 9
2. Find the GCD and the LCM for each of the following using the prime factorization method:
   a. 132 and 504
   b. 65 and 1690
   c. 900, 96, and 630
   d. 108 and 360
   e. 11 and 19
   f. 625, 750, and 1000
3. Find the GCD for each of the following using the Euclidean algorithm:
   a. 220 and 2924
   b. 14,595 and 10,856
   c. 122,368 and 123,152
4. Find the LCM for each of the following using any method:
   a. 24 and 36
   b. 72 and 90 and 96
   c. 90 and 105 and 315
   d. 9^{100} and 25^{100}
5. Find the LCM for each of the following pairs of numbers using Theorem 4-8 and the answers from problem 3 above:
   a. 220 and 2924
   b. 14,595 and 10,856
   c. 122,368 and 123,152
6. Use colored rods to find the GCD and the LCM of 6 and 10.
7. In Quinn's dormitory room, there are three snooze-alarm clocks, each of which is set at a different time. Clock A goes off every 15 min, clock B goes off every 40 min, and clock C goes off every 60 min. If all three clocks go off at 6:00 A.M., answer the following:
   a. How long will it be before the clocks go off together again after 6:00 A.M.?
   b. Would the answer to (a) be different if clock B went off every 15 min and clock A went off every 40 min?
8. At the Senior All-Night Party, a money chest contained enough money so that from 1 to 6 winners could share the money equally. The winners were to be chosen from those still in attendance at 4:00 A.M.; no one who had left early could win.
   a. What is the least amount of money that could be in the nonempty chest?
   b. If there were actually five winners, how much would each receive?
   c. If the prize money was to be given in $2 bills, how many bills were in the chest?
9. Midas has 120 gold coins and 144 silver coins. He wants to place his gold coins and his silver coins in stacks so that there are the same number of coins in each stack. What is the greatest number of coins that he can place in each stack?
10. Bill and Sue both work at night. Bill has every sixth night off and Sue has every eighth night off. If they are both off tonight, how many nights will it be before they are both off again?
11. By selling cookies at 24¢ each, José made enough money to buy several cans of pop costing 45¢ per can. If he had no money left over after buying the pop, what is the least number of cookies he could have sold?
12. Bijous I and II start their movies at 7:00 P.M. The movie at Bijou I takes 75 min, while the movie at Bijou II takes 90 min. If the shows run continuously, when will they start at the same time again?
13. Two bike riders ride around in a circular path. The first rider completes one round in 12 min and the second rider completes it in 18 min. If they both start at the same place and the same time and go in the same direction, after how many minutes will they meet again at the starting place?
14. Assume a and b are natural numbers and answer the following:
    a. If GCD(a, b) = 1, find LCM(a, b).
    b. Find GCD(a, a) and LCM(a, a).
    c. Find GCD(a^2, a) and LCM(a^2, a).
    d. If $a|b$, find GCD(a, b) and LCM(a, b).
    e. If a and b are two primes, find GCD(a, b) and LCM(a, b).
    f. What is the relationship between a and b if GCD(a, b) = a?
    g. What is the relationship between a and b if LCM(a, b) = a?
15. Classify each of the following as true or false. Justify your answers.
    a. If GCD(a, b) = 1, then a and b cannot both be even.
    b. If GCD(a, b) = 2, then both a and b are even.
    c. If a and b are even, then GCD(a, b) = 2.
    d. For all natural numbers a and b, LCM(a, b)|GCD(a, b).
    e. For all natural numbers a and b, LCM(a, b)|ab.
    f. GCD(a, b) $\leq a$
    g. LCM(a, b) $\geq a$
16. To find GCD(24, 20, 12), it is possible to find GCD(24, 20), which is 4, and then find GCD(4, 12), which is 4. Use this approach and the Euclidean algorithm to find
    a. GCD(120, 75, 105)
    b. GCD(34578, 4618, 4619)
17. a. Show that 97,219,988,751 and 4 are relatively prime.
    b. Show that 181,345,913 and 11 are relatively prime.
18. The radio station gave away a discount coupon for every fifth and sixth caller. Every twentieth caller received free concert tickets. Which caller was first to get both a coupon and a concert ticket?

19. Jackie spent the same amount of money on cassette tapes that she did on compact discs. If tapes cost $12 and CDs $16, what is the least amount she could have spent on each?
20. Larry and Mary bought a special 360-day joint membership to a tennis club. Larry will use the club every other day, and Mary will use the club every third day. They both use the club on the first day. How many days will neither person use the club in the 360 days?
21. At the Party Store, paper plates come in packages of 30, paper cups in packages of 15, and napkins in packages of 20. What is the least number of plates, cups, and napkins that can be purchased so that there is an equal number of each?
22. Determine how many complete revolutions gear 2 in the following must make before the arrows are lined up again:

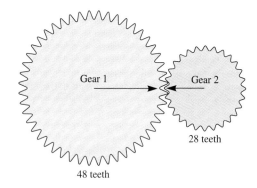

23. Determine how many complete revolutions each gear in the following must make before the arrows are lined up again:

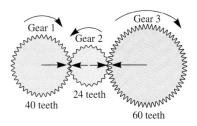

24. Venn diagrams can be used to show factors of two or more numbers. Draw Venn diagrams to show the common factors for each of the following sets of three numbers:
    a. 10, 15, 60
    b. 8, 16, 24
25. What are the factors of 4^{10}?
26. Find all natural numbers x such that $GCD(25, x) = 1$ and $1 \leq x \leq 25$.

Communication

27. Can two numbers have a greatest common multiple? Explain your answer.
28. Describe to a sixth-grade student the difference between a divisor and a multiple.
29. Is it true that $GCD(a, b, c) \cdot LCM(a, b, c) = abc$? Explain your answer.
30. A rectangular plot of land is 558 meters by 1212 meters. A surveyor needs to divide the plot into the largest possible square plots of the same size, being a whole number of meters long. What is the size of each square and how many square plots can be created? Explain your reasoning.
31. Suppose that $GCD(a, b, c) = 1$. Is it necessarily true that $GCD(a, b) = GCD(b, c) = 1$? Explain your reasoning.
32. Suppose $GCD(a, b) = GCD(b, c) = 2$. Does that always imply that $GCD(a, b, c) = 2$? Justify your answer.

Open-Ended

33. Examine three elementary school textbooks and report on how the introduction of the topics of GCD and LCM differ in the different textbooks.

Cooperative Learning

34. a. In your group, discuss whether the Euclidean algorithm for finding the GCD of two numbers should be introduced in middle school (to all students? to some?) Why or why not?
    b. If you decide that it should be introduced in middle school, discuss how it should be introduced. Report your group's decision to the class.

Review Problems

35. Find two whole numbers x and y such that $x \cdot y = 1,000,000$ and neither x nor y contains any zeros as digits.
36. Fill each blank space with a single digit that makes the corresponding statement true. Find all possible answers.
    a. $3 | 83\_51$
    b. $11 | 8\_691$
    c. $23 | 103\_6$
37. Is 3111 a prime? Prove your answer.
38. Find a number that has exactly six prime factors.
39. Produce the least positive number that is divisible by 2, 3, 4, 5, 6, 7, 8, 9, 10, and 11.
40. What is the greatest prime that must be used to determine if 2089 is prime?

TECHNOLOGY CORNER

1. Write a spreadsheet to generate the first 50 multiples of 3 and the first 50 multiples of 4. describe the intersection of the 2 sets.
2. Use a spreadsheet to find the factors of 2486. How far down do you need to copy the formula to be sure you have found all the divisors?

	A	B
1	1	= 2486/A1
2	2	
3	3	

3. Make a spreadsheet with four columns:
   Column A—the multiples of 6
   Column B—the multiples of 9
   Column C—the multiples of 12
   Column D—the multiples of 15

   a. What is the least number that appears in all four columns?
   b. Explain how to find this number without using a spreadsheet.

4. Edit the following Logo procedure for finding the GCD of two positive integers to fit your particular Logo software, and then use the procedure to find the GCD of the given numbers.

```
TO GCD :A :B
   IF :B = 0 OUTPUT :A
   OUTPUT GCD :B (REMAINDER :A :B)
END
```

   a. GCD (676, 221)
   b. GCD (10,764, 2300)

5. Use Theorem 4-8 and the preceding GCD procedure to write a Logo procedure LCM for finding the LCM of any two positive integers, :A and :B.

BRAIN TEASER

For any $n \times m$ rectangle such that $GCD(n, m) = 1$, find a rule for determining the number of unit squares (1×1) that a diagonal passes through. For example, in the drawings in Figure 4-27 the diagonal passes through 8 and 6 squares, respectively.

Figure 4-27

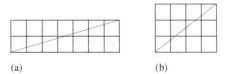

(a)   (b)

*Section 4-6 — Clock and Modular Arithmetic

In this section, we investigate clock arithmetic. Consider the following:

a. A doctor's prescription says to take a pill every 8 hr. If you take the first pill at 7:00 A.M., when should you take the next two pills?
b. Suppose you are following a bean soup recipe that calls for letting the beans soak for 12 hr. If you begin soaking them at 8:00 P.M., when should you take them out?
c. The odometer on a car gives the total miles traveled up to 99,999 miles and then starts counting from 0. If the odometer shows 99,124 miles, what will it show after a trip of 2,116 miles?

These situations involve the ability to solve arithmetic problems using clocks. Most people can solve these problems without thinking much about what they are doing. It is possible to use the clock in Figure 4-28(a) to determine that 8 hr after 7:00 A.M. is 3:00 P.M., and 8 hr after that is 11:00 P.M. Also, 12 hr after 8:00 P.M. is 8:00 A.M. We could record these additions on the clock as

$$7 \oplus 8 = 3, \quad 3 \oplus 8 = 11, \quad 8 \oplus 12 = 8,$$

where $\oplus$ indicates addition on a 12-hr clock.

Figure 4-28

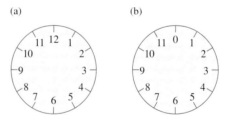

You probably noticed the special role of 12 when you found that $8 \oplus 12 = 8$. In 12-hr clock arithmetic, 12 acts like a 0 if you were adding in the set of whole numbers. For this reason, the 12 is often replaced by a 0, as in Figure 4-28(b). An addition table for the finite system based on the clock in Figure 4-28(b) is shown in Table 4-4.

Table 4-4

+	0	1	2	3	4	5	6	7	8	9	10	11
0	0	1	2	3	4	5	6	7	8	9	10	11
1	1	2	3	4	5	6	7	8	9	10	11	0
2	2	3	4	5	6	7	8	9	10	11	0	1
3	3	4	5	6	7	8	9	10	11	0	1	2
4	4	5	6	7	8	9	10	11	0	1	2	3
5	5	6	7	8	9	10	11	0	1	2	3	4
6	6	7	8	9	10	11	0	1	2	3	4	5
7	7	8	9	10	11	0	1	2	3	4	5	6
8	8	9	10	11	0	1	2	3	4	5	6	7
9	9	10	11	0	1	2	3	4	5	6	7	8
10	10	11	0	1	2	3	4	5	6	7	8	9
11	11	0	1	2	3	4	5	6	7	8	9	10

NOW TRY THIS 4-10

- Examine Table 4-4 to determine if the following properties hold for $\oplus$ on the set of numbers in the table:
  a. Commutative property of addition
  b. Associative property of addition
  c. Identity property of addition
  d. Inverse property of addition

When we allow numbers other than those on the 12-hr clock to be added, such as $8 \oplus 24 = 8$, we find that numbers such as 24, 36, 48, … act like 0. Likewise, the numbers 13, 25, 37, … act like the number 1. Similarly, we can generate classes of numbers that act like each of the numbers on the 12-hr clock. The members of any one class differ by multiples of 12. Consequently, to perform additions on a 12-hr clock we perform regular addition, divide by 12, and record the remainder as the answer. For example, we can find $11 \oplus 8$ and $8 \oplus 12$ as follows:

$11 + 8 = 19$. Next divide $19 \div 12$. The quotient is 1 with a remainder of 7, which is the answer.

$8 + 12 = 20$. Next divide $20 \div 12$. The quotient is 1 with a remainder of 8, which is the answer.

Whenever the sum of digits on a 12-hr clock exceeds 12, add the numbers normally and then obtain the remainder when the sum is divided by 12.

To perform other operations on the clock, such as $2 \ominus 9$, where $\ominus$ denotes clock subtraction, we could interpret it as the time 9 hr before 2 o'clock. Counting backward (counterclockwise) 9 units from 2 reveals that $2 \ominus 9 = 5$. If subtraction on the clock is defined in terms of addition, we have $2 \ominus 9 = x$ if, and only if, $2 = 9 \oplus x$. Consequently, $x = 5$.

Example 4-31 Perform each of the following computations on a 12-hr clock:
a. $8 \oplus 8$ b. $4 \ominus 12$ c. $4 \ominus 4$ d. $4 \ominus 8$

Solution a. $(8 + 8) \div 12$ has remainder 4. Hence, $8 \oplus 8 = 4$.
b. $4 \ominus 12 = 4$, since, by counting forward or backward 12 hr, you arrive at the original position.
c. $4 \ominus 4 = 12$. This should be clear from looking at the clock, but it can also be found by using the definition of subtraction in terms of addition.
d. $4 \ominus 8 = 8$ because $8 \oplus 8 = 4$.

Clock multiplication can be defined using repeated addition, as with whole numbers. For example, $2 \otimes 8 = 8 \oplus 8 = 4$, where $\otimes$ denotes clock multiplication. Similarly, $3 \otimes 5 = (5 \oplus 5) \oplus 5 = 10 \oplus 5 = 3$.

Clock division can be defined in terms of multiplication. For example, $8 \oslash 5 = x$, where $\oslash$ denotes clock division, if, and only if, $8 = 5 \otimes x$ for a unique x in the set $\{1, 2, 3, \ldots, 12\}$. Because $5 \otimes 4 = 8$, then $8 \oslash 5 = 4$.

Example 4-32 Perform the following operations on a 12-hr clock, if possible:

a. $3 \otimes 11$ **b.** $2 \ominus 7$ **c.** $3 \oslash 2$ **d.** $5 \oslash 12$

Solution
a. $3 \otimes 11 = (11 \oplus 11) \oplus 11 = 10 \oplus 11 = 9$.
b. $2 \ominus 7 = x$ if, and only if, $2 = 7 \oplus x$. Consequently, $x = 2$.
c. $3 \oslash 2 = x$ if, and only if, $3 = 2 \otimes x$. Multiplying each of the numbers 1, 2, 3, 4, ... , 12 by 2 shows that none of the multiplications yields 3. Thus the equation $3 = 2 \otimes x$ has no solution, and consequently, $3 \oslash 2$ is undefined.
d. $5 \oslash 12 = x$ if, and only if, $5 = 12 \otimes x$. However, $12 \otimes x = 12$ for every x in the set $\{1, 3, 4, ... , 12\}$. Thus $5 = 12 \otimes x$ has no solution on the clock; and therefore $5 \oslash 12$ is undefined.

Adding or subtracting 12 on a 12-hr clock gives the same result. Thus 12 behaves as 0 does in a base-ten addition or subtraction and is the additive identity for addition on the 12-hr clock. Similarly, on a 5-hr clock 5 behaves as 0 does.

Addition, subtraction, and multiplication on a 12-hr clock can be performed for any two numbers but, as shown in Example 4-32(d), not all divisions can be performed. Division by 12, the additive identity, on a 12-hr clock either can never be performed or is not meaningful, since it does not yield a unique answer. However, there are clocks on which all divisions can be performed, except by the corresponding additive identities. One such clock is a 5-hr clock, shown in Figure 4-29.

Figure 4-29

On this clock, $3 \oplus 4 = 2$, $2 \ominus 3 = 4$, $2 \otimes 4 = 3$, and $3 \oslash 4 = 2$. Since adding 5 to any number yields the original number, 5 is the additive identity for this 5-hr clock, as seen in Table 4-5(a). Consequently, you might suspect that division by 5 is not possible on a 5-hour clock. To determine which divisions are possible, consider Table 4-5(b), a multiplication table for 5-hr clock arithmetic. To find $1 \oslash 2$, we write $1 \oslash 2 = x$, which is equivalent to $1 = 2 \otimes x$. The second row of Table 4-5(b) shows that $2 \otimes 1 = 2$, $2 \otimes 2 = 4$, $2 \otimes 3 = 1$, $2 \otimes 4 = 3$, and $2 \otimes 5 = 5$. The solution of $1 = 2 \otimes x$ is $x = 3$, so $1 \oslash 2 = 3$. The information given in the second row of the table can be used to determine the following divisions:

$$2 \oslash 2 = 1 \text{ because } 2 = 2 \otimes 1$$
$$3 \oslash 2 = 4 \text{ because } 3 = 2 \otimes 4$$
$$4 \oslash 2 = 2 \text{ because } 4 = 2 \otimes 2$$
$$5 \oslash 2 = 5 \text{ because } 5 = 2 \otimes 5$$

Table 4-5

(a)

$\oplus$	1	2	3	4	5
1	2	3	4	5	1
2	3	4	5	1	2
3	4	5	1	2	3
4	5	1	2	3	4
5	1	2	3	4	5

(b)

$\otimes$	1	2	3	4	5
1	1	2	3	4	5
2	2	4	1	3	5
3	3	1	4	2	5
4	4	3	2	1	5
5	5	5	5	5	5

Because every element occurs in the second row, division by 2 is always possible. Similarly, division by all other numbers, except 5, is always possible. In the problem set, you are asked to perform arithmetic on different clocks and to investigate for which clocks all computations, except division by the additive identity, can be performed.

Modular Arithmetic

APRIL						
S	M	T	W	T	F	S
1	2	3	4	5	6	7
8	9	10	11	12	13	14
15	16	17	18	19	20	21
22	23	24	25	26	27	28
29	30					

Figure 4-30

Many of the concepts for clock arithmetic can be used to work problems that involve a calendar. On the calendar in Figure 4-30, the five Sundays have dates 1, 8, 15, 22, and 29. Any two of these dates for Sunday differ by a multiple of 7. The same property is true for any other day of the week. For example, the second and thirtieth days fall on the same day, since $30 - 2 = 28$ and 28 is a multiple of 7. We say that 30 is congruent to 2, modulo 7, and we write $30 \equiv 2 \pmod{7}$. Similarly, because 18 and 6 differ by a multiple of 12, we write $18 \equiv 6 \pmod{12}$. This leads to the following definition.

Definition of Modular Congruence

For integers a and b, ***a* is congruent to *b* modulo *m*,** written $a \equiv b \pmod{m}$, if, and only if, $a - b$ is a multiple of m, where m is a positive integer greater than 1.

REMARK This definition could be written as $a \equiv b \pmod{m}$ if, and only if, $m|(a - b)$, where m is a positive whole number greater than 1.

Notice that 18 and 25 are congruent modulo 7 and each number leaves the same remainder—4—upon division by 7. Indeed, $18 = 2 \cdot 7 + 4$ and $25 = 3 \cdot 7 + 4$. In general, we have the following property: *Two whole numbers are congruent modulo m if, and only if, their remainders on division by m are the same.* In the student page on page 238 from *Scott Foresman-Addison Wesley Middle School Math Course* 3, 1999, modular arithmetic modulo 7 is introduced through remainders. Answer the questions that follow the explanation on the student page.

Example 4-33 Tell why each of the following is true:

a. $23 \equiv 3 \pmod{10}$
b. $23 \equiv 3 \pmod{4}$
c. $23 \not\equiv 3 \pmod{7}$
d. $10 \equiv {}^{-}1 \pmod{11}$
e. $25 \equiv 5 \pmod{5}$

Solution **a.** $23 \equiv 3 \pmod{10}$ because $23 - 3$ is a multiple of 10.
b. $23 \equiv 3 \pmod{4}$ because $23 - 3$ is a multiple of 4.
c. $23 \not\equiv 3 \pmod{7}$ because $23 - 3$ is not a multiple of 7.
d. $10 \equiv {}^{-}1 \pmod{11}$ because $10 - ({}^{-}1) = 11$ is a multiple of 11.
e. $25 \equiv 5 \pmod{5}$ because $25 - 5 = 20$ is a multiple of 5.

Extend Key Ideas — Discrete Math

Modular Arithmetic

People keep track of time using months, days of the week, hours, and so on. In mathematics, modular arithmetic can be used to show these ways of counting. If today is Monday, then 7 days later it's Monday again. Days of the week use modulo 7 because there are seven days in a week.

Day	Sat.	Sun.	Mon.	Tues.	Wed.	Thurs.	Fri.
Day (Modulo 7)	1	2	3	4	5	6	0

If we apply the modulo 7 table above to a calendar for January 2011, you'll see how multiples of seven are always Fridays.

If you divide a date by 7 and there is a remainder of 1, then it's a Saturday.
Remainder 2 is a Sunday = 2 (modulo 7).
Remainder 3 is a Monday = 3 (modulo 7).
Remainder 4 is a Tuesday = 4 (modulo 7).

February 2011 has a total of 28 days and February 1 is on a Tuesday. You can determine which day is February 28 by using modulo 7 and letting Tuesday = 1 (modulo 7). 28 ÷ 7 = 4 with 0 remainder. So, the last day of February is on a Monday.

Day	Tues.	Wed.	Thurs.	Fri.	Sat.	Sun.	Mon.
Day (Modulo 7)	1	2	3	4	5	6	0

1. Use modular arithmetic to find what day February 14 falls on in 2011.
2. In 2011, March 1 is on a Tuesday; the month ends March 31. Make a modulo 7 table for March.

Use modular arithmetic to find the day of the week for the following dates.

3. March 9   4. March 22   5. March 31

Example 4-34 Find all integers x such that $x \equiv 1 \pmod{10}$.

Solution $x \equiv 1 \pmod{10}$ if, and only if, $x - 1 = 10k$, where k is any integer. Consequently, $x = 10k + 1$. Letting $k = 0, 1, 2, 3, \ldots$ yields the sequence $1, 11, 21, 31, 41, \ldots$. Likewise, letting $k = {}^-1, {}^-2, {}^-3, {}^-4, \ldots$ yields the negative integers ${}^-9, {}^-19, {}^-29, {}^-39, \ldots$. The two sequences can be combined to give the solution set

$$\{\ldots, {}^-39, {}^-29, {}^-19, {}^-9, 1, 11, 21, 31, 41, 51, \ldots\}.$$

The $\boxed{\text{INT} \div}$ button on a calculator can be used to work with modular arithmetic. If we press the following sequence of buttons, we see that $4325 \equiv 5 \pmod 9$ because the remainder when 4325 is divided by 9 is 5:

$$\boxed{4}\,\boxed{3}\,\boxed{2}\,\boxed{5}\,\boxed{\text{INT}\div}\,\boxed{9}\,\boxed{=},$$

and the display shows a remainder of 5.

Example 4-35 Heidi signed a promissory note that will become due in 90 days. She is worried that it will become due on a weekend. She signed the note on a Monday. On what day of the week will it be due?

Solution Because $90 = 7 \cdot 12 + 6$, we know that $90 \equiv 6 \pmod 7$. On a fraction calculator, you could enter $\boxed{9}\,\boxed{0}\,\boxed{\text{INT}\div}\,\boxed{7}\,\boxed{=}$, and a quotient of 12 with remainder 6 would be displayed. Therefore the note will come due 12 wk and 6 days after Monday, which is a Sunday.

Example 4-36
a. If it is now Monday, October 14, on what day of the week will October 14 fall next year if next year is not a leap year?
b. If Christmas falls on Thursday this year, on what day of the week will Christmas fall next year if next year is a leap year?

Solution
a. Because next year is not a leap year, we have 365 days in the year. Because $365 = 52 \cdot 7 + 1$, we have $365 \equiv 1 \pmod 7$. Thus 365 days after October 14 will be 52 wk and one day later. Thus October 14 will be on a Tuesday.
b. Because there are 366 days in a leap year, we have $366 \equiv 2 \pmod 7$. Thus Christmas will be 2 days after Thursday, on Saturday.

ONGOING ASSESSMENT 4-6

1. Dr. Harper prescribed some medicine for Camile. She is supposed to take a dose every 6 hr. If she takes her first dose at 8:00 A.M., when should she take her next dose?
2. Perform each of the following operations on a 12-hr clock, if possible:
   a. $7 \oplus 8$ b. $4 \oplus 10$ c. $3 \ominus 9$
   d. $4 \ominus 8$ e. $3 \otimes 9$ f. $4 \otimes 4$
   g. $1 \oslash 3$ h. $2 \oslash 5$
3. Perform each of the following operations on a 5-hr clock:
   a. $3 \oplus 4$ b. $3 \oplus 3$ c. $3 \otimes 4$
   d. $1 \otimes 4$ e. $4 \otimes 4$ f. $2 \otimes 3$
   g. $3 \ominus 4$ h. $1 \ominus 4$
4. a. Construct an addition table for a 7-hr clock.
   b. Using the addition table in (a), find $5 \ominus 6$ and $2 \ominus 5$.
   c. Using the addition table in (a), show that subtraction can always be performed on a 7-hr clock.
5. a. Construct a multiplication table for a 7-hr clock.
   b. Use the multiplication table in (a) to find $3 \oslash 5$ and $4 \oslash 6$.
   c. Use the multiplication table to find whether division by numbers different from 7 is always possible.
6. On a 12-hr clock, find each of the following:
   a. Additive inverse of 2   b. Additive inverse of 3
   c. $(^{-}2) \oplus (^{-}3)$ d. $^{-}(2 \oplus 3)$
   e. $(^{-}2) \ominus (^{-}3)$ f. $(^{-}2) \otimes (^{-}3)$
7. a. If April 23 falls on Tuesday, what are the dates of the other Tuesdays in April?
   b. If July 2 falls on Tuesday, list the dates of the Wednesdays in July.
   c. If September 3 falls on Monday, on what day of the week will it fall next year if next year is a leap year?
8. Fill in each of the following blanks so that the answer is nonnegative and the least possible number:
   a. $29 \equiv$ _____ (mod 5)
   b. $3498 \equiv$ _____ (mod 3)
   c. $3498 \equiv$ _____ (mod 11)
   d. $^{-}23 \equiv$ _____ (mod 10)
9. a. Find all x such that $x \equiv 0$ (mod 2).
   b. Find all x such that $x \equiv 1$ (mod 2).
   c. Find all x such that $x \equiv 3$ (mod 5).

Communication

10. Explain how the odometer on a car uses modular arithmetic. What is the mod?

Open-Ended

11. On a clock we define the additive inverse of a in the same way as the additive inverse was defined for integers. Having this definition in mind, list some similarities and some differences between the number system on the clock and the set of integers. Justify your answers.

Cooperative Learning

12. a. Have members of your group construct the multiplication tables for 3-hr, 4-hr, 6-hr, and 11-hr clocks.
    b. Compare your results. On which of the clocks in (a) can divisions by numbers other than the additive identity always be performed?
    c. How do the multiplication tables of clocks for which division can always be performed (except by an additive identity) differ from the multiplication tables of clocks for which division is not always meaningful?

BRAIN TEASER How many primes are in the following sequence?
9, 98, 987, 9876, ... , 987654321, 9876543219, 98765432198, ...

HINT FOR SOLVING THE PRELIMINARY PROBLEM

Consider all of the triples of three positive integers whose product is 36 and find the corresponding sum (e.g., $1 \cdot 3 \cdot 12 = 36$ and $1 + 3 + 12 = 16$). After you list all possible triples and the corresponding sums, it will be clear that the remark about the oldest child differentiates between two possible cases.

QUESTIONS FROM THE CLASSROOM

1. A fourth-grade student devised the following subtraction algorithm for subtracting $84 - 27$.

   Four minus seven equals negative three.
   $$\begin{array}{r} 84 \\ -\ 27 \\ \hline ^-3 \end{array}$$

   Eighty minus twenty equals sixty.
   $$\begin{array}{r} 84 \\ -\ 27 \\ \hline ^-3 \\ 60 \end{array}$$

   Sixty plus negative three equals fifty-seven.
   $$\begin{array}{r} 84 \\ -\ 27 \\ \hline ^-3 \\ +\ 60 \\ \hline 57 \end{array}$$

   Thus the answer is 57. What is your response as a teacher?

2. A seventh-grade student does not believe that $^-5 < {}^-2$. The student argues that a debt of \$5 is greater than a debt of \$2. How do you respond?

3. An eighth-grade student claims she can prove that subtraction of integers is commutative. She points out that if a and b are integers, then $a - b = a + {}^-b$. Since addition is commutative, so is subtraction. What is your response?

4. A student computes $^-8 - 2(^-3)$ by writing $^-10(^-3) = 30$. How would you help this student?

5. A student says that his father showed him a very simple method for dealing with expressions like $^-(a - b + 1)$ and $x - (2x - 3)$. The rule is, if there is a negative sign before the parentheses, change the signs of the expressions inside the parentheses. Thus, $^-(a - b + 1) = {}^-a + b - 1$ and $x - (2x - 3) = x - 2x + 3$. What is your response?

6. A student had the following picture of an integer and its opposite. Other students in the class objected, saying that ^-a should be to the left of 0. How do you respond?

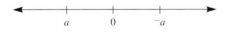

7. A student found that addition of integers can be performed by finding the sum or the difference of the absolute values of these integers. She would like to know if this is always true. How do you respond?

8. A student claims that $a|a$ and $a|a$ implies $a|(a - a)$, and hence, $a|0$. Is the student correct?

9. A student writes, "If $d \nmid a$ and $d \nmid b$, then $d \nmid (a + b)$." How do you respond?

10. Your seventh-grade class has just completed a unit on divisibility rules. One of the better students asks why divisibility by numbers other than 3 and 9 cannot be tested by dividing the sum of the digits by the tested number. How should you respond?

11. A student claims that a number with an even number of digits is divisible by 7 if, and only if, each of the numbers formed by pairing the digits into groups of two is divisible by 7. For example, 49,562,107 is divisible by 7, since each of the numbers 49, 56, 21, and 07 is divisible by 7. Is this true?

12. A sixth-grade student argues that there are infinitely many primes because "there is no end to numbers." How do you respond?

13. A student claims that a number is divisible by 21 if, and only if, it is divisible by 3 and by 7, and, in general, a number is divisible by $a \cdot b$ if, and only if, it is divisible by a and by b. What is your response?

14. A student claims that for any two integers a and b, $GCD(a, b)$ divides $LCM(a, b)$ and, hence, $GCD(a, b) < LCM(a, b)$. Is the student correct? Why or why not?

15. A student claims that there are infinitely many triples of positive integers x, y, and z, that make the equation $x^2 + y^2 = z^2$ true. How do you respond?

16. A student argues that 1 should be a prime because it has 1 and itself as divisors. How do you respond?

17. A student asks about the relation between least common multiple and least common denominator. How do you respond?

CHAPTER OUTLINE

I. Basic concepts of integers
  A. The set of **integers**, I, is $\{\ldots, {}^-3, {}^-2, {}^-1, 0, 1, 2, 3, \ldots\}$.
  B. The distance from any integer to 0 is called the **absolute value** of the integer. The absolute value of an integer x is denoted $|x|$.
  C. Operations with integers

  1. Addition: For any integers a and b,
  $$^-a + {}^-b = {}^-(a + b)$$

  2. Subtraction
   a. If a and b are any integers, then $a - b = n$ if, and only if, $a = b + n$.
   b. For all integers a and b, $a - b = a + {}^-b$.

3. **Multiplication:** For any integers a and b,
   a. $(^-a) \cdot (^-b) = ab$
   b. $(^-a) \cdot b = b \cdot (^-a) = {}^-(ab)$
4. **Division:** If a and b are any integers with $b \neq 0$, then $a \div b$ is the unique integer c, if it exists, such that $a = bc$.
5. **Order of operations:** When addition, subtraction, multiplication, and division appear without parentheses, multiplications and divisions are done first in the order of their appearance from left to right and then additions and subtractions are done in the order of their appearance from left to right. Any arithmetic in parentheses is done first.

II. The system of integers
   A. The set of integers, $I = \{\ldots, {}^-3, {}^-2, {}^-1, 0, 1, 2, 3, \ldots\}$, along with the operations of addition and multiplication, satisfy the following properties:

Property	+	×
Closure	Yes	Yes
Commutative	Yes	Yes
Associative	Yes	Yes
Identity	Yes, 0	Yes, 1
Inverse	Yes	No
Distributive Property of Multiplication over Addition		

   B. **Zero multiplication property of integers**
      For any integer a, $a \cdot 0 = 0 = 0 \cdot a$.
   C. **Addition property of equality:** For any integers a, b, and c, if $a = b$, then $a + c = b + c$.
   D. **Multiplication property of equality:** For any integers a, b, and c, if $a = b$, then $ac = bc$.
   E. **Substitution property:** Any number may be substituted for its equal.
   F. **Cancellation properties of equality**
      1. For any integers a, b, and c, if $a + c = b + c$, then $a = b$.
      2. For any integers a, b, and c, if $c \neq 0$ and $ac = bc$, then $a = b$.
   G. For all integers a, b, and c,
      1. $^-(^-a) = a$
      2. $a - (b - c) = a - b + c$
      3. $(a + b)(a - b) = a^2 - b^2$ (**difference-of-squares formula**)

III. Divisibility
   A. If a and b are any integers, then b **divides** a, denoted by $b|a$, if, and only if, there is a unique integer c such that $a = cb$.
   B. The following are basic divisibility theorems for integers a, b, and d:
      1. If $d|a$ and k is any integer, then $d|ka$.
      2. If $d|a$ and $d|b$, then $d|(a + b)$ and $d|(a - b)$.
      3. If $d|a$ and $d\!\not|\,b$, then $d\!\not|\,(a + b)$ and $d\!\not|\,(a - b)$.
   C. Divisibility tests
      1. An integer is divisible by 2, 5, or 10 if, and only if, its units digit is divisible by 2, 5, or 10, respectively.
      2. An integer is divisible by 4 if, and only if, the last 2 digits of the integer represent a number divisible by 4.
      3. An integer is divisible by 8 if, and only if, the last 3 digits of the integer represent a number divisible by 8.
      4. An integer is divisible by 3 or by 9 if, and only if, the sum of its digits is divisible by 3 or 9, respectively.
      5. An integer is divisible by 11 if, and only if, the sum of the digits in the places that are even powers of 10 minus the sum of the digits in the places that are odd powers of 10 is divisible by 11.
      6. An integer is divisible by 6 if, and only if, the integer is divisible by both 2 and 3.

IV. Prime and composite numbers
   A. Positive integers that have exactly two positive divisors are called **primes**. Integers greater than 1 and not primes are called **composites**.
   B. **Fundamental theorem of arithmetic:** Every composite number has one and only one prime factorization, aside from variation in the order of the prime factors.
   C. Criterion for determining if a given number n is prime: *If n is not divisible by any prime p such that $p^2 \leq n$, then n is prime.*
   D. If the prime factorization of a number is $p^n q^m$, where p and q are prime, then the number of divisors of n is $(n + 1)(m + 1)$.

V. Greatest common divisor and least common multiple
   A. The **greatest common divisor (GCD)** of two or more natural numbers is the greatest divisor, or factor, that the numbers have in common.
   B. **Euclidean algorithm:** If a and b are positive integers and $a \geq b$, then $\text{GCD}(a, b) = \text{GCD}(b, r)$, where r is the remainder when a is divided by b. The procedure of finding the GCD of two numbers a and b by using the above result repeatedly is the *Euclidean algorithm*.
   C. The **least common multiple (LCM)** of two or more natural numbers is the least positive multiple that the numbers have in common.
   D. $\text{GCD}(a, b) \cdot \text{LCM}(a, b) = ab$.
   E. If $\text{GCD}(a, b) = 1$, then a and b are **relatively prime**.

*VI. Modular arithmetic
   A. For any integers a and b, a **is congruent to** b **modulo** m if, and only if, $a - b$ is a multiple of m, where m is a positive integer greater than 1.
   B. Two integers are congruent modulo m if, and only if, their remainders upon division by m are the same.

CHAPTER REVIEW

1. Find the additive inverse of each of the following:
   a. 3   b. ^-a c. 0
   d. $x + y$ e. $^-x + y$ f. $(^-2)^5$ g. $^-2^5$
2. Perform each of the following operations:
   a. $(^-2 + {}^-8) + 3$ b. $^-2 - (^-5) + 5$
   c. $^-3(^-2) + 2$ d. $^-3(^-5 + 5)$
   e. $^-40 \div (^-5)$ f. $(^-25 \div 5)(^-3)$
3. For each of the following, find all integer values of x (if there are any) that make the given equation true:
   a. $^-x + 3 = 0$ b. $^-2x = 10$
   c. $0 \div (^-x) = 0$ d. $^-x \div 0 = {}^-1$
   e. $3x - 1 = {}^-124$ f. $^-2x + 3x = x$
4. Use a pattern approach to explain why $(^-2)(^-3) = 6$.
5. In each of the following chip models, the encircled chips are removed. Write the corresponding integer problem with its solution.

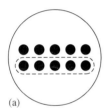
(a)

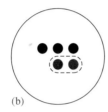
(b)

6. a. Show that $(x - y)(x + y) = x^2 - y^2$.
   b. Use the result in (a) to simplify $(^-2 - x)(^-2 + x)$.
7. Simplify each of the following expressions:
   a. ^-1x b. $(^-1)(x - y)$
   c. $2x - (1 - x)$ d. $(^-x)^2 + x^2$
   e. $(^-x)^3 + x^3$ f. $(^-3 - x)(3 + x)$
8. Factor each of the following expressions and then simplify, if possible:
   a. $x - 3x$ b. $x^2 + x$
   c. $x^2 - 36$ d. $81y^6 - 16x^4$
   e. $5 + 5x$ f. $(x - y)(x + 1) - (x - y)$
9. Classify each of the following as true or false (all letters represent integers). Justify your answers.
   a. $|x|$ always is positive.
   b. For all x and y, $|x + y| = |x| + |y|$.
   c. If $a < {}^-b$, then $a < 0$.
   d. For all x and y, $(x - y)^2 = (y - x)^2$.
   e. $(^-a)(^-b)$ is the additive inverse of ab.
10. Find a counterexample to disprove each of the following properties on the set of integers:
    a. Commutative property of division
    b. Associative property of subtraction
    c. Closure property for division
    d. Distributive property of division over subtraction
11. Classify each of the following as true or false:
    a. $8|4$ b. $0|4$ c. $4|0$
    d. If a number is divisible by 4 and by 6, then it is divisible by 24.
    e. If a number is not divisible by 12, then it is not divisible by 3.
12. Classify each of the following as true or false. If false, show a counterexample.
    a. If $7|x$ and $7 \nmid y$, then $7 \nmid xy$.
    b. If $d \nmid (a + b)$, then $d \nmid a$ and $d \nmid b$.
    c. If $16|10^4$, then $16|10^6$.
    d. If $d|(a + b)$ and $d \nmid a$, then $d \nmid b$.
    e. If $d|(x + y)$ and $d|x$, then $d|y$.
    f. If $4 \nmid x$ and $4 \nmid y$, then $4 \nmid xy$.
13. Test each of the following numbers for divisibility by 2, 3, 4, 5, 6, 8, 9, and 11:
    a. 83,160   b. 83,193
14. Assume that 10,007 is prime. Without actually dividing 10,024 by 17, prove that 10,024 is not divisible by 17.
15. Fill each blank with 1 digit to make each of the following true (find all the possible answers):
    a. $6|87\_4$
    b. $24|4\_856$
    c. $29|87\_\_4$
16. A student claims that the sum of 5 consecutive positive integers is always divisible by 5.
    a. Check the student's claim for a few cases.
    b. Prove or disprove the student's claim.
17. Determine whether each of the following numbers is prime or composite:
    a. 143   b. 223
18. How can you tell if a number is divisible by 24? Check 4152 for divisibility by 24.
19. Find the GCD for each of the following:
    a. 24 and 52
    b. 5767 and 4453
20. Find the LCM for each of the following:
    a. $2^3 \cdot 5^2 \cdot 7^3$, $2 \cdot 5^3 \cdot 7^2 \cdot 13$, and $2^4 \cdot 5 \cdot 7^4 \cdot 29$
    b. 278 and 279
21. Construct a number that has exactly five divisors. Explain your construction.
22. Find all divisors of 144.
23. Find the prime factorization of each of the following:
    a. 172   b. 288   c. 260   d. 111
24. Find the least positive number that is divisible by every positive integer less than or equal to 10.
25. Candy bars priced at 50¢ each were not selling, so the price was reduced. Then they all sold in one day for a total of $31.93. What was the reduced price for each candy bar?
26. Two bells ring at 8:00 A.M. For the remainder of the day, one bell rings every half hour and the other bell rings every 45 min. What time will it be when the bells ring together again?
27. If the GCD of two positive whole numbers is 1, what can you say about the LCM of the two numbers? Explain your reasoning.

28. If there were to be 9 boys and 6 girls at a party and the host wanted each to be given exactly the same number of candies that could be bought in packages containing 12 candies, what is the fewest number of packages that could be bought?
29. Jane and Ramon are running laps on a track. If they start at the same time and place and go in the same direction, with Jane running a lap in 5 min and Ramon running a lap in 3 min, how long will it take for them to be at the starting place at the same time if they continue to run at these speeds?
★30. The triplet 3, 5, 7 consists of consecutive odd integers that are all prime. Give a convincing argument that this is the only triplet of consecutive odd integers that are all prime.
★31. Prove the test for divisibility by 9 using a 3-digit number n such that $n = a \cdot 10^2 + b \cdot 10 + c$.
*32. The length of a week was probably inspired by the need for market days and religious holidays. The Romans, for example, once used an 8-day week. Assuming April still had 30 days but was based on an 8-day week, if the first day of the month was on Sunday and the extra day after Saturday was called Venaday, on what day would the last day of the month fall?
*33. In measuring angles of rotation that a light on a small island lighthouse sweeps, what mod system would be used and why?

SELECTED BIBLIOGRAPHY

Ball, J. "A Perfect Formula." *Mathematics Teaching in the Middle School* 2 (February 1997): 200.

Battista, M. "A Complete Model for Operations on Integers." *Arithmetic Teacher* 30 (May 1983): 26–31.

Bezuszka, S., and M. Kenney. "Even Perfect Numbers: (Update)². " *Mathematics Teacher* 90 (November 1997): 628–633.

Billstein, R. "Teach a Turtle to Add and Subtract." *The Computing Teacher* 14 (May 1987): 47–50.

Borlaug, V. "Building Equations Using M&M's." *Mathematics Teaching in the Middle School* 2 (February 1997): 290–292.

Christensen, E. "The Thinking of Students: Pythagorean Triples Served for Supper." *Mathematics Teaching in the Middle School* 3 (September 1997): 60–62.

Crowley, M., and K. Dunn. "On Multiplying Negative Numbers." *Mathematics Teacher* 78 (April 1985): 252–256.

Edwards, F. "Geometric Figures Make the LCM Obvious." *Arithmetic Teacher* 34 (March 1987): 17–18.

Ewbank, W. "LCM—Let's Put It in Its Place." *Arithmetic Teacher* 35 (November 1987): 45–47.

Florence, H., Ed. "Pythagorean Triples Revisited." *Mathematics Teaching in the Middle School* 3 (May 1998): 478–479.

Georgeson, J., and C. Laughlin. "March's Menu of Problems." *Mathematics Teaching in the Middle School* 2 (March–April 1997): 332–333.

Georgeson, J., and C. Laughlin. "December's Menu of Problems." *Mathematics Teaching in the Middle School* 4 (November–December 1998): 180–181.

Lott, J., Ed. "Menu Madness." *Student Math Notes*. Reston, VA: National Council of Teachers of Mathematics (May 1991).

Olson, M. "On the Ball." *Student Math Notes*. Reston, VA: National Council of Teachers of Mathematics (September 1990).

Peterson, J. "Fourteen Different Strategies for Multiplication of Integers, or Why (⁻1)(⁻1) = (+1)." *Arithmetic Teacher* 19 (May 1972): 396–403.

Pollack, P. "The Thinking of Students: My Application of the Pythagorean Theorem." *Mathematics Teaching in the Middle School* 1 (May 1996): 814–816.

Shultz, H. "The Postage-Stamp Problem, Number Theory, and the Programmable Calculator." *Mathematics Teacher* 92 (January 1999): 20–22.

Stallings, L., and P. Bullock. "Juniper Green." *Mathematics Teaching in the Middle School* 4 (April 1999): 438–440.

Tirman, A. "Pythagorean Triples." *Mathematics Teacher* 79 (November 1986): 652–655.

Wyatt, C. "Clock Beaters." *Arithmetic Teacher* 34 (September 1986): 20.

5

Rational Numbers as Fractions

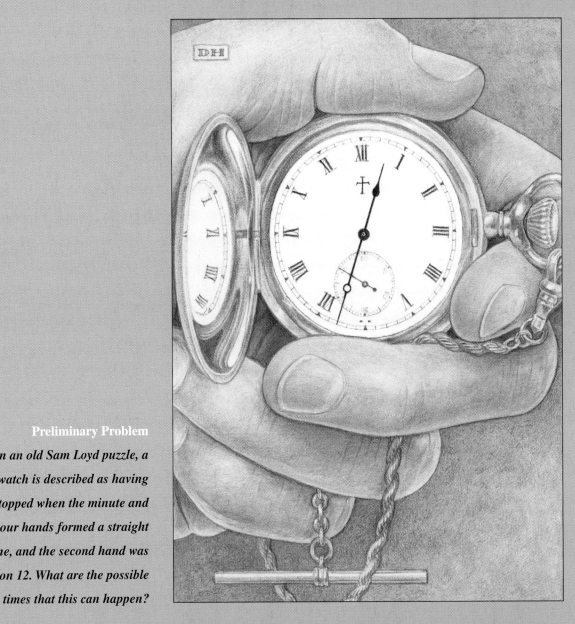

Preliminary Problem

In an old Sam Loyd puzzle, a watch is described as having stopped when the minute and hour hands formed a straight line, and the second hand was not on 12. What are the possible times that this can happen?

ntegers such as ⁻5 were invented to solve equations like $x + 5 = 0$. Similarly, a new type of number is needed to solve an equation like $2x = 1$. We can use *deductive reasoning* to know that a "new" number is needed. The needed number must be such that it is between 0 and 1 because $2 \cdot 0 = 0$, $2 \cdot 1 = 2$, and $0 < 1 < 2$. Thus, $2 \cdot 0 < 2x < 2 \cdot 1$, or $0 < x < 1$. There are no integers between 0 and 1 that meet these conditions. Therefore, similar to the development of new notation for negative integers, we need notation for this new number. If multiplication is to work with this new number as with multiplication of whole numbers and $2x = 1$, then $x + x = 1$. In other words, the number created must be added to itself to get 1. The number invented to solve the equation is called *one half* and is denoted by 1/2. It is an element of the set of numbers of the form a/b, where $b \neq 0$ and a and b are integers. Moreover, numbers of the form a/b are solutions to equations of the form $bx = a$. This set, denoted by Q, is the set of **rational numbers** and is defined as follows:

rational numbers

$$Q = \{a/b \mid a \text{ and } b \text{ are integers and } b \neq 0\}$$

Q is a subset of another set of numbers called *fractions*. Fractions are of the form a/b where $b \neq 0$ but a and b are not necessarily integers. For example, $1/\sqrt{2}$ is a fraction but not a rational number. (In this text we restrict ourselves to fractions where a and b are real numbers, but that restriction is not necessary.) The fact that $b \neq 0$ is always necessary because division by 0 is impossible.

Representations of rational numbers of the form a/b, where $b \neq 0$, are not unique. Again consider $2x = 1$ and multiply both sides of the equation by 2. Now we have $4x = 2$. The solution to the first equation was 1/2 and the solution to the second equation is 2/4. Because the equations are equivalent, the solutions should be equal. Similarly, $1/2 = 3/6$, $1/2 = 4/8$, and so on, so that we have a pattern:

$$\frac{1}{2} = \frac{2}{4} = \frac{3}{6} = \frac{4}{8} = \cdots = \frac{n}{2n}, \quad \text{where } n \text{ is any nonzero integer}$$

Hence, a single rational number has infinitely many representations, all considered to be equal. In general, we have the following property.

Fundamental Law of Fractions

$\frac{a}{b} = \frac{an}{bn}$, a, b, and n can be any numbers but $b \neq 0$ and $n \neq 0$.

Section 5-1 — The Set of Rational Numbers

numerator · denominator

In the rational number $\frac{a}{b}$, a is the **numerator** and b is the **denominator**. The rational number $\frac{a}{b}$ may also be represented as a/b or as $a \div b$. The word *fraction* is derived from the Latin word *fractus* meaning "to break." The word *numerator* comes from a Latin word meaning "numberer," and *denominator* comes from a Latin word meaning "namer." Table 5-1 shows several ways in which we use rational numbers.

Table 5-1 Uses of Rational Numbers

Use	Example
Division problem or solution to a multiplication problem	The solution to $2x = 3$ is $\frac{3}{2}$.
Partition, or part, of a whole	Joe received $\frac{1}{2}$ of Mary's salary each month for alimony.
Ratio	The ratio of Republicans to Democrats in the Senate is three to five.
Probability	When you toss a fair coin, the probability of getting heads is $\frac{1}{2}$.

HISTORICAL NOTE

The early Egyptian numeration system had symbols for fractions with numerators of 1. Most fractions with numerators other than 1 were expressed as a sum of different fractions with numerators of 1 $\left(\text{for example, } \frac{7}{12} = \frac{1}{3} + \frac{1}{4}\right)$.

Fractions with denominator 60 or powers of 60 were common in ancient Babylon about 2000 B.C., where 12,35 meant $12 + \frac{35}{60}$. The method was later adopted by the Greek astronomer Ptolemy (approximately A.D. 125). The same method was also used in Islamic and European countries and is presently used in the measurements of angles, where $13°19'\ 47''$ means $13 + \frac{19}{60} + \frac{47}{60^2}$ degrees.

The modern notation for fractions—a bar between numerator and denominator—is of Hindu origin. It came into general use in Europe in sixteenth-century books.

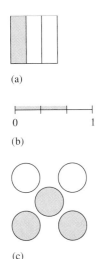

(a)

(b)

(c)

Figure 5-1

Figure 5-1 illustrates the use of rational numbers as part of a whole and as part of a given set. For example, in the area model in Figure 5-1(a), one part out of three congruent parts, or $\frac{1}{3}$ of the largest rectangle, is shaded. In Figure 5-1(b), two parts out of three congruent parts, or $\frac{2}{3}$ of the unit segment, are shaded. In Figure 5-1(c), three circles out of five congruent circles, or $\frac{3}{5}$ of the circles, are shaded.

Our early exposure to fractions, or rational numbers, usually takes the form of oral description rather than mathematical notation. We hear phrases such as "one half of a pizza," "one third of a cake," or "three fourths of a pie." We encounter such questions as "If three identical fruit bars are distributed equally among four friends, how much does each get?" The answer is that each receives $\frac{3}{4}$ of a bar. In the *Principles and Standards* for grades 3 through 5, we find that all students should

▲ *... develop understanding of fractions as parts of unit wholes, as parts of a collection, as locations on number lines, and as divisions of whole numbers; use models, benchmarks, and equivalent forms to judge the size of fractions; ...* (p. 148).

Rational numbers can be represented on a number line. Once the integers 0 and 1 are assigned to points on a line, every other rational number is assigned to a specific point. For example, to represent $\frac{3}{4}$ on the number line, we divide the segment from 0 to 1 into 4 segments of equal length. Then, starting from 0, we count 3 of these segments and stop at the mark corresponding to the right endpoint of the third segment to obtain the rational number $\frac{3}{4}$. Figure 5-2 shows the points that correspond to $\frac{3}{4}$, 1, $\frac{5}{4}$, 2, $\frac{^-3}{4}$, $^-1$, $\frac{^-5}{4}$, and $^-2$.

Figure 5-2

NOW TRY THIS 5-1

- Draw a Venn diagram to show the relationship among natural and whole numbers, integers, and rational numbers.

proper fraction A fraction $\frac{a}{b}$, where $0 \leq |a| < |b|$, is a **proper fraction.** For example, $\frac{4}{7}$ is a proper fraction,

improper fraction but $\frac{7}{4}, \frac{4}{4}$, and $\frac{^-9}{7}$ are not; $\frac{7}{4}$ is an **improper fraction.** In general $\frac{a}{b}$ is an improper fraction if $|a| \geq |b| > 0$.

Equal Fractions

That a rational number such as $\frac{1}{3}$ has many representations can be related to students through a concrete activity such as paperfolding. In Figure 5-3(a), one of three congruent parts, or $\frac{1}{3}$, is shaded. In Figure 5-3(b), each of the thirds has been folded in half so that now we have six sections, and two of six congruent parts, or $\frac{2}{6}$, are shaded. Thus both $\frac{1}{3}$ and $\frac{2}{6}$ represent exactly the same shaded portion. Although the symbols $\frac{1}{3}$ and $\frac{2}{6}$ do not look alike,

equivalent fractions they represent the same rational number and are **equivalent fractions** or **equal fractions.**
equal fractions Because they represent equal amounts, we write $\frac{1}{3} = \frac{2}{6}$ and say that $\frac{1}{3}$ equals $\frac{2}{6}$.

Figure 5-3

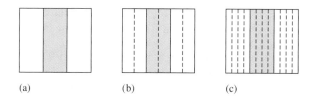

(a) (b) (c)

Figure 5-3(c) shows the rectangle with each of the original thirds folded into 4 equal parts with 4 of the 12 parts now shaded. Thus $\frac{1}{3}$ is equal to $\frac{4}{12}$ because the same portion of the model is shaded. Similarly, we could illustrate that $\frac{1}{3}, \frac{2}{6}, \frac{3}{9}, \frac{4}{12}, \frac{5}{15}, \ldots$ are equal.

The paperfolding example illustrates the Fundamental Law of Fractions mentioned earlier. It can be stated as follows: *The value of a fraction does not change if its numerator and denominator are multiplied by the same nonzero number.*

From the Fundamental Law of Fractions, $\frac{7}{-15} = \frac{-7}{15}$ because $\frac{7}{-15} = \frac{7 \cdot (^{-}1)}{-15 \cdot (^{-}1)} = \frac{-7}{15}$. Similarly, $\frac{a}{-b} = \frac{-a}{b}$. The form $\frac{-a}{b}$, where b is a positive number, is usually preferred for noncalculator calculations.

Example 5-1 Find a value for x so that $\frac{12}{42} = \frac{x}{210}$.

Solution By the Fundamental Law of Fractions, $\frac{12}{42} = \frac{12 \cdot 5}{42 \cdot 5} = \frac{60}{210}$. Hence, $\frac{x}{210} = \frac{60}{210}$, and $x = 60$.

Simplifying Fractions

simplifying fractions The Fundamental Law of Fractions justifies the process of **simplifying fractions.** Consider the following:
$$\frac{60}{210} = \frac{6 \cdot 10}{21 \cdot 10} = \frac{6}{21}.$$

Also,
$$\frac{6}{21} = \frac{2 \cdot 3}{7 \cdot 3} = \frac{2}{7}.$$

We can simplify $\frac{60}{210}$ because the numerator and denominator have a common factor of 10. We can simplify $\frac{6}{21}$ because 6 and 21 have a common factor of 3. However, we can not simplify $\frac{2}{7}$ because 2 and 7 have no common factor other than 1. The fraction $\frac{2}{7}$ is the **simplest (reduced) form** **simplest (or reduced) form** of $\frac{60}{210}$ because both 60 and 210 have been divided by their **lowest terms** greatest common divisor, 30. To write a fraction $\frac{a}{b}$ in simplest form, that is, in **lowest terms,** we divide both a and b by the GCD(a, b).

> **Definition of Simplest Form**
>
> A rational number $\frac{a}{b}$ is in simplest form if a and b have no common factor greater than 1, that is, if a and b are relatively prime.

An idea similar to the paper model in Figure 5-3 and that is used to find the lowest terms or simplest form of a fraction is illustrated on the following student page from *McDougal Littell Middle Grades Math Thematics, Book 1*, 1999.

We can use scientific/fraction calculators to simplify fractions. For example, to simplify $\frac{6}{12}$, we enter $\boxed{6}\ \boxed{/}\ \boxed{1}\ \boxed{2}$ and press $\boxed{\text{SIMP}}\ \boxed{=}$, and 3/6 appears on the screen. At this point, an indicator tells us that this is not in simplest form, so we press $\boxed{\text{SIMP}}\ \boxed{=}$ again to obtain 1/2. At any time, we can view the factor that was removed by pressing the $\boxed{x \circlearrowright y}$ key.

Example 5-2 Write each of the following in simplest form:

a. $\dfrac{28ab^2}{42a^2b^2}$ b. $\dfrac{(a+b)^2}{3a+3b}$ c. $\dfrac{x^2+x}{x+1}$ d. $\dfrac{3+x^2}{3x^2}$ e. $\dfrac{3+3x^2}{3x^2}$

Solution a. $\dfrac{28ab^2}{42a^2b^2} = \dfrac{2(14ab^2)}{3a(14ab^2)} = \dfrac{2}{3a}$

b. $\dfrac{(a+b)^2}{3a+3b} = \dfrac{(a+b)(a+b)}{3(a+b)} = \dfrac{a+b}{3}$

c. $\dfrac{x^2+x}{x+1} = \dfrac{x(x+1)}{x+1} = \dfrac{x(x+1)}{1(x+1)} = \dfrac{x}{1}$

d. $\dfrac{3+x^2}{3x^2}$ cannot be reduced further because $3+x^2$ and $3x^2$ have no factors in common except 1.

e. $\dfrac{3+3x^2}{3x^2} = \dfrac{3(1+x^2)}{3 \cdot x^2} = \dfrac{1+x^2}{x^2}$

Equality of Fractions

We can use several methods to show that two fractions such as $\dfrac{12}{42}$ and $\dfrac{10}{35}$ are equal.

1. Reduce both fractions to the same simplest form:

$$\dfrac{12}{42} = \dfrac{2^2 \cdot 3}{2 \cdot 3 \cdot 7} = \dfrac{2}{7} \quad \text{and} \quad \dfrac{10}{35} = \dfrac{5 \cdot 2}{5 \cdot 7} = \dfrac{2}{7}.$$

Thus
$$\dfrac{12}{42} = \dfrac{10}{35}.$$

2. Rewrite both fractions with the same least common denominator. Since LCM(42, 35) = 210, then

$$\dfrac{12}{42} = \dfrac{60}{210} \quad \text{and} \quad \dfrac{10}{35} = \dfrac{60}{210}.$$

Thus
$$\dfrac{12}{42} = \dfrac{10}{35}.$$

3. Rewrite both fractions with a common denominator (not necessarily the least). A common multiple of 42 and 35 may be found by finding the product 42 · 35, or 1470. Now,

$$\dfrac{12}{42} = \dfrac{420}{1470} \quad \text{and} \quad \dfrac{10}{35} = \dfrac{420}{1470}.$$

Section 3 Key Concepts

Equivalent Fractions (pp. 109, 111, 112)

Fractions that name the same part of a whole are equivalent.

Examples

$\frac{1}{2}$, $\frac{2}{4}$, and $\frac{3}{6}$, are equivalent fractions since

$\frac{1}{2}$ = ▢, $\frac{2}{4}$ = ▢, and $\frac{3}{6}$ = ▢

When given a fraction, you can find an equivalent fraction by multiplying or dividing the numerator and denominator by the same whole number other than 0.

Examples

$\frac{5}{6} = \frac{5 \cdot 3}{6 \cdot 3} = \frac{15}{18}$, so $\frac{5}{6}$ is equivalent to $\frac{15}{18}$.

$\frac{24}{28} = \frac{24 \div 4}{28 \div 4} = \frac{6}{7}$, so $\frac{24}{28}$ is equivalent to $\frac{6}{7}$.

Lowest Terms (p. 113)

In the example above, $\frac{6}{7}$ is in the lowest terms because 1 is the only whole number that will divide both 6 and 7 evenly.

Hence,
$$\frac{12}{42} = \frac{10}{35}.$$

The third method suggests a general algorithm for determining if two fractions $\frac{a}{b}$ and $\frac{c}{d}$ are equal. Rewrite both fractions with common denominator bd. That is,

$$\frac{a}{b} = \frac{ad}{bd} \quad \text{and} \quad \frac{c}{d} = \frac{bc}{bd}.$$

Because the denominators are the same, $\frac{ad}{bd} = \frac{bc}{bd}$ if, and only if, $ad = bc$. For example, $\frac{24}{36} = \frac{6}{9}$ because $24 \cdot 9 = 216 = 36 \cdot 6$. In general, the following property results.

Property

Two fractions $\frac{a}{b}$ and $\frac{c}{d}$ are equal if, and only if, $ad = bc$.

Using a calculator, we can determine if two fractions are equal by using the above property. Since both $\boxed{2} \times \boxed{2} \boxed{1} \boxed{9} \boxed{6} \boxed{=}$ and $\boxed{4} \times \boxed{1} \boxed{0} \boxed{9} \boxed{8} \boxed{=}$ yield a display of 4392, we see that $\frac{2}{4} = \frac{1098}{2196}$.

Ordering Rational Numbers

Children know that $\frac{7}{8} > \frac{5}{8}$ because if a pizza is divided into 8 parts, then 7 parts of a pizza is more than 5 parts. Similarly, $\frac{3}{7} < \frac{4}{7}$. Thus given two fractions with common positive denominators, the one with the greater numerator is the greater fraction. This can be written as follows.

Theorem 5-1

If a, b, and c are integers and $b > 0$, then $\frac{a}{b} > \frac{c}{b}$ if, and only if, $a > c$.

NOW TRY THIS 5-2

- Determine if Theorem 5-1 is true if $b < 0$.

Comparing fractions with unlike denominators can be accomplished by *rewriting the problem as a previously solved type of problem.* Using the common denominator bd, we can write the fractions $\frac{a}{b}$ and $\frac{c}{d}$ as $\frac{ad}{bd}$ and $\frac{bc}{bd}$. Because $b > 0$ and $d > 0$, $bd > 0$, we apply Theorem 5-1 to find that

$$\frac{ad}{bd} > \frac{bc}{bd} \quad \text{if, and only if,} \quad ad > bc.$$

Denseness of Rational Numbers

The set of rational numbers has a very special property that is not present for the set of whole numbers or for the set of integers. Consider $\frac{1}{2}$ and $\frac{2}{3}$. To find a rational number between $\frac{1}{2}$ and $\frac{2}{3}$, we first rewrite the fractions with a common denominator, as $\frac{3}{6}$ and $\frac{4}{6}$. Because there is no whole number between the numerators 3 and 4, we next find two fractions equal respectively to $\frac{1}{2}$ and $\frac{2}{3}$ with greater denominators. For example, $\frac{1}{2} = \frac{6}{12}$ and $\frac{2}{3} = \frac{8}{12}$, and $\frac{7}{12}$ is between the two fractions $\frac{6}{12}$ and $\frac{8}{12}$. So $\frac{7}{12}$ is between $\frac{1}{2}$ and $\frac{2}{3}$.

This property is generalized as follows.

Property

Given rational numbers $\frac{a}{b}$ and $\frac{c}{d}$, there is another rational number between these two numbers.

NOW TRY THIS 5-3

● Show that there are infinitely many rational numbers between any two rational numbers. ●

Example 5-3 Find two fractions between $\frac{1}{2}$ and $\frac{7}{18}$.

Solution Because $\frac{1}{2} = \frac{1 \cdot 9}{2 \cdot 9} = \frac{9}{18}$, we see that $\frac{8}{18}$, or $\frac{4}{9}$, is between $\frac{7}{18}$ and $\frac{9}{18}$. To find another fraction between the given fractions, we find two fractions equal to $\frac{7}{18}$ and $\frac{9}{18}$, respectively, but with greater denominators. For example, $\frac{7}{18} = \frac{14}{36}$ and $\frac{9}{18} = \frac{18}{36}$. We now see that $\frac{15}{36}, \frac{16}{36}$, and $\frac{17}{36}$ are all between $\frac{14}{36}$ and $\frac{18}{36}$.

ONGOING ASSESSMENT 5-1

1. Write a sentence illustrating the use of $\frac{7}{8}$ in each of the following ways:
   a. As a division problem
   b. As part of a whole
   c. As a ratio

2. For each of the following, write a fraction to represent the shaded portion:

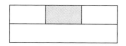
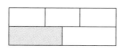

(a) (b)

(c) (d)

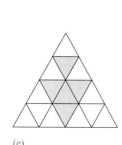
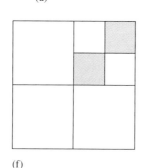

(e) (f)

3. For each of the following four squares, write a fraction to represent the shaded portion. What property of fractions does the diagram illustrate?

(a) (b) (c) (d)

4. Complete each of the following figures so that it shows $\frac{3}{5}$:

(a) (b)

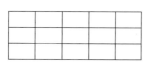

(c)

(d)

(e) (f)

5. Refer to the following figure and represent each of the following as a fraction:

   a. The dots inside the circle as a part of all the dots
   b. The dots inside the rectangle as a part of all the dots
   c. The dots in the intersection of the rectangle and the circle as a part of all the dots
   d. The dots outside the circle but inside the rectangle as part of all the dots

6. For each of the following, write three fractions equal to the given fraction:
   a. $\frac{2}{9}$ b. $\frac{^-2}{5}$ c. $\frac{0}{3}$ d. $\frac{a}{2}$

7. Find the simplest form for each of the following fractions:
   a. $\frac{156}{93}$ b. $\frac{27}{45}$ c. $\frac{^-65}{91}$
   d. $\frac{0}{68}$ e. $\frac{84^2}{91^2}$ f. $\frac{662}{703}$

8. Mr. Gonzales and Ms. Price gave the same test to their fifth-grade classes. In Mr. Gonzales's class, 20 out of 25 students passed the test, and in Ms. Price's class, 24 out of 30 students passed the test. One of Ms. Price's students heard about the results of the tests and claimed that the classes did equally well. Is the student right? Explain.

9. For each of the following, choose the expression in parentheses that equals or describes best the given fraction:
   a. $\frac{0}{0}$ (1, undefined, 0) b. $\frac{5}{0}$ (undefined, 5, 0)
   c. $\frac{0}{5}$ (undefined, 5, 0)
   d. $\frac{2+a}{a}$ (2, 3, cannot be simplified)
   e. $\frac{15+x}{3x}$ $\left(\frac{5+x}{x}, 5, \text{cannot be simplified}\right)$
   f. $\frac{2^6 + 2^5}{2^4 + 2^7}$ $\left(1, \frac{2}{3}, \text{cannot be simplified}\right)$

g. $\dfrac{2^{100} + 2^{98}}{2^{100} - 2^{98}} \left(2^{196}, \dfrac{5}{3}, \text{too large to simplify}\right)$

10. Find the simplest form for each of the following fractions:
    a. $\dfrac{x}{x}$
    b. $\dfrac{14x^2y}{63xy^2}$
    c. $\dfrac{a^2 + ab}{a + b}$
    d. $\dfrac{a}{3a + ab}$

11. Determine if the following pairs are equal:
    a. $\dfrac{3}{8}$ and $\dfrac{375}{1000}$
    b. $\dfrac{18}{54}$ and $\dfrac{23}{69}$
    c. $\dfrac{6}{10}$ and $\dfrac{600}{1000}$
    d. $\dfrac{17}{27}$ and $\dfrac{25}{45}$

12. Determine if the following pairs are equal by changing both to the same denominator:
    a. $\dfrac{10}{16}$ and $\dfrac{12}{18}$
    b. $\dfrac{3}{12}$ and $\dfrac{41}{154}$
    c. $\dfrac{3}{-12}$ and $\dfrac{-36}{144}$
    d. $\dfrac{-21}{86}$ and $\dfrac{-51}{215}$

13. A board is needed that is exactly $\dfrac{11}{32}$ in. wide to fill a hole. Can a board that is $\dfrac{3}{8}$ in. be shaved down to fit the hole? If so, how much must be shaved from the board?

14. Draw an area model to show that $\dfrac{3}{4} = \dfrac{6}{8}$.

15. If a fraction is equal to $\dfrac{3}{4}$ and the sum of the numerator and denominator is 84, what is the fraction?

16. The following two parking meters are next to each other with the times left as shown. Which meter has more time left on it? How much more?

Meter A   Meter B

17. Mr. Gomez filled his car's 16-gal gas tank. He took a short trip and used 6 gal of gas. Draw an arrow in the following figure to show what his gas gauge looked like after the trip:

18. Read each measurement as shown on the following ruler:

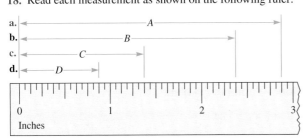

19. Solve for x in each of the following:
    a. $\dfrac{2}{3} = \dfrac{x}{16}$
    b. $\dfrac{3}{4} = \dfrac{-27}{x}$
    c. $\dfrac{3}{x} = \dfrac{3x}{x^2}$

20. a. If $\dfrac{a}{c} = \dfrac{b}{c}$, what must be true?
    b. If $\dfrac{a}{b} = \dfrac{a}{c}$, what must be true?

21. Let W be the set of whole numbers, N be the set of natural numbers, I be the set of integers, and Q the set of rational numbers. Write sentences to describe the relationship among $N, W, I,$ and Q.

22. In Amy's algebra class, 6 of the 31 students received A's on a test. The same test was given to Bren's class and 5 of the 23 students received A's. Which class had the higher rate of A's?

23. For each of the following pairs of fractions, replace the comma with the correct symbol ($<$, $=$, $>$) to make a true statement:
    a. $\dfrac{7}{8}, \dfrac{5}{6}$
    b. $2\dfrac{4}{5}, 2\dfrac{3}{6}$
    c. $\dfrac{-7}{8}, \dfrac{-4}{5}$
    d. $\dfrac{1}{-7}, \dfrac{1}{-8}$
    e. $\dfrac{2}{5}, \dfrac{4}{10}$
    f. $\dfrac{0}{7}, \dfrac{0}{17}$

24. Arrange each of the following in decreasing order:
    a. $\dfrac{11}{22}, \dfrac{11}{16}, \dfrac{11}{13}$
    b. $\dfrac{-1}{5}, \dfrac{-19}{36}, \dfrac{-17}{30}$

25. If $\dfrac{a}{b} < 1$ and $\dfrac{c}{d} > 0$, compare the size of $\dfrac{c}{d}$ with $\dfrac{a}{b} \cdot \dfrac{c}{d}$.

26. Show that the sequence $\dfrac{1}{2}, \dfrac{2}{3}, \dfrac{3}{4}, \dfrac{4}{5}, \dfrac{5}{6}, \dfrac{6}{7}, \ldots$ is an increasing sequence; that is, show that each term in the sequence is greater than the preceding one.

27. For each of the following, find two rational numbers between the given fractions:
    a. $\dfrac{3}{7}$ and $\dfrac{4}{7}$
    b. $\dfrac{-7}{9}$ and $\dfrac{-8}{9}$
    c. $\dfrac{5}{6}$ and $\dfrac{83}{100}$
    d. $\dfrac{-1}{3}$ and $\dfrac{3}{4}$

28. Consider the following number grid. The circled numbers form a rhombus (that is, all sides are the same length).

1	2	3	4	5	6	7	⑧	9	10
11	12	13	14	15	16	17	18	19	⑳
21	22	23	24	25	26	㉗	28	29	30
31	32	33	34	35	36	37	38	㊴	40
41	42	43	44	45	46	47	48	49	50

    a. If A is the sum of the four circled numbers and B is the sum of the four interior numbers, find A/B.
    b. Form a rhombus by circling the numbers 6, 18, 25, and 37. Compute A and B as in (a) and then find A/B.
    c. How do the answers in (a) and (b) compare? Why does this happen?

29. A scale on a map is 12 mi to the inch. What is the airline mileage between two cities that are 38 in. apart on the map?

30. Six ounces is what part of a pound? A ton?

Communication

31. Explain why there are 24 time zones on earth.
32. If $\frac{1}{3}$ of each of two classes is female, explain whether each class must contain the same number of females.
33. Consider the set of all fractions equal to $\frac{1}{2}$. If you take any ten of those fractions, add their numerators to obtain the numerator of a new fraction and add their denominators to obtain the denominator of a new fraction, how does the new fraction relate to $\frac{1}{2}$? Generalize what you found and explain.
34. Should fractions always be reduced to their simplest form? Why or why not?
35. How would you respond to each of the following students?
    a. Iris claims that if we have two positive rational numbers, the one with the greatest numerator is the greatest.
    b. Shirley claims that if we have two positive rational numbers, the one with the greatest denominator is the least.
36. If we were to take the set of fractions equivalent to $\frac{1}{3}$ and graph them as points on a coordinate system so that the numerator becomes the *x*-coordinate and the denominator becomes the *y*-coordinate for that point, explain what type of graph we would get.
37. Write an explanation of how to convert inches to yards and vice versa.

Open-Ended

38. List three types of measures that require rational numbers as the appropriate number of units in the measurements.
39. Some people have argued that the system of integers is more understandable than the system of positive rational numbers. If you could decide which should be taught first in school, which would you choose and why?

Cooperative Learning

40. Assume the tallest person in the class is 1 unit tall and do the following:
    a. Find rational numbers to represent other members of the class.
    b. Order the class members according to height.
    c. Make a number line using the rational numbers for each person ordered according to the heights of class members.
    d. Use equivalent fractions to determine if the ordering is correct.

TECHNOLOGY CORNER

Let *a*/*b* be any rational number. To write *a*/*b* in simplest form, first determine which is greater, *a* or *b*. Suppose *a* is the lesser of *a* and *b*. Create a spreadsheet in which the first column contains the whole numbers 1 through *a*. Use the formula *a*/A1 to create the second column and the formula *b*/A1 to create the third column. Fill down *a* rows of columns 2 and 3. Now consider the numbers in the cells of these columns. If in both columns whole numbers with no decimal parts appear in the same row as the quotients, then *a*/*b* may be simplified to *c*/*d*, where *c* and *d* are the respective whole numbers. To determine if *a*/*b* can be simplified further, replace *a* with *c* and *b* with *d* and repeat the process as many times as is necessary to obtain the simplest form of *a*/*b*. You have obtained the simplest form when there are no whole numbers in the same row in the entire second and third columns. Create a spreadsheet and try this method to write 60/210 in simplest form.

Section 5-2 — Addition and Subtraction of Rational Numbers

To determine how to add two fractions in general, or two rational numbers in particular, we examine the addition of two particular fractions using Polya's four-step problem-solving process.

Problem Solving   **Adding Rational Numbers Problem**

Determine how to add the rational numbers 2/3 and 1/4.

- **Understanding the Problem** We can model $\frac{2}{3}$ and $\frac{1}{4}$ as parts of a whole as seen in Figure 5-4, but we need a way to combine the two drawings to find the sum.

Figure 5-4

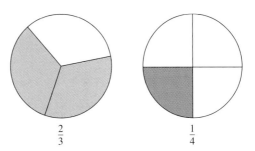

- **Devising a Plan** We use the strategy of *solving a simpler problem and consider adding rational numbers with the same denominators.* To find the sum, 2/5 + 1/5, we *use diagrams* as in Figure 5-5.

Figure 5-5

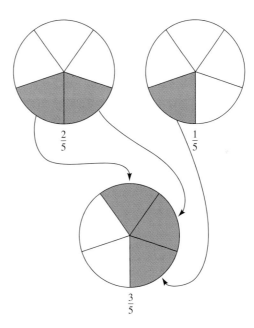

Thus 2/5 + 1/5 = 3/5. We see that if the rational numbers have the same denominators, then the addition can be accomplished by adding the numerators and keeping the denominators the same. Now we consider the original problem: 2/3 + 1/4. We can find the sum by writing each with a common denominator and completing the computation as above.

- **Carrying Out the Plan** From earlier work in the chapter, we know that 2/3 is equal to an infinite set of rational numbers, including 4/6, 6/9, 8/12, and so on. Also 1/4 is equal to an infinite set of rational numbers, including 2/8, 3/12, 4/16, and so on. By comparing the two sets of rational numbers, we see that 8/12 and 3/12 have the same denominator. One is 8 parts of 12 while the other is 3 parts of 12. Consequently, the sum is 2/3 + 1/4 = 8/12 + 3/12 = 11/12. Figure 5-6 illustrates the addition.

Figure 5-6

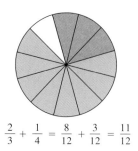

$$\frac{2}{3} + \frac{1}{4} = \frac{8}{12} + \frac{3}{12} = \frac{11}{12}$$

- **Looking Back** To add two rational numbers of unlike denominators, we considered equal rational numbers with like denominators. The common denominator for 2/3 and 1/4 was 12. This is also the least common denominator, or the LCM (3, 4). To add two fractions with unequal denominators such as 5/12 and 7/18, we could find equal fractions with the denominator as the LCM(12, 18), or 36. However, any common denominator will work as well, for example, 72 or even 12 · 18.

The generalization of the Adding Rational Numbers Problem on page 256 leads us to the following definition for addition of rational numbers with like denominators.

Definition of Addition of Rational Numbers

If $\frac{a}{b}$ and $\frac{c}{b}$ are rational numbers, then $\frac{a}{b} + \frac{c}{b} = \frac{a+c}{b}$.

By considering the sum $\frac{2}{3} + \frac{1}{4} = \frac{2 \cdot 4}{3 \cdot 4} + \frac{1 \cdot 3}{4 \cdot 3} = \frac{8}{12} + \frac{3}{12} = \frac{11}{12}$, we can generalize to the sum of two rational numbers with unlike denominators as in the following property.

Property

If $\frac{a}{b}$ and $\frac{c}{d}$ are any two rational numbers, then $\frac{a}{b} + \frac{c}{d} = \frac{ad + bc}{bd}$.

Example 5-4 Find each of the following sums:

a. $\frac{2}{15} + \frac{4}{21}$ **b.** $\frac{2}{-3} + \frac{1}{5}$ **c.** $\left(\frac{3}{4} + \frac{1}{5}\right) + \frac{1}{6}$ **d.** $\frac{3}{x} + \frac{4}{y}$

Solution **a.** Because LCM (15, 21) = 15 · 7, then

$$\frac{2}{15} + \frac{4}{21} = \frac{2 \cdot 7}{15 \cdot 7} + \frac{4 \cdot 5}{21 \cdot 5} = \frac{14}{105} + \frac{20}{105} = \frac{34}{105}$$

b. $\frac{2}{-3} + \frac{1}{5} = \frac{(2)(5) + (^-3)(1)}{(^-3)(5)} = \frac{10 + {}^-3}{{}^-15} = \frac{7}{{}^-15} = \frac{7(^-1)}{{}^-15(^-1)} = \frac{{}^-7}{15}$

c. $\dfrac{3}{4} + \dfrac{1}{5} = \dfrac{3 \cdot 5 + 4 \cdot 1}{4 \cdot 5} = \dfrac{19}{20}$. Hence, $\left(\dfrac{3}{4} + \dfrac{1}{5}\right) + \dfrac{1}{6} = \dfrac{19}{20} + \dfrac{1}{6}$
$= \dfrac{19 \cdot 6 + 20 \cdot 1}{20 \cdot 6} = \dfrac{134}{120}$ or $\dfrac{67}{60}$.

d. $\dfrac{3}{x} + \dfrac{4}{y} = \dfrac{3y}{xy} + \dfrac{4x}{xy} = \dfrac{3y + 4x}{xy}$

Mixed Numbers

mixed numbers

In everyday life, we often use **mixed numbers,** that is, numbers that are made up of an integer and a fractional part of an integer. For example, Figure 5-7 shows that the nail is $2\dfrac{3}{4}$ in. long. The mixed number $2\dfrac{3}{4}$ means $2 + \dfrac{3}{4}$. It is sometimes inferred that $2\dfrac{3}{4}$ means 2 times $\dfrac{3}{4}$, since xy means $x \cdot y$, but this is not correct. Also, the number $^-4\dfrac{3}{4}$ means $^-\left(4 + \dfrac{3}{4}\right)$, or $^-4 - \dfrac{3}{4}$, not $^-4 + \dfrac{3}{4}$.

Figure 5-7

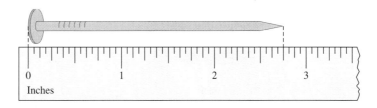

A mixed number is a rational number, and therefore it can always be written in the form $\dfrac{a}{b}$. For example,

$$2\dfrac{3}{4} = 2 + \dfrac{3}{4} = \dfrac{2}{1} + \dfrac{3}{4} = \dfrac{2 \cdot 4 + 1 \cdot 3}{1 \cdot 4} = \dfrac{8 + 3}{4} = \dfrac{11}{4}.$$

Example 5-5 Change each of the following mixed numbers to the form $\dfrac{a}{b}$, where a and b are integers:

a. $4\dfrac{1}{3}$ b. $^-3\dfrac{2}{5}$

Solution a. $4\dfrac{1}{3} = 4 + \dfrac{1}{3} = \dfrac{4}{1} + \dfrac{1}{3} = \dfrac{4 \cdot 3 + 1 \cdot 1}{1 \cdot 3} = \dfrac{12 + 1}{3} = \dfrac{13}{3}$

b. $^-3\dfrac{2}{5} = ^-\left(3 + \dfrac{2}{5}\right) = ^-\left(\dfrac{3}{1} + \dfrac{2}{5}\right) = ^-\left(\dfrac{3 \cdot 5 + 1 \cdot 2}{1 \cdot 5}\right) = \dfrac{^-17}{5}$

Example 5-6 Change $\dfrac{29}{5}$ to a mixed number.

Solution $\dfrac{29}{5} = \dfrac{5 \cdot 5 + 4}{5} = \dfrac{5 \cdot 5}{5} + \dfrac{4}{5} = 5 + \dfrac{4}{5} = 5\dfrac{4}{5}$

REMARK In elementary schools, problems like Example 5-6 are usually solved using division, as follows:

$$\begin{array}{r} 5 \\ 5\overline{)29} \\ \underline{25} \\ 4 \end{array}$$

Hence, $\dfrac{29}{5} = 5 + \dfrac{4}{5} = 5\dfrac{4}{5}$.

We can use scientific/fraction calculators to change improper fractions to mixed numbers. For example, if we enter $\boxed{2}\ \boxed{9}\ \boxed{/}\ \boxed{5}$ and press $\boxed{Ab/c}$, then 5 ⌴ 4/5 appears, which means $5\dfrac{4}{5}$.

Because mixed numbers are rational numbers, the methods of adding rationals can be extended to include mixed numbers. The following page from *Scott-Foresman Addison Wesley Middle School Math, Course 1,* 1999, shows a method of estimating sums of mixed numbers.

6-5 Adding Mixed Numbers

You'll Learn …
- to add mixed numbers

… How It's Used
Farmers add mixed numbers when working with rainfall data.

▶ **Lesson Link** In the last lesson, you estimated sums of mixed numbers. Now you'll learn how to find mixed number sums exactly. ◀

Explore Adding Mixed Numbers

Another Fine Mix!

Materials: Fraction Bars®

Adding Mixed Numbers

- Draw and label the whole number for the first mixed number.
- Next to that, draw and label the fraction for the first number.
- Next to that, draw and label the whole number for the second mixed number.
- Next to that, draw and label the fraction for the second number.
- Using a whole number and a fraction less than 1, describe the model.

$1\dfrac{3}{4} + 1\dfrac{1}{2} = 3\dfrac{1}{4}$

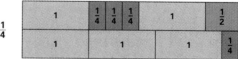

1. Model each problem.

  a. $1\dfrac{1}{4} + 2\dfrac{1}{2}$ **b.** $1\dfrac{2}{3} + 1\dfrac{3}{6}$ **c.** $1\dfrac{3}{8} + 2\dfrac{1}{8}$ **d.** $1\dfrac{3}{6} + 2\dfrac{1}{2}$

2. Does the whole number in the answer always equal the sum of the two whole numbers in the problem? Explain.

3. Is the sum of two mixed numbers always a mixed number? Explain.

We can also use scientific/fraction calculators to add mixed numbers. For example, to add $2\frac{4}{5} + 3\frac{5}{6}$, we enter $\boxed{2}\ \boxed{\text{Unit}}\ \boxed{4}\ \boxed{/}\ \boxed{5}\ \boxed{+}\ \boxed{3}\ \boxed{\text{Unit}}\ \boxed{5}\ \boxed{/}\ \boxed{6}\ \boxed{=}$, and the display reads 5 ⊔ 49/30. We then press $\boxed{Ab/c}$ to obtain 6 ⊔ 19/30, which means $6\frac{19}{30}$.

Properties of Addition for Rational Numbers

Rational numbers have the following properties for addition: closure, commutative, associative, additive identity, and additive inverse. To emphasize the additive inverse property of rational numbers, we state it explicitly, as follows.

> **Property**
>
> **Additive inverse property of rational numbers** For any rational number $\frac{a}{b}$, there exists a unique rational number $-\frac{a}{b}$, the additive inverse of $\frac{a}{b}$, such that
> $$\frac{a}{b} + \left(-\frac{a}{b}\right) = 0 = \left(-\frac{a}{b}\right) + \frac{a}{b}.$$

Another form of $-\frac{a}{b}$ can be found by considering the sum $\frac{a}{b} + \frac{-a}{b}$. Because

$$\frac{a}{b} + \frac{-a}{b} = \frac{a + -a}{b} = \frac{0}{b} = 0,$$

it follows that $-\frac{a}{b}$ and $\frac{-a}{b}$ are both additive inverses of $\frac{a}{b}$, so $-\frac{a}{b} = \frac{-a}{b}$.

Example 5-7 Find the additive inverses for each of the following:

a. $\frac{3}{5}$ b. $\frac{-5}{11}$ c. $4\frac{1}{2}$

Solution a. $-\frac{3}{5}$ or $\frac{-3}{5}$ b. $-\left(\frac{-5}{11}\right) = \frac{-(-5)}{11} = \frac{5}{11}$ c. $-4\frac{1}{2}$, or $\frac{-9}{2}$

Properties of the additive inverse for rational numbers are analogous to those of the additive inverse for integers, as shown in Table 5-2. As with the set of integers, the set of rational numbers also has the addition property of equality.

Table 5-2

	Integers	Rational Numbers
1.	$-(-a) = a$	$-\left(-\frac{a}{b}\right) = \frac{a}{b}$
2.	$-(a + b) = -a + -b$	$-\left(\frac{a}{b} + \frac{c}{d}\right) = \frac{-a}{b} + \frac{-c}{d}$

> **Property**
>
> **Addition property of equality** If $\frac{a}{b}$ and $\frac{c}{d}$ are any rational numbers such that $\frac{a}{b} = \frac{c}{d}$, and if $\frac{e}{f}$ is any rational number, then $\frac{a}{b} + \frac{e}{f} = \frac{c}{d} + \frac{e}{f}$.

Subtraction of Rational Numbers

In elementary school, subtraction of rational numbers is usually introduced by using a take-away model. If we have $\frac{6}{7}$ of a pizza and $\frac{2}{7}$ of the original pizza is taken away, $\frac{4}{7}$ of the pizza remains: that is, $\frac{6}{7} - \frac{2}{7} = \frac{6-2}{7} = \frac{4}{7}$. In general, subtraction of rational numbers with like denominators is determined as follows:

$$\frac{a}{b} - \frac{c}{b} = \frac{a-c}{b}$$

Subtraction of rational numbers, like subtraction of integers, can be defined in terms of addition as follows.

> **Definition of Subtraction of Rational Numbers**
>
> If $\frac{a}{b}$ and $\frac{c}{d}$ are any rational numbers, then $\frac{a}{b} - \frac{c}{d} = x$ if, and only if, $\frac{a}{b} + x = \frac{c}{d}$, where x is a rational number.

As with integers, we can see that subtraction of rational numbers can be performed by adding the additive inverses. The following theorem states this.

> **Theorem 5-2**
>
> If $\frac{a}{b}$ and $\frac{c}{d}$ are any rational numbers, then $\frac{a}{b} - \frac{c}{d} = \frac{a}{b} + \frac{^-c}{d}$.

Now, using the definition of addition of rational numbers, we obtain the following:

$$\frac{a}{b} - \frac{c}{d} = \frac{a}{b} + \frac{^-c}{d}$$
$$= \frac{ad + b(^-c)}{bd}$$
$$= \frac{ad + {}^-(bc)}{bd}$$
$$= \frac{ad - bc}{bd}$$

We summarize this result in the following theorem.

Theorem 5-3

If $\frac{a}{b}$ and $\frac{c}{d}$ are any rational numbers, then $\frac{a}{b} - \frac{c}{d} = \frac{ad - bc}{bd}$.

Example 5-8 Find each difference in the following:

a. $\frac{5}{8} - \frac{1}{4}$ **b.** $5\frac{1}{3} - 2\frac{3}{4}$

Solution **a.** One approach is to find the LCM for the fractions. Because LCM(8, 4) = 8, we have

$$\frac{5}{8} - \frac{1}{4} = \frac{5}{8} - \frac{2}{8} = \frac{3}{8}.$$

An alternative approach is as follows:

$$\frac{5}{8} - \frac{1}{4} = \frac{5 \cdot 4 - 8 \cdot 1}{8 \cdot 4} = \frac{5 \cdot 4 - 8 \cdot 1}{32} = \frac{12}{32}, \text{ or } \frac{3}{8}.$$

b. Two methods of solution are given:

$$5\frac{1}{3} = 5\frac{4}{12} = 4 + 1\frac{4}{12} = 4\frac{16}{12}$$
$$-2\frac{3}{4} = -2\frac{9}{12} = -2\frac{9}{12} \quad = -2\frac{9}{12}$$
$$\qquad\qquad\qquad\qquad\qquad\qquad\qquad 2\frac{7}{12}$$

$$5\frac{1}{3} - 2\frac{3}{4} = \frac{16}{3} - \frac{11}{4}$$
$$= \frac{16 \cdot 4 - 3 \cdot 11}{3 \cdot 4}$$
$$= \frac{64 - 33}{12}$$
$$= \frac{31}{12} \text{ or } 2\frac{7}{12}$$

Estimation with Rational Numbers

In the *Principles and Standards*, we find the following for grades 3–5:

> *At these grades, the emphasis should not be on developing general procedures to solve all decimal and fraction problems. Rather, students should generate solutions that are based on number sense and properties of the operations and that use a variety of models or representations. For example, in a fourth-grade class, students might work on this problem: Jamal invited seven of his friends to lunch on Saturday. He thinks that each of the eight people (his seven guests and himself) will eat one and a half sandwiches. How many sandwiches should he make? Students might draw a picture and count up the number of sandwiches, or they might use reasoning based on their knowledge of number and operations—for example, "That would be eight whole sandwiches and eight half sandwiches; since two halves make a whole sandwich, the eight halves will make four more sandwiches, so Jamal needs to make twelve sandwiches." (p. 155).*

Consider a student who added $\frac{3}{4}$ and $\frac{1}{2}$ and obtained $\frac{4}{6}$. An estimation of $\frac{3}{4} + \frac{1}{2}$ as a number greater than $\frac{1}{2} + \frac{1}{2}$ shows that the answer should be greater than 1 and that $\frac{4}{6}$ is unreasonable.

Sometimes it is desirable to round fractions to a convenient fraction, such as $\frac{1}{2}, \frac{1}{3}, \frac{1}{4}, \frac{1}{5}, \frac{2}{3}, \frac{3}{4}$, or 1. For example, if a student had 59 correct answers out of 80 questions, the student answered $\frac{59}{80}$ of the questions correctly, which is approximately $\frac{60}{80}$, or $\frac{3}{4}$. Intuitively, we know $\frac{60}{80}$ is greater than $\frac{59}{80}$. On a number line, the greater fraction is to the right of the lesser. The estimate $\frac{3}{4}$ for $\frac{59}{80}$ is a high estimate. In a similar way, we can estimate $\frac{31}{90}$ by $\frac{30}{90}$, or $\frac{1}{3}$. In this case, the estimate of $\frac{1}{3}$ is a low estimate because it is lower than the actual answer.

Example 5-9 A sixth-grade class is collecting cans to take to the recycling center. Becky's group brought the following amounts (in pounds). About how many pounds does her group have all together?

$$1\frac{1}{8}, 3\frac{4}{10}, 5\frac{7}{8}, \frac{6}{10}$$

Solution We can estimate the amount by using front-end estimation and then adjusting by using $0, \frac{1}{2}$, and 1 as reference points. The front-end estimate is $(1 + 3 + 5)$, or 9. The adjustment is $\left(0 + \frac{1}{2} + 1 + \frac{1}{2}\right)$, or 2. An adjusted estimate would be 11 lb.

Example 5-10 Estimate each of the following:

a. $\frac{27}{13} + \frac{10}{9}$ b. $3\frac{9}{10} + 2\frac{7}{8} + \frac{11}{12}$

Solution a. Because $\frac{27}{13}$ is more than 2 and $\frac{10}{9}$ is more than 1, an estimate is a number close to 3 but more than 3.

b. We first add the front-end parts to obtain $3 + 2$, or 5. Because each of the fractions, $\frac{9}{10}, \frac{7}{8}$, and $\frac{11}{12}$, is close to but less than 1, their sum is less than 3. The approximate answer is a number close to but less than 8.

ONGOING ASSESSMENT 5-2

1. Compute each of the following using any method:
   a. $\frac{1}{2} + \frac{2}{3}$
   b. $\frac{4}{12} - \frac{2}{3}$
   c. $\frac{5}{x} + \frac{-3}{y}$
   d. $\frac{-3}{2x^2y} + \frac{5}{6xy^2} + \frac{7}{x^2}$
   e. $\frac{5}{6} + 2\frac{1}{8}$
   f. $-4\frac{1}{2} - 3\frac{1}{6}$

2. Change each of the following fractions to mixed numbers:
   a. $\frac{56}{3}$
   b. $\frac{14}{5}$
   c. $-\frac{293}{100}$
   d. $-\frac{47}{8}$

3. Change each of the following mixed numbers to fractions in the form $\frac{a}{b}$, where a and b are integers:
   a. $6\frac{3}{4}$
   b. $7\frac{1}{2}$
   c. $-3\frac{5}{8}$
   d. $-4\frac{2}{3}$

4. Place the numbers 2, 5, 6, and 8 in the following boxes to make the equation true:

$$\frac{\Box}{\Box} + \frac{\Box}{\Box} = \frac{23}{24}$$

5. Approximate each of the following situations with a convenient fraction. Explain your reasoning. Tell whether your estimate is high or low.
   a. Giorgio had 15 base hits out of 46 times at bat.
   b. Ruth made 7 goals out of 41 shots.
   c. Laura answered 62 problems correctly out of 80.
   d. Jonathan made 9 baskets out of 19.

6. Use the information in the following table to answer each of the following questions:

Team	Games Played	Games Won
Ducks	22	10
Beavers	19	10
Tigers	28	9
Bears	23	8
Lions	27	7
Wildcats	25	6
Badgers	21	5

   a. Which team won just over $\frac{1}{2}$ of its games?
   b. Which team won just under $\frac{1}{2}$ of its games?
   c. Which team won just over $\frac{1}{3}$ of its games?
   d. Which team won just under $\frac{1}{3}$ of its games?
   e. Which team won just over $\frac{1}{4}$ of its games?
   f. Which team won just under $\frac{1}{4}$ of its games?

7. Sort the following fraction cards into the ovals by estimating in which oval the fraction belongs:

Sort these fraction cards: $\frac{1}{10}$, $\frac{4}{7}$, $\frac{8}{12}$, $\frac{1}{3}$, $\frac{7}{8}$, $\frac{2}{5}$, $\frac{3}{10}$, $\frac{13}{10}$, $\frac{1}{100}$, $\frac{9}{18}$

About 0   About $\frac{1}{2}$ About 1

8. Approximate each of the following fractions by 0, $\frac{1}{4}$, $\frac{1}{2}$, $\frac{3}{4}$, or 1. Tell whether your estimate is high or low.
   a. $\frac{19}{39}$ **b.** $\frac{3}{197}$ **c.** $\frac{150}{201}$ **d.** $\frac{8}{9}$
   e. $\frac{113}{110}$ **f.** $\frac{-2}{117}$ **g.** $\frac{150}{198}$ **h.** $\frac{999}{2000}$

9. Without actually finding the exact answer, state which of the numbers given in parentheses in the following is the best approximation for the given sum or difference:

   a. $\frac{6}{13} + \frac{7}{15} + \frac{11}{23} + \frac{17}{35}$ $(1, 2, 3, 3\frac{1}{2})$
   b. $\frac{30}{41} + \frac{1}{1000} + \frac{3}{2000}$ $(\frac{3}{8}, \frac{3}{4}, 1, 2)$
   c. $\frac{103}{300} + \frac{203}{601} - \frac{602}{897}$ $(1, \frac{1}{3}, \frac{2}{3}, 0)$
   d. $\frac{1}{100} - \frac{1}{101} + \frac{1}{102} - \frac{1}{103}$ $(\frac{1}{2}, 1, 0)$

10. Compute each of the following mentally:
   a. $1 - \frac{3}{4}$ **b.** $6 - \frac{7}{8}$
   c. $3\frac{3}{8} + 2\frac{1}{4} - 5\frac{5}{8}$ **d.** $2\frac{3}{5} + 4\frac{1}{10} + 3\frac{3}{10}$

11. The following ruler has regions marked M, A, T, H:

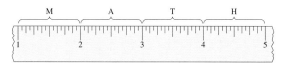

Use mental mathematics and estimation to determine which region each of the following falls into (for example, $\frac{12}{5}$ in. falls in region A).
   a. $\frac{20}{8}$ in.   **b.** $\frac{36}{8}$ in.
   c. $\frac{60}{16}$ in.   **d.** $\frac{18}{4}$ in.

12. A class consists of $\frac{2}{5}$ freshmen, $\frac{1}{4}$ sophomores, and $\frac{1}{10}$ juniors; the rest are seniors. What fraction of the class is seniors?

13. The Naturals Company sells its products in many countries. The following two circle graphs show the fractions of the company's earnings for 1990 and 2000. Based on this information, answer the following questions:
   a. In 1990, how much greater was the fraction of sales for Japan than for Canada?
   b. In 2000, how much less was the fraction of sales for England than for the United States?
   c. How much greater was the fraction of total sales for the United States in 2000 than in 1990?
   d. Is it true that the amount of sales in dollars in Australia was less in 1990 than in 2000? Why?

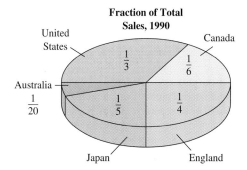

Fraction of Total Sales, 1990

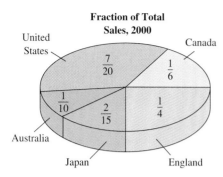

Fraction of Total Sales, 2000
- United States: $\frac{7}{20}$
- Canada: $\frac{1}{6}$
- England: $\frac{1}{4}$
- Japan: $\frac{2}{15}$
- Australia: $\frac{1}{10}$

14. A clerk sold three pieces of one type of ribbon to different customers. One piece was $\frac{1}{3}$ yd long, another was $2\frac{3}{4}$ yd long, and the third was $3\frac{1}{2}$ yd long. What was the total length of that type of ribbon sold?

15. A recipe requires $3\frac{1}{2}$ c of milk. Ran put in $1\frac{3}{4}$ c and emptied the container. How much more milk does he need to put in?

16. Martine bought $8\frac{3}{4}$ yd of fabric. She wants to make a skirt using $1\frac{7}{8}$ yd, pants using $2\frac{3}{8}$ yd, and a vest using $1\frac{2}{3}$ yd. How much fabric will be left over?

17. A $15\frac{3}{4}$-in. board is cut from a $38\frac{1}{4}$-in. board. The saw cut takes $\frac{3}{8}$ in. How much of the $38\frac{1}{4}$-in. board is left after cutting?

18. Students from Rattlesnake School formed four teams to collect cans for recycling during the months of April and May. The students received 10¢ for each 5 lb of cans. A record of their efforts follows:

Number of Pounds Collected

	Team 1	Team 2	Team 3	Team 4
April	$28\frac{3}{4}$	$32\frac{7}{8}$	$28\frac{1}{2}$	$35\frac{3}{16}$
May	$33\frac{1}{3}$	$28\frac{5}{12}$	$25\frac{3}{4}$	$41\frac{1}{2}$

   a. Which team collected the most for the 2-mo period? How much did they collect?
   b. What was the difference in the total amounts collected by the teams during the 2 mo?

19. Demonstrate by example that each of the following properties of rational numbers holds:
   a. Closure property of addition
   b. Commutative property of addition
   c. Associative property of addition

20. For each of the following sequences, discover a pattern and write three more terms of the sequence if the pattern continues. Which of the sequences are arithmetic, and which are not? Justify your answers.

   a. $\frac{1}{4}, \frac{1}{2}, \frac{3}{4}, 1, \frac{5}{4}, \ldots$ b. $\frac{1}{2}, \frac{2}{3}, \frac{3}{4}, \frac{4}{5}, \frac{5}{6}, \ldots$
   c. $\frac{2}{3}, \frac{5}{3}, \frac{8}{3}, \frac{11}{3}, \frac{14}{3}, \ldots$ d. $\frac{5}{4}, \frac{3}{4}, \frac{1}{4}, \frac{-1}{4}, \frac{-3}{4}, \ldots$

21. Find the nth term in each of the sequences in problem 20.

22. Use the following diagram from a 1994 United Nations *Human Development Report* to answer the following questions:
   a. In 1991, of the countries pictured, what fraction of the water was used by Japan and the United States together?
   b. Similarly, what fraction of the pesticides was used by Japan and the United States together?

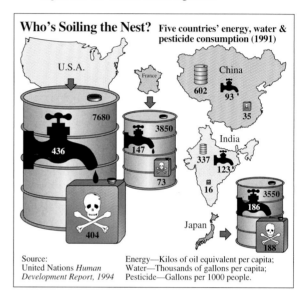

Who's Soiling the Nest? Five countries' energy, water & pesticide consumption (1991)

Source: United Nations *Human Development Report, 1994*
Energy—Kilos of oil equivalent per capita; Water—Thousands of gallons per capita; Pesticide—Gallons per 1000 people.

23. Insert five fractions between the numbers 1 and 2 so that the seven numbers (including 1 and 2) constitute an arithmetic sequence.

24. Let $f(x) = x + \frac{3}{4}$.
   a. Find the outputs if the inputs are the following:
      i. 0 ii. $\frac{4}{3}$ iii. $\frac{-3}{4}$
   b. For which inputs will the outputs be the following?
      i. 1 ii. $^-1$ iii. $\frac{1}{2}$

25. Let $f(x) = \frac{x+2}{x-1}$ and let the domain of the function be the set of all integers except 1. Find the following:
   a. $f(0)$ b. $f(^-2)$
   c. $f(^-5)$ d. $f(5)$

26. a. Check that each of the following is true:
   $\frac{1}{3} = \frac{1}{4} + \frac{1}{3 \cdot 4}$ $\frac{1}{4} = \frac{1}{5} + \frac{1}{4 \cdot 5}$
   $\frac{1}{5} = \frac{1}{6} + \frac{1}{5 \cdot 6}$
   b. Based on the examples in (a), write $\frac{1}{n}$ as a sum of two unit fractions, that is, as a sum of fractions with numerator 1.

Communication

27. Sally claims that it is easier to add two fractions if she adds the numerators and then adds the denominators. How can you help her?
28. Does each of the following properties hold for subtraction of rational numbers? Justify your answer.
    a. Closure   b. Commutative
    c. Associative d. Identity
    e. Inverse
29. Explain an error pattern in each of the following.
    a. $\frac{13}{35} = \frac{1}{5}, \quad \frac{27}{73} = \frac{2}{3}, \quad \frac{16}{64} = \frac{1}{4}$
    b. $\frac{4}{5} + \frac{2}{3} = \frac{6}{8}, \quad \frac{2}{5} + \frac{3}{4} = \frac{5}{9}, \quad \frac{7}{8} + \frac{1}{3} = \frac{8}{11}$
    c. $8\frac{3}{4} - 6\frac{1}{8} = 2\frac{2}{4}, \quad 5\frac{3}{8} - 2\frac{2}{3} = 3\frac{1}{5}, \quad 2\frac{2}{7} - 1\frac{1}{3} = 1\frac{1}{4}$
    d. $\frac{2}{3} \cdot 3 = \frac{6}{9}, \quad \frac{1}{4} \cdot 6 = \frac{6}{24}, \quad \frac{4}{5} \cdot 2 = \frac{8}{10}$

Open-Ended

30. Use the approximate population density (number of people per square mile) from the following table to answer the given questions:

Montana	Russia	United Kingdom	Nigeria	Bangladesh	United States
6	22	588	248	2028	68

   a. Explain why it is or is not feasible to add the numbers in the table for Montana and Russia to determine the population density of the combined country and state.
   b. Decide whether it is reasonable to decide that, based on the given data, the population of Bangladesh is approximately 355 times that of Montana.
   c. Using the data in the table, create and solve two questions that are reasonable for middle-school students to solve.

Cooperative Learning

31. Interview 10 people and ask them if and when they add and subtract fractions in their lives. Combine those responses with those of the rest of the class to get a view of how "ordinary" people must use computation of rational numbers in their daily lives.

Review Problems

32. Use the following graph and the marked points to answer the following:

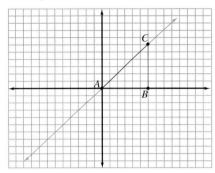

   a. Create a series of right triangles whose longest side (hypotenuse) lies along the slanted line. Use at least one marked point as a vertex of each triangle. Triangle ABC is given as an example.
   b. To get to point A from point C by traveling along the horizontal and vertical sides of the triangle, we must go down 6 units ($^-6$) and 6 units to the left ($^-6$). The ratio of directed segment lengths CB to BA is $^-6/(^-6)$, or 1. Find the ratio of directed segment lengths in the triangles created in (a).
   c. Make a conjecture about triangles created in the manner described in (a).
33. Simplify if possible.
    a. $\frac{14}{21}$ b. $\frac{117}{153}$ c. $\frac{5^2}{7^2}$
    d. $\frac{a^2 + a}{1 + a}$ e. $\frac{a^2 + 1}{a + 1}$
34. Determine if the fractions in each of the following pairs are equal:
    a. $\frac{a^2}{b}$ and $\frac{a^2 b^2}{b^3}$ b. $\frac{377}{400}$ and $\frac{378}{401}$
    c. $\frac{0}{10}$ and $\frac{0}{-10}$ d. $\frac{a}{b}$ and $\frac{a+1}{b+1}$, where $a \neq b$
35. a. What month of the year has the smallest fraction of days of the year?
    b. What fraction of days of the year occur before July 4?
    c. How many days are actually in a year? Express this as a mixed number and as an improper fraction.
36. If the same positive number is added to the numerator and denominator of a positive proper fraction, is the new fraction greater than, less than, or equal to the original fraction? Justify your answer.

BRAIN TEASER When Professor Sum was asked by Mr. Little how many students were in his classes, he answered, "All of them study either languages, physics, or not at all. One half of them study languages only, one fourth of them study French, one seventh of them study physics only, and 20 do not study at all." How many students does Professor Sum have?

Section 5-3 — Multiplication and Division of Rational Numbers

Multiplication of Rational Numbers

To motivate the definition of multiplication of rational numbers, we use the interpretation of multiplication as repeated addition. Using repeated addition, we can interpret $3 \cdot \left(\dfrac{3}{4}\right)$ as follows:

$$3 \cdot \left(\dfrac{3}{4}\right) = \dfrac{3}{4} + \dfrac{3}{4} + \dfrac{3}{4} = \dfrac{9}{4} = 2\dfrac{1}{4}.$$

The area model in Figure 5-8 shows this.

Figure 5-8

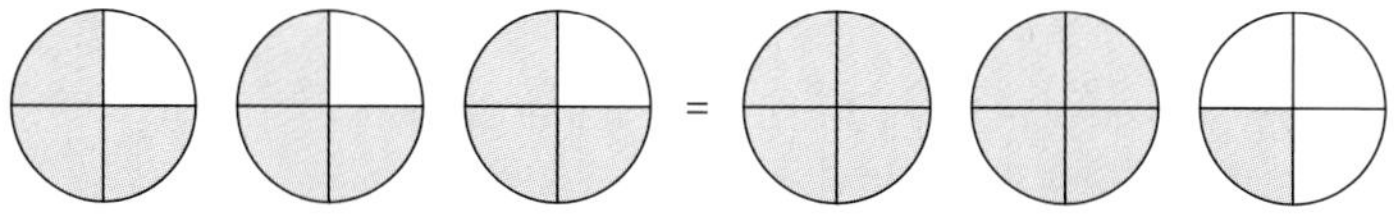

$$3 \cdot \dfrac{3}{4} = \dfrac{3}{4} + \dfrac{3}{4} + \dfrac{3}{4} = \dfrac{9}{4} \text{ or } 2\dfrac{1}{4}$$

If the commutative property of multiplication of rational numbers is true, then $3 \cdot \left(\dfrac{3}{4}\right) = \left(\dfrac{3}{4}\right) \cdot 3 = \dfrac{9}{4}.$

Next, we consider what happens when neither factor is an integer. If forests once covered about $\dfrac{3}{5}$ of Earth's land and only about $\dfrac{1}{2}$ of these forests remain, what fraction of Earth is covered with forests today? We can use an area model to find out.

Figure 5-9(a) shows a one-unit rectangle separated into fifths, with $\dfrac{3}{5}$ shaded. To find $\dfrac{1}{2}$ of $\dfrac{3}{5}$, we divide the shaded portion of the rectangle in Figure 5-9(a) into two equal parts and take one of those parts. The result would be the green portion of Figure 5-9(b). However, the green portion represents three parts out of 10, or $\dfrac{3}{10}$ of the one-unit rectangle. Thus

$$\dfrac{1}{2} \cdot \dfrac{3}{5} = \dfrac{3}{10} = \dfrac{1 \cdot 3}{2 \cdot 5}.$$

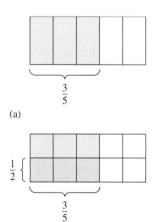

Figure 5-9

This discussion leads to the following definition of multiplication for rational numbers.

Definition of Multiplication of Rational Numbers

If $\dfrac{a}{b}$ and $\dfrac{c}{d}$ are any rational numbers, then $\dfrac{a}{b} \cdot \dfrac{c}{d} = \dfrac{a \cdot c}{b \cdot d}.$

Example 5-11 Answer the following:

If $\frac{5}{6}$ of the population of a certain city are college graduates and $\frac{7}{11}$ of the city's college graduates are female, what fraction of the population of that city is female and college graduates?

Solution The fraction should be 7/11 of 5/6 or $\frac{7}{11} \cdot \frac{5}{6} = \frac{7 \cdot 5}{11 \cdot 6} = \frac{35}{66}$.

The fraction of the population that is college-graduated females is $\frac{35}{66}$.

HISTORICAL NOTE

In the Middle Ages, mathematical skill was admired and supported by the monarchs. Leonardo of Pisa (1170–1230), known as Fibonacci, was the most prominent of the medieval mathematicians. In 1225, Fibonacci participated in a mathematical tournament before the Roman Emperor Frederic II, who came to Pisa with a group of mathematicians to test Fibonacci's immense reputation. One of the questions was to find a rational number that is a square, for example $\left(\frac{4}{9} = \left(\frac{2}{3}\right)^2\right)$, and that remains a square if it is decreased or increased by 5. Fibonacci found the number: $\frac{1681}{144}$, or $\left(\frac{41}{12}\right)^2$. When 5 is subtracted, it remains a square because $\frac{1681}{144} - 5 = \frac{961}{144} = \left(\frac{31}{12}\right)^2$, and when 5 is added, it remains a square because $\frac{1681}{144} + 5 = \frac{2401}{144} = \left(\frac{49}{12}\right)^2$.

Properties of Multiplication of Rational Numbers

Multiplication of rational numbers has properties analogous to the properties of addition of rational numbers. These include the following properties for multiplication: closure, commutative, associative, multiplicative identity, and multiplicative inverse. For emphasis, we list the last two properties.

Properties

Multiplicative identity of rational numbers The number 1 is the unique number such that for every rational number $\frac{a}{b}$,

$$1 \cdot \left(\frac{a}{b}\right) = \frac{a}{b} = \left(\frac{a}{b}\right) \cdot 1$$

Multiplicative inverse of rational numbers For any nonzero rational number $\frac{a}{b}$, $\frac{b}{a}$ is the unique rational number such that $\frac{a}{b} \cdot \frac{b}{a} = 1 = \frac{b}{a} \cdot \frac{a}{b}$. The multiplicative inverse of $\frac{a}{b}$ is also called the **reciprocal** of $\frac{a}{b}$.

reciprocal

Example 5-12 Find the multiplicative inverse of each of the following rational numbers:

a. $\frac{2}{3}$ b. $\frac{-2}{5}$ c. 4 d. 0 e. $6\frac{1}{2}$

Solution a. $\frac{3}{2}$

b. $\frac{5}{-2}$, or $\frac{-5}{2}$

c. Because $4 = \frac{4}{1}$, the multiplicative inverse of 4 is $\frac{1}{4}$.

d. Even though $0 = \frac{0}{1}$, $\frac{1}{0}$ is undefined; there is no multiplicative inverse of 0.

e. Because $6\frac{1}{2} = \frac{13}{2}$, the multiplicative inverse of $6\frac{1}{2}$ is $\frac{2}{13}$.

Multiplication and addition are connected through the distributive property of multiplication over addition. Also there is a multiplication property of equality for rational numbers and a multiplication property of zero similar to those for whole numbers and integers.

Properties

Distributive property of multiplication over addition for rational numbers If $\frac{a}{b}, \frac{c}{d}$, and $\frac{e}{f}$ are any rational numbers, then

$$\frac{a}{b}\left(\frac{c}{d} + \frac{e}{f}\right) = \left(\frac{a}{b} \cdot \frac{c}{d}\right) + \left(\frac{a}{b} \cdot \frac{e}{f}\right)$$

Multiplication property of equality for rational numbers If $\frac{a}{b}$ and $\frac{c}{d}$ are any rational numbers such that $\frac{a}{b} = \frac{c}{d}$, and $\frac{e}{f}$ is any rational number, then $\frac{a}{b} \cdot \frac{e}{f} = \frac{c}{d} \cdot \frac{e}{f}$.

Multiplication property of zero for rational numbers If $\frac{a}{b}$ is any rational number, then $\frac{a}{b} \cdot 0 = 0 = 0 \cdot \frac{a}{b}$.

Example 5-13 A bicycle is on sale at $\frac{3}{4}$ of its original price. If the sale price is $330, what was the original price?

Solution Let x be the original price. Then $\frac{3}{4}$ of the original price is $\frac{3}{4}x$. Because the sale price is $330, we have $\frac{3}{4}x = 330$. Solving for x gives

$$\frac{4}{3} \cdot \frac{3}{4}x = \frac{4}{3} \cdot 330$$

$$1 \cdot x = 440$$

$$x = 440.$$

Thus the original price was $440.

An alternative approach, which does not use algebra, follows. Because $\frac{3}{4}$ of the original price is $330, $\frac{1}{4}$ of the original price is $\frac{1}{3} \cdot 330$, or $110; thus $4 \cdot \frac{1}{4}$ of the original price is $4 \cdot 110$, or $440.

Problem Solving   Sonja's Deck

Sonja wants to build a square deck with the floor made out of 1-in.-by-6-in. boards. She wants the deck to be 30 boards wide. Boards come in lengths of 6, 8, 10, 12, 14, 16, 18, and 20 ft and sell for 32¢ per ft. How many boards of what length must Sonja order to make the floor? What is the minimum cost of the floor if she orders only complete boards?

• Understanding the Problem   In reality a 1-in.-by-6-in. board is $5\frac{1}{2}$ in. wide and $\frac{3}{4}$ in. thick. Only complete boards can be used. We need to find the minimum cost.

• Devising a Plan   Because the deck is a square and the deck is 30 boards wide, the length of the deck must be $30 \cdot \left(5\frac{1}{2}\right) = 165$ in. From this information, we can find the length of the boards Sonja needs and the cost.

• Carrying Out the Plan   Since the length of the deck is 165 in., we divide by 12 to convert to feet and obtain $13\frac{3}{4}$ ft. Hence, we need to order 30 of the 14-ft boards. This gives $30 \cdot 14$, or 420 ft at 32¢ per ft. Thus Sonja's bill would be 32¢ $\cdot$ 420 = $134.40. This is the minimum cost because boards shorter than 14 ft will not work, boards longer than 14 ft cost more, and two lengths cannot be cut from any board.

• Looking Back   We could vary this problem by changing the sizes of the deck or the boards. We could also work on *related problems* such as which size nail is needed to nail together three 1-in.-by-6-in. boards so that the nail would go through two boards and go $\frac{1}{2}$ in. into the third board. (Nail sizes increase by $\frac{1}{4}$ in. For example, a 2-penny nail is 1 in. long; a 3-penny nail is $1\frac{1}{4}$ in. long; a 4-penny nail is $1\frac{1}{2}$ in. long, and so forth.) How many nails would be needed to nail down the deck?

Extending the Notion of Exponents

Recall that a^m was defined for any number a and any natural number m as follows.

Definition of *a* to the *m*th Power

$a^m = \underbrace{a \cdot a \cdot a \cdot \ldots \cdot a}_{m \text{ factors}}$, where a is any rational number and m is any natural number.

From the definition, $a^3 \cdot a^2 = (a \cdot a \cdot a) \cdot (a \cdot a) = a^{3+2}$. In a similar way, it follows that

(1) $$a^m \cdot a^n = a^{m+n}$$

where a is any number and m and n are natural numbers. If (1) is true for all whole numbers m and n, then because $a^1 \cdot a^0 = a^{1+0} = a^1$, it makes sense for $a^0 = 1$. It is useful to give meaning to a^0 when $a \neq 0$ as follows:

(2) $$a^0 = 1, \quad \text{if } a \neq 0$$

These notions can be extended for rational number values of a. For example, consider the following:

$$\left(\frac{2}{3}\right)^4 = \frac{2}{3} \cdot \frac{2}{3} \cdot \frac{2}{3} \cdot \frac{2}{3}$$

$$\left(\frac{2}{3}\right)^2 \cdot \left(\frac{2}{3}\right)^3 = \left(\frac{2}{3} \cdot \frac{2}{3}\right) \cdot \left(\frac{2}{3} \cdot \frac{2}{3} \cdot \frac{2}{3}\right) = \left(\frac{2}{3}\right)^{2+3} = \left(\frac{2}{3}\right)^5$$

In general it can be shown that Equation (1) holds when a is any rational number and consequently Equation (2) remains true if a is a rational number and $a \neq 0$.

Exponents can also be extended to negative integers. Notice that as the exponents decrease by 1, the numbers on the right are divided by 10. Thus the pattern might be continued, as shown.

$$10^3 = 10 \cdot 10 \cdot 10$$
$$10^2 = 10 \cdot 10$$
$$10^1 = 10$$
$$10^0 = 1$$
$$10^{-1} = \frac{1}{10} = \frac{1}{10^1}$$
$$10^{-2} = \frac{1}{10} \cdot \frac{1}{10} = \frac{1}{10^2}$$
$$10^{-3} = \frac{1}{10^2} \cdot \frac{1}{10} = \frac{1}{10^3}$$

If the pattern is extended, then we would predict that $10^{-n} = \frac{1}{10^n}$. This is true, and in general, for any nonzero number a, $a^{-n} = \frac{1}{a^n}$.

Consider whether the property $a^m \cdot a^n = a^{m+n}$ can be extended to include all powers of a, where the exponents are integers. For example, is it true that $2^4 \cdot 2^{-3} = 2^{4+-3} = 2^1$? The definitions of 2^{-3} and the properties of nonnegative exponents ensure this is true, as shown next:

$$2^4 \cdot 2^{-3} = 2^4 \cdot \frac{1}{2^3} = \frac{2^4}{2^3} = \frac{2^1 \cdot 2^3}{2^3} = 2^1$$

Also, $2^{-4} \cdot 2^{-3} = 2^{-4+-3} = 2^{-7}$ is true because

$$2^{-4} \cdot 2^{-3} = \frac{1}{2^4} \cdot \frac{1}{2^3} = \frac{1 \cdot 1}{2^4 \cdot 2^3} = \frac{1}{2^{4+3}} = \frac{1}{2^7} = 2^{-7}$$

In general, with integer exponents, the following property holds.

Property

For any nonzero rational number a and any integers m and n, $a^m \cdot a^n = a^{m+n}$.

Other properties of exponents can be developed by using the properties of rational numbers. For example,

$$\frac{2^5}{2^3} = \frac{2^3 \cdot 2^2}{2^3} = 2^2 = 2^{5-3} \qquad \frac{2^5}{2^8} = \frac{2^5}{2^5 \cdot 2^3} = \frac{1}{2^3} = 2^{-3} = 2^{5-8}$$

With integer exponents, the following property holds.

Property

For any rational number a such that $a \neq 0$ and for any integers m and n, $\dfrac{a^m}{a^n} = a^{m-n}$.

At times, we may encounter an expression like $(2^4)^3$. This expression can be written as a single power of 2 as follows:

$$(2^4)^3 = 2^4 \cdot 2^4 \cdot 2^4 = 2^{4+4+4} = 2^{3 \cdot 4} = 2^{12}$$

In general, if a is a nonzero rational number and m and n are positive integers, then

$$(a^m)^n = \underbrace{a^m \cdot a^m \cdot a^m \cdot \ldots \cdot a^m}_{n \text{ factors}} = a^{\overbrace{m+m+\ldots+m}^{n \text{ terms}}} = a^{nm} = a^{mn}$$

Does this property hold for negative-integer exponents? For example, does $(2^3)^{-4} = 2^{(3)(-4)} = 2^{-12}$? The answer is yes because $(2^3)^{-4} = \dfrac{1}{(2^3)^4} = \dfrac{1}{2^{12}} = 2^{-12}$. Also, $(2^{-3})^4 = \left(\dfrac{1}{2^3}\right)^4 = \dfrac{1}{2^3} \cdot \dfrac{1}{2^3} \cdot \dfrac{1}{2^3} \cdot \dfrac{1}{2^3} = \dfrac{1^4}{(2^3)^4} = \dfrac{1}{2^{12}} = 2^{-12}$

Property

For any rational number $a \neq 0$ and any integers m and n,

$$(a^m)^n = a^{mn}.$$

Using the definitions and properties developed, we can derive additional properties. Notice, for example, that

$$\left(\frac{2}{3}\right)^4 = \frac{2}{3} \cdot \frac{2}{3} \cdot \frac{2}{3} \cdot \frac{2}{3} = \frac{2 \cdot 2 \cdot 2 \cdot 2}{3 \cdot 3 \cdot 3 \cdot 3} = \frac{2^4}{3^4}$$

This property can be generalized as follows.

Property

For any nonzero rational number $\frac{a}{b}$ and any integer m,

$$\left(\frac{a}{b}\right)^m = \frac{a^m}{b^m}.$$

Division of Rational Numbers

Recall that $6 \div 3$ means "How many 3s are there in 6?" We found that $6 \div 3 = 2$ because $3 \cdot 2 = 6$. Consider $3 \div \left(\frac{1}{2}\right)$, which is equivalent to finding how many halves there are in 3. We see from the area model in Figure 5-10 that there are 6 half pieces in the 3 whole pieces. We record this as $3 \div \left(\frac{1}{2}\right) = 6$. Also note that $\left(\frac{1}{2}\right) \cdot 6 = 3$.

Figure 5-10

Next, consider $\left(\frac{3}{4}\right) \div \left(\frac{1}{8}\right)$. This means "How many $\frac{1}{8}$s are in $\frac{3}{4}$?" Figure 5-11 shows that there are six $\frac{1}{8}$s in the shaded portion, which represents $\frac{3}{4}$ of the whole. Therefore $\left(\frac{3}{4}\right) \div \left(\frac{1}{8}\right) = 6$. Also, note that $\left(\frac{1}{8}\right) \cdot 6 = \frac{3}{4}$.

In the previous examples, we saw a relationship between division and multiplication of rational numbers. We can define division for rational numbers formally in terms of multiplication in the same way that we define division for integers.

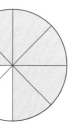

Figure 5-11

Definition of Division of Rational Numbers

If $\frac{a}{b}$ and $\frac{c}{d}$ are any rational numbers and $\frac{c}{d}$ is not zero, then $\frac{a}{b} \div \frac{c}{d} = x$ if, and only if, x is the unique rational number such that $\frac{c}{d} \cdot x = \frac{a}{b}$.

REMARK In the definition of division, $\frac{c}{d}$ is not zero because division by zero is impossible. Also, $\frac{c}{d} \neq 0$ implies that $c \neq 0$.

SECTION 5-3 Multiplication and Division of Rational Numbers

Algorithm for Division of Rational Numbers

To develop an algorithm for division of rational numbers, we consider what such a division might mean. For example,

$$\frac{2}{3} \div \frac{5}{7} = x \text{ implies } \frac{2}{3} = \frac{5}{7}x.$$

To find x, we multiply both sides of the equation by $\frac{7}{5}$. Thus,

$$\frac{7}{5} \cdot \frac{2}{3} = \frac{7}{5} \cdot \left(\frac{5}{7}x\right) = x.$$

A traditional justification of the division algorithm also follows. The algorithm for division of fractions is usually justified in the middle grades by using the Fundamental Law of Fractions, $\frac{a}{b} = \frac{ac}{bc}$, where a, b, and c are all fractions. For example,

$$\frac{2}{3} \div \frac{5}{7} = \frac{\frac{2}{3}}{\frac{5}{7}} = \frac{\frac{2}{3} \cdot \frac{7}{5}}{\frac{5}{7} \cdot \frac{7}{5}} = \frac{\frac{2}{3} \cdot \frac{7}{5}}{\frac{5}{7} \cdot \frac{7}{5}} = \frac{\frac{2}{3} \cdot \frac{7}{5}}{1} = \frac{2}{3} \cdot \frac{7}{5}.$$

Thus

$$\frac{2}{3} \div \frac{5}{7} = \frac{2}{3} \cdot \frac{7}{5}.$$

We summarize the algorithm as follows:

Algorithm for Division of Fractions

$$\frac{a}{b} \div \frac{c}{d} = \frac{a}{b} \cdot \frac{d}{c}, \text{ where } \frac{c}{d} \neq 0.$$

An alternative approach for developing an algorithm for division of fractions can be found by first dividing fractions that have equal denominators. For example, $\frac{9}{10} \div \frac{3}{10} = 9 \div 3$ and $\frac{15}{23} \div \frac{5}{23} = 15 \div 5$. These examples suggest that when two fractions with the same denominators are divided, the result can be obtained by dividing the numerator of the first fraction by the numerator of the second. To divide fractions with different denominators, we rename the fractions so that the denominators are equal. Thus

$$\frac{a}{b} \div \frac{c}{d} = \frac{ad}{bd} \div \frac{bc}{bd} = ad \div bc = \frac{ad}{bc}.$$

Example 5-14 A radio station provides 36 min for Public Service Announcements for every 24 hr of broadcasting.

a. What part of the broadcasting day is allotted to Public Service Announcements?
b. How many 3/4-min Public Service Announcements can be allowed in the 36 min?

Solution a. There are 60 min in an hour and 60 · 24 min in a day. Thus 36/(60 · 24), or 1/40, of the day is allotted for the announcements.
b. 36/(3/4) = 36(4/3), or 48, announcements are allowed.

Example 5-15 We have $35\frac{1}{2}$ yd of material available to make shirts. Each shirt requires $\frac{3}{8}$ yd of material.

a. How many shirts can be made?
b. How much material will be left over?

Solution a. We need to find the integer part of the answer to $35\frac{1}{2} \div \frac{3}{8}$. The division follows:

$$35\frac{1}{2} \div \frac{3}{8} = \frac{71}{2} \cdot \frac{8}{3} = \frac{284}{3} = 94\frac{2}{3}$$

Thus we can make 94 shirts.

b. Because the division in (a) was by $\frac{3}{8}$, the amount of material left over is $\frac{2}{3}$ of $\frac{3}{8}$, or $\frac{2}{3} \cdot \frac{3}{8}$, or $\frac{1}{4}$ yd.

Division of Rational Numbers Related to Exponents

From the definition of negative exponents, the above property, and division of fractions, we have

$$\left(\frac{a}{b}\right)^{-m} = \frac{1}{\left(\frac{a}{b}\right)^m} = \frac{1}{\frac{a^m}{b^m}} = \frac{b^m}{a^m} = \left(\frac{b}{a}\right)^m.$$

Consequently, $\left(\frac{a}{b}\right)^{-m} = \left(\frac{b}{a}\right)^m$.

A property similar to this holds for multiplication. For example,

$$(2 \cdot 3)^{-3} = \frac{1}{(2 \cdot 3)^3} = \frac{1}{2^3 \cdot 3^3} = \left(\frac{1}{2^3}\right) \cdot \left(\frac{1}{3^3}\right) = 2^{-3} \cdot 3^{-3}$$

and in general, it is true that $(a \cdot b)^m = a^m \cdot b^m$ if a and b are rational numbers and m is an integer.

The definitions and properties of exponents are summarized in the following list. For any rational numbers a and b and integers m and n (as long as 0^0 does not appear), we have the following:

1. $a^m = \underbrace{a \cdot a \cdot a \cdot \ldots \cdot a}_{m \text{ factors}}$, where m is a positive integer
2. $a^0 = 1$, where $a \neq 0$
3. $a^{-m} = \frac{1}{a^m}$, where $a \neq 0$
4. $a^m \cdot a^n = a^{m+n}$
5. $\frac{a^m}{a^n} = a^{m-n}$, where $a \neq 0$

6. $(a^m)^n = a^{mn}$

7. $\left(\dfrac{a}{b}\right)^m = \dfrac{a^m}{b^m}$, where $b \neq 0$

8. $(ab)^m = a^m \cdot b^m$

9. $\left(\dfrac{a}{b}\right)^{-m} = \left(\dfrac{b}{a}\right)^m$

Observe that all the properties of exponents refer to powers with either the same base or the same exponent. To evaluate expressions using exponents where different bases or powers are used, perform all the computations or rewrite the expressions in either the same base or the same exponent if possible. For example, $\dfrac{27^4}{81^3}$ can be rewritten as $\dfrac{27^4}{81^3} = \dfrac{(3^3)^4}{(3^4)^3} = \dfrac{3^{12}}{3^{12}} = 1$.

Example 5-16 Write each of the following in simplest form using positive exponents in the final answer:

a. $16^2 \cdot 8^{-3}$
b. $20^2 \div 2^4$
c. $(10^{-1} + 5 \cdot 10^{-2} + 3 \cdot 10^{-3}) \cdot 10^3$

Solution
a. $16^2 \cdot 8^{-3} = (2^4)^2 \cdot (2^3)^{-3} = 2^8 \cdot 2^{-9} = 2^{8 + ^-9} = 2^{-1} = \dfrac{1}{2}$

b. $\dfrac{20^2}{2^4} = \dfrac{(2^2 \cdot 5)^2}{2^4} = \dfrac{2^4 \cdot 5^2}{2^4} = 5^2$

c. $(10^{-1} + 5 \cdot 10^{-2} + 3 \cdot 10^{-3}) \cdot 10^3 = 10^{-1} \cdot 10^3 + 5 \cdot 10^{-2} \cdot 10^3 + 3 \cdot 10^{-3} \cdot 10^3$
$= 10^{-1+3} + 5 \cdot 10^{-2+3} + 3 \cdot 10^{-3+3}$
$= 10^2 + 5 \cdot 10^1 + 3 \cdot 10^0$
$= 153.$

BRAIN TEASER A castle in the faraway land of Aluossim was surrounded by four moats. One day, the castle was attacked and captured by a fierce tribe from the north. Guards were stationed at each bridge. Juan was allowed to take a number of bags of gold from the castle as he went into exile. However, the guard at the first bridge demanded half the bags of gold plus one more bag. Juan met this demand and proceeded to the next bridge. The guards at the second, third, and fourth bridges made identical demands, all of which the prince met. When Juan finally crossed all the bridges, a single bag of gold was left. With how many bags did Juan start?

ONGOING ASSESSMENT 5-3

1. In the following figures, a unit rectangle is used to illustrate the product of two fractions. Name the fractions and their products.

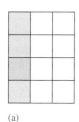
(a)

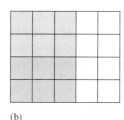
(b)

2. Use a rectangular region to illustrate each of the following products:
   a. $\dfrac{3}{4} \cdot \dfrac{1}{3}$ b. $\dfrac{1}{5} \cdot \dfrac{2}{3}$ c. $\dfrac{2}{5} \cdot \dfrac{1}{3}$

3. Find each of the following products. Write your answers in simplest form.
   a. $\dfrac{49}{65} \cdot \dfrac{26}{98}$ b. $\dfrac{a}{b} \cdot \dfrac{b^2}{a^2}$ c. $\dfrac{xy}{z} \cdot \dfrac{z^2 a}{x^3 y^2}$
   d. $2\dfrac{1}{3} \cdot 3\dfrac{3}{4}$ e. $\dfrac{22}{7} \cdot 4\dfrac{2}{3}$ f. $\dfrac{-5}{2} \cdot 2\dfrac{1}{2}$

4. Use the distributive property to find each product.
   a. $4\dfrac{1}{2} \cdot 2\dfrac{1}{3}$ $\left[\text{Hint: } \left(4 + \dfrac{1}{2}\right) \cdot \left(2 + \dfrac{1}{3}\right).\right]$
   b. $3\dfrac{1}{3} \cdot 2\dfrac{1}{2}$ c. $248\dfrac{2}{5} \cdot 100\dfrac{1}{8}$

5. Find the multiplicative inverse for each of the following:
   a. $\dfrac{-1}{3}$ b. $3\dfrac{1}{3}$
   c. $\dfrac{x}{y}$, if $x \neq 0$ and $y \neq 0$ d. $^-7$

6. Compute the following mentally. Find the exact answers.
   a. $3\dfrac{1}{4} \cdot 8$ b. $7\dfrac{1}{4} \cdot 4$ c. $9\dfrac{1}{5} \cdot 10$ d. $8 \cdot 2\dfrac{1}{4}$
   e. $3 \div \dfrac{1}{2}$ f. $3\dfrac{1}{2} \div \dfrac{1}{2}$ g. $3 \div \dfrac{1}{3}$ h. $4\dfrac{1}{2} \div 2$

7. Choose the number from among the numbers in parentheses that best approximates each of the following:
   a. $3\dfrac{11}{12} \cdot 5\dfrac{3}{100}$ (8, 20, 15, 16)
   b. $2\dfrac{1}{10} \cdot 7\dfrac{7}{8}$ (16, 14, 4, 3)
   c. $20\dfrac{2}{3} \div 9\dfrac{7}{8}$ $\left(2, 180, \dfrac{1}{2}, 10\right)$
   d. $\dfrac{1}{101} \div \dfrac{1}{103}$ $\left(0, 1, \dfrac{1}{2}, \dfrac{1}{4}\right)$

8. Estimate the following:
   a. $5\dfrac{4}{5} \cdot 3\dfrac{1}{10}$ b. $4\dfrac{10}{11} \cdot 5\dfrac{1}{8}$
   c. $\dfrac{20\dfrac{8}{9}}{3\dfrac{1}{12}}$ d. $\dfrac{12\dfrac{1}{3}}{1\dfrac{7}{8}}$

9. Without actually doing the computations, choose the number in parentheses that correctly describes each:
   a. $\dfrac{13}{14} \cdot \dfrac{17}{19}$ (greater than 1, less than 1)
   b. $3\dfrac{2}{7} \div 5\dfrac{1}{9}$ (greater than 1, less than 1)
   c. $4\dfrac{1}{3} \div 2\dfrac{3}{100}$ (greater than 2, less than 2)
   d. $16 \div 4\dfrac{3}{18}$ (greater than 4, less than 4)
   e. $16 \div 3\dfrac{8}{9}$ (greater than 4, less than 4)

10. A sewing project requires $6\dfrac{1}{8}$ yd of material that sells for 62¢ per yard and $3\dfrac{1}{4}$ yd that sells for 81¢ per yard. Choose from the following the best estimate for the cost of the project:
    a. Between $2 and $4   b. Between $4 and $6
    c. Between $6 and $8   d. Between $8 and $10

11. When you multiply a certain number by 3 and then subtract $\dfrac{7}{18}$, you get the same result as when you multiply the number by 2 and add $\dfrac{5}{12}$. What is the number?

12. Five-eighths of the students at Salem State College live in dormitories. If 6000 students at the college live in dormitories, how many students are there in the college?

13. Di Paloma University had a faculty reduction and lost $\dfrac{1}{5}$ of its faculty. If 320 faculty members were left after the reduction, how many members were there originally?

14. Alberto owns $\dfrac{5}{9}$ of the stock in the N.W. Tofu Company. His sister Renatta owns half as much stock as Alberto. What part of the stock is owned by neither Alberto nor Renatta?

15. A person has $29\dfrac{1}{2}$ yd of material available to make doll uniforms. Each uniform requires $\dfrac{3}{4}$ yd of material.
    a. How many uniforms can be made?
    b. How much material will be left over?

16. A suit is on sale for $180. What was the original price of the suit if the discount was $\dfrac{1}{4}$ of the original price?

17. Every employee's salary at the Sunrise Software Company increases each year by $\dfrac{1}{10}$ of that person's salary the previous year.
    a. If Martha's present annual salary is $100,000, what will her salary be in 2 yr?
    b. If Aaron's present salary is $99,000, what was his salary 1 yr ago?
    c. If Juanita's present salary is $363,000, what was her salary 2 yr ago?

18. Jasmine is reading a book. She has finished $\frac{3}{4}$ of the book and has 82 pages left to read. How many pages has she read?

19. John took all his money out of his bank savings account. He spent $50 on a radio and $\frac{3}{5}$ of what remained on presents. Half of what was left he put back in his checking account, and the remaining $35 he donated to charity. How much money did John originally have in his savings account?

20. Peter, Paul, and Mary start at the same time walking around a circular track in the same direction. Peter takes $\frac{1}{2}$ hr to walk around the track. Paul takes $\frac{5}{12}$ hr, and Mary takes $\frac{1}{3}$ hr.
   a. How many minutes does it take each person to walk around the track?
   b. How many times will each person go around the track before all three meet again at the starting line?

21. The formula for converting degrees Celsius (C) to degrees Fahrenheit (F) is $F = \left(\frac{9}{5}\right) \cdot C + 32$.
   a. If Samantha reads that the temperature is 32°C in Spain, what is the Fahrenheit temperature?
   b. If the temperature dropped to ⁻40°F in West Yellowstone, what is the temperature in degrees Celsius?

22. Glenn bought 175 shares of stock at $48\frac{1}{4}$ a share. A year later, he sold it at $35\frac{3}{8}$ a share. How much did Glenn lose on the transaction?

23. Al gives $\frac{1}{2}$ of his marbles to Bev. Bev gives $\frac{1}{2}$ of these to Carl. Carl gives $\frac{1}{2}$ of these to Dani. If Dani has four marbles, how many did Al have originally?

24. Believe it or not! Graham Greater supposedly averaged a hit every 1 1/2 sec in trapshooting 2264 targets in 1 hr. Is the arithmetic true?

25. The fastest centipede can travel at a rate of 19 17/25 in. per second. How far can it travel in one hr if the rate remains constant?

26. The normal brain weight for an African bull elephant is 9 1/4 lb. Approximately how much would be the weight of 13 of the brains of these elephants?

27. Write each of the following in simplest form using positive exponents in the final answer:
   a. $3^{-7} \cdot 3^{-6}$
   b. $3^7 \cdot 3^6$
   c. $5^{15} \div 5^4$
   d. $5^{15} \div 5^{-4}$
   e. $(^-5)^{-2}$
   f. $\frac{a^2}{a^{-3}}$, where $a \neq 0$
   g. $\frac{a}{a^{-1}}$, where $a \neq 0$
   h. $\frac{a^{-3}}{a^{-2}}$, where $a \neq 0$

28. Write each of the following in simplest form using positive exponents in the final answer:
   a. $\left(\frac{1}{2}\right)^3 \cdot \left(\frac{1}{2}\right)^7$
   b. $\left(\frac{1}{2}\right)^9 \div \left(\frac{1}{2}\right)^6$
   c. $\left(\frac{2}{3}\right)^5 \cdot \left(\frac{4}{9}\right)^2$
   d. $\left(\frac{3}{5}\right)^7 \div \left(\frac{3}{5}\right)^7$
   e. $\left(\frac{3}{5}\right)^7 \div \left(\frac{5}{3}\right)^4$
   f. $\left[\left(\frac{5}{6}\right)^7\right]^3$

29. If a and b are rational numbers, with $a \neq 0$ and $b \neq 0$, and if m and n are integers, which of the following are true and which are false? Justify your answers.
   a. $a^m \cdot b^n = (ab)^{m+n}$
   b. $a^m \cdot b^n = (ab)^{mn}$
   c. $a^m \cdot b^m = (ab)^{2m}$
   d. $a^0 = 0$
   e. $(a+b)^m = a^m + b^m$
   f. $(a+b)^{-m} = \frac{1}{a^m} + \frac{1}{b^n}$
   g. $a^{mn} = a^m \cdot a^n$
   h. $\left(\frac{a}{b}\right)^{-1} = \frac{b}{a}$

30. Solve for the integer n in each of the following:
   a. $2^n = 32$
   b. $n^2 = 36$
   c. $2^n \cdot 2^7 = 2^5$
   d. $2^n \cdot 2^7 = 8$
   e. $(2+n)^2 = 2^2 + n^2$
   f. $3^n = 27^5$

31. A human being has approximately 25 trillion ($25 \cdot 10^{12}$) red blood cells, each with an average radius of $4 \cdot 10^{-3}$ mm (millimeters).
   a. If these cells were placed end to end in a line, how long would the line be in millimeters?
   b. If 1 km is 10^6 mm, how long would the line be in kilometers?

32. Solve each of the following inequalities for x, where x is an integer:
   a. $3^x \leq 81$
   b. $4^x < 8$
   c. $3^{2x} > 27$
   d. $2^x > 1$

33. Determine which of the fractions in each of the following pairs is greater:
   a. $\left(\frac{1}{2}\right)^3$ or $\left(\frac{1}{2}\right)^4$
   b. $\left(\frac{3}{4}\right)^{10}$ or $\left(\frac{3}{4}\right)^8$
   c. $\left(\frac{4}{3}\right)^{10}$ or $\left(\frac{4}{3}\right)^8$
   d. $\left(\frac{3}{4}\right)^{10}$ or $\left(\frac{4}{5}\right)^{10}$
   e. $\left(\frac{4}{3}\right)^{10}$ or $\left(\frac{5}{4}\right)^{10}$
   f. $\left(\frac{3}{4}\right)^{100}$ or $\left(\frac{3}{4} \cdot \frac{9}{10}\right)^{100}$

34. Suppose the amount of bacteria in a certain culture is given as a function of time by $Q(t) = 10^{10}(6/5)^t$, where t is the time in seconds and $Q(t)$ is the amount of bacteria after t seconds. Find the following:
   a. The initial number of bacteria (that is, the number of bacteria at $t = 0$)
   b. The number of bacteria after 2 sec

35. Let $S = \frac{1}{2} + \frac{1}{2^2} + \frac{1}{2^3} + \ldots + \frac{1}{2^{64}}$.
   a. Use the distributive property of multiplication over addition to find an expression for $2S$.
   b. Show that $2S - S = S = 1 - \left(\frac{1}{2}\right)^{64}$.
   c. Find a simple expression for the sum
   $$\frac{1}{2} + \frac{1}{2^2} + \frac{1}{2^3} + \ldots + \frac{1}{2^n}.$$

36. In an arithmetic sequence, the first term is 1 and the hundredth term is 2. Find the following:
    a. The fiftieth term
    b. The sum of the first 50 terms
37. For each of the following sequences, (a) find a pattern and (b) write two more terms of the sequence, assuming the pattern continues. Which of the sequences are geometric? Justify your answers.
    i. $1, \frac{1}{2}, \frac{1}{4}, \frac{1}{8}, \frac{1}{16}, \ldots$
    ii. $1, \frac{^-1}{2}, \frac{1}{4}, \frac{^-1}{8}, \frac{1}{16}, \ldots$
    iii. $\frac{4}{3}, 1, \frac{3}{4}, \frac{9}{16}, \frac{27}{64}, \ldots$
    iv. $\frac{1}{3}, \frac{2}{3^2}, \frac{3}{3^3}, \frac{4}{3^4}, \ldots$
38. If $f(n) = \frac{3}{4} \cdot 2^n$, find the following:
    a. $f(0)$
    b. $f(5)$
    c. $f(^-5)$
    d. The greatest integer value of n for which $f(n) < \frac{3}{1400}$
39. If the nth term of a sequence is given by $a_n = 3 \cdot 2^{-n}$, answer the following:
    a. Find the first 5 terms.
    b. Show that the first 5 terms are in a geometric sequence.
    c. Find the first term that is less than $\frac{3}{1000}$.
40. In the following, determine which number is greater:
    a. 32^{50} or 4^{100}
    b. $(^-27)^{-15}$ or $(^-3)^{-75}$
41. There is a simple method for squaring any number that consists of a whole number and $\frac{1}{2}$. For example $\left(3\frac{1}{2}\right)^2$
    $= 3 \cdot 4 + \left(\frac{1}{2}\right)^2 = 12\frac{1}{4}; \left(4\frac{1}{2}\right)^2 = 4 \cdot 5 + \left(\frac{1}{2}\right)^2 = 20\frac{1}{4}; \left(5\frac{1}{2}\right)^2$
    $= 5 \cdot 6 + \left(\frac{1}{2}\right)^2 = 30\frac{1}{4}.$
    a. Write a statement for $\left(n + \frac{1}{2}\right)^2$ that generalizes these examples, where n is a whole number.
    ★b. Justify this procedure.
42. Let $f(x) = \frac{3x + 4}{4x - 5}$, where the domain is all rational numbers for which the function has a value.
    a. Find the outputs if the inputs are as follows:
       i. 0   ii. $\frac{2}{5}$ iii. $\frac{^-2}{5}$
    b. For which inputs will the outputs be the following?
       i. 0   ii. $\frac{2}{5}$ iii. $\frac{^-1}{2}$
    c. What value for x is not in the domain of the function?
43. Consider these products:
    First product: $\left(1 + \frac{1}{1}\right)\left(1 + \frac{1}{2}\right)$
    Second product: $\left(1 + \frac{1}{1}\right)\left(1 + \frac{1}{2}\right)\left(1 + \frac{1}{3}\right)$
    Third product: $\left(1 + \frac{1}{1}\right)\left(1 + \frac{1}{2}\right)\left(1 + \frac{1}{3}\right)\left(1 + \frac{1}{4}\right)$

   a. Calculate the value of each product. Based on the pattern in your answers, guess the value of the fourth product. Then check to determine if your guess is correct.
   b. Guess the value of the hundredth product.
   c. Find as simple an expression as possible for the nth product.
44. Show that the arithmetic mean of two rational numbers is between the two numbers; that is, for $0 < \frac{a}{b} < \frac{c}{d}$, prove that
    $$0 < \frac{a}{b} < \frac{1}{2}\left(\frac{a}{b} + \frac{c}{d}\right) < \frac{c}{d}.$$

Communication

45. Suppose you divide a natural number, n, by a positive rational number less than 1. Will the answer always be less than n, sometimes less than n, or never less than n? Why?
46. If the fractions represented by points C and D on the following number line are multiplied, what point best represents the product? Explain why.

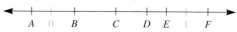

47. What are two reasonable estimates for $\frac{1}{7}$ of 39? Explain how you arrived at each estimate.
48. If the product of two numbers is 1 and one of the numbers is greater than 1, what do you know about the other number? Explain your answer.
49. Show that the following properties do *not* hold for the division of rational numbers:
    a. Commutative   b. Associative
    c. Identity   d. Inverse

Open-Ended

50. Would you use the problem in the following cartoon in your class? Why or why not? Solve the problem.

Cooperative Learning

51. Choose a brick building on your campus. Measure the height of one brick and the thickness of mortar between bricks. Estimate the height of the building and then calculate the height of the building. Were rational numbers used in your computations?

Review Problems

52. a. Maria noticed that every 30 sec the temperature of a chemical reaction in her lab decreased by the same number of degrees. Initially, she measured the temperature as 28°C and 5 min later as $^-12$°C. In a second experiment, she noticed that the temperature of the chemical reaction was initially $^-57$°C and was decreasing by 3°C every minute. If she started the two experiments at the same time, when were the temperatures of the experiments the same? (*Hint:* A spreadsheet may be used to do the calculations.)
    b. What was that temperature?

53. Perform each of the following computations. Leave your answers in simplest form.

    a. $\dfrac{^-3}{16} + \dfrac{7}{4}$ b. $\dfrac{1}{6} + \dfrac{^-4}{9} + \dfrac{5}{3}$

    c. $\dfrac{^-5}{2^3 \cdot 3^2} - \dfrac{^-5}{2 \cdot 3^3}$ d. $3\dfrac{4}{5} + 4\dfrac{5}{6}$

    e. $5\dfrac{1}{6} - 3\dfrac{5}{8}$ f. $^-4\dfrac{1}{3} - 5\dfrac{5}{12}$

54. Each student at Sussex Elementary School takes one foreign language. Two-thirds of the students take Spanish, $\dfrac{1}{9}$ take French, $\dfrac{1}{18}$ take German, and the rest take some other foreign language. If there are 720 students in the school, how many do not take Spanish, French, or German?

BRAIN TEASER

A woman's will decreed that her cats be shared among her three daughters as follows: $\dfrac{1}{2}$ of the cats to the eldest daughter, $\dfrac{1}{3}$ of the cats to the middle daughter, and $\dfrac{1}{9}$ of the cats to the youngest daughter. Since the woman had 17 cats, the daughters decided that they could not carry out their mother's wishes. The judge who held the will agreed to lend the daughters a cat so that they could share the cats as their mother wished. Now, $\dfrac{1}{2}$ of 18 is 9; $\dfrac{1}{3}$ of 18 is 6; and $\dfrac{1}{9}$ of 18 is 2. Since $9 + 6 + 2 = 17$, the daughters were able to divide the 17 cats and return the borrowed cat. They obviously did not need the extra cat to carry out their mother's bequest, but they could not divide 17 into halves, thirds, and ninths. Has the woman's will really been followed?

Section 5-4 — Proportional Reasoning

Ratios are encountered in everyday life. For example, there may be a 2-to-3 ratio of Democrats to Republicans on a certain legislative committee, a friend may be given a speeding ticket for driving 63 miles per hour, or eggs may cost 98¢ a dozen. Each of these

ratio   illustrates a **ratio**. A 1-to-2 ratio of males to females means that the number of males is $\dfrac{1}{2}$ the number of females; that is, there is one male for every two females. The ratio 1 to 2 can be written as $\dfrac{1}{2}$ or 1:2. In general, a ratio is denoted by $\dfrac{a}{b}$ or $a{:}b$, where $b \neq 0$.

Example 5-17 There were 7 males and 12 females in the Dew Drop Inn on Monday evening. In the Game Room next door were 14 males and 24 females.

  a. Express the number of males to females at the Inn as a ratio.
  b. Express the number of males to females at the Game Room as a ratio.

Solution a. The ratio is $\frac{7}{12}$.   b. The ratio is $\frac{14}{24}$.

In Example 5-17, the ratios $\frac{7}{12}$ and $\frac{14}{24}$ are equal and proportional to each other. In general, two ratios are **proportional** if, and only if, the fractions representing them are equal. Two equal ratios form a **proportion.**

For example, $\frac{2}{3}$ and $\frac{8}{12}$ form a proportion because $\frac{2}{3} = \frac{8}{12}$, since $2 \cdot 12 = 8 \cdot 3$. Also $\frac{3}{4} \neq \frac{4}{5}$ because $3 \cdot 5 \neq 4 \cdot 4$. In general, we have the following property.

> **Property**
>
> If $a, b, c,$ and d are all real numbers and $b \neq 0$ and $d \neq 0$, then
>
> $$\frac{a}{b} = \frac{c}{d} \quad \text{if, and only if,} \quad ad = bc.$$

This property can be justified by multiplying each side of the proportion by bd.

Frequently, one term in a proportion is missing, as in

$$\frac{3}{8} = \frac{x}{16}.$$

This equation is a proportion if, and only if,

$$3 \cdot 16 = 8 \cdot x$$
$$48 = 8 \cdot x$$
$$6 = x$$

Another way to solve the equation is to multiply both sides by 16, as follows:

$$\frac{3}{8} \cdot 16 = \frac{x}{16} \cdot 16$$
$$3 \cdot 2 = x$$
$$6 = x$$

It is important to remember that in the ratio $a \div b$, a and b do not have to be integers. For example, if in Eugene, Oregon, $\frac{7}{10}$ of the population exercises regularly, then $\frac{3}{10}$ of the population does not exercise regularly, and the ratio of those who do to those who do not is $\frac{7}{10} : \frac{3}{10}$. However, since ratios are usually expressed using natural numbers, the last ratio can be written as 7:3.

The following are examples of problems that use ratio and proportion.

Example 5-18 If there should be 3 tractors for every 4 farmers on a collective farm, how many tractors are needed for 44 farmers on the farm?

Solution We use the strategy of *setting up a table,* as shown in Table 5-3.

Table 5-3

Number of Tractors	3	x
Number of Farmers	4	44

The ratio of tractors to farmers should always be the same.

$$\begin{array}{l}\text{Tractors} \rightarrow \\ \text{Farmers} \rightarrow\end{array} \frac{3}{4} = \frac{x}{44}$$

$$3 \cdot 44 = 4 \cdot x$$
$$132 = 4x$$
$$33 = x$$

Thus 33 tractors are needed.

It is important to notice units of measure when we work with proportions. For example, if a turtle travels 5 in. every 10 sec, how many feet does it travel in 50 sec? If units of measure are ignored, we might set up the following proportion:

$$\frac{5 \text{ in.}}{10 \text{ sec}} = \frac{x \text{ ft}}{50 \text{ sec}}$$

This statement is incorrect. A correct statement must involve the same units in each ratio. We may write the following:

$$\frac{5 \text{ in.}}{10 \text{ sec}} = \frac{x \text{ in.}}{50 \text{ sec}}$$

This implies that $x = 25$. Consequently, since 12 in. = 1 ft, the turtle travels $\frac{25}{12}$ ft, or $2\frac{1}{12}$ ft, or 2 ft 1 in.

Example 5-19 Kai, Paulus, and Judy made $2520 for painting a house. Kai worked 30 hr, Paulus worked 50 hr, and Judy worked 60 hr. They divided the money in proportion to the number of hours worked. How much did each earn?

Solution If we denote the amount of money that Kai received by $30x$, then the amount of money Paulus received must be $50x$ because then, and only then, will the ratios of the amounts be the same as 30:50 as required. Similarly, Judy received $60x$. Because the total amount of money received is $30x + 50x + 60x$, we have

$$30x + 50x + 60x = 2520$$
$$140x = 2520$$
$$x = 18.$$

Hence,

$$\text{Kai received } 30x = 30 \cdot 18, \text{ or } \$540$$
$$\text{Paulus received } 50x = 50 \cdot 18, \text{ or } \$900$$
$$\text{Judy received } 60x = 60 \cdot 18, \text{ or } \$1080.$$

Dividing each of the amounts by 18 shows that the proportion is as required.

Consider the proportion $\frac{15}{30} = \frac{3}{6}$. Because the ratios in the proportion are equal fractions and because equal nonzero fractions have equal reciprocals, it follows that $\frac{30}{15} = \frac{6}{3}$. Also notice that the proportions are true because each results in $15 \cdot 6 = 30 \cdot 3$. In general, we have the following property.

Property

For any rational numbers $\frac{a}{b}$ and $\frac{c}{d}$, with $a \neq 0$ and $c \neq 0$, $\frac{a}{b} = \frac{c}{d}$ if, and only if, $\frac{b}{a} = \frac{d}{c}$.

Consider $\frac{15}{30} = \frac{3}{6}$ again. Notice that $\frac{15}{3} = \frac{30}{6}$; that is, the ratio of the numerators is proportional to the ratio of the corresponding denominators. In general we have the following property.

Property

For any rational numbers $\frac{a}{b}$ and $\frac{c}{d}$, with $b, c, d \neq 0$, $\frac{a}{b} = \frac{c}{d}$ if, and only if, $\frac{a}{c} = \frac{b}{d}$.

Problem Solving **Dog Cartoon Problem**

Analyze the statements in the cartoon on page 285. How did the author arrive at the ratio of 21 meals per day based on the information in the first two frames?

- **Understanding the Problem** The dog in the cartoon makes the following three statements:

  1. One year to a person is like seven years to a dog.
  2. One day to a person is like one week to a dog.
  3. I can eat 21 meals every day.

  We need to determine if the ratio in **3** follows from the ratios in **1** or **2** or both.

- **Devising a Plan** Each statement leads to a ratio, a proportion, or a bit of mathematics based on the resulting ratios and proportions. We need to translate each into its mathematical form and check the computations.

From the cartoon Mother Goose and Grimm

- **Carrying Out the Plan** The first frame of the cartoon leads to the following ratio:

$$\frac{1 \text{ year to a person}}{7 \text{ years to a dog}} \text{ or the ratio of person time to dog time is } 1/7$$

If we translate the next frame into a ratio, we have the following:

$$\frac{1 \text{ day to a person}}{1 \text{ week to a dog}}$$

which also implies that the ratio of person time to dog time is

$$\frac{1 \text{ day to a person}}{7 \text{ days to a dog}}$$

If a person eats 3 meals per day and if the ratio of person time to dog time carries over to meals, then we have the following:

$$\frac{3 \text{ meals for a person}}{x \text{ meals for a dog}} = \frac{1 \text{ day for a person}}{7 \text{ days for a dog}}.$$

Solving for x, we have $x = 21$ meals for a dog. Thus the ratio in the cartoon's last frame follows from the ratio in the first frames.

- **Looking Back** Each of the statements in the cartoon appears to be consistent with the other statements. Either frame 1 or frame 2 of the cartoon is not needed to analyze the problem.

The type of reasoning in the Dog Cartoon Problem is typical of reasoning found in newspapers. However, not all of such reasoning leads to a logical conclusion. The reader needs to consider carefully all statements given in an article to determine their consistency and evaluate the conclusions.

Scale Drawings

scale Ratio and proportions are used in scale drawings. For example, if the scale is 1:300 then the length of 1 cm in such a drawing represents 300 cm, or 3 m in true size. The **scale** is the ratio of the size of the drawing to the actual size of the object. The following example shows the use of scale drawings.

Example 5-20 The floor plan of the main floor of a house in Figure 5-12 is drawn in the scale of 1:300. Find the dimensions in meters of the living room.

Figure 5-12

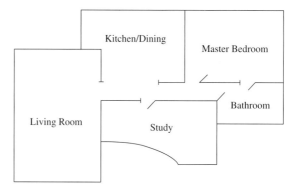

Solution In Figure 5-12, the dimensions of the living room measured with a centimeter ruler are approximately 3.7 cm by 2.5 cm. Because the scale is 1:300, 1 cm in the drawing represents 300 cm, or 3 m in true size. Hence, 3.7 cm represents $3.7 \cdot 3$, or 11.1 m, and 2.5 cm represents $2.5 \cdot 3$, or 7.5 m. Hence the dimensions of the living room are approximately 11.1 m by 7.5 m.

ONGOING ASSESSMENT 5-4

1. Answer the following regarding the English alphabet:
   a. Determine the ratio of vowels to consonants.
   b. Write a word that has a ratio of 2:3 of vowels to consonants.
2. Solve for x in each of the following proportions:
   a. $\dfrac{12}{x} = \dfrac{18}{45}$
   b. $\dfrac{x}{7} = \dfrac{^-10}{21}$
   c. $\dfrac{5}{7} = \dfrac{3x}{98}$
   d. $3\dfrac{1}{2}$ is to 5 as x is to 15.
3. There are approximately 2 lb of muscle for every 5 lb of body weight. For a 90-lb person, how much of the weight is muscle?
4. There are 5 adult drivers to each teenage driver in Aluossim. If there are 12,345 adult drivers in Aluossim, how many teenage drivers are there?
5. If 4 grapefruits sell for 79¢, how much do 6 grapefruits cost?
6. On a map, $\dfrac{1}{3}$ in. represents 5 mi. If New York and Aluossim are 18 in. apart on the map, what is the actual distance between them?
7. David read 40 pages of a book in 50 min. How many pages should he be able to read in 80 min if he reads at a constant rate?
8. A candle is 30 in. long. After burning for 12 min, the candle is 25 in. long. How long will it take for the whole candle to burn at the same rate?
9. Two numbers are in the ratio 3:4. Find the numbers if
   a. their sum is 98.
   b. their product is 768.
10. A rectangular yard has a width-to-length ratio of 5:9. If the distance around the yard is 2800 ft, what are the dimensions of the yard?
11. Gary, Bill, and Carmella invested in a corporation in the ratio of 2:4:5, respectively. If they divide the profit of $82,000 proportionally to their investment, how much will each receive?
12. Sheila and Dora worked $3\dfrac{1}{2}$ hr and $4\dfrac{1}{2}$ hr, respectively, on a programming project. They were paid $176 for the project. How much did each earn?
13. Vonna scored 75 goals in her soccer kicking practice. If her success-to-failure rate is 5:4, how many times did she attempt a goal?
14. The rise and span for a house roof are identified as follows. The pitch of a roof is the ratio of the rise to the half-span.
    a. If the rise is 10 ft and the span is 28 ft, what is the pitch?

b. If the span is 16 ft and the pitch is $\frac{3}{4}$, what is the rise?

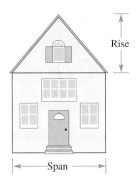

15. A grasshopper can jump 20 times its length. If jumping ability in humans were proportional to a grasshopper's, how far could a 6-ft-tall person jump?
16. Jim found out that after working for 9 mo he had earned 6 days of vacation time. How many days per year does he earn at this rate?
17. Gear ratios are used in industry. A gear ratio is the comparison of the number of teeth on two gears. When two gears are meshed, the revolutions per minute (rpm) are inversely proportional to the number of teeth; that is,

$$\frac{\text{rpm of large gear}}{\text{rpm of small gear}} = \frac{\text{number of teeth on small gear}}{\text{number of teeth on large gear}}.$$

   a. The rpm ratio of the large gear to the small gear is 4:6. If the small gear has 18 teeth, how many teeth does the large gear have?
   b. The large gear revolves at 200 rpm and has 60 teeth. How many teeth are there on the small gear that has an rpm of 600?
18. A Boeing 747 jet is approximately 230 ft long and has a wingspan of 195 ft. If a scale model of the plane is about 40 cm long, what is the model's wingspan?
19. Jennifer weighs 160 lb on Earth and 416 lb on Jupiter. Find Amy's weight on Jupiter if she weighs 120 lb on Earth.
20. **a.** If the ratio of boys to girls in a class is 2:3, what is the ratio of boys to all the students in the class? Why?
    b. If the ratio of boys to girls in a class is $m:n$, what is the ratio of boys to all the students in the class?
21. A recipe calls for 1 tsp of mustard seeds, 3 c of tomato sauce, 1 1/2 c of chopped scallions, and 3 1/4 c of beans. If one ingredient is altered as specified, how must the other ingredients be changed to keep the proportions the same? Explain your reasoning.
    a. 2 c of tomato sauce
    b. 1 c of chopped scallions
    c. 1 3/4 c of beans
22. The electrical resistance of a wire, measured in ohms (Ω), is equal to a constant times the length of the wire. If the electrical resistance of a 5-ft wire is 4.2 Ω, what is the resistance of 18 ft of the same wire?
23. At a particular time, the ratio of the height of an object that is perpendicular to the ground to the length of its shadow is the same for all objects. If a 30-ft tree casts a shadow of 12 ft, how tall is a tree that casts a shadow of 14 ft?
24. In a photograph of a father and his daughter, the daughter's height is 2.3 cm and the father's height is 5.8 cm. If the father is actually 188 cm tall, how tall is the daughter?
25. The following table shows several possible widths W and corresponding lengths L of a rectangle whose area is 10 ft².

Width (W) (Feet)	Length (L) (Feet)	Area (Square Feet)
0.5	20	$0.5 \cdot 20 = 10$
1	10	$1 \cdot 10 = 10$
2	5	$2 \cdot 5 = 10$
2.5	4	$2.5 \cdot 4 = 10$
4	2.5	$4 \cdot 2.5 = 10$
5	2	$5 \cdot 2 = 10$
10	1	$10 \cdot 1 = 10$
20	0.5	$20 \cdot 0.5 = 10$

   a. Use the values in the table and some additional values to graph the length L on the vertical axis versus the width W of the horizontal axis.
   b. What is the algebraic relationship between L and W?
   c. Write W as a function of L; that is, express W in terms of L.
   d. Write L as a function of W; that is, express L in terms of W.
26. **a.** In Room A of the University Center are one man and two women; in Room B are two men and four women; and in Room C are five men and ten women. If all the people in Rooms B and C go to Room A, what will be the ratio of men to women in Room A?
   ★**b.** Prove the following generalization of the proportions used in (a):

   If $\frac{a}{b} = \frac{c}{d} = \frac{e}{f}$, then $\frac{a}{b} = \frac{c}{d} = \frac{e}{f} = \frac{a+c+e}{b+d+f}$.

★27. Prove that if $\frac{a}{b} = \frac{c}{d}$, then the following are true:
   a. $\frac{a+b}{b} = \frac{c+d}{d}$ $\left(\text{Hint: } \frac{a}{b} + 1 = \frac{c}{d} + 1\right)$
   b. $\frac{a}{a+b} = \frac{c}{c+d}$
   c. $\frac{a-b}{a+b} = \frac{c-d}{c+d}$

Communication

28. Iris has found some dinosaur bones and a fossil footprint. The length of the footprint is 40 cm, the length of the thigh bone is 100 cm, and the length of the body is 700 cm.
    a. What is the ratio of the footprint's length to the dinosaur's length?
    b. Iris found a new track that she believes was made by the same species of dinosaur. If the footprint was 30 cm long and if the same ratio of foot length to body length holds, how long is the dinosaur?
    c. In the same area, Iris also found a 50-cm thigh bone. Do you think this thigh bone belonged to the same dinosaur that made the 30-cm footprint that Iris found? Why or why not?

29. Suppose a 10-in. pizza costs $4. For you to find the price x of a 14-in. pizza, is it correct to set up the proportion $\frac{x}{4} = \frac{14}{10}$? Why or why not?

30. The amount of gold in jewelry and other products is measured in karats (K), where 24K represents pure gold. The mark 14K on a chain indicates that the ratio between the mass of the gold in the chain and the mass of the chain is 14:24. If a gold ring is marked 18K and it weighs 0.4 oz, what is the value of the gold in the ring if pure gold is valued at $300 per oz? Explain your reasoning.

31. The approximate mass of the sun is $2.9 \cdot 10^{32}$ kg (kilograms) and the approximate mass of the hydrogen atom is $1.7 \cdot 10^{-29}$ kg. Is the mass of the hydrogen atom to your mass approximately the same as the ratio of your mass to the mass of the sun? Explain why or why not.

Open-Ended

32. List three real-world situations that involve ratio and proportion.

33. Boyle's law states that at a given temperature, the product of the volume V of a gas and the pressure P is a constant c as follows:

    $$PV = c$$

    a. If at a given temperature, a pressure of 48 lb/in² compresses a certain gas to a volume of 960 in³, what pressure would be necessary to compress the gas to a volume of 800 in³ at the same temperature?

    b. Find three other real-world situations in which the variables are related mathematically like the variables in Boyle's law. In each case, describe how the variables are related using ratio and proportion.

Cooperative Learning

34. Look at various elementary science and mathematics books and determine both if and how *dimensional analysis* is treated in those texts. How is dimensional analysis related to ratio and proportion?

35. Consider the stock market pages of a newspaper. Each person in a group should determine the profit-to-earnings ratio for at least five companies. Discuss with your group how you would rate stocks based on this indicator.

Review Problems

36. Find all values of x that satisfy the following equations or inequalities:
    a. $\frac{x}{3} = \frac{3x}{4}$
    b. $3^x = 243$
    c. $\frac{1}{3}x + \frac{3}{4}x = 49$
    d. $\frac{1}{3}x \leq 49 - \frac{3}{4}x$

37. Do all rational numbers have a multiplicative inverse?

38. Suppose the ratio between two sides of two squares is r. How do the areas of the two squares compare?

39. Find three rational numbers between 2 and 3 so that the numbers, including 2 and 3, form an arithmetic sequence.

40. If a geometric sequence has 4/9 as its seventh term, and the ratio of the sequence is 1/2, what is the first term?

41. Place values from the left to right in a number to form a geometric sequence. What is the ratio of the sequence?

42. Determine what, if anything, is wrong with each of the following:
    a. $2 = \frac{6}{3} = \frac{3+3}{3} = \frac{3}{3} + 3 = 1 + 3 = 4$
    b. $1 = \frac{4}{2+2} = \frac{4}{2} + \frac{4}{2} = 2 + 2 = 4$
    c. $\frac{ab + c}{a} = \frac{\cancel{a}b + c}{\cancel{a}} = b + c$
    d. $\frac{a^2 - b^2}{a - b} = \frac{a \cdot \cancel{a} - b \cdot \cancel{b}}{\cancel{a} - \cancel{b}} = a - b$
    e. $\frac{a + c}{b + c} = \frac{a + \cancel{c}}{b + \cancel{c}} = \frac{a}{b}$

HINT FOR SOLVING THE PRELIMINARY PROBLEM

The goal of the preliminary problem is to find all times on a watch when the hour and minute hands make a straight line. One such time is 6:00. Because the second hand was not on 12 when the watch stopped, neither 12:00 nor 6:00 are the desired answers. There are other times when the hands align to form a straight line. In order to find them, we have to consider carefully how the hands move in each part of an hour.

QUESTIONS FROM THE CLASSROOM

1. Is $\dfrac{0}{6}$ in simplest form? Why or why not?

2. A student says that taking one half of a number is the same as dividing the number by one half. Is this correct?

3. A student writes $\dfrac{15}{53} < \dfrac{1}{3}$ because $3 \cdot 15 < 53 \cdot 1$. Another student writes $\dfrac{15}{53} = \dfrac{1}{3}$. Where is the fallacy?

4. A student claims that the following is an arithmetic sequence. Is the student right?
$$\dfrac{1}{2}, \dfrac{2}{3}, \dfrac{3}{4}, \dfrac{4}{5}, \dfrac{5}{6}, \dfrac{6}{7}, \dfrac{7}{8}, \ldots$$

5. A student claims she found a new way to obtain a fraction between two positive fractions: If $\dfrac{a}{b}$ and $\dfrac{c}{d}$ are two positive fractions, then $\dfrac{a+c}{b+d}$ is between these fractions. Is she right?

6. A student claims that if $\dfrac{a}{b} = \dfrac{c}{d}$, then $\dfrac{a+c}{b+d} = \dfrac{a}{b} = \dfrac{c}{d}$. Is he right?

7. When working on the problem of simplifying
$$\dfrac{3}{4} \cdot \dfrac{1}{2} \cdot \dfrac{2}{3},$$
a student did the following:
$$\dfrac{3}{4} \cdot \dfrac{1}{2} \cdot \dfrac{2}{3} = \left(\dfrac{3 \cdot 1}{4 \cdot 2}\right)\left(\dfrac{3 \cdot 2}{4 \cdot 3}\right) = \dfrac{3}{8} \cdot \dfrac{6}{12} = \dfrac{19}{96}.$$
What was the error?

8. A student simplified the fraction $\dfrac{m+n}{p+n}$ to $\dfrac{m}{p}$. Is that student correct?

9. Without thinking, a student argued that a pizza cut into 12 pieces was more than a pizza cut into 6 pieces. How would you respond?

10. A student asks if adding the same very large number to both the numerator and the denominator of a fraction yields a quotient of 1. How do you respond?

11. The teacher asked the class to solve the equation
$$\dfrac{1}{4} + \dfrac{7}{4}\left(x + \dfrac{1}{5}\right) = x + \dfrac{6}{5}.$$
Nat wrote $\left(\dfrac{1}{4} + \dfrac{7}{4}\right)\left(x + \dfrac{1}{5}\right) = x + \dfrac{6}{5}$, solved the equation, and got the answer $x = \dfrac{4}{5}$, which is correct. The teacher told Nat he had obtained the correct answer by using an incorrect method. Nat in turn responded that his method will also work for the equation $\dfrac{3}{8} + \dfrac{1}{4}(x - 1) = x - \dfrac{11}{8}$ and for the equation $1 + \dfrac{1}{2}\left(x - \dfrac{1}{4}\right) = x + \dfrac{1}{4}$. How would you respond?

12. A student claimed that $\dfrac{9}{-11}$ is an improper fraction. How do you respond?

13. If $2^{-1} = \dfrac{1}{2}$, a student said that 3^{-2} should equal $\dfrac{2}{3}$. How do you respond?

CHAPTER OUTLINE

I. Fractions and rational numbers
   A. Numbers of the form $\dfrac{a}{b}$, where a and b are integers and $b \neq 0$, are called **rational numbers.**
   B. A rational number can be used as follows:
      1. A division problem or the solution to a multiplication problem
      2. A partition, or part, of a whole
      3. A ratio
      4. A probability
   C. **Fundamental Law of Fractions:** For any fraction $\dfrac{a}{b}$ and any number $c \neq 0$, $\dfrac{a}{b} = \dfrac{ac}{bc}$.
   D. Two fractions $\dfrac{a}{b}$ and $\dfrac{c}{d}$ are **equal** if, and only if, $ad = bc$.
   E. If $GCD(a, b) = 1$, then $\dfrac{a}{b}$ is said to be in **simplest form.**
   F. If $0 \leq |a| < |b|$, then $\dfrac{a}{b}$ is called a **proper fraction.** If $|a| \geq |b| > 0$, $\dfrac{a}{b}$ is an **improper fraction.**

II. Operations on rational numbers
   A. $\dfrac{a}{b} + \dfrac{c}{b} = \dfrac{a+c}{b}$
   B. $\dfrac{a}{b} + \dfrac{c}{d} = \dfrac{ad+bc}{bd}$
   C. $\dfrac{a}{b} - \dfrac{c}{d} = \dfrac{ad-bc}{bd}$

D. $\dfrac{a}{b} \cdot \dfrac{c}{d} = \dfrac{ac}{bd}$

E. $\dfrac{a}{b} \div \dfrac{c}{d} = \dfrac{a}{b} \cdot \dfrac{d}{c} = \dfrac{ad}{bc}$, where $c \neq 0$

III. Properties of rational numbers
  A.

	Addition	Subtraction	Multiplication	Division
Closure	Yes	Yes	Yes	Yes, except for division by 0
Commutative	Yes	No	Yes	No
Associative	Yes	No	Yes	No
Identity	Yes	No	Yes	No
Inverse	Yes	No	Yes, except 0	No

  B. Distributive property of multiplication over addition for rational numbers x, y, and z:

  $$x(y + z) = xy + xz.$$

  C. Denseness property: Between any two rational numbers, there is another rational number.

  D. Multiplication property of equality: If $\dfrac{a}{b}$ and $\dfrac{c}{d}$ are any rational numbers such that $\dfrac{a}{b} = \dfrac{c}{d}$, and $\dfrac{e}{f}$ is any rational number, then $\dfrac{a}{b} \cdot \dfrac{e}{f} = \dfrac{c}{d} \cdot \dfrac{e}{f}$.

  E. Multiplication property of zero for rational numbers:

  If $\dfrac{a}{b}$ is any rational number, then $\dfrac{a}{b} \cdot 0 = 0 = 0 \cdot \dfrac{a}{b}$.

IV. Exponents
  A. $a^m = \underbrace{a \cdot a \cdot a \cdot \ldots \cdot a}_{m \text{ factors}}$, where m is a positive integer and a is a rational number

  B. Properties of exponents involving rational numbers
    1. $a^0 = 1$, where $a \neq 0$
    2. $a^{-m} = \dfrac{1}{a^m}$, where $a \neq 0$
    3. $a^m \cdot a^n = a^{m+n}$
    4. $\dfrac{a^m}{a^n} = a^{m-n}$, where $a \neq 0$
    5. $(a^m)^n = a^{mn}$
    6. $\left(\dfrac{a}{b}\right)^m = \dfrac{a^m}{b^m}$, where $b \neq 0$
    7. $(ab)^m = a^m \cdot b^m$
    8. $\left(\dfrac{a}{b}\right)^{-m} = \left(\dfrac{b}{a}\right)^m$

V. Ratio and Proportion
  A. A fraction a/b is a **ratio.**
  B. A **proportion** is an equation of two ratios.
  C. Properties of proportions
    1. If $\dfrac{a}{b} = \dfrac{c}{d}$, then $\dfrac{b}{a} = \dfrac{d}{c}$, where $a \neq 0$ and $c \neq 0$.
    2. If $\dfrac{a}{b} = \dfrac{c}{d}$, then $\dfrac{a}{c} = \dfrac{b}{d}$, where $c \neq 0$.

CHAPTER REVIEW

1. For each of the following, draw a diagram illustrating the fraction:

   a. $\dfrac{3}{4}$ **b.** $\dfrac{2}{3}$ **c.** $\dfrac{3}{4} \cdot \dfrac{2}{3}$

2. Write three rational numbers equal to $\dfrac{5}{6}$.

3. Reduce each of the following rational numbers to simplest form:

   a. $\dfrac{24}{28}$ **b.** $\dfrac{ax^2}{bx}$

   c. $\dfrac{0}{17}$ **d.** $\dfrac{45}{81}$

   e. $\dfrac{b^2 + bx}{b + x}$ **f.** $\dfrac{16}{216}$

4. Replace the comma with $>$, $<$, or $=$ in each of the following pairs to make a true statement:

   a. $\dfrac{6}{10}, \dfrac{120}{200}$

   b. $\dfrac{-3}{4}, \dfrac{-5}{6}$

   c. $\left(\dfrac{4}{5}\right)^{10}, \left(\dfrac{4}{5}\right)^{20}$

   d. $\left(1 + \dfrac{1}{3}\right)^2, \left(1 + \dfrac{1}{3}\right)^3$

5. Find the additive and multiplicative inverses for each of the following:

   a. 3 **b.** $3\dfrac{1}{7}$ **c.** $\dfrac{5}{6}$ **d.** $-\dfrac{3}{4}$

6. Order the following numbers from least to greatest:

   $-1\dfrac{7}{8},\ 0,\ -2\dfrac{1}{3},\ \dfrac{69}{140},\ \dfrac{71}{140},\ \left(\dfrac{71}{140}\right)^{300},\ \dfrac{1}{2},\ \left(\dfrac{74}{73}\right)^{300}$

7. John has $54\frac{1}{4}$ yd of material. If he needs to cut the cloth into pieces that are $3\frac{1}{12}$ yd long, how many pieces can he cut? How much material will be left over?

8. Without actually performing the given operations, choose the most appropriate estimation (among the numbers in parentheses) for the following expressions:

   a. $\dfrac{30\frac{3}{8}}{4\frac{1}{9}} \cdot \dfrac{8\frac{1}{3}}{3\frac{8}{9}}$ (15, 20, 8)

   b. $\left(\dfrac{3}{800} + \dfrac{4}{5000} + \dfrac{15}{6}\right) \cdot 6$ (15, 0, 132)

   c. $\dfrac{1}{407} \div \dfrac{1}{1609}$ $\left(\dfrac{1}{4}, 4, 0\right)$

9. Justify the invert-and-multiply algorithm for division of rational numbers.

10. The ratio of boys to girls in Mr. Good's class is 3 to 5, the ratio of boys to girls in Ms. Garcia's is the same, and you know that there are 15 girls in Ms. Garcia's class. How many boys are in Ms. Garcia's class?

11. Find two rational numbers between $\dfrac{3}{4}$ and $\dfrac{4}{5}$.

12. Suppose the $\boxed{\div}$ button on your calculator is broken, but the $\boxed{1/x}$ button works. Explain how you could compute 504792/23.

13. Jim is starting a diet. When he arrived home, he ate $\dfrac{1}{3}$ of the half of pizza that was left from the previous night. The whole pizza contains approximately 2000 calories. How many calories did Jim consume?

14. If a person got heads on a flip of a fair coin one half the time and obtained 376 heads, how many times was the coin flipped?

15. If a person obtained 240 heads when flipping a coin 1000 times, what fraction of the time did the person obtain heads? Put the answer in simplest form.

16. If the University of New Mexico won 3/4 of its women's basketball games and 5/8 of its men's basketball games, explain whether it is reasonable to say that the university won 3/4 + 5/8 of its basketball games.

17. Explain why a negative rational number times a negative rational number is a positive rational number.

18. A student argues that the following fraction is not a rational number because it is not the quotient of two integers:

$$\dfrac{\frac{2}{3}}{\frac{3}{4}}$$

How would you respond?

19. If 2/3 of all students in the academy are female and 2/5 of those are blondes, what fraction describes the number of blond females in the academy?

20. Explain which is greater: $^-11/9$ or $^-12/10$.

21. Solve for x.
    a. $7^x = 343$
    b. $2^{-3x} = \dfrac{1}{512}$

22. In water (H_2O), the ratio of the weight of oxygen to the weight of hydrogen is approximately 8:1. How many ounces of hydrogen is in 1 lb of water?

23. Use the following scale drawing and a marked ruler to approximate each of the following:
    a. The widest distance across Crater Lake
    b. The aerial distance from Wizard Island to the closest point of Mount Scott

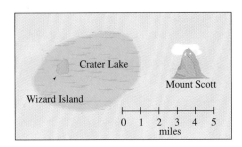

24. To estimate the number of fish in a lake, scientists use a tagging and recapturing technique. A number of fish are captured, tagged, and then released back into the lake. After a while, some fish are captured and the number of tagged fish is counted.

    Let T be the total number of fish captured, tagged, and released into the lake, n the number of fish in a recaptured sample, and t the number of fish found tagged in that sample. Finally let x be the number of fish in the lake. The assumption is that the ratio between tagged fish and the total number of fish in any sample is approximately the same and hence scientists assume $t/n = T/x$. Suppose 173 fish were captured, tagged, and released. Then 68 fish were recaptured and among them 21 were found to be tagged. Estimate the number of fish in the lake.

25. A manufacturer produces the same kind of computer chip in two plants. In the first plant, the ratio of defective chips to good chips is 15:100, and in the second plant, that ratio is 12:100. A buyer of a large number of chips is aware that some come from the first plant and some from the second. However, she is not aware of how many come from each. The buyer would like to know the ratio of defective chips to good chips in any given order. Can she determine that ratio? If so, explain how. If not, explain why not.

SELECTED BIBLIOGRAPHY

Bezuk, N. "Fractions in the Early Childhood Mathematics Curriculum." *Arithmetic Teacher* 35 (February 1988): 56–60.

Brinker, L. "Using Recipes and Ratio Tables." *Teaching Children Mathematics* 5 (December 1998): 218–224.

Caldwell, J. "Communicating about Fractions with Pattern Blocks." *Teaching Children Mathematics* 2 (November 1995): 156–161.

Collyer, S. "Adding Fractions." *Mathematics Teaching* 116 (September 1986): 9.

Conaway, B., and R. Midkiff. "Connecting Literature, Language, and Fractions." *Arithmetic Teacher* 41 (April 1994): 430–434.

Cramer, K., and N. Bezuk. "Multiplication of Fractions: Teaching for Understanding." *Arithmetic Teacher* 39 (November 1991): 34–37.

Curcio, F., and N. Bezuk. "Understanding Rational Numbers and Proportions." *Curriculum and Evaluation Standards for School Mathematics Addenda Series Grades 5–8.* Reston, VA: National Council of Teachers of Mathematics, 1994.

Edelman, L. "The Fractions of a Day." *Mathematics Teaching in the Middle School* 3 (November–December 1997): 192–195.

Edge, D. "Fractions and Panes." *Arithmetic Teacher* 34 (April 1987): 13–17.

Friel, J., and G. Gannon. " 'What If … ?' A Case in Point." *Mathematics Teacher* 88 (April 1995): 320–322.

Greeley, N., and T. R. Offerman. "Now & Then Dancing in Time and Space." *Mathematics Teaching in the Middle School* 4 (November–December 1998): 192–199.

Groff, P. "On My Mind: It Is Time to Question Fraction Teaching." *Mathematics Teaching in the Middle School* 1 (January–February 1996): 604–607.

Kennard, R. "Interpreting Fraction Form." *Mathematics Teaching* 112 (September 1985): 46–47.

Kieren, T., B. Davis, and R. Mason. "Fraction Flags: Learning from Children to Help Children Learn." *Mathematics Teaching in the Middle School* 2 (September–October 1996): 14–19.

Kuhns, C. "Half-Time Day." *Teaching Children Mathematics* 3 (January 1997): 218–221, 235.

Mack, N. "Making Connections to Understand Fractions." *Arithmetic Teacher* 40 (February 1993): 362–364.

Middleton, J., M. van den Heuvel-Panhuizen. "The Ratio Table." *Mathematics Teaching in the Middle School* 1 (January–March 1995): 282–288.

Middleton, J., M. van den Heuvel-Panhuizen, and J. Shew. "Using Bar Representations as a Model for Connecting Concepts of Rational Numbers." *Mathematics Teaching in the Middle School* 3 (January 1998): 302–312.

Nowlin, D. "Division with Fractions." *Mathematics Teaching in the Middle School* 2 (November–December 1996): 116–119.

Ott, J. "A Unified Approach to Multiplying Fractions." *Arithmetic Teacher* 37 (March 1990): 47–49.

Post, T. "Fractions and Other Rational Numbers." *Arithmetic Teacher* 37 (September 1989): 3, 28.

Post, T., and K. Cramer. "Research into Practice: Children's Strategies in Ordering Rational Numbers." *Arithmetic Teacher* 35 (October 1987): 33–35.

Quintero, A. "Helping Children Understand Ratios." *Arithmetic Teacher* 34 (April 1987): 17–21.

Reys, B. "Promoting Number Sense in the Middle Grades." *Mathematics Teaching in the Middle School* 1 (September–October, 1994): 114–120.

Van de Walle, J., and C. Thompson. "Fractions with Fraction Strips." *Arithmetic Teacher* 32 (December 1984): 48–52.

Warrington, M. "How Children Think about Division with Fractions." *Mathematics Teaching in the Middle School* 2 (May 1997): 390–394.

Warrington, M., and C. Kamii. "Multiplication with Fractions: A Piagetian, Constructivist Approach." *Mathematics Teaching in the Middle School* 3 (February 1998): 339–343.

Watanabe, T. "Ben's Understanding of One-Half." *Teaching Children Mathematics* 2 (April 1996): 460–464.

6

Decimals, Percents, and Real Numbers

Preliminary Problem

A basketball player has missed 23 of 137 free-throw attempts. She would like her shooting percentage for free throws to be at least 77% at the end of the season. She figures she will get 40 more attempts. Is it possible for her to achieve her goal? If so, how many of the 40 attempts does she have to put in the basket?

In the *Principles and Standards*, we find the following,

▲ *In grades 3–5, all students should recognize and generate equivalent forms of commonly used fractions, decimals, and percents* (p. 148).

Later in the *Principles and Standards* for grades 6–8 we find,

▲ *At the heart of flexibility in working with rational numbers is a solid understanding of different representations for fractions, decimals, and percents. In grades 3–5, students should have learned to generate and recognize equivalent forms of fractions, decimals, and percents, at least in some simple cases. In the middle grades, students should build on and extend this experience to become facile in using fractions, decimals, and percents meaningfully. Students can develop a deep understanding of rational numbers through experiences with a variety of models, such as fraction strips, number lines, 1010 grids, area models, and objects. These models offer students concrete representations of abstract ideas and support students' meaningful use of representations and their flexible movement among them to solve problems.*

As they solve problems in context, students also can consider the advantages and disadvantages of various representations of quantities. For example, students should understand not only that 15/100, 3/20, 0.15, and 15 percent are all representations of the same number but also that these representations may not be equally suitable to use in a particular context. For example, it is typical to represent a sales discount as 15%, the probability of winning a game as 3/20, a fraction of a dollar in writing a check as 15/100, and the amount of the 5 percent tax added to a purchase of $2.98 as $0.15 (pp. 215–216).

Although the Hindu-Arabic numeration system discussed in Chapter 3 was perfected around the sixth century, the extension of the system to decimals by the Dutch scientist Simon Stevin did not take place until about a thousand years later. The only significant improvement in the system since Stevin's time has been in notation. Even today there is no universally accepted form of writing a decimal. For example, in the United States, we write 6.75; in England, this number is written as 6 · 75; and in Germany and France, it is written 6,75.

Decimals play an important part in the mathematics education of students in grades 5 through 8. In the next section, we explore relationships between fractions and decimals and see how decimals are an extension of the base-ten system. Later in the chapter we consider operations on decimals, properties of decimals, percents and applications of percents, and finally real numbers.

HISTORICAL NOTE

In 1584 Simon Stevin (1548–1620), a quartermaster general in the Dutch army, wrote *La Thiende (The Tenth)*, a work that gave rules for computing with decimals. He not only stated rules for decimal computations but also suggested practical applications for decimals and recommended that his government adopt a system similar to the metric system. To show place value, Stevin used circled numerals between digits. For example, he wrote 0.4789 as 4 ① 7 ② 8 ③ 9 ④. Stevin also made contributions to military engineering and to physics in statics and hydrostatics.

Section 6-1 — Introduction to Decimals

decimal point

The word *decimal* comes from the Latin *decem*, meaning ten. Most people first encounter decimals when dealing with money. For example, where a sign that says a bike costs $128.95, the dot in $128.95 is the **decimal point.** Because 95¢ is $\frac{95}{100}$ of a dollar, we have $128.95 = 128 + \frac{95}{100}$ dollars. Because 95¢ is 9 dimes and 5 cents; one dime is $\frac{1}{10}$ of a dollar, and 1 cent is $\frac{1}{100}$ of a dollar, 95¢ is $9 \cdot \frac{1}{10} + 5 \cdot \frac{1}{100}$ of a dollar.

Consequently,

$$128.95 = 1 \cdot 10^2 + 2 \cdot 10 + 8 \cdot 1 + 9 \cdot \frac{1}{10} + 5 \cdot \frac{1}{10^2}.$$

The digits in 128.95 correspond to the following place values: 10^2, 10, 1, $\frac{1}{10}$, and $\frac{1}{10^2}$, respectively. Each term in the last sequence is $\frac{1}{10}$ of the previous term. Thus, 12.61843 represents

$$1 \cdot 10 + 2 \cdot 1 + \frac{6}{10^1} + \frac{1}{10^2} + \frac{8}{10^3} + \frac{4}{10^4} + \frac{3}{10^5}, \quad \text{or} \quad 12\frac{61,843}{100,000}.$$

The decimal 12.61843 is read "twelve and sixty-one thousand eight hundred forty-three hundred-thousandths." (The decimal point is read as "and.") Each place to the right of a decimal point may be named by its power of 10. For example, the places of 12.61843 can be named as shown in Table 6-1.

Table 6-1

1	2	.	6	1	8	4	3
Tens	Units	and	Tenths	Hundredths	Thousandths	Ten-thousandths	Hundred-thousandths

Decimals can be introduced with concrete materials. We can use a set of base-ten blocks and decide that 1 flat represents 1 unit, 1 long represents $\frac{1}{10}$, and 1 cube represents $\frac{1}{100}$, as in Figure 6-1(a). In this model, Figure 6-1(b) represents 1.23.

Figure 6-1

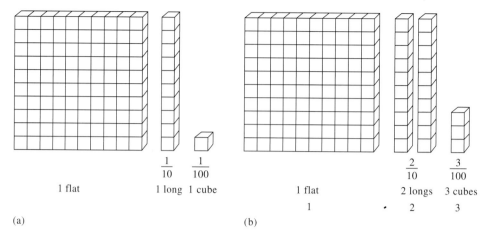

To represent a decimal such as 2.235, we can think of the block shown in Figure 6-2(a) as a unit. Then a flat represents $\frac{1}{10}$, a long represents $\frac{1}{100}$, and a cube represents $\frac{1}{1000}$. Using these objects, we show a representation of 2.235 in Figure 6-2(b).

Figure 6-2

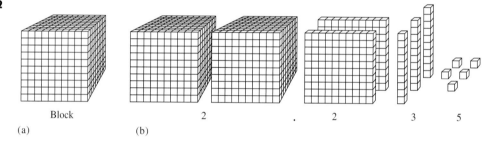

Table 6-2 shows other examples of decimals, their fractional meanings, and their common fraction notations.

Table 6-2

Decimal	Fractional Meaning	Common Fraction Notation
5.3	$5 + \frac{3}{10}$	$5\frac{3}{10}$, or $\frac{53}{10}$
0.02	$0 + \frac{0}{10} + \frac{2}{100}$	$\frac{2}{100}$
2.0103	$2 + \frac{0}{10} + \frac{1}{100} + \frac{0}{1000} + \frac{3}{10,000}$	$2\frac{103}{10,000}$, or $\frac{20,103}{10,000}$
$^-3.6$	$-\left(3 + \frac{6}{10}\right)$	$^-3\frac{6}{10}$, or $-\frac{36}{10}$

Decimals can also be written in expanded form using place value and negative exponents. Thus

$$12.61843 = 1 \cdot 10^1 + 2 \cdot 10^0 + 6 \cdot 10^{-1} + 1 \cdot 10^{-2} + 8 \cdot 10^{-3} + 4 \cdot 10^{-4} + 3 \cdot 10^{-5}$$

NOW TRY THIS 6-1

● Decimals can sometimes cause confusion. In the "Blondie" cartoon below, the writers probably intended the price of the prime rib to be half that of the steak. Is that what the cartoon says? Give other examples of this type of mistake involving dollars and cents.

Example 6-1 shows how to convert rational numbers whose denominators are powers of 10 to decimals.

Example 6-1 Convert each of the following to decimals:

a. $\dfrac{56}{100}$ **b.** $\dfrac{205}{10{,}000}$ **c.** $\dfrac{3.85}{1000}$

Solution **a.** $\dfrac{56}{100} = \dfrac{5 \cdot 10 + 6}{10^2} = \dfrac{5 \cdot 10}{10^2} + \dfrac{6}{10^2} = \dfrac{5}{10} + \dfrac{6}{10^2} = 0.56$

b. $\dfrac{205}{10{,}000} = \dfrac{2 \cdot 10^2 + 0 \cdot 10 + 5}{10^4} = \dfrac{2 \cdot 10^2}{10^4} + \dfrac{0 \cdot 10}{10^4} + \dfrac{5}{10^4}$

$= \dfrac{2}{10^2} + \dfrac{0}{10^3} + \dfrac{5}{10^4} = \dfrac{0}{10^1} + \dfrac{2}{10^2} + \dfrac{0}{10^3} + \dfrac{5}{10^4} = 0.0205$

c. $\dfrac{3.85}{1000} = \dfrac{3 + \dfrac{8}{10} + \dfrac{5}{10^2}}{10^3} = \dfrac{3}{10^3} + \dfrac{8}{10^4} + \dfrac{5}{10^5}$

$= 0 + \dfrac{0}{10^1} + \dfrac{0}{10^2} + \dfrac{3}{10^3} + \dfrac{8}{10^4} + \dfrac{5}{10^5} = 0.00385$

We reinforce the ideas in Example 6-1 through the use of a calculator. In Example 6-1(a), press [5] [6] [÷] [1] [0] [0] [=] and watch the display. Divide by 10 and look at the new placement of the decimal point. Once more, divide by 10 (which amounts to dividing the original number, 56, by 10,000) and note the placement of the decimal point. This leads to the following general rule for dividing a decimal by a power of 10:

To divide a decimal by 10^n, count n digits from right to left, annexing zeros if necessary, and insert the decimal point to the left of the nth digit.

The fractions in Example 6-1 are easy to convert to decimals because the denominators are powers of 10. If the denominator of a fraction is not a power of 10, as in $\frac{3}{5}$, we use the problem-solving strategy of *converting the problem to one we already know how to do*. First, we change $\frac{3}{5}$ to a fraction in which the denominator is a power of 10, and then we convert the fraction to a decimal.

$$\frac{3}{5} = \frac{3 \cdot 2}{5 \cdot 2} = \frac{6}{10} = 0.6$$

The reason for multiplying the numerator and the denominator by 2 is apparent when we observe that $10 = 2 \cdot 5$. In general, because $10^n = (2 \cdot 5)^n = 2^n \cdot 5^n$, the prime factorization of the denominator must be $2^n \cdot 5^n$ in order for the denominator of a rational number in simplest form to be 10^n. We use these ideas to write each fraction in Example 6-2 as a decimal.

Example 6-2 Express each of the following as decimals:

a. $\dfrac{7}{2^6}$ b. $\dfrac{1}{2^3 \cdot 5^4}$ c. $\dfrac{1}{125}$ d. $\dfrac{7}{250}$

Solution

a. $\dfrac{7}{2^6} = \dfrac{7 \cdot 5^6}{2^6 \cdot 5^6} = \dfrac{7 \cdot 15{,}625}{(2 \cdot 5)^6} = \dfrac{109{,}375}{10^6} = 0.109375$

b. $\dfrac{1}{2^3 \cdot 5^4} = \dfrac{1 \cdot 2^1}{2^3 \cdot 5^4 \cdot 2^1} = \dfrac{2}{2^4 \cdot 5^4} = \dfrac{2}{(2 \cdot 5)^4} = \dfrac{2}{10^4} = 0.0002$

c. $\dfrac{1}{125} = \dfrac{1}{5^3} = \dfrac{1 \cdot 2^3}{5^3 \cdot 2^3} = \dfrac{8}{(5 \cdot 2)^3} = \dfrac{8}{10^3} = 0.008$

d. $\dfrac{7}{250} = \dfrac{7}{2 \cdot 5^3} = \dfrac{7 \cdot 2^2}{(2 \cdot 5^3)2^2} = \dfrac{28}{(2 \cdot 5)^3} = \dfrac{28}{10^3} = 0.028$

A calculator can quickly convert some fractions to decimals. For example, to find $\dfrac{7}{2^6}$, press $\boxed{7}\ \boxed{\div}\ \boxed{2}\ \boxed{y^x}\ \boxed{6}\ \boxed{=}$; to convert $\dfrac{1}{125}$ to a decimal, press $\boxed{1}\ \boxed{\div}\ \boxed{1}\ \boxed{2}\ \boxed{5}\ \boxed{=}$, or press $\boxed{1}\ \boxed{2}\ \boxed{5}\ \boxed{1/x}\ \boxed{=}$. The display on some calculators may show $\boxed{8\ ^-03}$, which is the calculator's notation for $\dfrac{8}{10^3}$, or $8 \cdot 10^{-3}$. This notation, called *scientific notation*, is discussed in more detail later in this section.

terminating decimals   The answers in Example 6-2 are illustrations of **terminating decimals**—*decimals that can be written with only a finite number of places to the right of the decimal point*. If we attempt to rewrite $\dfrac{2}{11}$ as a terminating decimal using the method just developed, we first try to find a natural number b such that the following holds:

$$\frac{2}{11} = \frac{2b}{11b}, \quad \text{where } 11b \text{ is a power of 10.}$$

By the Fundamental Theorem of Arithmetic (discussed in Chapter 4), the only prime factors of a power of 10 are 2 and 5. Because $11b$ has 11 as a factor, we cannot write $11b$ as a power of 10, and therefore $\frac{2}{11}$ cannot be written as a terminating decimal. A similar argument using the Fundamental Theorem of Arithmetic holds in general, so we have the following result.

Theorem 6-1

A rational number $\frac{a}{b}$ in simplest form can be written as a terminating decimal if, and only if, the prime factorization of the denominator contains no primes other than 2 or 5.

Example 6-3 Which of the following fractions can be written as terminating decimals?

a. $\frac{7}{8}$ b. $\frac{11}{250}$ c. $\frac{21}{28}$ d. $\frac{37}{768}$

Solution
a. $\frac{7}{8} = \frac{7}{2^3}$. The denominator is 2^3, so $\frac{7}{8}$ can be written as a terminating decimal.

b. $\frac{11}{250} = \frac{11}{2 \cdot 5^3}$. The denominator is $2 \cdot 5^3$, so $\frac{11}{250}$ can be written as a terminating decimal.

c. $\frac{21}{28} = \frac{21}{2^2 \cdot 7} = \frac{3}{2^2}$. The denominator of the fraction in simplest form is 2^2, so $\frac{21}{28}$ can be written as a terminating decimal.

d. $\frac{37}{768} = \frac{37}{2^8 \cdot 3}$. This fraction is in simplest form and the denominator contains a factor of 3, so $\frac{37}{768}$ cannot be written as a terminating decimal.

REMARK As Example 6-3(c) shows, to determine whether a rational number $\frac{a}{b}$ can be represented as a terminating decimal, we consider the prime factorization of the denominator *only* if the fraction is in simplest form.

Ordering Terminating Decimals

To find which of two given decimals is greater, we could convert each to rational numbers in the form $\frac{a}{b}$, where a and b are integers, and determine which is greater. For example, because $0.36 = \frac{36}{100}$ and $0.9 = 0.90 = \frac{90}{100}$ and $\frac{90}{100} > \frac{36}{100}$, it follows that $0.9 > 0.36$. One could also tell that $0.9 > 0.36$ because $\$0.90$ is 90¢ and $\$0.36$ is 36¢ and 90¢ > 36¢. This suggests a way to order decimals without conversion to fractions. The steps we use to compare decimals are like those for comparing whole numbers:

1. Line up the numbers by place value.
2. Start at the left and find the first place where the face values are different.
3. Compare these digits. The number containing the greater face value in this place is the greater of the two original numbers.

For example, to compare 0.532 and 0.52 we could do the following:

0.532
0.52
↑

The digits in the hundredths place are different and $3 > 2$ so $0.532 > 0.52$.

Scientific Notation

Many calculators will display the decimals for the fractions $\frac{3}{45,689}$ or $\frac{5}{76,146}$ as $\boxed{6.5661319 \ -05}$ and $\boxed{6.566333 \ -05}$, respectively. The displays are in scientific notation. The first display is a notation for $6.5661319 \cdot 10^{-5}$ and the second for $6.566333 \cdot 10^{-5}$.

Scientists use scientific notation to handle either very small or very large numbers. For example, "the sun is 93,000,000 mi from Earth" is expressed as "the sun is $9.3 \cdot 10^7$ mi from Earth." A micron, a metric unit of measure that is 0.000001 m, is written $1 \cdot 10^{-6}$ m.

Definition of Scientific Notation

In **scientific notation,** a positive number is written as the product of a number greater than or equal to 1 and less than 10 and an integer power of 10.

The following numbers are in scientific notation:

$$8.3 \cdot 10^8, \quad 1.2 \cdot 10^{10}, \quad \text{and } 7.84 \cdot 10^{-6}$$

The numbers $0.43 \cdot 10^9$ and $12.3 \cdot 10^{-6}$ are not in scientific notation because 0.43 and 12.3 are not greater than or equal to 1 and less than 10. To write a number like 934.5 in scientific notation, we divide by 10^2 to get 9.345 and then multiply by 10^2 to retain the value of the original number:

$$934.5 = (934.5/10^2) \cdot 10^2 = 9.345 \cdot 10^2.$$

This amounts to moving the decimal point two places to the left (dividing by 10^2) and then multiplying by 10^2. Similarly to write 0.000078 in scientific notation, we first multiply by 10^5 to obtain 7.8 and then divide by 10^5 or multiply by 10^{-5} to keep the original value:

$$0.000078 = (0.000078 \cdot 10^5) \cdot 10^{-5} = 7.8 \cdot 10^{-5}.$$

This amounts to moving the decimal point five places to the right and multiplying by 10^{-5}.

Example 6-4 Write each of the following in scientific notation:

a. 413,682,000 b. 0.0000231 c. 83.7 d. 10,000,000

Solution
a. $413{,}682{,}000 = (413{,}682{,}000/10^8) \cdot 10^8 = 4.13682 \cdot 10^8$
b. $0.0000231 = (0.0000231 \cdot 10^5) \cdot 10^{-5} = 2.31 \cdot 10^{-5}$
c. $83.7 = (83.7/10^1) \cdot 10^1 = 8.37 \cdot 10^1$
d. $10{,}000{,}000 = (10{,}000{,}000/10^7) \cdot 10^7 = 1 \cdot 10^7$

Example 6-5 Convert each of the following to standard numerals:

a. $6.84 \cdot 10^{-5}$ b. $3.12 \cdot 10^7$

Solution
a. $6.84 \cdot 10^{-5} = 6.84 \cdot \left(\dfrac{1}{10^5}\right) = 0.0000684$
b. $3.12 \cdot 10^7 = 31{,}200{,}000$

Numbers in scientific notation are easy to manipulate using the laws of exponents. For example, $(5.6 \cdot 10^5)(6 \cdot 10^4)$ can be rewritten as $(5.6 \cdot 6)(10^5 \cdot 10^4) = 33.6 \cdot 10^9$, which is $3.36 \cdot 10^{10}$ in scientific notation. Also,

$$(2.35 \cdot 10^{-15})(2 \cdot 10^8) = (2.35 \cdot 2)(10^{-15} \cdot 10^8) = 4.7 \cdot 10^{-7}.$$

Calculators with an $\boxed{\text{EE}}$ key can be used to represent numbers in scientific notation. For example, to find $(5.2 \cdot 10^{16}) \cdot (9.37 \cdot 10^4)$, press

$\boxed{5}\,\boxed{.}\,\boxed{2}\,\boxed{\text{EE}}\,\boxed{1}\,\boxed{6}\,\boxed{\times}\,\boxed{9}\,\boxed{.}\,\boxed{3}\,\boxed{7}\,\boxed{\text{EE}}\,\boxed{4}\,\boxed{=}$.

ONGOING ASSESSMENT 6-1

1. Write each of the following in expanded form:
   a. 0.023
   b. 206.06
   c. 312.0103
   d. 0.000132

2. Rewrite each of the following as decimals:
   a. $4 \cdot 10^3 + 3 \cdot 10^2 + 5 \cdot 10 + 6 + 7 \cdot 10^{-1} + 8 \cdot 10^{-2}$
   b. $4 \cdot 10^3 + 6 \cdot 10^{-1} + 8 \cdot 10^{-3}$
   c. $4 \cdot 10^4 + 3 \cdot 10^{-2}$
   d. $2 \cdot 10^{-1} + 4 \cdot 10^{-4} + 7 \cdot 10^{-7}$

3. Write each of the following as numerals:
   a. Five hundred thirty-six and seventy-six ten-thousandths
   b. Three and eight thousandths
   c. Four hundred thirty-six millionths
   d. Five million and two tenths

4. Write each of the following terminating decimals as fractions:
   a. 0.436   b. 25.16   c. $^{-}316.027$
   d. 28.1902   e. $^{-}4.3$ f. $^{-}62.01$

5. Without performing the actual divisions, determine which of the following represent terminating decimals:
   a. $\dfrac{4}{5}$ b. $\dfrac{61}{2^2 \cdot 5}$
   c. $\dfrac{3}{6}$ d. $\dfrac{1}{2^5}$
   e. $\dfrac{36}{5^5}$ f. $\dfrac{133}{625}$
   g. $\dfrac{1}{3}$ h. $\dfrac{2}{25}$
   i. $\dfrac{1}{13}$ j. $\dfrac{26}{65}$

6. Where possible, write each of the numbers in problem 5 as terminating decimals.

7. Order each of the following decimals from greatest to least:
   a. 13.4919, 13.492, 13.49183, 13.49199
   b. $^{-}1.453$, $^{-}1.45$, $^{-}1.4053$, $^{-}1.493$

8. Convert each of the following to standard numerals:
   a. $3.2 \cdot 10^{-9}$ b. $3.2 \cdot 10^9$
   c. $4.2 \cdot 10^{-1}$ d. $6.2 \cdot 10^5$
9. Write the numerals in each of the following sentences in scientific notation:
   a. The diameter of Earth is about 12,700,000 m.
   b. The distance from Pluto to the sun is 5,797,000 km.
   c. Each year, about 50,000,000 cans are discarded in the United States.
10. Write the numerals in each of the following sentences in standard form:
    a. A computer requires $4.4 \cdot 10^{-6}$ sec to do an addition problem.
    b. There are about $1.99 \cdot 10^4$ km of coastline in the United States.
    c. Earth has existed for approximately $3 \cdot 10^9$ yr.
11. Write the results of each of the following in scientific notation:
    a. $(8 \cdot 10^{12}) \cdot (6 \cdot 10^{15})$
    b. $(16 \cdot 10^{12}) \div (4 \cdot 10^5)$
    c. $(5 \cdot 10^8) \cdot (6 \cdot 10^9) \div (15 \cdot 10^{15})$
12. Given any reduced rational number $\dfrac{a}{b}$ with $0 < a < b$, where b is of the form $2^m \cdot 5^n$ (m and n are whole numbers), determine a relationship between m and/or n and the number of digits in the terminating decimal. Justify your answer.
13. Which of the following numbers is the greatest: $100{,}000^3$, 1000^5, $100{,}000^2$? Justify your answer.
14. A newspaper reported that a stock gained 3 7/8 points (dollars) on a given day. How much did the stock increase in value in terms of dollars and cents on that day?
15. The five top swimmers in an event had the following times.

    Emily   64.54 seconds
    Molly   64.46 seconds
    Martha   63.59 seconds
    Kathy   64.02 seconds
    Rhonda   63.54 seconds

    List them in order of how they placed.

Communication

16. Explain how you would use base-ten blocks to represent two and three hundred forty-five thousandths.
17. Explain why in Theorem 6-1 the rational number must be in simplest form before examining the denominator.
18. In your own words, explain what a decimal point is and what it does.

Open-Ended

19. Determine how decimal notation is symbolized in different countries.
20. Examine three elementary school textbooks and report how the introductions of the topics of exponents, decimals, and scientific notation differ, if they do.

Cooperative Learning

21. Each member in a group should look up the size of some very small objects such as atoms, electrons, protons, and living organisms such as various bacteria and viruses.
    a. Order the objects in size from least to greatest.
    b. Compare your group's list with those of other groups in the class. What is the smallest object in all the lists?

Section 6-2 — Operations on Decimals

To develop an algorithm for addition of terminating decimals, consider the sum $2.16 + 1.73$. In elementary school, base-ten blocks are recommended to demonstrate such an addition problem. Figure 6-3 shows how the addition can be performed.

■ The computation in Figure 6-3 can also be approached by *changing it to a problem we already know how to solve,* that is, to a sum involving fractions. We then use the commutative and associative properties of addition to aid in the computation, as follows:

$$2.16 + 1.73 = \left(2 + \frac{1}{10} + \frac{6}{100}\right) + \left(1 + \frac{7}{10} + \frac{3}{100}\right)$$
$$= (2 + 1) + \left(\frac{1}{10} + \frac{7}{10}\right) + \left(\frac{6}{100} + \frac{3}{100}\right)$$
$$= 3 + \frac{8}{10} + \frac{9}{100}$$
$$= 3.89$$

Figure 6-3

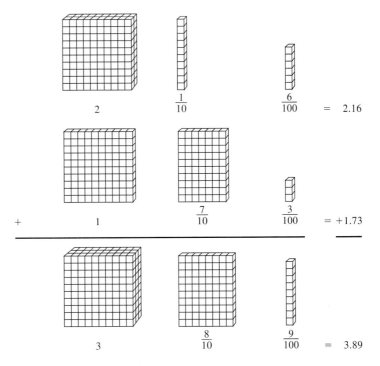

In the above computation, we see that we are adding units to units, tenths to tenths, and hundredths to hundredths. This can be accomplished most efficiently by keeping the numbers in their decimal forms, lining up the decimal points, and adding as if the numbers were whole numbers. This technique works for both addition and subtraction, as demonstrated on the student page on page 304 from *Scott Foresman-Addison Wesley Middle School Math, Course 1*, 1999. Try the problems on the bottom of the student page using this technique.

Multiplying Decimals

Just as we developed an algorithm for adding and subtracting terminating decimals by representing them as fractions, we can develop an algorithm for multiplication of decimals. Consider the product $4.62 \cdot 2.4$:

$$(4.62)(2.4) = \frac{462}{100} \cdot \frac{24}{10} = \frac{462}{10^2} \cdot \frac{24}{10^1} = \frac{462 \cdot 24}{10^2 \cdot 10^1} = \frac{11{,}088}{10^3} = 11.088.$$

The answer to this computation was obtained by multiplying the whole numbers 462 and 24 and then dividing the result by 10^3.

The algorithm for multiplying decimals can be stated as follows:

If there are n digits to the right of the decimal point in one number and m digits to the right of the decimal point in a second number, multiply the two numbers, ignoring the decimals, and then place the decimal point so that there are n + m digits to the right of the decimal point in the product.

> REMARK There are $n + m$ digits to the right of the decimal point in the product because $10^n \cdot 10^m = 10^{n+m}$.

CHAPTER 6 — Decimals, Percents, and Real Numbers

Learn | Adding and Subtracting Decimal Numbers

When you add, you must make sure you're adding tenths to tenths, hundredths to hundredths, and so on. To do this, line up the decimal points. Then add as if you were adding whole numbers.

Example 1

Add 1.7 and 2.49.

Estimate: 2 + 2 = 4.

Line up the decimal points.
↓

```
  1.7
+ 2.49
  4.19
```

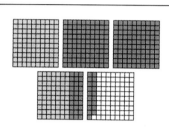

Use the same plan when subtracting decimals. Line up the decimal points, and then subtract as if you were subtracting whole numbers. Annex zeros if the second number has more digits after the decimal than the first.

Example 2

The United Kingdom uses a decimal currency based on the *pound* (£). Paul has £1.8. Edmund has only £1.38. How many more pounds does Paul have?

Subtract to find the difference.

Estimate to tenths: 1.8 − 1.4 = 0.4.

Line up the decimal points.
↓

```
  1.80     Annex zeros.
- 1.38
  0.42
```

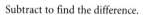
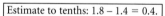

Remember
Annexing zeros to the right of a decimal number does not change the value of the number.
[Page 148]

Paul has £0.42 more than Edmund.

Try It

Find the sum or difference.

a. 4.631 + 3.986 **b.** 8.592 − 4.635 **c.** 5.6 + 1.973 **d.** 7.3 − 4.45

Example 6-6

Compute each of the following:

a. $(6.2)(1.43)$ **b.** $(0.02)(0.013)$ **c.** $(1000)(3.6)$

Solution

a.
$$
\begin{array}{r}
1.43 \\
\times\ 6.2 \\
\hline
286 \\
858 \\
\hline
8.866
\end{array}
$$
(2 digits after the decimal point)
(1 digit after the decimal point)
(3 digits after the decimal point)

b.
$$
\begin{array}{r}
0.013 \\
\times\ 0.02 \\
\hline
0.00026
\end{array}
$$

c.
$$
\begin{array}{r}
3.6 \\
\times\ 1000 \\
\hline
3600.0
\end{array}
$$

NOW TRY THIS 6-2

- Example 6.6(c) suggests that multiplication by 1000 or 10^3 results in moving the decimal point in the multiplicand 3 places to the right. Explain why this is true using expanded notation and the distributive property of multiplication over addition. In general, how does multiplication by 10^n, where n is a positive integer, affect the multiplicand? Why?

Dividing Decimals

Next we develop an algorithm for division of decimals. As with the other operations, we use fractions to represent a division that involves decimals and see if it suggests an algorithm to handle the division. Consider $75.45 \div 3$:

$$75.45 \div 3 = \frac{7545}{100} \div \frac{3}{1} = \frac{7545}{100} \cdot \frac{1}{3} = \frac{7545}{3} \cdot \frac{1}{100} = 2515 \cdot \frac{1}{100} = 25.15$$

This computation is shown using the algorithm on the student page on the following page from *Scott Foresman-Addison Wesley Math, Grade 5*, 1999.

From the student page, we see that when the divisor is a whole number, the division can be handled as with whole numbers and the decimal point placed directly over the decimal point in the dividend. Notice the role of estimation in the algorithm and that the division can be checked using the inverse operation of multiplication. When the divisor is not a whole number, as in $1.2032 \div 0.32$, we can obtain a whole-number divisor by expressing the quotient as a fraction and then multiplying the numerator and denominator of the fraction by 100. This corresponds to rewriting the division problem in form (a) as an equivalent problem in form (b), as follows:

a. $0.32\overline{)1.2032}$ **b.** $32\overline{)120.32}$

306 CHAPTER 6 *Decimals, Percents, and Real Numbers*

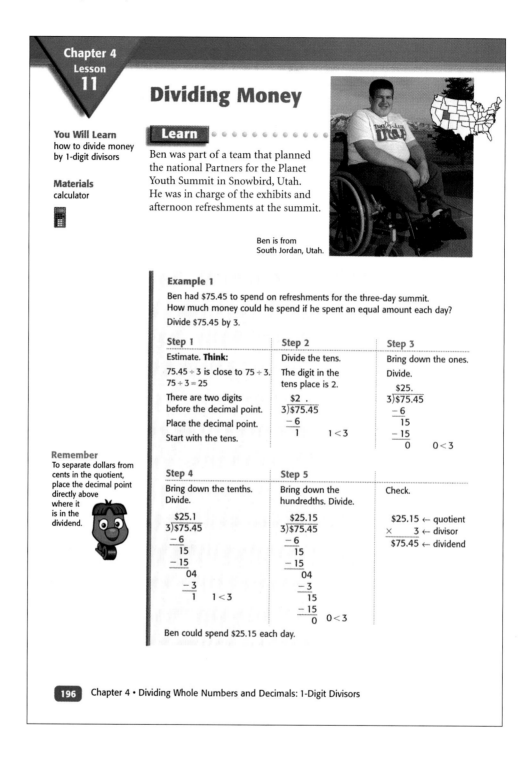

In elementary school texts, this process is usually described as "moving" the decimal point two places to the right in both the dividend and the divisor. This process is usually indicated with arrows, as shown on the following page.

$$\begin{array}{r}3.76\\0.32\overline{)1.2032}\\\underline{96}\\243\\\underline{224}\\192\\\underline{192}\\0\end{array}$$

Multiply divisor and dividend by 100.

Example 6-7 Compute each of the following:

a. $13.169 \div 0.13$ **b.** $9 \div 0.75$

Solution **a.**
$$\begin{array}{r}101.3\\0.13\overline{)13.169}\\\underline{13}\\16\\\underline{13}\\39\\\underline{39}\\0\end{array}$$

b.
$$\begin{array}{r}12\\0.75\overline{)9.00}\\\underline{75}\\150\\\underline{150}\\0\end{array}$$

In Example 6-7(b), we annexed two zeros in the dividend because $\dfrac{9}{0.75} = \dfrac{9 \cdot 100}{0.75 \cdot 100} = \dfrac{900}{75}$.

Example 6-8 An owner of a gasoline station must collect a gasoline tax of $0.11 on each gallon of gasoline sold. One week, the owner paid $1595 in gasoline taxes. The pump price of a gallon of gas that week was $1.35.

a. How many gallons of gas were sold during the week?
b. What was the revenue after taxes for the week?

Solution **a.** To find the number of gallons of gas sold during the week, we must divide the total gas tax bill by the amount of the tax per gallon:

$$\frac{1595}{0.11} = 14{,}500$$

Thus 14,500 gallons were sold.

b. To obtain the revenue after taxes, first determine the revenue before taxes. Then multiply the number of gallons sold by the cost per gallon:

$$(14{,}500)(\$1.35) = \$19{,}575$$

Next, subtract the gasoline taxes from the total revenue:

$$\$19{,}575 - \$1595 = \$17{,}980$$

Thus the revenue after gasoline taxes is $17,980.

NOW TRY THIS 6-3

- Use decimal division to help Hi, in the "Hi & Lois" cartoon, determine the best buy.

Mental Computation

Some of the tools used for mental computations with whole numbers can be used to perform mental computations with decimals, as seen in the following:

1. *Breaking and bridging*

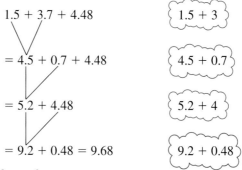

2. *Using compatible numbers*
   (Decimal numbers are compatible when they add up to a whole number.)

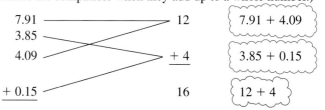

3. *Making compatible numbers*

$$\begin{aligned} 9.27 &= 9.25 + 0.02 \\ + 3.79 &= 3.75 + 0.04 \\ \hline &13.00 + 0.06 = 13.06 \end{aligned}$$

4. *Balancing with decimals in subtraction*

$$\begin{array}{rcrcr} 4.63 = & & 4.63 + 0.03 & = & 4.66 \\ -\,1.97 = & -& (1.97 + 0.03) & = & -2.00 \\ \hline & & & & 2.66 \end{array}$$

5. *Balancing with decimals in division*

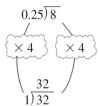

REMARK Balancing with decimals in division uses the property $\dfrac{a}{b} = \dfrac{a \cdot c}{b \cdot c}$.

Rounding Decimals

Frequently, it is not necessary to know the exact numerical answer to a question. For example, if we want to know the distance to the moon or the population of New York City, the approximate answers of 239,000 mi and 7,300,000 people, respectively, may be adequate.

Often a situation determines how you should round. For example, suppose a purchase came to $38.65 and the cashier used a calculator to figure out the sales tax by multiplying $0.06 \cdot 38.65$. The display showed 2.319. Because the display is between 2.31 and 2.32 and it is closer to 2.32, the cashier rounds up the sales tax to 2.32. Suppose a display of 8.7345649 needs to be reported to the nearest hundredth. The display is between 8.73 and 8.74 but is closer to 8.73, so we round it down to 8.73. Next suppose the number 6.8675 needs to be rounded to the nearest thousandth. Notice that 6.8675 is exactly halfway between 6.867 and 6.868. In such cases, it is common practice to round up and therefore the answer to the nearest thousandth is 6.868.

Example 6-9 Round each of the following numbers:

a. 7.456 to the nearest hundredth
b. 7.456 to the nearest tenth
c. 7.456 to the nearest unit
d. 7456 to the nearest thousand
e. 745 to the nearest ten
f. 74.56 to the nearest ten

Solution a. $7.456 \doteq 7.46$
b. $7.456 \doteq 7.5$
c. $7.456 \doteq 7$
d. $7456 \doteq 7000$
e. $745 \doteq 750$
f. $74.56 \doteq 70$

Rounding can also be done on some calculators using the [FIX] key. If you want the number 2.3669 to be rounded to thousandths, you enter [FIX] [3]. The display will show 0.000. If you then enter 2.3669 and press the [=] key, the display will show 2.367.

Estimating Decimal Computations Using Rounding

Rounded numbers can be useful for estimating answers to computations. For example, consider each of the following:

1. Karly goes to the grocery store to buy items that cost the following amounts. She estimates the total cost by rounding each amount to the nearest dollar and adding the rounded numbers.

$$
\begin{array}{rcl}
\$2.39 & \to & \$2 \\
0.89 & \to & 1 \\
6.13 & \to & 6 \\
4.75 & \to & 5 \\
+\,5.05 & \to & 5 \\
\hline
 & & \$19
\end{array}
$$

Thus Karly's estimate for her grocery bill is $19.

2. Karly's bill for car repairs was $72.80, and she has a coupon for $17.50 off. She can estimate her total cost by rounding each amount to the nearest ten dollars and subtracting.

$$
\begin{array}{rr}
\$72.80 & \$70 \\
-17.50 & -20 \\
\hline
 & \$50
\end{array}
$$

Thus an estimate for the repair bill is $50.

3. Karly sees a flash of lightning and hears the thunder 3.2 sec later. She knows that sound travels at 0.33 km/sec. She may estimate the distance she is from the lightning by rounding the time to the nearest unit and the speed to the nearest tenth and multiplying.

$$
\begin{array}{rcl}
0.33 & \to & 0.3 \\
\times\,3.2 & \to & \times\,3 \\
\hline
 & & 0.9
\end{array}
$$

Thus Karly estimates that she is approximately 0.9 km from the lightning.

An alternative approach is to recognize that $0.33 \doteq \frac{1}{3}$ and 3.2 is close to 3.3, so an approximation using compatible numbers is $\left(\frac{1}{3}\right) \cdot 3.3$, or 1.1 km.

4. Karly wants to estimate the cost per kilogram of a frozen turkey that sells for $17.94 and weighs 6.42 kg. She rounds and divides as follows:

$$
6.42 \overline{)17.94} \quad \to \quad 6 \overline{)18.00}^{\,3.00}
$$

Thus the turkey sells for approximately $3.00/kg.

When computations are performed with rounded numbers, the results may be significantly different from the actual answer. For example, suppose the distance, rounded to a tenth of a mile, along I-5 from Eugene to the first Albany exit is 42.6 mi, whereas the distance, rounded to the nearest tenth of a mile, from that exit to the first Salem exit is 22.4 mi. How far is it from Eugene to the first Salem exit? It seems that the answer is 42.6 + 22.4, or 65, mi. But how accurate is this answer? The distances might have been more accurately recorded as 42.55 and 22.35, when the sum would have been 64.9, or they may have been recorded as 42.64 and 22.44, when the sum would be 65.08, or 65.1 rounded to the nearest tenth. Thus the calculated sum of 65 mi could actually be 0.1 mi off in either direction. Similar errors may arise in other arithmetic operations.

NOW TRY THIS 6-4

- Other estimation strategies, such as front-end, clustering, and grouping to nice numbers, that you investigated with whole numbers also work with decimals.

- Take the grocery store bill in part 1 on the preceding page and use a front-end-with-adjustment strategy to estimate the bill.

ONGOING ASSESSMENT 6-2

1. If Maura went to the store and bought a chair for $17.95, a lawn rake for $13.59, a spade for $14.86, a lawn mower for $179.98, and two six-packs of mineral water for $2.43 each, what was the bill?
2. At 60°F, 1 qt of water weighs 2.082 lb. One cubic foot of water is 29.922 qt. What is the weight of a cubic foot of water to the nearest thousandth of a pound?
3. Complete the following magic square; that is, make the sum of every row, column, and diagonal the same:

8.2		
3.7	5.5	
	9.1	2.8

4. Keith bought 30 lb of nuts at $3.00/lb and 20 lb of nuts at $5.00/lb. If he wanted to buy 10 more pounds of a different kind of nut to make the average price per pound equal to $4.50, what price should he pay for the additional 10 lb?
5. A kilowatt hour means 1000 watts of electricity are being used continuously for 1 hr. The electric utility company in Laura's town charges $0.03715 for each kilowatt hour used. Laura heats her house with three electric wall heaters that use 1200 watts per hour each.
   a. How much does it cost to heat her house for one day?
   b. How many hours would a 75-watt light bulb have to stay on to result in $1 for electricity charges?
6. Automobile engines used to be measured in cubic inches but are now usually measured in cubic centimeters. If 2.54 cm is equivalent to 1 in., answer the following:
   a. Susan's 1963 Thunderbird has a 390 in.^{3} engine. Approximately how many cubic centimeters is this?
   b. Dan's 1991 Taurus has a 3000 cm^3 engine. Approximately how many cubic inches is this?
7. Florence Griffith-Joyner set a world record for the women's 100-m dash at the 1988 Summer Olympics in Seoul, South Korea. She covered the distance in 10.49 sec. If 1 m is equivalent to 39.37 in., express Griffith-Joyner's speed in terms of miles per hour.
8. Continue the following decimal patterns (assume each sequence is either arithmetic or geometric):
   a. 0.9, 1.8, 2.7, 3.6, 4.5, ____, ____, ____
   b. 0.3, 0.5, 0.7, 0.9, 1.1, ____, ____, ____
   c. 1, 0.5, 0.25, 0.125, ____, ____, ____
   d. 0.2, 1.5, 2.8, 4.1, 5.4, ____, ____, ____
9. A bank statement from a local bank shows that a checking account has a balance of $83.62. The balance recorded in the checkbook shows only $21.69. After checking the canceled checks against the record of these checks, the customer finds that the bank has not yet recorded six checks in the amounts of $3.21, $14.56, $12.44, $6.98, $9.51, and $7.49. Is the bank record correct? (Assume the person's checkbook records *are* correct.)

10. Round each of the following numbers as specified:
    a. 203.651 to the nearest hundred
    b. 203.651 to the nearest ten
    c. 203.651 to the nearest unit
    d. 203.651 to the nearest tenth
    e. 203.651 to the nearest hundredth
11. Jane's car travels 224 mi on 12 gal of gas. How many miles to the gallon does her car get, rounded to the nearest mile?
12. Audrey wants to buy some camera equipment to take pictures on her daughter's birthday. To estimate the total cost, she rounds each price to the nearest dollar and adds the rounded prices. What is her estimate for the items listed?

    | Camera | $54.56 |
    | Film | $ 4.50 |
    | Case | $17.85 |

13. Estimate the sum or difference in each of the following by using (i) rounding and (ii) front-end estimation. Then perform the computations to see how close your estimates are to the actual answers.

    a. 65.84   b. 89.47   c. 5.85   d. 223.75
        24.29   − 32.16   6.13   − 87.60
        12.18   9.10
      + 19.75   + 4.32

14. Mary Kim invested $964 in 18 shares of stock. A month later, she sold the 18 shares at $61.48 per share. She also invested in 350 shares of stock for a total of $27,422.50. She sold this stock for $85.35 a share and paid $495 in total commissions. What was Mary Kim's profit or loss on the transactions to the nearest dollar?
15. Luisa is traveling in Switzerland where the exchange rate is U.S. $1 = 1.59 Swiss francs for cash and U.S. $1 = 1.60 Swiss francs for traveler's checks.
    a. Luisa is exchanging $235 in cash. How many Swiss francs will she get?
    b. Luisa wants to buy a watch that costs 452.85 Swiss francs and hiking boots that cost 284.65 Swiss francs. What is the minimum number of dollars in cash that Luisa needs to exchange to purchase both?
    c. Luisa wants to buy a suit that costs 687.75 Swiss francs. She has traveler's checks in U.S. dollars in denominations of $100 and $20. What is the least amount of dollars in traveler's checks she needs to exchange? Explain your solution.
16. Find the least and the greatest possible products for each expression using one of the digits 1 through 9 exactly once in each part.
    a. ☐ . ☐ × ☐   b. ☐ . ☐ × ☐ . ☐
17. Some digits in the following number are covered by squares:

    4 ☐☐ 3 ☐ . ☐☐ 8 ☐

    If each of the digits 1 through 9 is used exactly once in the number, determine the number in each of these cases:
    a. The number is the greatest possible.
    b. The number is the least possible.

18. Iris worked a 40-hour week at $6.25/hr. Mentally compute her salary for the week and explain how you did it.
19. Mentally compute the number to fill in the blank in each of the following.
    a. $8.4 \cdot 6 = 4.2 \cdot$ ____   b. $10.2 \div 0.3 = 20.4 \div$ ____
    c. $a \cdot b = a/2 \cdot$ ____   d. $a \div b = 2a \div$ ____
20. Which of the following result in equal quotients?
    a. $7 \div 0.25$ b. $70 \div 2.5$
    c. $0.7 \div 0.25$ d. $700 \div 25$
21. Use estimation to place the decimal point in the correct position in each of the following products:
    a. $534 \cdot 0.34 = 18156$ b. $5.07 \cdot 29.3 = 148551$
22. Is subtracting 0.3 the same as (a) subtracting 0.30? (b) subtracting 0.03? Explain.
23. Use estimation to choose a decimal to multiply by 9 in order to get within 1 of 93. Explain how you made your choice and check your estimate.
24. a. Fill in the parentheses in each of the following, to write a true equation:

    $1 \cdot 2 + 0.25 = (\ \ )^2 \qquad 2 \cdot 3 + 0.25 = (\ \ )^2$

    Conjecture what the next two equations in this pattern will be.
    b. Do the computations to determine if your next two equations are correct.
    c. Generalize your answer in part (a) by filling in an appropriate expression in the parentheses $n \cdot (n + 1) + 0.25 = (\ \ )^2$.

Communication

25. How is multiplication of decimals like multiplication of whole numbers? How is it different?
26. On the second student page in this section there is a calculator symbol. Discuss whether you think a calculator is an appropriate tool for doing the mathematics on this page.
27. Why are estimation skills important in dividing decimals?
28. In the text, multiplication and division were done using both fractional and decimal forms. Discuss the advantages and disadvantages of each.
29. Explain why subtraction of terminating decimals can be accomplished by lining up the decimal points, subtracting as if the numbers were whole numbers, and then placing the decimal point in the difference.

Open-Ended

30. Find several examples of the use of decimals in the newspaper. Tell whether you think the numbers are exact or estimates. Also tell why you think decimals were used instead of fractions.
31. How could a calculator be used to develop or reinforce the understanding of multiplication of decimals?
32. Examine several elementary school textbooks and compare the ways that arithmetic operations involving decimals are introduced.

Cooperative Learning

33. In your group, decide on all the prerequisite skills that students need before learning to perform arithmetic operations on decimals.
34. You will need a calculator and a partner to play the game described below on the student page from McDougal Littell *Middle Grades MATH Thematics, Book 1*, 1999.

Review Problems

35. Write 14.0479 in expanded form.
36. Without dividing, determine which of the following represent terminating decimals:
    a. 24/36   b. 35/56
37. Write each of the following in scientific notation:
    a. 3,320,000   b. 0.0002367
38. Without using a calculator, write 35/56 as a decimal.

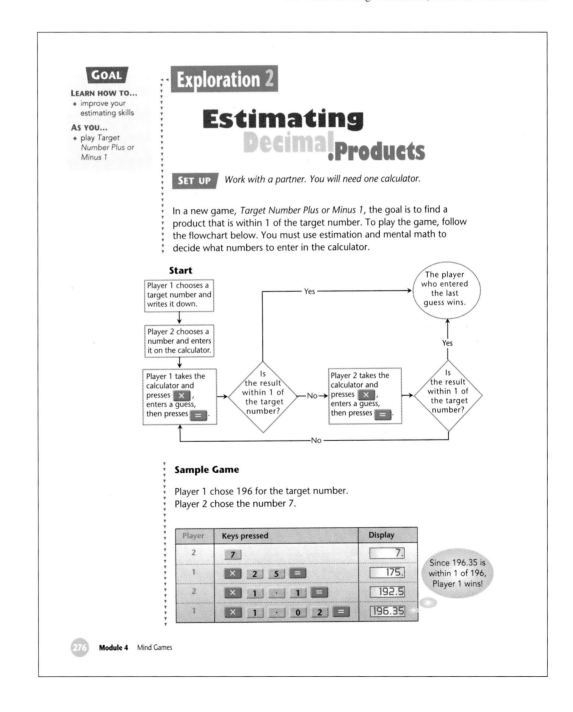

Section 6-3 — Nonterminating Decimals

Earlier in the text, procedures for converting some rational numbers to decimals were developed. For example, 7/8 can be written as a terminating decimal as follows:

$$\frac{7}{8} = \frac{7}{2^3} = \frac{7 \cdot 5^3}{2^3 \cdot 5^3} = \frac{875}{1000} = 0.875.$$

The decimal for 7/8 can also be found by division, as shown below:

$$\begin{array}{r}
0.875 \\
8\overline{)7.000} \\
\underline{6\ 4} \\
60 \\
\underline{56} \\
40 \\
\underline{40} \\
0
\end{array}$$

Repeating Decimals

If we use a calculator to find a decimal representation for 2/11, the calculator will display 0.1818181. It seems that the block of digits 18 repeats itself. To examine what digits, if any, the calculator did not display, consider the following division:

$$\begin{array}{r}
0.18 \\
11\overline{)2.00} \\
\underline{1\ 1} \\
90 \\
\underline{88} \\
2
\end{array}$$

At this point, if the division is continued, the division pattern would repeat, since the remainder 2 repeats the division. Thus the quotient is $0.181818\ldots$. A decimal of this type is a **repeating decimal,** and the repeating block of digits is the **repetend.** The repeating decimal is written as $0.\overline{18}$, where the bar indicates that the block of digits underneath is repeated infinitely.

repeating decimal
repetend

Example 6-10

Convert the following to decimals:

a. $\frac{1}{7}$ b. $\frac{2}{13}$

Solution If we use a calculator to divide 1 by 7 and 2 by 13, the display will show 0.1428571 and 0.1538461, respectively. It seems that the division pattern repeats. Thus $\frac{1}{7} = 0.\overline{142857}$ and $\frac{2}{13} = 0.\overline{153846}$.

To see why in Example 6-10 the division pattern repeated as predicted, consider the following divisions:

$$
\begin{array}{r}
\text{a.} \quad 0.142857 \\
7\overline{)1.000000} \\
\underline{7} \\
30 \\
\underline{28} \\
20 \\
\underline{14} \\
60 \\
\underline{56} \\
40 \\
\underline{35} \\
50 \\
\underline{49} \\
1
\end{array}
\qquad
\begin{array}{r}
\text{b.} \quad 0.153846 \\
13\overline{)2.000000} \\
\underline{1\,3} \\
70 \\
\underline{65} \\
50 \\
\underline{39} \\
110 \\
\underline{104} \\
60 \\
\underline{52} \\
80 \\
\underline{78} \\
2
\end{array}
$$

In $\frac{1}{7}$, the remainders obtained in the division are 3, 2, 6, 4, 5, and 1. These are all the possible nonzero remainders that can be obtained when dividing by 7. If we had obtained a remainder of 0, the decimal would terminate. Consequently, the seventh division cannot produce a new remainder. Whenever a remainder recurs, the process repeats itself. Using similar reasoning, we could predict that the repetend for $\frac{2}{13}$ could not be longer than 12, because there are only 12 possible nonzero remainders. However, one of the remainders could repeat sooner than that, which was actually the case in part (b) above. In general, if $\frac{a}{b}$ is any rational number in simplest form with $b > a$ and it does not represent a terminating decimal, the repetend has at most $b - 1$ digits. Therefore *a rational number may always be represented either as a terminating decimal or as a repeating decimal.*

NOW TRY THIS 6-5

- **a.** Write 1/9 as a decimal.
- **b.** Based on your answer in part (a), mentally compute the decimal representation for each of the following.
  (i) 2/9 (ii) 3/9
  (iii) 5/9 (iv) 8/9

Example 6-11 Use a calculator to convert $\frac{1}{17}$ to a repeating decimal.

Solution In using a calculator, if we press $\boxed{1}\,\boxed{\div}\,\boxed{1}\,\boxed{7}\,\boxed{=}$, we obtain the following, shown as part of a division problem:

$$
\begin{array}{r}
0.0588235 \\
17\overline{)1.}
\end{array}
$$

Without knowing whether the calculator has an internal round-off feature and with the calculator's having an 8-digit display, we find the greatest number of digits to be trusted in the quotient is 6 following the decimal point. (Why?) If we use those 6 places and multiply 0.058823 times 17, we may continue the operation as follows:

$$\boxed{.}\boxed{0}\boxed{5}\boxed{8}\boxed{8}\boxed{2}\boxed{3}\boxed{\times}\boxed{1}\boxed{7}\boxed{=}$$

We then obtain 0.999991, which we may place in the preceding division:

$$\begin{array}{r} 0.058823 \\ 17\overline{)1.000000} \\ \underline{999991} \\ 9 \end{array}$$

Next, we divide 9 by 17 to obtain 0.5294118. Again ignoring the rightmost digit, we continue as before, completing the division as follows, where the repeating pattern is apparent:

$$\begin{array}{r} 0.0588235294117647058235 \\ 17\overline{)1.0000000000000000000000} \\ \underline{999991} \\ 9000000 \\ \underline{8999987} \\ 13000000 \\ \underline{12999985} \\ 15 \end{array}$$

Thus $\dfrac{1}{17} = 0.0\overline{588235294117647}$, and the repetend is 16 digits long.

Writing a Repeating Decimal in the Form $\dfrac{a}{b}$ Where $a, b \in I$

We have already considered how to write terminating decimals in the form $\dfrac{a}{b}$, where $a, b \in I, b \neq 0$. For example,

$$0.55 = \dfrac{55}{10^2} = \dfrac{55}{100}.$$

To write $0.\overline{5}$ in a similar way, we see that because the repeating decimal has infinitely many digits, there is no single power of 10 that can be placed in the denominator. To overcome this difficulty, we must somehow eliminate the infinitely repeating part of the decimal. Our *subgoal* is to write an equation for n without the repeating part. Suppose $n = 0.\overline{5}$. It can be shown that $10(0.555\ldots) = 5.555\ldots = 5.\overline{5}$. Hence, $10n = 5.\overline{5}$. Using this information, we subtract the corresponding sides of the equations to obtain an equation whose solution can be written as a rational number in the form a/b, where a and b are integers and $b \neq 0$.

$$10n = 5.\overline{5}$$
$$\underline{n = 0.\overline{5}}$$
$$9n = 5$$
$$n = \dfrac{5}{9}$$

Thus $0.\overline{5} = \frac{5}{9}$. This result can be checked by performing the division $5 \div 9$. Performing the subtraction gives an equation that contains only integers. The repeating blocks "cancel" each other.

Suppose a decimal has a repetend of more than one digit, such as $0.\overline{235}$. To write it in the form $\frac{a}{b}$, we can reasonably multiply by 10^3, since there is a three-digit repetend. Let $n = 0.\overline{235}$. Our *subgoal* is again to write an equation for n without the repeating part of the decimal:

$$1000n = 235.\overline{235}$$
$$n = 0.\overline{235}$$
$$999n = 235$$
$$n = \frac{235}{999}$$

Hence, $0.\overline{235} = \frac{235}{999}$.

Notice that $0.\overline{5}$ repeats in one-digit blocks. Therefore, to write it in the form $\frac{a}{b}$, we first multiply by 10^1; $0.\overline{235}$ repeats in three-digit blocks, so we first multiply by 10^3. In general, *if the repetend is immediately to the right of the decimal point, first multiply by 10^n, where n is the number of digits in the repetend, and then continue as in the preceding cases.*

Now, suppose the repeating block does *not* occur immediately after the decimal point. For example, let $n = 2.3\overline{45}$. A strategy for solving this problem is to *change it to a related problem* we already know how to do; that is, change it to a problem where the repeating block immediately follows the decimal point. This is our new *subgoal*. To accomplish this, we multiply both sides by 10:

$$n = 2.3\overline{45}$$
$$10n = 23.\overline{45}$$

We now proceed as with previous problems. Because $10n = 23.\overline{45}$ and the number of digits in the repetend is 2, we multiply by 10^2 as follows:

$$100(10n) = 2345.\overline{45}$$

Thus

$$1000n = 2345.\overline{45}$$
$$10n = 23.\overline{45}$$
$$990n = 2322$$
$$n = \frac{2322}{990} \quad \text{or} \quad \frac{129}{55}.$$

Hence, $2.3\overline{45} = \frac{2322}{990}$, or $\frac{129}{55}$.

To find the $\frac{a}{b}$ form of $0.\overline{9}$, we proceed as follows. Let $n = 0.\overline{9}$, then $10n = 9.\overline{9}$. Next we subtract as above:

$$10n = 9.\overline{9}$$
$$n = 0.\overline{9}$$
$$9n = 9$$
$$n = 1$$

Hence, $0.\overline{9} = 1$. This approach to the problem may not be convincing. Another approach to show that $0.\overline{9}$ is really another name for 1 is shown next:

(1) $\dfrac{1}{3} = 0.33333333\ldots$ (2) $\dfrac{2}{3} = 0.66666666\ldots$

Adding Equations (1) and (2), we have $1 = 0.99999999\ldots$, or $0.\overline{9}$. This decimal represents the infinite sum $\dfrac{9}{10} + \dfrac{9}{10^2} + \dfrac{9}{10^3} + \ldots$. In more advanced mathematics courses, such sums are defined as the limits of finite sums.

NOW TRY THIS 6-6

- **a.** Based on the cartoon below, write Arlo's real age in years as a decimal. Is the decimal terminating or nonterminating? Why?
- **b.** Find Arlo's real age in years and months.
- **c.** Use the 74.9-year life expectancy and your age to determine what date of a one-year life it is for you.

Arlo and Janis *by Johnson*

ARLO & JANIS REPRINTED BY PERMISSION OF NEWSPAPER ENTERPRISE ASSOCIATION, INC. ALL RIGHTS RESERVED.

Ordering Repeating Decimals

We saw in Section 6-1 how to compare terminating decimals. To compare repeating decimals such as $1.\overline{3478}$ and $1.34\overline{7821}$, we use a similar procedure. We write the decimals one under the other, in their equivalent forms without the bars, and line up the decimal points, as follows:

$$1.34783478\ldots$$
$$1.34782178\ldots$$

The digits to the left of the decimal points and the first four digits after the decimal points are the same in each of the numbers. However, since the digit in the hundred-thousandths place of the top number, which is 3, is greater than the digit 2 in the hundred-thousandths place of the bottom number, $1.\overline{3478}$ is greater than $1.34\overline{7821}$.

It is easy to compare two fractions, such as $\frac{21}{43}$ and $\frac{37}{75}$, using a calculator. We convert each to a decimal and then compare the decimals.

$$\boxed{2}\,\boxed{1}\,\boxed{\div}\,\boxed{4}\,\boxed{3}\,\boxed{=} \rightarrow 0.4883721$$
$$\boxed{3}\,\boxed{7}\,\boxed{\div}\,\boxed{7}\,\boxed{5}\,\boxed{=} \rightarrow 0.4933333$$

Examining the digits in the hundredths place, we see that

$$\frac{37}{75} > \frac{21}{43}.$$

Example 6-12 Find a rational number in decimal form between $0.\overline{35}$ and $0.\overline{351}$.

Solution First, line up the decimals.

$$0.353535\ldots$$
$$0.351351\ldots$$

Then, to find a decimal between these two, observe that starting from the left, the first place at which the two numbers differ is the thousandths place. Clearly, one decimal between these two is 0.352. Others include 0.3514, $0.35\overline{15}$, and 0.35136. In fact, there are infinitely many others.

TECHNOLOGY CORNER

The following REPETEND Logo procedure displays the repetend when a fraction is represented by a nonterminating decimal. (To find a repetend in the decimal expansion of 1/17, type REPETEND 1 17.) Edit the procedure to fit your particular Logo software.

```
TO REPETEND :A :B         TO REPETEND1 :A :B :N
  REPETEND1 :A :B 1         PRINT QUOTIENT :A :B
END                         IF :N = :B + 1 [STOP]
                            REPETEND1 (REMAINDER :A :B)* 10 :B :N + 1
                          END
```

Run REPETEND to find the decimal expansions of 1/11, 1/17, and 1/29.

ONGOING ASSESSMENT 6-3

1. Find the decimal representation for each of the following:
   a. $\frac{4}{9}$ b. $\frac{2}{7}$ c. $\frac{3}{11}$ d. $\frac{1}{15}$
   e. $\frac{2}{75}$ f. $\frac{1}{99}$ g. $\frac{5}{6}$ h. $\frac{1}{13}$

2. Convert each of the following repeating decimals to $\frac{a}{b}$ form where a, b are integers, $b \neq 0$.
   a. $0.\overline{4}$ b. $0.\overline{6}$ c. $1.3\overline{9}$
   d. $0.5\overline{5}$ e. $-2.3\overline{4}$ f. $-0.0\overline{2}$

3. Order each of the following decimals from greatest to least:
   $-1.4\overline{54}, -1.\overline{454}, -1.\overline{45}, -1.45\overline{4}, -1.454$

4. Continue the following patterns for the arithmetic sequences.
   a. $0, 0.\overline{3}, 0.\overline{6}, 1, 1.\overline{3}$, ___, ___, ___
   b. $0, 0.5, 0.6, 0.75, 0.8, 0.8\overline{3}$, ___, ___, ___

5. Suppose $a = 0.\overline{32}$ and $b = 0.\overline{123}$.
   a. Find $a + b$ by adding from left to right. How many digits are in the repetend of the sum?
   b. Find $a + b$ if $a = 1.2\overline{34}$ and $b = 0.\overline{1234}$. Is the answer a rational number? How many digits are in the repetend?

6. Find repeating decimals for each of the following:
   a. $\frac{1}{13}$ b. $\frac{1}{21}$ c. $\frac{3}{19}$

7. Find three decimals between each of the two following pairs of decimals:
   a. 3.2 and 3.22   b. 462.24 and 462.243

8. Find the decimal halfway between the two following decimals:
   a. 0.4 and 0.5   b. 0.9 and 1.1

9. a. Find three rational numbers between 3/4 and $0.\overline{75}$.
   b. Find three rational numbers between 1/3 and $0.\overline{34}$.

10. a. What is the 21st digit in the decimal expansion of 3/7?
    b. What is the 5280th digit in the decimal expansion of 1/17?

11. a. Write each of the following as a fraction in the form a/b where a and b are integers, $b \neq 0$.
    (i) $0.\overline{1}$ (ii) $0.\overline{01}$ (iii) $0.\overline{001}$
    b. What fractions would you expect from $0.\overline{0001}$?
    c. Mentally compute the decimal equivalent for 1/90.

12. Use the fact that $0.\overline{1} = 1/9$ to mentally convert each of the following into fractions:
    a. $0.\overline{2}$ b. $0.\overline{3}$ c. $0.\overline{5}$ d. $2.\overline{7}$ e. $9.\overline{9}$

13. Use the fact that $0.\overline{01} = 1/99$ and $0.\overline{001} = 1/999$ to mentally convert each of the following into fractions:
    a. $0.\overline{05}$ b. $0.\overline{003}$ c. $3.\overline{25}$ d. $3.\overline{125}$

Communication

14. A friend claims that every finite decimal is equal to some infinite decimal. Is the claim true? Explain why or why not.

15. Some addition problems are easier to compute with fractions and some are easier to do with decimals. For example, $1/7 + 5/7$ is easier to compute than $0.\overline{142857} + 0.\overline{714285}$ and $0.4 + 0.25$ is easier to compute than $2/5 + 1/4$. Describe situations in which you think it would be easier to compute the additions with fractions than with decimals, and vice versa.

Open-Ended

16. Notice that $\frac{1}{7} = 0.\overline{142857}$, $2/7 = 0.\overline{285714}$, $3/7 = 0.\overline{428571}$, $4/7 = 0.\overline{571428}$, $5/7 = 0.\overline{714285}$, and $6/7 = 0.\overline{857142}$.
    a. Describe a common property that all of these repeating decimals share.
    b. Suppose you memorized the decimal form for $\frac{1}{7}$. How could you quickly find the answers for the decimal expansion of the rest of the above fractions? Describe as many ways as you can.
    c. Find the other fractions that behave like $\frac{1}{7}$. In what way is the behavior similar?
    d. Based on your answer in (a), describe shortcuts for writing $\frac{k}{14}$ as a repeating decimal for $k = 1, 2, 3, \ldots 11$.

Cooperative Learning

17. Choose a partner and play the following game. Write a repeating decimal of the form $0.\overline{abcdef}$. Tell your partner that the decimal is of that form but do not reveal the specific values for the digits. Your partner's objective is to find your repeating decimal. Your partner is allowed to ask you for the values of 6 digits that are at the 100th or larger places after the decimal point but not the digits in consecutive places. For example, your opponent may ask for the 100th, 200th, 300th, ... digits but may not ask for the 100th and 101st digit. Switch roles at least once. After playing the game, discuss in your group a strategy for asking your partner the least number of questions that will allow one to find the partner's repetend.

Review Problems

18. John is a payroll clerk for a small company. Last month, the employees' gross earnings (earnings before deductions) totaled $27,849.50. John deducted $1520.63 for social security, $723.30 for unemployment insurance, and $2843.62 for federal income tax. What was the employees' net pay (their earnings after deductions)?

19. The speed of light is approximately 186,000 mi/sec. It takes light from the nearest star, Alpha Centauri, approximately 4 yr to reach Earth. How many miles away is Alpha Centauri from Earth? Express the answer in scientific notation.

20. Find the product of 0.22 and 0.35 on a calculator. How does the placement of the decimal point in the answer on the calculator compare with the placement of the decimal point using the rule in this chapter? Explain.

Section 6-4 Percents

Percents are very useful in conveying information. People hear that there is a 60 percent chance of rain or that their savings accounts are drawing 6 percent interest. The word **percent** comes from the Latin phrase *per centum,* which means *per hundred.* For example, a bank that pays 6 percent simple interest on a savings account pays $6 for each $100 in the account for 1 yr; that is, it pays $\frac{6}{100}$ of whatever amount is in the account for 1 yr. The symbol, %, indicates percent. For example, we write 6% for $\frac{6}{100}$.

In general, we have the following definition.

> **Definition of Percent**
>
> $$n\% = \frac{n}{100}$$

Figure 6-4

Thus $n\%$ of a quantity is $\frac{n}{100}$ of the quantity. Therefore 1% is one hundredth of a whole and 100% represents the entire quantity, whereas 200% represents $\frac{200}{100}$, or 2 times, the given quantity. Percents can be illustrated by using a hundreds grid. For example, what percent of the grid is shaded in Figure 6-4? Because 30 out of the 100, or $\frac{30}{100}$, of the squares are shaded, we say that 30% of the grid is shaded.

Because $n\% = \frac{n}{100}$, to convert a number to a percent we write it as a fraction with denominator 100; the numerator gives the amount of the percent. For example, $\frac{3}{4} = \frac{3 \cdot 25}{4 \cdot 25} = \frac{75}{100}$. Hence, $\frac{3}{4} = 75\%$. Notice that to convert $\frac{75}{100}$ to a percent, we could have multiplied the fraction by 100. Thus $0.0002 = (100 \cdot 0.0002)\% = 0.02\%$. In general, to convert a number to a percent we multiply it by 100 and attach the % symbol.

Example 6-13 Write each of the following as a percent:

a. 0.03 b. $0.\overline{3}$ c. 1.2 d. 0.00042
e. 1 f. $\frac{3}{5}$ g. $\frac{2}{3}$ h. $2\frac{1}{7}$

Solution
a. $0.03 = 100 \cdot 0.03\% = 3\%$
b. $0.\overline{3} = 100 \cdot 0.\overline{3}\% = 33.\overline{3}\%$
c. $1.2 = 100 \cdot 1.2\% = 120\%$
d. $0.00042 = 100 \cdot 0.00042\% = 0.042\%$
e. $1 = 100 \cdot 1\% = 100\%$
f. $\frac{3}{5} = 100 \cdot \frac{3}{5}\% = \frac{300}{5}\% = 60\%$

g. $\frac{2}{3} = 100 \cdot \frac{2}{3}\% = \frac{200}{3}\% = 66.\overline{6}\%$ **h.** $2\frac{1}{7} = 100 \cdot 2\frac{1}{7}\% = \frac{1500}{7}\% = 214\frac{2}{7}\%$

A number can also be converted to a percent by using a *proportion*. For example, to write $\frac{3}{5}$ as a percent, find the value of n in the following proportion:

$$\frac{3}{5} = \frac{n}{100}$$

Solving the proportion, we obtain $\left(\frac{3}{5}\right) \cdot 100 = n$, or $n = 60$. Therefore $\frac{3}{5} = 60\%$.

Still another way to convert a number to a percent is to recall that $1 = 100\%$. Thus for example, $\frac{3}{4} = \frac{3}{4}$ of $1 = \frac{3}{4} \cdot 1 = \frac{3}{4} \cdot 100\% = 75\%$.

REMARK The % symbol is crucial in identifying the meaning of a number. For example, $\frac{1}{2}$ and $\frac{1}{2}\%$ are different numbers: $\frac{1}{2} = 50\%$, which is not equal to $\frac{1}{2}\%$. Similarly, 0.01 is different from 0.01%, which is 0.0001.

In our computations, it is sometimes useful to convert percents to decimals. This can be done by writing the percent as a fraction and then converting the fraction to a decimal.

Example 6-14 Write each of the following percents as a decimal:

a. 5%   **b.** 6.3%   **c.** 100%   **d.** 250%   **e.** $\frac{1}{3}\%$ **f.** $33\frac{1}{3}\%$

Solution
a. $5\% = \frac{5}{100} = 0.05$ **b.** $6.3\% = \frac{6.3}{100} = 0.063$
c. $100\% = \frac{100}{100} = 1$ **d.** $250\% = \frac{250}{100} = 2.50$
e. $\frac{1}{3}\% = \frac{\frac{1}{3}}{100} = \frac{0.\overline{3}}{100} = 0.00\overline{3}$ **f.** $33\frac{1}{3}\% = \frac{33\frac{1}{3}}{100} = \frac{33.\overline{3}}{100} = 0.\overline{3}$

Another approach to writing a percent as a decimal is first to convert 1% to a decimal. Because $1\% = \frac{1}{100} = 0.01$, we can conclude that $5\% = 5 \cdot 0.01 = 0.05$ and $6.3\% = 6.3 \cdot 0.01 = 0.063$.

NOW TRY THIS 6-7

- **a.** Investigate how your calculator handles percents and tell what the calculator does when the % key is pushed.
- **b.** Use your calculator to change 1/3 to a percent.

In the *Principles and Standards* we find the following under the Representation Standard for Grades 6–8.

▲ *Middle-grades students who are taught with this Standard in mind will learn to recognize, compare, and use an array of representational forms for fractions, decimals, percents and integers. They also will learn to use representational forms such as exponential and scientific notation when working with large and small numbers and to use a variety of graphical tools to represent and analyze data sets* (p. 280).

The *Principles and Standards* also states

▲ *Technnology should not be used as a replacement for basic understandings and intuitions; rather, it can and should be used to foster those understandings and intuitions. In mathematics-instruction programs, technology should be used widely and responsibly, with the goal of enriching students' learning of mathematics* (p. 25).

Applications Involving Percent

Application problems that involve percents usually take one of the following forms:

1. Finding a percent of a number
2. Finding what percent one number is of another
3. Finding a number when a percent of that number is known

Before we consider examples illustrating these forms, recall what it means to find a fraction "of" a number. For example, $\frac{2}{3}$ of 70 means $\frac{2}{3} \cdot 70$. Similarly, to find 40% of 70, we have $\frac{40}{100}$ of 70, which means $\frac{40}{100} \cdot 70$, or $0.40 \cdot 70 = 28$.

Example 6-15 A house that sells for \$92,000 requires a 20% down payment. What is the amount of the down payment?

Solution The down payment is 20% of \$92,000, or $0.20 \cdot \$92,000 = \$18,400$. Hence, the amount of the down payment is \$18,400.

Example 6-16 If Alberto has 45 correct answers on an 80-question test, what percent of his answers are correct?

Solution Alberto has $\frac{45}{80}$ of the answers correct. To find the percent of correct answers, we need to convert $\frac{45}{80}$ to a percent. We can do this by multiplying the fraction by 100 and attaching the % symbol as follows:

$$\frac{45}{80} = 100 \cdot \frac{45}{80}\%$$
$$= 56.25\%.$$

Thus 56.25% of the answers are correct.

The student page on page 324 from *Scott Foresman-Addison Wesley Middle School Math, Course 3,* 1999 demonstrates how students can work flexibly with equivalent fractions, decimals, and percents using a calculator.

324 CHAPTER 6 *Decimals, Percents, and Real Numbers*

6-1 Percents, Decimals, and Fractions

You'll Learn ...
■ to convert among fractions, decimals, and percents

... How It's Used
Advertisers use percents in many of the newspaper ads they create.

Vocabulary
percent

circle graph

▶ **Lesson Link** You have represented ratios as fractions and decimals. Now you will learn how to represent a ratio as a percent. ◀

Explore Fractions, Decimals, and Percents

Calculated Conversions

Materials: Calculator with [F↔D] key

Use your calculator to convert fractions, decimals, and percents.

1. Enter 3 [+] 4 [=] [F↔D]. What is displayed?
   Press [F↔D] again. What is displayed?
   Repeat for each fraction.

   a. $\frac{1}{100}$ b. $\frac{2}{5}$ c. $\frac{5}{4}$ d. $\frac{3}{3}$

2. Enter 0.3 [F↔D]. What is displayed?
   Repeat for each decimal.

   a. 1.75   b. 0.85   c. 0.003   d. 2.25

3. Enter 25 [%]. What is displayed?
   Press [F↔D]. What is displayed?
   Repeat for each percent.

   a. 30%   b. 45%   c. 3%   d. 99%

4. What patterns, if any, do you notice?

Learn Percents, Decimals, and Fractions

You find percents everywhere—40% off, 95% fat free, sales increase 10%. A **percent** is a ratio that compares a number to 100. *Percent* means "parts per hundred," "hundredths," or "out of every hundred."

We often see percents represented in a **circle graph**, sometimes called a pie chart.

274 Chapter 6 • Percent

An alternative solution uses proportion. Let n be the percent of correct answers and proceed as follows:

$$\frac{45}{80} = \frac{n}{100}$$

$$\frac{45}{80} \cdot 100 = n$$

$$n = \frac{4500}{80} = 56.25.$$

Example 6-17 Forty-two percent of the parents of the school children in the Paxson School District are employed at Di Paloma University. If the number of parents employed by the university is 168, how many parents are in the school district?

Solution Let n be the number of parents in the school district. Then 42% of n is 168. We translate this information into an equation and solve for n.

$$42\% \text{ of } n = 168$$

$$\frac{42}{100} \cdot n = 168$$

$$0.42 \cdot n = 168$$

$$n = \frac{168}{0.42} = 400.$$

There are 400 parents in the school district.

The problem can be solved using a proportion. Forty-two percent, or $\frac{42}{100}$, of the parents are employed at the university. If n is the total number of parents, then $168/n$ also represents the fraction of parents employed there. Thus

$$\frac{42}{100} = \frac{168}{n}$$

$$42n = 100 \cdot 168$$

$$n = \frac{16{,}800}{42} = 400.$$

We can also solve the problem as follows:

$$42\% \text{ of } n \text{ is } 168.$$

$$1\% \text{ of } n \text{ is } \frac{168}{42}.$$

$$100\% \text{ of } n \text{ is } 100\left(\frac{168}{42}\right).$$

Therefore,

$$n \text{ is } 100\left(\frac{168}{42}\right), \text{ or } 400.$$

Example 6-18 Kelly bought a bicycle and a year later sold it for 20% less than what she paid for it. If she sold the bike for $144, what did she pay for it?

Solution We are looking for the original price P that Kelly paid for the bike. We know that she sold the bike for $144 and that this included a 20% loss. Thus we can *write the following equation*:

$$\$144 = P - \text{Kelly's loss}.$$

Because Kelly's loss is 20% of P, we proceed as follows:

$$\$144 = P - 20\% \cdot P$$
$$\$144 = P - 0.20 \cdot P$$
$$\$144 = (1 - 0.20)P$$
$$\$144 = 0.80\, P$$
$$\$\frac{144}{0.80} = P$$
$$\$180 = P.$$

Thus, she paid $180 for the bike.

Example 6-19 Westerner's Clothing Store advertised a suit for 10% off, for a savings of $15. Later, the manager marked the suit at 30% off the original price. What is the amount of the current discount?

Solution A 10% discount amounts to a $15 savings. We could find the amount of the current discount if we knew the original price. Thus finding the original price becomes our *subgoal*. Because 10% of P is $15, we have the following:

$$10\% \cdot P = \$15$$
$$0.10 \cdot P = \$15$$
$$P = \$150.$$

To find the current discount, we calculate 30% of $150. Because $0.30 \cdot \$150 = \$45$, the amount of the 30% discount is $45.

In the *Looking Back* stage of problem solving, we check the answer and look for other ways to solve the problem. A different approach leads to a more efficient solution and confirms the answer. If 10% of the price is $15, then 30% of the price is 3 times $15, or $45.

Mental Math with Percents

Mental math may be helpful when working with percents. Two techniques follow:

1. *Using fraction equivalents*
   Knowing fraction equivalents for some percents can make some computations easier. Table 6-3 gives several fraction equivalents.

Table 6-3

Percent	25%	50%	75%	$33\frac{1}{3}\%$	$66\frac{2}{3}\%$	10%	1%
Fraction Equivalent	$\frac{1}{4}$	$\frac{1}{2}$	$\frac{3}{4}$	$\frac{1}{3}$	$\frac{2}{3}$	$\frac{1}{10}$	$\frac{1}{100}$

These equivalents can be used in such computations as the following:

$$50\% \text{ of } \$80 = \left(\frac{1}{2}\right)80 = \$40$$

$$66\frac{2}{3}\% \text{ of } 90 = \left(\frac{2}{3}\right)90 = 60$$

2. *Using a known percent*
   Frequently, we may not know a percent of something, but we know a close percent of it. For example, to find 55% of 62, we might do the following:

$$50\% \text{ of } 62 = \left(\frac{1}{2}\right)(62) = 31$$

$$5\% \text{ of } 62 = \left(\frac{1}{2}\right)(10\%)(62) = \left(\frac{1}{2}\right)(6.2) = 3.1$$

Adding, we see that 55% of 62 is $31 + 3.1 = 34.1$.

Estimations with Percents

Estimations with percents can be used to determine whether answers are reasonable. Following are two examples:

1. To estimate 27% of 598, note that 27% of 598 is a little more than 25% of 598, but 25% of 598 is approximately the same as 25% of 600, or $\frac{1}{4}$ of 600, or 150. Here, we have adjusted 27% downward and 598 upward, so 150 should be a reasonable estimate. A better estimate might be obtained by estimating 30% of 600 and then subtracting 3% of 600 to obtain 27% of 600, giving $180 - 18$, or 162.
2. To estimate 148% of 500, note that 148% of 500 should be slightly less than 150% of 500. 150% of 500 is $1.5(500) = 750$. Thus 148% of 500 should be a little less than 750.

Example 6-20 Laura wants to buy a blouse originally priced at $26.50 but now on sale at 40% off. She has $17 in her wallet and wonders if she has enough cash. How can she mentally find out?

Solution It is easier to find 40% of $25 (versus $26.50) mentally. One way is to find 10% of $25, which is $2.50. Now, 40% is 4 times that much, that is, $4 \cdot \$2.50$, or $10. Thus Laura estimates that the blouse will cost $26.50 - \$10$, or $16.50. Since the actual discount is greater than $10 (40% of 26.50 is greater than 40% of 25), Laura will have to pay less than $16.50 for the blouse and, hence, she has enough cash.

Sometimes it may not be clear which operations to perform with percent. The following example investigates this.

Example 6-21 Which of the following statements are true and which are false? Explain your answers.

   a. Leonardo got a 10% raise at the end of his first year on the job and a 10% raise after another year. His total raise was 20% of his original salary.
   b. Jung and Dina paid 45% of their first department store bill of $620 and 48% of the second department store bill of $380. They paid $45\% + 48\% = 93\%$ of the total bill of $1000.

c. Bill spent 25% of his salary on food and 40% on housing. Bill spent 25% + 40% = 65% of his salary on food and housing.
d. In Bordertown, 65% of the adult population works in town, 25% works across the border, and 15% is unemployed.
e. In Clean City, the fine for various polluting activities is a certain percentage of one's monthly income. The fine for smoking in public places is 40%, for driving a polluting car is 50%, and for littering is 30%. Mr. Schmutz committed all three polluting crimes in one day and paid a fine of 120% of his monthly salary.

Solution
a. In applications, percent has meaning only when it represents part of a quantity. For example, 10% of a quantity and another 10% of the same quantity is 20% of that quantity. In Leonardo's case, the first 10% raise was calculated based on his original salary and the second 10% raise was calculated on his new salary. Consequently, the percentages cannot be added, and the statement is false.
b. The answer does not make sense. Jung and Dina paid less than $\frac{1}{2}$ of each bill, so they could not have paid 93% (almost all) of the total. In fact, $\frac{1}{2}$ of one bill plus $\frac{1}{2}$ of the other bill is not $\frac{1}{2} + \frac{1}{2}$, or 1, the full amount of the total bill, because the bills are different.
c. Because the percentages are of the same quantity, the statement is true.
d. Because the percentages are of the same quantity, that is, the number of adults, we can add them: 65% + 25% + 15% = 105%. But 105% of the population accounts for more (5% more) than the town's population, which is impossible. Hence, the statement is false.
e. Again, the percentages are of the same quantity, that is, the individual's monthly income. Hence, we can add them: 120% of one's monthly income is a stiff fine, but possible.

ONGOING ASSESSMENT 6-4

1. Express each of the following as percents:
   a. 7.89   b. 0.032   c. 193.1   d. 0.2
   e. $\frac{5}{6}$ f. $\frac{3}{20}$ g. $\frac{1}{8}$ h. $\frac{3}{8}$
   i. $\frac{5}{8}$ j. $\frac{1}{6}$ k. $\frac{4}{5}$ l. $\frac{1}{40}$

2. Convert each of the following percents to decimals:
   a. 16%   b. $4\frac{1}{2}\%$ c. $\frac{1}{5}\%$ d. $\frac{2}{7}\%$
   e. $13\frac{2}{3}\%$ f. 125%   g. $\frac{1}{3}\%$ h. $\frac{1}{4}\%$

3. Fill in the following blanks to find other expressions for 4%:
   a. _____ for every 100
   b. _____ for every 50
   c. 1 for every _____
   d. 8 for every _____
   e. 0.5 for every _____

4. Different calculators compute percents in various ways. To investigate this, consider 5 · 6%.
   a. If the following sequence of keys is pressed, is the correct answer of 0.3 displayed on your calculator?
   $\boxed{5}\ \boxed{\times}\ \boxed{6}\ \boxed{\%}\ \boxed{=}$
   b. Press $\boxed{6}\ \boxed{\%}\ \boxed{\times}\ \boxed{5}\ \boxed{=}$. Is the answer 0.3?

5. Answer each of the following:
   a. What is 6% of 34?
   b. 17 is what percent of 34?
   c. 18 is 30% of what number?
   d. What is 7% of 49?
   e. 61.5 is what percent of 20.5?
   f. 16 is 40% of what number?

6. Marc had 84 boxes of candy to sell. He sold 75% of the boxes. How many did he sell?

7. Gail made $16,000 last year and received a 6% raise. How much does she make now?
8. Gail received a 7% raise last year. If her salary is now $27,285, what was her salary last year?
9. Joe sold 180 newspapers out of 200. Bill sold 85% of his 260 newspapers. Ron sold 212 newspapers, 80% of those he had.
   a. Who sold the most newspapers? How many?
   b. Who sold the greatest percentage of his newspapers? What percent?
   c. Who started with the greatest number of newspapers? How many?
10. If a dress that normally sells for $35 is on sale for $28, what is the "percent off"? (This could be called a *percent of decrease,* or a *discount.*)
11. A used car originally cost $1700. One year later, it was worth $1400. What is the percentage of depreciation?
12. On a certain day in Glacier Park, 728 eagles were counted. Five years later, 594 were counted. What was the percentage of decrease in the number of eagles counted?
13. Mort bought his house in 1975 for $59,000. It was recently appraised at $95,000. What is the *percent of increase* in value?
14. Xuan weighed 9 lb when he was born. At 6 mo, he weighed 18 lb. What was the percent of increase in Xuan's weight?
15. Sally bought a dress marked 20% off. If the regular price was $28.00, what was the sale price?
16. What is the sale price of a softball if the regular price is $6.80 and there is a 25% discount?
17. If a $\frac{1}{4}$-c serving of Crunchies breakfast food has 0.5% of the minimum daily requirement of vitamin C, how many cups would you have to eat to obtain the minimum daily requirement of vitamin C?
18. An airline ticket costs $320 without the tax. If the tax rate is 5%, what is the total bill for the airline ticket?
19. Bill got 52 correct answers on an 80-question test. What percent of the questions did he answer incorrectly?
20. A real estate broker receives 4% of an $80,000 sale. How much does the broker receive?
21. A survey reported that $66\frac{2}{3}$% of 1800 employees favored a new insurance program. How many employees favored the new program?
22. Mentally tell which can be represented by the greater percent: 325/500 or 600/1000. How can you tell?
23. a. How can an estimate of 10% of a number help you estimate 35% of the number?
    b. Mentally compute 35% of $8.00.
24. If 30 is 150% of a number, is the number greater than or less than 30? Why?
25. What is 40% of 50% of a number?
26. If you add 20% of a number to the number itself, what percent of the result would you have to subtract to get the original number back?

27. An advertisement reads that if you buy ten items you get 20% off your total purchase price. You need 8 items that cost $9.50 each.
    a. How much would 8 items cost? 10 items?
    b. Is it more economical to buy 8 items or 10 items?
28. Soda is advertised at 45¢ a can or $2.40 a six-pack. If 6 cans are to be purchased, what percent is saved by purchasing the six-pack?
29. John paid $330 for a new mountain bicycle to sell in his shop. He wants to price it so that he can offer a 10% discount and still make 20% of the price he paid for it. At what price should the bike be marked?
30. The price of a suit that sold for $200 was reduced by 25%. By what percent must the price of the suit be increased to bring the price back to $200?
31. The car Elsie bought 1 yr ago has depreciated by $1116.88, which is 12.13% of the price she paid for it. How much did she pay for the car, to the nearest cent?
32. Solve each of the following using mental mathematics:
    a. 15% of $22   b. 20% of $120
    c. 5% of $38   d. 25% of $98
33. If we build a 10 × 10 model with blocks, as shown in the following figure, and paint the entire model, what percent of the cubes will have each of the following?
    a. Four faces painted
    b. Three faces painted
    c. Two faces painted

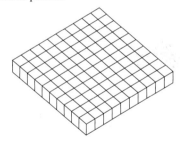

34. For people to be safe but still achieve a cardiovascular training effect, they should monitor their heart rates while exercising. The maximum heart rate can be approximated by subtracting your age from 220. You can achieve a safe training effect if you maintain your heart rate between 60% and 80% of that number for at least 20 min three times a week.
    a. Determine the range for your age.
    b. At the top of a long hill, Jeannie slows her bike and takes her pulse. She counts 41 beats in 15 sec.
        i. Express in decimal form the amount of time in seconds between successive beats.
        ii. Express the amount in terms of minutes.
35. A crew consists of 1 apprentice, 1 journeyman, and 1 master carpenter. The crew receives a check for $4200 for a job they just finished. A journeyman makes 200% of what an apprentice makes, and a master makes 150% of what a journeyman makes. How much does each person in the crew earn?

36. **a.** In an incoming freshman class of 500 students, only 20 claimed to be math majors. What percent of the freshman class is this?
    b. When the survey was repeated the next year, 5% of non-math majors had decided to switch and become math majors.
        i. How many math majors are there now?
        ii. What percent of the freshman class do they represent?
37. Ms. Price has received a 10% raise in salary in each of the last 2 yr. If her annual salary this year is $100,000, what was her salary 2 yr ago, rounded to the nearest penny?

Communication

38. Is 40% of 30 equal to 30% of 40? Explain why or why not.
39. Compute 20% of 120 in two ways.
40. Why is it possible to have an increase of 150% in a price but not a 150% decrease in price?
41. A student asks if 90% means 90 out of 100, how can she possibly score 90% on a test that has only 20 questions. How do you answer her?
42. Two equal amounts of money were invested in two different stocks. The value of the first stock increased by 15% the first year and then decreased by 15% the second year. The second stock decreased by 15% the first year and increased by 15% the second year. Was one investment better than the other? Explain your reasoning.

Open-Ended

43. Write and solve a word problem whose solution involves the following. If one of these tasks is impossible, explain why.
    a. Addition of percent
    b. Subtraction of percent
    c. Multiplication of percent
    d. Division of percent
    e. A percent whose decimal representation is raised to the second power
    f. A percent greater than 100
44. Look at newspapers and magazines for information given in percents.
    a. Based on your findings, write a problem that involves social science as well as mathematics.
    b. Write a clear solution to your problem in (a).

Cooperative Learning

45. Find the percentage of students in your class that engage in each of the following activities:
    a. Studying and doing homework
    b. Watching TV

Number of Hours per Week (h)	Percent
$h < 1$	
$1 \leq h < 3$	
$3 \leq h < 5$	
$5 \leq h < 10$	
$h \geq 10$	
Total	

Number of Hours per Week (h)	Percent
$h < 1$	
$1 \leq h < 5$	
$5 \leq h < 10$	
$h \geq 10$	
$h \geq 10$	
Total	

c. Did your totals add up to 100%? Why or why not?

Review Problems

46. **a.** Find $0.8 \div 0.32$ by converting these numbers to rational numbers in the form $\frac{a}{b}$ where a and b are integers, dividing the rational numbers, and then changing the quotient back to decimal form.
    b. Find $0.8 \div 0.32$ using decimal division and compare your answer with the answer from part (a).
47. Change $7.27\overline{1}$ to a rational number in the form $\frac{a}{b}$ where a and b are integers.
48. Find a repeating decimal between 0.2 and $0.\overline{2}$.
49. Order the following from least to greatest:
    $2.5, \quad 5/3, \quad 2.0\overline{5}, \quad 2.\overline{15}, \quad 7/3$
50. Show that $0.\overline{9} = 1$ in three ways.

BRAIN TEASER The crust of a certain pumpkin pie is 25% of the pie. By what percent should the amount of crust be reduced in order to make it constitute 20% of the pie?

TECHNOLOGY CORNER

We can use spreadsheets to solve mixture problems. For example, consider the problem of finding out how many liters of water must be added to 5 L of pure lemon juice to change its concentration from 100% to less than 30% lemon juice.

Six lemonade mixtures were prepared starting from 5 L of pure lemon juice and adding water in 2 L increments. At each step, the percent of lemon juice in the mixture was calculated. The results of the process are summarized in the spreadsheet in Figure 6-5. The formulas used to obtain the results in a particular column are given in row 12.

a. Explain how this spreadsheet can be used to help students solve the problem.
b. Explain the formulas in row 12.

Figure 6-5

	A	B	C	D
1	Liters of	Liters of	Total Liters	% Lemon Juice
2	Lemon Juice	Water Added	in Mixture	in Mixture
3	(L)	(L)	(L)	
4	5	0	5	100.00
5	5	2	7	71.43
6	5	4	9	55.56
7	5	6	11	45.45
8	5	8	13	38.46
9	5	10	15	33.33
10	5	12	17	29.41
11				
12	5	x	$5 + x$	$5/(5 + x)*100$
13		(where x is a multiple of 2)		

*Section 6-5 — Computing Interest

interest

principal · interest rate

simple interest

When a bank advertises a $5\frac{1}{2}$% interest rate on a savings account, the **interest** is the amount of money the bank will pay for using that money. The original amount deposited or borrowed is the **principal**. The percent used to determine the interest is the **interest rate.** Interest rates are given for specific periods of time, such as years, months, or days. Interest computed on the original principal is **simple interest.** For example, suppose we borrow $5000 from a company at a simple interest rate of 9% for 1 yr. The interest we owe on the loan for 1 yr is 9% of $5000, or $5000 · 0.09. In general, if a principal P is invested at an annual interest rate of r, then the simple interest after 1 yr is $Pr · 1$; after t years, it is Prt. Thus if I represents simple interest, we have

$$I = Prt.$$

amount/balance

The amount needed to pay off a $5000 loan at 9% simple interest is the $5000 borrowed plus the interest on the $5000, that is, $5000 + 5000 \cdot 0.09$, or $5450. In general, *an **amount** (or **balance**) A is equal to the principal P plus the interest I*, that is,

$$A = P + I = P + Prt = P(1 + rt).$$

Example 6-22 Vera opened a savings account that pays simple interest at the rate of $5\frac{1}{4}\%$ per year. If she deposits $2000 and makes no other deposits, find the interest and the final amount for the following time periods:

a. 1 yr   **b.** 90 days

Solution   **a.** To find the interest for 1 yr, we proceed as follows:

$$I = \$2000 \cdot 5\frac{1}{4}\% \cdot 1 = \$2000 \cdot 0.0525 \cdot 1 = \$105$$

Her final amount at the end of 1 yr is

$$\$2000 + \$105 = \$2105.$$

b. When the interest rate is annual and the interest period is given in days, we represent the time as a fractional part of a year by dividing the number of days by 365. Thus

$$I = \$2000 \cdot 5\frac{1}{4}\% \cdot \frac{90}{365}$$

$$= \$2000 \cdot 0.0525 \cdot \frac{90}{365} \doteq \$25.89.$$

Hence,

$$A \doteq \$2000 + \$25.89$$
$$A \doteq \$2025.89.$$

Thus Vera's amount after 90 days is approximately $2025.89.

Example 6-23 Find the annual interest rate if a principal of $10,000 increased to $10,900 at the end of 1 yr.

Solution   Let the annual interest be $x\%$. We know that $x\%$ of $10,000 is the increase. Because the increase is $10,900 − $10,000 = $900, we use the strategy of *writing an equation* for x as follows:

$$x\% \text{ of } 10{,}000 = 900$$

$$\frac{x}{100} \cdot 10{,}000 = 900$$

$$x = 9$$

Thus the interest is 9%. We can also solve this problem mentally by asking, "What percent of 10,000 is 900?" Because 1% of 10,000 is 100, to obtain 900, we take 9% of 10,000.

Compound Interest

compound interest

In business transactions, interest is sometimes calculated daily (365 times a year). In the case of savings, the earned interest is added daily to the principal, and each day the interest is earned on a different amount; that is, it is earned on the previous interest as well as the principal. When interest is computed in this way, it is called **compound interest.** Compounding usually is done annually (once a year), semiannually (twice a year), quarterly (4 times a year), or monthly (12 times a year). However, even when the interest is compounded, it is given as an annual rate. For example, if the annual rate is 6% compounded monthly, the interest per month is $\frac{6}{12}$%, or 0.5%. If it is compounded daily, the interest per day is $\frac{6}{365}$%. In general, *the interest rate per period is the annual interest rate divided by the number of periods in a year.*

We can use a spreadsheet to compare various compound interest rates. Work through the student page from *Scott Foresman-Addison Wesley Middle School Math, Course 3,* 1999 on the following page and answer the questions in the *TRY IT* and the *ON YOUR OWN* sections.

Example 6-24

If you invest $100 at 8% compounded quarterly, how much will you have in the account after 1 yr?

Solution The quarterly interest rate is $\frac{1}{4} \cdot 8\%$, or 2%. It seems that we would have to calculate the interest four times. But we can also reason as follows. If at the beginning of any of the four periods there are x dollars in the account, at the end of that period there will be

$$x + 2\% \text{ of } x = x + 0.02x$$
$$= x(1 + 0.02)$$
$$= x \cdot 1.02 \text{ dollars.}$$

Hence, to find the amount at the end of any period, we need only multiply the amount at the beginning of the period by 1.02. From Table 6-4, we see that the amount at the end of the fourth period is $100 \cdot 1.02^4$. On a scientific calculator, we can find the amount using $\boxed{1}\boxed{0}\boxed{0}\boxed{\times}\boxed{1}\boxed{.}\boxed{0}\boxed{2}\boxed{y^x}\boxed{4}\boxed{=}$. The calculator displays 108.24322. Thus the amount at the end of 1 yr is approximately $108.24.

Table 6-4

Period	Initial Amount	Final Amount
1	100	$100 \cdot 1.02$
2	$100 \cdot 1.02$	$(100 \cdot 1.02) \cdot 1.02$ or $100 \cdot 1.02^2$
3	$100 \cdot 1.02^2$	$(100 \cdot 1.02^2) \cdot 1.02$ or $100 \cdot 1.02^3$
4	$100 \cdot 1.02^3$	$(100 \cdot 1.02^3) \cdot 1.02$ or $100 \cdot 1.02^4$

In Example 6-24, finding the final amount at the end of the nth period amounts to finding the nth term of a geometric sequence whose first term is $100 \cdot 1.02$ (amount at the end of the first period) and whose ratio is 1.02. Thus, the amount at the end of the nth period is given by $(100 \cdot 1.02) \cdot (1.02)^{n-1} = 100 \cdot 1.02^n$. We can generalize this discussion. If the

TECHNOLOGY

Using a Spreadsheet • Compound Interest

Problem: Which investment strategy will cause your $100 investment to increase the most in 4 years: if you earn 4% interest compounded annually or if you earn 3.75% interest compounded monthly?

A spreadsheet can help you find the answer to this problem.

① Enter the following information in your spreadsheet as shown.

	A	B	C	D	E	F
1	Year	Amount	Annual Rate	Month	Amount	Monthly Rate
2	0	100		0	100	

② Enter the following formulas.
In cell C2, enter =.04.
In cell F2, enter =.0375/12.
In cell A3, enter =A2+1.
In cell B3, enter =B2+B2*C$2.
In cell D3, enter =D2+1.
In cell E3, enter =E2+E2*F$2.

	A	B	C	D	E	F
1	Year	Amount	Annual Rate	Month	Amount	Monthly Rate
2	0	100	0.04	0	100	0.003125
3	1	104		1	100.3125	

③ Select cells A3 to F50 and use the **Fill Down** command.

	A	B	C	D	E	F
1	Year	Amount	Annual Rate	Month	Amount	Monthly Rate
2	0	100	0.04	0	100	0.003125
3	1	104		1	100.3125	
4	2	108.16		2	100.6259	
5	3	112.4864		3	100.9404	
6	4	116.9858		4	101.2558	
50		48 657.0528			48 116.1562	

Solution: $100 at 4% interest compounded annually is $116.99; $100 at 3.75% interest compounded monthly is only $116.16.

TRY IT

Which investment strategy will cause your $100 investment to increase the most in 4 years: if you earn 5% interest compounded annually or if you earn 5% interest compounded quarterly (4 times a year)?

ON YOUR OWN

▶ Which is easier, computing compound interest by calculator or by using a spreadsheet? Explain.

▶ Why must you divide the interest rate by the number of compounding periods?

▶ Why do you have to enter formulas by using the "=" symbol?

principal is P and the interest rate per period is r, then the amount A after n periods is $P(1 + r) \cdot (1 + r)^{n-1}$ or $P(1 + r)^n$. Therefore, we have a formula for computing the amount at the end of the nth period, namely $A = P(1 + r)^n$.

Example 6-25 Suppose you deposit $1000 in a savings account that pays 6% interest compounded quarterly.

a. What is the balance at the end of 1 yr?
b. What is the *effective annual yield* on this investment; that is, what is the rate that would have been paid if the amount had been invested using simple interest?

Solution a. An annual interest rate of 6% earns $\frac{1}{4}$ of 6%, or an interest rate of $\frac{0.06}{4}$, in 1 quarter. Because there are 4 periods, we have the following:

$$A = 1000\left(1 + \frac{0.06}{4}\right)^4 \doteq \$1061.36$$

The balance at the end of 1 yr is approximately $1061.36.

b. Because the interest earned is $1061.36 − $1000.00 = $61.36, the effective annual yield can be computed by using the simple interest formula, $I = Prt$.

$$61.36 = 1000 \cdot r \cdot 1$$
$$\frac{61.36}{1000} = r$$
$$0.06136 = r$$
$$6.136\% = r$$

The effective annual yield is 6.136%.

Example 6-26 To save for their child's college education, a couple deposits $3000 into an account that pays 7% annual interest compounded daily. Find the amount in this account after 8 yr.

Solution The principal in the problem is $3000, the daily rate i is 0.07/365, and the number of compounding periods is 8 · 365, or 2920. Thus we have

$$A = \$3000\left(1 + \frac{0.07}{365}\right)^{2920} \doteq \$5251.74$$

Thus the amount in the account is approximately $5251.74.

ONGOING ASSESSMENT 6-5

You will need a calculator to do most of the following problems.

1. Complete the following compound-interest chart.

Compounding Period	Principal	Annual Rate	Length of Time (Years)	Interest Rate per Period	Number of Periods	Amount of Interest Paid
a. Semiannual	$1000	6%	2			
b. Quarterly	$1000	8%	3			
c. Monthly	$1000	10%	5			
d. Daily	$1000	12%	4			

2. Ms. Jackson borrowed $42,000 at 13% annual simple interest to buy her house. If she won the Irish Sweepstakes exactly 1 yr later and was able to repay the loan without penalty, how much interest would she owe?

3. Carolyn went on a shopping spree with her Bankamount card and made purchases totaling $125. If the interest rate is 1.5% per month on the unpaid balance and she does not pay this debt for 1 yr, how much interest will she owe at the end of the year?

4. A man collected $28,500 on a loan of $25,000 he made 4 yr ago. If he charged simple interest, what was the rate he charged?

5. Burger Queen will need $50,000 in 5 yr for a new addition. To meet this goal, the company deposits money in an account today that pays 9% annual interest compounded quarterly. Find the amount that should be invested to total $50,000 in 5 yr.

6. A company is expanding its line to include more products. To do so, it borrows $320,000 at 13.5% annual simple interest for a period of 18 mo. How much interest must the company pay?

7. To save for their retirement, a couple deposits $4000 in an account that pays 9% interest compounded quarterly. What will be the value of their investment after 20 yr?

8. A car company is offering car loans at a simple-interest rate of 9%. Find the interest charged to a customer who finances a car loan of $7200 for 3 yr.

9. Johnny and Carolyn have three savings plans, which accumulated the following amounts of interest for 1 yr.
   a. A passbook savings account that accumulated $53.90 on a principal of $980
   b. A certificate of deposit that accumulated $55.20 on a principal of $600
   c. A money-market certificate that accumulated $158.40 on a principal of $1200
   Which of these accounts paid the best interest rate for the year?

10. A hamburger costs $1.35 and the price continues to rise at a rate of 11% a year for the next 6 yr. What will the price of a hamburger be at the end of 6 yr?

11. If college tuition is $10,000 this year, what will it be 10 yr from now, assuming a constant inflation rate of 9% a year?

12. Sara invested money at a bank that paid 6.5% compounded quarterly. If she had $4650 at the end of 4 yr, what was her initial investment?

13. Adrien and Jarrell deposit $300 on January 1 in a holiday savings account that pays 1.1% per month interest and they withdraw the money on December 1 of the same year. What is the effective annual yield?

14. The number of trees in a rain forest decreases each month by 0.5%. If the forest has approximately $2.34 \cdot 10^9$ trees, how many trees will be left after 20 yr?

15. An amount of $3000 was deposited in a bank at a rate of 5% compounded quarterly for 3 yr. The rate then increased to 8% and was compounded quarterly for the next 3 yr. If no money was withdrawn, what was the balance at the end of this time? Explain your reasoning.

16. A money-market fund pays 14% annual interest compounded daily. What is the value of $10,000 invested in this fund after 15 yr? Explain your solution.

17. The New Age Savings Bank advertises 9% interest rates compounded daily, while the Pay More Bank pays 10.5% interest compounded annually. Which bank offers a better rate for a customer who plans to leave her money in for exactly 1 yr? Justify your answer.

18. A car is purchased for $15,000. If each year the car depreciates by 10% of its value the preceding year, what will its value be at the end of 3 yr? Explain your reasoning.

Communication

19. Because of a recession, the value of a new house depreciated 10% each year for 3 yr in a row. Then, for the next 3 yr, the value of the house increased 10% each year. Did the value of the house increase or decrease after 6 yr? Explain.

20. Determine the number of years (to the nearest tenth) it would take for any amount of money to double if it were deposited at a 10% interest rate compounded annually. Explain your reasoning.

21. A car parts manufacturer advertised three devices that could be installed in a car to save gas. The first could save 15% on fuel, the second 35%, and the third 50%. Explain whether you could conclude that when all three devices were installed, the savings on fuel would be 100%?

22. Each year a car's value depreciated 20% from the previous year. Mike claims that after 5 years the car would depreciate 100% and would not be worth anything. Is Mike correct? Explain why or why not. If not, find the actual percent the car would depreciate after 5 years.

Open-Ended

23. The effect of depreciation can be computed using a formula similar to the formula for compound interest.
    a. Assume depreciation is the same each month. Write a problem involving depreciation and solve it.
    b. Develop a general formula for depreciation defining what each variable in the formula stands for.

24. Find four large cities around the world and an approximate percentage rate of population growth for the countries in which the cities are located. Estimate the population in each of the four cities in 25 yr.

25. State different situations that do not involve money in which a formula like the one for compound interest is used. In each case, state a related problem and write its solution.

Cooperative Learning

26. The federal *Truth in Lending Act,* passed in 1969, requires lending institutions to quote an annual percentage rate (APR) that helps consumers compare the true cost of loans

regardless of how each lending institution computes the interest and adds on costs.
   a. Call different banks and ask for their APR on some loans and the meaning of APR.
   b. Based on your findings in (a), write a clear definition of APR.
   c. Use the information given by your credit card (you may need to call the bank) and compute the APR on cash advances. Is your answer the same as that given by the bank? Compare the APR for different credit cards.

Section 6-6 — Real Numbers

Every rational number can be expressed either as a repeating decimal or as a terminating decimal. The ancient Greeks discovered numbers that are not rational. Such numbers must have a decimal representation that neither terminates nor repeats. To find such decimals, we focus on the characteristics they must have:

1. There must be an infinite number of nonzero digits to the right of the decimal point.
2. There cannot be a repeating block of digits (a repetend).

One way to construct a nonterminating, nonrepeating decimal is to devise a pattern of infinite digits in such a way that there will definitely be no repeated block. Consider the number 0.1010010001... . If the pattern continues, the next groups of digits are four zeros followed by 1, five zeros followed by 1, and so on. It is possible to describe a pattern for this decimal, but there is no repeating block of digits. Because this decimal is nonterminating and nonrepeating, it cannot represent a rational number. Numbers that are not rational numbers are **irrational numbers.**

irrational numbers

π (pi)

In the mid-eighteenth century, it was proved that the ratio of the circumference of a circle to its diameter, symbolized by π **(pi)**, is an irrational number. The numbers $\frac{22}{7}$, 3.14, or 3.14159 are rational-number approximations of π. The value of π has been computed to billions of decimal places with no apparent pattern.

Square Roots

Irrational numbers occur in the study of area. For example, to find the area of a square, we use the formula $A = s^2$, where A is the area and s is the length of a side of the square. If a side of a square is 3 cm long, then the area of the square is 9 cm^2 (square centimeters). Conversely, we can use the formula to find the length of a side of a square, given its area. If the area of a square is 25 cm^2, then $s^2 = 25$, so $s = 5$ or $^-5$. Each of these solutions is a **square root** of 25. However, because lengths are always nonnegative, 5 is the only possible solution. The positive solution of $s^2 = 25$ (namely, 5) is the **principal square root** of 25 and is denoted by $\sqrt{25}$. Similarly, the principal square root of 2 is denoted by $\sqrt{2}$. Note that $\sqrt{16} \neq {^-4}$ because $^-4$ is not the principal square root of 16. Can you find $\sqrt{0}$?

square root
principal square root

Definition of the Principal Square Root

If a is any whole number, the **principal square root** of a is the nonnegative number b such that $b^2 = a$.

HISTORICAL NOTE

The discovery of irrational numbers by members of the Pythagorean Society (founded by Pythagoras) is one of the greatest events in the history of mathematics. This discovery was very disturbing to the Pythagoreans, who believed that everything depended on whole numbers, so they decided to keep the matter secret. One legend has it that Hippasus, a society member, was drowned because he relayed the secret to persons outside the society.

In 1525 Christoff Rudolff, a German mathematician, became the first to use the symbol $\sqrt{}$, for a radical or a root.

Example 6-27

Find the following:

a. The square roots of 144
b. The principal square root of 144
c. $\sqrt{\dfrac{4}{9}}$

Solution

a. The square roots of 144 are 12 and $^-12$.
b. The principal square root of 144 is 12.
c. $\sqrt{\dfrac{4}{9}} = \dfrac{2}{3}$.

Other Roots

We have seen that the positive solution to $x^2 = 25$ is denoted by $\sqrt{25}$. Similarly, the positive solution to $x^4 = 25$ is denoted $\sqrt[4]{25}$.

nth root
index

In general, if n is even the positive solution to $x^n = 25$ is $\sqrt[n]{25}$ and is the principal **nth root** of 25. The number n is the **index**. Note that in the expression $\sqrt{25}$, the index 2 is understood and not expressed. In general, the positive solution to $x^n = b$, where b is nonnegative, is $\sqrt[n]{b}$. Substituting $\sqrt[n]{b}$ for x in the equation $x^n = b$ gives the following:

$$(\sqrt[n]{b})^n = b.$$

If b is negative, $\sqrt[n]{b}$ may not be a real number. For example, consider $\sqrt[4]{-16}$. If $\sqrt[4]{-16} = x$, then $x^4 = {}^-16$. Because any nonzero real number raised to the fourth power is positive, there is no real-number solution to $x^4 = {}^-16$ and therefore $\sqrt[4]{-16}$ is not a real number. Similarly, if we are restricted to real numbers it is not possible to find *any* even root of a negative number. However, the value $^-2$ satisfies the equation $x^3 = {}^-8$. Hence, $\sqrt[3]{-8} = {}^-2$. *In general, the odd root of a negative number is a negative number.*

Because $\sqrt{a}$, if it exists, is positive by definition, $\sqrt{(^-3)^2} = \sqrt{9} = 3$ and not $^-3$. Many students think that $\sqrt{a^2}$ always equals a. This is true if $a \geq 0$, but false if $a < 0$. *In general*, $\sqrt{a^2} = |a|$. Similarly, $\sqrt[4]{a^4} = |a|$ and $\sqrt[6]{a^6} = |a|$, but $\sqrt[3]{a^3} = a$ (why?).

REMARK Notice that when n is even and $b > 0$, the equation $x^n = b$ has two real-number solutions, $\sqrt[n]{b}$ and $-\sqrt[n]{b}$. If n is odd, the equation has only one real-number solution, $\sqrt[n]{b}$ for any real number b.

Irrationality of Square Roots and Other Roots

Some square roots are rational numbers. Others, like $\sqrt{2}$, are irrational numbers. To see this, note that $1^2 = 1$ and $2^2 = 4$ and that there is no whole number s such that $s^2 = 2$. Is there a rational number $\frac{a}{b}$ such that $\left(\frac{a}{b}\right)^2 = 2$? We use the strategy of *indirect reasoning*. If we assume there is such a rational number, then the following must be true:

$$\left(\frac{a}{b}\right)^2 = 2$$
$$\frac{a^2}{b^2} = 2$$
$$a^2 = 2b^2$$

If $a^2 = 2b^2$, then by the Fundamental Theorem of Arithmetic, the prime factorizations of a^2 and $2b^2$ are the same. In particular, the prime 2 appears the same number of times in the prime factorization of a^2 as it does in the factorization of $2b^2$. Because $b^2 = b \cdot b$, then no matter how many times 2 appears in the prime factorization of b, it appears twice as many times in $b \cdot b$. Also, a^2 has an even number of 2s for the same reason b^2 does. In $2b^2$, another factor of 2 is introduced, resulting in an odd number of 2s in the prime factorization of $2b^2$ and, hence, of a^2. But 2 cannot appear both an odd number of times and an even number of times in the same prime factorization of a^2. We have a contradiction. This contradiction could have been caused only by the assumption that $\sqrt{2}$ is a rational number. Consequently, $\sqrt{2}$ must be an irrational number. We can use a similar argument to show that $\sqrt{3}$ is irrational or $\sqrt{n}$ is irrational, where n is a whole number but not the square of another whole number.

Many irrational numbers can be interpreted geometrically. For example, we can find a point on a number line to represent $\sqrt{2}$ by using the **Pythagorean theorem.** That is, if a and b are the lengths of the shorter sides (legs) of a right triangle and c is the length of the longer side (hypotenuse), then $a^2 + b^2 = c^2$, as shown in Figure 6-6.

Figure 6-6

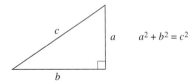

Figure 6-7 shows a segment one unit long constructed perpendicular to a number line at point P. Thus two sides of the triangle shown are each one unit long. If $a = b = 1$, then $c^2 = 2$ and $c = \sqrt{2}$. To find a point on the number line that corresponds to $\sqrt{2}$, we need to find a point Q on the number line such that the distance from 0 to Q is $\sqrt{2}$. Because $\sqrt{2}$ is the length of the hypotenuse, the point Q can be found by marking an arc with center 0 and radius c. The intersection of the positive number line with the arc is Q.

Figure 6-7

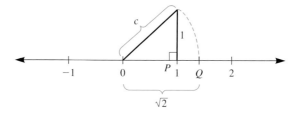

Similarly, other square roots can be constructed, as shown in Figure 6-8.

Figure 6-8

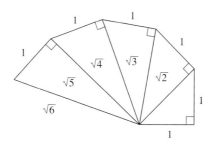

Estimating a Square Root

From Figure 6-7, we see that $\sqrt{2}$ must have a value between 1 and 2; that is, $1 < \sqrt{2} < 2$. To obtain a closer approximation of $\sqrt{2}$, we attempt to "squeeze" $\sqrt{2}$ between two numbers that are between 1 and 2. Because $(1.5)^2 = 2.25$ and $(1.4)^2 = 1.96$, it follows that $1.4 < \sqrt{2} < 1.5$. Because a^2 can be interpreted as the area of a square with side of length a, this discussion can be pictured geometrically, as in Figure 6-9.

Figure 6-9

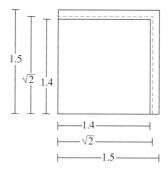

If we desire a more accurate approximation for $\sqrt{2}$, we can continue this squeezing process. We see that $(1.4)^2$, or 1.96, is closer to 2 than is $(1.5)^2$, or 2.25, so we choose numbers closer to 1.4 in order to find the next approximation. We find the following:

$$(1.42)^2 = 2.0164$$
$$(1.41)^2 = 1.9981$$

Thus $1.41 < \sqrt{2} < 1.42$. We can continue this process until we obtain the desired approximation. Note that if the calculator has a square-root key, we can obtain the approximation directly.

The System of Real Numbers

real numbers The set of **real numbers** R is the union of the set of rational numbers and the set of irrational numbers. Real numbers represented as decimals can be terminating, repeating, or nonterminating and nonrepeating.

Every integer is a rational number as well as a real number. Every rational number is a real number, but not every real number is rational, as has been shown with $\sqrt{2}$. The relationships among these sets of numbers are summarized in the Venn diagram in Figure 6-10, where the universe is the set of real numbers and the complement of the set of rationals is the set of irrational numbers.

Figure 6-10

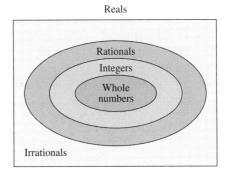

The concept of fractions can now be extended to include all numbers of the form $\frac{a}{b}$, where a and b are real numbers with $b \neq 0$, such as $\frac{\sqrt{3}}{5}$. Addition, subtraction, multiplication, and division are defined on the set of real numbers in such a way that all the properties of these operations on rationals still hold. The properties are summarized next.

Properties

Closure properties For real numbers a and b, $a + b$ and ab are unique real numbers.
Commutative properties For real numbers a and b, $a + b = b + a$ and $ab = ba$.
Associative properties For real numbers a, b, and c, $a + (b + c) = (a + b) + c$ and $a(bc) = (ab)c$.
Identity properties The number 0 is the unique additive identity and 1 is the unique multiplicative identity such that, for any real number a, $0 + a = a = a + 0$ and $1 \cdot a = a = a \cdot 1$.
Inverse properties (1) For every real number a, ^-a is its unique additive inverse; that is, $a + {^-a} = 0 = {^-a} + a$. (2) For every nonzero real number a, $\frac{1}{a}$ is its unique multiplicative inverse; that is, $a\left(\frac{1}{a}\right) = 1 = \left(\frac{1}{a}\right)a$.
Distributive property of multiplication over addition For real numbers a, b, and c, $a(b + c) = ab + ac$.
Denseness property For real numbers a and b, there exists a real number c such that $a < c < b$.

Radicals and Rational Exponents

Scientific calculators have a $\boxed{y^x}$ key with which we can find the values of expressions like $3.41^{2/3}$ and $4^{1/2}$. What does $4^{1/2}$ mean? By extending the properties of exponents previously developed for integer exponents, we have $4^{1/2} \cdot 4^{1/2} = 4^{1/2+1/2} = 4^1$. This implies that $(4^{1/2})^2 = 4$, or $4^{1/2}$, is a square root of 4. The number $4^{1/2}$ is assumed to be the principal square root of 4, that is, $4^{1/2} = \sqrt{4}$. In general, if x is a nonnegative real number, then $x^{1/2} = \sqrt{x}$. Similarly, $(x^{1/3})^3 = x^{(1/3) \cdot 3} = x^1$, and $x^{1/3} = \sqrt[3]{x}$. This discussion leads to the following:

1. $x^{1/n} = \sqrt[n]{x}$, where $\sqrt[n]{x}$ is meaningful.
2. $(x^m)^{1/n} = \sqrt[n]{x^m}$
3. $x^{m/n} = \sqrt[n]{x^m}$

More Properties of Exponents

The properties of integer exponents also hold for rational exponents. These properties are equivalent to the corresponding properties of radicals if the expressions involving radicals are meaningful.

Let r and s be any rational numbers, x and y be any real numbers, and n be any nonzero integer.

a. $x^{-r} = 1/x^r$.
b. $(xy)^r = x^r y^r$ implies $(xy)^{1/n} = x^{1/n} y^{1/n}$ and $\sqrt[n]{xy} = \sqrt[n]{x} \sqrt[n]{y}$.
c. $\left(\dfrac{x}{y}\right)^r = \dfrac{x^r}{y^r}$ implies $\left(\dfrac{x}{y}\right)^{1/n} = \dfrac{x^{1/n}}{y^{1/n}}$ and $\sqrt[n]{\dfrac{x}{y}} = \dfrac{\sqrt[n]{x}}{\sqrt[n]{y}}$.
d. $(x^r)^s = x^{rs}$ implies $(x^{1/n})^s = x^{s/n}$ and hence, $(\sqrt[n]{x})^s = \sqrt[n]{x^s}$.

REMARK The preceding properties can be used to write equivalent expressions for the roots of many numbers. For example, $\sqrt{96} = \sqrt{16 \cdot 6} = \sqrt{16} \cdot \sqrt{6} = 4 \cdot \sqrt{6}$. Similarly, $\sqrt[3]{54} = \sqrt[3]{27 \cdot 2} = \sqrt[3]{27} \cdot \sqrt[3]{2} = 3 \cdot \sqrt[3]{2}$.

Example 6-28 Simplify each of the following if possible.

a. $16^{1/4}$ b. $16^{5/4}$ c. $(-8)^{1/3}$ d. $125^{-4/3}$ e. $(-16)^{1/4}$

Solution
a. $16^{1/4} = (2^4)^{1/4} = 2^1 = 2$, or $16^{1/4} = \sqrt[4]{16} = 2$
b. $16^{5/4} = 16^{(1/4) \cdot 5} = (16^{1/4})^5 = 2^5 = 32$
c. $(-8)^{1/3} = ((-2)^3)^{1/3} = (-2)^1 = -2$ or $(-8)^{1/3} = \sqrt[3]{-8} = -2$
d. $125^{-4/3} = (5^3)^{-4/3} = 5^{-4} = 1/5^4 = 1/625$
e. Because every real number raised to the fourth power is positive, $\sqrt[4]{-16}$ is not a real number. Consequently, $(-16)^{1/4}$ is not a real number.

NOW TRY THIS 6-8

- Compute $\sqrt[8]{10}$ on a calculator using the following sequence of keys:

  $\boxed{10}\ \boxed{\sqrt{}}\ \boxed{\sqrt{}}\ \boxed{\sqrt{}}$

a. Explain why this approach works.
b. For what values of n can $\sqrt[n]{10}$ be compacted using only the $\boxed{\sqrt{}}$ key? Why?

ONGOING ASSESSMENT 6-6

1. Without using a radical sign, write an irrational number whose digits are twos and threes.
2. Use the Pythagorean theorem to find x.

   a. 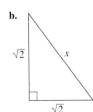   b. 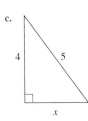   c. (triangle with legs 4 and x, hypotenuse 5)

3. Arrange the following real numbers in order from greatest to least:

   $$0.9,\ 0.\overline{9},\ 0.\overline{98},\ 0.9\overline{88},\ 0.9\overline{98},\ 0.\overline{898}$$

4. Determine which of the following represent irrational numbers:
   a. $\sqrt{51}$ b. $\sqrt{64}$ c. $\sqrt{324}$
   d. $\sqrt{325}$ e. $2 + 3\sqrt{2}$ f. $\sqrt{2} \div 5$

5. If possible find the square roots, correct to tenths, for each of the following without using a calculator:
   a. 225   b. 251   c. 169
   d. 512   e. $^-81$ f. 625

6. Find the approximate square roots for each of the following, rounded to hundredths, by using the squeezing method:
   a. 17   b. 7   c. 21
   d. 0.0120   e. 20.3   f. 1.64

7. Classify each of the following as true or false. If false, give a counterexample.
   a. The sum of any rational number and any irrational number is a rational number.
   b. The sum of any two irrational numbers is an irrational number.
   c. The product of any two irrational numbers is an irrational number.
   d. The difference of any two irrational numbers is an irrational number.

8. Find three irrational numbers between 1 and 3.
9. Find an irrational number between $0.\overline{53}$ and $0.\overline{54}$.
10. If R is the set of real numbers, Q is the set of rational numbers, I is the set of integers, W is the set of whole numbers, and S is the set of irrational numbers, find each of the following:
    a. $Q \cup S$ b. $Q \cap S$ c. $Q \cap R$
    d. $S \cap W$ e. $W \cup R$ f. $Q \cup R$

11. If the following letters correspond to the sets listed in problem 10, complete the following table by placing checkmarks in the appropriate columns.

	N	I	Q	R	S
a. 6.7			✓	✓	
b. 5					
c. $\sqrt{2}$					
d. $^-5$					
e. 3 1/7					

12. If the following letters correspond to the sets listed in problem 10, put a checkmark under each set of numbers for which a solution to the problem exists (N is the set of natural numbers):

	N	I	Q	R	S
a. $x^2 + 1 = 5$					
b. $2x - 1 = 32$					
c. $x^2 = 3$					
d. $x^2 = 4$					
e. $\sqrt{x} = ^-1$					
f. $\frac{3}{4}x = 4$					

13. Determine for what real values of x, if any, each of the following statements is true:
    a. $\sqrt{x} = 8$
    b. $\sqrt{x} = {}^-8$
    c. $\sqrt{-x} = 8$
    d. $\sqrt{-x} = {}^-8$
    e. $\sqrt{x} > 0$
    f. $\sqrt{x} < 0$
14. A diagonal brace is placed in a 4 ft × 5 ft rectangular gate. What is the length of the brace to the nearest tenth of a foot? (*Hint:* Use the Pythagorean theorem.)
15. Write each of the following square roots in the form $a\sqrt{b}$, where a and b are integers and b has the least value possible:
    a. $\sqrt{180}$
    b. $\sqrt{363}$
    c. $\sqrt{252}$
16. Write each of the following in the simplest form $a\sqrt[n]{b}$, where a and b are integers, $b > 0$, and b has the least value possible:
    a. $\sqrt[3]{-54}$
    b. $\sqrt[5]{96}$
    c. $\sqrt[3]{250}$
    d. $\sqrt[5]{-243}$
17. In each of the following geometric sequences, find the missing terms:
    a. 5, _, _, 10
    b. 2, _, _, _, 1
18. The following exponential function approximates the number of bacteria after t hours: $E(t) = 2^{10} \cdot 16^t$.
    a. What is the initial number of bacteria, that is, the number when $t = 0$?
    b. After $\frac{1}{4}$ hr, how many bacteria are there?
    c. After $\frac{1}{2}$ hr, how many bacteria are there?
19. Without using a calculator, determine which is greater in each of the following. Explain your reasoning.
    a. $\sqrt{3}$ or $\sqrt[3]{4}$
    b. $\sqrt[3]{3}$ or $\sqrt{2}$
20. Solve for x in the following, where x is a rational number:
    a. $3^x = 81$
    b. $4^x = 8$
    c. $128^{-x} = 16$
    d. $\left(\frac{4}{9}\right)^{3x} = \frac{32}{243}$
21. Classify each of the following numbers as rational or irrational:
    a. $\sqrt{2} - \frac{2}{\sqrt{2}}$
    b. $(\sqrt{2})^{-4}$
    c. $\frac{1}{1 + \sqrt{2}}$
    d. $\frac{4}{\sqrt{2}} - \sqrt{2}$
★22. Prove that $\sqrt{3}$ is irrational.
★23. Prove that if p is a prime number, then $\sqrt{p}$ is an irrational number.

Communication

24. Jim asked if $\sqrt{2}$ can be written as $\sqrt{2}/1$, why it isn't rational? How would you answer him?
25. Find the value of $\sqrt{3}$ on a calculator. Explain why this can't be the exact value of $\sqrt{3}$.
26. Is it true that $\sqrt{a + b} = \sqrt{a} + \sqrt{b}$? Explain.
27. Pi (π) is an irrational number. Could $\pi = \frac{22}{7}$? Why or why not?
28. Without using a calculator or doing any computation, determine if $\sqrt{13} = 3.60\overline{5}$. Explain why or why not.
29. Is $\sqrt{x^2 + y^2} = x + y$ for all values of x and y? Explain your reasoning.
30. Answer the following as being true sometimes, always, or never. Justify your answers.
    a. $\sqrt{a^2} = a$
    b. $\sqrt{({}^-x)^2} = {}^-x$
    c. $\sqrt{({}^-x)^2} = |x|$
    d. $\sqrt{(a + b)^2} = a + b$
    e. $\sqrt[4]{a^2} = \sqrt{a}$
31. Without using a calculator, arrange the following in increasing order. Explain your reasoning.
    $(4/25)^{-1/3}, (25/4)^{1/3}, (4/25)^{-1/4}$

Open-Ended

32. The sequence 1, 1.01, 1.001, 1.0001, ... is an infinite sequence of rational numbers.
    a. Write several other infinite sequences of rational numbers.
    b. Write several infinite sequences of irrational numbers.
33. a. Place five irrational numbers between 1/2 and 3/4.
    b. Write an infinite sequence of irrational numbers all of whose terms are between 1/2 and 3/4.

Cooperative Learning

34. Let each member of a group choose a number between 0 and 1 on a calculator and check what happens when the $\boxed{x^2}$ key is pressed in succession until it is clear that there is no reason to go on.
    a. Compare your answers and write a conjecture based on what you observe.
    b. Use other keys on the calculator in a similar way. Describe the process and state a corresponding conjecture.
    c. Why do you get the result you do?
35. A calculator displays the following: $(3.7)^{2.4} = 23.103838$. In your group, discuss the meaning of the expression $(3.7)^{2.4}$ in view of what you know about exponents. Compare your findings with those of other groups.

Review Problems

36. a. Human bones make up 0.18 of a person's body weight. How much do the bones of a 120-lb person weigh?
    b. Muscles make up about 0.4 of a person's body weight. How much do the muscles of a 120-lb person weigh?
37. Write each of the following decimals as rational numbers:
    a. 16.72
    b. 0.003
    c. ${}^-5.07$
    d. 0.123
38. Write a repeating decimal equal to each of the following without using more than one zero.
    a. 5
    b. 5.1
    c. $\frac{1}{2}$
39. Write 0.00024 as a fraction in simplest form.
40. Write $0.\overline{24}$ as a fraction in simplest form.
41. Write each of the following as a standard numeral:
    a. $2.08 \cdot 10^5$
    b. $3.8 \cdot 10^{-4}$

HINT FOR SOLVING THE PRELIMINARY PROBLEM

Since the player figures she will get at least 40 more attempts, she will end the season with at least 177 attempts. She has already made 137 − 23 = 114 shots. You need to determine if there is some number less than or equal to 40 such that when you add it to 114 and divide by 177 the percentage is greater than or equal to 77%.

QUESTIONS FROM THE CLASSROOM

1. A student claims that 0.36 is greater than 0.9 because 36 is greater than 9. How do you respond?
2. A student reports that ⁻438,340,000 cannot be written in scientific notation. How do you respond?
3. A student multiplies (6.5)(8.5) to obtain the following:

    $$\begin{array}{r} 8.5 \\ \times\ 6.5 \\ \hline 4\ 2\ 5 \\ 5\ 1\ 0 \\ \hline 5\ 5.2\ 5 \end{array}$$

    However, when the student multiplies $8\frac{1}{2} \cdot 6\frac{1}{2}$, she obtains the following:

    $$\begin{array}{r} 8\frac{1}{2} \\ \times\ 6\frac{1}{2} \\ \hline 4\frac{1}{4} \quad \left(\frac{1}{2} \cdot 8\frac{1}{2}\right) \\ 48 \quad (6 \cdot 8) \\ \hline 52\frac{1}{4} \end{array}$$

    How is this possible?

4. A student says that $3\frac{1}{4}\% = 0.03 + 0.25 = 0.28$. Is this correct? Why?
5. A student reports that it is impossible to mark a product up 150% because 100% of something is all there is. What is your response?
6. A student argues that a $p\%$ increase in salary followed by a $q\%$ decrease is equivalent to a $q\%$ decrease followed by a $p\%$ increase because of the commutative property of multiplication. How do you respond?
7. A student argues that 0.01% = 0.01 because in 0.01%, the percent is already written as a decimal. How do you respond?
8. A student tries to calculate $0.999^{10,000}$ on a calculator and finds the answer to be $4.65173346 \cdot 10^{-5}$. The student wonders how it could be that a number like 0.999 so close to 1 when raised to some power could result in a number close to 0. How do you respond?
9. Explain how you would respond to the following:
    a. A student claims that $\frac{9443}{9444}$ and $\frac{9444}{9445}$ are equal because both display 0.9998941 on his scientific calculator when the divisions are performed.
    b. Another student claims that the fractions are not equal and wants to know if there is any way the same calculator can determine which is greater.
10. Why is $\sqrt{25} \neq {}^-5$?
11. A student claims that $\sqrt{(^-5)^2} = {}^-5$ because $\sqrt{a^2} = a$. Is this correct?
12. Another student says that $\sqrt{(^-5)^2} = [(^-5)^2]^{1/2} = (^-5)^{2/2} = (^-5)^1 = {}^-5$. Is this correct?
13. A student claims that the equation $\sqrt{^-x} = 3$ has no solution, since the square root of a negative number does not exist. Why is this argument wrong?
14. A student claims that if the value of an item increases by 100% each year from its value the previous year and if the original price is d, then the value after n years will be $d \cdot 2^n$. Is the student correct? Why or why not?

CHAPTER REVIEW

1. **a.** On the number line, find the decimals that correspond to points A, B, and C.
   b. Indicate by D the point that corresponds to 0.09 and by E the point that corresponds to 0.15.

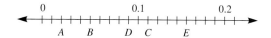

2. Write each of the following in expanded form:
   a. 32.012   **b.** 0.00103

3. Give a test to determine if a fraction can be written as a terminating decimal without one's actually performing the division. Explain why this test is valid.

4. A board is 442.4 cm long. How many shelves can be cut from it if each shelf is 55.3 cm long? (Disregard the width of the cuts.)

5. Write each of the following as a decimal:
   a. $\frac{4}{7}$ **b.** $\frac{1}{8}$ **c.** $\frac{2}{3}$ **d.** $\frac{5}{8}$

6. Write each of the following as a fraction in simplest form:
   a. 0.28   **b.** $^-5.07$ **c.** $0.\overline{3}$ **d.** $2.0\overline{8}$

7. Round each of the following numbers as specified:
   a. 307.625 to the nearest hundredth
   b. 307.625 to the nearest tenth
   c. 307.625 to the nearest unit
   d. 307.625 to the nearest hundred

8. Rewrite each of the following in scientific notation:
   a. 426,000   **b.** $324 \cdot 10^{-6}$
   c. 0.00000237   **d.** 0.325

9. Order each of the following decimals from greatest to least:
   $1.4\overline{519}$, $1.45\overline{19}$, 1.4519, $1.451\overline{9}$, $^-0.134$, $^-0.1\overline{3401}$, $0.1\overline{3401}$

10. Each of the following is a geometrical sequence. Find the missing terms.
    a. 5, _, 10   **b.** 1, _, _, _, 1/4

11. Write each of the following in scientific notation without using a calculator:
    a. 1783411.56   **b.** $347/10^8$ **c.** $49.3 \cdot 10^8$
    d. $29.4 \cdot 10^{12}/10^{-4}$ **e.** $0.47 \cdot 1000^{12}$ **f.** $3/5^9$

12. **a.** Find five decimals between 0.1 and 0.11 and order them from greatest to least.
    b. Find four decimals between 0 and 0.1 listed from least to greatest so that each decimal starting from the second is twice as large as the preceding one.
    c. Find four decimals between 0.1 and 0.2 and list them in increasing order so that the first one is halfway between 0.1 and 0.2, the second halfway between the first and 0.2, the third halfway between the second and 0.2, and similarly for the fourth one.

13. Answer each of the following:
    a. 6 is what percent of 24?
    b. What is 320% of 60?
    c. 17 is 30% of what number?
    d. 0.2 is what percent of 1?

14. Change each of the following to percents:
    a. $\frac{1}{8}$ **b.** $\frac{3}{40}$ **c.** 6.27
    d. 0.0123   **e.** $\frac{3}{2}$

15. Change each of the following percents to decimals:
    a. 60%   **b.** $\frac{2}{3}$%   **c.** 100%

16. Sandy received a dividend that equals 11% of the value of her investment. If her dividend was $1020.80, how much was her investment?

17. Five computers in a shipment of 150 were found to be defective. What percent of the computers were defective?

18. On a mathematics examination, a student missed 8 of 70 questions. What percent of the questions, rounded to the nearest tenth, did the student do correctly?

19. A microcomputer system costs $3450 at present. This is 60% of the cost 4 yr ago. What was the cost of the system 4 yr ago? Explain your reasoning.

20. If, on a purchase of one new suit, you are offered successive discounts of 5%, 10%, or 20% in any order you wish, what order should you choose?

21. Jane bought a bicycle and sold it for 30% more than she paid for it. She sold it for $104. How much did she pay for it?

*22. A company was offered a $30,000 loan at a 12.5% annual interest rate for 4 yr. Find the simple interest due on the loan at the end of 4 yr.

*23. A money-market fund pays 14% annual interest compounded quarterly. What is the value of a $10,000 investment after 3 yr?

24. Classify each of the following as rational or irrational (assume the patterns shown continue):
    a. 2.191199119991199991 19 …
    b. $\frac{1}{\sqrt{2}}$ **c.** $\frac{4}{9}$
    d. 0.0011001100110011 …
    e. 0.001100011000011 …

25. Write each of the following in the form $a\sqrt{b}$ or $a\sqrt[n]{b}$, where a and b are integers and b has the least value possible:
    a. $\sqrt{242}$ **b.** $\sqrt{288}$
    c. $\sqrt{360}$ **d.** $\sqrt[3]{162}$

26. Answer each of the following and explain your answers:
    a. Is the set of irrational numbers closed under addition?
    b. Is the set of irrational numbers closed under subtraction?
    c. Is the set of irrational numbers closed under multiplication?
    d. Is the set of irrational numbers closed under division?

27. Find an approximation for $\sqrt{23}$ correct to three decimal places without using the $\boxed{y^x}$ or the $\boxed{\sqrt{}}$ keys.

CHAPTER OUTLINE

I. Decimals
   A. Every rational number can be represented as a terminating or repeating decimal.
   B. A rational number $\frac{a}{b}$, in simplest form whose denominator is of the form $2^m \cdot 5^n$, where m and n are whole numbers, can be expressed as a **terminating decimal.**
   C. A **repeating decimal** is a decimal with a block of digits, called the **repetend,** that repeat infinitely many times.
   D. A number is in **scientific notation** if it is written as the product of a number n that is greater than or equal to 1 and an integral power of 10.

II. Percent and interest
   A. **Percent** means *per hundred.* Percent is written using the % symbol: $x\% = \frac{x}{100}$.
   *B. **Simple interest** is computed using the formula $I = Prt$, where I is the interest, P is the principal, r is the annual interest rate, and t is the time in years.
   *C. When **compound interest** is involved, we use the formula $A = P(1 + i)^n$, where A is the balance, P is the principal, i is the interest rate per period, and n is the number of periods.

III. Real numbers
   A. An **irrational number** is represented by a nonterminating, nonrepeating decimal.
   B. The set of **real numbers** is the set of all decimals, namely, the union of the set of rational numbers and the set of irrational numbers.
   C. If a is any whole number, then the **principal square root** of a, denoted by $\sqrt{a}$, is the nonnegative number b such that $b \cdot b = b^2 = a$.
   D. Square roots and nth roots can be found by using the **squeezing method.**

IV. Radicals and rational exponents
   A. $\sqrt[n]{x}$, or $x^{1/n}$, is the **nth root** of x and n is the **index.**
   B. The following properties hold for radicals if the expressions involving radicals are meaningful:
   a. $\sqrt[n]{xy} = \sqrt[n]{x} \cdot \sqrt[n]{y}$
   b. $\sqrt[n]{\frac{x}{y}} = \frac{\sqrt[n]{x}}{\sqrt[n]{y}}$
   c. $(\sqrt[n]{x})^m = \sqrt[n]{x^m}$

SELECTED BIBLIOGRAPHY

Bennett, A., and T. Nelson. "A Conceptual Model for Solving Percent Problems." *Mathematics Teaching in the Middle School* 1 (April, 1994): 20–25.

Coburn, T. "Percentage and the Hand Calculator." *Mathematics Teacher* 79 (May 1986): 361–367.

Oppenheimer, L., and R. Hunting. "Relating Fractions and Decimals: Listening to Students Talk." *Mathematics Teaching in the Middle School* 4 (February, 1999): 318–321.

Rossini, B. "Using Percent Problems to Promote Critical Thinking." *Mathematics Teacher* 81 (January 1988): 31–34.

Thompson, C., and V. Walker. "Connecting Decimals and Other Mathematical Content." *Teaching Children Mathematics* 2 (April 1996): 496–502.

Weibe, J. "Manipulating Percentages." *Mathematics Teacher* 79 (January 1986): 21, 23–26.

Williams, S., and J. Copley. "Using Calculators to Discover Patterns in Dividing Decimals." *Mathematics Teaching in the Middle School* 1 (April, 1994): 72–75.

Zawojewski, J. "Initial Decimal Concepts: Are They Really So Easy?" *Arithmetic Teacher* 30 (March, 1983): 52–56.

7

Probability

Preliminary Problem
It is your first day of teaching and your class has 40 students. A friend who does not know any children in your class bets you that at least two of them share a birthday (month and day). What are the friend's chances of winning the bet?

Probability, with its roots in gambling, is used in such areas as predicting sales, planning political campaigns, determining insurance premiums, making investment decisions, and testing experimental drugs. Some examples of uses of probability in everyday conversations include the following:

What is the probability that the Braves will win the World Series?
There is no chance you will get a raise.
There is a 50% chance of rain today.

In this chapter, we use tree diagrams and geometric probabilities (area models) to solve problems and to analyze games that involve spinners, cards, and dice. We introduce counting techniques and we discuss the role of simulations in probability.

Probability plays an important role in the *Principles and Standards,* as shown in the following:

In grades 6 through 8, all students should

- *understand and use appropriate terminology to describe complementary and mutually exclusive events;*
- *use proportionality and a basic understanding of probability to make and test conjectures about the results of experiments and simulations;*
- *compute probabilities for simple compound events, using such methods as organized lists, tree diagrams, and area models* (p. 248).

HISTORICAL NOTE

The first time probability appears to have been mentioned is in a 1477 commentary on Dante's *Divine Comedy,* but most historians think that it originated in an unfinished dice game. The French mathematician Blaise Pascal (1623–1665) received a letter from his friend Chevalier de Méré, a professional gambler, who asked how to divide the stakes if two players start, but fail to complete, a game consisting of five matches in which the winner is the one who wins three out of five matches. The players decided to divide the stakes according to their chances of winning the game. Pascal shared the problem with Pierre de Fermat (1601–1665) and together they solved the problem, which prompted the development of probability.

Since the work of the French mathematician Pierre Simon de Laplace (1749–1827), probability theory has become a major mathematical tool in science. In 1905 Albert Einstein (1879–1955) used the theory of probability in his work in physics.

Section 7-1 — How Probabilities Are Determined

experiment
outcome

Probabilities are ratios, expressed as fractions, decimals, or percents, determined by considering results or outcomes of experiments. An **experiment** is an activity whose results can be observed and recorded. Each of the possible results of an experiment is an **outcome.** If we toss a coin that cannot land on its edge, there are two distinct possible outcomes: heads (H) and tails (T).

350 CHAPTER 7 *Probability*

sample space

A set of all possible outcomes for an experiment is a **sample space.** In a single coin toss, the sample space S is given by $S = \{H, T\}$. The sample space can be modeled by a tree diagram, as shown in Figure 7-1. Each outcome of the experiment is designated by a separate branch in the tree diagram. The sample space S for rolling the standard die in Figure 7-2(a) is $S = \{1, 2, 3, 4, 5, 6\}$. Figure 7-2(b) gives a tree diagram for the sample space.

Figure 7-1

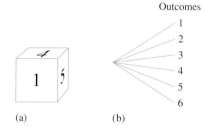

Figure 7-2 (a) (b)

event

Any subset of a sample space is an **event.** For example, the set of all even-numbered rolls $\{2, 4, 6\}$ is a subset of all possible rolls of a die $\{1, 2, 3, 4, 5, 6\}$ and is an event.

Example 7-1

Suppose an experiment consists of drawing one slip of paper from a jar containing 12 slips of paper, each with a different month of the year written on it. Find each of the following:

a. The sample space S for the experiment
b. The event A consisting of outcomes having a month beginning with J
c. The event B consisting of outcomes having the name of a month that has exactly 4 letters
d. The event C consisting of outcomes having a month that begins with M or N

Solution **a.** $S = \{$January, February, March, April, May, June, July, August, September, October, November, December$\}$
b. $A = \{$January, June, July$\}$
c. $B = \{$June, July$\}$
d. $C = \{$March, May, November$\}$

Determining Probabilities

Around 1900, the English statistician Karl Pearson tossed a coin 24,000 times and recorded 12,012 heads. During World War II, the Dane, John Kerrich, a prisoner of war, tossed a coin 10,000 times. A subset of his results is in Table 7-1. The *relative frequency* column on the right is obtained by dividing the number of heads by the number of tosses of the coin.

Table 7-1

Number of Tosses	Number of Heads	Relative Frequency (rounded)
10	4	0.400
50	25	0.500
100	44	0.440
500	255	0.510
1,000	502	0.502
5,000	2,533	0.507
8,000	4,034	0.504
10,000	5,067	0.507

As the number of Kerrich's tosses increased, he obtained heads close to half the time. The relative frequency for Pearson's 24,000 tosses gives a similar result of 12,012/24,000, or approximately $\frac{1}{2}$.

experimentally · empirically

When a probability is determined by observing outcomes of experiments, it is said to be determined **experimentally,** or **empirically.** The exact number of heads that occurs when a fair coin is tossed a few times cannot be predicted accurately. A *fair coin* is a coin that is just as likely to land "heads" as it is to land "tails." Probabilities only suggest what will happen in the "long run." When a fair coin is tossed many times and the fraction (or proportion) of heads is near $\frac{1}{2}$, we say that the probability of heads occurring is $\frac{1}{2}$ and write $P(H) = \frac{1}{2}$.

theoretical probabilities

We assign **theoretical probabilities** to the outcomes under ideal conditions. For example, we could argue that since an ideal coin is symmetric and has two sides, then each side should appear about the same number of times if the coin is tossed many times. Again we would conclude that

$$P(H) = P(T) = 1/2.$$

equally likely

When one outcome is just as likely as another, as in coin tossing, the outcomes are **equally likely.** In this text, by "probability" we mean *theoretical probability*. If an experiment is repeated many times, the experimental probability of the event's occurring should approach the theoretical probability of the event's occurring.

A *fair* die is a die that is just as likely to land showing any of the numerals 1 through 6. Its sample space S is given by $S = \{1, 2, 3, 4, 5, 6\}$, and $P(1) = P(2) = P(3) = P(4) = P(5) = P(6) = 1/6$. The probability of rolling an even number, that is, the probability of the event $E = \{2, 4, 6\}$, is 3/6, or 1/2. For a sample space with equally likely outcomes, the probability of an event A can be defined as follows.

Definition of Probability of an Event with Equally Likely Outcomes

For an experiment with sample space S and equally likely outcomes, the **probability of an event** A is given by

$$P(A) = \frac{\text{Number of elements of } A}{\text{Number of elements of } S} = \frac{n(A)}{n(S)}.$$

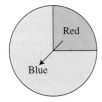

Figure 7-3

This definition applies only to a sample space that has equally likely outcomes. Applying the definition to a space with outcomes that are not equally likely leads to incorrect conclusions. For example, the sample space for spinning the spinner in Figure 7-3 is given by $S = \{\text{Red, Blue}\}$, but the outcome Blue is more likely to occur than is the outcome Red, so $P(\text{Red})$ is not equal to $\frac{1}{2}$ but to $\frac{90}{360}$, or $\frac{1}{4}$ (why?). If the spinner were spun 100 times, we could reasonably expect that about $\frac{1}{4}$, or 25, of the outcomes would be Red, whereas about $\frac{3}{4}$, or 75, of the outcomes would be Blue.

REMARK Because "landing heads" is the event $\{H\}$, the probability of heads occurring should be written $P(\{H\})$. However, for simplicity we write instead $P(H)$.

NOW TRY THIS 7-1

a. In an experiment of tossing a fair coin once, what is the sum of the probabilities of all the distinct outcomes in the sample space?
b. In an experiment of tossing a fair die once, what is the sum of the probabilities of all the distinct outcomes in the sample space?
c. Does the sum of the probabilities of all the distinct outcomes of any sample space always result in the same number? Why?

Example 7-2

at random

Let $S = \{1, 2, 3, 4, 5, \ldots, 25\}$. If a number is chosen **at random**, that is, with the same chance of being drawn as all other numbers in the set, calculate each of the following probabilities:

a. The event A that an even number is drawn
b. The event B that a number less than 10 and greater than 20 is drawn
c. The event C that a number less than 26 is drawn
d. The event D that a prime number is drawn
e. The event E that a number both even and prime is drawn

Solution Each of the 25 numbers in set S has an equal chance of being drawn.

a. $A = \{2, 4, 6, 8, 10, 12, 14, 16, 18, 20, 22, 24\}$, so $n(A) = 12$. Thus
$$P(A) = \frac{n(A)}{n(S)} = \frac{12}{25}.$$

b. $B = \emptyset$, so $n(B) = 0$. Thus $P(B) = \frac{0}{25} = 0$.

c. $C = S$ and $n(C) = 25$. Thus $P(C) = \frac{25}{25} = 1$.

d. $D = \{2, 3, 5, 7, 11, 13, 17, 19, 23\}$, so $n(D) = 9$. Thus
$$P(D) = \frac{n(D)}{n(S)} = \frac{9}{25}.$$

e. $E = \{2\}$, so $n(E) = 1$. Thus $P(E) = \frac{1}{25}$.

impossible event

In Example 7-2(b), event B is the empty set. An event such as B that has no outcomes in it is an **impossible event** *and has probability* 0. If the word *and* were replaced by *or* in Example 7-2(b), then event B would no longer be the empty set. In Example 7-2(c), event

SECTION 7-1   *How Probabilities Are Determined*   **353**

C consists of drawing a number less than 26 on a single draw. Because every number in *S* is less than 26, $P(C) = \frac{25}{25} = 1$. An event that has probability 1 is a **certain event.**

certain event

Because an event is a subset of a sample space, an event can have no more outcomes than in the sample space. In addition, an event can have no fewer than 0 outcomes. Thus, if *A* is any event, it occurs between 0% and 100% of the time, and we have the following:

$$0\% \leq P(A) \leq 100\% \quad \text{or} \quad 0 \leq P(A) \leq 1$$

Consider one spin of the wheel shown in Figure 7-4. For this experiment, $S = \{0, 1, 2, 3, 4, 5, 6, 7, 8, 9\}$. If *A* is the event of spinning a number in the set $\{0, 1, 2, 3, 4\}$ and *B* is the event of spinning a number in the set $\{5, 7\}$, then using the definition of probability for equally likely events, $P(A) = n(A)/n(S) = 5/10$ and $P(B) = n(B)/n(S) = 2/10$. The probability of an event can be found by adding the probabilities of the events representing the various outcomes in the set. For example, event $B = \{5, 7\}$ can be represented as the union of two disjoint events, that is spinning a 5 or spinning a 7. Then $P(B)$ can be found by adding the probabilities of each event.

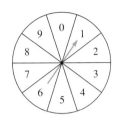

Figure 7-4

$$P(B) = P(5) + P(7) = 1/10 + 1/10 = 2/10$$

Likewise,

$$P(A) = 1/10 + 1/10 + 1/10 + 1/10 + 1/10 = 5/10.$$

These are special cases of the following property, which holds for all probabilities.

Property of Probability of an Event

The **probability of an event** is equal to the sum of the probabilities of the events representing all the outcomes in the original event.

Example 7-3   If we draw a card at random from an ordinary deck of playing cards, what is the probability that

a. the card is an ace?   **b.** the card is an ace or a queen?

Solution   **a.** There are 52 cards in a deck, 4 of which are aces. If event *A* is drawing an ace, then $A = \{$ 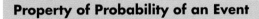 $\}$. We use the definition of probablilty for equally likely outcomes to compute the following:

$$P(A) = \frac{n(A)}{n(S)} = \frac{4}{52}$$

An alternative approach is to find the sum of the probabilities of obtaining each of the outcomes in the event, where the probability of drawing any single ace from the deck is $\frac{1}{52}$:

$$P(A) = \frac{1}{52} + \frac{1}{52} + \frac{1}{52} + \frac{1}{52} = \frac{4}{52}$$

b. The event E of getting an ace or a queen consists of 8 cards; 4 aces and 4 queens. Hence,

$$P(E) = \frac{n(E)}{n(S)} = \frac{8}{52}$$

Mutually Exclusive Events

Consider one spin of the wheel in Figure 7-4. For this experiment, we have $S = \{0, 1, 2, 3, 4, 5, 6, 7, 8, 9\}$. If $A = \{0, 1, 2, 3, 4\}$ and $B = \{5, 7\}$, then $A \cap B = \emptyset$. Two such events are **mutually exclusive** events. If event A occurs, then event B cannot occur, and we have the following definition.

mutually exclusive

> **Definition of Mutually Exclusive Events**
>
> Events A and B are **mutually exclusive** if $A \cap B = \emptyset$.

Each outcome in the space $S = \{0, 1, 2, 3, 4, 5, 6, 7, 8, 9\}$ is equally likely, with probability $\frac{1}{10}$. Thus, if we write the probability of A or B as $P(A \cup B)$, we have the following:

$$P(A \cup B) = \frac{n(A \cup B)}{n(S)} = \frac{7}{10} = \frac{5 + 2}{10} = \frac{5}{10} + \frac{2}{10}$$

$$= \frac{n(A)}{n(S)} + \frac{n(B)}{n(S)} = P(A) + P(B)$$

The result developed in this example is true for all mutually exclusive events. In general, we have the following property.

> **Property**
>
> If events A and B are mutually exclusive, then $P(A \cup B) = P(A) + P(B)$.

For a sample space with equally likely outcomes, this property follows immediately from the fact that if $A \cap B = \emptyset$, then $n(A \cup B) = n(A) + n(B)$.

Complementary Events

complements

If the weather forecaster tells us that the probability of rain is 25%, what is the probability that it will not rain? These two events—rain and not rain—are **complements** of each other. Therefore if the probability of rain is 25%, or 1/4, the probability it will not rain is $100\% - 25\% = 75\%$, or $1 - 1/4 = 3/4$. Notice that $P(\text{no rain}) = 1 - P(\text{rain})$. The two events rain and no rain are mutually exclusive because if one happens, the other cannot.

complementary events Two mutually exclusive events whose union is the sample space are **complementary events.** If A is an event, the complement of A, written $\overline{A}$, is also an event. For example, consider the event $A = \{2, 4\}$ of tossing a 2 or a 4 using a standard die. The complement of A is the set $\overline{A} = \{1, 3, 5, 6\}$. Because the sample space is $S = \{1, 2, 3, 4, 5, 6\}$, we have $P(A) = 2/6$ and $P(\overline{A}) = 4/6$. Notice that $P(\overline{A}) = 1 - P(A)$. This is true in general for any set A and its complement, $\overline{A}$.

Property

If A is an event and $\overline{A}$ is its complement, then

$$P(A) + P(\overline{A}) = 1,$$
$$P(\overline{A}) = 1 - P(A), \text{ or } P(A) = 1 - P(\overline{A}).$$

Non–Mutually Exclusive Events

Consider the spinner in Figure 7-5. Let E be the event of spinning an even number and T the event of spinning a number divisible by 3, as follows:

$$E = \{0, 2, 4, 6, 8\}$$
$$T = \{0, 3, 6, 9\}$$

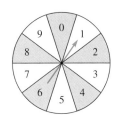

Figure 7-5

The event of spinning an even number and a number divisible by 3, denoted by $E \cap T$, is $\{0, 6\}$. Because $E \cap T = \{0, 6\}$, E and T are not mutually exclusive and $P(E \cup T) \neq P(E) + P(T)$. However, because each outcome is equally likely we can still use the definition of the probability of an event to compute the probability of E or T as follows:

$$P(E \cup T) = \frac{n(E \cup T)}{n(S)}$$

Because $E \cup T = \{0, 2, 4, 6, 8, 3, 9\}$, $n(E \cup T) = 7$. Also, $n(S) = 10$, and $P(E \cup T) = \frac{7}{10}$.

In general, we can also compute the probability of E or T by using notions of sets from Chapter 2. We know

$$n(E \cup T) = n(E) + n(T) - n(E \cap T).$$

Therefore, for a sample space with equally likely outcomes,

$$P(E \cup T) = \frac{n(E \cup T)}{n(S)}$$
$$= \frac{n(E) + n(T) - n(E \cap T)}{n(S)}$$
$$= \frac{n(E)}{n(S)} + \frac{n(T)}{n(S)} - \frac{n(E \cap T)}{n(S)}$$
$$= P(E) + P(T) - P(E \cap T).$$

This result, although proved for events in a sample space with equally likely outcomes, is true in general, and the properties of probability are summarized next.

Properties of Probability

1. $P(\emptyset) = 0$ (impossible event).
2. $P(S) = 1$, where S is the sample space (certain event).
3. For any event A, $0 \leq P(A) \leq 1$.
4. If A and B are events and $A \cap B = \emptyset$, then $P(A \cup B) = P(A) + P(B)$.
5. If A and B are events, then $P(A \cup B) = P(A) + P(B) - P(A \cap B)$.
6. If A is an event, then $P(\overline{A}) = 1 - P(A)$.

Example 7-4 A golf bag contains 2 red tees, 4 blue tees, and 5 white tees.

a. What is the probability of the event R that a tee drawn at random is red?
b. What is the probability of the event "not R," that is, that a tee drawn at random is not red?
c. What is the probability of the event that a tee drawn at random is either red (R) or blue (B), that is, $P(R \cup B)$.

Solution a. Because the bag contains a total of $2 + 4 + 5$, or 11, tees and 2 tees are red, $P(R) = \dfrac{2}{11}$.

b. The bag contains 11 tees and 9 are not red, so the probability of "not R" is $\dfrac{9}{11}$. Also, notice that $P(\overline{R}) = 1 - P(R) = 1 - \dfrac{2}{11} = \dfrac{9}{11}$.

c. The bag contains 2 red tees and 4 blue tees and $R \cap B = \emptyset$, so $P(R \cup B) = \dfrac{2}{11} + \dfrac{4}{11}$, or $\dfrac{6}{11}$.

Example 7-5 Find the probability of rolling a sum of 7 or 11 when rolling a fair pair of dice.

Solution To solve this problem, we use the strategy of *making a table,* as in Figure 7-6(a), to show all possible outcomes of tossing the dice. We know that there are 6 possible results from tossing the first die and 6 from tossing the second die, so by the Fundamental Counting Principle, there are $6 \cdot 6$, or 36, entries in the table. It may be easier to read the results when they are recorded as ordered pairs, as in Figure 7-6(b), where the first component represents the number on the first die and the second component represents the number on the second die. We find the possible sums in rolling the pair of dice as shown in Figure 7-6(c).

SECTION 7-1 *How Probabilities Are Determined*

Figure 7-6

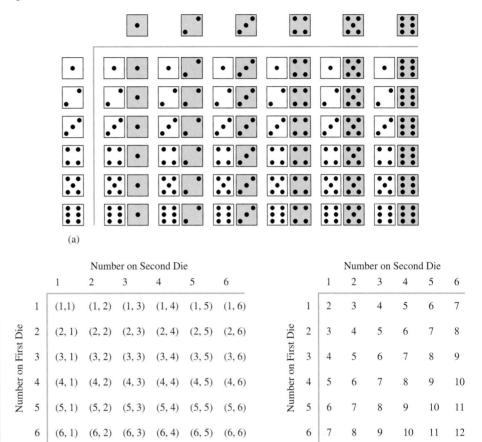

In Figure 7-6(c) we see that a sum of 7 appears six times; that is, in 6 ordered pairs. Hence, the event "a sum of 7" is the following subset of the set of ordered pairs in Figure 7-6(b):

$$\{(6, 1), (5, 2), (4, 3), (3, 4), (2, 5), (1, 6)\}.$$

Each outcome in this set is equally likely and hence P(a sum of 7) = 6/36. Similarly, P(a sum of 11) = 2/36. The probabilities of each of the elements in the sample space can be calculated in the same way and are displayed in Table 7-2 (the numbers in the "Outcome" row represent the sum of the numbers that appear on the two dice).

Table 7-2

Outcome	2	3	4	5	6	7	8	9	10	11	12
Probability	1/36	2/36	3/36	4/36	5/36	6/36	5/36	4/36	3/36	2/36	1/36

The probability of rolling a sum of 7 or 11 is given by P(sum of 7 or sum of 11) = P(sum of 7) + P(sum of 11) = 6/36 + 2/36 = 8/36.

Example 7-6 A fair pair of dice is rolled. Let E be the event of rolling a sum that is an even number and P the event of rolling a sum that is a prime number. Find the probability of rolling a sum that is even *or* prime, that is, $P(E \cup P)$.

Solution To solve this problem, we use Table 7-2. We know that events E and P are not mutually exclusive because $E = \{2, 4, 6, 8, 10, 12\}$, $P = \{2, 3, 5, 7, 11\}$, and $E \cap P = \{2\}$. One way to solve the problem is to note that $E \cup P = \{2, 4, 6, 8, 10, 12, 3, 5, 7, 11\}$. Therefore

$$P(E \cup P) = P(2) + P(4) + P(6) + P(8) + P(10) + P(12) + P(3) + P(5) + P(7) + P(11)$$
$$= 1/36 + 3/36 + 5/36 + 5/36 + 3/36 + 1/36 + 2/36 + 4/36 + 6/36 + 2/36$$
$$= 32/36.$$

Another approach is to use a property of probabilities (Property 5) established earlier:

$$P(E \cup P) = P(E) + P(P) - P(E \cap P)$$
$$= 18/36 + 15/36 - 1/36$$
$$= 32/36$$

A third approach to finding $P(E \cup P)$ is to find $P(\overline{E \cup P})$ and subtract this probability from 1. Because $E \cup P = \{2, 3, 4, 5, 6, 7, 8, 10, 11, 12\}$, then $\overline{E \cup P} = \{9\}$ and $P(\overline{E \cup P}) = 4/36$. Hence, $P(E \cup P) = 1 - P(\overline{E \cup P}) = 1 - 4/36 = 32/36$.

ONGOING ASSESSMENT 7-1

1. When a thumbtack is dropped, it will land point up (⊥) or point down (⋏). This experiment was repeated 80 times with the following results:
   Point up: 56 times   Point down: 24 times
   a. What is the experimental probability that the thumbtack will land point up?
   b. What is the experimental probability that the thumbtack will land point down?
   c. If you were to try this experiment another 80 times, would you get the same results? Why?
   d. Would you expect to get nearly the same results on a second trial? Why?

2. Consider the experiment of drawing a single card from a standard deck of cards and determine which of the following are sample spaces with equally likely outcomes:
   a. {face card, not face card}
   b. {club, diamond, heart, spade}
   c. {black, red}
   d. {king, queen, jack, ace, even card, odd card}

3. An experiment consists of selecting the last digit of a telephone number. Assume that each of the 10 digits is equally likely to appear as a last digit. List each of the following:
   a. The sample space
   b. The event consisting of outcomes that the digit is less than 5
   c. The event consisting of outcomes that the digit is odd
   d. The event consisting of outcomes that the digit is not 2
   e. Find the probability of each of the events in (b) through (d).

4. Each letter of the alphabet is written on a separate piece of paper and placed in a box and then one piece is drawn at random.
   a. What is the probability that the selected piece of paper has a vowel written on it?
   b. What is the probability that it has a consonant written on it?

5. The following spinner is spun:

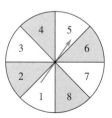

Find the probabilities of obtaining each of the following:
a. P(factor of 35)   b. P(multiple of 3)
c. P(even number)   d. P(6 or 2)

e. $P(11)$
   f. P(composite number)
   g. P(neither a prime nor a composite)
6. A card is selected from an ordinary deck of 52 cards. Find the probabilities for each of the following:
   a. A red card
   b. A face card
   c. A red card or a 10
   d. A queen
   e. Not a queen
   f. A face card or a club
   g. A face card and a club
   h. Not a face card and not a club
7. A drawer contains 6 black socks, 4 brown socks, and 2 green socks. Suppose one sock is drawn from the drawer and that it is equally likely that any one of the socks is drawn. Find the probabilities for each of the following:
   a. The sock is brown.
   b. The sock is either black or green.
   c. The sock is red.
   d. The sock is not black.
8. The drawer in the cartoon contains 6 pairs of black socks, 8 pairs of blue socks, and 3 pairs of white socks.
   a. A child reaches in the drawer without looking to pull out four socks. What is the probability that the child gets two socks of the same color?
   b. What is the least number of socks that one must pull out to guarantee getting at least one pair of each color? Justify your answer.
   c. A cat knocks one sock out of the drawer. What is the probability that the sock is not white?

9. Riena has six unmarked computer disks in a box, where each is dedicated to exactly one of English, mathematics, French, American history, chemistry, and computer science. Answer the following questions:
   a. If she chooses a computer disk at random, what is the probability she chooses the English disk?
   b. What is the probability that she chooses a disk that is neither math nor chemistry?
10. The following questions refer to a very popular dice game, craps, in which a player rolls two dice:
    a. Rolling a sum of 7 or 11 on the first roll of the dice is a win. What is the probability of winning on the first roll?
    b. Rolling a sum of 2, 3, or 12 on the first roll of the dice is a loss. What is the probability of losing on the first roll?
    c. Rolling a sum of 4, 5, 6, 8, 9, or 10 on the first roll is neither a win nor a loss. What is the probability of neither winning nor losing on the first roll?
    d. After rolling a sum of 4, 5, 6, 8, 9, or 10, a player must roll the same sum again before rolling a sum of 7. Which sum—4, 5, 6, 8, 9, or 10—has the highest probability of occurring again?
    e. What is the probability of rolling a sum of 1 on any roll of the dice?
    f. What is the probability of rolling a sum less than 13 on any roll of the dice?
    g. If the two dice are rolled 60 times, predict about how many times a sum of 7 will be rolled.
11. According to a weather report, there is a 30% chance it will rain tomorrow. What is the probability it will not rain tomorrow? Explain your answer.
12. A roulette wheel has 38 slots around the rim. The first 36 slots are numbered from 1 to 36. Half of these 36 slots are red, and the other half are black. The remaining 2 slots are numbered 0 and 00 and are green. As the roulette wheel is spun in one direction, a small ivory ball is rolled along the rim in the opposite direction. The ball has an equally likely chance of falling into any one of the 38 slots. Find each of the following:
    a. The probability that the ball lands in a black slot
    b. The probability that the ball lands on 0 or 00
    c. The probability that the ball does not land on a number from 1 through 12
    d. The probability that the ball lands on an odd number or on a green slot
13. If the roulette wheel in problem 12 is spun 190 times, predict about how many times the ball will land on 0 or 00.
14. Determine if each player has an equal probability of winning each of the following games:
    a. Toss a fair coin. If heads appears, I win; if tails appears, you lose.
    b. Toss a fair coin. If heads appears, I win; otherwise, you win.
    c. Toss a fair die numbered 1 through 6. If 1 appears, I win; if 6 appears, you win.

d. Toss a fair die numbered 1 through 6. If an even number appears, I win; if an odd number appears, you win.
e. Toss a fair die numbered 1 through 6. If a number greater than or equal to 3 appears, I win; otherwise, you win.
f. Toss two fair dice numbered 1 through 6. If a 1 appears on each die, I win; if a 6 appears on each die, you win.
g. Toss two fair dice numbered 1 through 6. If the sum is 3, I win; if the sum is 2, you win.
h. Toss two dice numbered 1 through 6; one die is red and one is white. If the number on the red die is greater than the number on the white die, I win; otherwise, you win.

15. Suppose a fair coin is tossed twice. Find the probability for each of the following:
    a. Exactly one head
    b. At least one head
    c. At most one head

16. In Sentinel High School, there are 350 freshmen, 320 sophomores, 310 juniors, and 400 seniors. If a student is chosen at random from the student body to represent the school, what is the probability that the chosen student is a freshman?

17. In each of the following, sketch a single spinner with the following characteristics:
    a. The outcomes are M, A, T, and H, each with equally likely probability.
    b. The outcomes are R, A, and T with $P(R) = 3/4$, $P(A) = 1/8$, and $P(T) = 1/8$.

18. In the game of "Between," two cards are dealt. You then pick a third card from the deck. To win, you must pick a card that has a value between the other two cards. The order of values is 2, 3, 4, 5, 6, 7, 8, 9, 10, J, Q, K, A, where the letters represent a jack, queen, king, and ace, respectively. Determine the probability of your winning if the first two cards dealt are the following:
    a. A 5 and a jack
    b. A 2 and a king
    c. A 5 and a 6

19. Calculators, watches, scoreboards, and many other devices display numbers using arrays like the following. The device lights up different parts of the array (any segment lettered a through g), to display any single digit 0 through 9. Suppose a digit is chosen at random. Determine the probability for each of the following. (*Hint:* Use a digital watch to see which segments are lit up for the different numbers.)

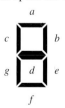

a. Segment a will be lit.
b. Segment b will be lit.
c. Segments e and b will be lit.
d. Segments e or b will be lit.

20. If A is the set of students taking algebra and C is the set of students taking chemistry at Central High School, describe in words what is meant by each of the following probabilities:
    a. $P(A \cup C)$
    b. $P(A \cap C)$
    c. $1 - P(C)$

21. If A and B are mutually exclusive and if $P(A) = 0.3$ and $P(B) = 0.4$, what is $P(A \cup B)$?

22. A calculus class is composed of 35 men and 45 women. There are 20 business majors, 30 biology majors, 10 computer science majors, and 20 mathematics majors. No person has a double major. If a single student is chosen from the class, what is the probability that the student is the following:
    a. Female
    b. A computer science major
    c. Not a mathematics major
    d. A computer science major or a mathematics major

23. A box contains 25% black balls and 75% white balls. The same number of black balls as was in the box is added (so the new number of black balls is twice the original number). A ball is now drawn from the box at random. What is the probability that it is black?

24. A box contains 5 white balls, 3 black balls, and 2 red balls.
    a. How many red balls must be added to the box so that the probability of drawing a red ball is 3/4?
    b. How many black balls must be added to the original box so that the probability of drawing a white ball is 1/4?

25. Suppose we have a box containing 5 white balls, 3 black balls, and 2 red balls. Is it possible to add the same number of each color balls to the box so that when a ball is drawn at random the probability that it is a black ball is the following (explain):
    a. $\dfrac{1}{3}$
    b. 0.32

Communication

26. Explain whether events A and B can be mutually exclusive if $P(A) = 0.8$ and $P(B) = 0.9$.

27. Bobbie says that when she shoots a free throw in basketball, she will either make it or miss it. Because there are only two outcomes and one of them is making a basket, Bobbie claims the probability of her making a free throw is 1/2. Explain whether Bobbie's reasoning is correct.

28. If the following spinner is spun 100 times, Joe claims that the number 4 will occur most often because the greatest area of the spinner is covered by the number 4. What would you tell Joe about his conjecture? What is the probability that a 4 will occur on any spin?

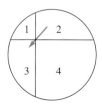

Open-Ended

29. Suppose you toss a die once. Name two events that could result that are mutually exclusive.
30. Select any book, go to the first complete paragraph in it, and count the number of words the paragraph has. Now count the number of words that start with a vowel. If the paragraph has fewer than 100 words, continue to count words until there are more than 100 words from where you started. What is the experimental probability that a word chosen at random from the book starts with a vowel? Open the book to any page and choose a paragraph with more than 100 words. Predict how many words will start with a vowel, and then count them to see how close you were.
31. List three real-world situations that do not involve weather or gambling where probability might be used.
32. For each of the following letters, describe an event, if possible, that has the approximate probability marked on the probability line:

Cooperative Learning

33. Form groups of three or four students. Each group has a pair of dice. Each player needs 18 markers and a sheet of paper with a gameboard drawn on it similar to the following:

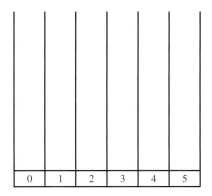

At the beginning of each game, each player places his or her markers on the boards in any arrangement above the numbers. Two players (or teams) then take turns rolling the dice. The result of each roll is the difference between the greater and the smaller of the two numbers. For example, if a 6 and a 4 were rolled, the difference would be 6 − 4, or 2. All players who have a marker on the number that represents the difference can remove one marker. Only one marker can be removed each roll. The first player to remove all of his or her markers is the winner. The game may be stopped if it is clear that there can be no winner.

   a. Play the game twice. What differences seem to occur most often? Least often?
   b. Roll the dice 20 times and record how often the various differences occur. Using this information, explain how you would distribute your 18 markers to win.
   c. Compute the theoretical probabilities for each possible difference. (Figure 7-6(a) might be useful.)
   d. Use your answers to (c) to explain how you would arrange the markers in order to win the game.

34. The following game is played with two players and a single fair die numbered 1 through 6. The die is rolled and one person receives a score that is the square of the number appearing on the die. The other person will receive a score of four times the value showing on the die. The person with the greatest score wins.

   a. Play the game several times to see if it appears to be a *fair game,* that is, a game in which each player has an equal chance of winning.
   b. Determine if this is a fair game. If it is not fair, who has the advantage? Explain how you arrived at your answer.

LABORATORY ACTIVITY

1. Suppose a paper cup is tossed in the air. The different ways it can land are shown here:

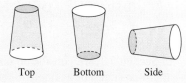

Toss a cup 100 times and record each result. From this information, calculate the experimental probability of each outcome. Do the outcomes appear to be equally likely? Using experimental probabilities, predict how many times the cup will land on its side if tossed 100 times.

2. Toss a fair coin 100 times and record the results. From this information, calculate the experimental probability of getting heads on a particular toss. Does the experimental result agree with the expected theoretical probability of $\frac{1}{2}$?

3. Hold a coin upright on its edge under your forefinger on a hard surface and then spin it with your other finger so that it spins before landing. Repeat this experiment 100 times and calculate the experimental probability of the coin's landing on heads on a particular spin. Compare your experimental probabilities with those in activity 2.

TECHNOLOGY CORNER

The following Logo procedures simulates flipping a coin.

```
TO TOSS
    OUTPUT PICK.ONE [HEADS TAILS]
END
TO PICK.ONE :FLIP
    OUTPUT ITEM (1 + RANDOM COUNT :FLIP) :FLIP
END
```

Edit the procedure to fit your particular Logo software and enter it into your computer. Then complete the following:

1. Execute `PRINT TOSS`.
2. To simulate tossing a coin 25 times, execute the following:
   `REPEAT 25 [PRINT TOSS]`
3. Using the results in number 2, determine the experimental probability for obtaining `HEADS`. For obtaining `TAILS`.
4. Edit the Logo procedure `TOSS` to make the list be `[HEADS HEADS TAILS TAILS TAILS]`. Now execute the following: `REPEAT 25 [PRINT TOSS]`

What is the experimental probability for obtaining `HEADS` now? For obtaining `TAILS` now?

Section 7-2 — Multistage Experiments with Tree Diagrams and Geometric Probabilities

In Section 7-1, we considered one-stage experiments, that is, experiments that were over after one step. For example, drawing one ball at random from the box containing a red, white, and green ball in Figure 7-7(a) is a one-stage experiment. A tree diagram for this experiment is shown in Figure 7-7(b).

Figure 7-7

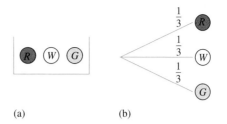

(a)   (b)

Next we consider a two-stage experiment. For example, a ball is drawn from the box in Figure 7-7(a) and its color is recorded. Then the ball is *replaced*, and a second ball is drawn and its color is recorded. A sample space for this experiment may be written as {RR, RW, RG, WR, WW, WG, GR, GW, GG}. A tree diagram for this experiment is given in Figure 7-8.

Figure 7-8

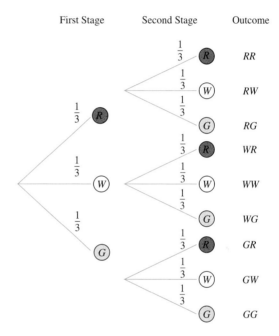

Each of the outcomes in the sample space is equally likely and there are nine total outcomes, so the probability of each outcome is 1/9.

The box in Figure 7-9(a) contains one colored ball and two white balls. If a ball is drawn at random and the color recorded, a tree diagram for the experiment might look like the one in Figure 7-9(b). Because each ball has the same chance of being drawn, we may combine the branches and obtain the tree diagram shown in Figure 7-9(c). Combining branches in this way is a common practice because it simplifies tree diagrams.

Figure 7-9

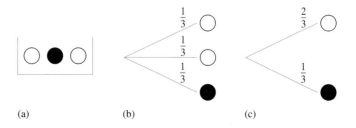

(a)   (b)   (c)

Suppose a ball is drawn at random from the box in Figure 7-9(a) and its color recorded. The ball is then *replaced,* and a second ball is drawn and its color recorded. The sample space for this two-stage experiment may be recorded using ordered pairs as {(●, ●), (●, ○), (○, ●), (○, ○)} or, more commonly, as {● ●, ● ○, ○ ●, ○ ○}, as shown in the tree diagram in Figure 7-10.

Figure 7-10

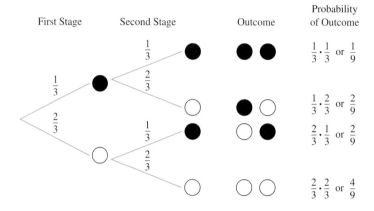

To assign the probability of the outcomes in this experiment, consider, for example, the path for the outcome ● ○. In the first stage, the probability of obtaining a colored ball is $\frac{1}{3}$. Then, the probability of obtaining a white ball in the second stage (second draw) is $\frac{2}{3}$. Thus we expect to obtain a colored ball on the first draw $\frac{1}{3}$ of the time and then on the second draw, to obtain a white ball $\frac{2}{3}$ of those times that we obtained a colored ball on the first draw, that is, $\frac{2}{3}$ of $\frac{1}{3}$, or $\frac{2}{3} \cdot \frac{1}{3}$. Observe that this product can be obtained by multiplying the probabilities along the branches used for the path leading to ● ○, that is, $\frac{1}{3} \cdot \frac{2}{3}$, or $\frac{2}{9}$. The probabilities shown in Figure 7-10 are obtained by following the paths leading to each of the four outcomes and multiplying the probabilities along the paths. This discussion yields the following property for tree diagrams.

Property: Multiplication Rule for Probabilities

For all multistage experiments, the probability of the outcome along any path is equal to the product of all the probabilities along the path.

REMARK The sum of the probabilities on all the branches from any point always equals 1, and the sum of the probabilities for the possible outcomes must also be 1.

Look again at the box pictured in Figure 7-9(a). This time, suppose two balls are drawn one by one *without replacement.* A tree diagram for this experiment, along with the set of possible outcomes, is shown in Figure 7-11.

Figure 7-11

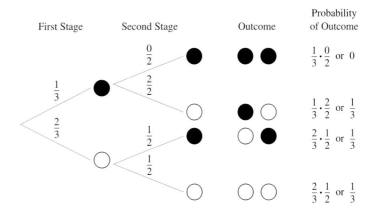

The denominators of the fractions along the second stage are all 2. Because the draws are made without replacement, only two balls remain for the second draw.

Consider event A, consisting of the outcomes for drawing exactly one colored ball in the two draws without replacement. This event is given by $A = \{\bullet \circ, \circ \bullet\}$. Since the outcome $\bullet \circ$ appears $\frac{1}{3}$ of the time, and the outcome $\circ \bullet$ appears $\frac{1}{3}$ of the time, then either $\bullet \circ$ or $\circ \bullet$ will appear $\frac{2}{3}$ of the time. Thus $P(A) = \frac{1}{3} + \frac{1}{3} = \frac{2}{3}$.

Event B, consisting of outcomes for drawing *at least* one colored ball, could be recorded as $B = \{\bullet \circ, \circ \bullet, \bullet \bullet\}$. Because $P(\bullet \circ) = \frac{1}{3}$, $P(\circ \bullet) = \frac{1}{3}$, and $P(\bullet \bullet) = 0$, then $P(B) = \frac{1}{3} + \frac{1}{3} + 0 = \frac{2}{3}$. Because $\overline{B} = \{\circ \circ\}$ and $P(\overline{B}) = \frac{1}{3}$, the probability of B could have been computed as follows: $P(B) = 1 - P(\overline{B}) = 1 - \frac{1}{3} = \frac{2}{3}$.

Example 7-7

Figure 7-12 shows a box with eleven letters. Some letters are repeated. Suppose four letters are drawn at random from the box one by one without replacement. What is the probability of the outcome *BABY,* with the letters chosen in exactly the order given?

Figure 7-12

PROBABILITY

Solution We do not need the entire tree diagram to find this probability because we are interested in only the branch leading to the outcome *BABY.* The portion needed is shown in Figure 7-13.

Figure 7-13

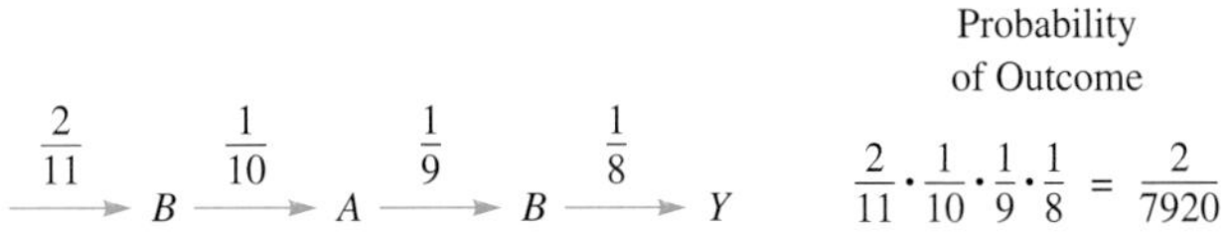

$\xrightarrow{\frac{2}{11}} B \xrightarrow{\frac{1}{10}} A \xrightarrow{\frac{1}{9}} B \xrightarrow{\frac{1}{8}} Y \qquad \frac{2}{11} \cdot \frac{1}{10} \cdot \frac{1}{9} \cdot \frac{1}{8} = \frac{2}{7920}$

Probability of Outcome

The probability of the first B is $\frac{2}{11}$ because there are two B's out of eleven letters. The probability of the second B is $\frac{1}{9}$ because there are nine letters left after one B and one A have been chosen. Then, $P(BABY)$ is $\frac{2}{7920}$, as shown.

In Example 7-7, suppose four letters are drawn one by one from the box and the letters are replaced after each drawing. In this case, the branch needed to find $P(BABY)$ in the order drawn is pictured in Figure 7-14. Then $P(BABY) = \left(\frac{2}{11}\right) \cdot \left(\frac{1}{11}\right) \cdot \left(\frac{2}{11}\right) \cdot \left(\frac{1}{11}\right)$, or $\frac{4}{14,641}$.

Figure 7-14

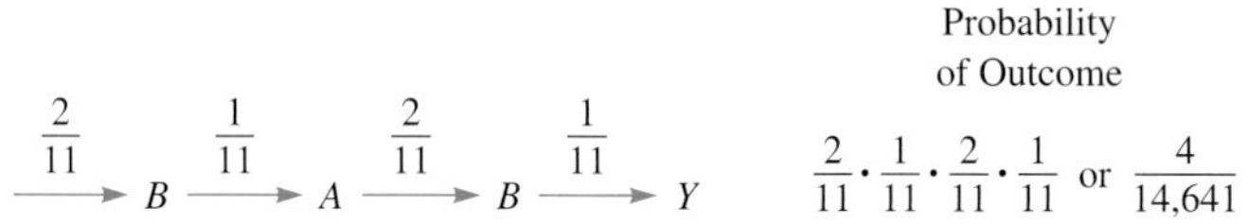

Example 7-8

Consider the three boxes in Figure 7-15. A letter is drawn from box 1 and placed in box 2. Then, a letter is drawn from box 2 and placed in box 3. Finally, a letter is drawn from box 3. What is the probability that the letter drawn from box 3 is B?

Figure 7-15

Box 1: AAB Box 2: AB Box 3: ABBB

Solution A tree diagram for this experiment is given in Figure 7-16. Notice that the denominators in the second stage are 3 rather than 2 because in this stage, there are now three letters in box 2. The denominators in the third stage are 5 because in this stage, there are five letters in box 3. To find the probability that a B is drawn from box 3, add the probabilities for the outcomes AAB, ABB, BAB, and BBB that make up this event.

Figure 7-16

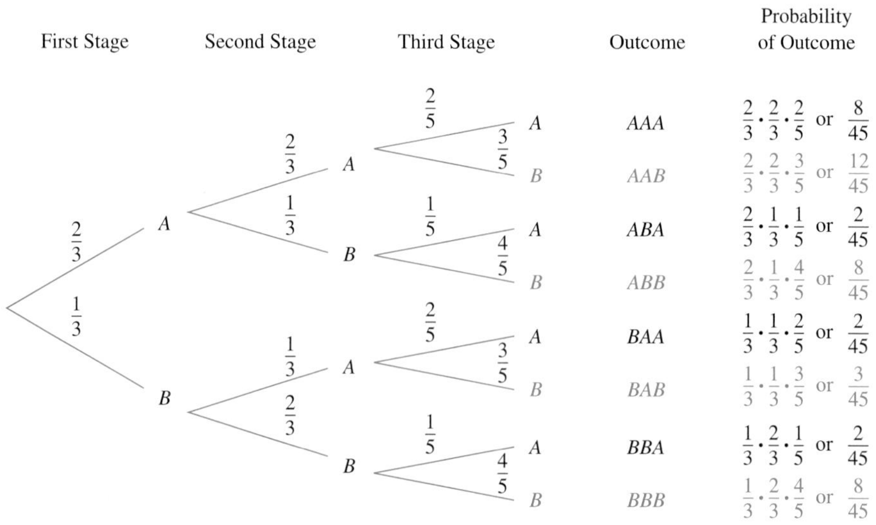

First Stage	Second Stage	Third Stage	Outcome	Probability of Outcome
A (2/3)	A (2/3)	A (2/5)	AAA	$\frac{2}{3} \cdot \frac{2}{3} \cdot \frac{2}{5}$ or $\frac{8}{45}$
		B (3/5)	AAB	$\frac{2}{3} \cdot \frac{2}{3} \cdot \frac{3}{5}$ or $\frac{12}{45}$
	B (1/3)	A (1/5)	ABA	$\frac{2}{3} \cdot \frac{1}{3} \cdot \frac{1}{5}$ or $\frac{2}{45}$
		B (4/5)	ABB	$\frac{2}{3} \cdot \frac{1}{3} \cdot \frac{4}{5}$ or $\frac{8}{45}$
B (1/3)	A (1/3)	A (2/5)	BAA	$\frac{1}{3} \cdot \frac{1}{3} \cdot \frac{2}{5}$ or $\frac{2}{45}$
		B (3/5)	BAB	$\frac{1}{3} \cdot \frac{1}{3} \cdot \frac{3}{5}$ or $\frac{3}{45}$
	B (2/3)	A (1/5)	BBA	$\frac{1}{3} \cdot \frac{2}{3} \cdot \frac{1}{5}$ or $\frac{2}{45}$
		B (4/5)	BBB	$\frac{1}{3} \cdot \frac{2}{3} \cdot \frac{4}{5}$ or $\frac{8}{45}$

Thus the probability of obtaining a B on the draw from box 3 in this experiment is $\frac{12}{45} + \frac{8}{45} + \frac{3}{45} + \frac{8}{45} = \frac{31}{45}$.

NOW TRY THIS 7-2

- Suppose that in Example 7-8, it is known that the letter A was drawn on the first draw. What is the probability that
  a. the last letter drawn is a B?
  b. the last letter drawn is an A?
  c. the last 2 letters drawn will match, that is, 2 A's or 2 B's?

Modeling Games

We can use models to analyze games that involve probability. Consider the following game, which Arthur and Guinevere play: There are two colored marbles and one white marble in a box. Guinevere mixes the marbles, and Arthur draws two marbles at random without replacement. If the two marbles match, Arthur wins; otherwise, Guinevere wins. Does each player have an equal chance of winning? We *develop a model* to analyze the game. One possible model is a tree diagram, as shown in Figure 7-17.

Figure 7-17

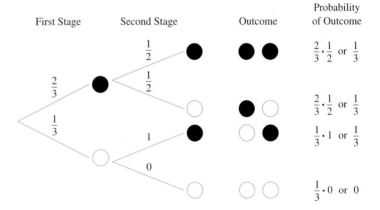

The probability that the marbles are the same color is $\frac{1}{3} + 0$, or $\frac{1}{3}$, and the probability that they are not the same color is $\frac{1}{3} + \frac{1}{3}$, or $\frac{2}{3}$. Because $\frac{1}{3} \neq \frac{2}{3}$, the players do not have the same chance of winning.

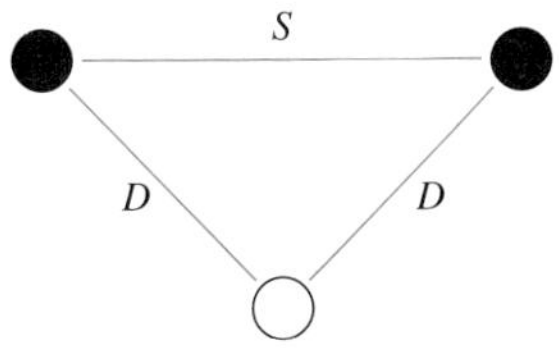

An alternative model for analyzing this game is given in Figure 7-18, where the colored and white marbles are shown along with the possible ways of drawing two marbles. Each line segment in the diagram represents one pair of marbles that could be drawn. S indicates that the marbles in the pair are the same color, and D indicates that the marbles

Figure 7-18

are different colors. Because there are two *D*'s in Figure 7-18, we see that the probability of drawing two different-colored marbles is $\frac{2}{3}$. Likewise, the probability of drawing two marbles of the same color is $\frac{1}{3}$. Because $\frac{2}{3} \neq \frac{1}{3}$, the players do not have an equal chance of winning. Will adding another white marble give each player an equal chance of winning? With two white and two colored marbles, we have the model in Figure 7-19. Therefore $P(D) = \frac{4}{6}$, or $\frac{2}{3}$, and $P(S) = \frac{2}{6}$, or $\frac{1}{3}$. We see that adding a white marble does not change the probabilities.

Figure 7-19

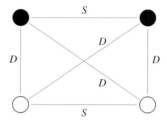

Next, consider a game with the same rules but using three colored marbles and one white marble. Figure 7-20 shows a model for this situation.

Figure 7-20

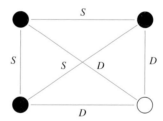

Thus the probability of drawing two marbles of the same color is $\frac{3}{6}$, and the probability of drawing two marbles of different colors is $\frac{3}{6}$. Finally, we have a game in which each player has an equal chance of winning.

NOW TRY THIS 7-3

● Referring to the games described above, answer the following:

a. Does each player have an equal chance of winning if only one white marble and one colored marble are used and the marble is replaced after the first draw?

b. Find games that involve different numbers of marbles in which each player has an equal chance of winning.

c. Find a pattern for the numbers of colored and white marbles that allow each player to have an equal chance of winning. ●

Problem Solving   A String-Tying Game

In a party game, a child is handed six strings, as shown in Figure 7-21(a). Another child ties the top ends two at a time, forming three separate knots, and the bottom ends, forming three separate knots, as in Figure 7-21(b). If the strings form one closed ring, as in Figure 7-21(c), the child tying the knots wins a prize. What is the probability that the child wins a prize on the first try?

Figure 7-21

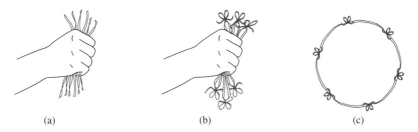

(a)   (b)   (c)

- **Understanding the Problem** The problem is to determine the probability that one closed ring will be formed. One closed ring means that all six pieces are joined end to end to form one, and only one, ring, as shown in Figure 7-21(c).

- **Devising a Plan** Figure 7-22(a) shows what happens when the ends of the strings of one set are tied in pairs at the top. Notice that no matter in what order those ends are tied, the result appears as in Figure 7-22(a).

Figure 7-22

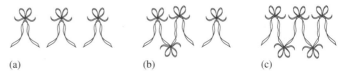

(a)   (b)   (c)

Then, the other ends are tied in a three-stage experiment. If we pick any string in the first stage, then there are five choices for its mate. Four of these choices are favorable choices for forming a ring. Thus the probability of forming a favorable first tie is $\frac{4}{5}$. Figure 7-22(b) shows a favorable tie at the first stage.

For any one of the remaining four strings, there are three choices for its mate. Two of these choices are favorable ones. Thus the probability of forming a favorable second tie is $\frac{2}{3}$. Figure 7-22(c) shows a favorable tie at the second stage.

Now, two ends remain. Since nothing can go wrong at the third stage, the probability of making a favorable tie is 1. If we use the probabilities completed at each stage and a single branch of a tree diagram, we can calculate the probability of performing three successful ties in a row and hence the probability of forming one closed ring.

- **Carrying Out the Plan** If we let S represent a successful tie at each stage, then the branch of the tree with which we are concerned is the one shown in Figure 7-23.

Figure 7-23

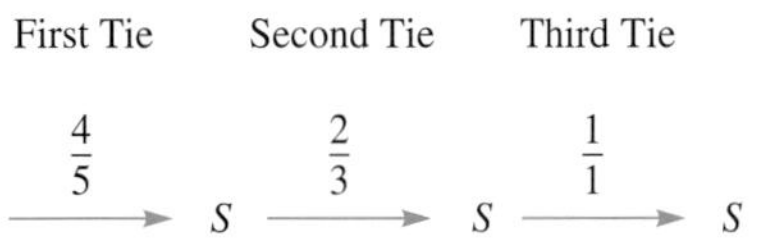

First Tie   Second Tie   Third Tie

$$\xrightarrow{\frac{4}{5}} S \xrightarrow{\frac{2}{3}} S \xrightarrow{\frac{1}{1}} S$$

Thus the probability of forming one ring is $P(\text{ring}) = \frac{4}{5} \cdot \frac{2}{3} \cdot \frac{1}{1} = \frac{8}{15} = 0.5\overline{3}$.

- **Looking Back** The probability that a child will form a ring on the first try is $\frac{8}{15}$. A class might simulate this problem several times with strings to see how the fraction of successes compares with the theoretical probability of $\frac{8}{15}$.

  Related problems that could be posed for solution include the following:

  1. If a child fails to get a ring ten times in a row, the child may not play again. What is the probability of such a streak of bad luck?
  2. If the number of strings is reduced to three and the rule is that an upper end must be tied to a lower end, what is the probability of a single ring?
  3. If the number of strings is three, but an upper end can be tied to either an upper or a lower end, what is the probability of a single ring?
  4. What is the probability of forming three rings in the original problem?
  5. What is the probability of forming two rings in the original problem?

independent

A concept related to multistage experiments is that of independent events. When the outcome of one event has no influence on the outcome of a second event, the events are called **independent**. For example, if two coins are flipped and event E_1 is obtaining a head on the first coin and E_2 is obtaining a tail on the second coin, then E_1 and E_2 are independent events because one event has no influence on the second. Notice that $P(E_1) = \frac{1}{2}$, $P(E_2) = \frac{1}{2}$, and $P(E_1 \cap E_2) = \frac{1}{4}$. So in this case, $P(E_1 \cap E_2) = P(E_1) \cdot P(E_2)$.

Next, consider two boxes: box 1 contains 2 white and 2 black balls, and box 2 contains 2 white balls and 3 black balls. Let B_1 be the event of drawing a black ball from box 1 and B_2 the event of drawing a black ball from box 2. Notice that the events are independent. Suppose a ball is drawn from each box and we are interested in the probability that each ball is black; that is, $P(B_1 \cap B_2)$. We know that $P(B_1) = \frac{2}{4}$ and $P(B_2) = \frac{3}{5}$. Is $P(B_1 \cap B_2) = P(B_1) \cdot P(B_2)$ in this case as well? We can answer this question by computing $P(B_1 \cap B_2)$ using a familiar approach. If we consider all the black balls to be different and all the white balls to be different, there are $4 \cdot 5$ different pairs of balls. (Why?) Among these pairs, we are interested in pairs consisting only of black balls. There are $2 \cdot 3$ such pairs. (Why?) Hence,

$$P(B_1 \cap B_2) = \frac{2 \cdot 3}{4 \cdot 5} = \frac{2}{4} \cdot \frac{3}{5} = \frac{1}{2} \cdot \frac{3}{5}$$

Thus we see that for the above independent events B_1 and B_2, we have $P(B_1 \cap B_2) = P(B_1) \cdot P(B_2)$. The property can be generalized as follows:

Property

For any independent events E_1 and E_2,

$$P(E_1 \cap E_2) = P(E_1) \cdot P(E_2)$$

The student page below from *Scott Foresman-Addison Wesley Middle School MATH Course 2,* 1999, gives examples of problems that involve independent events. Answer questions 2 through 9 on the student page.

12-7 Exercises and Applications

Practice and Apply

1. **Getting Started** Five red cubes and five green cubes are in a bag. Follow the steps to find *P*(green, green)—the probability that you pull two green cubes out of the bag in a row.

   a. What is the probability that the first cube taken from the bag is green?

   b. After a green cube is taken out of the bag, how many green cubes are left? How many cubes are left all together?

   c. What is the probability that the second cube taken from the bag is green?

   d. Find the product of your answers from **a** and **c**. Write your answer in lowest terms.

Tell whether the events are dependent or independent.

2. One tossed coin landing heads and the next landing tails

3. Rolling two sixes in a row on a number cube

4. Being the tallest person in your class one year, then being the tallest again the next year

Exercises 5–7 refer to rolling a number cube, then spinning the spinner shown. Find each probability.

5. *P*(rolling a 2, spinning an A)

6. *P*(rolling an even number, spinning a vowel)

7. *P*(rolling a number less than 3, spinning a consonant)

8. **Social Studies** During Hanukkah, children play with a *dreidel.* The dreidel has four sides, with the Hebrew letters that correspond to the letters N, G, S, and H. The children spin the dreidel like a top, and the letter that comes up determines the result for each turn.

   a. Are the spins of a dreidel dependent or independent events?

   b. What is the probability of spinning 2 Hs in a row?

9. **Science** Suppose the weather report says there is a 25% chance of rain for the next two days.

   a. If the events are independent, what is the probability that it rains *both* days?

   b. Do you think these events are actually independent? Explain why or why not.

Geometric Probability

A probability model that uses geometric shapes is called an *area model*. When area models are used to determine probabilities geometrically, outcomes are associated with points chosen at random in a geometric region that represents the sample space. For example, suppose we throw darts at a square target 2 units long on a side and divided into four congruent triangles, as shown in Figure 7-24. If the dart must hit the target somewhere and if all spots can be hit with equal probability, what is the probability that the dart will land in the shaded region? The entire target, which has an area of 4 square units, represents the sample space. The shaded area is the event of a successful toss. The area of the shaded part is $\frac{1}{4}$ of the sample space. Thus the probability of the dart's landing in the shaded region is the ratio of the area of the event to the area of the sample space, or $\frac{1}{4}$.

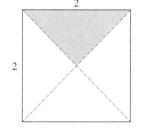

Figure 7-24

Other geometric probability problems are investigated in the problem set. The following problem is solved by using both tree diagrams and a geometric approach.

Problem Solving   A Quiz-Show Game

On a quiz show, a contestant stands at the entrance to a maze that opens into two rooms, as shown in Figure 7-25. The master of ceremonies' assistant is to place a new car in one room and a donkey in the other. The contestant must walk through the maze into one of the rooms and will win whatever is in that room. If the contestant makes each decision in the maze at random, in which room should the assistant place the car to give the contestant the best chance to win?

Figure 7-25

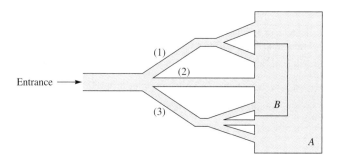

- **Understanding the Problem**   The contestant must first choose one of the paths marked 1, 2, or 3 and then choose other paths as she proceeds through the maze. To determine the room the contestant is most likely to choose, the assistant must be able to determine the probability of the contestant's reaching each room.

- **Devising a Plan**   One way to determine where the car should be placed is to *model the choices with a tree diagram* and to compute the probabilities along the branches of the tree.

- **Carrying Out the Plan**   A tree diagram for the maze is shown in Figure 7-26, along with the possible outcomes and the probabilities of each branch.

Figure 7-26

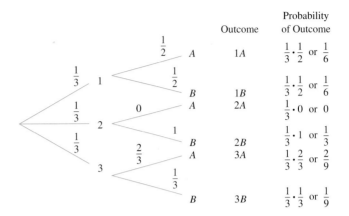

The probability that room A is chosen is $\frac{1}{6} + 0 + \frac{2}{9} = \frac{7}{18}$. Hence the probability that room B is chosen is $1 - \frac{7}{18} = \frac{11}{18}$. Thus room B has the greater probability of being chosen. This is where the car should be placed for the contestant to have the best chance of winning it.

• **Looking Back** An alternative model for this problem and for many probability problems is an area model. The rectangle in Figure 7-27(a) represents the first three choices that the contestant can make. Because each choice is equally likely, each is represented by an equal area. If the contestant chooses the upper path, then rooms A and B have an equal chance of being chosen. If she chooses the middle path, then only room B can be entered. If she chooses the lower path, then room A is entered $\frac{2}{3}$ of the time. This can be expressed in terms of the area model shown in Figure 7-27(b). Dividing the rectangle into pieces of equal area, we obtain the model in Figure 7-27(c), in which the area representing room B is shaded. Because the area representing room B is greater than the area representing room A, room B has the greater probability of being chosen. If we want, Figure 7-27(c) can enable us to find the probability of choosing room B. Because the shaded area consists of 11 rectangles out of a total of 18 rectangles, the probability of choosing room B is $\frac{11}{18}$. We can vary the problem by changing the maze or by changing the locations of the rooms.

Figure 7-27

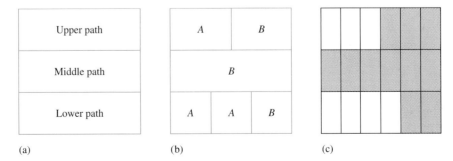

ONGOING ASSESSMENT 7-2

1. **a.** A box contains 3 white balls and 2 red balls. A ball is drawn at random from the box and not replaced. Then a second ball is drawn from the box. Draw a tree diagram for this experiment and find the probability that the two balls are of different colors.

   b. Suppose that a ball is drawn at random from the box in part (a), its color is recorded, and then it is put back in the box. Draw a tree diagram for this experiment and find the probability that the two balls are of different colors.

2. Suppose an experiment consists of spinning X and then spinning Y, as follows:

   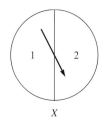 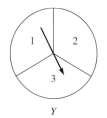

   Find the following:
   a. The sample space S for the experiment
   b. The event A consisting of outcomes from spinning an even number followed by an even number
   c. The event B consisting of outcomes from spinning at least one 2
   d. The event C consisting of outcomes from spinning exactly one 2

3. A box contains six letters, shown as follows. What is the probability of the outcome *DAN* in that order if three letters are drawn one by one
   a. with replacement?
   b. without replacement?

RANDOM

4. Following are three boxes containing letters:

   a. From box 1, three letters are drawn one by one without replacement and recorded in order. What is the probability that the outcome is *HAT*?
   b. From box 1, three letters are drawn one by one with replacement and recorded in order. What is the probability that the outcome is *HAT*?
   c. One letter is drawn at random from box 1, then another from box 2, and then another from box 3, with the results recorded in order. What is the probability that the outcome is *HAT*?
   d. If a box is chosen at random and then a letter is drawn at random from the box, what is the probability that the outcome is A?

5. An executive committee consisted of ten members: four women and six men. Three members were selected at random to be sent to a meeting in Hawaii. A blindfolded woman drew three of the ten names from a hat. All three names drawn were women's. What was the probability of such luck?

6. Two boxes with letters follow. You are to choose a box and draw three letters at random, one by one, without replacement. If the outcome is SOS, you win a prize.

   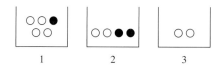

   a. Which box should you choose?
   b. Which box would you choose if the letters are to be drawn with replacement?

7. Following are three boxes containing balls. Draw a ball from box 1 and place it in box 2. Then draw a ball from box 2 and place it in box 3. Finally, draw a ball from box 3.

   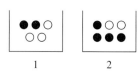

   a. What is the probability that the last ball, drawn from box 3, is white?
   b. What is the probability that the last ball drawn is colored?

8. An assembly line has two inspectors. The probability that the first inspector will miss a defective item is 0.05. If the defective item passes the first inspector, the probability that the second inspector will miss it is 0.01. What is the probability that a defective item will pass by both inspectors?

9. Following are two boxes containing colored and white balls. A ball is drawn at random from box 1. Then a ball is drawn at random from box 2, and the colors of balls from both boxes are recorded in order.

Find each of the following:
a. The probability of two white balls
b. The probability of at least one colored ball
c. The probability of at most one colored ball
d. The probability of ●○ or ○●

10. A penny, a nickel, a dime, and a quarter are tossed. What is the probability of obtaining at least three heads?

11. Assume the probability is $\frac{1}{2}$ that a child born is a boy. What is the probability that if a family is going to have four children, they will all be boys?

12. Brittany is going to ascend a four-step staircase. At any time, she is just as likely to stride up one step or two steps. Find the probability that she will ascend the four steps in
a. two strides.   b. three strides.   c. four strides.

13. A box contains five slips of paper. Each slip has one of the numbers 4, 6, 7, 8, or 9 written on it. There are two players for the game. The first player reaches into the box and draws two slips and adds the two numbers. If the sum is even, the player wins. If the sum is odd, the player loses.
a. What is the probability that the first player wins?
b. Does the probability change if the two numbers are multiplied? Explain.

14. Suppose we spin the following spinner with the first spin giving the numerator and the second spin giving the denominator of a fraction. What is the probability that the fraction will be greater than 1 1/2?

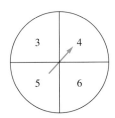

15. The following shows the numbers of symbols on each of the three dials of a standard slot machine:

Symbol	Dial 1	Dial 2	Dial 3
Bar	1	3	1
Bell	1	3	3
Plum	5	1	5
Orange	3	6	7
Cherry	7	7	0
Lemon	3	0	4
Total	20	20	20

Find the probability for each of the following:
a. three plums   b. three oranges
c. three lemons   d. No plums

16. You play a game in which you first choose one of the two spinners shown below. You then spin your spinner and a second person spins the other spinner. The one with the greater number wins.

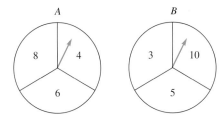

a. Which spinner should you choose? Why?
b. Notice that the sum of the numbers on each spinner is 18. Design two spinners with unequal sums so that choosing the spinner with the least sum will give the player a greater probability of winning.

17. An experiment consists of spinning the spinner shown below and then flipping a coin with sides numbered 1 and 2.

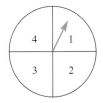

What is the probability that
a. the number on the spinner will be greater than the number on the coin?
b. the outcome will consist of two consecutive integers?

18. If a person takes a five-question true-false test, what is the probability that the score is 100% correct if the person guesses on every question?

19. Rattlesnake and Paxson Colleges play four games against each other in a chess tournament. Rob Fisher, the chess whiz from Paxson, withdrew from the tournament, so the probabilities that Rattlesnake and Paxson will win each game are $\frac{2}{3}$ and $\frac{1}{3}$, respectively. Determine the following probabilities:
a. Paxson loses all four games.
b. The match is a draw with each school winning two games.

20. The combinations on the lockers at the high school consist of three numbers, each ranging from 0 to 39. If a combination is chosen at random, what is the probability that the first two numbers are multiples of 9 and the third number is a multiple of 4?

21. The following box contains the 11 letters shown. The letters are drawn one by one without replacement, and the results are recorded in order. Find the probability of the outcome MISSISSIPPI.

| MIIIPPSSSS |

22. Consider the following dart board: (Assume that all quadrilaterals are squares and that the x's represent equal measures.)

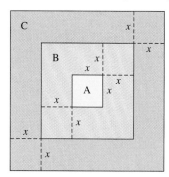

If a dart may hit any point on the board with equal probability, what is the probability that it will land in
   a. section A?   b. section B?   c. section C?

23. In the following square dart board, suppose a dart is equally likely to land in any region of the board.

Points are given as follows:

Region	Points
A	10
B	8
C	6
D	4
E	2

   a. What is the total area of the board?
   b. What is the probability of a dart's landing in each region of the board?
   c. If two darts are tossed, what is the probability of scoring 20 points?
   d. What is the probability that the dart will land in neither D nor E?

24. The land area of Earth is approximately 57,500,000 mi². The water area of Earth is approximately 139,600,000 mi². If a meteor lands at random on the planet, what is the probability, to the nearest tenth, that it will hit water?

25. An electric clock is stopped by a power failure. What is the probability that the second hand is stopped between the 3 and the 4?

26. A husband and wife discover that there is a 10% probability of their passing on a hereditary disease to one of their children. If they plan to have three children, what is the probability that at least one child will inherit the disease?

27. Let $A = \{x \mid {-1} < x < 1\}$ and $B = \{x \mid {-3} < x < 2\}$. If a real number is picked at random from set B, what is the probability it will be in set A?

28. At a certain hospital, 40 patients have lung cancer, 30 patients smoke, and 25 have lung cancer and smoke. Suppose the hospital contains 200 patients. If a patient chosen at random is known to smoke, what is the probability that the patient has lung cancer?

29. There are 40 employees in a certain firm. We know that 28 of these employees are males, 2 of these males are secretaries, and 10 secretaries are employed by the firm. What is the probability that an employee chosen at random is a secretary, given that the person is a male?

30. In a certain population of caribou, the probability of an animal's being sickly is $\frac{1}{20}$. If a caribou is sickly, the probability of its being eaten by wolves is $\frac{1}{3}$. If a caribou is not sickly, the probability of its being eaten by wolves is $\frac{1}{150}$. If a caribou is chosen at random from the herd, what is the probability that it will be eaten by wolves?

31. Four blue socks, 4 white socks, and 4 gray socks are mixed in a drawer. You pull out two socks, one at a time, without looking.
   a. Draw a tree diagram along with the possible outcomes and the probabilities of each branch.
   b. What is the probability of getting a pair of socks of the same color?
   c. What is the probability of getting two gray socks?
   d. Suppose that, instead of pulling out two socks, you pull out 4 socks. What is now the probability of getting two socks of the same color?

32. Solve the Quiz-Show Game on page 372 by replacing Figure 7-25 with the following maze:

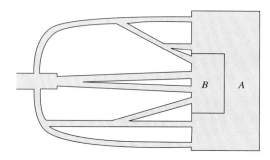

★33. Carolyn will win a large prize if she wins two tennis games in a row out of three games. She is to play alternately against Billie and Bobby. She may choose to play Billie-Bobby-Billie or Bobby-Billie-Bobby. She wins against Billie 50% of the time and against Bobby 80% of the time. Which alternative should she choose, and why?

★34. Jane has two tennis serves, a hard serve and a soft serve. Her hard serve is in (a good serve) 50% of the time, and her soft serve is in (good) 75% of the time. If her hard serve is in, she wins 75% of her points. If her soft serve is in, she wins 50% of her points. Since she is allowed to re-serve one time if her first serve is out, what should her serving strategy be? That is, should she serve hard followed by soft; both hard; soft followed by hard; or both soft?

Communication

35. Jim rolled a fair die three times and obtained a 3 every time. He concluded that on the next roll, a 3 is more likely to occur than the other numbers. Explain whether this is true.
36. If you had 2 witnesses, 1 seeing the red car and 1 not, the question would make more sense. If the only witness saw a blond-haired, blue-eyed driver of a red care, the police wouldn't look at anybody without the red car at first.
37. You are given 3 white balls, 1 red ball, and 2 identical boxes. You are asked to distribute the balls in the boxes in any way you like. You then are asked to select a box (after the boxes have been shuffled) and to pick a ball at random from that box. If the ball is red, you win a prize. How should you distribute the balls in the boxes to maximize your chances of winning? Justify your reasoning.

Open-Ended

38. Make up a game in which the players have an equal chance of winning and that involves rolling two regular dice.
39. How can the faces of two cubes be numbered so that when they are rolled, the resulting sum is a number 1 to 12 inclusive and each sum has the same probability?
40. Use graph paper to design a dart board such that the probability of hitting a certain part of the board is $\frac{3}{5}$. Explain your reasoning.

Cooperative Learning

41. Play the following game in groups of two. One player chooses one of four equally likely outcomes from the sample space {*HH, HT, TH, TT*}, obtained by tossing a fair coin twice. The other player then chooses one of the other outcomes. A coin is flipped until either player's choice appears. For example, the first player chooses *TT* and the second player chooses *HT*. If the first two flips yield *TH*, then no one wins and the game continues. If, after five flips, the string *THHHT* appears, the second player is the winner because the sequence *HT* finally appeared. Play the game ten times. Does each player appear to have the same chance of winning? Analyze the game forthe case in which the first player chooses *TT* and the second *HT*, and explain whether the game is fair. (*Hint:* Find the probability that the first player wins the game by showing that the first player will win if and only if "tails" appears on the first and on the second flip.)

42. Consider the three spinners *A*, *B*, and *C* shown in the following figure:

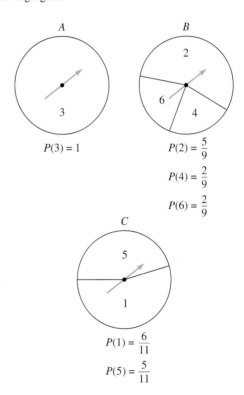

a. Suppose there are only two players and that the first player chooses a spinner and then the second player chooses a different spinner and each person spins his or her spinner with the highest number winning. Play the game several times to get a feeling for it. Determine if each player appears to have the same chances of winning the game. If not, which spinner should you choose in order to win?
b. This time play the same game with three players. If each player must choose a different spinner, is the winning strategy the same as it was in (a)? Why or why not?

Review Problems

43. Match the following phrase to the probability that describes it:

    a. A certain event
    b. An impossible event
    c. A very likely event
    d. An unlikely event
    e. A 50% chance

    (i) $\dfrac{1}{1000}$
    (ii) $\dfrac{999}{1000}$
    (iii) 0
    (iv) $\dfrac{1}{2}$
    (v) 1

44. A date in the month of April is chosen at random. Find the probability of the date's being each of the following:

    a. April 7
    b. April 31
    c. Before April 20

BRAIN TEASER Bradley Efron, a Stanford University statistician, designed a set of nonstandard dice whose faces are numbered as shown in the accompanying figure. The dice are to be used in a game in which each player chooses a die and then rolls it. Whoever rolls the greatest number is the winner. What strategy should you use so that you have the best chance of winning this game?

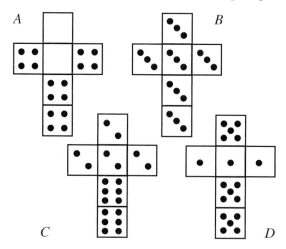

Section 7-3 — Using Simulations in Probability

Students can use simulations to study phenomena too complex to analyze by other means. Using simulations, students can estimate, rather than determine, probabilities analytically.

Suppose we want to simulate the results of tossing a coin 100 times. We could do this using random digits, as in Table 7-3. Random-digit tables are lists of digits selected at random, often by a computer or calculator. To simulate the coin toss, pick a number at random to start and then read across the table, letting an even digit represent heads and an odd digit represent tails. Continue this process for 100 digits. The simulated probability of heads is the ratio of the number of even digits found (heads) to 100. For example, if we choose the top two rows in Table 7-3 and use the first 100 numbers, we find that there are 44 even numbers. Because even numbers represent heads, the simulated probability of tossing a head is $P(H) = 44/100$. Notice that we did not obtain the theoretical probability of 1/2. However, if the number of random digits chosen was much greater, then the simulated probability should approach the theoretical probability.

Similarly, to simulate the probability of a couple's having two girls (GG) in an expected family, we could use the random-digit table with an even digit representing a girl and an odd digit representing a boy. Because there are two children, we need to consider pairs of digits. If we examine 100 pairs, then the simulated probability of GG will be the number of pairs of even digits divided by 100, the total number of pairs considered.

Table 7-3 Random Digits

36422	93239	76046	81114	77412	86557	19549	98473	15221	87856
78496	47197	37961	67568	14861	61077	85210	51264	49975	71785
95384	59596	05081	39968	80495	00192	94679	18307	16265	48888
37957	89199	10816	24260	52302	69592	55019	94127	71721	70673
31422	27529	95051	83157	96377	33723	52902	51302	86370	50452
07443	15346	40653	84238	24430	88834	77318	07486	33950	61598
41348	86255	92715	96656	49693	99286	83447	20215	16040	41085
12398	95111	45663	55020	57159	58010	43162	98878	73337	35571
77229	92095	44305	09285	73256	02968	31129	66588	48126	52700
61175	53014	60304	13976	96312	42442	96713	43940	92516	81421
16825	27482	97858	05642	88047	68960	52991	67703	29805	42701
84656	03089	05166	67571	25545	26603	40243	55482	38341	97782
03872	31767	23729	89523	73654	24626	78393	77172	41328	95633
40488	70426	04034	46618	55102	93408	10965	69744	80766	14889
98322	25528	43808	05935	78338	77881	90139	72375	50624	91385
13366	52764	02407	14202	74172	58770	65348	24115	44277	96735
86711	27764	86789	43800	87582	09298	17880	75507	35217	08352
53886	50358	62738	91783	71944	90221	79403	75139	09102	77826
99348	21186	42266	01531	44325	61042	13453	61917	90426	12437
49985	08787	59448	82680	52929	19077	98518	06251	58451	91140
49807	32863	69984	20102	09523	47827	08374	79849	19352	62726
46569	00365	23591	44317	55054	99835	20633	66215	46668	53587
09988	44203	43532	54538	16619	45444	11957	69184	98398	96508
32916	00567	82881	59753	54761	39404	90756	91760	18698	42852
93285	32297	27254	27198	99093	97821	46277	10439	30389	45372
03222	39951	12738	50303	25017	84207	52123	88637	19369	58289
87002	61789	96250	99337	14144	00027	43542	87030	14773	73087
68840	94259	01961	42552	91843	33855	00824	48733	81297	80411
88323	28828	64765	08244	53077	50897	91937	08871	91517	19668
55170	71062	64159	79364	53088	21536	39451	95649	65256	23950

NOW TRY THIS 7-4

- **a.** Use the random-digit table to estimate the probability that in a family of three, there are two girls and one boy.
- **b.** Determine the theoretical probability of the family's having two girls and one boy and compare the answer to the simulated probability in (a).
- **c.** Should the answers in (a) and (b) always be exactly the same? Why? How can you make sure that the answers are approximately the same?

The following page from *Scott Foresman-Addison Wesley Middle School MATH Course 2,* 1999, shows how middle-school students learn to use simulations to solve probability problems. Work the problem on the bottom of the student page using a graphing calculator and the random-digit table for the random-digit generator.

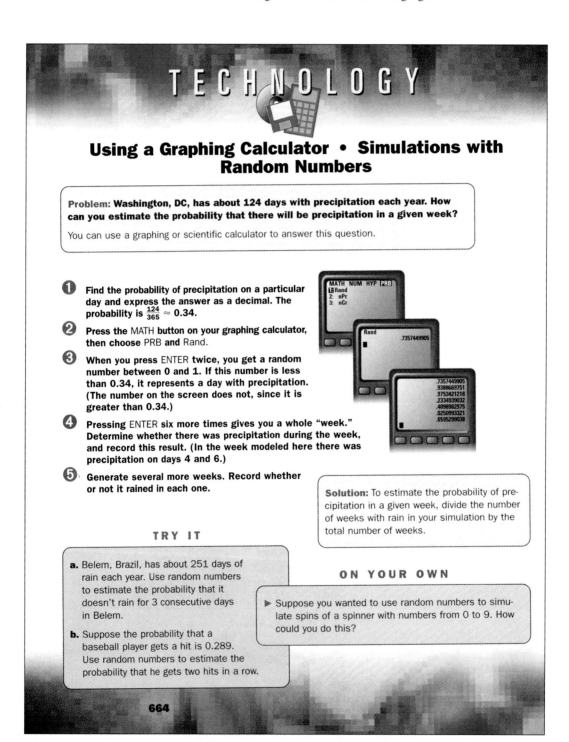

Example 7-9

A baseball player, Reggie, has a batting average of 0.400; that is, his probability of getting a hit on any particular time at bat is 0.400. Estimate the probability that he will get at least one hit in his next three times at bat.

Solution We use a random-digit table to simulate this example. We choose a starting point and place the random digits in groups of three. Because Reggie's probability of getting a hit on any particular time at bat is 0.400, we could use the occurrence of four numbers from 0 through 9 to represent a hit. Suppose a hit is represented by the digits 0, 1, 2, and 3. At least one hit is obtained in three times at bat if, in any sequence of three digits, a 0, 1, 2, or 3 appears. Data for 50 trials are given next:

780	862	760	580	783	720	590	506	021	366
848	118	073	077	042	254	063	667	374	153
377	883	573	683	780	115	662	591	685	274
279	652	754	909	754	892	310	673	964	351
803	034	799	915	059	006	774	640	298	961

We see that a 0, 1, 2, or 3 appears in 42 out of the 50 trials. Thus an estimate for the probability of at least one hit on Reggie's next three times at bat is $\frac{42}{50}$. Try to determine the theoretical probability for this experiment.

From a random sample, we can deduce information about the population from which the sample was taken. To see how this can be done, consider Example 7-10.

Example 7-10

To determine the number of fish in a certain pond, suppose we capture 300 fish, mark them, and throw them back into the pond. Suppose that the next day, 200 fish are caught and 20 of these are already marked. These 200 fish are then thrown back into the pond. Estimate how many fish are in the pond.

Solution Because 20 of the 200 fish are marked, we assume that $\frac{20}{200}$, or $\frac{1}{10}$, of the fish are marked. Thus $\frac{1}{10}$ of the population is marked. If n represents the population, then $\frac{1}{10}n = 300$ and $n = 300 \cdot 10 = 3000$. Hence, an estimate for the fish population of the pond is 3000 fish.

The following Peanuts cartoon suggests a simulation problem concerning chocolate chip cookies.

Example 7-11 Suppose Lucy makes enough batter for exactly 100 chocolate chip cookies and mixes 100 chocolate chips into the batter. If the chips are distributed at random and Charlie Brown chooses a cookie at random from the 100 cookies, estimate the probability that it will contain exactly one chocolate chip.

Solution We can use a simulation to estimate the probability of choosing a cookie with exactly one chocolate chip. We construct a 10×10 grid, as shown in Figure 7-28(a), to represent the 100 cookies Lucy made. Each square (cookie) can be associated with some ordered pair, where the first component is for the horizontal scale and the second is for the vertical scale. For example, the squares (0, 2) and (5, 3) are pictured in Figure 7-28(a). Using the random-digit table, close your eyes and then take a pencil and point to one number to start. Look at the number and the number immediately following it. Consider these numbers an ordered pair and continue on until you obtain 100 ordered pairs to represent the 100 cookies. For example, suppose we start at a 3 and the numbers following 3 are as follows:

$$39968 \qquad 80495 \qquad 00192 \ldots$$

Then the ordered pairs would be given as (3, 9), (9, 6), (8, 8), (0, 4), and so on. Use each pair of numbers as the coordinates for the square (cookie) and place a tally on the grid to represent each chip, as shown in Figure 7-28(b). We estimate the probability that a cookie has exactly one chip by counting the number of squares with exactly one tally and dividing by 100.

Figure 7-28

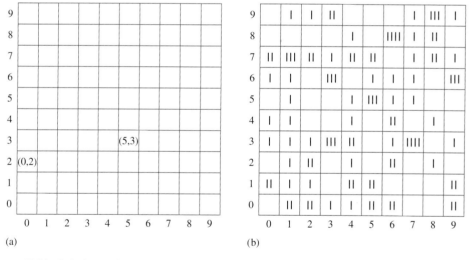

Table 7-4 shows the results of one simulation. Thus the estimate for the probability of Charlie Brown receiving a cookie with exactly one chip is $\frac{34}{100}$.

Table 7-4

Number of Chips	Number of Cookies
0	38
1	34
2	20
3	6
≥ 4	2

Try a simulation on your own and compare your results with the preceding ones and with the results given in Table 7-5, obtained by theoretical methods.

Table 7-5

Number of Chips	Number of Cookies
0	36.8
1	36.8
2	18.4
3	6.1
≥ 4	1.9

ONGOING ASSESSMENT 7-3

1. How might you use a deck of cards to simulate the birth of boys and girls?
2. The weather forecast for Pelican, Alaska, is for a 90% chance of rain on any given day.
   a. How might you simulate the probability of rain in Pelican on any given day?
   b. Use your simulation from (a) to estimate the probability of rain in Pelican for seven days in a row.
   c. What is the theoretical probability of not having rain for seven days in a row in Pelican?
3. How might you use a random-digit table to simulate each of the following?
   a. Tossing a single die
   b. Choosing three people at random from a group of 20 people
   c. Spinning the spinner, where the probability of each color is as shown in the following figure:

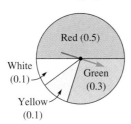

4. A school has 500 students. The principal is to pick 30 students at random from the school to go to the Rose Bowl. How can this be done by using a random-digit table?
5. In a certain city, the probability that it will rain on a certain day is 0.8 if it rained the day before. The probability that it will be dry on a certain day is 0.3 if it was dry the day before. It is now Sunday, and it is raining. Use the random-digit table to simulate the weather for the rest of the week.
6. It is reported that 15% of people who come into contact with a person infected with strep throat contract the disease. How might you use the random-digit table to simulate the probability that at least one child in a three-child family will catch the disease, given that each child has come into contact with the infected person?
7. Pick a block of two digits from the random-digit table. What is the probability that the block picked is less than 30?
8. An estimate of the fish population of a certain pond was found by catching 200 fish and marking and returning them to the pond. The next day, 300 fish were caught, of which 50 had been marked the previous day. Estimate the fish population of the pond.
9. Suppose that in the World Series, the two teams are evenly matched. The two teams play until one team wins four games, and no ties are possible.
   a. What is the maximum number of games that could be played?
   b. Use simulation to approximate the probabilities that the series will end in (i) four games and in (ii) seven games.
10. Assume Carmen Smith, a basketball player, makes free throws with 80% probability of success and is placed in a one-and-one situation where she is given a second foul shot only if the first shot goes through the basket. Simulate the 25 attempts from the foul line in one-and-one situations to determine how many times we would expect Carmen to score 0 points, 1 point, and 2 points.

Communication

11. In an attempt to reduce the growth of its population, China instituted a policy limiting a family to one child. Rural Chinese suggested revising the policy to limit families to one son. Assuming the suggested policy is adopted and that any birth is as likely to produce a boy as a girl, explain how to use simulation to answer the following:
    a. What would be the average family size?
    b. What would be the ratio of newborn boys to newborn girls?

12. Consider a "walk" on the following grid starting out at the origin 0 and "walking" one unit (block) north, and at each intersection turning left with probability $\frac{1}{2}$, turning right with probability $\frac{1}{6}$, and moving straight with probability $\frac{1}{3}$. Explain how to simulate the "walk" using a regular six-sided die.

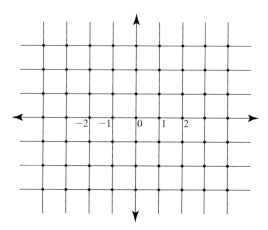

Open-Ended

13. What is the probability that in a group of five people chosen at random, at least two will have birthdays in the same month? Design a simulation for this problem and try your simulation 10 times.
14. The probability of the home team's winning a basketball game is 80%. Describe a simulation of the probability that the home team will win three home games in a row.
15. Montana duck hunters are all perfect shots. Ten Montana hunters are in a duck blind when 10 ducks fly over. All 10 hunters pick a duck at random to shoot at, and all 10 hunters fire at the same time. How many ducks could be expected to escape, on the average, if this experiment were repeated a large number of times? How could this problem be simulated?

Cooperative Learning

16. The sixth-grade class decided that the ideal number of children in a family is four: two boys and two girls.
    a. As a group, design a simulation to determine the probability of two boys and two girls in a family of four.
    b. Have each person in your group try the simulation 25 times and compare the probabilities.
    c. Combine the results of all the members of your group and use this information to find a simulated probability.
    d. Compute the theoretical probability of having two boys and two girls in a family of four and compare your answer to the simulated probability.
17. Have you ever wondered how you would score on a 10-item true-false test if you guessed at every answer?
    a. Simulate your score by tossing a coin 10 times with heads representing true and tails representing false. Check your answers by using the following key and score yourself:

1	2	3	4	5	6	7	8	9	10
F	T	F	F	T	F	T	T	T	F

   b. Combine your results with those of others in your group to find the average (mean) number of correct answers for your group.
   c. How many items would you expect to get correct if the number of items were 30 instead of 10?
   d. What is the theoretical probability of your getting all the answers correct on a 10-item true-false test if all answers were chosen by flipping a coin?
18. a. Estimate how many cards you would expect to have to turn over on the average in an ordinary playing deck before an ace appeared.
    b. Have each person in your group either try the actual experiment or simulate it for 10 trials and find the average number of aces.
    c. Combine the results of all the members of your group and find the average number of cards that needs to be drawn. How does this average compare to your individual estimate?
★19. A cereal company places a coupon bearing a number from 1 to 9 in each box of cereal. If the numbers are distributed at random in the boxes, estimate the number of boxes, on the average, you would have to purchase in order to obtain all nine numbers. Explain how the random-digit table could be used to estimate the number of coupons. Each person in the group should simulate 10 trials, and the results in the group should be combined to find the estimate.

Review Problems

20. In a two-person game, four coins are tossed. If exactly two heads come up, you win. If anything else comes up, you lose. Does each player have an equal chance of winning the game? Explain why or why not.
21. A single card is drawn from an ordinary deck. What is the probability of obtaining each of the following?
    a. A club
    b. A queen and a spade
    c. Not a queen
    d. Not a heart
    e. A spade or a heart
    f. The 6 of diamonds
    g. A queen or a spade
    h. Either red or black
22. From a sack containing 7 red marbles, 8 blue marbles, and 4 white marbles, marbles are drawn at random for several experiments. Determine the probability of each of the following events:
    a. One marble drawn at random is either red or blue.
    b. The first draw is red and the second is blue, where one marble is drawn at random, its color is recorded, the marble is replaced, and another marble is drawn.
    c. The event in (b) where the first marble is not replaced.

TECHNOLOGY CORNER

Edit the following Logo procedure for your software to simulate the rolling of a die.

```
TO ROLL
    OUTPUT (1 + RANDOM 6)
END
```

Execute the procedure with the following line:

```
REPEAT 100 [PRINT ROLL]
```

Count the number of times each digit is printed to estimate the probabilities of obtaining a 1, 2, 3, 4, 5, or 6 when a die is tossed. (a) How close is this approximation to the theoretical probability for each of those events? (b) How would you change the procedure to simulate tossing two dice?

TECHNOLOGY CORNER

RANDINT(

1. On some graphing calculators, if we choose the MATH menu and then select PRB, which stands for PROBABILITY, we find **RANDINT(**, the random integer feature. RANDINT(generates a random integer within a specified range. It requires two inputs that are the upper and lower boundaries for the integers. For example, RANDINT(1, 10) generates a random integer from 1 through 10.
   a. How could you use RANDINT(to simulate tossing a single die?
   b. How could you use RANDINT(to simulate the sum of the numbers when tossing 2 dice?

RAND

int

2. Some graphing calculators have a **RAND** function, a random-digit generator. RAND generates and returns a random number greater than 0 and less than 1. For example, RAND might produce the numbers .5956605, .049599836, or .876572691. To have RAND produce random numbers from 1 to 10 as in (1), we enter int (10 *RAND) + 1. The **int** (greatest integer) feature is found in the MATH menu under NUM. The feature int returns the greatest integer less than or equal to a number.
   a. How could you use RAND to simulate tossing a single die?
   b. How could you use RAND to simulate tossing the sum of the numbers when tossing two dice?

3. Use one of these random features to simulate tossing two dice 30 times. Based on your simulation, what is the probability that a sum of 7 will occur?

Section 7-4 — Odds and Expected Value

Computing Odds

odds in favor

People talk about the *odds in favor of* and the *odds against* a particular event's happening. When the **odds in favor** of the president's being reelected are 4 to 1, this refers to how likely the president is to win the election relative to how likely the president is to lose. The probability of the president's winning is four times the probability of losing. If W represents the event the president wins the election and L represents the event the president loses, then $P(W) = 4P(L)$ or as a proportion, we have

$$\frac{P(W)}{P(L)} = \frac{4}{1}, \text{ or } 4:1.$$

Because W and L are complements of each other, $L = \overline{W}$, we have

$$\frac{P(W)}{P(\overline{W})} = \frac{P(W)}{1 - P(W)} = \frac{4}{1}, \text{ or } 4:1.$$

The **odds against** the president's winning are how likely the president is to lose relative to how likely the president is to win. Using the information above, we have

$$\frac{P(L)}{P(W)} = \frac{1}{4}, \text{ or } 1:4.$$

Because $L = \overline{W}$, we have

$$\frac{P(\overline{W})}{P(W)} = \frac{1 - P(W)}{P(W)} = \frac{1}{4}, \text{ or } 1:4.$$

Formally, odds are defined as follows.

Definition of Odds

Let $P(A)$ be the probability that A occurs and $P(\overline{A})$ be the probability that A does not occur. Then the **odds in favor** of an event A are

$$\frac{P(A)}{P(\overline{A})}, \quad \text{or} \quad \frac{P(A)}{1 - P(A)},$$

and the **odds against** an event A are

$$\frac{P(\overline{A})}{P(A)}, \quad \text{or} \quad \frac{1 - P(A)}{P(A)}.$$

When odds are calculated for equally likely outcomes, the denominators of the probabilities divide out. Thus alternative definitions for odds in case of *equally likely* outcomes are as follows:

$$\text{Odds in favor} = \frac{\text{Number of favorable outcomes}}{\text{Number of unfavorable outcomes}}$$

$$\text{Odds against} = \frac{\text{Number of unfavorable outcomes}}{\text{Number of favorable outcomes}}$$

When you roll a die, the number of favorable ways of rolling a 4 in one throw of a die is 1, and the number of unfavorable ways is 5. Thus the odds in favor of rolling a 4 are 1 to 5.

Example 7-12 For each of the following, find the odds in favor of the event's occurring:

a. Rolling a number less than 5 on a die
b. Tossing heads on a fair coin

c. Drawing an ace from an ordinary 52-card deck
d. Drawing a heart from an ordinary 52-card deck

Solution a. The probability of rolling a number less than 5 is $\frac{4}{6}$; the probability of rolling a number not less than 5 is $\frac{2}{6}$. The odds in favor of rolling a number less than 5 are $\left(\frac{4}{6}\right) \div \left(\frac{2}{6}\right)$, or 4:2, or 2:1.

b. $P(H) = \frac{1}{2}$ and $P(\overline{H}) = \frac{1}{2}$. The odds in favor of getting heads are $\left(\frac{1}{2}\right) \div \left(\frac{1}{2}\right)$, or 1:1.

c. The probability of drawing an ace is $\frac{4}{52}$, and the probability of not drawing an ace is $\frac{48}{52}$. The odds in favor of drawing an ace are $\left(\frac{4}{52}\right) \div \left(\frac{48}{52}\right)$, or 4:48, or 1:12.

d. The probability of drawing a heart is $\frac{13}{52}$, or $\frac{1}{4}$, and the probability of not drawing a heart is $\frac{39}{52}$, or $\frac{3}{4}$. The odds in favor of drawing a heart are $\left(\frac{13}{52}\right) \div \left(\frac{39}{52}\right) = \frac{13}{39}$, or 13:39, or 1:3.

NOW TRY THIS 7-5

- In Example 7-12(a), there are four ways to roll a number less than 5 on a die (favorable outcomes) and two ways of not rolling a number less than 5 (unfavorable outcomes), so the odds in favor of rolling a number less than 5 are 4:2, or 2:1. Work the other three parts of Example 7-12 using this approach.

Given the probability of an event, it is possible to find the odds in favor of (or against) the event and vice versa. For example, if the odds in favor of an event A are 5:1, then the following proportion holds:

$$\frac{P(A)}{1 - P(A)} = \frac{5}{1}$$
$$P(A) = 5[1 - P(A)]$$
$$6P(A) = 5$$
$$P(A) = \frac{5}{6}$$

The probability $\frac{5}{6}$ is a ratio. The exact number of favorable outcomes and the exact total of all outcomes are not necessarily known.

Example 7-13 In the following cartoon, find the probability of making totally black copies if the odds are 3 to 1 against making totally black copies:

TODAY'S ODDS	
Makes totally black copies	3-1
Makes copies with wavy black lines	4-1
Misfeeds	2-1
Gives no change	3-1
Gives double change	8-1
Mystery light appears and 2 or more of the above occur	5-1

Cable

Solution If the odds against making totally black copies are 3 to 1, B represents the event of making a totally black copy and $\bar{B}$ represents not making a totally black copy. We have

$$\frac{P(\bar{B})}{1 - P(\bar{B})} = \frac{3}{1}$$

$$P(\bar{B}) = 3(1 - P(\bar{B}))$$

$$P(\bar{B}) = \frac{3}{4}$$

$$P(B) = 1 - P(\bar{B}), \text{ or } 1 - \frac{3}{4}$$

$$P(B) = \frac{1}{4}.$$

Expected Value

Racetracks use odds for betting purposes. If the odds against Fast Jack are 3:1, this means the track will pay $3 for every $1 you bet. If Fast Jack wins, then for a $5 bet, the track will return your $5 plus $15 more, or $20. The 3:1 odds means the track expects Fast Jack to lose 3 out of 4 times in this situation. If the odds at racetracks were accurate, bettors would receive an even return for their money, that is, bettors would not expect to win or lose money in the long run. For example, if the stated odds of 3:1 against Fast Jack were accurate, then the probability of Fast Jack's losing the race would be 3/4 and for Fast Jack's winning 1/4. If we compute the expected average winnings (expected value) over the long run, the gain is $3 for every $1 bet for a win and a loss of $1 otherwise. The expected value, E, is computed as follows:

$$E = 3(1/4) + {}^-1(3/4) = 0$$

Therefore the expected value is $0. If the racetrack gave accurate odds, it could not stay in business because it could not cover its expenses and make a profit. This is why the track overestimates the horses' chances of winning by about 20%.

Consider the spinner in Figure 7-29, with the payoff in each sector of the circle. Using area models, we can assign the following probabilities to each region:

$$P(\$1.00) = \frac{1}{2} \qquad P(\$2.00) = \frac{1}{4} \qquad P(\$3.00) = \frac{1}{8} \qquad P(\$4.00) = \frac{1}{8}$$

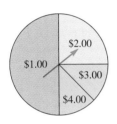

Figure 7-29

Should the owner of this spinner expect to make money over an extended period of time if the charge is $2.00 per spin?

To determine the average payoff over the long run, we find the product of the probability of landing on the payoff and the payoff itself and then find the sum of the products. This computation is given by

$$E = (1/2)1 + (1/4)2 + (1/8)3 + (1/8)4 = 1.875.$$

The owner can expect to pay out about $1.88 per spin. This is less than the $2.00 charge, so the owner should make a profit if the spinner is used many times. The sum of the products in this example, $1.875, is the **expected value,** or *mathematical expectation,* of the experiment of spinning the wheel in Figure 7-29 once. The owner's expected average earnings are $2.00 − $1.875 = $0.125 per spin and the player's expected average earnings (loss) are $^-$$.125.

expected value

The expected value is an average of winnings over the long run. Expected value can be used to predict the average result of an experiment when it is repeated many times. But *an expected value cannot be used to determine the outcome of any single experiment.*

> ### Definition of Expected Value
>
> If, in an experiment, the possible outcomes are numbers $a_1, a_2, \ldots, a_n$, occurring with probabilities $p_1, p_2, \ldots, p_n$, respectively, then the **expected value** (mathematical expectation) E is given by the equation
>
> $$E = a_1 \cdot p_1 + a_2 \cdot p_2 + a_3 \cdot p_3 + \ldots + a_n \cdot p_n.$$

fair game When payoffs are involved and the net winning of a game of chance is $0, the game is a **fair game.** Needless to say, gambling casinos and lotteries make sure that the games are not fair.

> **REMARK** In problem 34, of Ongoing Assessment 7-1, a *fair game* has been defined as a game in which each player has an equal chance of winning the game. If there are two players and each has the same probability of winning a given number of dollars from the other, it follows that the net winning for each player is 0.

Example 7-14 Suppose you pay $5.00 to play the following game. Two coins are tossed. You receive $10 if two heads occur, $5 if exactly one head occurs, and nothing if no heads appear. Is this a fair game? That is, are the net winnings $0?

Solution Before we determine the average payoff, recall that $P(HH) = \frac{1}{4}$, $P(HT \text{ or } TH) = \frac{1}{2}$, and $P(TT) = \frac{1}{4}$. To find the expected value, we perform the following computation:

$$E = \left(\frac{1}{4}\right) \cdot (\$10) + \left(\frac{1}{2}\right) \cdot (\$5) + \left(\frac{1}{4}\right) \cdot (0) = \$5$$

Because the price of playing is equal to the average payoff, the net winnings are $0. This is a fair game.

Problem Solving   A Coin-Tossing Game

Al and Betsy played a coin-tossing game in which a fair coin was tossed until a total of either three heads or three tails occurred. Al was to win when a total of three heads were tossed, and Betsy was to win when a total of three tails were tossed. Each bet $50 on the game. If the coin was lost when Al had two heads and Betsy had one tail, how should the stakes be fairly split if the game is not continued?

- **Understanding the Problem** Al and Betsy each bet $50 on a coin-tossing game in which a fair coin was to be tossed five times. Al was to win when a total of three heads was obtained; Betsy was to win when a total of three tails was obtained. When Al had two heads and Betsy had one tail, the coin was lost. The problem is how to split the stakes fairly.

  Different people could have different interpretations of what "splitting fairly" means. Possibly, though, the best is to split the pot in proportion to the probabilities of each player's winning the game when play was halted. We must calculate the expected value for each player and split the pot accordingly.

- **Devising a Plan** A third head would make Al the winner, whereas Betsy needs two more tails to win. A *tree diagram* that simulates the completion of the game allows us to find the probability of each player's winning the game. Once we find the probabilities, all we need do is multiply the probabilities by the amount of the pot, $100, to determine each player's fair share.

- **Carrying Out the Plan** The tree diagram in Figure 7-30 shows the possibilities for game winners if the game is completed. We can find the probabilities of each player's winning as follows:

$$P(\text{Betsy wins}) = \frac{1}{2} \cdot \frac{1}{2} = \frac{1}{4}$$

$$P(\text{Al wins}) = 1 - \frac{1}{4} = \frac{3}{4}$$

Figure 7-30

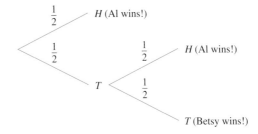

Hence, the fair way to split the stakes is for Al to receive $\frac{3}{4}$ of $100, or $75, and Betsy should receive $\frac{1}{4}$ of $100, or $25.

- **Looking Back** The problem could be made even more interesting by assuming that the coin is not fair so that the probability is not $\frac{1}{2}$ for each branch in the tree diagram. Other possibilities arise if the players have unequal amounts of money in the pot or if more tosses are required in order to win.

ONGOING ASSESSMENT 7-4

1. a. What are the odds in favor of drawing a face card from an ordinary deck of playing cards?
   b. What are the odds against drawing a face card?
2. On a single roll of a pair of dice, what are the odds against rolling a sum of 7?
3. If the probability of a boy's being born is $\frac{1}{2}$, and a family plans to have four children, what are the odds against having all boys?
4. Diane tossed a coin nine times and got nine tails. Assume that Diane's coin is fair and answer each of the following questions:
   a. What is the probability of tossing a tail on the tenth toss?
   b. What is the probability of tossing ten more tails in a row?
   c. What are the odds against tossing ten more tails in a row?
5. If the odds against Deborah's winning first prize in a chess tournament are 3 to 5, what is the probability that she will win first prize?
6. What are the odds in favor of tossing at least two heads if a fair coin is tossed three times?
7. If the probability of rain for the day is 60%, what are the odds against its raining?
8. On an American roulette wheel, half of the slots numbered 1 through 36 are red and half are black. Two slots, numbered 0 and 00, are green. What are the odds against a red slot's coming up on any spin of the wheel?
9. On a tote board at a race track, the odds for Gameylegs are listed as 26:1. Tote boards list the odds that the horse will lose the race. If this is the case, what is the probability of Gameylegs's winning the race?
10. You pay $2.00 to play a game in which two dice are rolled. If a sum of 7 appears, you win $10; otherwise, you lose $2.00. If you intend to play this game for a long time, should you expect to make money, lose money, or come out about even? Explain.
11. On a roulette wheel are 36 slots numbered 1 through 36 and 2 slots numbered 0 and 00. You can bet on a single number. If the ball lands on your number, you receive 35 chips plus the chip you played.
    a. What is the probability that the ball will land on 17?
    b. What are the odds against the ball landing on 17?
    c. If each chip is worth $1, what is the expected payoff for a player who plays the number 17 for a long time?

392 CHAPTER 7 *Probability*

12. Suppose five quarters, five dimes, five nickels, and ten pennies are in a box. One coin is selected at random. What is the expected value of this experiment?
13. If the odds in favor of Fast Leg's winning a horse race are 5 to 2 and the first prize is $14,000, what is the expected value of Fast Leg's winning?
14. Sweepstakes are required by law to display the odds of winning as well as the payoffs. That information is sufficient to calculate the expected value of a sweepstakes. Suppose that mailing the sweepstakes costs $0.33 and the odds in favor of winning the various prizes are as follows:

Odds	Prize	Quantity
1 to 20,000,000	$1,000,000	1
1 to 20,000,000	$100,000	10
1 to 1,000,000	$1,000	100

    a. What is the expected value of the sweeptakes for any individual?
    b. At what postage rate would the drawing be fair?
15. Suppose it costs $8 to roll a pair of dice. You get paid the sum of the numbers in dollars that appear on the dice.
    a. What is the expected value of the game?
    b. Is it a fair game?

Communication

16. Explain the difference between odds and probability.
17. A prominent newspaper reported that the odds of getting AIDS in June 1991 were 68,000 to 1. Explain why you believe or disbelieve this report.
18. A game involves tossing two coins. A player wins $1.00 if both tosses result in heads. What should you pay to play this game in order to make it a fair game? Explain your answer.

Open-Ended

19. An insurance company sells a policy that pays $50,000 in case of accidental death. According to company figures, the rate of accidental death is 47 per 100,000 population. What annual premium should the company charge for this coverage? Explain how much profit the company will make under your plan, how you determined the amount of profit needed for the company, and how the annual premium was computed.

20. Write a game-type problem about odds and payoffs so that the odds in favor of an event are 2:3 and the game is a fair game.

Cooperative Learning

21. As a group, design a game that involves cards, dice, or spinners.
    a. Write the rules so that any person who wants to play can understand the game.
    b. Write a description explaining whether the game is fair and how you arrived at your conclusion.
    c. Calculate the odds of each player's winning.
    d. If betting is involved, discuss expected values.
    e. Exchange a game with another group and compare your analysis of their game with their analysis of your group's game.

Review Problems

22. Refer to the following spinners and write the sample space for each of the following experiments:
    a. Spin spinner 1 once.
    b. Spin spinner 2 once.
    c. Spin spinner 1 once and then spin spinner 2 once.
    d. Spin spinner 2 once and then roll a die.
    e. Spin spinner 1 twice.
    f. Spin spinner 2 twice.

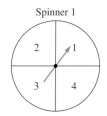
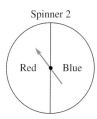

23. Draw a spinner with two sections, red and blue, such that the probability of getting (Blue, Blue) on two spins is $\frac{25}{36}$.
24. Find the probability of getting two vowels when someone draws two letters from the English alphabet with replacement.

Section 7-5 — Methods of Counting

Permutations of Unlike Objects

permutation An arrangement of things in a definite order with no repetitions is a **permutation**. For example, RAT, RTA, ART, ATR, TRA, and TAR are all different arrangements of the three letters R, A, and T. Notice that order is important and there are no repetitions. Determining the number of possible arrangements of the three letters without making a list can be done using the *Fundamental Counting Principle.* Because there are three ways to choose the first letter, two ways to choose the second letter, and one way to choose the third letter, there are $3 \cdot 2 \cdot 1$, or 6 ways to arrange the letters. It is common to record the number of permutations of three objects taken three at a time as $_3P_3$. Therefore, $_3P_3 = 6$.

Consider how many ways the owner of an ice cream parlor can display ten flavors in a row along the front of the display case. The first position can be filled in ten ways, the second position in nine ways, the third position in eight ways, and so on. By the Fundamental Counting Principle, there are $10 \cdot 9 \cdot 8 \cdot 7 \cdot 6 \cdot 5 \cdot 4 \cdot 3 \cdot 2 \cdot 1$, or 3,628,800, ways to display the flavors. If there were 16 flavors, there would be $16 \cdot 15 \cdot 14 \cdot 13 \cdot \ldots \cdot 3 \cdot 2 \cdot 1$ ways to arrange them. In general, *if there are n objects, then the number of possible ways to arrange the objects in a row is the product of all the natural numbers from n to 1, inclusive.* This expression is called **n factorial** and is denoted by **n!**, as shown next.

n factorial (*n*!)

$$n! = n \cdot (n-1) \cdot (n-2) \cdot \ldots \cdot 3 \cdot 2 \cdot 1$$

For example, $5! = 5 \cdot 4 \cdot 3 \cdot 2 \cdot 1$, $3! = 3 \cdot 2 \cdot 1$, and $1! = 1$. Using factorial notation is helpful in counting and probability problems.

Many calculators have a factorial key such as $\boxed{x!}$. To use this key, enter a whole number and then press the factorial key. For example, to compute 5!, press $\boxed{5}$ $\boxed{x!}$ and 120 will appear on the display.

Consider the set of people in a small club, {Al, Betty, Carl, Dan}. For them to elect a president and a secretary, order is important and no repetitions are possible. How many ways are there to elect a committee of two in which one person is president and the other secretary? One way to answer the question is to agree that the choice "Al, Betty" denotes Al as president and Betty as secretary, while the choice "Betty, Al" indicates that Betty is president and Al is secretary. Thus order is important and no repetitions are possible. Consequently, counting the number of possibilities is a permutation problem. Since there are four ways of choosing a president and then three ways of choosing a secretary, by the Fundamental Counting Principle, there are $4 \cdot 3$, or 12, ways of choosing a president and a secretary. Choosing two officers from a club of four is a permutation of four people chosen two at a time. The number of possible permutations of four objects taken two at a time, denoted by $_4P_2$, may be counted using the Fundamental Counting Principle, as seen in Figure 7-31. Therefore we have $_4P_2 = 4 \cdot 3$, or 12.

Figure 7-31

NOW TRY THIS 7-6

a. Write $_nP_2$, $_nP_3$, and $_nP_4$ in terms of *n*.

b. Based on your answers in part (a), write $_nP_r$ in terms of *n* and *r*.

The number of permutations can be written in terms of factorials. Consider the number of permutations of 20 objects three at a time:

$$_{20}P_3 = 20 \cdot 19 \cdot 18$$
$$= \frac{20 \cdot 19 \cdot 18 \,(17 \cdot \ldots \cdot 3 \cdot 2 \cdot 1)}{(17 \cdot \ldots \cdot 3 \cdot 2 \cdot 1)}$$
$$= \frac{20!}{17!}$$
$$= \frac{20!}{(20-3)!}.$$

This can be generalized as follows:

$$_nP_r = \frac{n!}{(n-r)!}.$$

$_nP_n$ is the number of permutations of n objects chosen n at a time—that is, the number of ways of rearranging n objects in a row. We have seen that this number is $n!$. If we use the formula for $_nP_r$ to compute $_nP_n$, we obtain

$$_nP_r = \frac{n!}{(n-n)!} = \frac{n!}{0!}.$$

Consequently, $n! = n!/0!$. To make this equation true, we define $0!$ to be 1.

Many calculators, especially graphing calculators, can calculate the number of permutations of n objects taken r at a time. This feature is usually denoted by $\boxed{_nP_r}$. To use this key, enter the value of n, then press $\boxed{_nP_r}$, followed by the value of r. If you then press $\boxed{=}$ or $\boxed{\text{ENTER}}$, the number of permutations is displayed.

NOW TRY THIS 7-7

a. Try to use a factorial key $\boxed{x!}$ on your calculator to compute $\dfrac{100!}{98!}$. What happens? Why?

b. Without using a calculator, use the definition of factorials to compute the expression in part (a).

Example 7-15

a. A baseball team has nine players. Find the number of ways the manager can arrange the batting order.

b. Find the number of ways of choosing three initials from the alphabet if none of the letters can be repeated.

Solution

a. Because there are nine ways to choose the first batter, eight ways to choose the second batter, and so on, there are $9 \cdot 8 \cdot 7 \cdot \ldots \cdot 2 \cdot 1 = 9!$, or 362,880, ways of arranging the batting order. Using the formula for permutations, we have $_9P_9 = 9!/0! = 362{,}880$.

b. There are 26 ways of choosing the first letter, 25 ways of choosing the second letter, and 24 ways of choosing the third letter. Hence, there are $26 \cdot 25 \cdot 24$, or 15,600, ways of choosing the 3 letters. Using the formula for permutations, we have

$$_{26}P_3 = \frac{26!}{23!} = \frac{26 \cdot 25 \cdot 24 \cdot (23 \cdot 22 \cdot 21 \cdot \ldots \cdot 1)}{23 \cdot 22 \cdot 21 \cdot \ldots \cdot 1}$$
$$= 26 \cdot 25 \cdot 24$$
$$= 15{,}600$$

Permutations Involving Like Objects

In the previous counting examples, each object to be counted was distinct. Suppose we wanted to rearrange the letters in the word *ZOO*. How many choices would we have? A tree diagram, as in Figure 7-32, suggests that there might be $3 \cdot 2 \cdot 1 = 3!$, or 6, possibilities. However, looking at the list of possibilities shows that *ZOO*, *OZO*, and *OOZ* each appears twice because the *O*'s are not different. We need to determine how to remove the duplication in arrangements such as this where some objects are the same. To eliminate the duplication, we divide the number of arrangements shown by the number of ways the two *O*'s can be rearranged, which is 2!. Consequently, there are $\frac{3!}{2!}$, or 3, ways of arranging the letters in *ZOO*. The arrangements are *ZOO*, *OZO*, and *OOZ*.

Figure 7-32

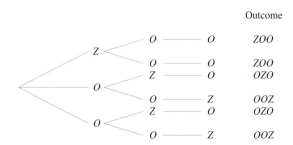

Permutations of Like Objects

If a set contains n elements, of which r_1 are of one kind, r_2 are of another kind, and so on through r_k, then the number of different arrangements of all n elements is equal to

$$\frac{n!}{r_1! \cdot r_2! \cdot r_3! \cdot \ldots \cdot r_k!}.$$

Example 7-16 Find the number of rearrangements of the letters in each of the following words:

a. bubble   **b.** statistics

Solution **a.** There are six letters with b repeated three times. Hence, the number of arrangements is

$$\frac{6!}{3!} = 6 \cdot 5 \cdot 4 = 120.$$

b. There are ten letters in the word *statistics,* with three *s*'s, three *t*'s, and two *i*'s duplicated in the word. Hence, the number of arrangement is

$$\frac{10!}{3! \cdot 3! \cdot 2!} = \frac{10 \cdot 9 \cdot 8 \cdot 7 \cdot 6 \cdot 5 \cdot 4 \cdot 3 \cdot 2 \cdot 1}{3 \cdot 2 \cdot 1 \cdot 3 \cdot 2 \cdot 1 \cdot 2 \cdot 1} = 50{,}400.$$

Combinations

combination

Reconsider the club {Al, Betty, Carl, Dan}. Suppose a two-person committee is selected with no chair. In this case, order is not important, and an Al-Betty choice is the same as a Betty-Al choice. An arrangement of objects in which the order makes no difference is called a **combination.** A comparison of the results of electing a president and a secretary for the club and the results of simply selecting a two-person committee are shown in Figure 7-33. We see that the number of combinations is the number of permutations divided by 2, or

$$\frac{4 \cdot 3}{2} = 6.$$

Because each two-person choice can be arranged in 2!, or 2, ways, we divide the number of permutations by 2.

Figure 7-33

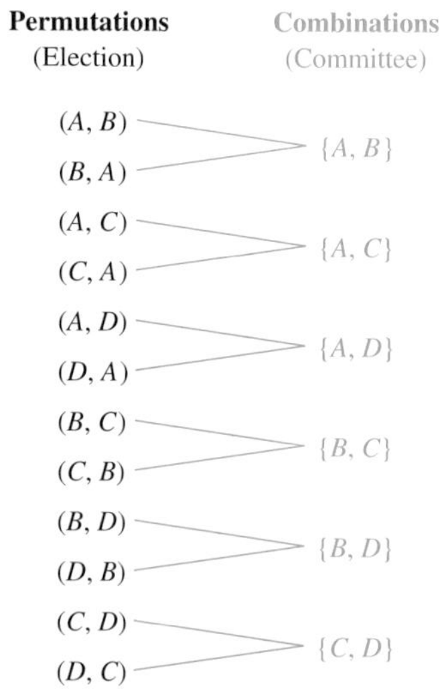

Permutations (Election)	Combinations (Committee)
(A, B) (B, A)	{A, B}
(A, C) (C, A)	{A, C}
(A, D) (D, A)	{A, D}
(B, C) (C, B)	{B, C}
(B, D) (D, B)	{B, D}
(C, D) (D, C)	{C, D}

In how many ways can a committee of three people be selected from the club {Al, Betty, Carl, Dan}? To solve this problem, we proceed as we did above and find the number of ways to select three people from a group of four for three offices, say president, vice president, and secretary (a permutation problem) and then use this result to see how many combinations of people are possible for the committee. Figure 7-34 shows a partial list for both problems.

Figure 7-34

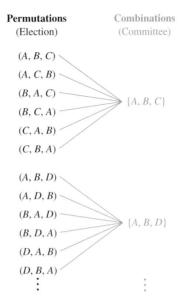

By the Fundamental Counting Principle, if order is important the number of ways to choose three people from the list of four is 4 · 3 · 2, or 24. However, with each triple chosen, there are 3!, or 6, ways to rearrange the triple, as seen in Figure 7-34. Therefore there are 3! times as many permutations as combinations. To find the number of combinations, we divide the number of permutations, 24, by 3!, or 6, to obtain 4. The four committees are {A, B, C}, {A, B, D}, {B, C, D}, and {A, C, D}.

In general, we use the following rule to count combinations: *To find the number of combinations possible in a counting problem, first use the fundamental counting principle to find the number of permutations and then divide by the number of ways in which each choice can be arranged.*

Symbolically, the number of combinations of n objects taken r at a time is denoted by $_nC_r$. From the preceding rule, we develop the following formula:

$$_nC_r = \frac{_nP_r}{_rP_r} = \frac{\frac{n!}{(n-r)!}}{r!} = \frac{n!}{r!(n-r)!}$$

It is not necessary to memorize this formula; we can always find the number of combinations by using the reasoning developed in the committee example.

Example 7-17 The Library of Science Book Club offers three free books from a list of 42. If you circle three choices from a list of 42 numbers on a postcard, how many possible choices are there?

Solution By the Fundamental Counting Principle, there are 42 · 41 · 40 ways to choose the three free books. Because each set of three circled numbers could be rearranged 3 · 2 · 1 different ways, there is an extra factor of 3! in the original $42 \cdot 41 \cdot 40$ ways. Therefore the number of combinations possible for three books is

$$\frac{42 \cdot 41 \cdot 40}{3!} = 11{,}480.$$

Example 7-18 At the beginning of the second quarter of a mathematics class for elementary-school teachers, each of the class's 25 students shook hands with each of the other students exactly once. How many handshakes took place?

Solution Since the handshake between persons A and B is the same as that between persons B and A, this is a problem of choosing combinations of 25 people two at a time. There are

$$_{24}C_2 = \frac{25 \cdot 24}{2!} = 300$$

different handshakes.

Example 7-19 Given a class of 12 girls and 10 boys, answer each of the following:

a. In how many ways can a committee of five consisting of 3 girls and 2 boys be chosen?
b. What is the probability that a committee of five, chosen at random from the class, consists of 3 girls and 2 boys?
c. How many of the possible committees of five have no boys?
d. What is the probability that a committee of five, chosen at random from the class, consists only of girls?

Solution **a.** Based on the information in the problem, we do not assign special functions to members on a committee and, hence, the order of the children on a committee does not matter. From 12 girls we can choose 3 girls in $_{12}C_3$ ways. Each of these choices can be paired with $_{10}C_2$ combinations of boys. Hence the total number of committees is

$$_{12}C_3 \cdot {_{10}C_2} = \frac{12 \cdot 11 \cdot 10}{3!} \cdot \frac{10 \cdot 9}{2} = 9900$$

b. The total number of committees of 5 is $_{22}C_5$. Using part (a), we find the probability that a committee of five will consist of 3 girls and 2 boys to be

$$\frac{_{12}C_3 \cdot {_{10}C_2}}{_{22}C_5} = 0.37594$$

c. The number of ways to choose 5 girls from the 12 girls in the class is

$$_{12}C_5 = \frac{12 \cdot 11 \cdot 10 \cdot 9 \cdot 8}{1 \cdot 2 \cdot 3 \cdot 4 \cdot 5} = 792$$

d. $\dfrac{_{12}C_5}{_{22}C_5} = \dfrac{792}{26334} \doteq 0.030075$

Problem Solving A True-False Test Problem

In the following "Peanuts" cartoon, suppose Peppermint Patty took a six-question true-false test. If she answered each question true or false at random, what is the probability that she answered 50% of the questions correctly?

- **Understanding the Problem** A score of 50% indicates that Peppermint Patty answered $\frac{1}{2}$ of the six questions, or three questions, correctly. She answered the questions true or false at random, so the probability that she answered a given question correctly is $\frac{1}{2}$. We are asked to determine the probability that Patty answered exactly three of the questions correctly.

- **Devising a Plan** We do not know which three questions Patty missed. She could have missed any three out of six on the test. Suppose she answered questions 2, 4, and 5 incorrectly. In this case, she would have answered questions 1, 3, and 6 correctly. We can compute the probability of this set of answers by *using the branch of a tree diagram,* as in Figure 7-35, where C represents a correct answer and I represents an incorrect answer.

Figure 7-35

Question:	1	2	3	4	5	6	Probability of Outcome
	$\xrightarrow{\frac{1}{2}} C$	$\xrightarrow{\frac{1}{2}} I$	$\xrightarrow{\frac{1}{2}} C$	$\xrightarrow{\frac{1}{2}} I$	$\xrightarrow{\frac{1}{2}} I$	$\xrightarrow{\frac{1}{2}} C$	$\left(\frac{1}{2}\right)^6$

Multiplying the probabilities along the branches, we obtain $\left(\frac{1}{2}\right)^6$ as the probability of answering questions 1 through 6 in the following way: $C\,I\,C\,I\,I\,C$. There are other ways to answer exactly three questions correctly: for example, $C\,C\,C\,I\,I\,I$. The probability of answering questions 1 through 6 in this way is also $\left(\frac{1}{2}\right)^6$. The number of ways to answer the questions is simply the number of ways of arranging three C's and three I's in a row, which is also the number of ways of choosing three correct questions out of six, that is, $_6C_3$. Because all these arrangements give Patty a score of 50%, the desired probability is the sum of the probabilities for each arrangement.

- **Carrying Out the Plan** There are $_6C_3$, or 20, sets of answers similar to the one in Figure 7-35, with three correct and three incorrect answers. The product of the probabilities for each of these sets of answers is $\left(\frac{1}{2}\right)^6$, so the sum of the probabilities for all 20 sets is $20 \cdot \left(\frac{1}{2}\right)^6$, or approximately 0.3125. Thus Peppermint Patty has a probability of 0.3125 of obtaining a score of exactly 50% on the test.

- **Looking Back** It seems paradoxical to learn that the probability of obtaining a score of 50% on a six-question true-false test is not close to $\frac{1}{2}$. As an extension of the problem, suppose a passing score is a score of at least 70%. Now what is the probability that Peppermint Patty will pass? What is the probability of her obtaining a score of at least 50% on the test? If the test is a six-question multiple-choice test with five alternative answers for each question, what is the probability of obtaining a score of at least 50% by random guessing?

Problem Solving **Matching Letters to Envelopes**

Stephen placed three letters in envelopes while he was having a telephone conversation. He addressed the envelopes and sealed them without checking if each letter was in the correct envelope. What is the probability that each of the letters was inserted correctly?

- **Understanding the Problem** Stephen sealed three letters in addressed envelopes without checking to see if each was in the correct envelope. We are to determine the probability that each of the three letters was placed correctly. This probability could be found if we knew the sample space, or at least how many elements are in the sample space.

- **Devising a Plan** To aid in solving the problem, we represent the respective letters as a, b, and c and the respective envelopes as A, B, and C. For example, a correctly placed letter a would be in envelope A. To construct the sample space, we use the strategy of *making a table*. The table should show all the possible permutations of letters in envelopes. Once the table is completed, we can determine the probability that each letter is placed correctly.

- **Carrying Out the Plan** Table 7-6 is constructed by using the envelope labels A, B, and C as headings and listing all possibilities of letters a, b, and c below the headings. Case 1 is the only case out of 6 in which each of the envelopes is labeled correctly, so the probability that each envelope is labeled correctly is $\frac{1}{6}$.

Table 7-6

Addresses

Letters	A	B	C
1	a	b	c
2	a	c	b
3	b	a	c
4	b	c	a
5	c	a	b
6	c	b	a

• **Looking Back** Is the probability of having each letter placed incorrectly the same as the probability of having each letter placed correctly? A first guess might be that the probabilities are the same, but that is not true. Why?

We also could have used a counting argument to solve the problem. Given an envelope, there is only one correct letter to place in the envelope. Thus there is one correct way to place the letters in the envelopes. By the Fundamental Counting Principle, there are $3 \cdot 2 \cdot 1$ ways of choosing the letters to place in the envelopes, so the probability of having the letters correctly placed is $\frac{1}{6}$.

ONGOING ASSESSMENT 7-5

1. The eighth-grade class at a grade school has 16 girls and 14 boys. How many different boy-girl dates can be arranged?
2. If a coin is tossed five times, in how many different ways can the sequence of heads and tails appear?
3. The telephone prefix for a university is 243. The prefix is followed by four digits. How many telephones are possible before a new prefix is needed?
4. Radio stations in the United States have call letters that begin with either *K* or *W*. Some have three letters; others have four letters. How many sets of three-letter call letters are possible? How many sets of four-letter call letters are possible?
5. Carlin's Pizza House offers 3 kinds of salads, 15 kinds of pizza, and 4 kinds of desserts. How many different three-course meals can be ordered?
6. Decide whether each of the following is true or false:
   a. $6! = 6 \cdot 5!$
   b. $3! + 3! = 6!$
   c. $\frac{6!}{3!} = 2!$
   d. $\frac{6!}{3} = 2!$
   e. $\frac{6!}{5!} = 6$
   f. $\frac{6!}{4!2!} = 15$
   g. $n!(n+1) = (n+1)!$
7. In how many ways can the letters in the word *SCRAMBLE* be rearranged?
8. How many two-person committees can be formed from a group of six people?
9. Find the number of ways to rearrange the letters in the following words:
   a. *OHIO*
   b. *ALABAMA*
   c. *ILLINOIS*
   d. *MISSISSIPPI*
   e. *TENNESSEE*
10. In a car race, there are six Chevrolets, four Fords, and two Pontiacs. In how many ways can the 12 cars finish if we consider only the makes of the cars?
11. Assume a class has 30 members.
    a. In how many ways can a president, vice president, and a secretary be selected?
    b. How many committees of three people can be chosen?
12. A basketball coach was criticized in the newspaper for not trying out every combination of players. If the team roster has 12 players, how many five-player combinations are possible?
13. A five-volume numbered set of books is placed randomly on a shelf. What is the probability that the books will be numbered in the correct order from left to right?
14. Take 10 points in a plane, no three of them on a line. How many straight lines can be drawn if each line is drawn through a pair of points?
15. Sally has four red flags, three green flags, and two white flags. How many nine-flag signals can she run up a flag-pole?
16. Find the number of shortest paths from point *A* to point *B* along the edges of the cubes in each of the following. (For example, in (a) one shortest path is *A-C-D-B*.)

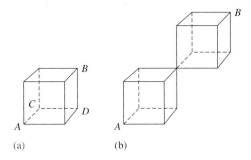

(a) (b)

17. At a party, 28 handshakes took place. Each person shook hands exactly once with each of the others present. How many people were at the party?
18. A committee of three people is selected at random from a set consisting of seven Americans, five French people, and three English people.
    a. What is the probability that the committee consists of all Americans?
    b. What is the probability that the committee has no Americans?
19. How many different five-card hands can be dealt from a standard deck of 52 playing cards?

20. License plates in a certain state have three letters followed by three digits. How many different plates are possible if no repetitions of letters or numbers are allowed?
21. In a certain lottery game, 54 numbers are randomly mixed and six are selected. A person must pick all six numbers to win. Order is not important. What is the probability of winning?
22. Social Security numbers are in the form ###-##-####, where each symbol represents a number 0 through 9. How many Social Security numbers are possible using this format?
23. The probability of a basketball player's making a free throw successfully at any time in a game is $\frac{2}{3}$. If the player attempts ten free throws in a game, what is the probability that exactly six are made?
24. A fair die is rolled eight times. What is the probability of getting
    a. 1 on each of the eight rolls?
    b. 6 exactly twice in the eight rolls?
    c. 6 at least once in the eight rolls?
25. Two fair dice are rolled five times and the sum of the numbers that come up is recorded. Find the probability of getting
    a. a sum of 7 on each of the five rolls.
    b. a sum of 7 exactly twice in the five rolls.
26. From a group of 10 boys and 12 girls, a committee of 4 students is chosen at random. What is the probability that
    a. all four members on the committee will be girls?
    b. there will be at least one girl on the committee?
    c. all four members of the committee will be boys?
27. From a group of 20 Britons, 21 Italians, and 4 Danes, a committee of 8 people is chosen at random. Find the probability that (Express your answers using notation for combinations.)
    a. the committee will consist of 2 Britons, 4 Italians, and 2 Danes.
    b. the committee will have no Britons.
    c. there will be at least one Briton on the committee.
    d. all members of the committee will be Britons.

Communication

28. The terms *fundamental counting principle, permutations,* and *combinations* are all used to work with counting problems. In your own words, explain how all these terms are related and how they are used.
29. Explain why $0! = 1$.
30. a. A bicycle lock has three reels, each of which contains the numbers 0 through 9. To open the lock, you must enter the numbers in the correct order, such as 369 or 455, where one number is chosen from each reel. How many different possibilities are there for the numbers to open the lock? Explain how you arrived at your answer.
    b. These kinds of locks are called *combination* locks. Explain why this is probably not a good name for these locks for someone who has studied counting problems.
31. Dan and Jessica study together. One day the math teacher assigned the following problems:
    Three debate teams with 5, 7, and 8 members, respectively, are to sit in a row of 20 seats. Each team is to sit together.
    (i) *How many seating arrangements are possible?*
    (ii) *In how many ways can this be carried out if two particular people on the third team insist on not sitting next to each other?*
    Dan had no idea how to approach the problems, but Jessica quickly wrote the answers:
    (i) $3! \cdot 5! \cdot 7! \cdot 8!$
    (ii) $3! \cdot 5! \cdot 7! \cdot (8! - 7! \cdot 2)$
    She explained part (i) by saying, "First I think about placing the 3 teams. They can be arranged in 3! ways. Then, in each of these seating arrangements, the first team can be rearranged in 5! ways, the second team in 7! ways, and the third team in 8! ways. By the fundamental counting principle, I multiply the numbers and obtain the answer I wrote."
    Dan was not entirely sure that her approach was correct, but he was even more puzzled by the answer to part (ii). Jessica said, "That's easy! The only thing I need to change is the number of seating arrangements for the third team. So instead of 8! ways of rearranging 8 people, I find the number of ways when two particular people do want to sit next to each other and then I subtract that number from 8!. If I consider the two particular people as one unit, then, with the other 6 people, I have 7 units, which can be rearranged in 7! ways. However, for each of these arrangements, I can rearrange the two particular people in 2!, or 2, ways, for a total of $7! \cdot 2$ arrangements. That's where the number $8! - 7! \cdot 2$ comes from." Dan felt that, based on what Jessica did in part (i), the answer for part (ii) should be $3! \cdot 5! \cdot 7! \cdot 8! - 7! \cdot 2$
    a. Is Jessica's solution to part (i) correct? Explain why or why not.
    b. Whose solution in part (ii) is correct? Explain why.
32. a. Ten people are to be seated on 10 chairs in a line. Among them is a family of 3 that does not want to be separated. How many different seating arrangements are possible? Explain how you arrived at your answer.
    b. How many possible seating arrangements are there in part (a) in which the family members do not sit all together? Explain how you arrived at your answer.
33. In how many ways can five couples be seated in a row of ten chairs if no couple is separated? Explain how you arrived at your answer.

Open-Ended

34. Suppose the Department of Motor Vehicles uses only six spaces and the numbers 0 through 9 to create its license plates.
    a. How many license plates are possible?

b. Based on the 2000 census, determine whether there are any states in which the answer in (a) might provide enough license plates?

c. If you were in charge of making license plates for the state of California, describe the method you would use to ensure you would have enough license plates.

Cooperative Learning

35. The following triangular array of numbers is a part of **Pascal's triangle**:

```
                    1              (0)
                  1   1            (1)
                1   2   1          (2)
              1   3   3   1        (3)
            1   4   6   4   1      (4)
          1   5  10  10   5   1    (5)
        1   6  15  20  15   6   1  (6)
```
Row

a. In your group, decide how the triangle was constructed and complete the next two rows.

b. Describe at least three number patterns in Pascal's triangle.
c. Find the sum of the numbers in each row. Predict the sum of the numbers in row ten.
d. The entries in row two are just $_2C_0$, $_2C_1$, and $_2C_2$. Have different members of your group investigate whether a similar pattern holds for other rows in Pascal's triangle.
e. Describe how you could use combinations to find any entry in Pascal's triangle.

Review Problems

36. Two cards are drawn at random without replacement from a deck of 52 cards. What is the probability that
   a. at least one card is an ace?
   b. exactly one card is red?
37. If two regular dice are tossed, what is the probability of tossing a sum greater than 10?
38. Two coins are tossed. You win $5.00 if both coins are heads and $3.00 if both coins are tails and lose $4.00 if the coins do not match. What is the expected value of this game? Is this a fair game?

BRAIN TEASER An airplane can complete its flight if at least $\frac{1}{2}$ of its engines are working. If the probability that an engine fails is 0.01 and all engine failures do not depend on each other, what is the probability of a successful flight if the plane has

a. two engines?   b. four engines?

HINT FOR SOLVING THE PRELIMINARY PROBLEM

Find the probability of the complementary event that all the birthdays are different, and then subtract the result from 1. Proceed as follows: The first child's birthday might fall on any day of the year. The probability that the second child's birthday is different is $\frac{364}{365}$. The probability that the third child's birthday is different from the first two is $\frac{363}{365}$. Hence, the probability that the second child's birthday is different from the first, and the third child's birthday is different from the first and the second is $\frac{364}{365} \cdot \frac{363}{365}$. Continue in this way to find the probability that all 40 children have different birthdays.

QUESTIONS FROM THE CLASSROOM

1. A student claims that if a fair coin is tossed and comes up heads five times in a row, then, according to the law of averages, the probability of tails on the next toss is greater than the probability of heads. What is your reply?
2. A student observes the following spinner and claims that the color red has the highest probability of appearing, since there are two red areas on the spinner. What is your reply?

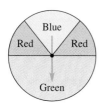

3. A student tosses a coin three times, and tails appears each time. The student concludes that the coin is not fair. What is your response?
4. An experiment consists of tossing a coin twice. The student reasons that there are three possible outcomes: two heads, one head and one tail, or two tails. Thus $P(HH) = \frac{1}{3}$. What is your reply?
5. In response to the question, "If a fair die is rolled twice, what is the probability of rolling a pair of fives?" a student replies, "One third, because $\frac{1}{6} + \frac{1}{6} = \frac{1}{3}$." How do you respond?
6. A student wonders why probabilities cannot be negative. What is your response?
7. A student claims that "if the probability of an event is $\frac{3}{5}$, then there are three ways the event can occur and only five elements in the sample space." How do you respond?
8. A student claims that if the odds in favor of winning a game are $a:b$, then out of every $a + b$ games she would win a games. Hence the probability of winning the game is $\frac{a}{a+b}$. Is the student's reasoning correct? Why or why not?
9. A student does not understand the meaning of $_4P_0$. The student wants to know how we can consider permutations of four objects chosen zero at a time. How do you respond?
10. A student wants to know why, if we can define $0!$ as 1, we cannot define $\frac{1}{0}$ as 1. How do you respond?
11. A student is not sure when to add and when to multiply probabilities. How do you respond?

CHAPTER OUTLINE

I. Probability
   A. Probabilities can be determined **experimentally (empirically)** or **theoretically.**
   B. A **sample space** is the set of all possible outcomes of an **experiment.**
   C. An **event** is a subset of a sample space.
   D. Outcomes are **equally likely** if each outcome is as likely to occur as another.
   E. If all outcomes of an experiment are *equally likely*, the **probability of an event A** from sample space S is given by
   $$P(A) = \frac{n(A)}{n(S)}.$$
   F. An **impossible event** is an event with a probability of zero. An impossible event can never occur.
   G. A **certain event** is an event with a probability of 1. A certain event is sure to happen.
   H. Two events are **mutually exclusive** if, and only if, exactly one of the events can occur at any given time—that is, if, and only if, the events are disjoint.
   I. The probability of the **complement of an event** is given by $P(\overline{A}) = 1 - P(A)$, where A is the event and $\overline{A}$ is its complement.
   J. **Multiplication rule for probabilities** For all **multistage experiments,** the probability of the outcome along any path of a tree diagram is equal to the product of all the probabilities along the path.
   K. If events E_1 and E_2 are independent—that is, the outcome of one does not depend on the outcome of the other—then $P(E_1 \text{ and } E_2) = P(E_1) \cdot P(E_2)$.
   L. **Simulations** can play an important part in probability. Fair coins, dice, spinners, and random-digit tables are useful in performing simulations.

II. Odds and expected value
   A. The **odds in favor** of an event A are given by
   $$\frac{P(A)}{P(\overline{A})} = \frac{P(A)}{1 - P(A)}.$$
   B. The **odds against** an event A are given by
   $$\frac{P(\overline{A})}{P(A)} = \frac{1 - P(A)}{P(A)}.$$

C. If, in an experiment, the possible outcomes are numbers $a_1, a_2, \ldots, a_n$, occurring with probabilities $p_1, p_2, \ldots, p_n$, respectively, then the **expected value** E is defined as

$$E = a_1 \cdot p_1 + a_2 \cdot p_2 + a_3 \cdot p_3 + \ldots + a_n \cdot p_n.$$

D. A **fair game** is a game in which the expected net winnings or expected value is $0.

III. Counting principles
   A. **Fundamental counting principle** If an event M can occur in m ways and, after it has occurred, event N can occur in n ways, then event M followed by event N can occur in $m \cdot n$ ways.
   B. **Permutations** are arrangements in which order is important:

   $$_nP_r = \frac{n!}{(n-r)!}$$

C. The expression **$n!$**, called **n factorial**, represents the product of all the natural numbers less than or equal to n. 0! is defined as 1.

D. **Permutations of like objects** If a set contains n elements, of which r_1 are of one kind, r_2 are of another kind, and so on through r_k, then the number of different arrangements of all n elements is equal to

$$\frac{n!}{r_1! \cdot r_2! \cdot r_3! \cdot \ldots \cdot r_k!}.$$

E. **Combinations** are arrangements in which order is *not* important. To find the number of combinations possible, first use the fundamental counting principle to find the number of permutations and then divide by the number of ways in which each choice can be arranged:

$$_nC_r = \frac{_nP_r}{_rP_r}$$

CHAPTER REVIEW

1. A coin is flipped 3 times and heads (H) or tails (T) are recorded.
   a. List all the elements in the sample space.
   b. List the elements in the event "at least two heads appear."
   c. Find the probability that the event in part (b) occurs.
2. Suppose the names of the days of the week are placed in a box and one name is drawn at random.
   a. List the sample space for this experiment.
   b. List the event consisting of outcomes that the day drawn starts with the letter T.
   c. What is the probability of drawing a day that starts with T?
3. If you have a jar of 1000 jelly beans and you know that $P(\text{Blue}) = \frac{4}{5}$ and $P(\text{Red}) = \frac{1}{8}$, list several things you can say about the beans in the jar.
4. In the 1960 presidential election, John F. Kennedy received 34,226,731 votes and Richard M. Nixon received 34,108,157. If a 1960 voter is chosen at random, answer the following:
   a. What is the probability that the person voted for Kennedy?
   b. What is the probability that the person voted for Nixon?
   c. What are the odds that a person chosen at random did not vote for Nixon?
5. A box contains three red balls, five black balls, and four white balls. Suppose one ball is drawn at random. Find the probability of each of the following events:
   a. A black ball is drawn.
   b. A black or a white ball is drawn.
   c. Neither a red nor a white ball is drawn.
   d. A red ball is not drawn.
   e. A black ball and a white ball are drawn.
   f. A black or white or red ball is drawn.
6. One card is selected at random from an ordinary set of 52 cards. Find the probability of each of the following events:
   a. A club is drawn.
   b. A spade and a 5 are drawn.
   c. A heart or a face card is drawn.
   d. A jack is not drawn.
7. A box contains five colored balls and four white balls. If three balls are drawn one by one, find the probability that they are all white if the draws are made as follows:
   a. With replacement   b. Without replacement
8. Consider the following two boxes. If a letter is drawn from box 1 and placed into box 2 and then a letter is drawn from box 2, what is the probability that the letter is an L?

9. Use the following boxes for a two-stage experiment. First select a box at random and then select a letter at random from the box. What is the probability of drawing an A?

10. Consider the following boxes. Draw a ball from box 1 and put it into box 2. Then draw a ball from box 2 and put it into box 3. Finally, draw a ball from box 3. Construct a tree diagram for this experiment and calculate the probability that the last ball chosen is colored.

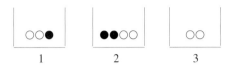

11. What are the odds in favor of drawing a jack when one card is drawn from an ordinary deck of playing cards?
12. A die is rolled once. What are the odds against rolling a prime number?
13. If the odds in favor of a certain event are 3 to 5, what is the probability that the event will occur?
14. A game consists of rolling two dice. Rolling double ones pays $7.20. Rolling double sixes pays $3.60. Any other roll pays nothing. What is the expected value for this game?
15. A total of 3000 tickets have been sold for a drawing. If one ticket is drawn for a single prize of $1000, what is a fair price for a ticket?
16. In a special raffle, a ticket costs $2. You mark any 4 digits on a card (repetition and 0 are allowed). If you select the winning number you win $15,000. What is the expected value?
17. How many four-digit numbers can be formed if the first digit cannot be zero and the last digit must be 2?
18. A club consists of ten members. In how many different ways can a group of three people be selected to go on a European trip?
19. Find the number of ways that four flags can be displayed on a flagpole, one above the other, if ten different flags are available.
20. Five women live together in an apartment. Two have blue eyes. If two of the women are chosen at random, what is the probability that they both have blue eyes?
21. Five horses (Applefarm, Bandy, Cash, Deadbeat, and Egglegs) run in a race.
    a. In how many ways can the first-, second-, and third-place horses be determined?
    b. Find the probability that Deadbeat finishes first and Bandy finishes second in the race.
    c. Find the probability that the first-, second-, and third-place horses are Deadbeat, Egglegs, and Cash, in that order.
22. Al and Ruby each roll an ordinary die once. What is the probability that the number of Ruby's roll is greater than the number of Al's roll?
23. Amy has a quiz on which she is to answer any three of the five questions. If she is equally well versed on all questions and chooses three questions at random, what is the probability that question 1 is not chosen?
24. On a certain street are three traffic lights. At any given time, the probability that a light is green is 0.3. What is the probability that a person will hit all three lights when they are green?
25. A three-stage rocket has the following probabilities for failure. The probability for failure at stage one is $\frac{1}{6}$; at stage two, $\frac{1}{8}$; and at stage three, $\frac{1}{10}$. What is the probability of a successful flight, given that the first stage was successful?
26. How could each of the following be simulated by using a random-digit table?
    a. Tossing a fair die
    b. Picking three months at random from the 12 months of the year
    c. Spinning the spinner shown

27. If a dart is thrown at the following tangram dart board and we assume the dart lands at random on the board, what is the probability of its landing in each of the following areas?
    a. Area A
    b. Area B
    c. Area C

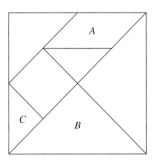

28. The points M, N, O, P, and Q in the following figure represent exits on a highway (the numbers represent miles). An accident occurs at random between points M and Q. What is the probability that it has occurred between N and O?

SELECTED BIBLIOGRAPHY

Bright, G. "Teaching Mathematics with Technology: Probability Simulations." *Arithmetic Teacher* 36 (May 1989): 16–18.

Brulag, D. "Choice and Chance in Life: The Game of 'Skunk.'" *Mathematics Teaching in the Middle School* 1 (April 1994): 28–33.

English, L. "Problem Solving with Combinations." *Arithmetic Teacher* 40 (October 1992): 72–77.

Erickson, D., M. Frank, and R. Kelley. "WITPO (What Is the Probability Of)." *Mathematics Teacher* 84 (April 1991): 258–264.

Fennell, F. "Implementing the *Standards:* Probability." *Arithmetic Teacher* 38 (December 1990): 18–22.

Freda, A. "Roll the Dice; An Introduction to Probability." *Mathematics Teaching in the Middle School* 4 (October 1998): 85–89.

Hatfield, L. "Explorations with Chance." *Mathematics Teacher* 85 (April 1992): 280–282.

Kader, G., and M. Perry. "Push-Penny: What Is Your Expected Score?" *Mathematics Teaching in the Middle School* 3 (February 1998): 370–377.

Kanold, C. "Teaching Probability Theory Modeling Real Problems." *Mathematics Teacher* 85 (April 1994): 232–235.

Lappan, G., et al. "Area Models for Probability." *Mathematics Teacher* 80 (November 1987): 650–654.

Lappan, G., and M. Winter. "Probability Simulation in Middle School." *Mathematics Teacher* 73 (September 1980): 446–449.

Lawrence, A. "From *The Giver* to *The Twenty-One Balloons:* Explorations with Probability." *Mathematics Teaching in the Middle School* 4 (May 1999): 504–509.

Litwiller, B., and D. Duncan. "Combinatorics Connections: Playoff Series and Pascal's Triangle." *Mathematics Teacher* 85 (October 1992): 532–535.

Litwiller, B., and D. Duncan. "Matching Garage Door Openers." *Mathematics Teacher* 85 (March 1992): 217–219.

Martin, H., and J. Zawojewski. "Dealing with Data and Chance: An Illustration from the Middle School Addendum to the Standards." *Arithmetic Teacher* 41 (December 1993): 220–223.

May, E. "Are Seven-Game Baseball Playoffs Fairer?" *Mathematics Teacher* 85 (October 1992): 528–531.

National Council of Teachers of Mathematics. *Dealing with Data and Chance. Grades 5–8 Addenda Book* (Reston, VA: NCTM) 1994.

Quinn, R. "Having Fun with Baseball Statistics." *Mathematics Teaching in the Middle School* 1 (May 1996): 780–785.

Shaughnessy, M. "Probability and Statistics." *Mathematics Teacher* 86 (March 1993): 244–248.

Shaughnessy, J., and T. Dick. "Monty's Dilemma: Should You Stick or Switch?" *Mathematics Teacher* 84 (April 1991): 252–256.

Shulte, A. "Learning Probability Concepts in Elementary School Mathematics." *Arithmetic Teacher* 34 (January 1987): 32–33.

Shultz, H., and B. Leonard. "Probability and Intuition." *Mathematics Teacher* 82 (January 1989): 52–53.

Van Zoest, L., and R. Walker. "Racing to Understand Probability." *Mathematics Teaching in the Middle School* 3 (October 1997): 162–170.

Walton, K. "Probability, Computer Simulation, and Mathematics." *Mathematics Teacher* 83 (January 1990): 22–25.

Wiest, L., and R. Quinn. "Exploring Probability through an Evens-Odds Dice Game." *Mathematics Teaching in the Middle School* 4 (March 1999): 358–362.

Woodword, E., and M. Woodword. "Expected Value and the Wheel of Fortune Game." *Mathematics Teacher* 87 (January 1994): 13–17.

8

Statistics: An Introduction

Preliminary Problem

At a birthday party for the chair of the mathematics department, the honoree would not tell the group his age but agreed to give some hints. He computed and announced that the mean age of his seven guests at the party was 21. When 29-year-old Jill arrived at the party, the honoree announced that the mean age of the eight people was now 22. Jack, another 29-year-old, arrived next. The honoree then added his age to the set of ages of the other nine people and announced that the mean was now 27. How old was the math department chair?

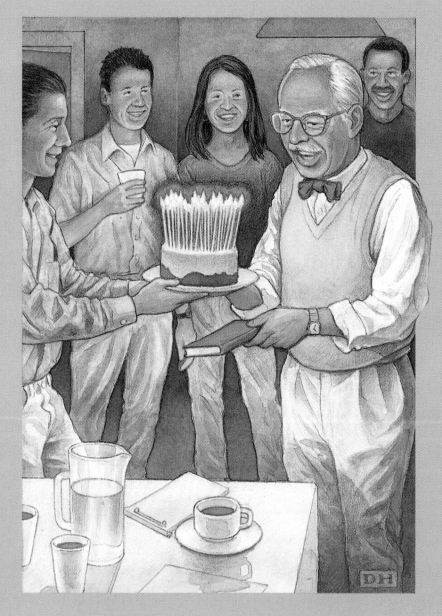

In the *Principles and Standards* we find the following.

A fundamental idea in prekindergarten through grade 2 is that data can be organized or ordered and that this "picture" of the data provides information about the phenomenon or question. In grades 3–5, students should develop skill in representing their data, often using bar graphs, tables, or line plots. They should learn what different numbers, symbols, and points mean. Recognizing that some numbers represent the values of the data and others represent the frequency with which those values occur is a big step. As students begin to understand ways of representing data, they will be ready to compare two or more data sets. Books, newspapers, the World Wide Web, and other media are full of displays of data, and by the upper elementary grades, students ought to learn to read and understand these displays. Students in grades 6–8 should begin to compare the effectiveness of various types of displays in organizing the data for further analysis or in presenting the data clearly to an audience. As students deal with larger or more-complex data sets, they can reorder data and represent data in graphs quickly, using technology so that they can focus on analyzing the data and understanding what they mean (p. 49).

We also find the following.

Beginning in grades 3–5 and continuing in the middle grades, the emphasis should shift from analyzing and describing one set of data to comparing two or more sets (Konold forthcoming). As they move through the middle grades into high school, students will need new tools, including histograms, stem-and-leaf plots, box plots, and scatterplots, to identify similarities and differences among data sets. Students also need tools to investigate association and trends in bivariate data, including scatterplots and fitted lines in grades 6–8 and residuals and correlation in grades 9–12 (p. 50).

For many years, the word *statistics* referred to numerical information about state or political territories. The word itself comes from the Latin *statisticus,* meaning "of the state." We now live in an information age and the study of statistics is more important than ever before. In today's world, much of statistics involves making sense of data.

In the early grades, students explore the basic ideas of statistics by collecting data, organizing the data pictorially, and then interpreting information from their displays. These ideas of gathering, representing, and analyzing data are expanded in the later grades. In this chapter we deal with representations of data and also key statistical concepts including measures of central tendency and of variation. We also address some uses and misuses of statistics.

HISTORICAL NOTE

The seventeenth-century work of John Graunt (1620–1674) and the eighteenth-century work of Adolph Quetelet (1796–1874) involved making predictions on the collection of data. Graunt dealt with birth and death records, while Quetelet dealt with crime and mortality rates. Florence Nightingale (1820–1910) worked with mortality tables during the Crimean War to get British hospitals changed to improve care. Other notables who worked with data collection and analysis include Sir Francis Galton (1822–1911) and Gregor Mendel (1822–1884). In the twentieth century, work continued by Ronald Fisher (1890–1962) in genetics and Andrei Nikolaevich Kolmogorov (1903–1987), who also was chairman of the Commission for Mathematical Education under the Presidium of the Academy of Sciences of the U.S.S.R. Also working in this century have been John Tukey (1915–), who developed many of the current graphical representations used to depict statistics, including stem-and-leaf plots, and Gertrude Mary Cox (1900–1978), who wrote *Experimental Designs* in 1950, a classic textbook on design and analysis of replicated experiments.

CHAPTER 8 Statistics: An Introduction

Section 8-1 Statistical Graphs

data

Visual illustrations are an important part of statistics. Such illustrations or graphs take many forms: pictographs, circle graphs, pie charts, line plots, scatterplots, stem-and-leaf plots, box-and-whisker plots, frequency tables, histograms, bar graphs, and frequency polygons or line graphs. A *graph* is a picture that displays **data.** Graphs are used to try to tell a story. In the "Herman" cartoon, we see a graph being used to display some particular data to make a point to an audience. What message do you think the presenter is trying to get across? What labels might appear on the vertical and horizontal axes?

"That's the last time I go on vacation."

NOW TRY THIS 8-1

- Explain how the graph in Figure 8-1 can be used to tell a story about the water level in a tub when a person prepares and takes a bath.

Figure 8-1

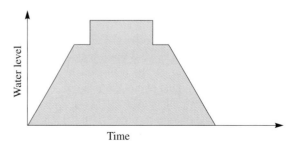

Pictographs

pictograph

One type of graph that children often construct and that we see in newspapers and magazines is a **pictograph.** In a pictograph, a symbol or an icon is used to represent a quantity of items. A *key* that tells what the symbol represents is usually presented. Pictographs are used frequently to show comparisons of outputs, as in Figure 8-2. A major disadvantage of pictographs is evident in Figure 8-2(a). The month of September contains a partial bundle of newspapers. It is impossible to tell from the graph the weight of that bundle with any accuracy.

Figure 8-2

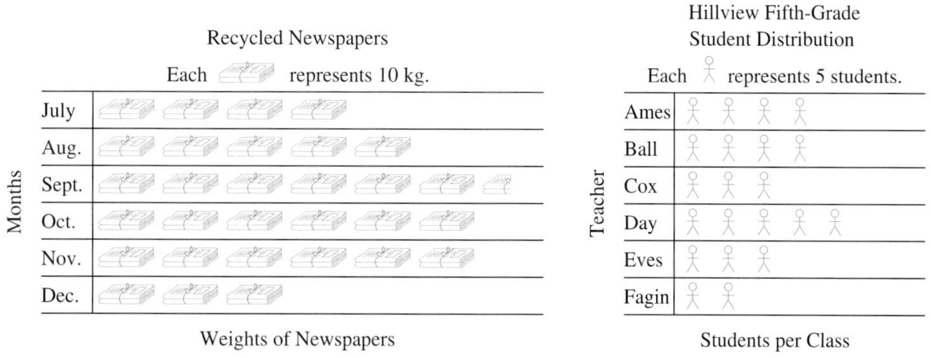

Line Plots

line plot

Next we examine a **line plot.** A line plot is somewhat like a pictograph, but no numerical values are lost in the graph. Line plots provide a quick, simple way of organizing numerical data. Typically, we use them when there is only one group of data with fewer than 50 values.

Suppose the 30 students in Abel's class received the following test scores:

| 82 | 97 | 70 | 72 | 83 | 75 | 76 | 84 | 76 | 88 | 80 | 81 | 81 | 52 | 82 |
| 82 | 73 | 98 | 83 | 72 | 84 | 84 | 76 | 85 | 86 | 78 | 97 | 97 | 82 | 77 |

A line plot for the class scores consists of a horizontal number line on which each score is denoted by an x above the corresponding number-line value, as shown in Figure 8-3. The number of x's above each score indicates how many times each score occurred.

Figure 8-3

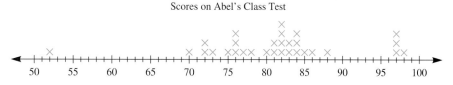

Figure 8-3 yields information about Abel's exam scores. For example, 3 students scored 76 and 4 scored greater than 90. We also see that the low score was 52, the high score was 98, and the most frequent score was 82. Several features of the data become more obvious when line plots are used. For example, outliers, clusters, and gaps are apparent. An **outlier** is a data point whose value is significantly larger or smaller than other values, such as the score of 52 in Figure 8-3. (Outliers are discussed in greater detail in the next section.)

outlier

412 CHAPTER 8 *Statistics: An Introduction*

cluster A **cluster** is an isolated group of points, such as the one located at the scores 97 and 98. A
gap **gap** is a large space between points, such as the one between 88 and 97.

If a line plot is constructed on grid paper, then shading in the squares with x's and adding a vertical axis depicting the scale allows the formation of a *bar graph,* as in Figure 8-4. (Bar graphs are discussed in more detail later in this section.)

Figure 8-4

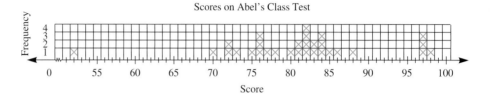

Stem-and-Leaf Plots

stem-and-leaf plot The **stem-and-leaf plot** is closely related to the line plot except that the number line is usually vertical and digits are used rather than x's. A stem-and-leaf plot of test scores for Abel's class is shown in Figure 8-5.

Figure 8-5

Scores on Abel's Class Test

```
5 | 2
6 |
7 | 0223566678
8 | 0112222334444568      9 | 7 represents 97
9 | 7778
```

stems The numbers on the left side of the vertical line are the **stems,** and the numbers on the
leaves right side are the **leaves.** In Figure 8-5, the stems are the tens digits of the scores on the test and the leaves are the units digits. In this case the legend, "9 | 7 represents 97," shows how to read the plot. We now construct a stem-and-leaf plot using the data in Table 8-1, which lists the presidents of the United States and their ages at death.

Table 8-1

President	Age at Death	President	Age at Death	President	Age at Death
George Washington	67	Millard Fillmore	74	Theodore Roosevelt	60
John Adams	90	Franklin Pierce	64	William Taft	72
Thomas Jefferson	83	James Buchanan	77	Woodrow Wilson	67
James Madison	85	Abraham Lincoln	56	Warren Harding	57
James Monroe	73	Andrew Johnson	66	Calvin Coolidge	60
John Q. Adams	80	Ulysses Grant	63	Herbert Hoover	90
Andrew Jackson	78	Rutherford Hayes	70	Franklin Roosevelt	63
Martin Van Buren	79	James Garfield	49	Harry Truman	88
William H. Harrison	68	Chester Arthur	57	Dwight Eisenhower	78
John Tyler	71	Grover Cleveland	71	John Kennedy	46
James K. Polk	53	Benjamin Harrison	67	Lyndon Johnson	64
Zachary Taylor	65	William McKinley	58	Richard Nixon	81

The death ages range from the 40s to the 90s. Thus we concentrate on numbers from 40 to 99. We choose the tens digits of the numbers as the stems. The leaves are the units digits. The plot is formed by placing the stem digits in a column from least to greatest on the left side of a vertical line, as shown in Figure 8-6(a). The leaves (which represent the units digits of the ages) are given on the right side of the vertical line, in whichever row contains their stem, as shown in Figure 8-6(b).

Figure 8-6

Stem	Leaf
4	
5	
6	
7	
8	
9	

(a)

Ages of Presidents at Death

4	96
5	36787
6	785463707034
7	3891470128
8	35081
9	00

4 | 9 represents 49 years old

(b)

In Figure 8-6(b), the top row has 4 as a stem and 9 and 6 as leaves. These numbers represent the ages 49 and 46, the ages at death of James Garfield and John Kennedy, respectively. *The graph should be titled and accompanied by a legend telling how to interpret the symbols used in it.*

In some sense, the data in Figure 8-6(b) are still not orderly because the numbers within each leaf are not in order from least to greatest on a given row. To make an **ordered stem-and-leaf plot,** we arrange the leaves on their rows from least to greatest, starting at the left, as in Figure 8-7.

ordered stem-and-leaf plot

Figure 8-7

Ages of Presidents at Death

4	69
5	36778
6	003344567778
7	0112347889
8	01358
9	00

4 | 9 represents 49 years old

There is no unique way to construct stem-and-leaf plots. Smaller numbers are usually placed at the top so that when the plot is turned counterclockwise 90°, it resembles a bar graph or a histogram (discussed later in this section). Important advantages of stem-and-leaf plots are that they can be created by hand rather easily and they do not become unmanageable when the number of values becomes large. Moreover, no original values are lost in a stem-and-leaf plot. For example, we can still tell that the youngest age at death was 46 and that exactly two presidents died when they were 90. A disadvantage of stem-and-leaf plots is that we do lose some information; for example, we know from the plot that a president died at age 88, but we do not know which one.

Following is a summary of how to construct a stem-and-leaf plot.

1. Find the high and low values of the data.
2. Decide on the stems.
3. List the stems in a column from least to greatest.
4. Use each piece of data to create leaves to the right of the stems on the appropriate rows.
5. If the plot is to be ordered, list the leaves in order from least to greatest.
6. Add a legend identifying the values represented by the stems and leaves.
7. Add a title explaining what the graph is about.

The following page from *Scott Foresman-Addison Wesley Math,* Grade 5, 1999, shows another example of the construction of a stem-and-leaf plot. This plot uses the tens digits for the stems and the units digits for the leaves. Answer the questions in the *Talk About It* and *Check* sections.

Chapter 1 Lesson 3

Reading Stem-and-Leaf Plots

Learn

You Will Learn
how to use a stem-and-leaf plot to answer questions about data

Vocabulary

stem-and-leaf plot
a graph for organizing data

stem
with 2-digit data, the part that shows tens

leaf
with 2-digit data, the part that shows ones

Although Frank and his uncle live in different cities, they don't let the distance come between them.

Frank jotted down the number of minutes each time they chatted online in August. Are most of their chats less than a half-hour?

Minutes per Chat
25 20 23 15 18 32 15 22 8 31
5 23 11 21 30 16 10 28 12 45

Frank from Forrest City, Arkansas and his uncle use their computers to keep in touch.

One way to organize Frank's list is in a **stem-and-leaf plot**. This kind of graph organizes data visually. It makes data easy to understand.

Math Tip
This row shows the data 15, 18, 15, 11, 16, 10, 12.

The tens digit of a number is the stem. ↓
The ones digit of a number is the leaf. ↓

```
4 | 5        4 | 5 represents 45.
3 | 2 1 0
2 | 5 0 3 2 3 1 8
1 | 5 8 5 1 6 0 2
0 | 8 5
```

The shape of the leaves shows most chats are greater than 9 minutes and less than 30 minutes.

Frank and his uncle usually chat for less than a half-hour.

Talk About It

Why is zero used as a stem in the stem-and-leaf plot?

Check

Use the data to answer **1** and **2**.

Number of Calls to the Homework Line in 10 Days

25	26	9	24	18
6	12	7	18	20

Stem-and-Leaf Plot

```
2 | 5 6 4 0
1 | 8 2 8
0 | 9 6 7
```

1. Which numbers are stems? Leaves? What does each stand for?

2. Reasoning On how many days were more than 15 calls made?

14 Chapter 1 • Data, Graphs, and Facts Review

If two sets of related data with a similar number of data values are to be compared, a *back-to-back stem-and-leaf plot* can be used. In this case, two plots are made: one with leaves to the right, and one with leaves to the left. For example, if Abel gave the same test to two classes, he might prepare a back-to-back stem-and-leaf plot, as shown in Figure 8-8.

Figure 8-8

Abel's Test Class Scores

Second-period Class			Fifth-period Class
20	5	2
531	6	24
99987542	7	1257
875420	8	4456999
0 | 5 | represents	1	9	2457
a score of 50		10	0	| 5 | 2 represents a score of 52

NOW TRY THIS 8-2

- In the stem-and-leaf plots in Figure 8-8, which class do you think did better on the test? Why?

Example 8-1

Group the presidents in Table 8-1 into two groups, the first consisting of George Washington to Ulysses Grant and the second consisting of Rutherford Hayes to Richard Nixon.

a. Create back-to-back stem-and-leaf plots of the two groups and see if there appears to be a difference in ages at death between the two groups.
b. Which group of presidents seems to have lived longer?

Solution a. Because the ages at death vary from 46 to 90, the stems vary from 4 to 9. In Figure 8-9, the first 18 presidents are listed on the left and the remaining 18 on the right.

Figure 8-9

Ages of Presidents at Death

Early Presidents			Later Presidents
	4	96
63	5	787
364587	6	707034
741983	7	0128
3 | 8 | represents	053	8	81
83 years old	0	9	0	| 6 | 7 represents 67 years old

b. The early presidents seem, on average, to have lived longer because the ages at the high end, especially in the 70s and 80s, come more often from the early presidents. The ages at the lower end come more often from the later presidents. For the stems in the 50s and 60s, the numbers of leaves are about equal.

A stem-and-leaf plot shows how wide a range of values the data cover, where the values are concentrated, whether the data has any symmetry, where gaps in the data are, and whether any data points are decidedly different from the rest of the data.

Frequency Tables

classes The stem-and-leaf plot in Figure 8-7 naturally groups scores into intervals or **classes.** For the data in Figure 8-7, the following classes are used: 40–49, 50–59, 60–69, 70–79, 80–89, and 90–99. Each class has interval size 10; that is, 10 different scores can fall within the interval 40 and 49. Students often incorrectly report the interval size as 9 because $49 - 40 = 9$.

grouped frequency table A slightly different way to display data is to use a grouped frequency table. A **grouped frequency table** shows how many times a certain piece of data occurs. For example, consider the data in Table 8-1 for the ages of the presidents at death. These results may be summarized in the grouped frequency table in Table 8-2.

Table 8-2

Ages at Death	Tally	Frequency
40–49	‖	2
50–59	⊮	5
60–69	⊮ ⊮ ‖	12
70–79	⊮ ⊮	10
80–89	⊮	5
90–99	‖	2
	Total	36

Figure 8-7 contains more information than does Table 8-2 because the actual death ages are not available in the table. Although Table 8-2 shows that 12 ages appear in the interval 60 through 69, it does not show the particular ages in the interval. As the interval size increases, more information is lost. Choices of interval size may vary. Classes should be chosen to accommodate all the data and each item should fit into only one class; that is, the classes should not overlap. Data from frequency tables can be graphed, as will be shown below.

Histograms and Bar Graphs

histogram The data in Table 8-2 can be pictured graphically using a **histogram,** a graph closely related to a stem-and-leaf plot. Figure 8-10(a) shows a histogram of the frequencies in Table 8-2. A histogram is made up of adjoining rectangles, or bars. In this case, the death ages are shown on the horizontal axis and the numbers along the vertical axis give the scale for the frequency. The frequencies of the death ages are shown by the bars, which are all the same width. The scale on the vertical axis must be uniform. In addition, all histograms should have the axes labeled and should include a title identifying the graph's content.

Figure 8-10

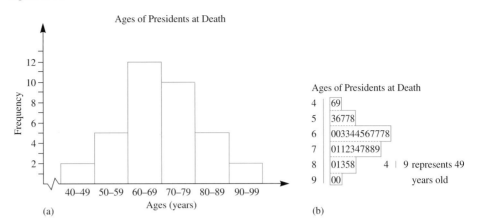

Histograms can be made easily from single-sided stem-and-leaf plots. For example, if we take the stem-and-leaf plot in Figure 8-7 and enclose each row (set of leaves) in a bar, as in Figure 8-10(b), we have what looks like a histogram. We can make Figure 8-10(b) resemble Figure 8-10(a) by rotating the graph 90° counterclockwise. Histograms show gaps and clusters just as stem-and-leaf plots do. However, with a histogram we cannot retrieve data as we can in a stem-and-leaf plot. Another disadvantage of a histogram is that it is often necessary to estimate the heights of the bars.

bar graph A **bar graph** typically has spaces between the bars. A bar graph showing the heights in centimeters of five students is given in Figure 8-11.

Figure 8-11

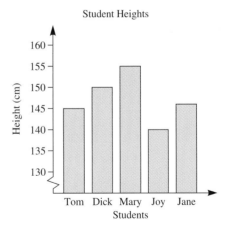

The break in the vertical axis, denoted by a squiggle, indicates that part of the scale has been omitted; therefore the scale is not accurate from 0 to 130. The height of each bar represents the height in centimeters of each student named on the horizontal axis. Each space between bars is usually one half the width of the bars.

double-bar graph **Double-bar graphs** can be used to make comparisons in data. For example, the data in the back-to-back stem-and-leaf plot of Figure 8-9 can be pictured as shown in Figure 8-12. The dark-colored bars represent the later presidents, while the light-colored bars represent the earlier presidents.

Figure 8-12

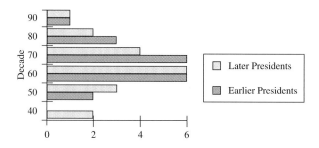

Line Graphs

line graph Another graphical form used to present data is a line graph. A **line graph** typically shows the trend of a variable over time. Time is usually marked on the horizontal axis with the variable being considered marked on the vertical axis. An example is seen in Figure 8-13, where Sanna's weight over 10 years is depicted. Observe that consecutive data points are connected by line segments.

Figure 8-13

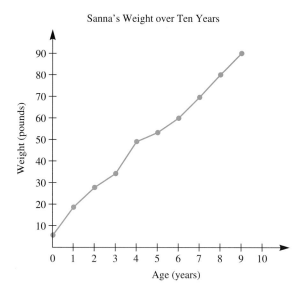

Scatterplots

scatterplot A relationship between two sets of data is sometimes determined by using a **scatterplot.** Figure 8-14(a) shows a scatterplot depicting the relation between the number of hours studied and quiz scores. The highest score is a 10 and the lowest is 1. We see that three students studied 4 hours for the test and that the mode is 5 hours.

Frequently, not all of the points on a scatterplot fall on a particular line, but on some **trend line** scatterplots the points fall near a **trend line.** A trend line is used to make predictions. If a trend line slopes up from left to right as in Figure 8-14(b), then we say there is a *positive* *correlation.* From the trend line in Figure 8-14(b), we would predict that students who studied 7 hours typically scored 6.

Figure 8-14

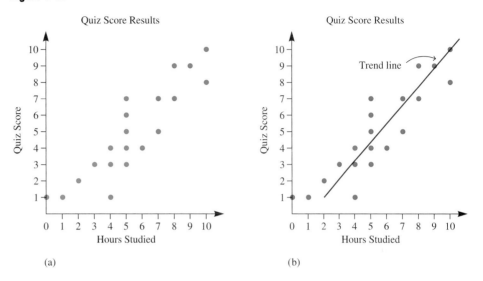

In Figure 8-14, note that the student who studied 4 hours and received a score of 1 did no better than the one who did not study. Although there is a possible correlation between studying and scoring well on a quiz, the example illustrates that studying for a long time does not guarantee a good quiz score. We can use scatterplots and trend lines to make predictions but cannot deduce cause and effect based on them.

If the trend line slopes downward to the right, we can also make predictions; we say there is a *negative correlation.* If the points do not approximately fall on any line, we say there is *no correlation.* Scatterplots also show clusters of points and outliers. Examples of various correlations are given in Figure 8-15.

Figure 8-15

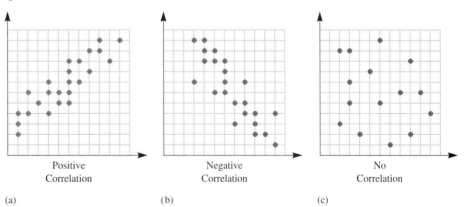

Circle Graphs (Pie Charts)

circle graph · pie chart

Another type of graph used to represent data is the circle graph. A **circle graph,** or **pie chart,** consists of a circular region partitioned into disjoint sections, with each section representing a part or percentage of the whole. A circle graph shows how parts are related to the whole. An example of a circle graph is given in Figure 8-16.

Figure 8-16

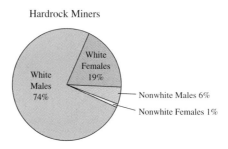

Example 8-2 shows how a circle graph can be constructed from given data.

Example 8-2 Construct a circle graph for the information in Table 8-3, which is based on information taken from a U.S. Bureau of the Census Report, 1994.

Table 8-3

Age	Number of U.S. People (nearest million)
Under 5	20
5–19	54
20–29	39
30–44	62
45–65	48
Over 65	32
Total	255

Solution The entire circle represents the total 255 million people. The measure of the central angle (an angle whose vertex is at the center of the circle) of each sector of the graph is proportional to the fraction or percent of the population the section represents. For example, the measure of the angle for the sector for the under-5 group is 20/255, or approximately 8% of the circle. Because the entire circle is 360°, then 8% of 360°, which is approximately 29°, should be devoted to the under-5 group. Similarly, we can compute the number of degrees for each age group, as shown in Table 8-4. (What formulas could be used in a spreadsheet to create columns 3 and 4?)

Table 8-4

Age	Ratio	Approximate Percent	Approximate Degrees
Under 5	20/255	8	29
5–19	54/255	21	76
20–29	39/255	15	54
30–44	62/255	24	86
45–65	48/255	19	68
Over 65	32/255	13	47
Total	255/255	100	360

The percents and degrees in Table 8-4 are only approximate. A compass can be used to draw a circle and a protractor can be used to draw the sectors in the circle graph, as shown in Figure 8-17.

Figure 8-17

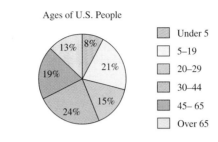

Choosing a Data Display

Choosing an appropriate data display is not always easy. Each type of graph is suitable for presenting certain kinds of data. In this section you have seen pictographs, line plots, stem-and-leaf plots, histograms, bar graphs, line graphs, scatterplots, and circle graphs.

NOW TRY THIS 8-3

- **a.** Which graph in Figure 8-18 displays the data more effectively? Why?

Figure 8-18

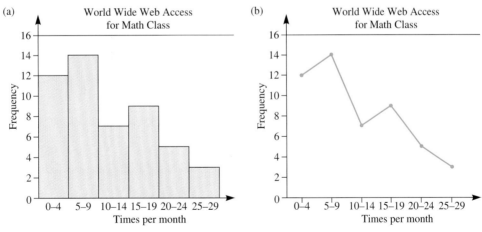

- **b.** To show each of the following, which graph is the best choice: line graph, bar graph, or circle graph? Why?
    (i) The percentage of a college student's budget devoted to housing, clothing, food, tuition and books, taxes
    (ii) Showing the change in the cost of living over the past 12 months

In Section 8-2, you will see another choice for data display, the box-and-whisker plot, as shown below. The following student page from *McDougal Littell's Middle Grades MATH Thematics,* Book 3, 1999, makes suggestions for choosing an appropriate data display.

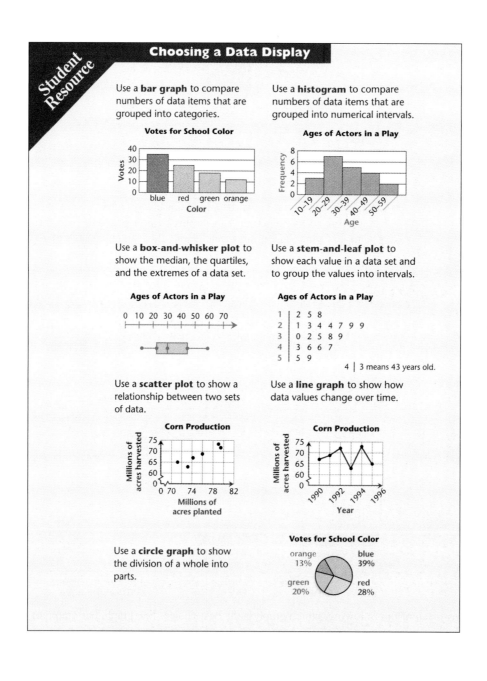

ONGOING ASSESSMENT 8-1

1. Make a pictograph to represent the data in the following table. Use 🥛 to represent 10 glasses of lemonade sold.

    Glasses of Lemonade Sold

Day	Tally	Frequency
Monday	𝍬 𝍬 𝍬	15
Tuesday	𝍬 𝍬 𝍬 𝍬	20
Wednesday	𝍬 𝍬 𝍬 𝍬 𝍬 𝍬	30
Thursday	𝍬	5
Friday	𝍬 𝍬	10

2. The following pictograph shows the approximate number of people who speak the six most common languages on Earth.
    a. About how many people speak Spanish?
    b. About how many people speak English?
    c. About how many more people speak Mandarin than Arabic?

    Number of People Speaking the Six Most Common Languages

    | Arabic | ● ◗ |
    | English | ● ● ● ◖ |
    | Hindi | ● ● ◗ |
    | Mandarin | ● ● ● ● ● ● ◖ |
    | Russian | ● ● ◖ |
    | Spanish | ● ● ◗ |

    Each ● represents 100 million people.

3. Following are the ages of the 30 students from Washington School who participated in the city track meet. Draw a line plot to represent these data.

    | 10 | 10 | 11 | 10 | 13 | 8 | 10 | 13 | 14 | 9 |
    | 14 | 13 | 10 | 14 | 11 | 9 | 13 | 10 | 11 | 12 |
    | 11 | 12 | 14 | 13 | 12 | 8 | 13 | 14 | 9 | 14 |

4. The following stem-and-leaf plot gives the weight in pounds of all 15 students in the Algebra 1 class at East Junior High:
    a. Write the weights of the 15 students.
    b. What is the weight of the lightest student in the class?
    c. What is the weight of the heaviest student in the class?

    Weights of Students in East Junior High Algebra 1 Class

    ```
     7 | 24
     8 | 112578
     9 | 2478
    10 | 3           10 | 3 represents
    11 |                      103 lb
    12 | 35
    ```

5. Draw a histogram based on the stem-and-leaf plot in problem 4 above.
6. Toss a coin 30 times.
    a. Construct a line plot for the data.
    b. Draw a histogram for the data.
7. The following figure shows a bar graph of the rainfall in centimeters during the last school year.

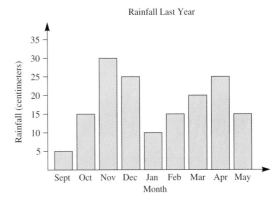

Rainfall Last Year

a. Which month had the most rainfall, and how much did it have?
b. How much total rain fell in October, December, and January?

8. HKM Company employs 40 people of the following ages:

    | 34 | 58 | 21 | 63 | 48 | 52 | 24 | 52 | 37 | 23 |
    | 23 | 34 | 45 | 46 | 23 | 26 | 21 | 18 | 41 | 27 |
    | 23 | 45 | 32 | 63 | 20 | 19 | 21 | 23 | 54 | 62 |
    | 41 | 32 | 26 | 41 | 25 | 18 | 23 | 34 | 29 | 26 |

    a. Draw a stem-and-leaf plot for the data.
    b. Are more employees in their 40s or in their 50s?
    c. How many employees are less than 30 years old?
    d. What percentage of the people are 50 years or older?

9. Given the following bar graph, estimate the length of the following rivers:
    a. Mississippi   b. Columbia

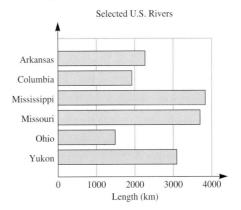

Selected U.S. Rivers

10. Draw a bar graph to represent the data in the following table:

Distances from the Sun

Planet	Distance (millions of miles)
Earth	93
Mars	142
Mercury	36
Venus	67

11. Five coins are tossed 64 times. A distribution for the number of heads obtained is shown in the following table. Draw a histogram for the data.

Number of Heads	0	1	2	3	4	5
Frequency	2	10	20	20	10	2

12. The following table shows the grade distribution for the final examination in the mathematics course for elementary teachers. Draw a circle graph for the data.

Grade	Frequency
A	4
B	10
C	37
D	8
F	1

13. The following are the amounts (to the nearest dollar) paid by 25 students for textbooks during the fall term:

```
35  42  37  60  50
42  50  16  58  39
33  39  23  53  51
48  41  49  62  40
45  37  62  30  23
```

a. Draw an ordered stem-and-leaf plot to illustrate the data.
b. Construct a grouped frequency table for the data, starting the first class at $15.00 with intervals of $5.00 each.
c. Draw a histogram of the data.

14. The following horizontal bar graph gives the top speeds of several animals:

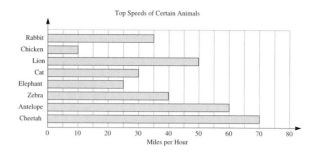

a. Which is the slowest animal shown?
b. How fast can a chicken run?
c. Which animal can run twice as fast as a rabbit?
d. Can a lion outrun a zebra?

15. The following bar graph shows the life expectancies for men and women:

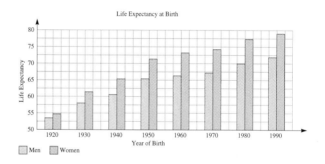

a. Whose life expectancy has changed the most since 1920?
b. In 1920, about how much longer was a woman expected to live than a man?
c. In 1990, about how much longer was a woman expected to live than a man?

16. The following graph shows how the value of a car depreciates each year. This graph allows us to find the trade-in value of a car for each of 5 yr. The percents given in the graph are based on the selling price of the new car.

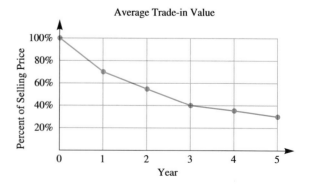

a. What is the approximate trade-in value of a $12,000 car after 1 yr?
b. How much has a $20,000 car depreciated after 5 yr?
c. What is the approximate trade-in value of a $20,000 car after 4 yr?
d. Dani wants to trade in her car before it loses half its value. When should she do this?

17. Use the following circle graph to answer the following questions:

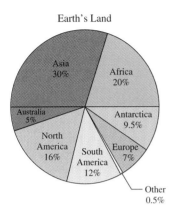

a. Which is the largest continent?
b. Which continent is about twice the size of Antarctica?
c. How does Africa compare in size to Asia?
d. Which two continents make up about half of Earth's surface?
e. What is the ratio of the size of Australia to North America?
f. If Europe has approximately 4.1 million mi^2 of land, what is the total area of the land on Earth?

18. A list of presidents, with the number of children for each, follows:

1. Washington, 0
2. J. Adams, 5
3. Jefferson, 6
4. Madison, 0
5. Monroe, 2
6. J. Q. Adams, 4
7. Jackson, 0
8. Van Buren, 4
9. W. H. Harrison, 10
10. Tyler, 14
11. Polk, 0
12. Taylor, 6
13. Fillmore, 2
14. Pierce, 3
15. Buchanan, 0
16. Lincoln, 4
17. A. Johnson, 5
18. Grant, 4
19. Hayes, 8
20. Garfield, 7
21. Arthur, 3
22. Cleveland, 5
23. B. Harrison, 3
24. McKinley, 2
25. T. Roosevelt, 6
26. Taft, 3
27. Wilson, 3
28. Harding, 0
29. Coolidge, 2
30. Hoover, 2
31. F. D. Roosevelt, 6
32. Truman, 1
33. Eisenhower, 2
34. Kennedy, 3
35. L. B. Johnson, 2
36. Nixon, 2
37. Ford, 4
38. Carter, 3
39. Reagan, 4
40. Bush, 5
41. Clinton, 1

a. Construct a line plot for these data.
b. Make a frequency table for these data.
c. What is the most frequent number of children?

19. Coach Lewis kept track of the basketball team's jumping records for a 10-year period, as follows:

Year	1991	1992	1993	1994	1995	1996
Record (nearest in.)	65	67	67	68	70	74

Year	1997	1998	1999	2000
Record (nearest in.)	77	78	80	81

a. Draw a scatterplot for the data.
b. What kind of correlation is there for these data?

20. Refer to the following scatterplot regarding movie attendance in a certain city.

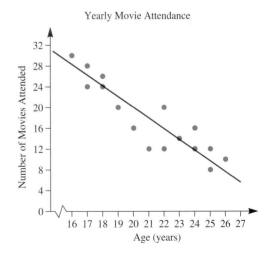

a. What type of correlation exists for these data?
b. About how many movies does an average 25-year-old attend?
c. From the data in the scatterplot, conjecture how old you think a person is who attends 16 movies a year.

21. The following table depicts the number of deaths in males in the United States from Acquired Immunodeficiency Syndrome (AIDS) by age in 1985 and in 1995.

Age (in years)	1985	1995
13–19	28	229
20–29	1505	8434
30–39	3585	25,904
40–49	1636	16,335
50–59	596	4749
60–69	159	1580

a. Choose and construct a graph to display the data.
b. Are any patterns of difference evident in the comparison of the two groups of data?

Communication

22. a. Discuss when a pictograph might be more appropriate than a circle graph.
    b. Discuss when a circle graph might be more appropriate than a bar graph or a line graph.
    c. Discuss when a line graph would be more appropriate than a bar graph.
    d. Give an example of a set of data for which a stem-and-leaf plot would be more informative than a histogram.
23. Explain whether a circle graph would change if the number of data in each category were doubled.
24. Explain why the sum of the percents in a circle graph should always be 100%. How could it happen that the sum is close to 100%?
25. The federal budget for one year is typically depicted with one type of visual representation. Which one is used and why?
26. Tell whether it would be more appropriate to use a bar graph or a line graph for each of the following. In each case, draw the appropriate graph.

a. U.S. Population   b. Continents of the World

Year	Population
1920	105,710,620
1930	122,775,046
1940	131,669,275
1950	150,697,361
1960	179,323,175
1970	203,302,031
1980	226,542,203
1990	248,765,170

Continent	Area in Square Miles (mi²)
Africa	11,694,000
Antarctica	5,100,000
Asia	16,968,000
Australia	2,966,000
Europe	4,066,000
North America	9,363,000
South America	6,886,000

27. Discuss the trend in U.S. car sales from 1983 to 1997 based on the information in the circle graphs given below.

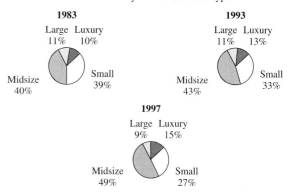

Data taken from *The World Almanac and Book of Facts 1999*, World Almanac Books, 1999.

28. The following graphs give the temperatures for a certain day. Which graph is more helpful for guessing the actual temperature at 10:00 A.M.? Why?

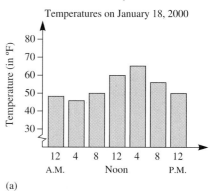

(a)

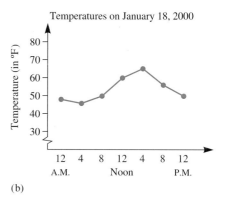

(b)

Open-Ended

29. Find six recent examples of different types of visual representations of data in your newspaper. Explain whether you think the representations are appropriate.

30. Choose a topic, describe how you would go about collecting data on the topic, and then explain how you would display your data in a graph. Tell why you chose the particular graph.

31. Look at the cartoon below and estimate the percentages for each of the 4 parts of the circle graph. Based on your estimates, determine the number of degrees that would be used to construct each angle in the circle graph. Construct a circle graph based on your estimates and computations and compare it to the one in the cartoon.

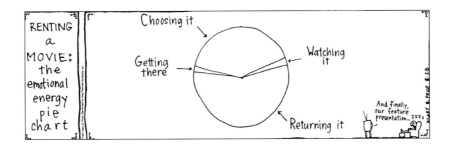

32. Study the graph below and draw some conclusions about the death rates for the selected causes.

U.S. Death Rates for Selected Causes

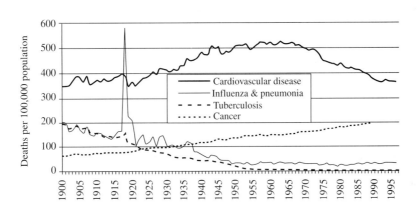

Data taken from *The World Almanac and Book of Facts 1999*, World Almanac Books, 1999.

Cooperative Learning

33. In groups, collect data on head circumference in centimeters of class members. Make a frequency table of all class data, choose an appropriate graph, and display the data. If you were a seller of hats, what is the most common size head for which you would purchase hats?

Section 8-2 — Measures of Central Tendency and Variation

The media present us with a variety of data and statistics. For example, we find in the *World Almanac* that the average person's lifetime includes 6 yr of eating, 4 yr of cleaning, 2 yr of trying to return telephone calls to people who never seem to be in, 6 mo waiting at stop lights, 1 yr looking for misplaced objects, and 8 mo opening junk mail. In the previous section, we examined data by looking at graphs to display the overall distribution of values. In this section, we describe specific aspects of data by using a few carefully chosen numbers. Two important aspects of data are its *center* and its *spread*. The mean, median, and mode are **measures of central tendency** that describe where data are centered. Each of these measures is a single number that describes the data. However, each does it slightly differently. The *range, variance,* and *standard deviation* introduced later in this section describe the spread of data.

measures of central tendency

A word that is often used in statistics is *average*. For example, suppose that, as the cartoon below suggests, the average number of children in a family is 1.5. What does this mean? How can the average number of children be 1.5?

"Bob and Ruth! Come on in .... Have you met Russell and Bill, our 1.5 children?"

To explore more about averages, examine the following set of data for three teachers, each of whom claims that his or her class scored better *on the average* than the other two classes did:

$$\begin{array}{rl} \text{Mr. Smith:} & 62, 94, 95, 98, 98 \\ \text{Mr. Jones:} & 62, 62, 98, 99, 100 \\ \text{Ms. Rivera:} & 40, 62, 85, 99, 99 \end{array}$$

All of these teachers are correct in their assertions because each has used a different number to characterize the scores in his or her class. In the following, we examine how each teacher can justify the claim.

Computing Means

arithmetic mean
average · mean

The number commonly used to characterize a set of data is the **arithmetic mean,** frequently called the **average,** or the **mean.** To find the mean of scores for each of the teachers given previously, we find the sum of the scores in each case and divide by 5, the number of scores.

$$\text{Mean (Smith):} \quad \frac{62 + 94 + 95 + 98 + 98}{5} = \frac{447}{5} = 89.4$$

$$\text{Mean (Jones):} \quad \frac{62 + 62 + 98 + 99 + 100}{5} = \frac{421}{5} = 84.2$$

$$\text{Mean (Rivera):} \quad \frac{40 + 62 + 85 + 99 + 99}{5} = \frac{385}{5} = 77$$

In terms of the mean, Mr. Smith's class scored better than the others. In general, we define the *arithmetic mean* as follows.

Definition of Mean

The **arithmetic mean** of the numbers $x_1, x_2, \ldots, x_n$, denoted by $\bar{x}$ and read "x bar," is given by

$$\bar{x} = \frac{x_1 + x_2 + x_3 + \cdots + x_n}{n}.$$

Understanding the Mean as a Balance Point

Because the mean is the most widely used measure of central tendency, we provide a model for thinking about it. Suppose a student at a rural school reports that the mean number of pets for the six students in a group is 5. Do we know anything about the distribution of these pets? One way to have a mean of 5 is that all six students have exactly five pets, as shown in the line plot in Figure 8-19(a).

Figure 8-19

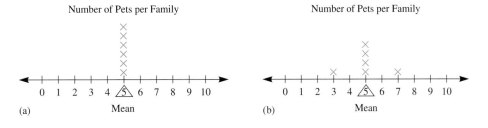

If we change the line plot as shown in Figure 8-19(b), the mean is still 5. Notice that the new line plot could be obtained by moving one value from Figure 8-19(a) 2 units to the right and then balancing this by moving one value 2 units to the left. We can think of the mean as a *balance point*.

Consider Figure 8-20, which shows the number of children for each family in a group. The mean of 5 is the balance point where the sum of the total distances from the mean to the data points above the mean equals the sum of the total distances from the mean to the data points below the mean. The sum of the distances above the mean is 3 + 5, or 8. The sum of the distances below the mean is 1 + 2 + 2 + 3, or 8. In this case, we see that the data are centered about the mean, but the mean does not belong to the set of data.

Figure 8-20

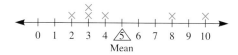

Number of Children per Family

NOW TRY THIS 8-4

a. A litter of six puppies was born with a mean average weight of 7 pounds. List two possibilities for the weights of the pups.

b. Could the mean average of a set of scores ever be equal to the greatest score? the least score? Explain your answer.

Computing Medians

median   The value exactly in the middle of an ordered set of numbers is the **median**. To find the median for the teachers' scores, we arrange each of their scores in increasing or decreasing order and pick the middle score. Intuitively, we know that half the scores are greater than the median and half are less.

$$\text{Median (Smith):} \quad 62, 94, \boxed{95}, 98, 98 \qquad \text{median} = 95$$
$$\text{Median (Jones):} \quad 62, 62, \boxed{98}, 99, 100 \qquad \text{median} = 98$$
$$\text{Median (Rivera):} \quad 40, 62, \boxed{85}, 99, 99 \qquad \text{median} = 85$$

In terms of the median, Mr. Jones's class scored better than the others.

With an odd number of scores, as in the present example, the median is the middle score. With an even number of scores, however, the median is defined as the mean of the middle two scores. Thus to find the median, we add the middle two scores and divide by 2. For example, the median of the scores

$$64, 68, \boxed{70, 74}, 82, 90$$

is given by

$$\frac{70 + 74}{2}, \text{ or } 72.$$

In general, to find the median for a set of n numbers, proceed as follows.

1. Arrange the numbers in order from least to greatest.
2. **a.** If n is odd, the median is the middle number.
   b. If n is even, the median is the mean of the two middle numbers.

Finding Modes

mode The **mode** of a set of data is the number that appears most frequently, if there is one. In some distributions, no number appears more than once. In other distributions, there may be more than one mode. For example, the set of scores 64, 79, 80, 82, 90 has no mode. (Some would say that this set of data has five modes, but because no one score appears more than another, we prefer to say there is no mode.)

bimodal The set of scores 64, 75, 75, 82, 90, 90, 98 is **bimodal** (two modes) because both 75 and 90 are modes. It is possible for a set of data to have too many modes for this type of number to be useful in describing the data.

For the three classes listed previously, if the mode is used as the criterion, Ms. Rivera's class scored better than the others.

Mode (Smith): 62, 94, 95, 98, 98 mode = 98
Mode (Jones): 62, 62, 98, 99, 100 mode = 62
Mode (Rivera): 40, 62, 85, 99, 99 mode = 99

Example 8-3 Find (a) the mean, (b) the median, and (c) the mode for the following collection of data:

60 60 70 95 95 100

Solution **a.** $\bar{x} = \dfrac{60 + 60 + 70 + 95 + 95 + 100}{6} = \dfrac{480}{6} = 80$

b. The median is $\dfrac{70 + 95}{2}$, or 82.5.

c. The set of data is bimodal and has both 60 and 95 as modes.

Example 8-4 When the data values are all the same, the mean, median, and mode are all the same. Describe a situation in which not all the data points are the same and the mean, median, and mode are still the same.

Solution Answers may vary. For example, one set of data is 92, 94, 94, 94, 96, in which the mean, median, and mode are all 94.

> **NOW TRY THIS 8-5**
>
> **a.** Suppose the average number of children per family for the employees of the university in a certain city is 2.58. Could this be a mean? median? mode? Explain why.
>
> **b.** Answer the same questions as above if the average number of children was reported to be 2.5.

Choosing the Most Appropriate Average

Although the *mean* is the number most commonly used to describe a set of data, it may not always be the most appropriate choice.

Example 8-5 Suppose a company employs 20 people. The president of the company earns $200,000, the vice president earns $75,000, and 18 employees earn $10,000 each. Is the mean the best number to choose to represent the "average" salary for the company?

Solution The mean salary for this company is

$$\frac{\$200{,}000 + \$75{,}000 + 18(\$10{,}000)}{20} = \frac{\$455{,}000}{20} = \$22{,}750.$$

In this case, the mean salary of $22,750 is not representative. Either the median or mode, both of which are $10,000, would describe the typical salary better. Notice that *the mean is affected by extreme values.*

In most cases, the *median* is not affected by extreme values. The median, however, can also be misleading, as shown in the following example.

Example 8-6 Suppose nine students make the following scores on a test:

$$30, 35, 40, 40, 92, 92, 93, 98, 99$$

Is the median the best "average" to represent the set of scores?

Solution The median score is 92. From that score, one might infer that the individuals all scored very well, yet 92 is certainly not a typical score. In this case, the mean of approximately 69 might be more appropriate than the mode. However, with the spread of the scores, neither is very appropriate for this distribution.

The *mode*, too, can be misleading in describing a set of data with very few items that occur frequently as shown in the following example.

Example 8-7 Is the mode an appropriate "average" for the following test scores?

$$40, 42, 50, 62, 63, 65, 98, 98$$

Solution The mode of the set of scores is 98 because this score occurs most frequently. The score of 98 is not representative of the set of data because of the large spread of scores.

The choice of which number to use to represent a particular set of data is not always easy. In the example involving the three teachers, each teacher chose the number that best suited his or her claim. The type of number used should always be specified.

Problem Solving   **The Missing Grades**

Students of Dr. Van Horn were asked to keep track of their own grades. One day, Dr. Van Horn asked the students to report their grades. One student had lost the papers but claims to remember the grades on four of six assignments: 100, 82, 74, and 60. In addition, the student remembered that the mean of all six papers was 69, and the other two papers had identical grades. What were the grades on the other two homework papers?

- **Understanding the Problem** The student had scores of 100, 82, 74, and 60 on four of six papers. The mean of all six papers was 69, and two identical scores were missing. The missing scores must be less than 60; otherwise, from observation, the mean could not be less than three of the four known scores and greater than the fourth one given.

- **Devising a Plan** To find the missing grades, we use the strategy of *writing an equation* for x. The mean is obtained by finding the sum of the scores and then dividing by the number of scores, which is 6. So if we let x stand for each of the two missing grades, we have

$$69 = \frac{100 + 82 + 74 + 60 + x + x}{6}.$$

- **Carrying Out the Plan** We now solve the equation as follows:

$$69 = \frac{100 + 82 + 74 + 60 + x + x}{6}$$

$$69 = \frac{316 + 2x}{6}$$

$$49 = x$$

Since the solution to the equation is $x = 49$, each of the two missing scores was 49.

- **Looking Back** The answer of 49 seems reasonable, since the mean of 69 is less than three of the four given scores. We can check this by computing the mean of the scores 100, 82, 74, 60, 49, 49 and showing that it is 69.

Measures of Dispersion

The mean, median, and mode provide limited information about the whole distribution of data. For example, if you sit in a sauna for 30 minutes and then a freezer for 30 minutes, an average temperature of your surroundings for that hour might sound comfortable. To tell

range

how much the data are scattered, we develop measures of *spread* or *dispersion*. Perhaps the easiest way to measure spread is the **range,** the difference between the greatest and the least values in a data set. For example, the range in the set of data 1, 3, 7, 8, 10 is $10 - 1 = 9$. However, just because the range of two sets of data is the same, the data do not have to have the same dispersion. For example, the data set 1, 10, 10, 10, 10 also has a range of 9 and is spread quite differently from the first collection of data. For this reason we need other measures of dispersion besides the range.

Another measure of dispersion is the **interquartile range (IQR).** The IQR is the range of the middle half of the data. Consider the following set of thirteen test scores:

20 25 40 50 50 60 70 75 80 80 90 100 100.

The range for this set of scores is $100 - 20 = 80$. The median score for this set of data is 70. We mark this location with a vertical bar between the 7 and the 0 as shown below.

20 25 40 50 50 60 7|0 75 80 80 90 100 100

vertical bar

Next, we consider only the data values to the left of the **vertical bar** and draw another vertical bar where the median of those values is located:

20 25 40 | 50 50 60

lower quartile · first quartile (Q_1)
upper quartile (Q_3)

The score of $45 = (40 + 50)/2$ is the median of the scores less than the median of all scores and is the **lower quartile.** The lower quartile is often called the **first quartile** and is denoted by Q_1. One quarter or 25% of the scores lie at or below Q_1. Similarly, we can find the upper, or third, quartile (Q_3), which is $(80 + 90)/2$, or 85. The **upper quartile (Q_3)** is the median of the scores greater than the median of all scores. Three quarters or 75% of the scores lie at or below Q_3. Thus we have divided the scores into four groups of three scores each:

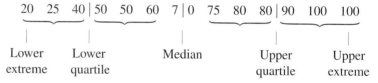

interquartile range (IQR)

The **interquartile range (IQR)** is the difference between the upper quartile and the lower quartile. In this case, $IQR = 85 - 45 = 40$. The IQR is itself another useful measure of variation because it is less influenced by extreme values. The IQR contains the middle 50% of the values.

Box-and-Whisker Plots

box-and-whisker plot

A **box-and-whisker plot** is a way to display data visually and draw informal conclusions. Box-and-whisker plots show only certain data. These plots are visual representations of the *five-number summary*. The five numbers are the median, the upper and lower quartiles, and the least and greatest values in the distribution. The center, the spread, and the overall range are immediately evident by looking at the plot.

To construct a box-and-whisker plot we need data along with its median, upper and lower quartiles, and extremes marked with vertical bars. To construct the box, we connect the vertical bars at the quartiles to form a box. We draw segments from each end of the box to the extreme values to form the whiskers. The box-and-whisker plot can be either vertical or horizontal. A vertical version of the box-and-whisker plot for the given data is shown in Figure 8-21.

SECTION 8-2 *Measures of Central Tendency and Variation* 435

Figure 8-21

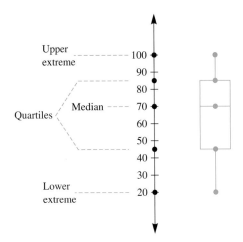

The box-and-whisker plot gives a fairly clear picture of the spread of the data. If we look at the graph in Figure 8-21, we can see that the median is 70, the maximum value is 100, the minimum value is 20, and the upper and lower quartiles are 45 and 85. The median is above the center of the box and so there are more scores above than below it.

Example 8-8 What are the minimum and maximum values, the median, and the lower and upper quartiles of the box plot shown in Figure 8-22?

Figure 8-22

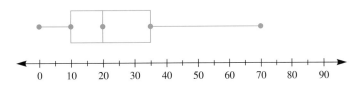

Solution The minimum value is 0, the maximum value is 70, the median is 20, the lower quartile is 10, and the upper quartile is 35.

Another example of the construction of a box-and-whisker plot is given on the student page on page 436 from *Scott Foresman-Addison Wesley Middle Grade Math,* Course 3, 1999. Read the examples and answer the questions in the "Try It" section.

Outliers

An *outlier* is a value that is widely separated from the rest of a group of data. For example, in a set of scores such as

$$91 \quad 92 \quad 92 \quad 93 \quad 93 \quad 93 \quad 94$$

all data are grouped close together and no values are widely separated. However, in a set of scores such as

$$21 \quad 92 \quad 92 \quad 93 \quad 93 \quad 93 \quad 95 \quad 150$$

DID YOU KNOW?

Why is it called a box-and-whisker plot? Because it looks like a box with whiskers!

Suppose a student scores 97, 80, 85, 72, 65, 94, and 76, on seven quizzes.

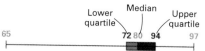

The **lower quartile**, 72, is the median of the lower half of the data set.

The **upper quartile**, 94, is the median of the upper half of the data set.

The whiskers show the range to be 65 to 97.

The median is 80.

The box shows that the middle half of the grades are between 72 and 94.

Math Quiz Grades

Example 1

Below is a box-and-whisker plot showing the gross earnings of the top 50 movies in 1994. Give the range and median. What does the whisker left of the median show?

1994 Movie Earnings ($ millions)

The range is $26.4 million to $298.9 million.   Look at the whiskers.

The median is $50.9 million.   Look at the middle line in the box.

The whisker left of the median shows that one-quarter of the movies grossed between $26.4 million and $38.8 million.

Try It

One hundred 12 to 15 year olds were asked how many times they had been in a music store within the past 30 days. The results are shown in the box-and-whisker plot.

Number of Times Teens Browse Music Stores

a. What is the range of the data?   b. What is the median?

c. What are the lower and upper quartiles?

d. What does the whisker to the right of the median show?

both 21 and 150 are widely separated from the rest of the data. These values are potential outliers. The upper and lower extreme values are not necessarily outliers. In data such as

$$75 \quad 90 \quad 91 \quad 92 \quad 92 \quad 93 \quad 93$$

outlier   it is not easy to decide, so we develop a convention for determining outliers. *An **outlier** is any value that is more than 1.5 times the interquartile range above the upper quartile or below the lower quartile.* Statisticians sometimes use values different from 1.5 to determine outliers.

It is common practice to indicate outliers with asterisks. Whiskers are then drawn to the extreme points that are not outliers. To investigate how this works, consider Example 8-9.

Example 8-9   Draw a box-and-whisker plot of the data in Table 8-5 and identify possible outliers.

Table 8-5   Final Medal Standings for Top 20 Countries—1996 Summer Olympics

Country	Medals	
United States	101	
Germany	65	
China	63	
Australia	50	
France	41	
Italy	37	$Q_3 = 39$
South Korea	35	
Cuba	27	
Ukraine	25	
Canada	23	
Hungary	22	Median (Q_2) = 22.5
Romania	21	
Netherlands	20	
Poland	19	
Spain	17	$Q_1 = 17$
Bulgaria	17	
Brazil	15	
Great Britain	15	
Belarus	15	
Japan	15	

Data taken from *The World Almanac and Book of Facts 1999*, World Almanac Books, 1999.

Solution The extreme scores are 101 and 15, the median is 22.5, $Q_1 = 17$, and $Q_3 = 39$. The IQR is $39 - 17$, or 22. Outliers are scores that are greater than $39 + 1.5(22)$, or 72, or less than $17 - 1.5(22)$, or $^-16$. Therefore, in this data 101 is the only outlier. A box-and-whisker plot is given in Figure 8-23. Notice that the whisker stops at the extreme point 15 on the lower end and at 65 on the upper end. The outlier is indicated with an asterisk.

Figure 8-23

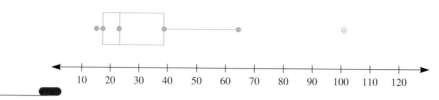

Comparing Sets of Data

Box-and-whisker plots are used primarily for large sets of data or for comparing several distributions. The stem-and-leaf plot is usually a much clearer display for a single distribution. Several box-and-whisker plots drawn below the same number line give us the easiest comparison of medians, extreme scores, and the quartiles for the sets of data. As an example, we construct box-and-whisker plots comparing the data in Table 8-6.

Table 8-6 Median Earnings of Males and Females, 1960–1992 (full-time workers age 15 and older)

Year	Women	Men
1960	$ 3,257	$ 5,368
1970	5,323	8,966
1980	11,197	18,612
1983	13,915	21,881
1984	14,780	23,218
1985	15,624	24,195
1986	16,232	25,256
1987	16,911	25,946
1988	17,606	26,656
1989	18,769	27,331
1990	19,822	27,678
1991	20,553	29,421
1992	21,440	30,358

Data taken from *Information Please Almanac Atlas and Yearbook 1995,* 48th Edition. Boston: Houghton Mifflin Company, 1995.

Before constructing horizontal box-and-whisker plots, we find the five important values for each group of data. These values are given in Table 8-7.

Table 8-7

Value	Women	Men
Maximum	$21,440.00	$30,358.00
Upper quartile	19,295.50	27,504.50
Median	16,232.00	25,256.00
Lower quartile	12,556.00	20,246.50
Minimum	3,257.00	5,368.00

In this example, the IQR for men is $7,258.00. The lower cutoff point for outliers is $20,246.50 − 1.5(7,258.00), or $9,359.50. Two salaries, $5,368.00 and $8,966.00, are less than the cutoff point and thus are outliers. Checking reveals there are no outliers in the women's salaries.

Next we draw the horizontal scale and construct the box-and-whisker plots for the women and men using the data in Table 8-7 as shown in Figure 8-24.

Figure 8-24

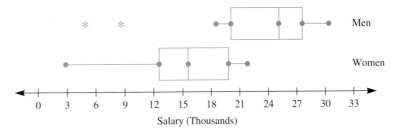

From Figure 8-24, we can see that the length of the box (IQR) for men is longer than the box (IQR) for women. This implies that the salaries for men have varied more over time than have those of women. The salaries for men have been higher than those for women, since the extreme values, median, and quartiles for the men are greater than those for women. Also, approximately 60% of the median salaries for men are greater than those for the median of women over the time period.

Although we cannot spot clusters or gaps in box-and-whisker plots as we can with stem-and-leaf or line plots, we can more easily compare data from different sets. With box-and-whisker plots, we do not need to have sets of data that are approximately the same size, as we did for stem-and-leaf plots. To compare data from two or more sets using their box-and-whisker plots, we first study the boxes to see if they are located in approximately the same places. Next, we consider the lengths of the boxes to see if the variability of the data is about the same. We also check whether the median, the quartiles, and the extreme values in one set are greater than those in another set. If they are, the data in the first set are greater than those of the other set, no matter how we compare them. If they are not, we can continue to study the data for other similarities and differences.

> **NOW TRY THIS 8-6**
>
> - Using only the data from 1983 through 1992 in Table 8-6 decide whether there are outliers. How do you think the two box-and-whisker plots would compare for this data with the comparison made in Figure 8-24?

Variance and Standard Deviation

Suppose Professors Abel and Babel each taught a section of a graduate statistics course and each had six students. Both professors gave the same final exam. The results, along with the means for each group of scores, are given in Table 8-8, with stem-and-leaf plots in Figure 8-25(a) and (b), respectively. As the stem-and-leaf plots show, the sets of data are very different. The first is more spread out, or varies more, than the second. However, each set has 60 as the mean. Each median also equals 60. Although the mean and the median for these two groups are the same, the two distributions of scores are very different.

Table 8-8

Abel's Class Scores	Babel's Class Scores
100	70
80	70
70	60
50	60
50	60
10	40
$\bar{x} = \frac{360}{6} = 60$	$\bar{x} = \frac{360}{6} = 60$

Figure 8-25

```
       Professor Abel's Class Scores
  1 | 0
  2 |
  3 |
  4 |                                    Professor Babel's Class Scores
  5 | 00
  6 |                                  4 | 0
  7 | 0        7 | 0 represents        5 |
  8 | 0          a score of 70         6 | 000
  9 |                                  7 | 00        7 | 0 represents
 10 | 0                                                 a score of 70
       (a)                                  (b)
```

As we have seen, there are many ways to measure the spread of data. The simplest way is to find the range. The range for Professor Abel's class is $100 - 10$, or 90. The range for Professor Babel's class is $70 - 40$, or 30. If we use the range as our measure of dispersion, we see that Abel's class is much more spread out than Babel's class. If we use the interquartile range, the IQR for Abel's class is 30 and for Babel's class is 10. Again these measures of dispersion show more of a spread for Abel's class than for Babel's.

Two other commonly used measures of dispersion are the **variance** and the **standard deviation.** These measures are based on how far the scores are from the mean. To find out how far each value differs from the mean, we subtract each value in the data from the mean to obtain the deviation. Some of these deviations may be positive, and others may be negative. Because the mean is the balance point, the total of the deviations above the mean equals the total of the deviations below the mean. (The mean of the deviations is 0 because the sum of the deviations is 0.) Squaring the deviations makes them all positive. The mean of the squared deviations is the *variance*. Because the variance involves squaring the deviations, it does not have the same units of measurement as the original observations. For example, lengths measured in feet have a variance measured in square feet. To obtain the same units as the original observations, we take the square root of the variance and obtain the *standard deviation*.

The steps involved in calculating the variance v and standard deviation s of n numbers are as follows.

1. Find the mean of the numbers.
2. Subtract the mean from each number.
3. Square each difference found in step 2.
4. Find the sum of the squares in step 3.
5. Divide by n to obtain the variance, v.
6. Find the square root of v to obtain the standard deviation, s.

These six steps can be summarized for the numbers $x_1, x_2, x_3, \ldots, x_n$ as follows, where $\bar{x}$ is the mean of these numbers:

$$s = \sqrt{v} = \sqrt{\frac{(x_1 - \bar{x})^2 + (x_2 - \bar{x})^2 + (x_3 - \bar{x})^2 + \cdots + (x_n - \bar{x})^2}{n}}$$

REMARK In some textbooks, this formula involves division by $n - 1$ instead of by n. Division by $n - 1$ is more useful for advanced work in statistics.

The variances and standard deviations for the final exam data from the classes of Professors Abel and Babel are calculated by using Tables 8-9 and 8-10, respectively.

Table 8-9 Abel's Scores

x	$x - \bar{x}$	$(x - \bar{x})^2$
100	40	1600
80	20	400
70	10	100
50	-10	100
50	-10	100
10	-50	2500
Totals 360	0	4800

$\bar{x} = \dfrac{360}{6} = 60$

$v = \dfrac{4800}{6} = 800$

$s = \sqrt{800} \doteq 28.3$

Table 8-10 Babel's Scores

x	$x - \bar{x}$	$(x - \bar{x})^2$
70	10	100
70	10	100
60	0	0
60	0	0
60	0	0
40	-20	400
Totals 360	0	600

$\bar{x} = \dfrac{360}{6} = 60$

$v = \dfrac{600}{6} = 100$

$s = \sqrt{100} = 10$

Example 8-10 Professor Abel gave two group exams. Exam A had grades of 0, 0, 0, 100, 100, 100, and exam B had grades of 50, 50, 50, 50, 50, 50. Find the following for each exam:

a. Mean   b. Median   c. Standard deviation

Solution
a. The means for exams A and B are each 50.
b. The medians for the exams are each 50.
c. The standard deviations for exams A and B are as follows:

$$s_A = \sqrt{\frac{3(0-50)^2 + 3(100-50)^2}{6}} = 50$$

$$s_B = \sqrt{\frac{6(50-50)^2}{6}} = 0$$

BRAIN TEASER The speeds of racing cars were timed after 3 mi, $4\frac{1}{2}$ mi, and 6 mi. One driver averaged 140 mph for the first 3 mi, 168 mph for the next $1\frac{1}{2}$ mi, and 210 mph for the last $1\frac{1}{2}$ mi. What was the driver's mean speed for the total 6-mi run?

ONGOING ASSESSMENT 8-2

1. Calculate the mean, the median, and the mode for each of the following data sets:
   a. 2, 8, 7, 8, 5, 8, 10, 5
   b. 10, 12, 12, 14, 20, 16, 12, 14, 11
   c. 18, 22, 22, 17, 30, 18, 12
   d. 82, 80, 63, 75, 92, 80, 92, 90, 80, 80
   e. 5, 5, 5, 5, 5, 10

2. a. If each of six students scored 80 on a test, find each of the following for the set of six scores:
      i. Mean   ii. Median   iii. Mode
   b. Make up another set of six scores that are not all the same but in which the mean, median, and mode are all 80.

3. The mean score on a set of 20 tests is 75. What is the sum of the 20 test scores?

4. The tram at a ski area has a capacity of 50 people with a load limit of 7500 lb. What is the mean weight of the passengers if the tram is loaded to capacity?

5. The mean for a set of 28 scores is 80. Suppose two more students take the test and score 60 and 50. What is the new mean?

6. The names and ages for each person in a family of five follow:

Name	Dick	Jane	Kirk	Jean	Scott
Age	40	36	8	6	2

   a. What is the mean age?
   b. Find the mean of the ages 5 years from now.
   c. Find the mean 10 years from now.
   d. Describe the relationships among the means found in (a), (b), and (c).

7. Suppose you own a hat shop and decide to order hats in only *one* size for the coming season. To decide which size to order, you look at last year's sales figures, which are itemized according to size. Should you find the mean, median, or mode for the data? Why?

8. A table showing Jon's fall quarter grades follows. Find his grade point average for the term (A = 4, B = 3, C = 2, D = 1, F = 0).

Course	Credits	Grade
Math	5	B
English	3	A
Physics	5	C
German	3	D
Handball	1	A

9. If the mean weight of seven tackles on a team is 230 lb and the mean weight of the four backfield members is 190 lb, what is the mean weight of the 11-person team?

10. If 99 people had a mean income of $12,000, how much is the mean income increased by the addition of a single income of $200,000?

11. The following table gives the annual salaries of the 40 dancers of a certain troupe.
    a. Find the mean annual salary for the troupe.
    b. Find the median annual salary.
    c. Find the mode.

Salary	Number of Dancers
$ 18,000	2
22,000	4
26,000	4
35,000	3
38,000	12
44,000	8
50,000	4
80,000	2
150,000	1

12. Refer to the following chart. In a gymnastics competition, each competitor receives six scores. The highest and lowest scores are eliminated, and the official score is the mean of the four remaining scores.

Gymnast	Scores					
Balance Beam						
Meta	9.2	9.2	9.1	9.3	9.8	9.6
Lisa	9.3	9.1	9.4	9.6	9.9	9.4
Olga	9.4	9.5	9.6	9.6	9.9	9.6
Uneven Bars						
Meta	9.2	9.1	9.3	9.2	9.4	9.5
Lisa	10.0	9.8	9.9	9.7	9.9	9.8
Olga	9.4	9.6	9.5	9.4	9.4	9.4
Floor Exercises						
Meta	9.7	9.8	9.4	9.8	9.8	9.7
Lisa	10.0	9.9	9.8	10.0	9.7	10.0
Olga	9.4	9.3	9.6	9.4	9.5	9.4

a. If the only events in the competition are the balance beam, the uneven bars, and the floor exercise, find the winner of each event.
b. Find the overall winner of the competition if the overall winner is the person with the highest combined official scores.

13. Maria needed 8 gal of gas to fill her car's gas tank. The mileage odometer read 42,800 mi. When the odometer read 43,030, Maria filled the tank with 12 gal for a trip she was taking. At the end of the trip, she filled the tank with 18 gal and the odometer read 43,390 mi. How many miles per gallon did she get for the entire trip?

14. The youngest person in a company is 24 years old. The range of ages is 34 yr. How old is the oldest person in the company?

15. Choose the data set(s) that fit the descriptions given in each of the following:
    a. The mean is 6.
       The range is 6.
       Set A: 3, 5, 7, 9
       Set B: 2, 4, 6, 8
       Set C: 2, 3, 4, 15
    b. The mean is 11.
       The median is 11.
       The mode is 11.
       Set A: 9, 10, 10, 11, 12, 12, 13
       Set B: 11, 11, 11, 11, 11, 11, 11
       Set C: 9, 11, 11, 11, 11, 12, 12
    c. The mean is 3.
       The median is 3.
       It has no mode.
       Set A: 0, $2\frac{1}{2}$, $6\frac{1}{2}$
       Set B: 3, 3, 3, 3
       Set C: 1, 2, 4, 5
    d. The box-and-whisker plot shown below.

    Set A: 2, 3, 4, 4, 6, 6, 7, 15
    Set B: 2, 3, 6, 6, 8, 9, 12, 14, 15
    Set C: 2, 4.5, 8, 13, 15
    Set D: 2, 3, 6, 6, 8, 9, 10, 11, 15

16. To receive an A in a class, Willie needs at least a mean of 90 on five exams. Willie's grades on the first four exams were 84, 95, 86, and 94. What minimum score does he need on the fifth exam to receive an A in the class?

17. Ginny's median score on three tests was 90. Her mean score was 92 and her range was 6. What were her three test scores?

18. The mean of five numbers is 6. If one of the five numbers is removed, the mean becomes 7. What is the value of the number that was removed?

19. a. Find the mean and the median of the following arithmetic sequences:
  i. 1, 3, 5, 7, 9
  ii. 1, 3, 5, 7, 9, ... , 199
  iii. 7, 10, 13, 16, ... , 607
  b. Based on your answers in (a), make a conjecture about the mean and the median of any arithmetic sequence.

20. Construct a box-and-whisker plot for the following gas mileages per gallon of various company cars:

   22 18 14 28 30 12 38 22
   30 39 20 18 14 16 10

21. Following are box-and-whisker plots comparing the ticket prices of two performing arts theaters:
  a. What is the median ticket price for each theater?
  b. Which theater has the greatest range of prices?
  c. What is the highest ticket price at either theater?
  d. Make some statements comparing the ticket prices at the two theaters.

22. Construct a box-and-whisker plot for the following set of test scores. Indicate outliers, if any, with asterisks.

   20 95 40 70 90 70 80 80 90 95

23. The following table shows the heights in feet of the tallest ten buildings in Los Angeles and in Minneapolis:

Los Angeles	Minneapolis
858	950
750	775
735	668
699	579
625	561
620	447
578	440
571	416
534	403
516	366

  a. Draw horizontal box-and-whisker plots to compare the data.
  b. Are there any outliers in this data? If so, which values are they?
  c. Based on your box-and-whisker plots from (a), make some comparisons of the heights of the buildings in the two cities.

24. The following shows box-and-whisker plots for the piano practice times in hours per week for Tom and Dick. Make some comparisons for their practice times.

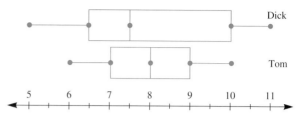

25. What is the standard deviation of the heights of seven trapeze artists if their heights are 175 cm, 182 cm, 190 cm, 180 cm, 192 cm, 172 cm, and 190 cm?

26. a. If all the numbers in a set are equal, what is the standard deviation?
  b. If the standard deviation of a set of numbers is zero, must all the numbers in the set be equal?

27. In a Math 131 class at DiPaloma University, the grades on the first exam were as follows:

   96 71 43 77 75 76 61
   83 71 58 97 76 74 91
   74 71 77 83 87 93 79

  a. Find the mean.   **b.** Find the median.
  c. Find the mode.   **d.** Find the IQR.
  e. Find the variance of the scores.
  f. Find the standard deviation of the scores.

28. In a school system, teachers start at a salary of $25,200 and have a top salary of $51,800. The teachers' union is bargaining with the school district for next year's salary increment.
  a. If every teacher is given a $1000 raise, what happens to each of the following?
     i. Mean   ii. Median
     iii. Extremes   iv. Quartiles
     v. Standard deviation   vi. IQR
  b. If every teacher received a 5% raise, what does this do to the following?
     i. Mean   ii. Standard deviation

★**29.** Show that the following formula for variance is equivalent to the one given in the text:

$$v = \frac{x_1^2 + x_2^2 + \ldots + x_n^2}{n} - \bar{x}^2$$

Communication

30. If you were considering ages and wanted one number to represent the age at which a person can get a driver's license, which "average" would you use and why?

31. A movie chain conducts a popcorn poll in which each person entering a theater and buying a box of popcorn is asked a yes-no question. Which "average" do you think is used to report the result and why?

32. When a government agency reports the rainfall for a state for a year, which "average" do you think they use and why?

33. Carl had scores of 90, 95, 85, and 90 on his first four tests.
  a. Find the median, mean, and mode.
  b. Carl scored a 20 on his fifth exam. Which of the three averages would Carl want the instructor to use to compute his grade? Why?
  c. Which measure is affected the most by an extreme score?

34. The mean of the five numbers given below is 50:

20   35   50   60   85

  a. Add four numbers to the list so that the mean of the nine numbers is still 50.
  b. Explain how you could choose the four numbers to add to the list so that the mean did not change.
  c. How does the mean of the four numbers you added to the list compare to the original mean of 50? Why?

35. Sue drives 5 mi at 30 mph and then 5 mi at 50 mph. Is the mean speed for the trip 40 mph? Why or why not?

Open-Ended

36. In 1991, students from countries around the world took the International Assessment of Educational Progress test. The table below gives the average mathematics scores for 15 countries along with the number of days that students in each country spend in school each year.

Country	Days of school	Math score
Canada	188	62
France	174	64
Hungary	177	68
Ireland	173	61
Israel	215	63
Italy	204	64
Jordan	191	40
Korea	222	73
Scotland	191	61
Slovenia	190	57
Soviet Union	198	70
Spain	188	55
Switzerland	207	71
Taiwan	222	73
United States	178	55

  (a) Make a conjecture about the relationship between the country's mathematics score and the number of days spent in school.
  (b) Test your conjecture by using a box-and-whisker plot to compare students who spend 190 days or fewer in school with those who spend more than 190 days in school.
  (c) Explain how your graph supports or disproves your conjecture.
  (d) Draw a scatterplot for the data and see if there seems to be a correlation between mathematics scores and days in school. Is this the same message your box-and-whisker plot gave you? Which plot would you use to support your conjecture?

37. Use the data in the following table to compare the number of people living in the United States from 1800 through 1890 and from 1900 through 1990. Use any form of graphical representation to make the comparison and explain why you chose the representation that you did.

Year	Number (thousands)	Year	Number (thousands)
1800	5,308	1900	76,212
1810	7,239	1910	92,228
1820	9,638	1920	106,021
1830	12,866	1930	123,202
1840	17,068	1940	132,164
1850	23,191	1950	151,325
1860	31,443	1960	179,323
1870	38,558	1970	203,302
1880	50,189	1980	226,542
1890	62,979	1990	248,765

Data taken from *The World Almanac and Book of Facts 1999*, World Almanac Books, 1999.

Cooperative Learning

38. In small groups, determine a method of finding the number and types of graphs and statistical representations used in at least two newspapers in your campus library. Based on your findings, write a report defending which type(s) of representation should be emphasized in a journalistic statistics class.

Review Problems

39. Given the following double-bar graph, make some comparisons of the number of men and women in the labor force over the years.

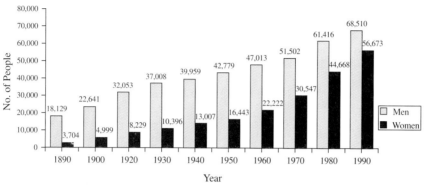

Data taken from *The World Almanac and Book of Facts 1999,* World Almanac Books, 1999.

40. Consider the following circle graph. What is the number of degrees in each sector of the graph?

41. Given the following bar graph, answer the following:
    a. Which mountain is the highest? Approximately how high is it?
    b. Which mountains are higher than 6000 m?

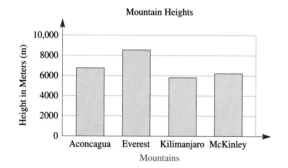

42. Following are raw test scores from a history test:

86	85	87	96	55
90	94	82	68	77
88	89	85	74	90
72	80	76	88	73
64	79	73	85	93

    a. Construct an ordered stem-and-leaf plot for the given data.
    b. Construct a grouped frequency table for these scores with intervals of 5, starting the first class at 55.
    c. Draw a histogram of the data.
    d. If a circle graph of the grouped data in (b) were drawn, how many degrees would be in the section representing the 85 through 89 interval?

LABORATORY ACTIVITY

We can model the mean as a measure of central tendency using a strip of cardboard that is 1 in. wide and 1 ft long with holes 1 in. apart and $\frac{1}{8}$ in. from the edge, as in Figure 8-26.

Figure 8-26

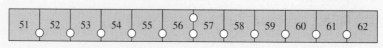

Use string and tape to suspend the strip from a desk with the string tied through a hole punched between 56 and 57. Then use paper clips of equal size to investigate means, as in Figure 8-27.

Figure 8-27

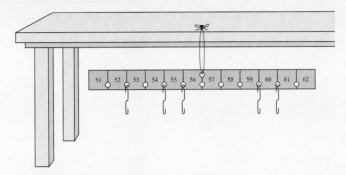

1. **a.** If paper clips are hung in the holes at 51, 53, and 60, where should an additional clip be hung in order to achieve a balance?
   b. If paper clips are hung at 51, 54, and 60, where should two additional paper clips be hung to achieve a balance?
   c. If paper clips are hung at 51, 53, 54, and 55, where should four additional clips be hung to achieve a balance?
2. Find the mean of the data of all the numbers in each part in (1) and compare this answer with the number in the center of the strip.
3. Would any of the means in (2) change if we hung an additional paper clip in the center of the strip?
4. Find the median and mode for 51, 53, 54, 56, 58, 58, and 59.
5. Hang paper clips in each hole in (4). If a number appears more than once, hang that number of paper clips in the hole. Is there a balance around the median? Is there a balance around the mode?
6. Under what conditions do you think there would be a balance around (a) the mean, (b) the median, and (c) the mode? Test your conjecture using the cardboard strip.

TECHNOLOGY CORNER

Use a graphing calculator or create a spreadsheet to find the variance and standard deviation of the set of scores 32, 41, 47, 53, and 57. If you use a spreadsheet, make column A be the set of scores, column B the score minus the mean, and column C the square of the difference of the score and the mean. Compute the variance and the standard deviation.

Section 8-3 — Abuses of Statistics

Statistics are frequently abused. Benjamin Disraeli (1804–1881), an English prime minister, once remarked, "There are three kinds of lies: lies, damned lies, and statistics." People sometimes deliberately use statistics to mislead others. This can be seen in advertising. More often, however, the misuse of statistics is the result of misinterpreting what the statistics mean. For example, if we were told that the "average" depth of water in a lily pond was 2 ft, most of us would presume that a heron could stand up in any part of the pond. That this is not necessarily the case, as seen in the following cartoon.

Consider an advertisement reporting that of the people responding to a recent survey, 98% of those tested said that Buffepain is the most effective pain reliever of headaches and arthritis. To certify that the statistics are not being misused, the following information should have been reported:

1. The number of people surveyed
2. The number of people who responded
3. How the people who participated in the survey were chosen
4. The number and type of pain relievers tested

Without the information listed, the following situations are possible, all of which could cause the advertisement to be misleading:

1. Suppose 1,000,000 people nationwide were sent the survey, and only 50 responded. This would mean that there was only a 0.005% response, which would certainly cause us to mistrust the ad.
2. Of the 50 responding in (1), suppose 49 responses were affirmative. The 98% claim is true, but 999,950 people did not respond.
3. Suppose all the people who received the survey were chosen from a town in which the major industry was the manufacture of Buffepain. It is very doubtful that the survey would represent an unbiased sample.
4. Suppose only two "pain relievers" were tested: Buffepain, whose active ingredient is 100% aspirin, and a placebo containing only powdered sugar.

This is not to say that advertisements of this type are all misleading or dishonest, but simply that statistics are only as honest as their users.

A different type of misuse of statistics involves graphs. Among the things to look for in a graph are the following. If they are not there, then the graph may be misleading.

1. Title
2. Labels on both axes of a line or bar chart and on all sections of a pie chart
3. Source of the data
4. Key to a pictograph
5. Uniform size of symbols in a pictograph
6. Scale: Does it start with zero? If not, is there a break shown?
7. Scale: Are the numbers equally spaced?

To see an example of a misleading use of graphs, consider how graphs can be used to distort data or exaggerate certain pieces of information. Graphs using a break in the vertical axis can be used to create different visual impressions, which are sometimes misleading. For example, consider the two graphs in Figure 8-28, which represent the number of girls trying out for basketball at each of three middle schools. As we can see, the graph in Figure 8-28(a) portrays a different picture from the one in Figure 8-28(b).

Figure 8-28

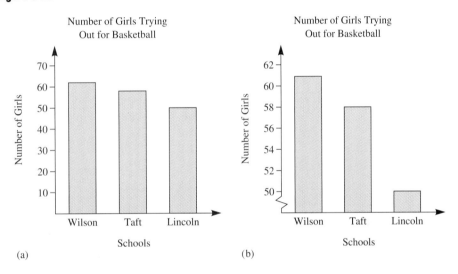

A line graph, histogram, or bar graph can be altered by changing the scale of the graph. For example, consider the data in Table 8-11 for the number of graduates from a community college for the years 1996 through 2000.

Table 8-11

Year	1996	1997	1998	1999	2000
Number of Graduates	140	180	200	210	160

The graphs in Figure 8-29(a) and (b) represent the same data, but different scales are used in each. The statistics presented are the same, but these graphs do not convey the same psychological message. In Figure 8-29(b), the spacing of the years on the horizontal axis of the graph is more spread out and that for the numbers on the vertical axis is more condensed than in Figure 8-29(a). Both of these changes minimize the variability of the data. A college administrator might use a graph like the one in Figure 8-29(b) to convince people that the college was not in serious enrollment trouble.

Figure 8-29

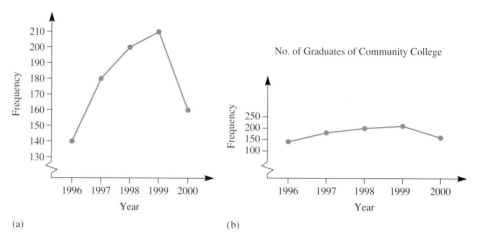

(a) (b)

Another error that frequently occurs in all arenas is the use of continuous curve graphs, as in Figure 8-29, to depict data that are discrete (a finite number of data values). In Figure 8-29, it may or may not make sense to discuss the enrollment at 1999.5, yet the way in which the graph is constructed leads us to believe that such a value exists.

Other ways to distort graphs include omitting a scale, as in Figure 8-30(a). The scale is given in Figure 8-30(b).

Figure 8-30

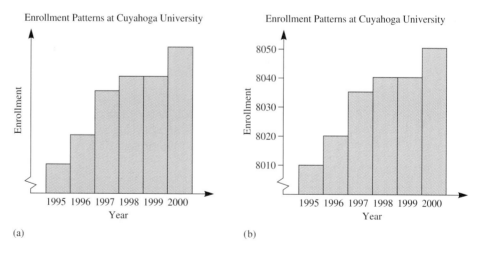

(a) (b)

Other graphs can also be misleading. Suppose, for example, that the number of boxes of cereal sold by Sugar Plops last year was 2 million and the number of boxes of cereal sold by Korn Krisps was 8 million. The Korn Krisps executives prepared the graph in Figure 8-31 to demonstrate the data. The Sugar Plops people objected. Do you see why?

The graph in Figure 8-31 clearly distorts the data, since the figure for Korn Krisps is both four times as high and four times as wide as the bar for Sugar Plops. Thus the area of the box face representing Korn Krisps is 16 times the comparable area representing Sugar Plops, rather than four times the area, as would be justified by the original data.

Figure 8-31

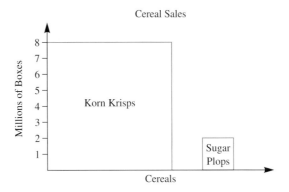

Figure 8-32 shows how the comparison of Sugar Plops and Korn Krisps cereals might look if the figures were made three-dimensional. The figure for Korn Krisps has a volume 64 times the volume of the Sugar Plops figure.

Figure 8-32

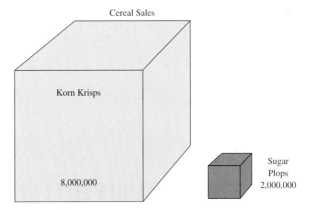

Circle graphs easily become distorted when attempts are made to depict them as three-dimensional. Many graphs of this type do not acknowledge either the variable thickness of the depiction or the distortion due to perspective. Observe that the 27% sector pictured in Figure 8-33 looks far greater than the 23% sector, although they should be very nearly the same size.

Figure 8-33

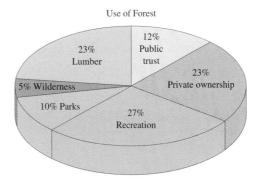

The final examples of the misuses of statistics involve misleading uses of mean, median, and mode. All these are "averages" and can be used to suit a person's purposes. As discussed in Section 8-2 in the example involving the teachers Smith, Jones, and Rivera, each teacher had reported that his or her class had done better than the other two. Each of the teachers used a different number to represent the test scores.

As another example, company administrators wishing to portray to prospective employees a rosy salary picture may find a mean salary of $38,000 for line workers as well as upper management in the schedule of salaries. At the same time, a union that is bargaining for salaries may include part-time employees as well as line workers and will exclude management personnel in order to present a mean salary of $29,000 at the bargaining table. The important thing to watch for when a mean is reported is disparate cases in the reference group. If the sample is small, then a few extremely high or low scores can have a great influence on the mean.

Suppose Figure 8-34 shows the salaries of both management and line workers of the company. If the median is being used as the average, then the median might be $33,500, which is representative of neither major group of employees. The bimodal distribution means the median is nonrepresentative of the distribution.

Figure 8-34

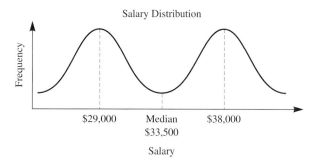

To conclude the comments on the misuse of statistics, consider a quote from Darrell Huff's book *How to Lie with Statistics* (p. 8):

So it is with much that you read and hear. Averages and relationships and trends and graphs are not always what they seem. There may be more in them than meets the eye, and there may be a great deal less.

The secret language of statistics, so appealing in a fact-minded culture, is employed to sensationalize, inflate, confuse, and oversimplify. Statistical methods and statistical terms are necessary in reporting the mass data of social and economic trends, business conditions, "opinion" polls, census. But without writers who use the words with honesty and understanding and readers who know what they mean, the result can be semantic "nonsense."

ONGOING ASSESSMENT 8-3

This entire set of assessment items is appropriate for communication and cooperative learning. Many items are open-ended and several lend themselves to further investigation.

1. Discuss whether the following claims could be misleading. Explain why and how.
    a. A car manufacturer claims its car is quieter than a glider.
    b. A motorcycle manufacturer claims that more than 95% of its cycles that were sold in the United States in the last 15 yr are still on the road.

c. A company claims its fruit juice has 10% more fruit solids than is required by U.S. government standards. (The government requires 10% fruit solids.)
d. A brand of bread claims to be 40% fresher.
e. A used-car dealer claims that a car she is trying to sell will get up to 30 mpg.
f. Sudso claims that its detergent will leave your clothes brighter.
g. A sugarless gum company claims that eight of every ten dentists responding to the survey recommend sugarless gum.
h. Most accidents occur in the home. Therefore, to be safer, you should stay out of your house as much as possible.
i. More than 95% of the people who fly to a certain city do so on Airline A. Therefore most people prefer Airline A to other airlines.

2. The city of Podunk advertised that its temperature was the ideal temperature in the country because its mean temperature was 25°C. What possible misconceptions could people draw from this advertisement?

3. Jenny averaged 70 on her quizzes during the first part of the quarter and 80 on her quizzes during the second part of the quarter. When she found out that her final average for the quarter was not 75, she went to argue with her teacher. Give a possible explanation for Jenny's misunderstanding.

4. Suppose the following circle graphs are used to illustrate the fact that the number of elementary teaching majors at teachers' colleges has doubled between 1990 and 2000, while the percent of male elementary teaching majors has stayed the same. What is misleading about the way the graphs are constructed?

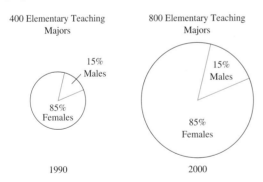

5. What is wrong with the following line graph?

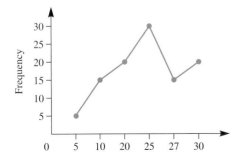

6. Can you draw any valid conclusions about a set of data in which the mean is less than the median?

7. A student read that nine out of ten pickup trucks sold in the last 10 yr are still on the road. She concluded that the average life of a pickup is around 10 yr. Is she correct?

8. General Cooster once asked a person by the side of a river if the river was too deep to ride his horse across. The person responded that the average depth was 2 ft. If General Cooster rode out across the river, what assumptions did he make on the basis of the person's information?

9. Doug's Dog Food Company wanted to impress the public with the magnitude of the company's growth. Sales of Doug's Dog Food had doubled from 1999 to 2000, so the company displayed the following graph, in which the radius of the base and the height of the 2000 can are double those of the 1999 can. What does the graph really show with respect to the growth of the company? (*Hint:* The volume of a cylinder is given by $V = \pi r^2 h$, where r is the radius of the base and h is the height.)

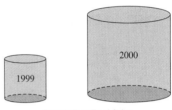

Doug's Dog Food Sales

10. Explain what is wrong with the following graph:

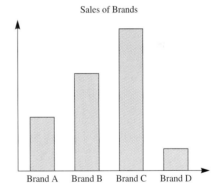

11. Refer to the following pictograph:

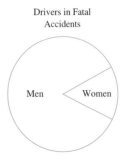

Ms. McNulty claims that on the basis of this information, we can conclude that men are worse drivers than women. Discuss whether you can reach that conclusion from the pictograph or you need more information. If more information is needed, what would you like to know?

12. The following graph was prepared to compare prices of washing machines at three stores:

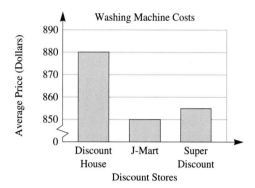

Which of the following statements is true? Explain why or why not.
   a. Prices vary widely at the three stores.
   b. The price at Discount House is four times as great as that at J-Mart.
   c. The prices at J-Mart and Super Discount differ by less than $10.

13. The following table gives the number of accidents per year on a certain highway for a 5-yr period:

Year	1996	1997	1998	1999	2000
Number of Accidents	24	26	30	32	38

   a. Draw a bar graph to convince people that the number of accidents is on the rise and that something should be done about it.
   b. Draw a bar graph to show that the rate of accidents is almost constant, and that nothing needs to be done.

14. Write a list of scores for which the mean and median are not representative of the list.

15. The following graph depicts the mean center of population of the United States and shows how the center has shifted from 1790 to 1990. Based solely on this graph, could you conclude that the population of the West Coast has increased since 1790?

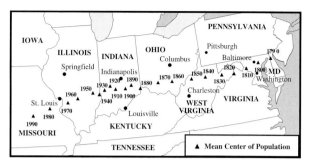

Source: U.S. Bureau of the Census, *1990 Census of Population and Housing, Population and Housing Unit Counts, United States* (1990 CPH-2-1).

16. The data in the graph of problem 15 roughly follows a line. If the mean center of the population continues to move westward in the next 200 years, where would you expect it to be at the end of that period? At the end of 400 years?

17. A prospective homeowner considered the dropping interest rates for house loans in the early 1990s and decided to wait until the year 2000 to buy. Explain what type of statistics might be used in making this decision and whether you consider the prospective homeowner's decision to be wise.

18. A student made 99 on a quiz and was ecstatic over the grade. What other information might you need in order to decide if the student was justified in being happy?

19. A school administrator reports to you, a school board member, that the "average" number of students in a class in a school is 32. What other information might you need in order to predict whether any single classroom was overcrowded?

20. Consider a state such as Montana that has both mountains and prairies. What numbers might you report to depict the "average" height above sea level of such a state? Why?

21. Is it possible for a state or country to have a mean sea level height that is negative? If so, what might such a region look like?

22. Describe how you might pick a random sample of adults that is representative of the members of your town.

23. A very large and successful manufacturer of computer chips released data in the early 1990s stating that approximately 65% of its chips were defective when they came off the assembly line. Give some reasons why this statistic could be accurate and yet the company could still be successful.

Review Problems

24. On the English 100 exam, the scores were as follows:

$$\begin{array}{cccc} 43 & 91 & 73 & 65 \\ 56 & 77 & 84 & 91 \\ 82 & 65 & 98 & 65 \end{array}$$

   a. Find the mean.
   b. Find the median.
   c. Find the mode.
   d. Find the variance.
   e. Find the standard deviation.

25. If the mean of a set of 36 scores is 27 and two more scores of 40 and 42 are added, what is the new mean?
26. On a certain exam, Tony corrected 10 papers and found the mean for his group to be 70. Alice corrected the remaining 20 papers and found that the mean for her group was 80. What is the mean of the combined group of 30 students?
27. Following are the men's gold-medal times for the 100-m run in the Olympic games from 1896 to 1996 rounded to the nearest tenth. Construct an ordered stem-and-leaf plot for the data.

Year	Time (Seconds)
1896	12.0
1900	11.0
1904	11.0
1908	10.8
1912	10.8
1920	10.8
1924	10.6
1928	10.8
1932	10.3
1936	10.3
1948	10.3
1952	10.4
1956	10.5
1960	10.2
1964	10.0
1968	10.0
1972	10.1
1976	10.1
1980	10.3
1984	10.0
1988	9.9
1992	9.7
1996	9.8

28. Following are the record swimming times of the women's 100-m freestyle and 100-m butterfly in the Olympics from 1960 to 1996. Draw box-and-whisker plots of the two sets of data using the same number line to compare them.

Year	Time— 100-m Freestyle (Seconds)	Time— 100-m Butterfly (Seconds)
1960	61.20	69.50
1964	59.50	64.70
1968	60.00	65.50
1972	58.59	63.34
1976	55.65	60.13
1980	54.79	60.42
1984	55.92	59.26
1988	54.93	59.00
1992	54.64	58.62
1996	54.50	59.13

HINT FOR SOLVING THE PRELIMINARY PROBLEM

We know that the sum of the ages of the first seven guests divided by 7 is 21. Hence, we know the sum of the ages of the first seven people is 7 · 21 (Why?). If we add another age of 29 to the sum, we must divide by 8 to find the new mean. If we add still another age of 29 to the previous total, we must divide by 9 to find the new mean. Since we do not know the age of the chair, we might use a variable for his age and use algebra to solve the equation involving finding the mean for the 10 ages.

QUESTIONS FROM THE CLASSROOM

1. A student asks, "If the average income of each of ten people is $10,000 and one person gets a raise of $10,000, is the median, the mean, or the mode changed and, if so, by how much?"
2. A student asks for an example of when the mode is the best average. What is your response?
3. A student says that a stem-and-leaf plot is always the best way to present data. How do you respond?
4. Suppose the class takes a test and the following averages are obtained: mean, 80; median, 90; mode, 70. Tom, who scored 80, would like to know if he did better than half the class. What is your response?
5. A student wants to know the advantages of presenting data in graphical form rather than in tabular form. What is your response? What are the disadvantages?
6. A student asks if it is possible to find the mode for data in a grouped frequency table. What is your response?
7. A student asks if she can draw any conclusions about a set of data if she knows that the mean for the data is less than the median. How do you answer?
8. A student asks if it is possible to have a standard deviation of $^-5$. How do you respond?
9. Mel's mean on ten tests for the quarter was 89. She complained to the teacher that she should be given an A because she missed the cutoff of 90 by only a single point. Did she really miss an A by only a single point?
10. A student claims that bar graphs can be used to give the same information as line graphs, so no one should have to learn how to do line graphs. What is your response?
11. A student asks, "Is it always true that in any set of data, there must be at least one data point in the data that is less than or equal to the mean and at least one data point greater than or equal to the mean?"
12. A student asks if the precision with which manufacturers must calibrate their tools is at all related to statistics. How do you respond?
13. A student wants to know if there is such a thing as average deviation and if anyone ever uses it. How do you relate average deviation to standard deviation and its use?

CHAPTER OUTLINE

I. Descriptive statistics
   Information can be summarized in each of the following forms:
   1. **Pictographs**
   2. **Line plots**
   3. **Stem-and-leaf plots**
   4. **Frequency tables**
   5. **Histograms**
   6. **Bar graphs**
   7. **Line graphs**
   8. **Circle graphs** or **pie charts**
   9. **Box-and-whisker plots**
II. Measures of central tendency
   A. The **mean** of n given numbers is the sum of the numbers divided by n.
   B. The **median** of a set of numbers is the middle number if the numbers are arranged in numerical order; if there is no middle number, the median is the mean of the two middle numbers.
   C. The **mode** of a set of numbers is the number or numbers that occur most frequently in the set.
III. Measures of variation
   A. The **range** is the difference between the greatest and least numbers in the data.
   B. The **variance** of a data set is found by subtracting the mean from each value, squaring each of these differences, finding the sum of these squares, and dividing by n, where n is the number of observations.
   C. The **standard deviation** is equal to the square root of the variance.
   D. **Box-and-whisker plots** focus attention on the median, the quartiles, and the extremes and invite comparisons among them.
      1. The **lower quartile** is the median of the subset of data less than the median of all the values in the data set.
      2. The **upper quartile** is the median of the subset of data greater than the median of all the values in the data set.
      3. The **interquartile range (IQR)** is calculated as the difference between the upper quartile and the lower quartile.
      4. An **outlier** is any value more than 1.5 IQR above the upper quartile or more than 1.5 IQR below the lower quartile.
   E. **Scatterplots** are graphs of ordered pairs that allow us to examine the relationship (correlation) between two sets of data.

CHAPTER REVIEW

1. Suppose you read that "the average family in Rattlesnake Gulch has 2.41 children." What average is being used to describe the data? Explain your answer. Suppose the sentence had said 2.5? Then what are the possibilities?
2. At Bug's Bar-B-Q restaurant, the average weekly wage for full-time workers is $150. There are ten part-time employees whose average weekly salary is $50 and the total weekly payroll is $3950. How many full-time employees are there?
3. Find the mean, the median, and the mode for each of the following groups of data:
   a. 10, 50, 30, 40, 10, 60, 10
   b. 5, 8, 6, 3, 5, 4, 3, 6, 1, 9
4. Find the range, variance, and standard deviation for each set of scores in problem 3.
5. The mass, in kilograms, of each child in Ms. Rider's class follows:

   40 49 43 48 46 42 49 39 47 49
   42 41 42 39 41 40 45 43 44 42

   a. Make a line plot for the data.
   b. Make an ordered stem-and-leaf plot for the data.
   c. Make a frequency table for the data.
   d. Make a bar graph of the data.
6. The grades on a test for 30 students follow:

   96 73 61 76 77 84
   78 98 98 80 67 82
   61 75 79 90 73 80
   85 63 86 100 94 77
   86 84 91 62 77 64

   a. Make a grouped frequency table for these scores, using four classes and starting the first class at 61.
   b. Draw a histogram of the grouped data.
7. The budget for the Wegetem Crime Co. is $2,000,000. Draw a circle graph to indicate how the company spends its money where $600,000 is spent on bribes, $400,000 for legal fees, $300,000 for bail money, $300,000 for contracts, and $400,000 for public relations. Indicate percentages on your graph.
8. What, if anything, is wrong with the following bar graph?

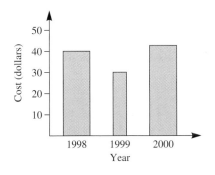

Monthly Health Club Costs

9. The mean salary of 24 people is $9000. How much will one additional salary of $80,000 increase the mean salary?
10. A cheetah can run 70 mph, a lion can run 50 mph, and a human can run 28 mph. Draw a bar graph to represent these data.
11. The life expectancies at birth for males and females are given in the following table:

Year	Male	Female	Year	Male	Female
1970	67.1	74.7	1987	71.5	78.4
1971	67.4	75.0	1988	71.5	78.3
1972	67.4	75.1	1989	71.7	78.5
1973	67.6	75.3	1990	71.8	78.8
1974	68.2	75.9	1991	72.0	78.9
1975	68.8	76.6	1992	72.1	78.9
1976	69.1	76.8	1993	72.1	78.9
1977	69.5	77.2	1994	72.4	79.0
1978	69.6	77.3	1997	73.6	79.2
1979	70.0	77.8	1996	73.1	79.1
1980	70.0	77.5	1995	72.5	78.9
1981	70.4	77.8			
1982	70.9	78.1			
1983	71.0	78.1			
1984	71.2	78.2			
1985	71.2	78.2			
1986	71.3	78.3			

   a. Draw back-to-back ordered stem-and-leaf plots to compare the data.
   b. Draw box-and-whisker plots to compare the data.
12. Larry and Marc took the same courses last quarter. Each bet that he would receive the better grades. Their courses and grades are as follows:

Course	Larry's Grades	Marc's Grades
Math (4 credits)	A	C
Chemistry (4 credits)	A	C
English (3 credits)	B	B
Psychology (3 credits)	C	A
Tennis (1 credit)	C	A

   Marc claimed that the results constituted a tie, since both received 2 A's, 1 B, and 2 C's. Larry said that he won the bet because he had the higher grade-point average for the quarter. Who is correct? (Allow 4 points for an A, 3 points for a B, 2 points for a C, 1 point for a D, and 0 points for an F.)

13. Following are the lengths in yards of the nine holes of the University Golf Course:

$$\begin{array}{ccc} 160 & 360 & 330 \\ 350 & 180 & 460 \\ 480 & 450 & 380 \end{array}$$

Find each of the following measures with respect to the lengths of the holes:
a. Median
b. Mode
c. Mean
d. Standard deviation

14. The speeds in miles per hour of 30 cars were checked by radar. The data are as follows:

$$\begin{array}{cccccccccc} 62 & 67 & 69 & 72 & 75 & 60 & 58 & 86 & 74 & 68 \\ 56 & 67 & 82 & 88 & 90 & 54 & 67 & 65 & 64 & 68 \\ 74 & 65 & 58 & 75 & 67 & 65 & 66 & 64 & 45 & 64 \end{array}$$

a. Find the median.
b. Find the upper and lower quartiles.
c. Draw a box-and-whisker plot for the data and indicate outliers (if any) with asterisks.
d. What percentage of the scores is in the interquartile range?
e. If every person driving faster than 70 mph received a ticket, what percentage of the drivers received speeding tickets?
f. Is the median in the center of the box? Why or why not?

15. The following scattergram was developed with information obtained from the girls trying out for the high-school basketball team:

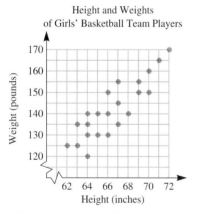

a. What kind of correlation exists between the heights and weights that are listed?
b. What is the weight of the girl who is 72 in. tall?
c. How tall is the girl who weighs 145 lb?
d. What is the mode of the heights?
e. What is the range of the weights?

16. The Nielsen Television Index rating of 30 means that an estimated 30% of American televisions are tuned to the show with that rating. The ratings are based on the preferences of a scientifically selected sample of 1200 homes.
a. Discuss possible ways in which viewers could bias this sample.
b. How could networks attempt to bias the results?

17. List and give examples of several ways to misuse statistics graphically.

18. Explain whether you think it is reasonable for a ski resort to advertise excellent skiing because the runs have 64 in. of snow at the top of the hill and 23 in. at the bottom.

19. The following bar graph shows the number of students per computer in the United States from 1983 to 1996.

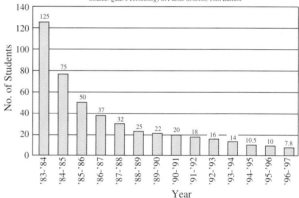

Data taken from *The World Almanac and Book of Facts 1999*, World Almanac Books, 1999.

a. Predict the results for the years 97–98, 98–99, and 99–00.
b. Between what two years did the greatest decrease in the number of students per computer occur?
c. Is a bar graph the most appropriate graph to use for this data? Why?
d. How could you distort this data to make it appear that the change has been very dramatic over the years?

20. The following are the United States Infant Mortality Rates for males and females from 1980 to 1996.
   a. Draw a back-to-back stem-and-leaf plot for the data.
   b. What can you say about the mortality rates based on your graph?
   c. What other kinds of graphs might you use to show comparisons you made in part (b)? Choose one and graph it.

U.S. Infant Mortality Rates 1980–1996*

Year	Male Deaths	Female Deaths	Year	Male Deaths	Female Deaths
1980	13.9	11.2	1989	10.8	8.8
1981	13.1	10.7	1990	10.3	8.1
1982	12.8	10.2	1991	10.0	7.8
1983	12.3	10.0	1992	9.4	7.6
1984	11.9	9.6	1993	9.3	7.4
1985	11.9	9.3	1994	8.8	7.2
1986	11.5	9.1	1995	8.3	6.8
1987	11.2	8.9	1996	8.0	6.6
1988	11.0	8.9			

*The rates are per 1000 live births.

1999 Data taken from *The World Almanac and Book of Facts 1999*, World Almanac Books, 1, 1999.

SELECTED BIBLIOGRAPHY

Barnes, S., K. D. Michalowicz, and D. Lee. "Now & Then: Some Moments in the History of Statistics: The Measurement of Uncertainty." *Mathematics Teaching in the Middle School* 1 (November-December 1994): 211–217.

Bloom, S. "Data Buddies: Primary-Grade Mathematicians Explore Data." *Teaching Children Mathematics* 1 (October 1994): 80–86.

Botula, M., and M. Ford. "All About Us: Connecting Statistics with Real Life." *Teaching Children Mathematics* 4 (September 1997): 14–19.

Brosnan, P. "Implementing Data Analysis in a Sixth-Grade Classroom." *Mathematics Teaching in the Middle School* 1 (January-February 1996): 622–628.

Browning, C., D. E. Channell, and R. A. Meyer. "Preparing Teachers to Present Techniques of Exploratory Data Analysis." *Mathematics Teaching in the Middle School* 1 (September-October 1994): 166–172.

Bryan, E. "Exploring Data with Box Plots." *Mathematics Teacher* 81 (November 1988): 658–663.

Burrill, G. "Statistics and Probability." *Mathematics Teacher* 83 (February 1990): 113–118.

Davis, G. "Using Data Analysis to Explore Class Enrollment." *Mathematics Teacher* 83 (February 1990): 104–106.

Dixon, J., and C. Falba. "Graphing in the Information Age: Using Data from the World Wide Web." *Mathematics Teaching in the Middle School* 2 (March-April 1997): 298–304.

Friel, S., and W. O'Connor. "Sticks to the Roof of Your Mouth?" *Mathematics Teaching in the Middle School* 4 (March 1999): 404–411.

Goldman, P. "Teaching Arithmetic Averaging: An Activity Approach." *Arithmetic Teacher* 37 (March 1990): 38–43.

Grummer, D. "Plotting Margo's Party." *Teaching Children Mathematics* 2 (November 1995): 176–179.

Hitch, C., and G. Armstrong. "Daily Activities for Data Analysis." *Arithmetic Teacher* 41 (January 1994): 242–245.

Hofstetter, E., and L. Sgroi. "Data with Snap, Crackle, and Pop." *Mathematics Teaching in the Middle School* 1 (March-April 1996): 760–764.

Huff, D. *How to Lie with Statistics.* New York: Norton, 1954.

Isaacs, A., and C. Kelso. "Pictures, Tables, Graphs, and Questions: Statistical Processes." *Teaching Children Mathematics* 2 (February 1996): 340–345.

Kader, G. "Means and MADs." *Mathematics Teaching in the Middle School* 4 (March 1999): 398–403.

Karp, K. "Telling Tales: Creating Graphs Using Multicultural Literature." *Teaching Children Mathematics* 1 (October 1994): 87–91.

Korithoski, T. P., and P. A. Korithoski. "Mean or Meaningless?" *Arithmetic Teacher* 41 (December 1993): 194–197.

Landwehr, J., and A. Watkins. *Exploring Data.* Palo Alto, Calif.: Dale Seymour Publishing, 1994.

Litton, N. "Graphing from A to Z." *Teaching Children Mathematics* 2 (December 1995): 220–223.

McClain, K. "Reflecting on Students' Understanding of Data." *Mathematics Teaching in the Middle School* 4 (March 1999): 374–380.

Morita, J. "Capture and Recapture Your Students' Interest in Statistics." *Mathematics Teaching in the Middle School* 4 (March 1999): 412–418.

O'Keefe, J. "The Human Scatterplot." *Mathematics Teaching in the Middle School* 3 (November-December 1997): 208–212.

Quinn, R. "Having Fun with Baseball Statistics." *Mathematics Teaching in the Middle School* 1 (May 1996): 780–785.

Rosenberg, M. "Learn about Statistics—Math League Baseball." *Arithmetic Teacher* 41 (April 1994): 459–461.

Rubink, W., and S. Taube. "Mathematics Connections from Biology: 'Killer' Bees Come to Life in the Classroom." *Mathematics Teaching in the Middle School* 4 (March 1999): 350–356.

Russell, S., and J. Mokros. "What Do Children Understand about Average?" *Teaching Children Mathematics* 2 (February 1996): 360–364.

Taylor, L., and J. A. Nichols. "Graphing Calculators Aren't Just for High School Students." *Mathematics Teaching in the Middle School* 1 (November-December 1994): 190–196.

Uccellini, J. "Teaching the Mean Meaningfully." *Mathematics Teaching in the Middle School* 2 (November-December 1996): 112–115.

Wilson, M., and C. Krapfl. "Exploring Mean, Median, and Mode with a Spreadsheet." *Mathematics Teaching in the Middle School* 1 (September–October 1995): 490–495.

9

Introductory Geometry

Preliminary Problem
Jon, a tile designer, was working to come up with convex tiles that had only acute angles. Could he find more than one type of shape for his tiles?

Geometry has taken on a very different meaning since antiquity when the word itself came from two Greek words meaning "earth measure." Even in the time of Euclid (ca. 300 B.C.), geometry was developing into a mathematical field that was removed from its physical beginnings. Today, we still consider geometry of the plane but recognize that as only one aspect of the mathematical field known as geometry.

HISTORICAL NOTE

Little is known of Euclid of Alexandria (ca. 300 B.C.), although legend has it that he studied geometry for its beauty and logic. Euclid is best known for *The Elements,* a work so systematic and encompassing that many earlier mathematical works were simply discarded and lost to all future generations. *The Elements,* composed of 13 books, included not only geometry but arithmetic and topics in algebra. Euclid set up a *deductive system* by starting with a set of statements that he assumed to be true and showing that geometric discoveries followed logically from these assumptions.

From knowing that the world is not flat, to measurement between planets, to measurement between cells, to Mandelbrot's work with fractals, geometry has changed as our knowledge base has expanded. In schools, too, geometry has changed. We no longer consider geometry as consisting solely of theorems and proofs; we no longer consider geometry in isolation from the rest of mathematics. As a result of these changes, the notion of what constitutes geometry for prospective elementary teachers has changed.

According to the *Principles and Standards,*

The study of geometry in grades 3–5 requires thinking and doing. As students sort, build, draw, model, trace, measure, and construct, their capacity to visualize geometric relationships will develop. ... This exploration requires access to a variety of tools, such as graph paper, rulers, pattern blocks, geoboards, and geometric solids, and is greatly enhanced by electronic tools that support exploration, such as dynamic geometry software (p. 165).

Section 9-1 — Basic Notions

The fundamental building blocks of geometry are *points, lines,* and *planes.* Ironically the building blocks are *undefined terms* in order to avoid circular definitions. An example of a circular definition and the frustration involved in starting without some basic undefined notions is given in the following "B.C." cartoon.

Reprinted by permission of Johnny Hart and Creators Syndicate, Inc.

Because other geometric concepts are developed from such undefined terms, we present an intuitive notion of these terms in Table 9-1.

Table 9-1

Term	Illustration	Symbolism
Point	(compass with points B, A, C)	Point A Point B Point C
Line	ℓ — Centerline of road (line m through points A and B)	line ℓ line m, line AB, $\overleftrightarrow{AB}$, or $\overleftrightarrow{BA}$
Plane	(table with plane α; plane γ containing points A, B, C)	Plane α Plane ABC or Plane γ

Linear Notions

In geometry, a line has no thickness and it extends forever in two directions. *It is determined by two points.* When we discuss situations that involve points and lines, we rely on such undefined relations as "contains," "belongs to," or "is on," and "is between." The meanings of these words are usually understood from the context. For example, Table 9-2 illustrates many common terms and intuitive meanings.

Table 9-2

Term	Illustration and Symbolism
Collinear points are points on the same line. (Any two points are collinear but not every three points have to be collinear.)	Line ℓ contains points A, B, and C. Points A, B, and C belong to line ℓ. Points A, B, and C are collinear. Points A, B, and D are not collinear.
Point B **is between** points A and C on line ℓ.	
A **line segment**, or **segment**, is a subset of a line that contains two points of the line and all points between those two points.	$\overline{AB}$ or $\overline{BA}$
A **ray** is a subset of a line that contains one point and all points on the line on one side of the point.	$\overrightarrow{AB}$

Planar Notions

A plane has no thickness and it extends indefinitely in two directions. *A plane is determined by three points that are not all on the same line.* In other words, given three noncollinear points, a unique plane is determined. Table 9-3 illustrates intuitive planar notions.

Table 9-3

Term	Illustration and Symbolism
Points in the same plane are **coplanar.** **Noncoplanar points** cannot be placed in the same plane. **Skew lines** are noncoplanar. **Intersecting lines** are two coplanar lines with exactly one point in common. **Concurrent lines** are lines that contain the same point.	Points D, E, and G are coplanar. Points D, E, F, and G are noncoplanar. Lines DE and EG are coplanar. Lines GF and DE are skew lines. Lines DE and GE are intersecting lines; they intersect at point E. Lines DE, EG, and EF are concurrent.
Two distinct coplanar lines m and n that have no points in common are **parallel lines.**	m is parallel to n, written $m \| n$.

NOW TRY THIS 9-1

- **a.** How many different lines can be drawn through two points?
- **b.** Can skew lines be parallel? Why?
- **c.** On a globe, a "line" is a great circle, that is, a circle the same size as the equator. How many different lines can be drawn through two different points?

Example 9-1 Answer each of the following:

a. Why is a tripod considered "level?"
b. How many planes can be determined by a line and a point not on the line?
c. How many planes do two intersecting lines determine?

Solution
a. The three feet of a tripod determine three points. The three points essentially form a level surface, or a plane, through those three points.
b. Any two distinct points determine a line. Using any two points on the line and the point not on the line, we have three noncollinear points that determine exactly one plane. There is no other plane that satisfies the required conditions. (Why?)
c. Two intersecting lines have one point in common and each line contains at least one other point not on the other line. Thus, we have three noncollinear points that determine exactly one plane. Any plane that contains the two lines must contain the points used here, so the plane is unique.

Properties of points, lines, and planes are summarized in the following:

Properties of Points, Lines, and Planes

1. There is exactly one line that contains any two distinct points.
2. If two points lie in a plane, then the line containing the points lies in the plane.
3. If two distinct planes intersect, then their intersection is a line.
4. There is exactly one plane that contains any three distinct noncollinear points.
5. A line and a point not on the line determine a plane.
6. Two parallel lines determine a plane.
7. Two intersecting lines determine a plane.

Problem Solving **Lines Through Points**

Given 15 points, no three of which are collinear, how many different lines can be drawn through the 15 points?

• **Understanding the Problem** We are asked to find the number of lines that are determined by 15 points, no 3 of which are collinear. Because two points determine exactly one line, we must consider ways to find out how many different lines are determined by the 15 points.

• **Devising a Plan** We use the strategy of *examining related simpler cases* of the problem in order to think through the original problem. Figure 9-1 shows that 3 non-collinear points determine 3 lines, 4 determine 6 lines, 5 determine 10 lines, and 6 determine 15 lines.

Figure 9-1

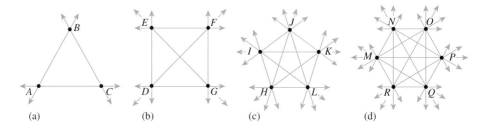

(a)   (b)   (c)   (d)

Examining Figure 9-1(d), we see that through any one of the six points, we can draw only five lines to connect to the other five points. If five lines are drawn through each point, there are 6 times 5, or 30, lines. But each of the 30 lines is counted twice. (Why?) Thus, there are $\frac{6 \cdot 5}{2}$, or 15, lines in the figure. We use this information to determine the solution for 15 points.

• **Carrying Out the Plan** Each of the 15 points can be paired with all points other than itself to determine a line; that is, each point can be paired with the other 14 points to determine 14 lines. If we do this and account for counting each line twice, we see that there should be $\frac{15 \cdot 14}{2}$, or 105, lines.

• **Looking Back** Using this reasoning, we conclude that the number of lines determined by n points, no 3 of which are collinear, is $\frac{n(n-1)}{2}$. (Why?)

An alternative solution to this problem uses the notion of combinations found in Chapter 7. The number of ways that n points, no 3 of which are collinear, can be chosen two at a time to form lines is $_nC_2$, or $\frac{n(n-1)}{2}$.

Other Planar Notions

Two distinct planes either intersect in a line or are parallel. In Figure 9-2(a), planes α and β are parallel; that is, they have no points in common. Figure 9-2(b) shows planes that intersect in $\overleftrightarrow{AB}$.

Figure 9-2

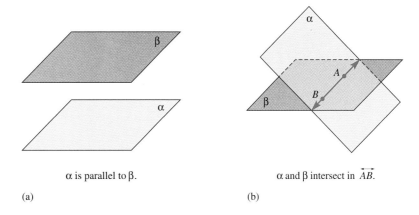

α is parallel to β.

(a)

α and β intersect in $\overleftrightarrow{AB}$.

(b)

A line and a plane can be related in one of three ways. If a line and a plane have no points in common, the line is parallel to the plane, as in Figure 9-3(a). If two points of a line are in the plane, then the entire line containing the points is contained in the plane, as in Figure 9-3(b). In Figure 9-3(b), the line AB separates plane α into two **half-planes** denoted as AB-C and AB-D. Plane α is the union of the three mutually disjoint sets: the two half-planes AB-C and AB-D, and line AB. If a line intersects a plane but is not contained in the plane, it intersects the plane at only one point, as in Figure 9-3(c).

half-plane

Figure 9-3

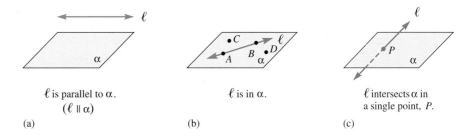

ℓ is parallel to α.
($\ell \parallel α$)

(a)

ℓ is in α.

(b)

ℓ intersects α in a single point, P.

(c)

Angles

angle
side · vertex

When two rays share a common endpoint, an **angle** is formed, as shown in Figure 9-4(a). The rays of an angle are the **sides** of the angle, and the common endpoint is the **vertex** of the angle. An angle can be named by three different points: the vertex and a point on each ray, with the vertex always listed between the other two points. Thus the angle in Figure 9-4(a) may be named ∠CBA or ∠ABC. When there is no risk of confusion, it is customary simply to name an angle by its vertex, by a number, or by a lowercase Greek letter. The angle in Figure 9-4(a) therefore also can be named ∠B or ∠1. In Figure 9-4(b), however, more than one angle has vertex P, namely, ∠QPR, ∠RPS, and ∠QPS. Thus the notation ∠P is inadequate for naming any one of the angles α, β, or ∠QPS. Angles, such as ∠QPR (or α) and ∠RPS (or β) in Figure 9-4(b) are **adjacent angles,** which share a common vertex and a common side and do not have overlapping interiors.

adjacent angles

Figure 9-4

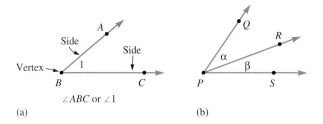

(a) ∠ABC or ∠1   (b)

Angle Measurement

degree An angle is measured according to the amount of "opening" between its sides. The **degree** is commonly used to measure angles. A complete rotation about a point has a measure of 360°. One degree is then $\frac{1}{360}$ of a complete rotation. Figure 9-5 shows that ∠BAC has a measure of 30 degrees, written $m(\angle BAC) = 30°$. The measuring device pictured in the figure is a **protractor**. A degree is subdivided into 60 equal parts—**minutes**—and each minute is further divided into 60 equal parts—**seconds**. The measurement 29 degrees, 47 minutes, 13 seconds is written 29°47′13″.

protractor • minutes seconds

Figure 9-5

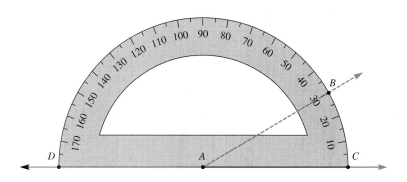

HISTORICAL NOTE

In 1634 Pierre Herigone used a symbol for an angle. It was not until 1923 that the Mathematical Association of America recommended ∠ as the standard symbol for angle in the United States. The use of 360° to measure angles seems to date to the Babylonian culture (4000–3000 B.C.), with minutes and seconds coming from Latin translations of Arabic translations of Babylonian sexagesimal (base 60) fractions.

Example 9-2

a. In Figure 9-6, find the measure of ∠BAC if $m(\angle 1) = 47°45'$ and $m(\angle 2) = 29°58'$.
b. Express $47°45'$ as a number of degrees.

Figure 9-6

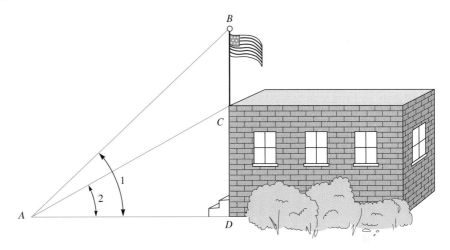

Solution

a. $m(\angle BAC) = 47°45' - 29°58'$
$= 46°(60 + 45)' - 29°58'$
$= 46°105' - 29°58'$
$= (46 - 29)° + (105 - 58)'$
$= 17°47'$

b. $47°45' = 47\dfrac{45°}{60} = 47.75°$

Types of Angles

We can create different types of angles by paper folding, especially with wax paper. Consider the folds shown in Figure 9-7(a) and (b). A piece of paper is folded in half and then reopened. If any point on the fold line labeled ℓ is chosen as the vertex, then the measure of the angle pictured is 180°. If the paper is refolded and folded once more, as shown in Figure 9-7(c), and then is reopened, as shown in Figure 9-7(d), four angles of the same size are created. Each angle has measure 90° and is a right angle. The symbol ⌐ denotes a right angle.

Figure 9-7

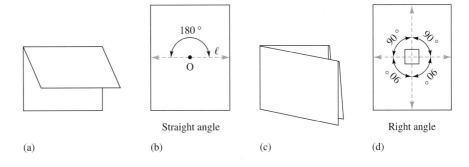

(a)　　(b) Straight angle　　(c)　　(d) Right angle

If the paper is folded as shown in Figure 9-8 and reopened, then angles α and β are formed, with measures that are less than 90° and greater than 90°, respectively. (Note that β has measure less than 180°.) Angle α is an *acute* angle while β is an *obtuse* angle.

Figure 9-8

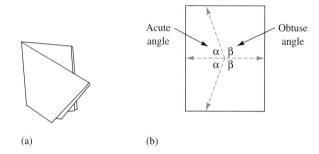

(a) (b)

The types of planar angles just discovered are shown in Figure 9-9, along with their definitions.

Figure 9-9

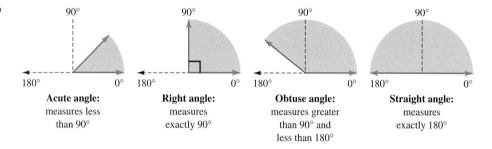

Acute angle: measures less than 90°

Right angle: measures exactly 90°

Obtuse angle: measures greater than 90° and less than 180°

Straight angle: measures exactly 180°

On the student page on page 471 from *Scott Foresman-Addison Wesley Middle School Math, Course 3*, 1999, we find a way to use paper folding to create angles of various measures.

Perpendicular Lines

perpendicular lines

When two lines intersect so that the angles formed are right angles, as in Figure 9-10, the lines are **perpendicular lines.** In Figure 9-10, lines *m* and *n* are perpendicular, and we write $m \perp n$. Two intersecting segments, two intersecting rays, or one segment and one ray that intersect are perpendicular if they lie on perpendicular lines. For example, in Figure 9-10, $\overline{AB} \perp \overline{BC}$, $\overrightarrow{BA} \perp \overline{BC}$, and $\overleftrightarrow{AB} \perp \overline{BC}$.

Figure 9-10

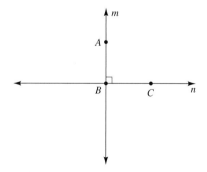

SECTION 9-1 *Basic Notions* **471**

8-4 Lines and Angles

You'll Learn ...
- to draw, measure, and identify angles

... How It's Used
Lighting technicians know how to set a dramatic mood by staging lights at various angles.

▶ **Lesson Link** You have learned how to use various units of measurement. Now you will learn about lines and angles. ◀

An **angle** is formed when two lines (or line segments, or rays) meet at one point, called the **vertex**. Angles are usually measured in degrees.

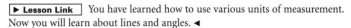

There are 360° in a complete circle.

Explore | Angles

The Angle in Creases

Materials: Sheet of wax paper, Compass, Scissors

1. Use your compass to draw the largest circle possible on the sheet of wax paper. Cut out the circle.

2. Fold the circle in half by lining up the edge of your circle. Fold the circle in half again.

3. Fold the circle in half two more times.

4. Each time you folded the circle, you made an angle. What part of the circle was located at the vertex of each angle?

5. The full circle shows 360°. When you fold the full circle in half, you make a 180° angle. What is the resulting angle measure if you fold this in half?

6. What is the number pattern that results each time an angle is folded in half? Write an algebraic equation for this.

7. If each quarter of the circle is 90°, what is the angle measure of three-quarters of a circle? How did you arrive at your answer?

8. On your circle, label as many creases as you can with the appropriate angle measure.

9. Use your circle as a measuring tool to approximate the measure of angles you can find in your classroom.

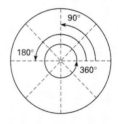

Vocabulary
angle
vertex
line
line segment
endpoints
ray
straight angle
right angle
acute angle
obtuse angle
complementary
supplementary
congruent
angle bisector

410 Chapter 8 • Geometry and Measurement

A Line Perpendicular to a Plane

If a line and a plane intersect, they can be perpendicular. For example, consider Figure 9-11, where planes β and γ represent two walls intersecting along $\overleftrightarrow{AB}$. The edge $\overleftrightarrow{AB}$ is perpendicular to the floor. Also, every line in the plane of the floor (plane α) passing through point *A* is perpendicular to $\overleftrightarrow{AB}$. This discussion should help you understand what we mean by "a line perpendicular to a plane."

Figure 9-11

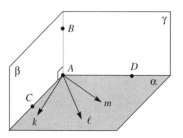

line perpendicular to a plane A **line perpendicular to a plane** is a line that is perpendicular to every line in the plane through its intersection with the plane.

ONGOING ASSESSMENT 9-1

1. For the following figure, answer the following:
   a. Name two pairs of skew lines.
   b. Are $\overleftrightarrow{BD}$ and $\overleftrightarrow{FH}$ parallel, skew, or intersecting lines?
   c. Are $\overleftrightarrow{BD}$ and $\overleftrightarrow{GH}$ parallel?
   d. Find the intersection of $\overleftrightarrow{BD}$ and plane *EFG*.
   e. Find the intersection of $\overleftrightarrow{BH}$ and plane *DCG*.
   f. Name two pairs of perpendicular planes.
   g. Name two lines that are perpendicular to plane *EFH*.

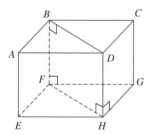

2. A **dihedral angle** is the union of two half-planes and the common line defining the half-planes. A dihedral angle is measured by any of the associated planar angles such as ∠*OPD* in the following figure. (Note that $\overrightarrow{PO} \perp \overleftrightarrow{AC}$ and $\overrightarrow{PD} \perp \overleftrightarrow{AC}$.)

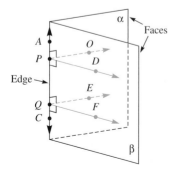

Dihedral angle

Using the figure in problem 1, do the following:
   a. Name a planar angle that could be used to measure dihedral angle *E-FH-B*.
   b. What is the measure of dihedral angle *D-HG-F*?

3. Use the following drawing of one of the Great Pyramids of Egypt to find the following:
   a. The intersection of $\overline{AD}$ and $\overline{CE}$
   b. The intersection of planes *ABC*, *ACE*, and *BCE*
   c. The intersection of $\overleftrightarrow{AD}$ and $\overleftrightarrow{CA}$
   d. A pair of skew lines
   e. A pair of parallel lines
   f. A plane not determined by one of the triangular faces or by the base

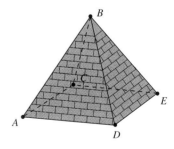

4. Determine how many pairs of adjacent angles are in the following figure:

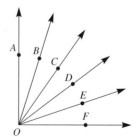

5. Identify a physical model for each of the following:
   a. Perpendicular lines
   b. An acute angle
   c. An obtuse angle
6. Find the measure of each of the following angles:
   a. ∠EAB
   b. ∠EAD
   c. ∠GAF
   d. ∠CAF

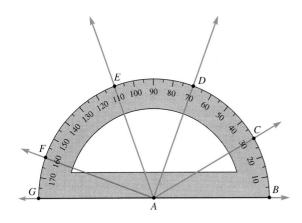

7. Use a protractor to find the measure of each of the following pictured angles:

Paper scissors
(a)

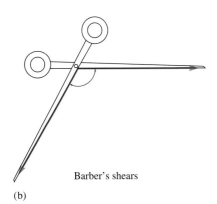

Barber's shears
(b)

8. a. Perform each of the following operations. Leave your answers in simplest form.

   i. $18°35'29'' + 22°55'41''$
   ii. $93°38'14'' - 13°49'27''$

   b. Express each of the following in degrees, minutes, and seconds, without decimals:
      i. $0.9°$
      ii. $15.13°$

9. Consider a correctly set clock that starts ticking at noon and answer the following:
   a. Find the measure of the angle swept by the hour hand by the time it reaches
      i. 3 P.M.
      ii. 12:25 P.M.
      iii. 6:50 P.M.
   b. Find the exact angle between the minute and the hour hands at 1:15 P.M.
   c. At what time between 12 noon and 1 P.M. will the angle between the hands be 180°?

10. Mario was studying right angles and wondered if during his math class the minute and hour hands of the clock formed a right angle. If his class meets from 2:00 P.M. to 2:50 P.M., is a right angle formed? If it is, figure out to the nearest minute when the hands form the right angle.

11. Determine how many rays are formed by each of the following:
    a. Three collinear points
    b. Four collinear points
    c. Five collinear points
    d. n collinear points

12. Refer to the following table.
    a. Sketch the possible intersections of the given number of lines. Three sketches are given for you.

Number of Intersection Points

		0	1	2	3	4	5
Number of Lines	2		✕	Not possible	Not possible	Not possible	Not possible
	3					Not possible	Not possible
	4				✳		✳
	5						
	6						

b. Given n lines, find a formula for determining the greatest possible number of intersection points.

13. Trace each of the following drawings. In your tracings, use dashed lines for segments that would not be seen and solid lines for segments that would be seen. (Different people may see different perspectives.)

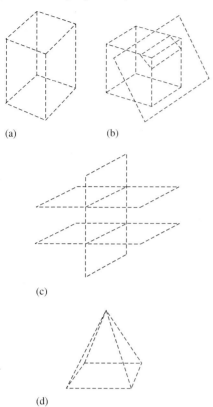

14. Write Logo procedures to draw each of the following:
    a. A procedure called ANGLE with input :SIZE to draw a variable-sized angle
    b. A procedure called SEGMENT with input :LENGTH to draw a variable-sized segment
    c. A procedure called PERPENDICULAR with inputs :LENGTH1 and :LENGTH2 to draw two variable-sized perpendicular segments
    d. A procedure called PARALLEL with inputs :LENGTH1 and :LENGTH2 to draw two variable-sized parallel segments
15. Draw pictures illustrating a real-world example of the following:
    a. Three planes intersecting in a common line
    b. Three planes intersecting in a common point.
16. In each of the following pairs, determine whether the symbols name the same geometric figure.
    a. $\overline{AB}$ and $\overline{BA}$ b. $\overrightarrow{AB}$ and $\overrightarrow{BA}$ c. $\overleftrightarrow{AB}$ and $\overleftrightarrow{BA}$.
17. a. How many planes are determined by 3 noncollinear points?
    b. How many planes are determined by four points, no three of which are collinear?
    c. How many planes are determined by *n* points, no three of which are collinear?

Communication

18. Forest rangers use degree measures to identify directions and locate critical spots such as fires. In the following drawing, a forest ranger at tower *A* observes smoke at a bearing of 149° (clockwise from the north), while another forest ranger at tower *B* observes the same source of smoke at a bearing of 250° (clockwise from the north).

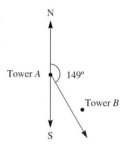

    a. Choose two locations for *A* and *B* and use a protractor and a straightedge to locate the source of the smoke.
    b. Explain how the forest rangers could find the location of the fire.
    c. Describe other situations in which location can be determined by similar methods.
19. a. Is it possible for a line to be perpendicular to one line in a plane but not perpendicular to the plane? Explain.
    b. Is it possible for a line to be perpendicular to two distinct lines in a plane and yet not be perpendicular to the plane? Explain.
    c. If a line not in a given plane is perpendicular to two distinct lines in the plane, is the line necessarily perpendicular to the plane? Explain.
20. Is it possible to locate four points in a plane such that the number of lines determined by the points is not exactly 1, 4, or 6? Explain.

Cooperative Learning

21. Each member of your group should use a protractor to make a triangle out of cardboard that has one angle measuring 30° and another 50°. Answer the following and compare your solutions with other members of your group:
    a. Show how to use the triangle (without a protractor) to draw an angle with measure 40°.
    b. Is there more than one way to draw an angle as in (a) using the triangle? Explain.
    c. What other angles can be drawn with the triangle? Why?

Open-Ended

22. Within the classroom, identify a physical object with the following shapes:
    a. Parallel lines   b. Parallel planes
    c. Skew lines   d. Right angles

23. On a sheet of dot paper or on a geoboard like the one shown, create the following shapes:

    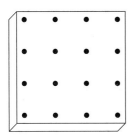

    a. Right angle
    b. Acute angle
    c. Obtuse angle
    d. Adjacent angles
    e. Parallel lines
    f. Intersecting lines

TECHNOLOGY CORNER

Use Logo to predict the outcomes by sketching the corresponding figure when each of the following is executed. Check your predictions on the computer.

a. RT 900 FD 50
b. LT 900 FD 50
c. REPEAT 4 [FD 50 RT 90]
d. REPEAT 5 [FD 60 RT 72]
e. REPEAT 5 [FD 60 RT 144]

LABORATORY ACTIVITY

Many geometric ideas can be experienced through paper folding, as in Figure 9-12. Consider the following construction of perpendicular lines and answer the questions that follow:

Figure 9-12

Fold one corner of a sheet of paper over and crease it along the fold.

Fold any part of the crease onto itself.

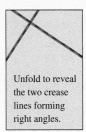
Unfold to reveal the two crease lines forming right angles.

a. Explain why the two crease lines are perpendicular.
b. Use a sheet of paper and follow the above instructions to create two pairs of perpendicular lines.
c. Use paper folding to create angles with the following measures:
    i. 45°   ii. 135°   iii. 22°30′

(*Mathematics Through Paper Folding* by Olsen and *Geometric Exercises in Paper Foldings* by Row are listed in the bibliography as resource books.)

Section 9-2 — Polygons

With a pencil, draw a path on a piece of paper without lifting the pencil and without retracing any part of the path except single points. The resulting drawing is restricted to the plane of the paper and not lifting the pencil implies that there are no breaks in the drawing. The drawing is *connected* and is a **curve.**

curve

In Table 9-4 each curve is in the plane and is connected, but there are some obvious differences among the curves. Table 9-4 shows sample curves and their classifications.

Table 9-4

Simple	Closed	Polygon	Convex	Concave
 ⌒	 ∞			
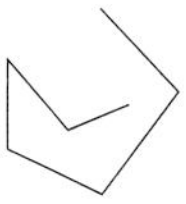 ⌵	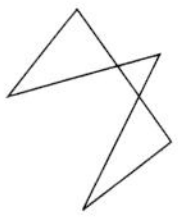 ✦			
 ϟ	 ϟ			 ϟ
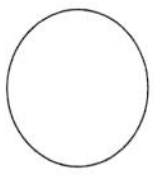 ○	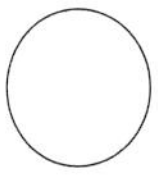 ○		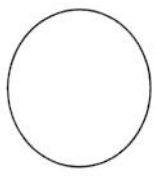 ○	
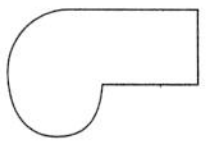 ⌐	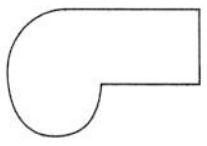 ⌐			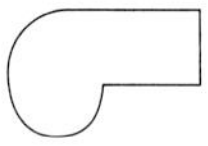 ⌐
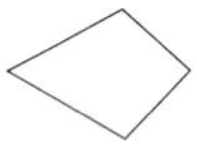 ◇	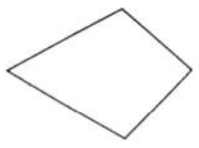 ◇	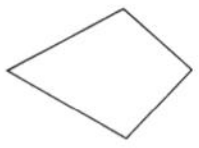 ◇	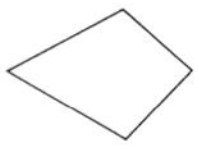 ◇	
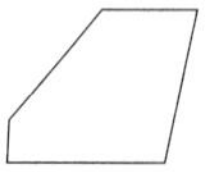 ▱	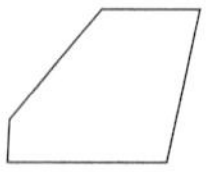 ▱	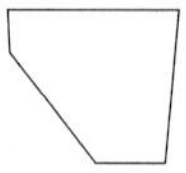 ▱	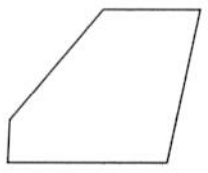 ▱	
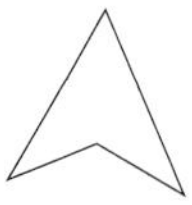 ⌂	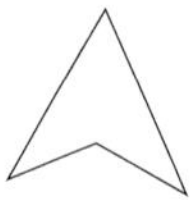 ⌂	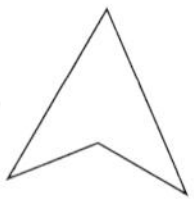 ⌂		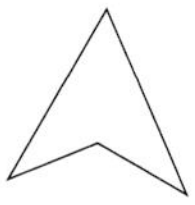 ⌂

simple A **simple** curve does not cross itself, except that if you draw it with a pencil, the starting
closed and stopping points may be the same. A **closed** curve can be drawn starting and stopping at
the same point. A curve can be classified as simple, nonsimple, closed, nonclosed, and so
polygon on. **Polygons** are simple and closed and have *sides* that are segments. A point where two
convex sides of a polygon meet is a *vertex*. **Convex** curves are simple, closed, and have no indentations. More precisely, the segment connecting any two points *in the interior of the curve*
concave *is wholly contained in the interior of the curve*. **Concave** curves are simple, closed, and not
convex; that is, they have an indentation.

NOW TRY THIS 9-2

- Draw a curve that is neither simple nor closed and explain why.

As in Figure 9-13(a), every simple closed curve separates the plane into three disjoint
subsets: the interior of the curve, the exterior of the curve, and the curve itself. Of specific
polygonal regions interest to us are polygons and their interiors, together called **polygonal regions.** Figure
9-13(b) shows a polygonal region.

Figure 9-13

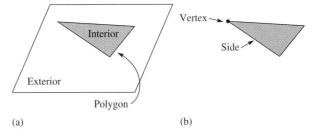

(a)   (b)

Whether a point is inside or outside a curve is not always obvious. This is explored in
the next investigation.

NOW TRY THIS 9-3

- Determine whether point *X* is inside or outside the simple closed curve of Figure 9-14. Explain your reasoning so that it can be generalized to other simple closed curves.

Figure 9-14

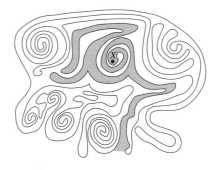

More About Polygons

Polygons are classified according to the number of sides or vertices they have. For example, consider the polygons listed in Table 9-5.

Table 9-5

Polygon	Number of Sides or Vertices
Triangle	3
Quadrilateral	4
Pentagon	5
Hexagon	6
Heptagon	7
Octagon	8
Nonagon	9
Decagon	10
n-gon	n

A polygon is referred to by the capital letters that represent its consecutive vertices, such as shown in Figure 9-15(a). *ABCD* or *CDAB* (but not *BCAD*). Any two sides of a polygon having a common vertex determine an **interior angle,** or **angle, of the polygon,** such as ∠1 of polygon *ABCD* in Figure 9-15(a). An **exterior angle of a polygon** is determined by a side of the polygon and the extension of a contiguous side of the polygon. An example is ∠2 in Figure 9-15(b). Any line segment connecting nonconsecutive vertices of a polygon such as $\overline{AC}$ in Figure 9-15(a) is a **diagonal** of polygon *ABCD*.

interior angle, or angle, of a polygon · **exterior angle of a polygon**

diagonal

Figure 9-15

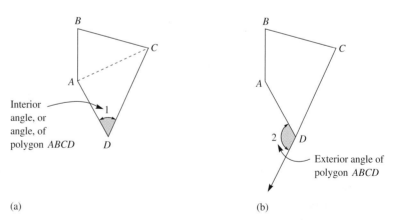

Congruent Segments and Angles

congruent parts Most modern industries operate on the notion of creating **congruent parts,** parts that are of the same size and shape. For example, the specifications for all cars of a particular model are the same, and all parts produced for that model are basically the same.

congruent segments Usually congruent figures refer to figures in a plane. For example, two line **segments** are **congruent** (≅) if a tracing of one line segment can be fitted exactly on top of the other.

congruent angles If $\overline{AB}$ is congruent to $\overline{CD}$, we write $\overline{AB} \cong \overline{CD}$. Two **angles** are **congruent** if they have the same measure. Congruent segments and congruent angles are shown in Figure 9-16(a) and (b), respectively.

Figure 9-16

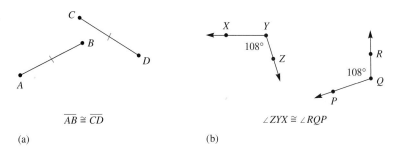

$\overline{AB} \cong \overline{CD}$

(a)

∠ZYX ≅ ∠RQP

(b)

Regular Polygons

regular polygons

Polygons in which all the interior angles are congruent and all the sides are congruent are **regular polygons.** A regular polygon is both *equiangular* and *equilateral*. A regular triangle is an equilateral triangle. A regular pentagon and a regular hexagon are illustrated in Figure 9-17. The congruent sides and congruent angles are marked.

Figure 9-17

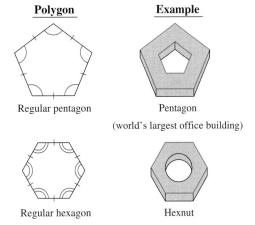

Polygon	Example
Regular pentagon	Pentagon (world's largest office building)
Regular hexagon	Hexnut

Triangles and Quadrilaterals

Triangles may be classified according to their angle measures, as shown in Table 9-6. Triangles and quadrilaterals may also be classified as shown in Table 9-6.

Table 9-6

Definition	Illustration	Example
A triangle containing one right angle is a **right triangle.**		

continued

Table 9-6 *continued*

Definition	Illustration	Example
A triangle in which all the angles are acute is an **acute triangle.**		
A triangle containing one obtuse angle is an **obtuse triangle.**		
A triangle with no congruent sides is a **scalene triangle.**		
A triangle with at least two congruent sides is an **isosceles triangle.**		
A triangle with three congruent sides is an **equilateral triangle.**		
A **trapezoid** is a quadrilateral with at least one pair of parallel sides.		
A **kite** is a quadrilateral with at least two distinct pairs of consecutive congruent sides.		
An **isosceles trapezoid** is a trapezoid with one pair of congruent base angles. (Equivalently, an isosceles trapezoid is a trapezoid with two congruent nonadjacent sides.)		
A **parallelogram** is a quadrilateral in which each pair of opposite sides is parallel.		
A **rectangle** is a parallelogram with a right angle. (Equivalently, a rectangle is a quadrilateral with four right angles.)		
A **rhombus** is a parallelogram with all sides congruent. (Equivalently, a rhombus is a quadrilateral with all sides congruent.)		
A **square** is a rectangle with all sides congruent. (Equivalently, a square is a quadrilateral with four right angles and four congruent sides.)		

Some texts give different definitions for a trapezoid. Many elementary texts define a trapezoid as a quadrilateral with *exactly* one pair of parallel sides. Note the definition of a trapezoid on the following student page from *Scott Foresman-Addison Wesley Middle School Math Course 2*, 1999. An excellent teaching aid, a *tangram,* is used on this page.

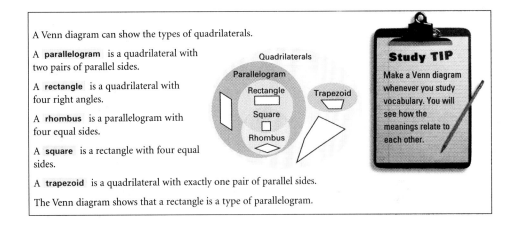

A Venn diagram can show the types of quadrilaterals.

A **parallelogram** is a quadrilateral with two pairs of parallel sides.

A **rectangle** is a quadrilateral with four right angles.

A **rhombus** is a parallelogram with four equal sides.

A **square** is a rectangle with four equal sides.

A **trapezoid** is a quadrilateral with exactly one pair of parallel sides.

The Venn diagram shows that a rectangle is a type of parallelogram.

Study TIP

Make a Venn diagram whenever you study vocabulary. You will see how the meanings relate to each other.

Hierarchy Among Polygons

Every triangle is a polygon, and every equilateral triangle is also isosceles. However, not every isosceles triangle is equilateral. Using set concepts, we can say that the set of all triangles is a proper subset of the set of all polygons. Also, the set of all equilateral triangles is a proper subset of the set of all isosceles triangles. This hierarchy is shown in Figure 9-18, where more general terms appear above more specific ones. Compare Figure 9-18 with the Venn diagram of the student page. How do they differ?

Figure 9-18

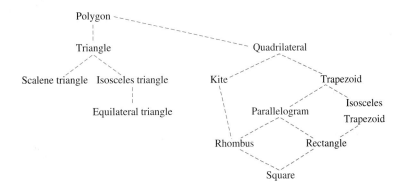

NOW TRY THIS 9-4

● Use the definitions in Table 9-6 to experiment with several drawings to decide which of the following are true:

1. An equilateral triangle is isosceles.
2. A square is a regular quadrilateral.
3. If one angle of a rhombus is a right angle, then all the angles of the rhombus are right angles.
4. A square is a rhombus with a right angle.
5. All the angles of a rectangle are right angles.
6. A rectangle is an isosceles trapezoid.
7. Some isosceles trapezoids are kites.
8. If a kite has a right angle, then it must be a square.

ONGOING ASSESSMENT 9-2

1. Determine for each of the following which of the figures labeled (1) through (10) can be classified under the given term:
   a. Simple closed curve
   b. Polygon
   c. Convex polygon
   d. Concave polygon

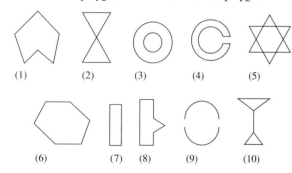

2. What is the maximum number of intersection points between a quadrilateral and a triangle (where no sides of the polygons are on the same line)?
3. What type of polygon must have a diagonal such that part of the diagonal falls outside of the polygon?
4. Which of the following figures are convex, and which are concave? Why?

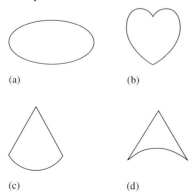

5. If possible, draw the following triangles. If it is not possible, state why.
   a. An obtuse scalene triangle
   b. An acute scalene triangle
   c. A right scalene triangle
   d. An obtuse equilateral triangle
   e. A right equilateral triangle
   f. An obtuse isosceles triangle
   g. An acute isosceles triangle
   h. A right isosceles triangle

6. Determine how many diagonals each of the following has:
   a. Decagon
   b. 20-gon
   c. 100-gon

7. Identify each of the following triangles as scalene, isosceles, or equilateral:

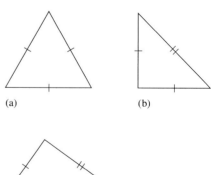

8. Describe regions (a) and (b) in the following Venn diagram:

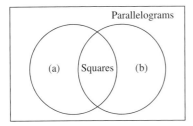

9. Use the labeled points in the following drawing to answer the questions:

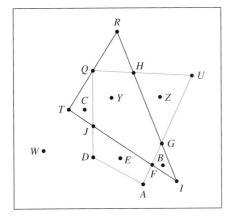

   a. Which points belong to triangle *TRI*?
   b. Which points belong to the interior of the quadrilateral *QUAD*?
   c. Which points belong to the exterior of triangle *TRI*?
   d. Which points belong to triangle *TRI* and quadrilateral *QUAD*?
   e. Which points belong to the intersection of the interiors of triangle *TRI* and quadrilateral *QUAD*?

10. Write a Logo program to draw each of the following:
    a. A square
    b. A rectangle

Communication

11. a. Fold a rectangular piece of paper to create a square. Describe your procedure in writing and orally with a classmate. Explain why your approach creates a square.
    b. Crease the square in (a) so that the two diagonals are shown. Use paper folding to show that the diagonals of a square are congruent and perpendicular and bisect each other. Describe your procedure and explain why it works.

Open-Ended

12. On a geoboard or dot paper, construct each of the following:
    a. A scalene triangle   b. A square
    c. A trapezoid   d. A convex hexagon

    e. A concave quadrilateral
    f. A parallelogram

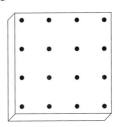

Cooperative Learning

13. Work with a partner. One of you should construct a figure on a geoboard or draw it on a piece of paper and identify it. Do not show the figure to your partner but tell your partner sufficient properties of the figure to identify it. Have your partner identify the figure you constructed. Your partner earns 1 point if the figure is correctly identified and 2 points if a figure is found that has all the required attributes but is different from the one you drew. Each of you should take the same number of turns. Try this with each of the following types of figures:
    a. Scalene triangle
    b. Isosceles triangle
    c. Square
    d. Parallelogram
    e. Trapezoid
    f. Rectangle
    g. Regular polygon
    h. Rhombus
    i. Isosceles trapezoid
    j. A kite that is not a rhombus

14. a. Compare the information in the Venn diagram on the student page on page 481 in this section to the quadrilateral hierarchy of Figure 9-18.
    b. Work with partners to create a Venn diagram showing the triangle portion of Figure 9-18.

15. a. Investigate the meaning and uses of Reuleaux triangles.
    b. Explain the similarities and differences between a Reuleaux triangle and an equilateral triangle.

16. a. With partners, decide how you might define a triangle on a globe.
    b. With partners, a globe, string, and a ruler, decide if you think a square can exist on a globe.

Review Problems

17. If three distinct rays with the same vertex are drawn as shown in the following figure, then three angles are formed: $\angle AOB$, $\angle AOC$, and $\angle BOC$:

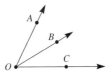

What is the maximum number of angles (measuring less than 180°) formed by using
a. 10 distinct noncollinear rays with the same vertex?
b. n distinct noncollinear rays with the same vertex?
18. Determine the possible intersection sets of a line and an angle.
19. Classify the following as true or false. If false, tell why.
a. A ray has two endpoints.
b. For any points M and N, $\overrightarrow{MN} = \overleftarrow{NM}$.
c. Skew lines are coplanar.
d. $\overleftrightarrow{MN} = \overrightarrow{NM}$
e. A line segment contains an infinite number of points.
f. If two distinct planes intersect, their intersection is a line segment.

TECHNOLOGY CORNER

Use Logo to create each of the following. In each case, give the sequence of commands used to create the figure.

a. A figure similar to that in Figure 9-19
b. A figure that resembles Figure 9-19 but that is made of ten squares

Figure 9-19

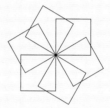

Section 9-3 — More About Angles

In Figure 9-20, two lines intersect and form the angles marked 1, 2, 3, and 4.

Figure 9-20

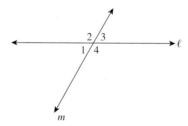

vertical angles

Vertical angles are pairs of angles such as ∠1 and ∠3 and appear any time two lines intersect. Another pair of vertical angles in Figure 9-20 is ∠2 and ∠4.

Other pairs of angles appear frequently enough that it is convenient to refer to them by specific names. Table 9-7 shows several types of angles.

Table 9-7

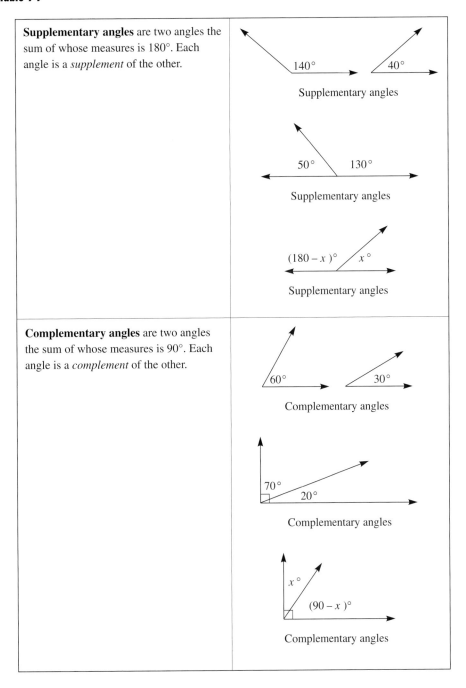

Supplementary angles are two angles the sum of whose measures is 180°. Each angle is a *supplement* of the other.	
Complementary angles are two angles the sum of whose measures is 90°. Each angle is a *complement* of the other.	

Angles are also formed when a line intersects two distinct lines. Any line that intersects a pair of lines is a **transversal** of those lines. In Figure 9-21(a), line *p* is a transversal of lines *m* and *n*. Angles formed by these lines are named according to their placement in relation to the transversal and the two given lines. They are listed in Table 9-8.

transversal

Figure 9-21

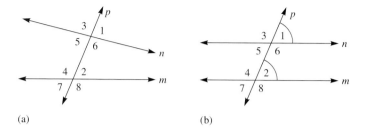

(a)   (b)

Table 9-8

Interior angles	∠2, ∠4, ∠5, ∠6
Exterior angles	∠3, ∠1, ∠7, ∠8
Alternate interior angles	∠5 and ∠2, ∠4 and ∠6
Alternate exterior angles	∠1 and ∠7, ∠3 and ∠8
Corresponding angles	∠3 and ∠4, ∠5 and ∠7, ∠1 and ∠2, ∠6 and ∠8

Suppose corresponding angles such as ∠1 and ∠2 in Figure 9-21(b) are congruent. With this assumption, and because ∠1 and ∠5 are congruent vertical angles, we know that the pair of alternate interior angles ∠2 and ∠5 are also congruent. Similarly, each pair of corresponding angles, alternate interior angles, and alternate exterior angles are congruent.

If we examine Figure 9-21(b) further, we see that lines m and n appear to be parallel when ∠1 is congruent to ∠2. Conversely, if the lines are parallel, the sets of angles mentioned previously are congruent. This is true and is summarized in the following theorem, which we state without proof.

Theorem 9-1

If any two distinct coplanar lines are cut by a transversal, then a pair of corresponding angles, alternate interior angles, or alternate exterior angles are congruent if, and only if, the lines are parallel.

Constructing Parallel Lines

A method commonly used by architects to construct a line ℓ through a given point P parallel to a given line m is shown in Figure 9-22. Place the side $\overline{AB}$ of triangle ABC on line m, as shown in Figure 9-22(a). Next, place a ruler on side $\overline{AC}$. Keeping the ruler stationary, slide triangle ABC along the ruler's edge until its side $\overline{AB}$ (marked $\overline{A'B'}$) contains point P, as in Figure 9-22(b). Use the side $\overline{A'B'}$ to draw the line ℓ through P parallel to m.

To show that the construction produces parallel lines, notice that when triangle ABC slides, the measures of its angles are unchanged. The angles of triangle ABC and triangle $A'B'C'$ in Figure 9-22(b) are correspondingly congruent angles. ∠A and ∠A' are corresponding angles formed by m and ℓ and the transversal $\overline{EF}$. Because corresponding angles are congruent, Theorem 9-1 implies that $\ell \| m$. In Chapter 10, we show how to construct parallel lines using only a compass and straightedge.

Figure 9-22

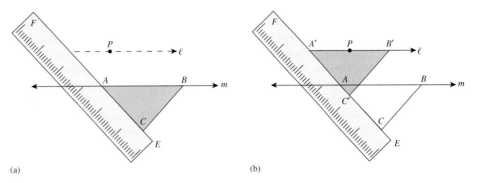

(a)   (b)

The Sum of the Measures of the Angles of a Triangle

The sum of the measures of the angles in a triangle can be shown intuitively to be 180°. We show this by using a torn triangle, as shown in Figure 9-23. Angles 1, 2, and 3 of triangle *ABC* in Figure 9-23(a) are torn as pictured and then replaced as shown in Figure 9-23(b). The three angles seem to form a single line ℓ.

Figure 9-23

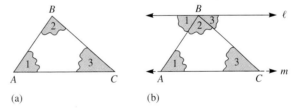

(a)   (b)

The angle measures seem to add up to a straight angle, which measures 180°. This conclusion, which is based on observation, is an example of *inductive reasoning* discussed in Chapter 1. In contrast, *deductive reasoning* shows that a statement is true by using the given information, previously defined and undefined terms, theorems or statements assumed to be true, and logic. A conclusion based on *deductive reasoning* must be true if the hypothesis is true.

Next, we use deductive reasoning to show that the sum of the measures of the interior angles in every triangle is 180°. In Figure 9-23(b), the line ℓ appears to be parallel to *m*. This suggests drawing a line ℓ parallel to $\overline{AC}$ through vertex *B* of triangle *ABC*, as in Figure 9-24(b), and showing that the angles formed are congruent to the interior angles of the triangle.

Figure 9-24

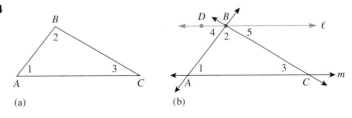

(a)   (b)

In Figure 9-24(b), ℓ and $\overleftrightarrow{AC}$ are parallel, with transversals $\overleftrightarrow{AB}$ and $\overleftrightarrow{BC}$, so it follows that alternate interior angles are congruent. Consequently, $m(\angle 1) = m(\angle 4)$ and $m(\angle 3) = m(\angle 5)$. Thus $m(\angle 1) + m(\angle 2) + m(\angle 3) = m(\angle 4) + m(\angle 2) + m(\angle 5) = 180°$. So, $m(\angle 1) + m(\angle 2) + m(\angle 3) = 180°$. From this, we obtain the following theorem.

Theorem 9-2

The sum of the measures of the interior angles of a triangle is 180°.

NOW TRY THIS 9-5

- An alternative way to show that the sum of the measures of the interior angles of a triangle is 180° is suggested in Figure 9-25. We start at vertex A in Figure 9-25(a) facing B, walk all the way around the triangle, and stop at the same position and facing in the same direction as when we started. This trip is shown in Figure 9-25(b).

Figure 9-25

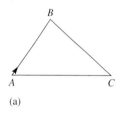
(a)

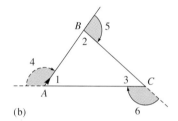
(b)

a. Write an argument for the sum of the measures of the exterior angles of the triangle to be 360°.

b. Use the fact that the sum of the measures of the exterior angles of a triangle is 360° to write an argument that the sum of the measures of the interior angles of the triangle is 180°.

c. Generalize the sum of the measures of the exterior angles of any convex polygon. Explain your reasoning. ●

Now Try This 9-5 can be summarized in Theorem 9-3.

Theorem 9-3

The sum of the measures of the exterior angles (one at each vertex) of a convex polygon is 360°.

Example 9-3

In the framework for a bridge, shown in Figure 9-26(a), *ABCD* is a parallelogram. If ∠*ADC* of the parallelogram measures 50°, what are the measures of the other angles of the parallelogram?

Figure 9-26

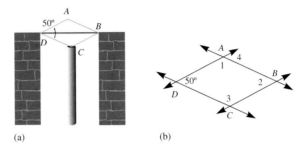

(a) (b)

Solution Refer to Figure 9-26(b). We draw the lines containing the sides of parallelogram *ABCD*. ∠*ADC* has a measure of 50°, and ∠4 and ∠*ADC* are corresponding angles formed by parallel lines $\overleftrightarrow{AB}$ and $\overleftrightarrow{CD}$ cut by transversal $\overleftrightarrow{AD}$. So it follows that $m(\angle 4) = 50°$. Because ∠1 and ∠4 are supplementary, $m(\angle 1) = 180° - 50° = 130°$. Using similar reasoning, we find that $m(\angle 2) = 50°$ and $m(\angle 3) = 130°$.

Example 9-4

In Figure 9-27, $m \| n$ and k is a transversal. Explain why $m(\angle 1) + m(\angle 2) = 180°$.

Figure 9-27

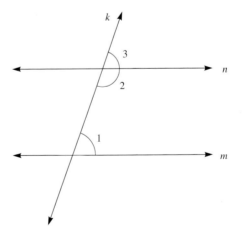

Solution Because ∠1 and ∠3 are corresponding angles and $m \| n$, $m(\angle 1) = m(\angle 3)$. Also because ∠2 and ∠3 are supplementary angles, $m(\angle 2) + m(\angle 3) = 180°$. Substituting $m(\angle 1)$ for $m(\angle 3)$, we have $m(\angle 2) + m(\angle 1) = 180°$.

NOW TRY THIS 9-6

When a plane flies directly from New York to London and then to Nairobi and back to New York, it flies along the sides of a spherical triangle, as shown in Figure 9-28(a), approximately following the surface of Earth. On a sphere, the shortest path between two points is along an arc of a great circle. A great circle is obtained when a plane through the center of the sphere intersects the sphere. (An example of a great circle is the equator.) To obtain a great circle through two given points on the sphere, we need consider only a plane through the two points and the center of the sphere. A great circle through points A and B is shown in Figure 9-28(b). A *spherical triangle* consists of three points on the sphere and arcs of large circles (the shortest path) connecting the points.

Figure 9-28

(a)

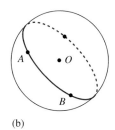
(b)

a. How many great circles go through the North Pole and the South Pole on the globe?

b. Describe a spherical triangle that has one vertex at the North Pole and the other two vertices on the equator.

c. Is there a spherical triangle with three right angles? If so, find one.

d. What can be said about the sum of the measures of the angles of a spherical triangle.

The Sum of the Measures of the Interior Angles of a Convex Polygon with *n* Sides

We can find the sum of the measures of all the interior angles in any convex *n*-gon. We first illustrate the approach for a convex pentagon. Extend each side of the pentagon as shown in Figure 9-29 so that at each vertex, an exterior angle is formed. Because at each vertex an interior angle and the corresponding exterior angle are supplementary, the sum of the measures of all the exterior and interior angles of the pentagon is $5 \cdot 180°$. By Theorem 9-3 the sum of the measures of the exterior angles is $360°$. Thus, the sum of the measures of the interior angles is $5 \cdot 180° - 360°$, or $540°$. The same reasoning applies to any convex *n*-gon. The sum of the measures of an interior angle and the corresponding exterior angle at

Figure 9-29

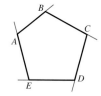

each vertex is 180°. Because there are n vertices, the sum of the measures of all the interior and exterior angles is $n \cdot 180°$. The sum of the measures of the exterior angles is 360°, so the sum of the measures of the interior angles is $n \cdot 180° - 360°$, or $(n - 2)180°$.

In a regular n-gon, all n interior angles are congruent and the sum of their measures is $180n - 360$, so the measure of a single angle is $\frac{180n - 360}{n}$, or $\frac{(n - 2)180°}{n}$.

The results from this discussion are summarized by the following theorem.

Theorem 9-4

a. The sum of the measures of the interior angles of any convex polygon with n sides is $180n - 360$, or $(n - 2)180°$.

b. The measure of a single interior angle of a regular n-gon is $\frac{180n - 360}{n}$, or $\frac{(n - 2)180°}{n}$.

NOW TRY THIS 9-7

- In this investigation, we consider an alternative way to justify Theorem 9-4.

  We can use the fact that the sum of the measures of a triangle's interior angles is 180° to find the sum of the measures of the interior angles of a quadrilateral by dividing the quadrilateral into two triangles. Because the sum of the measures of the interior angles in each triangle is 180°, the sum of the measures of the interior angles in the quadrilateral is $2 \cdot 180°$, or 360°. Use this approach to find the sum of the measures of the interior angles for any convex n-gon. Is your result the same as in Theorem 9-4?

Example 9-5

a. Find the measure of each angle of a regular decagon.

b. Find the number of sides of a regular polygon, each of whose angles has a measure of 175°.

Solution

a. Because a decagon has ten sides, the sum of the measures of the angles of a decagon is $10 \cdot 180 - 360$, or 1440°. A regular decagon has ten angles, all of which are congruent, so each one has a measure of $\frac{1440°}{10}$, or 144°. As an alternative solution, each exterior angle is $\frac{360°}{10}$, or 36°. Hence, each interior angle is $180 - 36$, or 144°.

b. Each interior angle of the regular polygon is 175°. Thus the measure of each exterior angle of the polygon is $180° - 175°$, or 5°. Because the sum of the measures of all exterior angles of a convex polygon is 360°, the number of exterior angles is $\frac{360}{5}$, or 72. Hence, the number of sides is 72.

BRAIN TEASER Find the sum of the measures of ∠1, ∠2, ∠3, ∠4, and ∠5 in any five-pointed star like the one in Figure 9-30.

Figure 9-30

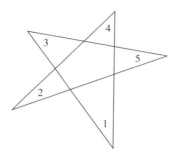

ONGOING ASSESSMENT 9-3

1. If five lines all meet in a single point, how many pairs of vertical angles are formed?
2. Find the measure of the third angle in each of the following triangles:

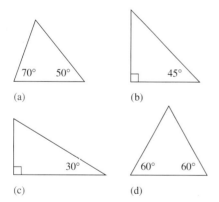

3. For each of the following figures, determine whether *m* and *n* are parallel lines. Justify your answers.

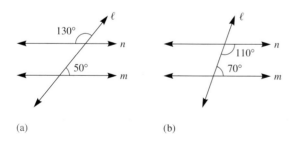

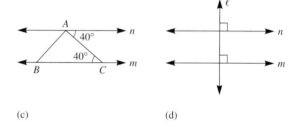

4. Two angles are complementary and the ratio of their measures is 7:2. What are the angle measures?
5. a. In a regular polygon, the measure of each angle is 162°. How many sides does the polygon have?
   b. Find the measure of each of the angles of a regular dodecagon.
6. In the following figure, $\overleftrightarrow{DE} \parallel \overleftrightarrow{BC}$, $\overleftrightarrow{EF} \parallel \overleftrightarrow{AB}$, and $\overleftrightarrow{DF} \parallel \overleftrightarrow{AC}$. Also, $m(\angle 1) = 45°$ and $m(\angle 2) = 65°$. Find each of the following values:
   a. $m(\angle 3)$ b. $m(\angle D)$
   c. $m(\angle E)$ d. $m(\angle F)$

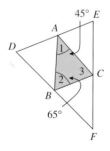

7. In the following figures, find the measures of the angles marked *x* and *y*:

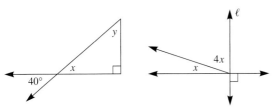

(a)　　　(b)

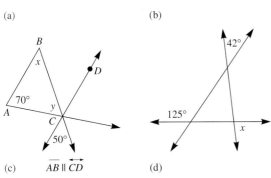

(c)　　$\overline{AB} \parallel \overleftrightarrow{CD}$　　(d)

8. **a.** Determine the measure of an angle whose measure is twice that of its complement.
   b. If two angles of a triangle are complementary, what is the measure of the third angle?

9. Find the sum of the measures of the marked angles in each of the following figures:

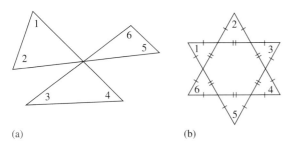

(a)　　　(b)

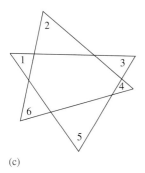

(c)

10. Calculate the measure of each angle of a pentagon, where the measures of the angles form an arithmetic sequence and the least measure is 60°.

11. Two sides of a regular octagon are extended as shown in the following figure. Find the measure of ∠1.

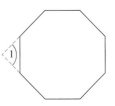

12. Find the measure of angle *x* in the following figure:

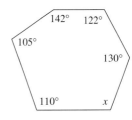

13. Find the measures of angles 1, 2, and 3 given that *TRAP* is a trapezoid with $\overline{TR} \parallel \overline{PA}$.

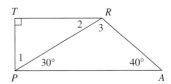

14. Home plate on a baseball field has three right angles and two other congruent angles. Refer to the following figure and find the measures of each of these two other congruent angles:

15. Refer to the following figure and answer (a) and (b):

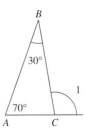

   a. Find m(∠1).
   b. ∠1 is an exterior angle of △*ABC*. Use your answer in (a) to make a conjecture concerning the measure of an exterior angle of a triangle. Justify your conjecture.

16. Refer to the following figure and answer (a) and (b):

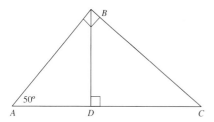

   a. If $m(\angle ABC) = 90°$ and $\overline{BD} \perp \overline{AC}$ and $m(\angle A) = 50°$, find the measure of all the angles of $\triangle ABC$, $\triangle ADB$, and $\triangle CDB$.
   b. If $m(\angle A) = \alpha$ in (a), find the measures of all the angles of $\triangle ABC$, $\triangle ADB$, and $\triangle CDB$ in terms of α.

Communication

17. Explain how you might find a measure of an angle of a staircase that will describe the staircase's steepness.
18. a. If one angle of a triangle is obtuse, can another also be obtuse? Why or why not?
    b. If one angle in a triangle is acute, can the other two angles also be acute? Why or why not?
    c. Can a triangle have two right angles? Why or why not?
    d. If a triangle has one acute angle, is the triangle necessarily acute? Why or why not?
19. In the following figure, A is a point not on line ℓ. Discuss whether it is possible to have two distinct perpendicular segments from A to ℓ in a plane.

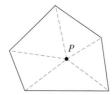

20. a. Explain how to find the sum of the measures of the interior angles of any convex pentagon by choosing any point P in the interior and constructing triangles, as shown in the following figure:

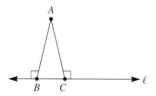

   b. Using the method suggested by the diagram in (a), explain how to find the sum of the measures of the angles of any convex n-gon. Is your answer in (b) the same as the one already obtained in this section, $n \cdot 180 - 360$?

21. In the following figure, the legs of the ladder are congruent. If the ladder makes an angle of $120°$ with the ground, what is x? Explain your reasoning.

22. Explain how, through paper folding, you would show each of the following:
    a. In an equilateral triangle, all the interior angles are congruent.
    b. An isosceles trapezoid has two pairs of congruent angles.
23. a. Explain how to find the sum of the measures of the interior angles of a quadrilateral like the following:

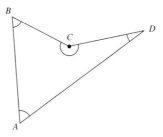

   b. Conjecture whether the formula for the sum of the measures of the angles of a convex polygon is true for nonconvex polygons.
   c. Justify your conjecture in (b) for pentagons and hexagons and explain why you think your conjecture is true in general.
24. Regular hexagons have been used to tile floors. Can a floor be tiled using only regular pentagons? Why or why not?
25. Study the following figure. Notice that you can determine a parallelogram if you know the length of two sides, `:L` and `:W`, and the measure of one angle, `:A`.

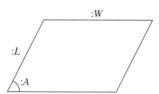

   a. Write a procedure called PARALLELOGRAM with inputs `:L`, `:A`, and `:W` that draws such a parallelogram.
   b. Write a procedure called RECTANGLE that calls the PARALLELOGRAM procedure to draw a rectangle.

c. Write a procedure called RHOMBUS that calls the PARALLELOGRAM procedure to draw a rhombus.
d. How could a square of size 50 be generated by the PARALLELOGRAM procedure?
e. How could a square of size 50 be generated by the RHOMBUS procedure?

Open-Ended

26. Draw three different nonconvex polygons. When you walk around a polygon, at each vertex you need to turn either right (clockwise) or left (counter-clockwise). A turn to the left is measured by a positive number of degrees and a turn to the right by a negative number of degrees. Find the sum of the measures of the turn angles of the polygons you drew. Assume you start at a vertex facing in the direction of a side, walk around the polygon, and end up at the same vertex facing in the same direction as when you started.

27. In Now Try This 9-6, you may have found that on a sphere a triangle with three right angles is possible. List other geometric properties of figures on a sphere that are different from corresponding properties in the plane.

Cooperative Learning

28. In $\triangle ABC$, $\overrightarrow{AD}$ and $\overrightarrow{BD}$ are *angle bisectors,* that is, they divide the angles at A and B into congruent angles.
    a. If the measures of $\angle A$ and $\angle B$ are known, then $m(\angle D)$ can be found. How?

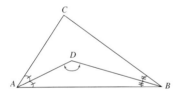

   b. Suppose the measures of $\angle A$ and $\angle B$ are not known but that of $\angle C$ is. Can $m(\angle D)$ be found? To answer this question, assign each member of your group a triangle with different angles but with the same measure for $\angle C$. Each person should compute $m(\angle D)$ for his or her triangle. Use the results to make a conjecture related to the previous question.
   c. Discuss a strategy for answering the question in (b) and write a solution to be distributed to the entire class.

29. Each person in your group is to draw a large triangle like $\triangle ABC$ in the following figure and cut it out. Obtain the crease $\overline{BB'}$ by folding the triangle at B so that A falls on some point A' on $\overline{AC}$. Next, unfold and fold the top B so that B falls on B'. Then fold vertices A and C to match point B', as shown in the following figures:

a. Why is $\overline{BB'}$ perpendicular to $\overline{AC}$?
b. What theorem does the folded figure illustrate? Why?
c. The folded figure seems to be a rectangle. Explain why.
d. What is the length of the base of the rectangle in terms of the base $\overline{AC}$ of $\triangle ABC$? Why?

Review Problems

30. In each of the following, find the required properties. If this is not possible, explain why.
    a. Two properties that hold true for all rectangles but not for all rhombuses
    b. Two properties that hold true for all squares but not for all isosceles trapezoids
    c. Two properties that hold true for all parallelograms but not for all squares

31. Can a circle be thought of as an "infinite-sided" polygon? Explain your answer.

32. Sort the shapes below according to the following attributes:
    a. Number of parallel sides
    b. Number of right angles
    c. Number of congruent sides
    d. Polygons with congruent diagonals

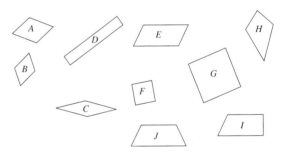

33. Use the figures in problem 32 to identify properties characteristic of different classes of figures. For example, "Congruent opposite sides describe a parallelogram."

Section 9-4 — Geometry in Three Dimensions

Simple Closed Surfaces

A visit to the grocery store exposes us to many three-dimensional objects that have simple closed surfaces. Figure 9-31 shows some examples.

Figure 9-31

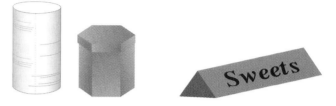

simple closed surface
sphere
center
solid
polyhedron
face · vertices
edges

A **simple closed surface** has exactly one interior, has no holes, and is hollow. An example is a sphere, as shown in Figure 9-32(c). A **sphere** is defined as the set of all points at a given distance from a given point, the **center.** The set of all points on a simple closed surface with all interior points is a **solid.** Figures 9-32(a), (b), (c), and (d) are examples of simple closed surfaces; (e) and (f) are not. A **polyhedron** (polyhedra is the plural) is a simple closed surface made up of polygonal regions, or **faces.** The vertices of the polygonal regions are the **vertices** of the polyhedron, and the sides of each polygonal region are the **edges** of the polyhedron. Figures 9-32(a) and (b) are examples of polyhedra, but (c), (d), (e), and (f) are not.

Figure 9-32

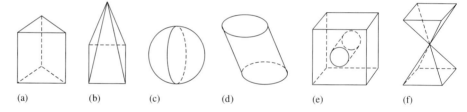

prism

A **prism** is a polyhedron in which two congruent faces lie in parallel planes and the other faces are bounded by parallelograms. Figure 9-33 shows four different prisms. The upper and lower parallel faces of a prism are the **bases** of the prism. A prism usually is named after its bases, as the figure suggests.

bases

Figure 9-33

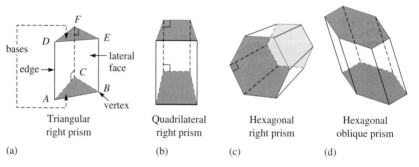

(a) Triangular right prism
(b) Quadrilateral right prism
(c) Hexagonal right prism
(d) Hexagonal oblique prism

lateral face
right prism
oblique prism

The faces other than the bases are the **lateral faces** of a prism. If the lateral faces of a prism are all bounded by rectangles, the prism is a **right prism,** as in Figure 9-33(a)-(c). Figure 9-33(d) is an **oblique prism** because some of its lateral faces are *not* bounded by rectangles.

Students often have trouble drawing three-dimensional figures. Figure 9-34 gives an example of how to draw a right pentagonal prism.

Figure 9-34

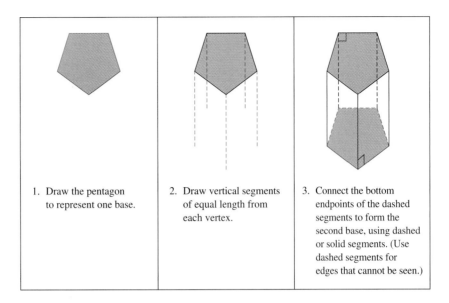

pyramid

base · apex

A **pyramid** is a polyhedron determined by a polygon and a point not in the plane of the polygon. The pyramid consists of the triangular regions determined by the point and each pair of consecutive vertices of the polygon and the polygonal region determined by the polygon. The polygonal region is the **base** of the pyramid, and the point is the **apex.** As with a prism, the faces other than the base are **lateral faces.** Pyramids are classified according to their bases, as shown in Figure 9-35.

Figure 9-35

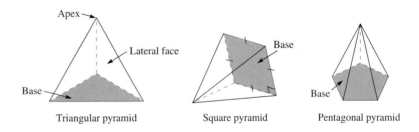

To draw a pyramid, follow the steps in Figure 9-36.

Figure 9-36

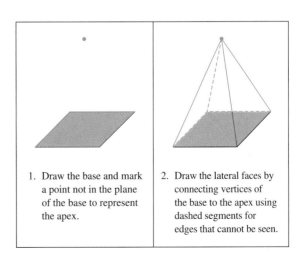

Regular Polyhedra

convex polyhedron

A polyhedron is a **convex polyhedron** if, and only if, the segment connecting any two points in the interior of the polyhedron is itself in the interior. Figure 9-37 shows a concave polyhedron (that is, one that is caved in).

regular polyhedron

A **regular polyhedron** is a convex polyhedron whose faces are congruent regular polygonal regions such that the number of edges that meet at each vertex is the same for all the vertices of the polyhedron.

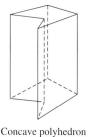

Concave polyhedron

Figure 9-37

HISTORICAL NOTE

The regular solid polyhedra are known as the **Platonic solids,** after the Greek philosopher Plato (ca. 350 B.C.). Plato attached a mystical significance to the five regular polyhedra, associating them with what he believed were the four elements (earth, air, fire, water) and the universe. Plato suggested that the smallest particles of earth have the form of a cube, those of air an octahedron, those of fire a tetrahedron, those of water an icosahedron, and those of the universe a dodecahedron.

Regular polyhedra as in Figure 9-38 have fascinated mathematicians for centuries. At least three of them were identified by the Pythagoreans (ca. 500 B.C.). Two others were known to the followers of Plato (ca. 350 B.C.). Three of the five polyhedra occur in nature in the form of crystals of sodium sulphantimoniate, sodium chloride (common salt), and chrome alum. The other two do not occur in crystalline form but have been observed as skeletons of microscopic sea animals called radiolaria.

Figure 9-38

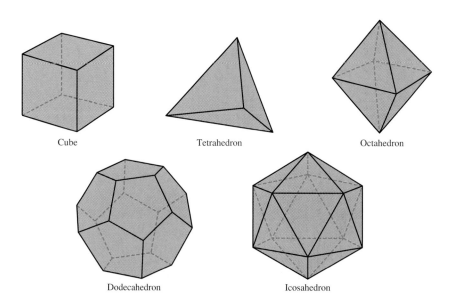

> **Problem Solving** Regular Polyhedra?

How many regular polyhedra are there?

- **Understanding the Problem** Each face of a regular polyhedron is congruent to each of the other faces of that polyhedron, and each face is bounded by a regular polygon. We are to find the number of different regular polyhedra.

- **Devising a Plan** The sum of the measures of all the angles at a vertex of a regular polyhedron must be less than 360°. (Do you see intuitively why this is true?) We next examine the measures of the interior angles of regular polygons to determine which of the polygons could be faces of a regular polyhedron. Then we try to determine how many types of polyhedra there are.

- **Carrying Out the Plan** We determine the size of an angle of some regular polygons as shown in Table 9-9. Could a regular heptagon be a face of a regular polyhedron? At least three figures must fit together at a vertex to make a polyhedron. (Why?) If three angles of a regular heptagon were together at one vertex, then the sum of the measures of these angles would be $\frac{3 \cdot 900°}{7}$, or $\frac{2700°}{7}$, which is greater than 360°. Similarly, more than three angles cannot be used at a vertex. Thus a heptagon cannot be used to make a regular polyhedron.

Table 9-9

Polygon	Measure of an Interior Angle
Triangle	60°
Square	90°
Pentagon	108°
Hexagon	120°
Heptagon	$\left(\frac{900}{7}\right)°$

The measure of an interior angle of a regular polygon increases as the number of sides of the polygon increases. (Why?) Thus any polygon with more than six sides has an interior angle greater than 120°. So if three angles were to fit together at a vertex, the sum of the measures of the angles would be greater than 360°. This means that the only polygons that might be used to make regular polyhedra are equilateral triangles, squares, regular pentagons, and regular hexagons. Consider the possibilities given in Table 9-10.

Notice that we were not able to use six equilateral triangles to make a polyhedron because $6(60°) = 360°$ and the triangles would lie in a plane. Similarly, we could not use four squares or any hexagons. We also could not use more than three pentagons because if we did, the sum of the measures of the angles would be more than 360°.

Table 9-10

Polygon	Measure of an Interior Angle	Number of Polygons at a Vertex	Sum of the Angles at the Vertex	Polyhedron Formed	Model
Triangle	60°	3	180°	Tetrahedron	

continued

Table 9-10 *continued*

Polygon	Measure of an Interior Angle	Number of Polygons at a Vertex	Sum of the Angles at the Vertex	Polyhedron Formed	Model
Triangle	60°	4	240°	**Octahedron**	
Triangle	60°	5	300°	**Icosahedron**	
Square	90°	3	270°	**Cube**	
Pentagon	108°	3	324°	**Dodecahedron**	

semiregular polyhedra

- **Looking Back** Interested readers may want to investigate **semiregular polyhedra.** These are also formed by using regular polygons as faces, but the regular polygons used need not have the same number of sides. For example, a semiregular polyhedron might have squares and regular octagons as its faces.

The patterns in Figure 9-39, called *nets,* can be used to construct the five regular polyhedra. It is left as an exercise to determine other patterns for constructing the regular polyhedra.

Figure 9-39

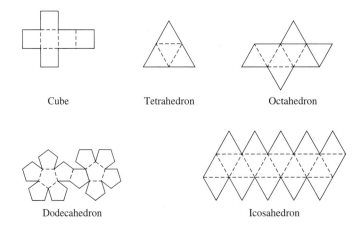

Cube   Tetrahedron   Octahedron

Dodecahedron   Icosahedron

NOW TRY THIS 9-8

- A simple relationship among the number of faces, the number of edges, and the number of vertices of any polyhedron was discovered by the French mathematician and philosopher René Descartes (1596–1650) and rediscovered by the Swiss mathematician Leonhard Euler (1707–1783).

Table 9-11 suggests a relationship among the numbers of vertices (V), edges (E), and faces (F). This relationship is known as **Euler's formula.**

a. State a relationship suggested by the table.
b. Try your relationship on other polyhedra.

Table 9-11

Name	V	F	E
Tetrahedron	4	4	6
Cube	8	6	12
Octahedron	6	8	12
Dodecahedron	20	12	30
Icosahedron	12	20	30

HISTORICAL NOTE

Leonhard Euler went blind in 1766 and for the remaining 17 years of his life continued to do mathematics by dictating to a secretary and by writing formulas in chalk on a slate for his secretary to copy down. He published 530 papers in his lifetime and left enough work to supply the *Proceedings of the St. Petersburg Academy* for the next 47 years.

Cylinders and Cones

A cylinder is an example of a simple closed surface that is not a polyhedron. Consider line segment $\overline{AB}$ and a line ℓ as shown in Figure 9-40. When $\overline{AB}$ moves so that it always remains parallel to a given line ℓ and points A and B trace simple closed planar curves other than polygons, the surface generated by $\overline{AB}$, along with the simple closed curves and their interiors, form a **cylinder.** The simple closed curves traced by A and B, along with their interiors, are the **bases** of the cylinder and the remaining points constitute the *lateral surface of the cylinder.*

cylinder
bases

Figure 9-40

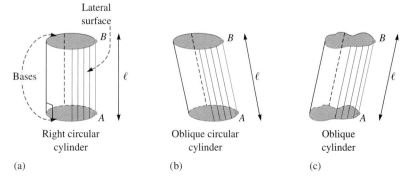

(a) Right circular cylinder
(b) Oblique circular cylinder
(c) Oblique cylinder

circular cylinder   If a base of a cylinder is a circular region, as in Figure 9-40 (a) and (b), the cylinder is a
right cylinder   **circular cylinder.** If the line segment forming a cylinder is perpendicular to a base, the cylin-
oblique cylinder   der is a **right cylinder.** Cylinders that are not right cylinders are **oblique cylinders.** The cylin-
der in Figure 9-40(a) is a right cylinder; those in Figures 9-40(b) and (c) are oblique cylinders.
   Suppose we have a simple closed curve, other than a polygon, in a plane and a point *P*
not in the plane of the curve. The union of line segments connecting point *P* to each point
cone   of a simple closed curve, the simple closed curve, and the interior of the curve is a **cone.**
vertex   Cones are pictured in Figure 9-41. Point *P* is the **vertex** of the cone. The points of the cone
not in the base constitute the *lateral surface of the cone.* A line segment from vertex *P* per-
altitude · **right circular cone**   pendicular to the plane of the base is the **altitude.** A **right circular cone,** such as the one
in Figure 9-41(a), is a cone whose altitude intersects the base (a circular region) at the cen-
ter of the circle. Figure 9-41(b) illustrates an oblique cone, and Figure 9-41(c) illustrates an
oblique circular cone   **oblique circular cone.**

Figure 9-41

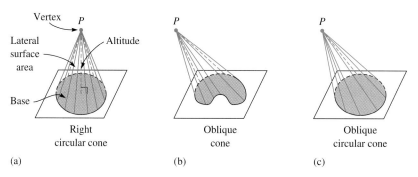

ONGOING ASSESSMENT 9-4

1. Identify each of the following polyhedra. If a polyhedron can be described in more than one way, give as many names as possible.

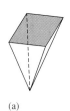   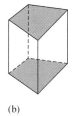   

(a)   (b)   (c)

2. Given the following tetrahedron, name the following:
   a. Vertices
   b. Edges
   c. Faces
   d. Intersection of face *DRW* and edge $\overline{RA}$
   e. Intersection of face *DRW* and face *DAW*

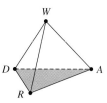

3. Identify five shapes of containers that can be found in a grocery store.
4. Determine for each of the following the minimum number of faces possible:
   a. Prism
   b. Pyramid
   c. Polyhedron
5. Classify each of the following as true or false:
   a. If the lateral faces of a prism are rectangles, it is a right prism.
   b. Every pyramid is a prism.

c. Some pyramids are polyhedra.
   d. The bases of a prism lie in perpendicular planes.
   e. The bases of all cones are circles.
   f. A cylinder has only one base.
   g. All lateral faces of an oblique prism are rectangular regions.
   h. All regular polyhedra are convex.
6. If possible, sketch each of the following:
   a. An oblique square prism
   b. An oblique square pyramid
   c. A noncircular right cone
   d. A noncircular cone that is not right
7. For each of the following, draw a prism and a pyramid that have the given region as a base:
   a. Triangle
   b. Pentagon
   c. Regular hexagon
8. Two prisms are sketched on dot paper, as in the following figure. Complete the drawings by using dashed segments for the hidden edges.

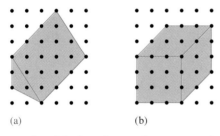

9. Name each polyhedron that can be constructed using the following nets:

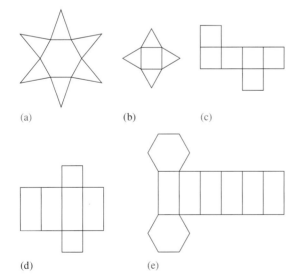

10. The figure on the left in each of the following represents a card attached to a wire as shown. Match each figure on the left with what it would look like if you were to revolve it by spinning the wire between your fingers.

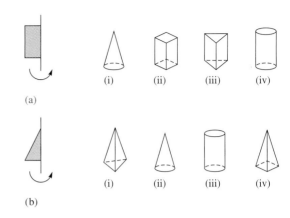

11. Which of the following three-dimensional figures could be used to make the shadow shown in (a)? in (b)?

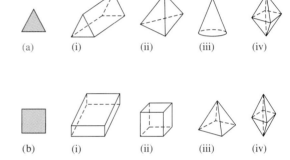

12. A diagonal of a prism is any segment determined by two vertices that do not lie in the same face, as shown in the following figure.

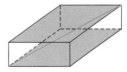

Complete the following table showing the total number of diagonals for various prisms:

Prism	Vertices per Base	Diagonals per Vertex	Total Number of Diagonals
Quadrilateral	4	1	4
Pentagonal	5		
Hexagonal			
Heptagonal			
Octagonal			
.			
.			
.			
n-gonal			

13. Consider a jar with a lid, as illustrated in the following figure. The jar is half filled with water. In drawings (a) and (b), sketch the water.

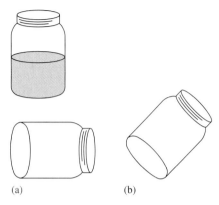

14. On the left of each of the following figures is a net for a three-dimensional object. On the right are several objects. Which object will the net fold to make?

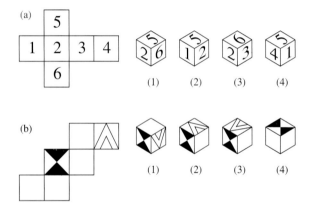

15. Sketch the intersection of each of the following with the plane shown:

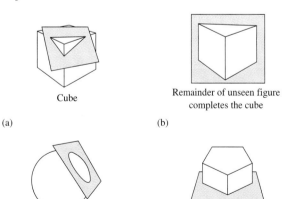

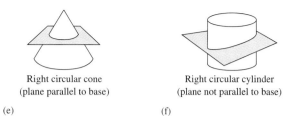

(e) Right circular cone (plane parallel to base)

(f) Right circular cylinder (plane not parallel to base)

16. For each of the following three-dimensional figures, draw all possible cross-sections when the three-dimensional figure is sliced by a plane.
    a. Cube
    b. Cylinder

17. For each of the following figures, find $V + F - E$, where V, E, and F stand, respectively, for the number of vertices, edges, and faces:

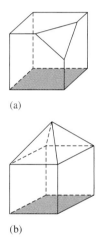

18. Complete the following table for each of the polyhedra described in the table:

Polyhedron	Vertices	Faces	Edges
a.		8	12
b.	20	30	
c.	6		15

State the relationship suggested by the table.

19. Answer each of the following questions about a pyramid and a prism, each having an n-gon as a base:
    a. How many faces does each have?
    b. How many vertices does each have?
    c. How many edges does each have?
    d. Use your answers to (a), (b), and (c) to verify Euler's formula for all pyramids and all prisms.

Communication

20. How many possible pairs of bases does a rectangular prism have? Explain.

21. A circle can be approximated by a "many-sided" polygon. Use this notion to describe the relationship between each of the following:
    a. A pyramid and a cone
    b. A prism and a cylinder
22. Can either or both of the following be drawings of a quadrilateral pyramid? If yes, where would you be standing in each case? Explain why.

(a)   (b)

Open-Ended

23. When a box in the shape of a right prism, like the one in the following figure, is cut by a plane halfway between the opposite sides and parallel to these sides, that plane is a *plane of symmetry*. If a mirror is placed at the plane of symmetry, the reflection of the front part of the box will look just like the back part. Draw several space figures and find the number of planes of symmetry for each. Summarize your results in a table. Can you identify any figures with infinitely many planes of symmetry?

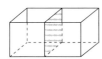

Cooperative Learning

24. In a two-person game, draw a three-dimensional figure without showing it to your partner. Tell your partner the shape of all possible cross-sections of your figure sufficient to identify the figure. If your partner can identify your figure, 1 point is earned by your partner. If a figure is identified that has all the cross-sections listed but that figure is not your figure, 2 points are earned by your partner. Each of you should take an equal number of turns.
25. Some of the following nets can be folded into a cube. Have each person in your group draw all the nets that can be folded into a cube. Share your findings with the group and decide how many such different nets there are. Discuss what "different" means in this case. Finally, compare your group's answers with those of other groups.

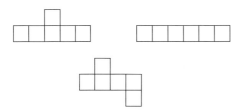

26. Choose different triples of points on the edges of each of the following figures. Then find the cross-section of where a plane through each triple of points intersects the figure. (The following figures show only a few examples. What are the possible figures that can be obtained in this manner?)

(a) Cube

(b) Tetrahedron

Review Problems

27. If two angles of one triangle are congruent to two angles of another triangle, must the third angles of both be congruent? Why or why not?
28. Triangles *ABC* and *CDE* are equilateral triangles with points *A*, *C*, and *E* collinear. Find the measure of ∠*BCD*.

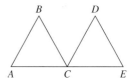

29. What is the measure of each angle in a regular nonagon?
30. Classify the following as true or false. If false, tell why.
    a. Every rhombus is a parallelogram.
    b. Every polygon has at least three sides.
    c. Triangles can have at most two acute angles.
31. Assume that lines ℓ, *m*, and *n* as shown are in the same plane. What can you conclude? Why?

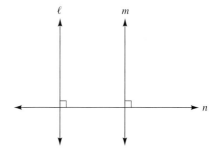

TECHNOLOGY CORNER

Use technology to draw a staircase similar to that in Figure 9-42.

Figure 9-42

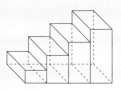

BRAIN TEASER

A rectangular region can be rolled to form the lateral surface of a right circular cylinder. What shape of paper is needed to make an oblique circular cylinder? (See "Making a Better Beer Glass" by A. Hoffer.)

LABORATORY ACTIVITY

Consider a structure made of cubes with the front view, the side view looking from the right, and the top view as in Figure 9-43. Build the structure.

Figure 9-43

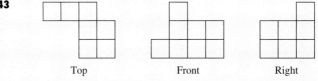

Top   Front   Right

*Section 9-5 — Networks

In the 1700s, the people of Königsberg, Germany, used to enjoy walking over the bridges of the Pregel River. There were two islands in the river and seven bridges over it, as shown in Figure 9-44. These walks eventually led to the following problem.

Königsberg Bridge Problem

Is it possible to walk across all the bridges so that each bridge is crossed exactly once on the same walk?

Figure 9-44

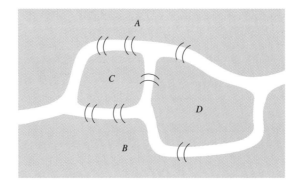

There is no restriction on where to start the walk or where to finish. Leonhard Euler became interested in this problem and solved it in 1736. He made the problem much simpler by representing the land masses, islands, and bridges in a **network,** as shown in the pink portion of Figure 9-45.

network

Figure 9-45

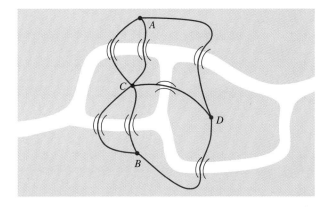

vertices · arcs

traversable

Euler circuit

The points in a network are **vertices,** and the curves are **arcs.** Using a network diagram, we can restate the Königsberg bridge problem as follows: *Is there a path through the network beginning at some vertex and ending at the same or another vertex such that each arc is traversed exactly once?* A network having such a path is **traversable;** that is, each arc is passed through exactly once. A network that is traversable in such a way that the starting point and the stopping point are the same is an **Euler circuit.**

We can walk around an ordinary city block, as illustrated in Figure 9-46(a), and because the starting point is the same as the stopping point, the network is an Euler circuit. We need not start at a particular point, and, in general, we can traverse any simple closed curve. Now consider walking around two city blocks and down the street that runs between them, as in Figure 9-46(b). To traverse this network, it is necessary to start at vertex B or C. Starting at points other than B or C might suggest that the figure is not traversable, but this is not the case, as shown in Figure 9-46(b). If we start at B, we end at C and vice versa. Note that it is permissible to pass through a vertex more than once, but an arc may be traversed only once. Vertices B and C are endpoints of three arcs, and each of the other vertices are endpoints of two arcs.

Figure 9-46

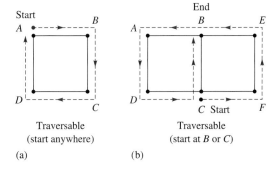

508 CHAPTER 9 *Introductory Geometry*

A traversable network is the type of network, or route, that a highway inspector would like to have if given the responsibility of checking out all the roads in a highway system. The inspector needs to traverse each road (arc) in the system but would save time by not having to make repeat journeys during an inspection tour. It would be feasible for the inspector to go through any town (vertex) more than once on the route. Consider the networks in Figure 9-47. Is it possible for the highway inspector to do the job with these networks without traversing a road twice?

Figure 9-47

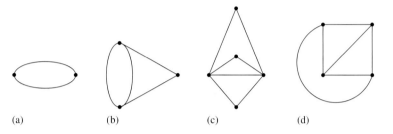

(a)　　　(b)　　　(c)　　　(d)

The first three networks in Figure 9-47 are traversable; the fourth network, (d), is not. Notice that the number of arcs *meeting at each vertex* in networks (a) and (c) is even. Any such vertex is an **even vertex.** If the number of arcs *meeting at a vertex* is odd, it is an **odd vertex.** In network (b), only the odd vertices will work as starting or stopping points. In network (d), which is not traversable, all the vertices are odd. If a network is traversable, each arrival at a vertex other than a starting or a stopping point requires a departure. Thus *each vertex that is not a starting or stopping point must be even.* The starting and stopping vertices in a traversable network may be even or odd, as seen in Figure 9-47(a) and (b), respectively. Which networks, if any, form an Euler circuit?

even vertex · odd vertex

Properties of a Network

In general, networks have the following properties.

1. If a network has all even vertices, it is traversable. Any vertex can be a starting point, and the same vertex must be the stopping point. Thus the network is an Euler circuit.
2. If a network has two odd vertices, it is traversable. One odd vertex must be the starting point, and the other odd vertex must be the stopping point.

NOW TRY THIS 9-9

a. Is there a traversable network with more than two odd vertices? Why or why not?

b. Is there a network with exactly one odd vertex?

Example 9-6   a. Which of the networks in Figure 9-48 are traversable?
b. Which of the networks in Figure 9-48 are Euler circuits?

Figure 9-48

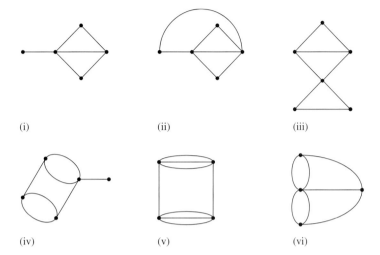

(i)   (ii)   (iii)

(iv)   (v)   (vi)

Solution   a. Networks in (ii) and (v) have all even vertices and therefore are traversable. Networks in (i) and (iii) have exactly two odd vertices and are traversable. Networks in (iv) and (vi) have four odd vertices and are not traversable.
b. Networks (ii) and (v) are Euler circuits.

The network in Figure 9-48(vi) represents the Königsberg bridge problem. It has four odd vertices and consequently is not traversable. Hence, no walk configuration is possible to solve the problem.

A problem similar to the highway inspector problem involves a traveling salesperson. Such a person might have to travel networks comparable to those of the highway inspector. However, the salesperson is interested only in visiting each town (vertex) once, not necessarily in following each road. It is not known for which networks this can be accomplished. Can you find a route for the traveling salesperson for each network in Figure 9-48?

A different type of application of network problems is discussed in Example 9-7.

Example 9-7   Look at the floor plan of the house shown in Figure 9-49. Is it possible for a security guard to go through all the rooms of the house and pass through each door exactly once?

Figure 9-49

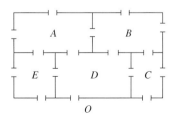

510 CHAPTER 9 *Introductory Geometry*

Solution Represent the floor plan as a network, as in Figure 9-50. Designate the rooms and the outside as vertices and the paths through the doors as arcs. The network has more than two odd vertices, namely, *A, B, D,* and *O*. Thus the network is not traversable, and it is impossible to go through all the rooms and pass through each door exactly once.

Figure 9-50

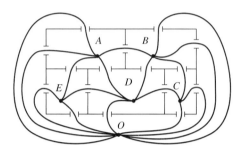

ONGOING ASSESSMENT 9-5

1. Which of the following networks are traversable? If the network is traversable, draw an appropriate path through it, labeling the starting and stopping vertices. Indicate which networks are Euler circuits.

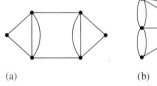

(a)   (b)

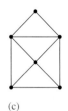

(c)   (d)   (e)

  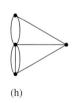

(f)   (g)   (h)

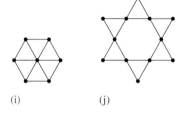

(i)   (j)

2. Which of the networks in problem 1 can be traveled efficiently by a traveling salesperson, with no vertex visited more than once?

3. A city contains one river, three islands, and ten bridges, as shown in the following figure. Is it possible to take a walk around the city by starting at any land area, returning after visiting every part of the city, and crossing each bridge exactly once? If so, show such a path both on the original figure and on the corresponding network.

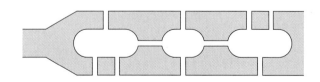

4. Refer to the following floor plans:

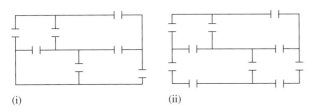

(i)   (ii)

   a. Draw a network that corresponds to each floor plan.
   b. Determine whether a person could pass through each room of each house by passing through each door exactly once. If it is possible, draw the path of such a trip.
5. Refer to the following floor plans:

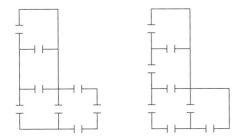

   Can a person walk through each door once and only once and also go through both of the following houses in a single path? If it is possible, draw the path of such a trip.
6. The following drawing represents the floor plan of an art museum. All tours begin and end at the entry. If possible, design a tour route that will allow a person to see every room but not go through any room twice.

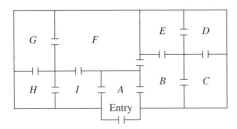

7. Each network in problem 1 separates the plane into several subsets. If R is the number of interior and exterior regions of the plane, V is the number of vertices, and A is the number of arcs, complete the following chart using each of the networks. (The first one is done for you.)

Network	R	V	A	$R + V - A$
(a)	6	6	10	2

8. Molly is making her first trip to the United States and would like to tour the eight states pictured in the following figure. She would like to plan her trip so that she can cross each border between neighboring states exactly once—that is, the Washington-Oregon border, the Washington-Idaho border, and so on.

   Is such a trip possible? If so, does it make any difference in which state she starts her trip?

Communication

9. The following network is not an Euler circuit:

   a. Add two arcs to the network so that the resulting network will be an Euler circuit.
   b. Add exactly one arc so that the resulting network will be an Euler circuit.
   c. Explain a real-life application in which the answer in (b) is useful.
10. If you were commissioned to build an eighth bridge to make the Königsberg bridge problem traversable, where would you build your bridge? Is there more than one location where you could build it? Explain why.

Open-Ended

11. **a.** How would a postal worker's route differ from that of a traveling salesperson or a highway inspector?
    b. Consider a 10-square block area of a city where a postal worker has to cover each side of each street. Design an efficient way to do this route.
12. Using as few vertices and arcs as possible, draw a network that is not traversable.
13. Draw a network that is not an Euler circuit and then add the least number of edges possible so that the new network will be an Euler circuit.
14. One application of Euler circuits is the checking of parking meters. List other real-life applications that could involve the use of Euler circuits. In each case, give a concrete example and describe the corresponding Euler circuit.

Cooperative Learning

15. *Traveler's Dodecahedron* is a puzzle invented in 1857 by the Irish mathematician William Rowen Hamilton. It consists of a wooden dodecahedron (a polyhedron with 12 regular pentagons as faces) with a peg at each vertex of the dodecahedron. The 20 vertices are labeled with the names of different cities around the world. The solver of the puzzle is to find a path that starts at some city, travels along the edges, goes through each of the remaining cities exactly once, and returns to the starting city. The path traveled is to be marked by a string connecting the pegs.

   a. Find a solution to the puzzle, first on the following network and then on the following dodecahedron. Compare your answer with those of other members of your group.

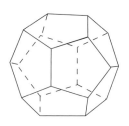 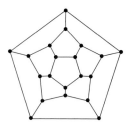

   b. Play the following game with a partner. Draw a polyhedron that can be traversed in the way described earlier in this problem. Also draw a two-dimensional network for the polyhedron similar to the one shown in (a). Ask your partner to answer the question posed in (a) for the new polyhedron and the accompanying network. Then switch roles. The person who draws the polyhedron with the greatest number of vertices wins.

LABORATORY ACTIVITY

1. Take a strip of paper like the one in Figure 9-51(i). Give one end a half-twist and join the ends by taping them, as in Figure 9-51(ii). The surface obtained is a Möbius strip.

Figure 9-51

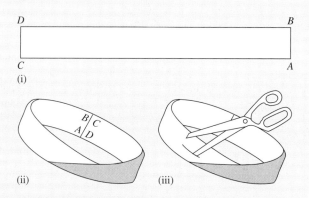

   a. Use a pencil to shade one side of a Möbius strip. What do you discover?
   b. Imagine cutting a Möbius strip all around midway between the edges, as in Figure 9-51(iii). What do you predict will happen? Now do the cutting. What is the result?
   c. Imagine cutting a Möbius strip one third of the way from an edge and parallel to the edge all the way through until you return to the starting point. Predict the result. Then do the cutting. Was your prediction correct?
   d. Imagine cutting around a Möbius strip one fourth of the way from an edge. Predict the result. Then do the cutting. How does the result compare with the result of the experiment in (c)?

2. **a.** Take a strip of paper and give it two half-twists (one full twist). Then join the ends together. Answer the questions in part 1.
   b. Repeat the experiment in (a), using three half-twists.
   c. Repeat the experiment in (a), using four half-twists. What do you find for odd-numbered twists? Even-numbered twists?
3. Take two strips of paper and tape each of them in a circular shape. Join them as shown in Figure 9-52.

Figure 9-52

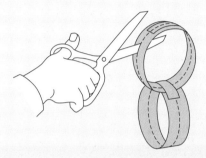

   a. What happens if you cut completely around the middle of each strip as shown?
   b. Repeat part (a) if both strips are Möbius strips. Does it make any difference if the half-twists are in opposite directions?
   c. Repeat part (a) if one strip is a Möbius strip and the other is not. What happens when you cut the strips?

BRAIN TEASER Given three buildings A, B, and C, as shown in Figure 9-53, and three utility centers for electricity (E), gas (G), and water (W), determine whether it is possible to connect each of the three buildings to each of the three utility centers without crossing lines.

Figure 9-53

HINT FOR SOLVING THE PRELIMINARY PROBLEM

This problem involves looking at the sizes of angles in a convex polygon. Think about the measures of the exterior angles of the polygon. If the polygon has n acute angles, what are the bounds on the sum of the measures of the corresponding exterior angles? What is the maximum number that n can be? Based on the answer to the last question, what types of convex polygons can be used for the tiles?

QUESTIONS FROM THE CLASSROOM

1. Henry claims that a line segment has a finite number of points because it has two endpoints. How do you respond?
2. A student claims that if any two planes that do not intersect are parallel, then any two lines that do not intersect should also be parallel. How do you respond?
3. A student says that it is actually impossible to measure an angle, since each angle is the union of two rays that extend infinitely and therefore continue forever. What is your response?
4. Maggie claims that to make the measure of an angle greater, you just extend the rays. How do you respond?
5. A student asks whether a polygon whose sides are congruent is necessarily a regular polygon and whether a polygon with all angles congruent is necessarily a regular polygon. How do you answer?
6. A student thinks that a square is the only regular polygon with all right angles. The student asks if this is true and if so, why. How do you answer?
7. A student says that a line is parallel to itself. How do you reply?
8. A student asks how to find the shortest path between two points A and B on a right circular cylinder. How do you respond?
9. One student says, "My sister's high-school geometry book talked about equal angles. Why don't we use the term 'equal angles' instead of 'congruent angles'?" How do you reply?
10. A student says there can be only 360 different rays emanating from a point, since there are only 360° in a circle. How do you respond?
11. Jodi identifies the following figure (a) as a rectangle and the figure (b) as a square. She claims that the figure (b) is not a rectangle because it is a square. How do you respond?

(a)   (b)

12. Millie claims that a rhombus is regular because all of its sides are congruent. How do you respond?
13. A student asks what exactly a Euclidean geometry is. How do you answer?
14. A student asks if geometry is worth studying if lines have different meanings in different settings. What do you say?

CHAPTER OUTLINE

I. Basic geometric notions
   A. Points, lines, and planes
      1. **Points, lines,** and **planes** are basic, but undefined, terms.
      2. **Collinear points** are points that belong to the same line.
      3. **Segments** and **rays** are important subsets of lines.
      4. **Coplanar points** are points that lie in the same plane. **Coplanar lines** are defined similarly.
      5. Two lines with exactly one point in common are **intersecting lines.**
      6. **Concurrent lines** are lines that contain a common point.
      7. Two distinct coplanar lines with no points in common are **parallel.**
      8. **Skew lines** are lines that cannot be contained in the same plane.
      9. **Parallel planes** are planes with no points in common.
      10. **Space** is the set of all points.
      11. An **angle** is the union of two rays with a common endpoint.
      12. Angles are classified according to size as **acute, obtuse, right,** or **straight.**
      13. Two lines that meet to form a right angle are **perpendicular.**
      14. A **dihedral angle** is the union of two half-planes and the common line defining the half-planes.
   B. Plane figures
      1. A **closed curve** is a curve that, when traced, has the same starting and stopping points and may cross itself at individual points.
      2. A **simple curve** is a curve that does not cross itself when traced, although the starting and stopping points may be the same.
      3. A **polygon** is a simple closed curve with segments as sides.
         a. A **diagonal** is any line segment connecting two nonconsecutive vertices of a polygon.
         b. A **convex polygon** is one such that if any two points of the polygonal region are connected by a segment, the segment is a subset of the polygonal region.
         c. A **concave polygon** is a nonconvex polygon.
         d. A **regular polygon** is a polygon in which all the interior angles are congruent and all the sides are congruent.
      4. A **polygonal region** is the union of a polygon and its interior.
      5. Triangles are classified according to the lengths of their sides as **scalene, isosceles,** or **equilateral** and according to the measures of their angles as **acute, obtuse,** or **right.**
      6. Quadrilaterals with special properties are **trapezoids, parallelograms, rectangles, kites, isosceles trapezoids, rhombuses,** and **squares.**

II. Theorems involving angles
   A. **Supplements** of the same angle, or of congruent angles, are congruent.
   B. **Complements** of the same angle, or of congruent angles, are congruent.
   C. **Vertical angles** formed by intersecting lines are congruent.
   D. If any two distinct coplanar lines are cut by a transversal, then a pair of **corresponding angles, alternate interior angles,** or **alternate exterior angles** are congruent if, and only if, the lines are parallel.
   E. The sum of the measures of the angles of a triangle is $180°$.
   F. The sum of the measures of the interior angles of any convex polygon with n sides is $180n - 360$, or $180(n - 2)$. One interior angle of a regular n-gon measures $\frac{180n - 360}{n}$, or $\frac{180(n - 2)}{n}$.
   G. The sum of the measures of the exterior angles of any convex polygon is $360°$.

III. Three-dimensional figures
   A. A **polyhedron** is a simple closed surface formed by polygonal regions.
   B. Three-dimensional figures with special properties are **prisms, pyramids, regular polyhedra, cylinders, cones,** and **spheres.**

*IV. Networks
   A. A **network** is a collection of points or **vertices** and a collection of curves or **arcs.**
   B. A vertex of a network is called an **even vertex** if the number of arcs meeting at the vertex is even. A vertex is an **odd vertex** if the number of arcs meeting at a vertex is odd.
   C. A network is called **traversable** if there is a path through the network such that each arc is passed through exactly once.
      1. If all the vertices of a network are even, then the network is traversable. Any vertex can be a starting point, and the same vertex must be the stopping point.
      2. If a network has two odd vertices, it is traversable. One odd vertex must be the starting point, and the other must be the stopping point.
      3. If a network has more than two odd vertices, it is not traversable.
      4. No network has exactly one odd vertex.
   D. A network that is traversable by starting and ending at the same point is an **Euler circuit.**

CHAPTER REVIEW

1. Refer to the following line m:

   a. List three names for the line.
   b. Name two rays on m that have endpoint B.
   c. Find a simpler name for the intersection of rays $\overrightarrow{AB}$ and $\overrightarrow{BA}$.
   d. Find a simpler name for the intersection of rays $\overrightarrow{BA}$ and $\overrightarrow{AC}$.

2. In the following figure, $\overleftrightarrow{PQ}$ is perpendicular to α:

   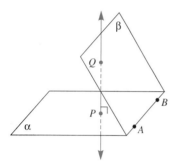

   a. Name a pair of skew lines.
   b. Using only the letters in the figure, name as many planes as possible that are perpendicular to α.
   c. What is the intersection of planes APQ and β?
   d. Is there a single plane containing A, B, P, and Q? Explain your answer.

3. Draw each of the following curves:
   a. A simple closed curve
   b. A closed curve that is not simple
   c. A concave hexagon
   d. A convex decagon

4. a. Can a triangle have two obtuse angles? Justify your answer.
   b. Can a parallelogram have four acute angles? Justify your answer.

5. In a certain triangle, the measure of one angle is twice the measure of the smallest angle. The measure of the third angle is seven times greater than the measure of the smallest angle. Find the measures of each of the angles in the triangle.

6. a. Explain how to derive an expression for the sum of the measures of the interior angles in a convex n-gon.
   b. In a certain regular polygon, the measure of each angle is 176°. How many sides does the polygon have?

7. Sketch each of the following:
   a. Three planes that intersect in a point
   b. A plane and a cone that intersect in a circle
   c. A plane and a cylinder that intersect in a segment
   d. Two pyramids that intersect in a triangle

8. Sketch drawings to illustrate different possible intersections of a square pyramid and a plane.

9. In a periscope, a pair of mirrors are parallel. If the dotted line in the following figure represents a path of light and $m(\angle 1) = m(\angle 2) = 45°$, find $m(\angle 3)$ and $m(\angle 4)$:

   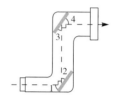

10. Find $6°48'59'' + 28°19'36''$. Write your answer in simplest terms.

11. In the figure, ℓ is parallel to m, and $m(\angle 1) = 60°$. Find each of the following:
    a. $m(\angle 3)$
    b. $m(\angle 6)$
    c. $m(\angle 8)$

    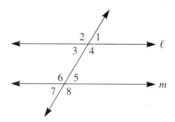

12. Draw a large triangle ABC and tear off $\angle B$. Fit the torn-off angle as $\angle BAD$, as shown in the following figure.

    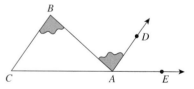

    a. Why is $\overrightarrow{AD} \parallel \overleftrightarrow{BC}$?
    b. Why is $\angle DAE \cong \angle C$?
    c. Use parts (a) and (b) to show that the sum of the measures of the interior angles of a triangle is 180°.

13. If a cube intersects a plane, what possible figures can be obtained by the intersection? Sketch the planes and figures obtained in each case.

14. If a pyramid has an octagon for a base, how many lateral faces does it have?

15. If ABC is a right triangle and $m(\angle A) = 42°$, what is the measure of the other acute angle?

16. In the following, dot the unseen segments.

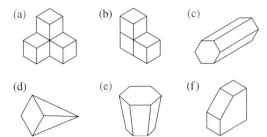

(a) (b) (c)
(d) (e) (f)

17. If there are *n* points in a plane, no three of which are collinear, how many triangles could be found with those points as vertices?
18. If one is drawing a two-dimensional representation of a large $n \times n \times n$ cube, will the edges of the square faces appear parallel as they do in a normal square on a piece of paper? Why or why not?
19. Explain whether or not a cylinder has to have a circular base.
20. Draw a curve that is neither simple nor closed.
*21. **a.** Which of the following networks are traversable?
   b. Find a corresponding path for the networks that are traversable.

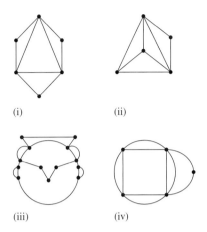

(i) (ii)
(iii) (iv)

SELECTED BIBLIOGRAPHY

Andrews, A. G. "Solving Geometric Problems by Using Unity Blocks." *Teaching Children Mathematics* 5 (February 1999): 318–323.

Bartels, B. H. "Truss(t)ing Triangles." *Mathematics Teaching in the Middle School.* 3 (March–April 1998): 394–396.

Bledsoe, G. "Guessing Geometric Shapes." *Mathematics Teacher* 80 (March 1987): 178–180.

Bright, G., and J. Harvey. "Games, Geometry, and Teaching." *Mathematics Teacher* 81 (April 1988): 250–259.

Camblin, B. "The Mathematics in Your Note Paper." *Mathematics Teaching in the Middle School.* 4 (November–December 1998): 168–169.

Carroll, W. "Cross Sections of Clay Solids." *Arithmetic Teacher* 35 (March 1988): 6–11.

Carroll, W. "Middle School Students' Reasoning about Geometric Situations." *Mathematics Teaching in the Middle School.* 3 (March–April 1998): 398–403.

Carroll, W. "Polygon Capture: A Geometry Game." *Mathematics Teaching in the Middle School.* 4 (October 1998): 90–94.

Greeley, N., and T. R. Offerman. "Now & Then Measuring Angles in Physical Therapy." *Mathematics Teaching in the Middle School.* 2 (March–April 1997): 338–344.

Hannibal, M. A. "Young's Children's Developing Understanding of Geometric Shapes." *Teaching Children Mathematics* 5 (February 1999): 353–357.

Hoffer, A. "Making a Better Beer Glass." *Mathematics Teacher* 75 (May 1982): 378–379.

Oberdorf, C. D., and J. Taylor-Cox. "Shape Up." *Teaching Children Mathematics* 5 (February 1999): 340–345.

Olson, A. T. *Mathematics through Paper Folding.* Washington, D.C.: NCTM (1975).

Olson, J. "What Shapes Can You Make?" *Teaching Children Mathematics* 5 (February 1999): 330–331.

Posamentier, A. "Geometry: A Remedy for the Malaise of Middle School Mathematics." *Mathematics Teacher* 82 (December 1989): 678–680.

Row, T. S. *Geometric Exercises in Paper Folding.* Dover Publications, Inc. (1966).

Schifter, D. "Learning Geometry: Some Insights Drawn from Teacher Writing." *Teaching Children Mathematics* 5 (February 1999): 360–366.

Sundberg, S. "A Plethora of Polyhedra." *Mathematics Teaching in the Middle School.* 3 (March–April 1998): 388–391.

Thiessen, D., and M. Matthias. "Selected Children's Books for Geometry." *Arithmetic Teacher* 37 (December 1989): 47–51.

van Hiele, P. M. "Developing Geometric Thinking through Activities That Begin with Play." *Teaching Children Mathematics* 5 (February 1999): 310–316.

Wheatley, G. H., and A. M. Reynolds. "'Image Maker': Developing Spatial Sense." *Teaching Children Mathematics* 5 (February 1999): 374–378.

Williams, C. G. "Sorting Activities for Polygons." *Mathematics Teaching in the Middle School.* 3 (March–April 1998): 444–445.

Woodward, E., and Brown, R. "Polydrons and Three-Dimensional Geometry." *Arithmetic Teacher* 41 (April 1994): 451–458.

10

Constructions, Congruence, and Similarity

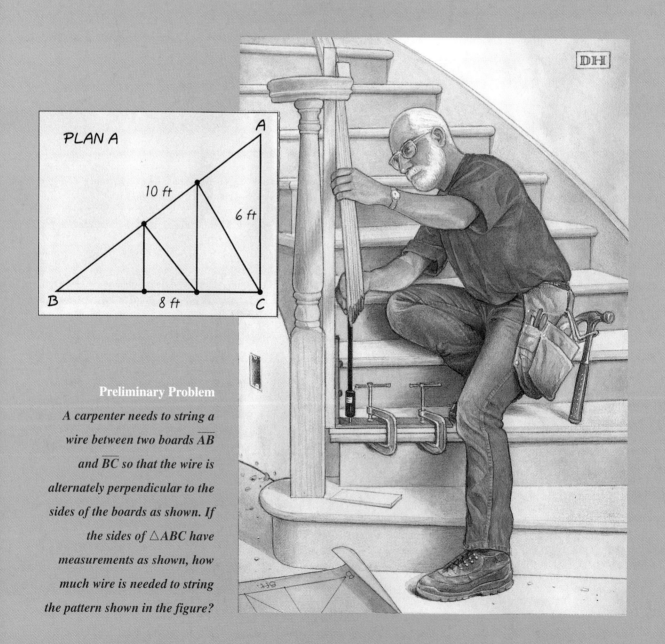

Preliminary Problem
A carpenter needs to string a wire between two boards $\overline{AB}$ and $\overline{BC}$ so that the wire is alternately perpendicular to the sides of the boards as shown. If the sides of $\triangle ABC$ have measurements as shown, how much wire is needed to string the pattern shown in the figure?

SECTION 10-1 *Congruence Through Constructions* **519**

The *Principles and Standards* recommend that:

In grades 6–8, all students should

- create and critique inductive and deductive arguments concerning geometric ideas and relationships, such as congruence, similarity, and the Pythagorean theorem.
- use coordinate geometry to represent and examine the properties of geometric shapes;
- use coordinate geometry to examine special geometric shapes, such as regular polygons or those with pairs of parallel or perpendicular sides.

…

- draw geometric objects with specified properties, such as side lengths or angle measures … ; (p. 232).

In this chapter we introduce, through constructions and visualization, the concepts of congruence and similarity. The last section develops equations of lines, and investigates systems of equations both geometrically and algebraically.

Section 10-1 — Congruence Through Constructions

similar
congruent

In mathematics, the word **similar** (∼) describes objects that have the same shape but not necessarily the same size, while the word **congruent** (≅) describes objects that have the same size as well as the same shape. Whenever two figures are congruent, they are also similar. However, the converse is not true (why?). Examples of similar and congruent objects are seen in Figure 10-1.

Figure 10-1

Symmetry Work 22
(a)

(b)

Figure 10-1(a) shows sets of congruent fish (and birds) in an Escher print. Figure 10-1(b) contains both congruent and similar equilateral triangles. The smaller triangles are congruent. They also are similar to the large one that contains them. In addition, Figure 10-1(b) depicts an example of a **rep-tile**, a figure that is used to construct a larger similar figure. In Figure 10-1(b), one of the smaller equilateral triangles is a rep-tile.

rep-tile

Before studying congruent and similar figures, we first consider some notation and review definitions from Chapter 9. For example, *any two line segments are congruent if they have the same length, while two angles are congruent if they have the same measure*. The length of line segment $\overline{AB}$ is denoted by AB. Symbolically, we may write the following about congruent segments and angles.

> **Definition of Congruent Segments and Angles**
>
> $\overline{AB} \cong \overline{CD}$ if, and only if, $AB = CD$
> $\angle ABC \cong \angle DEF$ if, and only if, $m(\angle ABC) = m(\angle DEF)$

HISTORICAL NOTE

The straight line and circle were considered the basic geometric figures by the Greeks and the straightedge and compass are their physical analogs. It is believed that the Greek philosopher Plato (c. 427–347 B.C.) rejected the use of mechanical devices other than the straightedge and compass for geometric constructions because use of other tools emphasized practicality rather than "ideas," which he regarded as more important.

Geometric Constructions

Ancient Greek mathematicians constructed geometric figures with a straightedge (no markings on it) and a collapsible compass. Figure 10-2 shows a modern compass. It can be used to mark off and duplicate lengths and to construct circles or arcs with a radius of a given measure. To draw a circle when given the radius PQ of a circle, we follow the steps illustrated in Figure 10-2.

Figure 10-2 Constructing a circle given its radius

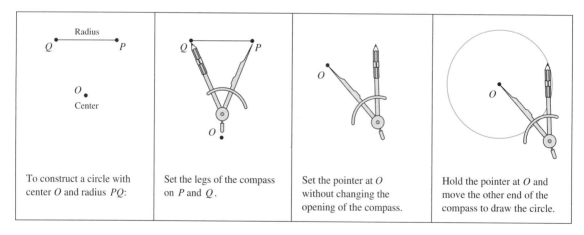

The figure formed in Figure 10-2 is a circle with center O and is referred to as circle O. Any circle is similar to any other circle because they have the same shape. However, *any circle is congruent to another circle if, and only if, the radii of the two circles are congruent.*

arc
center of arc

An **arc** of a circle is any part of the circle that can be drawn without lifting a pencil. (The entire circle could be considered an arc.) The **center of an arc** is the center of the circle that contains the arc. If there is no danger of ambiguity in a discussion, the endpoints of the arc are used to name the arc, otherwise three points are used in the naming. In Figure 10-3, $\widehat{ACB}$ is a **minor arc** and $\widehat{ADB}$ is a **major arc.** If the major arc and the minor arc of a circle are the same size, each is a **semicircle.** We will use the convention that if only two

minor arc • major arc
semicircle

points are used in the arc name, then the minor arc is intended. Thus, in Figure 10-3, $\overset{\frown}{AB}$ is another name for $\overset{\frown}{ACB}$.

Figure 10-3

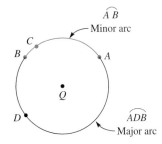

Constructing Segments

There are many ways to construct a segment congruent to a given segment $\overline{AB}$. A natural approach is to use a ruler, measure $\overline{AB}$, and then draw a congruent segment. A different way is to trace $\overline{AB}$ onto a piece of paper. A third method is to use a straightedge and a compass as in Figure 10-4.

Figure 10-4 Constructing a line segment congruent to a given segment

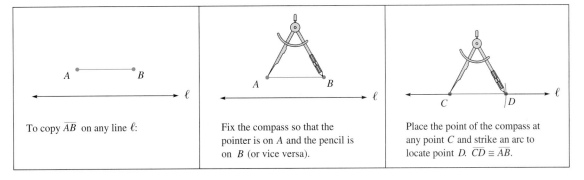

Triangle Congruence

Informally, two figures are congruent if it is possible to fit one figure onto the other so that all matching parts coincide. In Figure 10-5, $\triangle ABC$ and $\triangle A'B'C'$ have corresponding congruent parts. Tick marks are used to show the congruent segments and angles in the triangles. If we were to trace $\triangle ABC$ in Figure 10-5 and put the tracing over $\triangle A'B'C'$ so that the tracing of A is over A', the tracing of B is over B', and the tracing of C is over C', $\triangle ABC$ would coincide with $\triangle A'B'C'$. This suggests the following definition of congruent triangles.

Figure 10-5

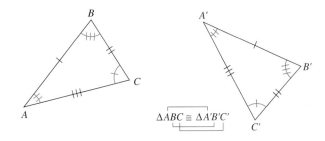

522 CHAPTER 10 *Constructions, Congruence, and Similarity*

Definition of Congruent Triangles

$\triangle ABC$ is congruent to $\triangle A'B'C'$, written $\triangle ABC \cong \triangle A'B'C'$, if $\angle A \cong \angle A'$, $\angle B \cong \angle B'$, $\angle C \cong \angle C'$, $\overline{AB} \cong \overline{A'B'}$, $\overline{BC} \cong \overline{B'C'}$, and $\overline{AC} \cong \overline{A'C'}$.

REMARK The statement "Corresponding parts of congruent triangles are congruent" is sometimes abbreviated as CPCTC.

NOW TRY THIS 10-1

- In Figure 10-5, $\triangle ABC \cong \triangle A'B'C'$. List all possible ways that the congruence can be symbolized.

Example 10-1 Write an appropriate symbolic congruence for each of the pairs of congruent triangles in Figure 10-6.

Figure 10-6

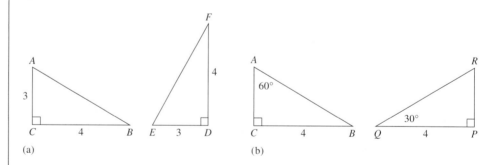

(a)   (b)

Solution
a. Vertex C corresponds to D because the angles at C and D are right angles. Also, because $\overline{CB} \cong \overline{DF}$ and C corresponds to D, B corresponds to F. Consequently, A corresponds to E. Thus $\triangle ABC \cong \triangle EFD$.
b. Vertex C corresponds to P because the angles at C and P are both right angles. To establish the other correspondences, we first find the missing angles in the triangles. We see that $m(\angle B) = 90° - 60° = 30°$ and $m(\angle R) = 90° - 30° = 60°$. Consequently, A corresponds to R because $m(\angle A) = m(\angle R) = 60°$, and B corresponds to Q because $m(\angle B) = m(\angle Q) = 30°$. Thus $\triangle ABC \cong \triangle RQP$.

Side, Side, Side Property (SSS)

In an automotive assembly line, the entire production process is designed in such a way that the same bodies of the same model cars are essentially congruent to each other. Calibration experts work to ensure that car parts are interchangeable so that the same part fits on all basic models of the same car. For the cars to be congruent, the parts must be congruent. In the assembly line of automotive production, decisions have to be made about the minimal set of items to consider for eventual congruency. In considering congruence of figures in geometry, we apply the same process.

If three sides and three angles of one triangle are congruent to the corresponding three sides and three angles of another triangle, then we can conclude from the definition of congruent triangles that the triangles are congruent. However, do we need to know that all six parts of one triangle are congruent to the corresponding parts of the second triangle in order to conclude that the triangles are congruent?

Consider the triangle formed by attaching three segments, as in Figure 10-7. Such a triangle is *rigid;* that is, its size and shape cannot be changed. Because of this property, a manufacturer can make duplicates if the lengths of the sides are known.

Figure 10-7

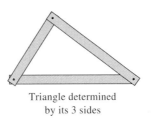

Triangle determined
by its 3 sides

Many bridges or other structures that have exposed frameworks demonstrate the practical use of the rigidity of triangles, as seen in Figure 10-8.

Figure 10-8

Because a triangle is completely determined by its three sides, we have the following property.

> **Property**
>
> **Side, Side, Side (SSS)** If the three sides of one triangle are congruent, respectively, to the three sides of a second triangle, then the triangles are congruent.

Example 10-2 For each part in Figure 10-9, use SSS to explain why the given triangles are congruent:

Figure 10-9

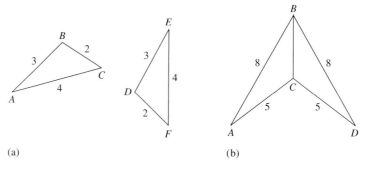

(a)   (b)

Solution a. $\triangle ABC \cong \triangle EDF$ by SSS because $\overline{AB} \cong \overline{ED}$, $\overline{BC} \cong \overline{DF}$, and $\overline{AC} \cong \overline{EF}$.
b. $\triangle ABC \cong \triangle DBC$ by SSS because $\overline{AB} \cong \overline{DB}$, $\overline{AC} \cong \overline{DC}$, and $\overline{BC} \cong \overline{BC}$.

Constructing a Triangle Given Three Sides

Using the SSS property, we can construct a triangle $A'B'C'$ congruent to a given triangle ABC if we know the lengths of the three sides. We can do this on a geometry drawing utility or with a compass and straightedge, as in Figure 10-10.

Figure 10-10

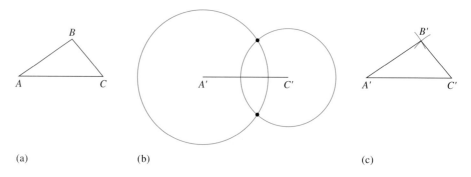

(a)   (b)   (c)

First, we construct a segment congruent to one of the three segments. For example, we may construct $\overline{A'C'}$ so that it is congruent to $\overline{AC}$. To complete the triangle construction, we must locate the other vertex, B'. The distance from A' to B' is AB. All points at a distance AB from A' are on a circle with center at A' and radius of length AB. Similarly, B' must be on a circle with center C' and radius of length BC. The only possible locations for B' are at the points where the two circles intersect. Either point is acceptable. Usually, a picture of the construction shows only one possibility and the construction uses only arcs, as pictured in Figure 10-10(c).

REMARK Starting the construction with a segment $\overline{A'B'}$ congruent to $\overline{AB}$ or with $\overline{B'C'}$ congruent to $\overline{BC}$ would also result in triangles congruent to $\triangle ABC$.

NOW TRY THIS 10-2

● From using the SSS construction as in Figure 10-10, we could think that given any three segments, we could construct a triangle whose sides are congruent to the given segments. To determine if this is true, cut at least ten pieces of straws of different lengths. Make all the possible triangles and answer the following questions:

a. Could a triangle be constructed from each of the three pieces of straws?

b. If three pieces were exactly the same length, what type of triangle could be constructed?

c. If the length of one piece of straw is the exact sum of the lengths of the other two pieces, can a triangle be constructed from the three pieces? Why or why not?

d. If one piece of straw is longer than the two other pieces put together, can a triangle be constructed from the three pieces? Why or why not? ●

TECHNOLOGY CORNER

Use the *Geometer's Sketchpad* or another geometry drawing utility to do the following:

a. Construct any triangle ABC.
b. Measure each of its sides.
c. Draw a segment $\overline{A'B'}$ congruent to $\overline{AB}$.
d. Draw a circle with center A' and radius of length AC.
e. Draw a circle with center B' and radius BC.
f. Label a point of intersection of the two circles as C'.
g. Draw $\overline{B'C'}$ and $\overline{A'C'}$.
h. Measure $\overline{B'C'}$ and $\overline{A'C'}$. Compare these lengths to BC and AC.
i. Measure and compare the following pairs of angles: $\angle ABC$ and $\angle A'B'C'$, $\angle BCA$ and $\angle B'C'A'$, and $\angle CAB$ and $\angle C'A'B'$.
j. What can you say about the two triangles?

Both the previous *Now Try This* and Technology Corner lend credence to the following property.

Property

Triangle inequality The sum of the measures of any two sides of a triangle must be greater than the measure of the third side.

Constructing Congruent Angles

We use the SSS notion of congruent triangles to construct an angle congruent to a given angle $\angle B$ by making $\angle B$ a part of an isosceles triangle and then reproducing this triangle, as in Figure 10-11.

Figure 10-11 shows how to construct an angle congruent to a given angle. However, with the compass and straightedge alone, it is impossible in general to construct an angle if given only its measure. For example, an angle of measure $20°$ cannot be constructed with a compass and a straightedge only. Instead, a protractor or some other measuring tool must be used. A geometry drawing utility can also be used to construct angles, but such constructions are approximations (why?).

Figure 10-11 Copying an angle

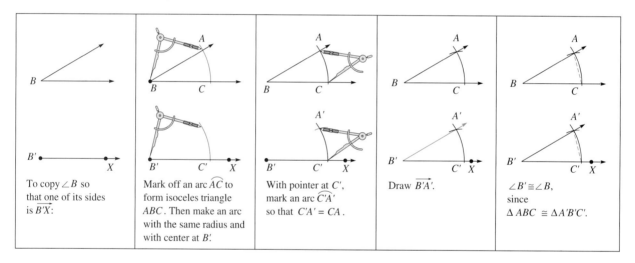

Side, Angle, Side Property (SAS)

We have seen that, given three segments, no more than one triangle can be constructed. Could more than one triangle be constructed from only two segments? Consider Figure 10-12(b), which shows three different triangles with sides congruent to the segments given in Figure 10-12(a). The length of the third side depends on the measure of the angle **included angle** between the other two sides. This angle is the **included angle**.

Figure 10-12

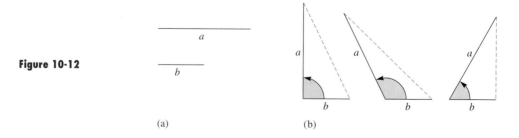

SECTION 10-1 *Congruence Through Constructions* 527

It appears that if we knew the lengths of two sides and the measure of the angle included between them, we could construct a unique triangle. This is true, and we can express the rule as the **Side, Angle, Side (SAS)** property.

Side, Angle, Side (SAS)

> **Property**
>
> **Side, Angle, Side (SAS)** If two sides and the included angle of one triangle are congruent to two sides and the included angle of another triangle, respectively, then the two triangles are congruent.

Example 10-3 For each part of Figure 10-13, use SAS to show that the given pair of triangles are congruent.

Figure 10-13

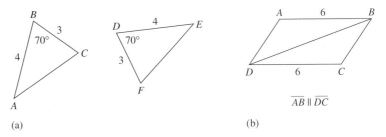

Solution
a. $\triangle ABC \cong \triangle EDF$ by SAS because $\overline{AB} \cong \overline{ED}$, $\angle B \cong \angle D$, and $\overline{BC} \cong \overline{DF}$.
b. Because $\overline{AB} \cong \overline{CD}$ and $\overline{DB} \cong \overline{BD}$, we need either another side or another angle to show that the triangles are congruent. We know nothing about the sides except that $\overline{AB} \parallel \overline{DC}$. Since parallel segments $\overline{AB}$ and $\overline{DC}$ are cut by transversal $\overline{BD}$, we have alternate interior angles $\angle ABD$ and $\angle BDC$ congruent. Now $\triangle ABD \cong \triangle CDB$ by SAS.

Figure 10-14

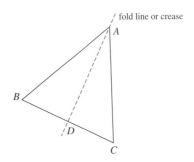

The SAS property of triangle congruence allows us to investigate a number of properties of triangles. For example, in Figure 10-14, consider isosceles triangle ABC with $\overline{AB} \cong \overline{AC}$. We use *paper folding* to construct the angle bisector of $\angle A$. For that purpose we fold a crease through vertex A so that vertex B "falls" on vertex C (this can be done because $\overline{AB} \cong \overline{AC}$). If point D is the intersection of the crease and $\overline{BC}$, then our folding assures that

∠BAD ≅ ∠CAD. Thus, by SAS, △ABD ≅ △ACD. Because corresponding parts must be congruent we have: ∠B ≅ ∠C, $\overline{BD}$ ≅ $\overline{CD}$, ∠BDA ≅ ∠CDA. Because the last pair of angles are supplementary, it follows that each is a right angle. Consequently, $\overline{AD}$ is perpendicular to $\overline{BC}$ and bisects $\overline{BC}$. A line that is perpendicular to a segment and bisects it, is the **perpendicular bisector** of the segment.

perpendicular bisector
altitude

An **altitude** of a triangle is the perpendicular segment from a vertex of the triangle to the line containing the opposite side of the triangle. These findings are summarized in Theorem 10-1.

Theorem 10-1

The following holds for every isosceles triangle:

a. The angles opposite the congruent sides are congruent. (Base angles of an isosceles triangle are congruent.)
b. The angle bisector of an angle formed by two congruent sides is an altitude of the triangle as well as the perpendicular bisector of the third side of the triangle.

In Figure 10-14, notice that if point A is equidistant from the endpoints B and C, then A is on the perpendicular bisector of $\overline{BC}$. The converse of this statement is also true. The statement and its converse are given in Theorem 10-2.

Theorem 10-2

a. Any point equidistant from the endpoints of a segment is on the perpendicular bisector of the segment.
b. Any point on the perpendicular bisector of a segment is equidistant from the endpoints of the segment.

REMARK Constructions with paper folding are an alternative to constructions with pencil, straightedge, and compass and are especially suitable for elementary and middle school students.

Construction of the Perpendicular Bisector of a Segment

Theorem 10-2 can be used to construct the perpendicular bisector of a segment by constructing any two points equidistant from the endpoints of the segment. Each point is a vertex of an isosceles triangle, and the two points determine the perpendicular bisector of the segment. In Figure 10-15 we have constructed point P equidistant from A and B by drawing any intersecting arcs with the same radius—one with center at A, and the other with center at B. Point Q is constructed similarly with two intersecting arcs. By Theorem 10-2 (a), each point is on the perpendicular bisector of $\overline{AB}$. Because two points determine a unique line, $\overleftrightarrow{PQ}$ must be the perpendicular bisector of $\overline{AB}$.

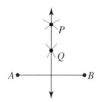

Figure 10-15

NOW TRY THIS 10-3

- In Figure 10-15 we constructed the perpendicular bisector of $\overline{AB}$ by drawing four arcs. An alternative construction of the perpendicular bisector of $\overline{AB}$ can be accomplished by drawing one large enough arc (or circle) with center at A and another one with the same radius and centered at B. The two points where the arcs intersect determine the perpendicular bisector of the segment. Draw any segment and construct its perpendicular bisector using only two arcs.

Construction of a Circle Circumscribed about a Triangle

circumscribed In Figure 10-16(a), a circle is **circumscribed** about a given $\triangle ABC$; that is, every vertex of the triangle is on the circle. How can such a circle be constructed?

To discover an appropriate construction, imagine that we know the location of center O of the circle as in Figure 10-16(b). The properties of the center of a circle should enable us to find its location. Because O is the center of the circle circumscribed about $\triangle ABC$, $OA = OC = OB$. The fact that $OA = OC$ implies that O is equidistant from the endpoints of segment $\overline{AC}$. Hence, by Theorem 10-2, O is on the perpendicular bisector, k, of $\overline{AC}$ as shown in Figure 10-16(c). Similarly, because $OC = OB$, O is on the perpendicular bisector, p, of $\overline{BC}$. Because O is on k and on p, it is the point of intersection of the two perpendicular bisectors. Thus, given $\triangle ABC$, we can construct the center of the circumscribed circle by constructing perpendicular bisectors of any two sides of the triangle. The point where the perpendicular bisectors intersect is the center of the required circle. Thus, the required circle is the circle with the center at O and radius OA (or OC or OB). The construction is shown in Figure 10-16(c). The center of the circle circumscribed about the triangle is the **circumcenter** of the triangle.

Figure 10-16

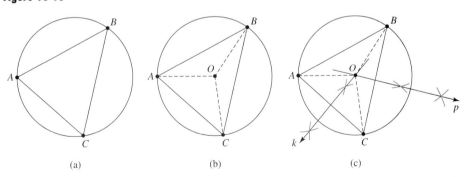

Constructions Involving Two Sides and an Angle of a Triangle

Figure 10-17 shows how to construct a triangle congruent to $\triangle ABC$ by using two sides $\overline{AB}$ and $\overline{AC}$ and the included angle, $\angle A$, formed by these sides. First, a ray with an arbitrary endpoint A' is drawn, and $\overline{A'C'}$ is constructed congruent to $\overline{AC}$. Then, $\angle A'$ is constructed so that $\angle A' \cong \angle A$ and B' is marked on the side of $\angle A'$ not containing C' so that $\overline{A'B'} \cong \overline{AB}$. Connecting B' and C' completes $\triangle A'B'C'$ so that $\triangle A'B'C' \cong \triangle ABC$.

Figure 10-17

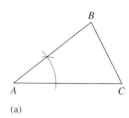
(a)

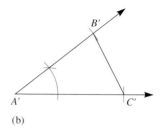
(b)

NOW TRY THIS 10-4

- If, in two triangles, two sides and an angle not included between these sides are congruent, respectively, determine whether the triangles must be congruent.

  a. If they are congruent, explain why you believe this is true.

  b. If they are not congruent, state as few additional conditions as possible that can be placed on the sides or angles to make the triangles congruent.

ONGOING ASSESSMENT 10-1

1. **a.** Use any tool to draw triangle ABC in which $\overline{BC}$ is greater than $\overline{AC}$. Measure the angles opposite $\overline{BC}$ and $\overline{AC}$. Compare the angle measures. What did you find?
   b. Based on your finding in (a), make a conjecture concerning the lengths of sides and the measures of angles of a triangle.

2. Use any tools to construct each of the following, if possible:
   a. A segment congruent to $\overline{AB}$ and an angle congruent to $\angle CAB$

   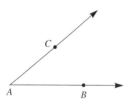

   b. A triangle with sides of lengths 2 cm, 3 cm, and 4 cm
   c. A triangle with sides of lengths 4 cm, 3 cm, and 5 cm (what kind of triangle is it?)
   d. A triangle with sides 4 cm, 5 cm, and 10 cm
   e. An equilateral triangle with sides 5 cm
   f. A triangle with sides 6 cm and 7 cm and an included angle of measure 75°
   g. A triangle with sides 6 cm and 7 cm and a nonincluded angle of measure 40°
   h. A triangle with sides 6 cm and 6 cm and a nonincluded angle of measure 40°
   i. A right triangle with legs 4 cm and 8 cm (the legs include the right angle)

3. For each of the conditions in problem 2(b) through (i), does the given information determine a unique triangle? Explain why or why not.

4. How many different triangles can be constructed with toothpicks by connecting the toothpicks only at their ends if each triangle can contain at most five toothpicks per side?

5. For each of the following, determine whether the given conditions are sufficient to prove that $\triangle PQR \cong \triangle MNO$. Justify your answers.
   a. $\overline{PQ} \cong \overline{MN}, \overline{PR} \cong \overline{MO}, \angle P \cong \angle M$
   b. $\overline{PQ} \cong \overline{MN}, \overline{PR} \cong \overline{MO}, \overline{QR} \cong \overline{NO}$
   c. $\overline{PQ} \cong \overline{MN}, \overline{PR} \cong \overline{MO}, \angle Q \cong \angle N$

6. A rancher designed a wooden gate as illustrated in the following figure. Explain the purpose of the diagonal boards on the gate.

   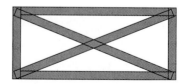

7. A rural homeowner had his television antenna held in place by three guy wires, as shown in the following figure. If the distances to each of the stakes from the base of the antenna

are the same, what is true about the lengths of the wires? Why?

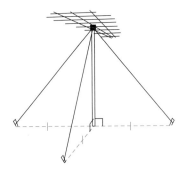

8. A group of students on a hiking trip wants to find the distance *AB* across a pond (see the following figure). One student suggests choosing any point *C*, connecting it with *B*, and then finding point *D* such that $\angle DCB \cong \angle ACB$ and $\overline{DC} \cong \overline{AC}$. How and why does this help in finding the distance *AB*?

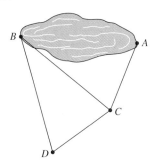

9. Using only a compass and a straightedge, perform each of the following:
   a. Reproduce $\angle A$ shown in the following figure:

   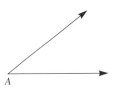

   b. Construct an equilateral triangle with the following side $\overline{AB}$.

   c. Construct a 60° angle.
   d. Construct an isosceles triangle with $\angle A$ (see the following figure) as the angle included between the two congruent sides:

   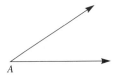

10. Refer to the following figure and, using only a compass and a straightedge, perform each of the following:

    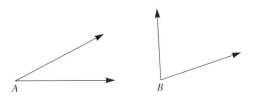

    a. Construct $\angle C$ so that $m(\angle C) = m(\angle A) + m(\angle B)$.
    b. Using the angles in (a), construct $\angle C$ so that $m(\angle C) = m(\angle B) - m(\angle A)$.

11. An equilateral triangle *ABC* is congruent to itself.
    a. Write all possible true correspondences between the triangle and itself.
    b. Use one of your answers in (a) to show that an equilateral triangular is also equiangular.

12. a. Draw a circle and use any method to mark off six points that are equally spaced on the circumference of the circle, as shown in the following figure:

    b. Connect each of the points to point *O* and connect the points in order around the circle to form a regular hexagon.
    c. Explain why all of the triangles are congruent.
    d. If the radius of the circle is *r*, what is the length of the side of the hexagon? Justify your answer.
    e. Construct any circle, then use your answer to part (d) to inscribe a regular hexagon in the circle; that is, the sides of the hexagon are congruent and all the vertices of the hexagons are on the circle. Describe your construction in words.
    f. Construct any circle. Use the construction in part (e) to construct an equilateral triangle inscribed in the circle.
    *g. Use the methods of this problem to construct a regular hexagon on a geometric drawing utility.

13. Suppose polygon *ABCD* is any square with diagonals $\overline{AC}$ and $\overline{BD}$ intersecting in point *F*, as shown in this figure:

    a. What is the relationship between point *F* and the diagonals $\overline{BD}$ and $\overline{AC}$? Why?
    b. What are the measures of angles *BFA* and *AFD*? Why?

14. a. If the diagonals of a quadrilateral bisect each other, what kind of quadrilateral must it be? Why?
    b. If the diagonals of a quadrilateral bisect each other and are congruent to each other, what type of quadrilateral can it be?
    c. If the diagonals of a quadrilateral are perpendicular bisectors of each other, what type of quadrilateral can it be?
15. Construct several noncongruent rhombuses and several noncongruent parallelograms that are not rhombuses. In each case, construct the diagonals.
    a. Based on your observations, what is true about the angles formed by the diagonals of a rhombus that is not necessarily true about the angles formed by the diagonals of a parallelogram that is not a rhombus?
    b. Justify your conjecture in (a).
16. What kind of figure is a quadrilateral in which both pairs of opposite sides are congruent? Why?
17. Write a definition for congruent arcs.
18. What minimum amount of information must you know in order to be sure that two cubes are congruent?
19. What is the least number of congruent parts that one must know in order to be sure that square pyramids are congruent?
20. How many different one-to-one correspondences could be listed between the following:
    a. Vertices of two triangles
    b. Vertices of two quadrilaterals
    c. Vertices of two *n*-gons
21. In a pair of right triangles, suppose two legs of one are congruent respectively to two legs of the other. Explain whether the triangles are congruent and why.
22. If two triangles are congruent, what can be said about their perimeters? Why?
23. Construct an obtuse triangle similar to the one shown below. Then, using only a compass and a straightedge, construct the circle that circumscribes the triangle.

24. For which of the following figures is it possible to find a circle that circumscribes the figure? If it is possible to find such a circle, draw the figure and construct the circumscribing circle.
    a. A right triangle
    b. A rectangle
    c. A rhombus that is not a square
    d. An isosceles trapezoid
    e. A parallelogram that is not a rectangle
    f. A regular hexagon

Communication

25. Explain whether you think the SSS and SAS properties can be used to determine if two triangles are similar. Use drawings constructed using a geometry utility or other tools to make your argument.
26. Write arguments to convince the class that Theorems 10-1 and 10-2 are true.
27. In the following figure congruent segments are shown with tick marks.

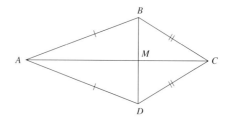

    a. Argue that the diagonal $\overline{AC}$ bisects $\angle A$ and $\angle C$.
    b. Let *M* be the point on which the diagonals of kite *ABCD* intersect. Measure $\angle AMD$ and make a conjecture concerning the angle between the diagonals of a kite. Justify your conjecture.
    c. Show that $\overline{BM} \cong \overline{MD}$.

Open-Ended

28. a. Find at least five examples of congruent objects.
    b. Find at least five examples of similar objects that are not congruent.
29. Design a quilt pattern that involves rep-tiles or find a pattern and describe the rep-tiles involved.
30. Consider the $\triangle AFB$ and $\triangle CED$ shown on the geoboard below to verify that segments $\overline{AB}$ and $\overline{CD}$ constructed are
    a. congruent
    ★ b. perpendicular to each other (consider the angles in $\triangle BGC$).

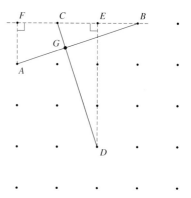

31. Using the following shapes, answer the questions below.
    a. Parallelogram   b. Rectangle   c. Kite
    d. Rhombus   e. Square
       i. For each shape, list sufficient properties to define that shape. For example, if given the words *isosceles triangle,* you might say that it is a triangle with at least two sides congruent.

ii. Now by using your answer in (i) for each shape, derive other properties of the shape. For example, you could show that the base angles are congruent in the isosceles triangle.

32. Describe a minimal set of conditions that can be used to argue that two quadrilaterals are congruent.

Cooperative Learning

33. Make a set of drawings of congruent figures and write an argument to convince a group of fellow students that the correspondence among vertices is necessary for determining that two triangles are congruent.

TECHNOLOGY CORNER

1. Write a Logo procedure to draw a variable-sized equilateral triangle.
2. Edit the following program to fit your particular Logo software for constructing a triangle when two sides and the included angle are given; type it into your computer and answer the questions below

```
TO SAS :SIDE1 :ANGLE :SIDE2
  DRAW
  BACK :SIDE1
  RIGHT :ANGLE
  FORWARD :SIDE2
  HOME
END
```

  a. If SAS 50 190 60 were executed, what would be the result?
  b. Is it possible in reality to draw a triangle with sides of 50 and 60 units and an included angle of 190°? Why?
  c. What line could be added to the procedure to correct the "bug" you encountered in (b)?

Section 10-2 — Other Congruence Properties

Angle, Side, Angle (ASA)

Triangles can be determined to be congruent by SSS and SAS. Can a triangle be constructed congruent to a given triangle by using two angles and a side? Figure 10-18 shows the construction of a triangle $A'B'C'$ such that $\overline{A'C'} \cong \overline{AC}$, $\angle A' \cong \angle A$, and $\angle C' \cong \angle C$. It seems that $\triangle A'B'C' \cong \triangle ABC$. This construction illustrates the **Angle, Side, Angle (ASA)** property of congruence.

Figure 10-18

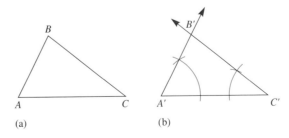

(a)   (b)

534 CHAPTER 10 *Constructions, Congruence, and Similarity*

Property

Angle, Side, Angle (ASA) If two angles and the included side of one triangle are congruent to two angles and the included side of another triangle, respectively, then the triangles are congruent.

In Figure 10-19 △*ABC* and △*DEF* have two pairs of angles congruent and a pair of sides congruent. Notice that ∠*A* ≅ ∠*D* and ∠*B* ≅ ∠*E*, which implies that ∠*C* ≅ ∠*F* because the measure of each is 180° − (70 + 40)°. Then, since $\overline{AC} \cong \overline{DF}$, we have △*ABC* ≅ △*DEF* by ASA.

Figure 10-19

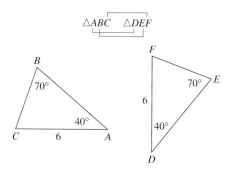

In general, by using the ASA property, we can justify the following property.

Property

Angle, Angle, Side (AAS) If two angles and a corresponding side of one triangle are congruent to two angles and a corresponding side of another triangle, respectively, then the two triangles are congruent.

Example 10-4 Show that the triangles in each part of Figure 10-20 are congruent.

Figure 10-20

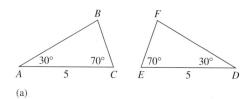

(a)

SECTION 10-2 *Other Congruence Properties* **535**

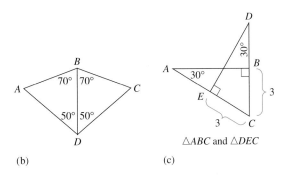

(b)   (c)

Solution
a. $\angle A \cong \angle D$, $\overline{AC} \cong \overline{DE}$, and $\angle C \cong \angle E$. Consequently, by ASA, $\triangle ABC \cong \triangle DFE$.

b. $\angle ABD \cong \angle CBD$, $\overline{BD} \cong \overline{BD}$, and $\angle ADB \cong \angle CDB$. Consequently, by ASA, $\triangle ABD \cong \triangle CBD$.

c. $\angle A \cong \angle D$, $\angle ABC \cong \angle DEC$, and $\overline{BC} \cong \overline{EC}$. Consequently, by AAS, $\triangle ABC \cong \triangle DEC$.

NOW TRY THIS 10-5

- Draw a parallelogram and one of its diagonals. Use the definition of a parallelogram given in Chapter 9 and ASA congruence property to prove that opposite sides of a parallelogram are congruent.

Example 10-5

a. Using the definition of a parallelogram and the property that opposite sides in a parallelogram are congruent (see Now Try This 10-5), prove that the diagonals of a parallelogram bisect each other; that is, in Figure 10-21(a), show that $AO = OC$ and $BO = OD$.

b. Draw a line through the point O where the diagonals of a parallelogram intersect as in Figure 10-21(b). The line intersects the opposite sides of the parallelogram at points P and Q. Prove that $\overline{OP} \cong \overline{OQ}$.

Figure 10-21

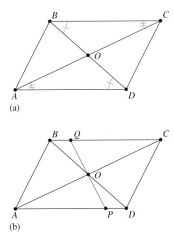

Solution **a.** In Figure 10-21(a) we need to show that $\overline{BO} \cong \overline{DO}$ and $\overline{AO} \cong \overline{CO}$. These segments appear in two pairs of triangles that seem to be congruent. One pair is $\triangle AOD$ and $\triangle COB$. To prove that these triangles are congruent we notice that $\overline{AD} \cong \overline{CB}$ (see Now Try This 10-5). We have no information about the corresponding sides, however we know that $\overline{BC} \parallel \overline{AD}$. Consequently, $\angle CBO \cong \angle ODA$, as these are alternate interior angles formed by the parallel lines and the transversal $\overleftrightarrow{BD}$. Similarly, using the transversal $\overleftrightarrow{AC}$ we have $\angle OAD \cong \angle OCB$ (congruent angles have been similarly marked in Figure 10-21(a)). Now $\triangle AOD \cong \triangle COB$ by ASA. Because corresponding parts in these triangles are congruent we have $\overline{AO} \cong \overline{OC}$ and $\overline{BO} \cong \overline{OD}$.

b. As in part (a), there are two pairs of triangles that have segments $\overline{OQ}$ and $\overline{OP}$ as sides; one such pair is $\triangle AOP$ and $\triangle COQ$. In these triangles, $\overline{AO} \cong \overline{OC}$ (as was proved in part (a)). As in part (a) $\angle OAP \cong \angle OCQ$. Also, $\angle AOP \cong \angle COQ$ because these are both vertical angles. Thus $\triangle AOP \cong \triangle COQ$ (why?) and $\overline{OP} \cong \overline{OQ}$ because these are corresponding parts in these triangles.

NOW TRY THIS 10-6

- If the diagonals of a quadrilateral bisect each other, must the quadrilateral be a parallelogram? Explain why or why not.

Using properties of congruent triangles, we can deduce various properties of quadrilaterals. Table 10-1 summarizes the definitions and lists some properties of six quadrilaterals. These and other properties of quadrilaterals are further investigated in Ongoing Assessment 10-2.

Table 10-1

Quadrilateral and Its Definition	Properties of the Quadrilateral
Trapezoid: A quadrilateral with at least one pair of parallel sides	Consecutive angles between parallel sides are supplementary.
Parallelogram: A quadrilateral in which each pair of opposite sides is parallel	**a.** A parallelogram has all the properties of a trapezoid. **b.** Opposite sides are congruent. **c.** Opposite angles are congruent. **d.** Diagonals bisect each other.

Table 10-1 Continued

Quadrilateral and Its Definition	Properties of the Quadrilateral
Rectangle: A parallelogram with a right angle	a. A rectangle has all the properties of a parallelogram. b. All the angles of a rectangle are right angles. c. A quadrilateral in which all the angles are right angles is a rectangle. d. The diagonals of a rectangle are congruent and bisect each other.
Kite: A quadrilateral with two distinct pairs of congruent adjacent sides.	a. Lines containing the diagonals are perpendicular to each other. b. A line containing one diagonal is a bisector of the other. c. One diagonal bisects nonconsecutive angles.
Rhombus: A parallelogram with all sides congruent	a. A rhombus has all the properties of a parallelogram. b. A quadrilateral in which all the sides are congruent is a rhombus. c. The diagonals of a rhombus are perpendicular and bisect each other. d. Each diagonal bisects opposite angles.
Square: A rectangle with all sides congruent	A square has all the properties of a parallelogram, a rectangle, and a rhombus.

ONGOING ASSESSMENT 10-2

1. Use any tools to construct each of the following, if possible:
   a. A triangle with angles measuring 60° and 70° and an included side of 8 in.
   b. A triangle with angles measuring 60° and 70° and a non-included side of 8 cm on a side of the 60° angle
   c. A right triangle with one acute angle measuring 75° and a leg of 5 cm on a side of the 75° angle
   d. A triangle with angles measuring 30°, 70°, and 80°
2. For each of the conditions in problem 1(a) through (d), is it possible to construct two noncongruent triangles? Explain why or why not.

3. For each of the following, determine whether the given conditions are sufficient to prove that $\triangle PQR \cong \triangle MNO$. Justify your answers.
   a. $\angle Q \cong \angle N$, $\angle P \cong \angle M$, $\overline{PQ} \cong \overline{MN}$
   b. $\angle R \cong \angle O$, $\angle P \cong \angle M$, $\overline{QR} \cong \overline{NO}$
   c. $\overline{PQ} \cong \overline{MN}$, $\overline{PR} \cong \overline{MO}$, $\angle N \cong \angle Q$
   d. $\angle P \cong \angle M$, $\angle Q \cong \angle N$, $\angle R \cong \angle O$

4. A parallel ruler, shown as follows, can be used to draw parallel lines. The distance between the parallel segments $\overline{AB}$ and $\overline{DC}$ can vary. The ruler is constructed so that the distance between A and B equals the distance between D and C. The distance between A and C is the same as the distance between B and D. Explain why $\overline{AB}$ and $\overline{DC}$ are always parallel.

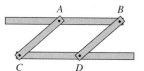
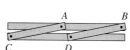

5. In each of the following, choose as many of the words *parallelogram, rectangle, rhombus, trapezoid, kite,* or *square* so that the resulting sentence is true. If none of the words makes the sentence true, answer "none" and justify your answer.
   a. A quadrilateral is a ____ if, and only if, its diagonals bisect each other.
   b. A quadrilateral is a ____ if, and only if, its diagonals are congruent.
   c. A quadrilateral is a ____ if, and only if, its diagonals are perpendicular.
   d. A quadrilateral is a ____ if, and only if, its diagonals are congruent and bisect each other.
   e. A quadrilateral is a ____ if, and only if, its diagonals are perpendicular and bisect each other.
   f. A quadrilateral is a ____ if, and only if, its diagonals are congruent and perpendicular and they bisect each other.
   g. A quadrilateral is a ____ if, and only if, a pair of opposite sides is parallel and congruent.

6. Use a geometry utility or other tools to create several trapezoids that have a pair of nonparallel sides congruent. Measure all angles and make a conjecture about the relationships among pairs of angles.

7. In both the ASA and the AAS properties of congruence, if two angles of one triangle are congruent respectively to two angles of another triangle, what must be true about the third angles of the triangles? Justify your answer.

8. a. For two right triangles, give a minimal set of conditions based on ASA and AAS to argue that the two triangles are congruent.
   b. Using your answer in (a), write two theorems that can be used to show that right triangles are congruent.

9. Classify each of the following statements as true or false. If the statement is false, provide a counterexample.
   a. The diagonals of a square are perpendicular bisectors of each other.
   b. If all sides of a quadrilateral are congruent, the quadrilateral is a rhombus.
   c. If a rhombus is a square, it must also be a rectangle.
   d. An isosceles trapezoid can be a rectangle.
   e. A square is a trapezoid.
   f. A trapezoid is a parallelogram.
   g. A parallelogram is a trapezoid.
   h. No rectangle is a rhombus.
   i. No trapezoid is a square.
   j. Some squares are trapezoids.

10. a. Construct quadrilaterals having exactly one, two, and four right angles.
    b. Can a quadrilateral have exactly three right angles? Why?
    c. Can a parallelogram have exactly two right angles? Why?

11. Each fourth grader is given a protractor, two 30-cm sticks, and two 20-cm sticks and is asked to form a quadrilateral with a 75° angle. Sketch all possibilities.

12. The game of Triominoes has equilateral-triangular playing pieces with numbers at each vertex, shown as follows:

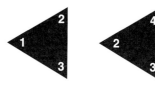

If two pieces are placed together as shown in the following figure, explain what type of quadrilateral is formed:

13. A **sector** of a circle is a pie-shaped section bounded by two radii and an arc. What is a minimal set of conditions for determining that two sectors of the same circle are congruent?

14. Draw two quadrilaterals such that two angles and an included side in one quadrilateral are congruent respectively to two angles and an included side in the other quadrilateral. However, the quadrilaterals are not congruent.

15. a. In an isosceles trapezoid, make a conjecture concerning the lengths of the sides which are not bases.

b. Make a conjecture concerning the diagonals of an isosceles trapezoid.
   c. Justify your conjectures in (a) and (b).
16. Two angles of the trapezoid shown below are 45° each. Suppose the lengths of the parallel sides of the trapezoid are a cm and b cm. Find the height of the trapezoid (the distance between the parallel sides) in terms of a and b.

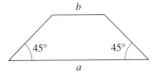

17. Using a straightedge and a compass, construct any convex kite. Then construct a second convex kite that is not congruent to the first but whose sides are congruent to the corresponding sides of the first kite.
18. **a.** What type of figure is formed by joining the midpoints of a rectangle (see the following figure)?

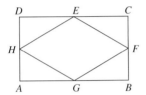

   b. What type of figure is formed by joining the midpoints of the sides of a parallelogram?
   ★**c.** Justify your answer to part (b).
19. What information is needed to determine congruency for each of the following?
   a. Two squares
   b. Two rectangles
   c. Two parallelograms
20. Describe a set of minimal conditions to determine if two regular polygons are congruent.
21. Suppose polygon $ABCD$ shown in the following figure is any parallelogram:

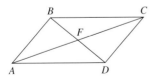

   Use congruent triangles to justify each of the following:
   a. $\angle A \cong \angle C$ and $\angle B \cong \angle D$ (opposite angles are congruent).
   b. $\overline{BC} \cong \overline{AD}$ and $\overline{AB} \cong \overline{CD}$ (opposite sides are congruent).
   c. $\overline{BF} \cong \overline{DF}$ and $\overline{AF} \cong \overline{CF}$ (the diagonals bisect each other).
   d. $\angle DAB$ and $\angle ABC$ are supplementary.

*22. **a.** Write a Logo procedure called RHOMBUS with inputs :SIDE and :ANGLE in which the first variable is the length of the side of the rhombus and the second is the measure of an interior angle of the rhombus.
   b. Execute RHOMBUS 80 50 and RHOMBUS 80 130. What is the relationship between the two figures? Why?
   c. Write a Logo procedure called SQ.RHOM with input :SIDE that will draw a square by calling on the RHOMBUS procedure.

Communication

23. Stan is standing on the bank of a river wearing a baseball cap. Standing erect and looking directly at the other bank, he pulls the bill of his cap down until it just obscures his vision of the opposite bank. He then turns around, being careful not to disturb the cap, and picks out a spot that is just obscured by the bill of his cap. He then paces off the distance to this spot and claims that the distance across the river is approximately equal to the distance he paced. Is Stan's claim true? Why?
24. Most ironing boards are collapsible for storage and can be adjusted to fit the height of the person using them. The surface of the board, though, remains parallel to the floor regardless of the height. Explain how to construct the legs of an ironing board to ensure the surface is always parallel to the floor.

Open-Ended

25. **a.** Go to a wallpaper store and examine the pattern books. Determine whether all rolls of a specific pattern of wallpaper are congruent.
   b. On an individual roll of wallpaper, determine the length of wallpaper before a pattern is repeated.
   c. If you were wallpapering a room, explain how congruence or noncongruence could save you money.

Cooperative Learning

26. **a.** Record the definitions of *trapezoid* and *kite* given in different grade 6–8 and secondary-school geometry textbooks.
   b. Compare the definitions found with those in this text and with those other groups found.
   c. Defend the use of one definition over another.

Review Problems

27. In the following regular pentagon, use the existing vertices to find all the triangles congruent to △ABC. Show that the triangles actually are congruent.

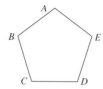

28. If possible, construct a triangle that has the following three segments a, b, and c as its sides; if not possible, explain why not.

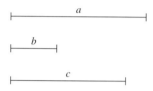

29. Construct an equilateral triangle whose sides are congruent to the following segment:

30. For each of the following pairs of triangles, determine whether the given conditions are sufficient to show that the triangles are congruent. If the triangles are congruent, tell which property can be used to verify this fact.

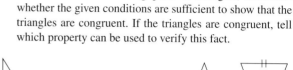

(a) (b)

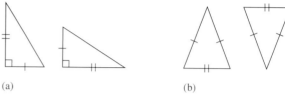

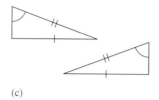

(c)

TECHNOLOGY CORNER

Use a geometry drawing utility to test the truth of the following statement. "If one pair of sides of a quadrilateral is congruent and parallel, then the quadrilateral is a parallelogram."

Section 10-3 — Other Constructions

We use the definition of a rhombus and the following properties (also listed in Table 10-1) to accomplish basic compass-and-straightedge constructions:

1. A rhombus is a parallelogram in which all the sides are congruent.
2. A quadrilateral in which all the sides are congruent is a rhombus.
3. Each diagonal of a rhombus bisects the opposite angles.
4. The diagonals of a rhombus are perpendicular.
5. The diagonals of a rhombus bisect each other.

Constructing Parallel Lines

To construct a line parallel to a given line ℓ through a point P not on ℓ, as in the leftmost panel of Figure 10-22, our strategy is to construct a rhombus (using property 2 listed above) with one of its vertices at P and one of its sides on line ℓ. Because the opposite sides of a rhombus are parallel, one of the sides through P will be parallel to ℓ. This construction is shown in Figure 10-22.

Figure 10-22 Constructing parallel lines (rhombus method)

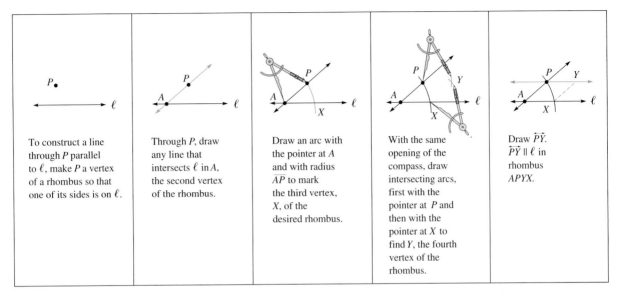

Figure 10-23 shows another way to do the construction. If congruent corresponding angles are formed by a transversal cutting two lines, then the lines are parallel. Thus the first step is to draw a transversal through P that intersects ℓ. The angle marked α is formed by the transversal and line ℓ. By constructing an angle with a vertex at P congruent to α, we create congruent corresponding angles; therefore $\overleftrightarrow{PQ} \parallel \ell$.

Figure 10-23 Constructing parallel lines (corresponding angle method)

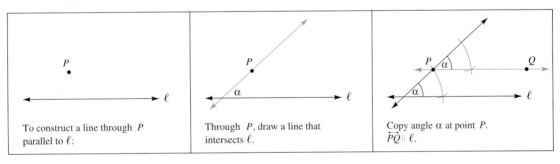

NOW TRY THIS 10-7

● Parallel lines are frequently constructed using either a ruler and one triangle or two triangles. If a ruler and a triangle are used, the ruler is left fixed and the triangle is slid so that one side of the triangle touches the ruler at all times. In Figure 10-24 the hypotenuses of the right triangles are all parallel (also the legs not on the ruler are all parallel). How can this method be used to construct a line through a given point parallel to a given line? ●

Figure 10-24

Paper folding can be used to construct parallel lines. For example, in Figure 10-25(a) if we wish to construct a line m parallel to line p through point Q, we can fold a perpendicular to line p so that the fold line does not contain point Q, as shown in Figure 10-25(b). Then by marking the image of point Q and connecting point Q and its image, Q', we have $\overline{QQ'}$ parallel to line p (why?).

Figure 10-25

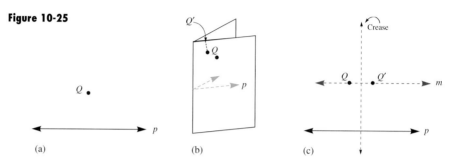

Constructing Angle Bisectors

angle bisector Another construction based on a property of a rhombus is the construction of an **angle bisector,** a ray that separates an angle into two congruent angles. The diagonal of a rhombus with vertex A bisects $\angle A$, as shown in Figure 10-26.

Figure 10-26 Bisecting an angle

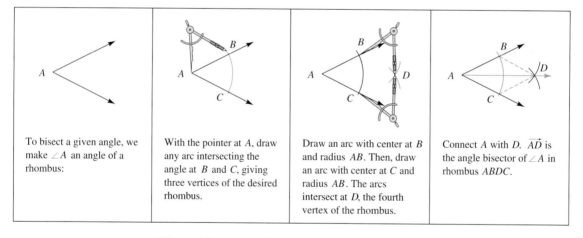

We can bisect an angle by folding a line through the vertex so that one side of the angle folds onto the other side. For example, in Figure 10-27 we bisect $\angle ABC$ by folding and creasing the paper through the vertex B so that $\overrightarrow{BC}$ coincides with $\overrightarrow{BA}$.

Figure 10-27

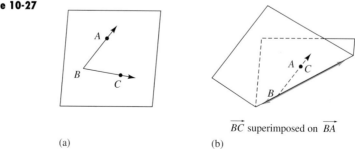

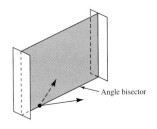

Figure 10-28

A Mira is a plastic device that acts as a reflector so that the image of an object can be seen behind the Mira. The drawing edge of the Mira acts as a folding line on paper. Any construction demonstrated in this text using paper folding can also be done with a Mira. To construct the bisector of an angle with a Mira, we place the drawing edge of the Mira on the vertex of the angle and reflect one side of the angle onto the other, as shown in Figure 10-28.

Constructing Perpendicular Lines

To construct a line through P perpendicular to line ℓ, where P is not a point on ℓ, as in Figure 10-29, recall that the diagonals of a rhombus are perpendicular to each other. If we construct a rhombus with a vertex at P and two vertices A and B on ℓ, as in Figure 10-29, the segment connecting the fourth vertex Q to P is perpendicular to ℓ because $\overline{AB}$ and $\overline{PQ}$ are diagonals of the rhombus.

Figure 10-29 Constructing a perpendicular to a line from a point not on a line

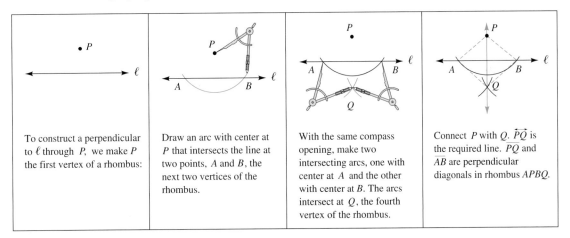

In Section 10-1 we saw how to construct the perpendicular bisector of a segment using a property of a perpendicular bisector stated in Theorem 10-2. Here we show how a property of a rhombus can also be used for constructing the perpendicular bisector of a segment.

To construct the perpendicular bisector of a line segment, as in Figure 10-30, we use the fact that the diagonals of a rhombus are perpendicular bisectors of each other.

Figure 10-30 Bisecting a line segment

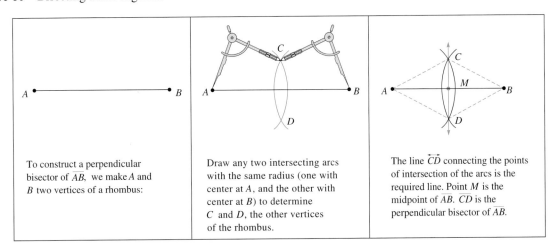

The construction yields a rhombus such that the original segment is one of the diagonals of the rhombus and the other diagonal is the perpendicular bisector, as in Figure 10-30.

Constructing a perpendicular to a line ℓ at a point M on ℓ is based on the same property of a rhombus just used; that is, the diagonals of a rhombus are perpendicular bisectors of each other. Observe in Figure 10-30 that $\overline{CD}$ is a perpendicular to $\overline{AB}$ through M. Thus we construct a rhombus whose diagonals intersect at point M, as in Figure 10-31.

Figure 10-31 Constructing a perpendicular to a line from a point on the line

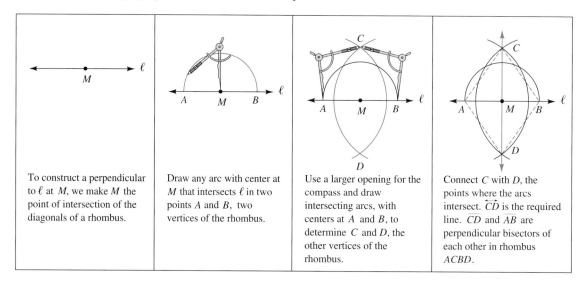

Perpendicularity constructions can also be completed by means of paper folding or by using a Mira. To use paper folding to construct a perpendicular to a given line ℓ at a point P on the line, we fold the line onto itself, as shown in Figure 10-32(a). The fold line is perpendicular to ℓ. To perform the construction with a Mira, we place the Mira with the drawing edge on P, as shown in Figure 10-32(b), so that ℓ is reflected onto itself. The line along the drawing edge is the required perpendicular.

Figure 10-32

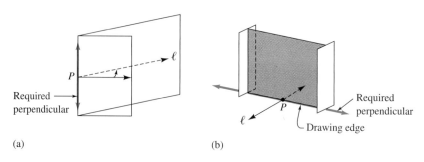

Constructing perpendiculars is useful in locating altitudes of a triangle. The construction of altitudes is described in Example 10-6.

Example 10-6 Given triangle *ABC*, construct an altitude from vertex *A* in each part of Figure 10-33.

Figure 10-33

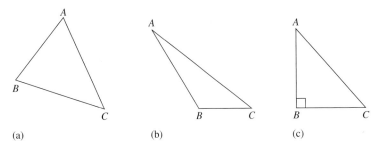

(a) (b) (c)

Solution **a.** An altitude is the perpendicular from a vertex to the line containing the opposite side of a triangle, so we need to construct a perpendicular from point *A* to the line containing $\overline{BC}$. Such a construction is shown in Figure 10-34. $\overline{AD}$ is the required altitude.

Figure 10-34

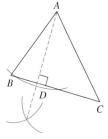

b. The construction of the altitude from vertex *A* is shown in Figure 10-35. Notice that the required altitude $\overline{AD}$ does not intersect the interior of $\triangle ABC$.

Figure 10-35

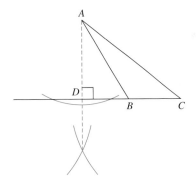

c. Triangle *ABC* is a right triangle. The altitude from vertex *A* is the side $\overline{AB}$. No construction is required.

TECHNOLOGY CORNER

Use a geometry drawing utility to draw all the altitudes of a triangle. Make conjectures about the altitudes of each of the following types of triangles: acute, right, and obtuse.

Properties of Angle Bisectors

Consider the angle bisector in Figure 10-36. It seems that any point P on the angle bisector is equidistant from the sides of the angle; that is, $PD = PE$. (The distance from a point to a line is the length of the perpendicular from the point to the line.)

Figure 10-36

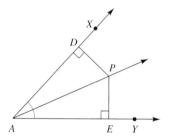

To justify this, we find two congruent triangles that have these segments as corresponding sides. The only triangles pictured are $\triangle ADP$ and $\triangle AEP$. Because $\overrightarrow{AP}$ is the angle bisector, $\angle DAP \cong \angle EAP$. Also, $\angle PDA$ and $\angle PEA$ are right angles and are thus congruent. $\overline{AP}$ is congruent to itself, so $\triangle PDA \cong \triangle PEA$ by AAS. Thus $\overline{PD} \cong \overline{PE}$ because they are corresponding parts of congruent triangles PDA and PEA. Consequently, we have the following theorem.

Theorem 10-3

a. Any point P on an angle bisector is equidistant from the sides of the angle.
b. Any point that is equidistant from the sides of an angle is on the angle bisector of the angle.

Notice that we have proved only Theorem 10-3(a). You are asked to justify part (b) in the following Now Try This.

NOW TRY THIS 10-8

- Use a geometry drawing utility or any other tool to demonstrate that if a point is in the interior of an angle and is equidistant from the sides of the angle, the point must be on the angle bisector of that angle.

Example 10-7

In the interior of $\triangle ABC$ in Figure 10-37 find the point P that is equidistant from the three sides of the triangle and construct the congruent segments through P perpendicular to each side.

Figure 10-37

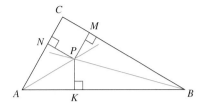

Solution To discover an appropriate construction imagine that we know the location of *P* in Figure 10-37. Then $\overline{PK}$, $\overline{PM}$, and $\overline{PN}$ are perpendicular to the corresponding sides and $PK = PM = PN$. Because *P* is equidistant from the sides of $\angle A$ it must be on the angle bisector of *A*. Similarly *P* must be on the angle bisector of $\angle B$. Thus to find the point *P* we construct two angle bisectors of $\triangle ABC$. The intersection of the angle bisectors is the required point. We then construct through *P* perpendiculars to the three lines. The points of intersection of the perpendiculars with the corresponding sides determine the equal distances *PM*, *PK*, and *PN*. The construction is shown in Figure 10-38.

Figure 10-38

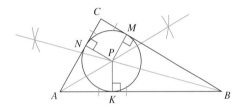

tangent

inscribed
incenter

Notice that if in Figure 10-38 we draw a circle with the center *P* and radius *PM* (or *PK* or *PN*) the circle seems to fit exactly inside the triangle. In fact each line "touches" the circle at one point only. We say that the sides of the triangle are **tangent** to the circle. A line is *tangent to a circle if it intersects the circle in one and only one point*. It is possible to show that a line perpendicular to a radius at the endpoint of the radius, which is not the center, is tangent to the circle. Thus the sides of the $\triangle ABC$ in Figure 10-38 are tangent to the circle. Such a circle is said to be **inscribed** in the triangle. The center of the inscribed circle is called the **incenter** of the triangle.

ONGOING ASSESSMENT 10-3

1. Refer to the following figure and use a compass and a straightedge to construct a line *m* through *P* parallel to ℓ, using each of the following:
   a. Alternate interior angles.
   b. Alternate exterior angles.

2. For each of the following items compare these methods of constructions: paper folding, Mira, compass and straightedge, and geometric drawing utility (if available). Give advantages and disadvantages of each method.
   a. Bisector of $\angle A$
   b. Perpendicular bisector of $\overline{AB}$
   c. Perpendicular from point *P* to line *m*

3. Construction companies avoid vandalism at night typically by hanging expensive pieces of equipment from the boom of a crane, as shown in the following figure:

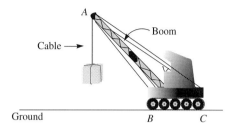

   a. If you consider a triangle with two vertices A and B as marked in the above figure and the intersection of a line through the cable holding the equipment and the ground as the third vertex, what type of triangle is formed?
   b. If you consider the triangle formed by points A, B, and C, describe where the altitude containing vertex A of the triangle is.

4. Construct the perpendicular bisectors of each of the following triangles. Use any desired method.

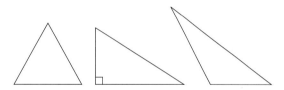

   a. Make a conjecture about the perpendicular bisectors of the sides of an acute triangle.
   b. Make a conjecture about the perpendicular bisectors of the sides of a right triangle.
   c. Make a conjecture about the perpendicular bisectors of the sides of an obtuse triangle.
   d. For each of the three triangles construct the circle that circumscribes the triangle.

5. a. Given triangle ABC, as in the following figure, construct a point P that is equidistant from the three vertices of the triangle. Explain why your construction is correct.

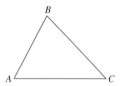

   b. Repeat (a) for the following obtuse triangle:

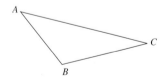

6. Draw an acute triangle and an obtuse triangle and in each case construct the circle inscribed in the triangle.

7. For which of the following figures is it possible to construct a circle that is inscribed in the figure (each side of the figure must be tangent to the circle)? If it is possible to find such a circle, draw the figure and construct the inscribed circle using only a straightedge and a compass; if not, explain how you decided that it is impossible to find an inscribed circle.
   a. A rectangle   b. A rhombus
   c. A kite that is not a rhombus
   d. A square   e. A regular hexagon

8. A **chord** of a circle is a segment with endpoints on the circle.
   a. Construct a circle, several chords, and a perpendicular bisector of each chord. Make a conjecture concerning the perpendicular bisector of a chord and the center of the circle.
   b. Justify your conjecture in (a).
   c. Given a circle with an unmarked center, find the center of the circle.

9. Given $\overline{AB}$ in the following figure, construct a square with $\overline{AB}$ as a side:

10. Suppose you are "charged" 10¢ each time you use your straightedge to draw a line segment and 10¢ each time you use your compass to draw an arc. Using only a compass and a straightedge, determine the cheapest way to construct a square. Explain your reasoning.

11. Given A, B, and C as vertices, use a compass and a straightedge to construct a parallelogram:

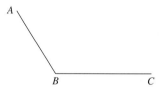

12. In the following concave quadrilateral $APBQ$, $\overline{PQ}$ and $\overline{AB}$ are the diagonals: $\overline{AP} \cong \overline{BP}$, $\overline{AQ} \cong \overline{BQ}$, and $\overline{PQ}$ has been extended until it intersects $\overline{AB}$ at C:

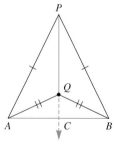

   a. Make a conjecture concerning $\overrightarrow{PQ}$ and $\overline{AB}$.
   b. Justify your conjecture in (a).
   c. Make conjectures concerning the relationships between $\overrightarrow{PQ}$ and $\angle APB$ and between $\overrightarrow{QC}$ and $\angle AQB$.
   d. Justify your conjectures in (c).

13. Use definitions of quadrilaterals to classify quadrilateral *APBQ* in problem 12.
14. Using any tools, construct each of the following, if possible. If the construction is not possible, explain why.
    a. A square, given one side
    b. A square, given one diagonal
    c. A rectangle, given one diagonal
    d. A parallelogram, given two of its adjacent sides
    e. A rhombus, given two of its diagonals
    f. A triangle with two obtuse angles
    g. A parallelogram with exactly three right angles
15. Using only a compass and a straightedge, construct angles with each of the following measures:
    a. 30°   b. 15°   c. 45°
    d. 75°   e. 105°
16. Given *AB* in the following figure, use a compass and a straightedge to construct the perpendicular bisector of $\overline{AB}$. You are not allowed to put any marks below $\overline{AB}$.

17. a. Explain why the hypotenuses in Figure 10-24 are all parallel.
    b. Use the "sliding triangle" method described in Figure 10-24 to construct a line through a point *P* parallel to a line ℓ.
18. Draw a line ℓ and a point *P* not on the line and use the sliding triangle method described in Figure 10-24 to construct a perpendicular to ℓ through *P* using a straightedge and a right triangle.
19. Draw a convex quadrilateral similar to the one shown below and construct each of the following points, if possible. If it is not possible, explain why. Describe each construction in words and explain why it produces the required point.
    a. The point that is equidistant from $\overline{AB}$ and $\overline{AD}$ and also from points *B* and *C*
    b. The point that is equidistant from *AB*, *AD*, and *BC*
    c. The point that is equidistant from *A*, *B*, and *C*
    d. The point that is equidistant from *A*, *B*, *C*, and *D*

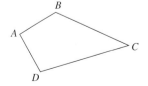

20. Describe how to inscribe a square in any circle. Explain why your construction produces a square.
21. Use problem 20 to describe how to inscribe a regular octagon in a circle.
22. Use variables to write a Logo procedure to draw each:
    a. The angle bisector of a variable-sized angle
    b. The perpendicular bisector of a variable-sized segment
    c. A set of two variable-sized parallel segments

Communication

23. Given an angle and a roll of tape, describe how you might construct the bisector of the angle.
24. Write to a curriculum developer explaining whether or not the geometry curriculum in grades 5–8 should include construction problems that use only a compass and straightedge.

Open-Ended

25. Explain whether you think there is a single perpendicular from a point to a line on a sphere if a line is defined as a great circle of the sphere (a circle obtained by intersecting the sphere with a plane through its center).
26. Explain whether you think there are parallel lines on a sphere.
27. Explain whether your geometry drawing utility allows the construction of regular polygons.

Cooperative Learning

28. In your group, perform the following paper-folding construction and answer the questions that follow. Let *P* and *Q* be two opposite vertices of a rectangular piece of paper as shown in Figure (i). Fold *P* onto *Q* so that a crease is formed. This results in Figure (ii) where *A* and *B* are the endpoints of the crease. Next crease again so that point *A* folds onto point *B*. This results in Figure (iii). Next, unfold to obtain two creases as in Figure (iv).

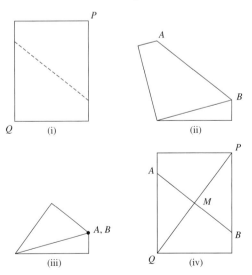

a. Individually write an explanation of why every point on $\overline{AB}$ is equidistant from the endpoints of $\overline{PQ}$. Compare

the explanations in your group and prepare one explanation to be presented to the class.
b. Discuss in your group whether *APBQ* is a rhombus.

Review Problems

29. In the following figure $\overleftrightarrow{AB} \parallel \overleftrightarrow{ED}$ and $\overline{BC} \cong \overline{CE}$. Explain why $DE = AB$.

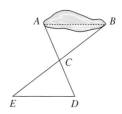

30. Draw $\triangle ABC$. Then construct $\triangle PQR$ congruent to $\triangle ABC$ using each of the following combinations:
    a. Two sides of $\triangle ABC$ and an angle included between these sides
    b. The three sides of $\triangle ABC$
    c. Two angles and a side included between these angles
31. *ABCD* is a trapezoid with $\overline{AB} \cong \overline{CD}$. Show that
    a. $\angle A \cong \angle D$ (through *C* draw a line parallel to $\overline{AB}$ and show that the triangle that is created is isosceles).
    b. $\overline{BD} \cong \overline{CA}$.
32. In two right triangles, $\triangle ABC$ and $\triangle DEF$, if $\angle A$ and $\angle D$ are congruent and $\overline{AC}$ and $\overline{DF}$ are congruent, what can you conclude about the two triangles. Why?

TECHNOLOGY CORNER

Use a geometry drawing utility to do each of the following:

1. a. Draw a circle. Mark its center and draw any angle whose vertex is the center and whose sides intersect the circle, as in Figure 10-39. Such an angle is a **central angle** of the circle. The measure of a central angle and its intercepted arc are considered to be the same measure.

central angle

Figure 10-39

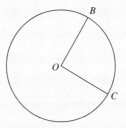

inscribed angle

b. Choose any point *P* on the circumference of the circle that is not *B* and not *C*, and not in the interior of $\angle BOC$. Draw $\angle BPC$ and find its measure. $\angle BPC$ is an **inscribed angle** of the circle.
c. Make a conjecture about the measures of $\angle BOC$ and $\angle BPC$.

2. Draw an inscribed triangle in a circle in such a way that one side of the triangle is a diameter of the circle. Find the measure of the angle that does not have the diameter as a side. Make a conjecture about the measure of any such angle.

Section 10-4 — Similar Triangles and Similar Figures

When a germ is examined under a microscope or when a slide is projected on a screen, the shapes in each case remain the same, but the sizes are altered. *Two figures that have the same shape but not necessarily the same size are* **similar.** For example, on the following student page from *Scott Foresman-Addison Wesley Middle School MATH Course 2, 1999,*

Matching sides and angles of similar figures are called **corresponding sides** and **corresponding angles**. The measures of these sides and angles are related in a special way.

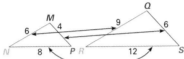

Remember
Congruent angles have equal measures.
[Page 214]

DEFINITION OF SIMILARITY
Figures are similar if their corresponding angles are congruent and the lengths of their corresponding sides have equal ratios.

The ratio of corresponding side lengths is the **scale factor** of the figures. It has the same meaning as the map and model scales you worked with earlier. The scale factor from $\triangle MNP$ to $\triangle QRS$ is $\frac{3}{2}$, since $\frac{9}{6} = \frac{6}{4} = \frac{12}{8} = \frac{3}{2}$.

Example 1

A movie uses a model skyscraper that is similar to a real one. Find the scale factor from the actual skyscraper to the model.

In similar figures, the ratio of any pair of corresponding side lengths can be used to find the scale factor. You can use the buildings' heights *or* widths to find the scale factor.

Using heights:
$$\frac{\text{model height}}{\text{actual height}} = \frac{4}{480} = \frac{1}{120}$$

Using widths:
$$\frac{\text{model width}}{\text{actual width}} = \frac{1}{120}$$

The scale factor from the actual skyscraper to the model is $\frac{1}{120}$.

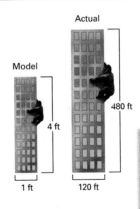

▶ **Science Link**
The largest ape in the world is the gorilla. An adult male usually weighs between 300 and 400 pounds.

The statement $\triangle ABC \sim \triangle EFD$ says that $\triangle ABC$ is similar to $\triangle EFD$. The order of the letters shows the corresponding parts. When you write a similarity statement, be sure to list the parts in the right order.

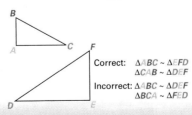

Correct: $\triangle ABC \sim \triangle EFD$
$\triangle CAB \sim \triangle DEF$
Incorrect: $\triangle ABC \sim \triangle DEF$
$\triangle BCA \sim \triangle FED$

7-8 • Creating and Exploring Similar Figures

552 CHAPTER 10 Constructions, Congruence, and Similarity

we see a skyscraper and its model. The ratio of the corresponding side lengths is the **scale factor**. On the student page the scale factor is $\frac{1}{120}$.

It seems that in any enlargement or reduction, such as that on the student page, the results will be similar; that is, the corresponding angle measures remain the same and the corresponding sides are proportional. In particular, the observations about similar figures (having corresponding angles congruent and corresponding sides being proportional) extend to triangles, as given in the following definition.

> **Definition of Similar Triangles**
>
> $\triangle ABC$ is similar to $\triangle DEF$, written $\triangle ABC \sim \triangle DEF$, if, and only if, $\angle A \cong \angle D$, $\angle B \cong \angle E$, $\angle C \cong \angle F$, and $\frac{AB}{DE} = \frac{AC}{DF} = \frac{BC}{EF}$.

Example 10-8 Given the pairs of similar triangles in Figure 10-40, find a one-to-one correspondence among the vertices of the triangles such that the corresponding angles are congruent. Then write the proportion for the corresponding sides that follows from the definition.

Figure 10-40

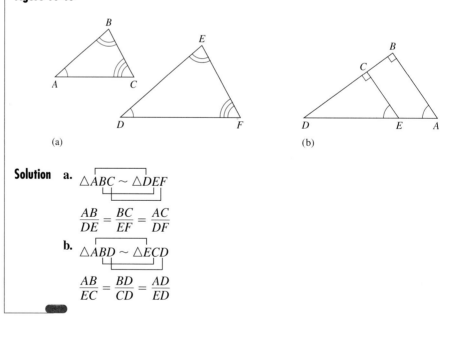

Solution a. $\triangle ABC \sim \triangle DEF$

$$\frac{AB}{DE} = \frac{BC}{EF} = \frac{AC}{DF}$$

b. $\triangle ABD \sim \triangle ECD$

$$\frac{AB}{EC} = \frac{BD}{CD} = \frac{AD}{ED}$$

Angle, Angle Property (AA)

As with congruent triangles, minimal conditions may be used to determine when two triangles are similar. For example, suppose two triangles each have angles with measures of 50°, 30°, and 100°, but the side opposite the 100° angle is 5 units long in one of the triangles and 1 unit long in the other.

SECTION 10-4 *Similar Triangles and Similar Figures* 553

Angle, Angle, Angle (AAA)

Angle, Angle (AA)

The triangles appear to have the same shape, as shown in Figure 10-41. The figure suggests that if the angles of the two triangles are congruent, then the sides are proportional and the triangles are similar. This statement is true in general; it is the **Angle, Angle, Angle** property of similarity for triangles, abbreviated **AAA**. But does it give a minimal set of conditions for similarity of triangles? We know that given the measures of any two angles of a triangle, the measure of the third angle can be found. Hence, if two angles in one triangle are congruent to two angles in another triangle, then the third angles must also be congruent. Consequently, we have the **Angle, Angle (AA)** property.

Figure 10-41

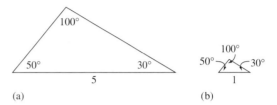

(a)   (b)

Property of Similar Triangles: Angle, Angle (AA)

Angle, Angle (AA) If two angles of one triangle are congruent, respectively, to two angles of a second triangle, then the triangles are similar.

Example 10-9 For each part of Figure 10-42, find a pair of similar triangles.

Figure 10-42

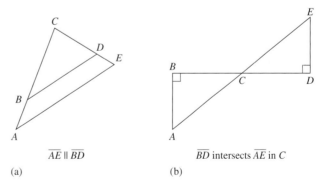

$\overline{AE} \parallel \overline{BD}$ $\overline{BD}$ intersects $\overline{AE}$ in C
(a)   (b)

Solution a. Because $\overline{AE} \parallel \overline{BD}$, congruent corresponding angles are formed by a transversal cutting the parallel segments. Thus $\angle CBD \cong \angle CAE$ and $\angle CDB \cong \angle CEA$. Also, $\angle C \cong \angle C$, so $\triangle CBD \sim \triangle CAE$ by AA.

b. $\angle B \cong \angle D$ because both are right triangles. Also, $\angle ACB \cong \angle ECD$ because they are vertical angles. Thus $\triangle ACB \sim \triangle ECD$ by AA.

NOW TRY THIS 10-9

● Congruency of corresponding angles is sufficient to prove that two triangles are similar. Is the same condition sufficient to prove other polygons are similar to each other? Explain your answer. ●

Example 10-10 In Figure 10-43 find x.

Figure 10-43

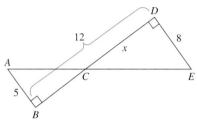

$\triangle ABC \sim \triangle EDC$

Solution $\triangle ABC \sim \triangle EDC$, so

$$\frac{AB}{ED} = \frac{AC}{EC} = \frac{BC}{DC}.$$

Now, $AB = 5$, $ED = 8$, and $CD = x$, so $BC = 12 - x$. Thus

$$\frac{5}{8} = \frac{12 - x}{x}$$

$$5x = 8(12 - x)$$

$$5x = 96 - 8x$$

$$13x = 96$$

$$x = \frac{96}{13}.$$

Problem Solving   **Mysterious Triangles**

For her art project, Rosa needed a wooden triangle that had one 18-in. side. The angles of the triangle needed to have measures 36°, 84°, and 60°. Rosa had a carpenter make such a triangle. A few months later, she needed another triangle with the same specifications. The first carpenter was unavailable, so she contacted a different one. When the second triangle arrived, she was surprised that it was not congruent to the first one. Explain how this is possible.

• **Understanding the Problem**   We are to explain how two triangles with one 18-in. side and angles of measures 36°, 84°, and 60° could be different sizes. To explain how this is possible, we construct two noncongruent triangles with Rosa's specifications.

• **Devising a Plan**   First, we need to construct a triangle with the given side and angles. Then, because the second triangle has the same angles, it must be similar to the first by AAA. Consequently, we try to construct a triangle similar to the first but not congruent to it.

• **Carrying Out the Plan** In Figure 10-44, $\overline{AB}$ represents the 18-in. side. We then use a protractor to construct $\angle A$ measuring 60° and $\angle B$ measuring 36°. Hence, $m(\angle C) = 84°$ (why?). Next, we are to construct a noncongruent triangle similar to $\triangle ABC$ with one side as long as $\overline{AB}$. Because this may seem difficult, we consider *a simpler but related problem*. We drop the condition that the triangle must have a side of length AB and consider constructing a triangle similar to $\triangle ABC$. This can be conveniently achieved by using one of the existing vertices of $\triangle ABC$ and drawing a side parallel to the opposite side. For example, in Figure 10-44 $\overline{C_1B_1} \parallel \overline{CB}$ and $\triangle ACB \sim \triangle AC_1B_1$ (why?). There are infinitely many such triangles AC_1B_1. We imagine sliding $\overrightarrow{BC}$ along $\overrightarrow{AB}$ and $\overrightarrow{AC}$ so that $\overline{B_1C_1} \parallel \overline{BC}$ and until $AC_1 = AB$. Consequently, $\triangle ACB$ and $\triangle AC_1B_1$ are two noncongruent similar triangles such that $AC_1 = AB$ and therefore there are two noncongruent triangles with Rosa's specifications.

Figure 10-44

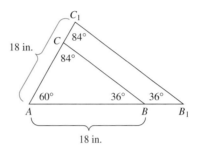

• **Looking Back** If Rosa gave each carpenter the measurements of two sides of the desired triangles and all the angles, could she then be sure to obtain a unique triangle? If she told the carpenter where the 18-in. side should be (for example, between the 36° and 60° angles), would that ensure her new triangle would be congruent to the old one?

Properties of Proportion

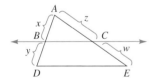

Figure 10-45

Similar triangles give rise to various properties that involve proportions. For example, in Figure 10-45 if $\overline{BC} \parallel \overline{DE}$, then $\dfrac{AB}{BD} = \dfrac{AC}{CE}$. This can be justified as follows: $\overline{BC} \parallel \overline{DE}$, so $\triangle ADE \sim \triangle ABC$ (why?). Consequently, $\dfrac{AD}{AB} = \dfrac{AE}{AC}$, which may be written as follows:

$$\frac{x+y}{x} = \frac{z+w}{z}$$

$$\frac{x}{x} + \frac{y}{x} = \frac{z}{z} + \frac{w}{z}$$

$$1 + \frac{y}{x} = 1 + \frac{w}{z}$$

$$\frac{y}{x} = \frac{w}{z}$$

$$\frac{x}{y} = \frac{z}{w}$$

This result is summarized in the following theorem.

Theorem 10-4

If a line parallel to one side of a triangle intersects the other sides, then it divides those sides into proportional segments.

The converse of Theorem 10-4 is also true; that is, if in Figure 10-45 we know that $\frac{AB}{BD} = \frac{AC}{CE}$, then we can conclude that $\overline{BC} \parallel \overline{DE}$. We summarize this result in the following theorem.

Theorem 10-5

If a line divides two sides of a triangle into proportional segments, then the line is parallel to the third side.

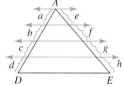

Figure 10-46

Similarly, if lines parallel to $\overline{DE}$ intersect $\triangle ADE$, as shown in Figure 10-46, so that $a = b = c = d$, it can be shown that $e = f = g = h$. This result is stated in the following theorem.

Theorem 10-6

If parallel lines cut off congruent segments on one transversal, then they cut off congruent segments on any transversal.

NOW TRY THIS 10-10

● In Figure 10-47, a graph shows similar triangles used to depict fuel economy standards for automobiles. Explain whether you believe the sides of the triangles are depicted in a way that misleads the reader. ●

Figure 10-47

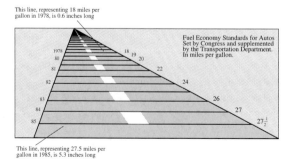

Theorem 10-6 can be used to divide a given segment into any number of congruent parts. For example, using only a compass and a straightedge, we can divide segment $\overline{AB}$ in Figure 10-48 into three congruent parts by making the construction resemble Figure 10-46.

Figure 10-48 Separating a segment into congruent parts

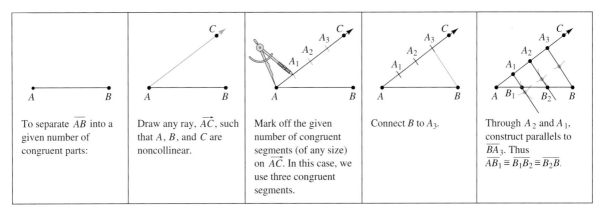

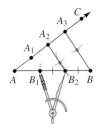

Figure 10-49

REMARK It is necessary to construct only $\overline{A_2B_2}$. We can then use a compass to mark off point B_1, such that $B_1B_2 = BB_2$, as in Figure 10-49.

Midsegments of Triangles and Quadrilaterals

The segment connecting the midpoints of two sides of a triangle or two adjacent sides of a quadrilateral is a **midsegment.** In Figure 10-50, M and N are midpoints of $\overline{AB}$ and $\overline{BC}$, respectively, and $\overline{MN}$ is a midsegment. Because $\dfrac{MB}{MA} = \dfrac{BN}{CN} = 1$ by Theorem 10-5, $\overline{MN} \parallel \overline{AC}$.

How is MN related to AC? Notice that $\triangle MBN \sim \triangle ABC$ (why?). Therefore

$$\frac{MN}{AC} = \frac{MB}{AB} = \frac{MB}{2MB} = \frac{1}{2}$$

Consequently we have the following theorem.

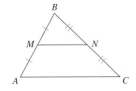

Figure 10-50

midsegment

Theorem 10-7

The Midsegment Theorem The segment connecting the midpoints of two sides of a triangle is parallel to the third side and half as long.

Using Theorem 10-7 and Theorem 10-6 the following theorem follows (why?).

Theorem 10-8

If a line bisects one side of a triangle and is parallel to a second side, then it bisects the third side and therefore is a midsegment.

Example 10-11

a. In the quadrilateral ABCD in Figure 10-51, M, N, P, and Q are the midpoints of the sides. What kind of quadrilateral is MNPQ?

b. What kind of quadrilateral is MNPQ if the diagonals of ABCD are congruent?

Figure 10-51

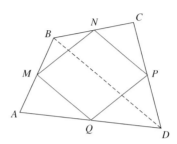

Solution

a. Careful drawings of several quadrilaterals and the corresponding midpoints of their sides suggest that MNPQ is a parallelogram. To prove this conjecture notice that $\overline{NP}$ is a midsegment in $\triangle BCD$ and consequently $\overline{NP} \parallel \overline{BD}$ and $NP = \frac{1}{2}BD$. Similarly in $\triangle ABD$, $\overline{MQ}$ is a midsegment and therefore $\overline{MQ} \parallel \overline{BD}$ Consequently $\overline{MQ} \parallel \overline{NP}$. In a similar way we could show that $\overline{MN} \parallel \overline{QP}$ (consider midsegments in $\triangle ABC$ and in $\triangle ADC$) and therefore by the definition of a parallelogram, MNPQ is a parallelogram.

b. The Midsegment Theorem also tells us that $NP = \frac{1}{2}BD$ and $MN = \frac{1}{2}AC$. Because the diagonals are congruent $MN = NP$ and hence the parallelogram MNPQ is a rhombus.

Indirect Measurements

Similar triangles have long been used to make indirect measurements. Thales of Miletus (ca. 600 B.C.) is believed to have determined the height of the Great Pyramid of Egypt by using ratios involving shadows, similar to those pictured in Figure 10-52. The sun is so far away that it should make approximately congruent angles at B and B'. Because the angles at C and C' are right angles, $\triangle ABC \sim \triangle A'B'C'$. Hence

$$\frac{AC}{A'C'} = \frac{BC}{B'C'}$$

Figure 10-52

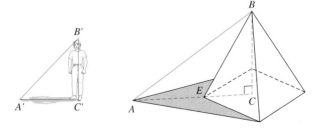

And because $AC = AE + EC$, the following proportion is obtained:

$$\frac{AE + EC}{A'C'} = \frac{BC}{B'C'}$$

The person's height and shadow can be measured. Also, the length AE of the shadow of the pyramid can be measured, and EC can be found because the base of the pyramid is a square. Each term of the proportion except the height of the pyramid is known. Thus the height BC of the pyramid can be found by solving the proportion.

Example 10-12

On a sunny day, a tall tree casts a 40-m shadow. At the same time, a meter stick held vertically casts a 2.5-m shadow. How tall is the tree?

Solution In Figure 10-53 the triangles are similar by AA because the tree and the stick both meet the ground at right angles and the angles formed by the sun's rays are congruent (because the shadows are measured at the same time).

$$\frac{x}{40} = \frac{1}{2.5}$$
$$2.5x = 40$$
$$x = 16$$

The tree is 16 m tall.

Figure 10-53

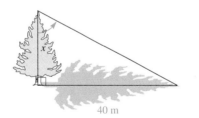

BRAIN TEASER

Two neighbors, Smith and Wheeler, plan to erect flagpoles in their yards. Smith wants a 10-ft pole, and Wheeler wants a 15-ft pole. To keep the poles straight while the concrete bases harden, guy wires are to be tied from the tops of the flagpoles to a fence post on the property lines and to the bases of the flagpoles, as shown in Figure 10-54. How high should the fence post be and how far apart should they erect flagpoles for this scheme to work?

Figure 10-54

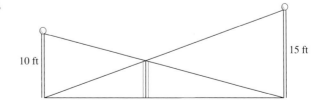

LABORATORY ACTIVITY

The device pictured in Figure 10-55 is a pantograph. It is used to draw enlarged versions of figures. Well-made adjustable pantographs are available from drafting or art supply stores, but you can make a crude one from wooden lath or other material.

In Figure 10-55 the red dots represent either brads or nuts and bolts. The strips are made of lath or cardboard and are rigid. A pointer at *D* is used to trace along an original figure, which causes the pencil at *F* to draw an enlarged version of the figure. Make or obtain a pantograph and experiment with enlarging figures. Explain how and why this works.

Figure 10-55

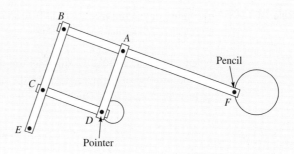

ONGOING ASSESSMENT 10-4

1. Which of the following are always similar? Why?
   a. Any two equilateral triangles
   b. Any two squares
   c. Any two rectangles
   d. Any two rhombuses
   e. Any two circles
   f. Any two regular polygons
   g. Any two regular polygons with the same number of sides
2. Use grid paper to draw figures that have sides three times as large as the ones in the following figure and that are colored similarly.

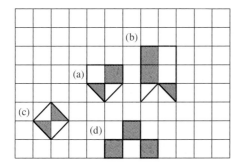

3. a. Construct a triangle with sides of lengths 4 cm, 6 cm, and 8 cm.
   b. Construct another triangle with sides of lengths 2 cm, 3 cm, and 4 cm.
   c. Make a conjecture about the similarity of triangles that have proportional sides only.
4. a. Construct a triangle with sides of lengths 4 cm and 6 cm and an included angle measuring 60°.
   b. Construct a triangle with sides of lengths 2 cm and 3 cm and an included angle measuring 60°.
   c. Make a conjecture about the similarity of triangles that have two sides proportional and congruent included angles.
5. a. Sketch two nonsimilar polygons for which corresponding angles are congruent.
   b. Sketch two nonsimilar polygons for which corresponding sides are proportional.
6. Examine several examples of similar polygons and make a conjecture concerning the ratio of their perimeters.

7. **a.** Which of the following pairs of triangles are similar? If they are similar, explain why.

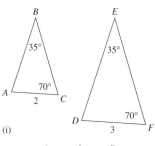
(i)

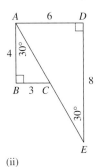
(ii)

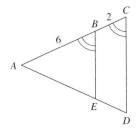
(iii)

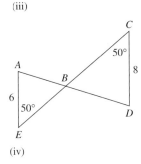
(iv)

b. For each pair of similar triangles, find the scale factor of the sides of the triangles.

8. Assume that in the following figures the triangles in each part are similar and find the measures of the unknown sides.

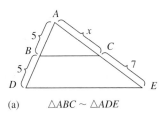
(a)　　△ABC ~ △ADE

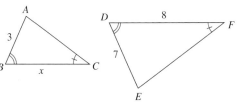
(b)　　△ABC ~ △EDF

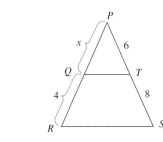
(c)　△PQT ~ △PRS

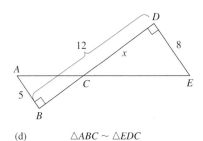
(d)　　△ABC ~ △EDC

9. Given the following figure, use a compass and a straightedge to separate $\overline{AB}$ into five congruent pieces.

10. In the following right triangle ABC, $\overline{CD} \perp \overline{AB}$:

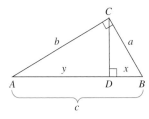

a. Find three pairs of similar triangles. Justify your answers.
b. Write the corresponding proportions for each set of similar triangles.
c. Use part (b) and the above figure to show that $a^2 = xc$. Also argue that $b^2 = yc$.
d. Use part (c) to show that $a^2 + b^2 = c^2$. State this result in words using legs and hypotenuse of a right triangle.

11. In the cartoon above, if a smaller map were obtained and its scale were half the size of the original scale, would the distance to be traveled be different? Why?

12. In the following figure, find the distance *AB* across the pond using the similar triangles shown:

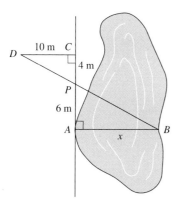

13. To find the height of a tree, a group of Girl Scouts devised the following method. A girl walks toward the tree along its shadow until the shadow of the top of her head coincides with the shadow of the top of the tree. If the girl is 150 cm tall, her distance to the foot of the tree is 15 m, and the length of her shadow is 3 m, how tall is the tree?

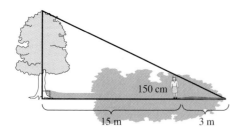

14. The angle bisector of one of the angles in an isosceles triangle is constructed. This angle bisector partitions the original triangle into two isosceles triangles.
    a. What are the angle measures of the original triangle? (There are two possibilities.)
    b. Which, if any, triangles are congruent? similar? Explain your answers.

15. Samantha wants to know how far above the ground the top of a leaning flagpole is. At high noon, when the sun is directly overhead, the shadow cast by the pole is 7 ft long. Samantha holds a plumb bob with a string 3 ft long up to the flagpole and determines that the point of the plumb bob touches the ground 13 in. from the base of the flagpole. How far above the ground is the top of the pole?

16. Dian has prepared a report on the gorilla, an endangered species, for her Earth Week information booth. For her backdrop, she wants to project a life-sized image of a gorilla on a screen. She has a slide showing a gorilla standing erect. In the slide, the image of the gorilla is $\frac{3}{4}$ in. tall. Dian's research shows that adult gorillas often reach 6 ft in height. If the bulb in the slide projector is 3 in. from the slide, where should the projector be placed so that the gorilla appears life-size on the screen?

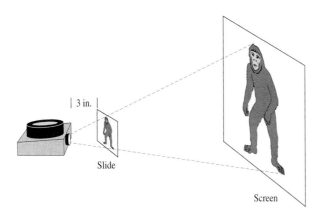

17. **a.** In the accompanying figure, $ABCD$ is a trapezoid. M is the midpoint of $\overline{AB}$. Through M, a line parallel to the bases has been drawn, intersecting $\overline{CD}$ at N. (i) Explain why N must be the midpoint of $\overline{CD}$ and (ii) express MN in terms of a and b, the lengths of the parallel sides of the trapezoid.

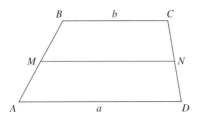

b. Denote MN by c. Use your answer to part (a) to show that b, c, a form an arithmetic sequence.

★**c.** In the trapezoid $ABCD$, the lengths of the bases are a and b as shown. Side $\overline{AB}$ has been divided into 9 congruent segments. Through the endpoints of the segments, lines parallel to $\overline{AD}$ have been drawn. In this way, 8 new segments connecting the sides $\overline{AB}$ and $\overline{CD}$ have been created. Show that the sequence of 10 terms, starting with b, proceeding with the lengths of the parallel segments, and ending with a, is an arithmetic sequence and find the sum of the sequence in terms of a and b.

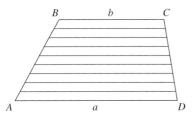

18. The midpoints M, N, P, Q of the sides of a quadrilateral $ABCD$ have been connected and an interior quadrilateral is obtained. We have shown that $MNPQ$ is a parallelogram. What is the most you can say about the kind of parallelogram $MNPQ$ is if $ABCD$ is

    a. a rhombus?
    b. a kite?
    c. an isosceles trapezoid?
    d. a quadrilateral that is neither a rhombus nor a kite but whose diagonals are perpendicular to each other?

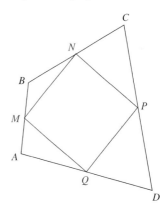

19. Explain why all pairs of regular octagons are similar.
20. If the ratio of corresponding sides of two similar triangles is $1/k$, what is the ratio of their perimeters? Why?
21. If you took cross-sections of a typical ice cream cone parallel to the circular opening where the ice cream is usually placed, explain whether the cross-sections would be similar.
22. Must all cross-sections of a circular cylinder be similar? Why? Draw a sketch to illustrate your answer.
23. A toy maker wants to cut the plastic rectangle $EFGH$ in the figure into 4 right triangles and a rectangle. Given the measurements shown, determine the length of $\overline{CE}$.

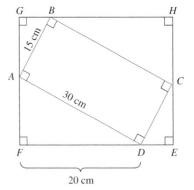

24. **a.** Write a procedure called RECTANGLE that draws a rectangle of variable size with inputs :LEN and :WID. Then write a procedure called SIM.RECT that draws a rectangle whose sides are twice as long as those of the rectangle drawn by RECTANGLE when the same inputs are used for :LEN and :WID.

    b. Write a procedure called SIM.RECTANGLE that draws a rectangle similar to the one drawn by RECTANGLE, with a scale factor called :SCALE that affects the size of the rectangle.

c. Write a procedure called `PARALLELOGRAM` that draws a parallelogram of variable size with inputs `:LEN`, `:WID`, and `:ANGLE`. Then write a procedure called `SIM.PAR` that generates similar parallelograms.

Communication

25. Do you think any two cubes are similar? Why or why not?
26. Architects frequently use scale drawings to construct models of projects. Are the models similar to the finished products? Why or why not?
27. Assuming the lines on an ordinary piece of notebook paper are parallel and equidistant, describe a method for using the paper to divide a piece of licorice evenly among 2, 3, 4, or 5 children. Explain why it works.

Open-Ended

28. Build two similar towers out of blocks.
    a. What is the ratio of the heights of the towers?
    b. What is the ratio of the perimeters of the bases of the towers?

Cooperative Learning

29. A building was to be built on a triangular piece of property. The architect was given the approximate measurements of the angles of the triangular lot as 54°, 39°, and 87° and the lengths of two of the sides as 100 m and 80 m. When the architect began the design on drafting paper, she drew a triangle to scale with the corresponding measures and found that the lot was considerably smaller than she had been led to believe. It appeared that the proposed building would not fit. The surveyor was called. He confirmed each of the measurements and could not see a problem with the size. Neither the architect nor surveyor could understand the reason for the other's opinion.
    a. Have one person in your group play the part of the architect and explain why she felt she was correct.
    b. Have one person in the group explain the reason for the miscommunication.
    c. Have the group suggest a way to provide an accurate description of the lot.

Review Problems

30. If a person holds a mirror at arm's length and looks into it, is the image seen congruent to the original? Why or why not?
31. Given the following base of an isosceles triangle and the altitude to that base, construct the triangle:

    _____
    Base

    _____
    Altitude

32. Given the following length of a side of the triangle, construct an altitude of an equilateral triangle.

    _____

33. Write a paragraph describing how you could construct an isosceles right triangle when given the length of the hypotenuse of the 45°-45°-90° triangle.
34. Use a compass and a straightedge to draw a pair of obtuse vertical angles and the angle bisector of one of these angles. Extend the angle bisector. Does the extended angle bisector bisect the other vertical angle? Justify your answer.

TECHNOLOGY CORNER

In 1975 Benoit Mandelbrot invented the word *fractal* to describe certain irregular and fragmented shapes. These shapes are such that if one looks at a small part of the shape, the small part resembles the larger shape. These shapes are somewhat like rep-tiles, although the smaller structures of fractals are not necessarily identical to the larger structure. (Recall that with rep-tiles, the smaller structures are similar and are identical in all aspects except size.) Fractal geometry was introduced for the purpose of modeling natural phenomena such as irregular coastlines, arteries and veins, the branching structure of plants, the thermal agitation of molecules in a fluid, and sponges. Figure 10-56 shows an example of a fractal—a computer-generated picture of the Mandelbrot set known as the "Tail of the Seahorse."

Figure 10-56

Earlier, in 1906, Helge von Koch came up with a curve that has infinite perimeter. To visualize this curve, we construct in Figure 10-57 a sequence of polygons $S_1, S_2, S_3, \ldots$ as follows:

a. S_1 is an equilateral triangle.
b. S_2 is obtained from S_1 by dividing each side of the triangle into 3 congruent parts and constructing on the middle part an equilateral triangle with the base removed.
c. S_3 is obtained from S_2 like S_2 was obtained from S_1.

We continue in a similar way to obtain the other polygons of the sequence in Figure 10-57(d) and (e). These polygons come closer and closer to a curve, called the *snowflake curve,* which is another example of a fractal. Edit the following SNOWFLAKE to fit your Logo software, type it into your computer, and display the polygons shown in Figure 10-57.

If S_1 has perimeter 3 units, what is the perimeter of S_2 and S_3? What do you think happens to the perimeter of the snowflakes as the number of sides of the polygons increases?

Figure 10-57

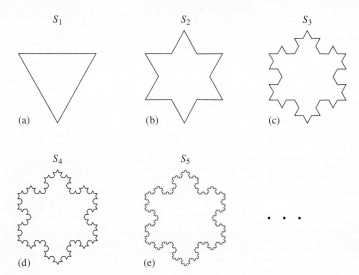

```
TO SNOWFLAKE :LEVEL
  CS PU SETXY 90 52
  MAKE "SIZE 180
  MAKE "N 5 — :LEVEL
  MAKE "LEVEL 1
  REPEAT :N[MAKE "LEVEL :LEVEL *3]
  REPEAT 3 [DRAW.IT :SIZE :LEVEL RT 120]
END

TO DRAW.IT :SIZE :LEVEL
  IF :SIZE < :LEVEL THEN FD :SIZE STOP
  DRAW.IT :SIZE/3 :LEVEL LT 60
  DRAW.IT :SIZE/3 :LEVEL RT 120
  DRAW.IT :SIZE/3 :LEVEL LT 60
  DRAW.IT :SIZE/3 :LEVEL
END
```

BRAIN TEASER A particular kaleidoscope is a right prism with an equilateral triangle as a base. A beam of light is reflected at a 60° angle from a point *P* on a side of the triangular base, as shown in Figure 10-58. The beam is reflected in the plane of the base to the different mirrored surfaces and continues bouncing off at 60° angles. Find the length of the path of the reflected light when it reaches the point at which it originated.

Figure 10-58

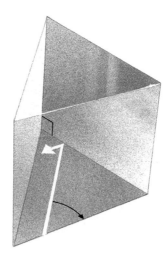

Section 10-5 — Lines in a Cartesian Coordinate System

The Cartesian coordinate system (named for René Descartes) enables us to study geometry using algebra and to interpret algebraic phenomena geometrically. A Cartesian coordinate system is constructed by placing two number lines perpendicular to each other, as shown in Figure 10-59. The intersection point of the two lines is the **origin,** the horizontal line is the **x-axis,** and the vertical line is the **y-axis.** The location of any point P can be described by an ordered pair of numbers.

If a perpendicular from P to the x-axis intersects at a point with coordinate a and a perpendicular from P to the y-axis intersects at a point with coordinate b, point P has coordinates (a, b). The first component in the ordered pair (a, b) is the **abscissa,** or **x-coordinate,** of P. The second component is the **ordinate,** or **y-coordinate,** of P. To each point in the plane, there corresponds an ordered pair (a, b) and vice versa. Hence, there is a one-to-one correspondence between all the points in the plane and all the ordered pairs of real numbers.

Figure 10-59

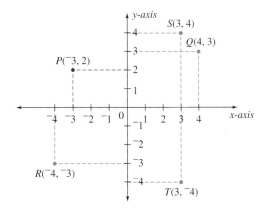

For example, in Figure 10-59 the x-coordinate of P is $^-3$ and the y-coordinate of P is 2, so P has coordinate $(^-3, 2)$. Similarly, R has coordinates $(^-4, ^-3)$, written as $R(^-4, ^-3)$.

Equations of Vertical and Horizontal Lines

Every point on the x-axis has a y-coordinate of zero. Thus the x-axis can be described as the set of all points (x, y) such that $y = 0$. This set of points on the x-axis has equation $y = 0$. Similarly, the y-axis can be described as the set of all points (x, y) such that $x = 0$ and y is an arbitrary real number. Thus $x = 0$ is the equation of the y-axis. If we plot the set of all points that satisfy a given condition, the resulting picture on the Cartesian coordinate system is called the **graph** of the set.

Example 10-13 Sketch the graph for each of the following:

a. $x = 2$ **b.** $y = 3$ **c.** $x < 2$ and $y = 3$

Solution **a.** The equation $x = 2$ represents the set of all points (x, y) for which $x = 2$ and y is any real number. This set is the line perpendicular to the x-axis at $(2, 0)$, as in Figure 10-60(a).
b. The equation $y = 3$ represents the set of all points (x, y) for which $y = 3$ and x is any real number. This set is the line perpendicular to the y-axis at $(0, 3)$, as in Figure 10-60(b).

Figure 10-60

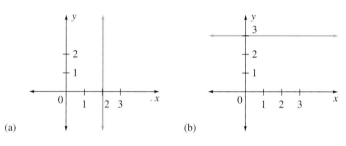

c. Together, the statements represent the set of all points (x, y) for which $x < 2$, but y is always 3. The set describes part of a line, as shown in Figure 10-61. Note that the hollow dot at $(2, 3)$ indicates that this point is not included in the solution.

Figure 10-61

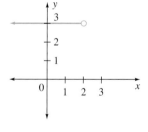

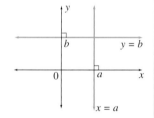

Figure 10-62

Table 10-2

Number of Term	Term
1	4
2	7
3	10
4	13
.	.
.	.
.	.
x	$3x + 1$

In Example 10-13 we found the graphs of the equations $x = 2$ and $y = 3$. In general, the graph of the equation $x = a$, where a is some real number, is a line perpendicular to the x-axis through the point with coordinates $(a, 0)$, as shown in Figure 10-62. Similarly, the graph of the equation $y = b$ is a line perpendicular to the y-axis through the point with coordinates $(0, b)$.

Equations of Lines

Consider the arithmetic sequence 4, 7, 10, 13, ... in Table 10-2 whose xth term is $3x + 1$. If the number of the term is the x-coordinate and the corresponding term the y-coordinate, the set of points appear to lie on a line that is parallel to neither the x-axis nor the y-axis, as in Figure 10-63.

Figure 10-63

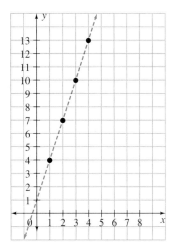

Do all points corresponding to arithmetic sequences lie along lines? To help answer this question, we consider the following sequences in Table 10-3.

Table 10-3

Number of Term x	$1 \cdot x$	$2 \cdot x$	$\frac{1}{2} \cdot x$	$(^-1) \cdot x$	$(^-2) \cdot x$
1	1	2	$\frac{1}{2}$	$^-1$	$^-2$
2	2	4	1	$^-2$	$^-4$
3	3	6	$\frac{3}{2}$	$^-3$	$^-6$
4	4	8	2	$^-4$	$^-8$
5	5	10	$\frac{5}{2}$	$^-5$	$^-10$
6	6	12	3	$^-6$	$^-12$
.	.	.			
.	.	.			
.	.	.			
x	x	$2x$	$\frac{1}{2}x$	^-x	^-2x

In Figure 10-64, the sets of ordered pairs $(x, y,)$ are plotted on a graph so that the number of the term is the *x*-coordinate and the corresponding term appearing in a particular column in Table 10-3, the *y*-coordinate. The sets of points appear to determine straight lines.

All five lines in Figure 10-64 have equations of the form $y = mx$, where *m* takes the values 2, 1, $\frac{1}{2}$, $^-1$, and $^-2$. If all points along a given dashed line are connected, then all the points on that line satisfy the corresponding equation. The number *m* is a measure of

slope steepness and is called the **slope** of the line whose equation is $y = mx$. The graph goes up from left to right (increases) if m is positive, and it goes down from left to right (decreases) if m is negative.

Figure 10-64

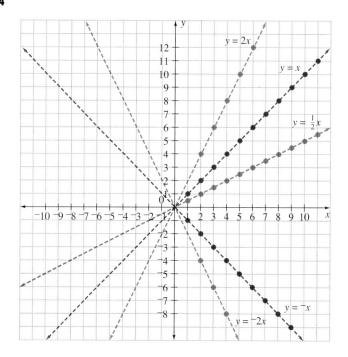

NOW TRY THIS 10-11

- In the equation $y = mx$, if m is 0, what happens to the line? What happens as m continues to increase? What happens when m continues to decrease?

All lines in Figure 10-64 pass through the origin. This is true for any line whose equation is $y = mx$. If $x = 0$, then $y = m \cdot 0 = 0$ and $(0, 0)$ is a point on the graph of $y = mx$. Conversely, it is possible to show that any nonvertical line passing through the origin has an equation of the form $y = mx$ for some value of m.

Example 10-14 Find the equation of the line that contains $(0, 0)$ and $(2, 3)$.

Solution The line goes through the origin; therefore its equation has the form $y = mx$. To find the equation of the line, we must find the value of m. The line contains $(2, 3)$, so we substitute 2 for x and 3 for y in the equation $y = mx$ to obtain $3 = m \cdot 2$, and thus $m = \frac{3}{2}$.

Hence, the required equation is $y = \frac{3}{2}x$.

Next, we consider equations of the form $y = mx + b$, where b is a real number. To do this, we examine the graphs of $y = x + 2$ and $y = x$. Given the graph of $y = x$, we can obtain the graph of $y = x + 2$ by "raising" each point on the first graph by 2 units. This is because for a certain value of x, the corresponding y value is 2 units greater. This is shown in Figure 10-65(a). Similarly, to sketch the graph of $y = x - 2$, we first draw the graph of $y = x$ and then lower each point vertically by 2 units, as shown in Figure 10-65(b).

Figure 10-65

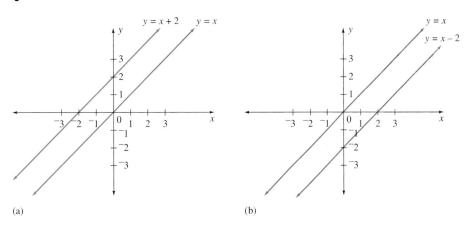

(a)   (b)

The graphs of $y = x + 2$ and $y = x - 2$ are straight lines. Moreover, the lines whose equations are $y = x$, $y = x + 2$, and $y = x - 2$ are parallel. In general, for a given value of m, the graph of $y = mx + b$ is a straight line through $(0, b)$ and parallel to the line whose equation is $y = mx$.

Further, the graph of the line $y = mx + b$, where $b > 0$, can be obtained from the graph of $y = mx$ by sliding $y = mx$ up b units, as shown in Figure 10-66. If $b < 0$, $y = mx$ must be slid down $|b|$ units. In general, *any two parallel lines have the same slope, or are vertical lines with no slope.*

The graph of $y = mx + b$ in Figure 10-66 crosses the y-axis at point $P(0, b)$. The value of y at the point of intersection of any line with the y-axis is the **y-intercept**. Thus b is the y-intercept of $y = mx + b$, and this form of the equation of a straight line is the **slope-intercept form**. Similarly, the value of x at the point of intersection of a line with the x-axis is the **x-intercept**.

y-intercept

slope-intercept form

x-intercept

Figure 10-66

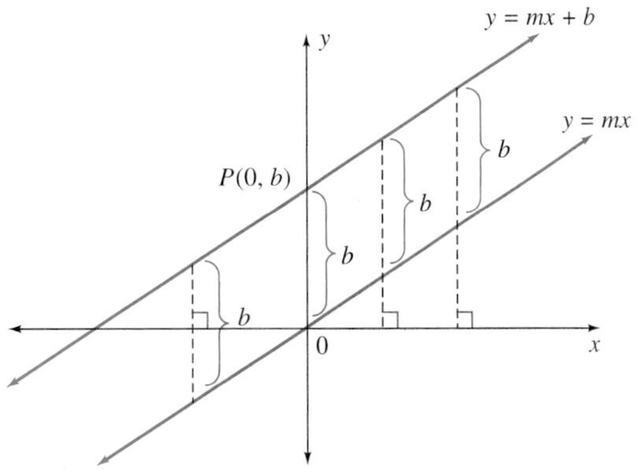

Example 10-15 Given the equation $y - 3x = {}^-6$, do the following:

a. Find the slope of the line.
b. Find the y-intercept.
c. Find the x-intercept.
d. Sketch the graph of the equation.

Solution a. To write the equation in the form $y = mx + b$, we add $3x$ to both sides of the given equation to obtain $y = 3x + ({}^-6)$. Hence, the slope is 3.
b. The form $y = 3x + ({}^-6)$ shows that $b = {}^-6$, which is the y-intercept. (The y-intercept can also be found directly by substituting $x = 0$ in the equation and finding the corresponding value of y.)
c. The x-intercept is the x-coordinate of the point where the graph intersects the x-axis. At that point, $y = 0$. Substituting 0 for y in $y = 3x - 6$ gives 2 as the x-intercept.
d. The y-intercept and the x-intercept are located at $(0, {}^-6)$ and $(2, 0)$, respectively, on the line. We plot these points and draw the line through them to obtain the desired graph in Figure 10-67. Note that any two points of the line can be used to sketch the graph because any two points determine a line.

Figure 10-67

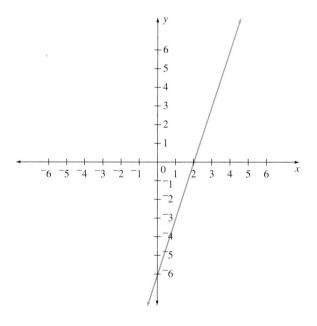

The equation $y = b$ can be written in slope-intercept form as $y = 0 \cdot x + b$. Consequently, its slope is 0 and its y-intercept is b. This should not be surprising. Because the line is parallel to the x-axis, its steepness, or slope, should be 0. Any vertical line has equation $x = a$ for some real number a. This equation cannot be written in slope-intercept form. In general, *every straight line has an equation of either the form $y = mx + b$ or $x = a$.* Any equation that can be put in one of these forms is a **linear equation**.

linear equation

SECTION 10-5 *Lines in a Cartesian Coordinate System* **573**

> **Equation of a Line**
>
> Every line has an equation of either the form $y = mx + b$ or $x = a$, where m is the slope and b is the y-intercept.

Using Similar Triangles to Determine Slope

We have defined the slope of a line with equation $y = mx + b$ to be m. The slope is a measure of steepness of a line. A different way to discuss the steepness of a line is to consider how much the line "rises" in relation to how much it "runs." In Figure 10-68, line k is steeper than line ℓ. In other words, line k rises higher than line ℓ for the same horizontal run.

Figure 10-68

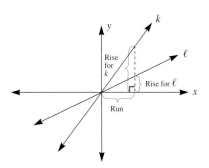

slope We express the steepness, or **slope,** as the ratio $\frac{\text{rise}}{\text{run}}$. In Figure 10-69, right triangles have been constructed and shaded on several lines. In each triangle, the horizontal side is the run and the vertical side is the rise.

Figure 10-69

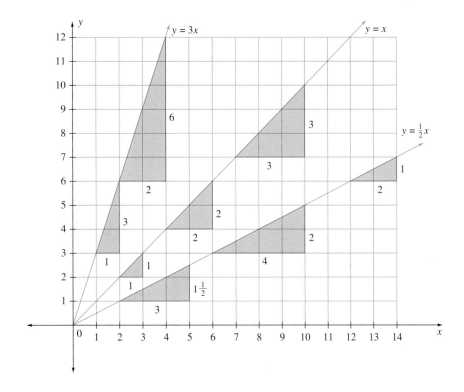

The slope of each line in Figure 10-69 can be calculated as the rise over the run in any of the shaded triangles, with hypotenuse (the side opposite the right angle) along the given line. To test this fact, notice that

$$\text{for } y = \frac{1}{2}x, \quad m = \frac{\text{rise}}{\text{run}} = \frac{1\frac{1}{2}}{3} = \frac{2}{4} = \frac{1}{2}$$

$$\text{for } y = x, \quad m = \frac{\text{rise}}{\text{run}} = \frac{1}{1} = \frac{2}{2} = \frac{3}{3}$$

$$\text{for } y = 3x, \quad m = \frac{3}{1} = \frac{6}{2}.$$

Figure 10-70 illustrates the situation generally when a line is inclined upward from the left to the right.

Figure 10-70

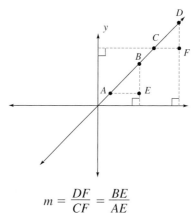

$$m = \frac{DF}{CF} = \frac{BE}{AE}$$

Using the previous notion, we find that the slope of a line $\overleftrightarrow{AB}$ is the change in y-coordinates divided by the corresponding change in x-coordinates of any two points on $\overleftrightarrow{AB}$. The difference $x_2 - x_1$ is the *run*, and the difference $y_2 - y_1$ is the *rise*. Thus the slope is often defined as "rise over run," or $\frac{\text{rise}}{\text{run}}$. The slope formula can be interpreted with coordinates, as shown in Figure 10-71. The ratio $\frac{y_2 - y_1}{x_2 - x_1}$ is always the same, regardless of which two points on a given nonvertical line are chosen.

Figure 10-71

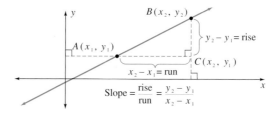

The discussion of slope is summarized in the following formula.

Slope Formula

Given two points $A(x_1, y_1)$ and $B(x_2, y_2)$ with $x_1 \neq x_2$, the slope m of the line $\overleftrightarrow{AB}$ is

$$m = \frac{y_2 - y_1}{x_2 - x_1} = \frac{\text{rise}}{\text{run}}.$$

REMARK By multiplying both the numerator and the denominator on the right side of the slope formula by $^-1$, we obtain

$$m = \frac{y_2 - y_1}{x_2 - x_1} = \frac{(y_2 - y_1)(-1)}{(x_2 - x_1)(-1)} = \frac{y_1 - y_2}{x_1 - x_2}.$$

This shows that while it does not matter which point is named (x_1, y_1) and which is named (x_2, y_2), *the order of the coordinates in the subtraction must be the same.*

NOW TRY THIS 10-12

a. Use the slope formula to find the slope of any horizontal line.
b. What happens when we attempt to use the slope formula for a vertical line? What is your conclusion about the slope of a vertical line?

When a line is inclined downward from the left to the right, the slope is negative. This is illustrated in Figure 10-72, where the graph of the line $y = {}^-2x$ is shown. The slope of line $y = {}^-2x$ can be calculated as $\frac{\text{rise}}{\text{run}} = \frac{{}^-4}{2} = \frac{{}^-2}{1}$.

Figure 10-72

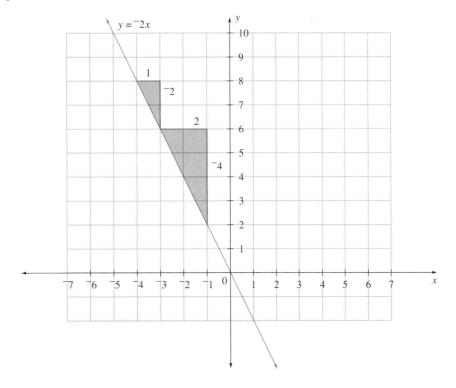

Example 10-16

a. Given $A(3, 1)$ and $B(5, 4)$, find the slope of $\overleftrightarrow{AB}$.
b. Find the slope of the line passing through the points $A(^-3, 4)$ and $B(^-1, 0)$.

Solution
a. $m = \dfrac{4 - 1}{5 - 3} = \dfrac{3}{2}$, or $\dfrac{1 - 4}{3 - 5} = \dfrac{^-3}{^-2} = \dfrac{3}{2}$

b. $m = \dfrac{4 - 0}{^-3 - (^-1)} = \dfrac{4}{^-2} = ^-2$, or $\dfrac{0 - 4}{^-1 - (^-3)} = \dfrac{^-4}{2} = ^-2$

Given two points on a nonvertical line, we can use the slope formula to find the slope of the line and its equation. This is demonstrated in the following example.

Example 10-17

In Figure 10-73, the points $(^-4, 0)$ and $(1, 4)$ are on the line ℓ. Find:

a. the slope of the line
b. the equation of the line

Figure 10-73

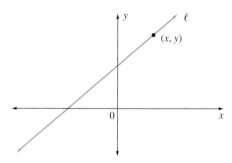

Solution
a. $m = \dfrac{4 - 0}{1 - (^-4)} = \dfrac{4}{5}$.

b. Point (x, y) represents any point on the line different from $(^-4, 0)$ if and only if the slope determined by points $(^-4, 0)$ and (x, y) is $\dfrac{4}{5}$, the slope of the line we found in part (a). Thus,

$$\dfrac{y - 0}{x - (^-4)} = \dfrac{4}{5},$$

$$y = \dfrac{4}{5}(x + 4).$$

The slope intercept from the equation of the line is: $y = \dfrac{4}{5}x + \dfrac{16}{5}$.

Systems of Linear Equations

The mathematical descriptions of many problems involve more than one equation, each having more than one unknown. To solve such problems, we must find a common solution to the equations, if it exists. An example is given on the following student page from *Scott Foresman-Addison Wesley Middle School MATH Course 3*, 1999.

Another way to find the solution of a system of linear equations is to graph both equations and find a point that lies on both lines. Then determine the x and y values for that point.

Example 2

Glen needs to decide whether to purchase hang-gliding equipment. If he rents equipment, he will pay $75 for each flight. If he purchases equipment for $200, he will pay $25 for each flight. For what number of flights would each option cost the same?

Let x = number of flights. Let y = total cost.

$y = 75x$ This is the total cost if he rents equipment.

$y = 25x + 200$ This is the total cost if he buys equipment.

Find two or three solutions for each equation and graph each line. Find the point where the lines intersect.

x	$y = 75x$
0	0
2	150

x	$y = 25x + 200$
0	200
2	250

The lines intersect at (4, 300). Check the coordinates for x and y in both equations.

$y = 75x$ $y = 25x + 200$

$300 \stackrel{?}{=} 75(4)$ $300 \stackrel{?}{=} 25(4) + 200$

$300 = 300$ ✓ $300 = 300$ ✓

If Glen makes 4 flights, the total cost would be the same—$300.

Test Prep

You may be able to find a solution of a system using number sense. If you see that Glen pays $50 more per flight if he rents, you might realize that he pays $200 more for 4 flights.

Check Your Understanding

1. How do you know whether an ordered pair is a solution of a system of two equations?
2. How can you find the solution of a system of equations from a graph?

Example 10-18   May Chin ordered lunch for herself and several friends by phone without checking prices. She paid $7.00 for five soyburgers and four orders of fries, and another time she paid $7.00 for four soyburgers and six orders of fries. Set up a system of equations with two unknowns representing the prices of a soyburger and an order of fries, respectively.

Solution   Let x be the price in dollars of a soyburger and y be the price of an order of fries. Five soyburgers cost $5x$ dollars, and four orders of fries cost $4y$ dollars. Because May paid $7.00 for the order, we have $5x + 4y = 7$. Similarly, $4x + 6y = 7$.

An ordered pair satisfying both equations is a point that belongs to each of the lines. Figure 10-74 shows the graphs of $5x + 2y = 6$ and $x - 4y = {}^-1$. The two lines appear to intersect at $\left(1, \frac{1}{2}\right)$. Thus $\left(1, \frac{1}{2}\right)$ appears to be the solution of the given system of equations. This solution can be checked by substituting 1 for x and $\frac{1}{2}$ for y in each equation. Because two distinct lines intersect in only one point, $\left(1, \frac{1}{2}\right)$ is the only solution to the system.

Figure 10-74

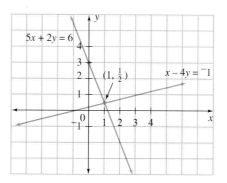

Substitution Method

Drawbacks to estimating graphically a solution to a system of equations include an inability to read noninteger real-number coordinates of points. However, there are algebraic methods for solving systems of linear equations. Consider, for example, the system $y = x + 3$ and $y = 2x - 1$. By substitution, $x + 3 = 2x - 1$, an equation with one unknown. Solving for x gives $3 + 1 = 2x - x$, and, hence, $4 = x$. Substituting 4 for x in either equation gives $y = 7$. Thus $(4, 7)$ is the solution to the given system. As before, this solution can be checked by substituting the obtained values for x and y in the original equation. This method for solving a system of linear equations is the **substitution method.**

substitution method

Example 10-19 Solve the following system:

$$3x - 4y = 5$$
$$2x + 5y = 1$$

Solution First, rewrite each equation, expressing y in terms of x.

$$y = \frac{3x - 5}{4} \quad \text{and} \quad y = \frac{1 - 2x}{5}$$

Then equate the expressions for y and solve the resulting equation for x.

$$\frac{3x - 5}{4} = \frac{1 - 2x}{5}$$
$$5(3x - 5) = 4(1 - 2x)$$
$$15x - 25 = 4 - 8x$$
$$23x = 29$$
$$x = \frac{29}{23}$$

Substituting $\frac{29}{23}$ for x in $y = \frac{3x - 5}{4}$ gives $y = \frac{-7}{23}$. Hence, $x = \frac{29}{23}$ and $y = \frac{-7}{23}$. This can be checked by substituting the values for x and y in the original equations.

REMARK Sometimes it is more convenient to solve a system of equations by expressing x in terms of y in one of the equations and substituting the obtained expression for x in the other equation.

Elimination Method

elimination method The **elimination method** for solving two equations with two unknowns is based on eliminating one of the variables by adding or subtracting the original or equivalent equations. For example, consider the following system:

$$x - y = {}^-3$$
$$x + y = 7$$

By adding the two equations, we can eliminate the variable y. The resulting equation can then be solved for x.

$$x - y = {}^-3$$
$$\underline{x + y = 7}$$
$$2x = 4$$
$$x = 2$$

Substituting 2 for x in the first equation (either equation may be used) gives $y = 5$. Checking this result shows that $x = 2$ and $y = 5$, or $(2, 5)$, is the solution to the system.

Often, another operation is required before equations are added so that an unknown can be eliminated. For example, consider the following system:

$$3x + 2y = 5$$
$$5x - 4y = 3$$

Adding the equations does not eliminate either unknown. However, if the first equation contained $4y$ rather than $2y$, the variable y could be eliminated by adding. To obtain $4y$ in the first equation, we multiply both sides of the equation by 2 to obtain the equivalent equation $6x + 4y = 10$. Adding the equations in the equivalent system gives the following:

$$6x + 4y = 10$$
$$\underline{5x - 4y = 3}$$
$$11x = 13$$
$$x = \frac{13}{11}$$

To find the corresponding value of y, we substitute $\frac{13}{11}$ for x in either of the original equations and solve for y, or we use the elimination method again and solve for y.

An alternative method is to eliminate the *x*-values from the original system by multiplying the first equation by 5 and the second by $^-3$ (or the first by $^-5$ and the second by 3). Then we add the two equations and solve for *y*.

$$15x + 10y = 25$$
$$^-15x + 12y = ^-9$$
$$22y = 16$$
$$y = \frac{16}{22}, \quad \text{or} \quad \frac{8}{11}$$

Consequently, $\left(\frac{13}{11}, \frac{8}{11}\right)$ is the solution of the original system. This solution, as always, should be checked by substitution in the *original* equations.

Solutions to Systems of Linear Equations

All examples thus far have had unique solutions. However, other situations may arise. Geometrically, a system of two linear equations can be characterized as follows:

1. The system has a unique solution if, and only if, the graphs of the equations intersect in a single point.
2. The system has no solution if, and only if, the equations represent parallel lines.
3. The system has infinitely many solutions if, and only if, the equations represent the same line.

NOW TRY THIS 10-13

- Find all the solutions (if any) of each of the following systems by graphing the equations in each system. Then justify your answers by an algebraic approach.

  a. $x - y = 1$
  $2x - y = 5$

  b. $2x - y = 1$
  $2y - 4x = 3$

  c. $2x - 3y = 1$
  $6y - 4x = 2$

Fitting a Line to Data

In many practical situations, a relationship between two variables comes from collected data such as from population or business surveys. When the data is graphed, there may not be a single line that goes through all of the points, but the points may appear to approximate, or "follow," a straight line. In such cases, it is useful to find the equation of what seems to be the **best-fitting line**. Knowing the equation of such a line enables us to predict an outcome without actually performing the experiment.

best-fitting line

REMARK Some data may be scattered in a way that does not approximate a straight line and hence it may be impossible to find the best-fitting line for the data.

There are several approaches to define, and hence find, the best-fitting line. We take a graphical approach as follows:

1. Choose a line that seems to follow the given points so that there are about an equal number of points below the line as above the line.
2. Determine two convenient points on the line and approximate the *x*- and *y*-coordinates of these points.
3. Use the points in (2) to determine the equation of the line.

Example 10-20 A shirt manufacturer noticed that the number of units sold depends on the price charged. The data in Table 10-4 shows the number of units sold for a given price per unit.

a. Find the equation of a line that seems to fit the data best.
b. Use the equation in (a) to predict the number of units that will be sold if the price per unit is $60.

Table 10-4

Price per Unit (In dollars)	Number of Units Sold (Thousands)
50	200
44	250
41	300
33	380
31	400
24.5	450
20	500
14.5	550

Solution **a.** Figure 10-75(a) shows the graph of the data displayed in Table 10-4. Figure 10-75(b) shows a line that seems to fit the data so that approximately the same number of points are below the line as above the line. We choose the points (50, 200) and (20, 500), which are on the line in Figure 10-75(b).

Figure 10-75

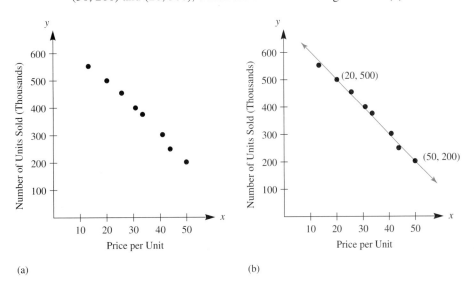

(a)

(b)

To find the equation of the line, we need to find m and b in the equation $y = mx + b$ (alternatively, we could use the point-slope form of the line). Substituting the points (50, 200) and (20, 500) into this equation, we obtain the following:

$$200 = 50m + b$$
$$500 = 20m + b$$

One way to solve the equations is to express b in terms of m for each equation:

$$b = 200 - 50m$$
$$b = 500 - 20m$$

We equate the expressions for b and solve for m:

$$200 - 50m = 500 - 20m$$
$$200 - 500 = 50m - 20m$$
$$^{-}300 = 30m$$
$$m = {}^{-}10$$

Substituting this value for m, we obtain

$$b = 200 - 50(^{-}10) = 700.$$

Consequently, the equation of the fitted line is $y = {}^{-}10x + 700$.

b. Using the equation in (a), substitute $x = 60$ to obtain $y = {}^{-}10(60) + 700$, or $y = 100$. Thus we predict that 100,000 units will be sold if the price per unit is \$60.

ONGOING ASSESSMENT 10-5

1. The graph of $y = mx$ is given in the following figure. Sketch the graphs for each of the following on the same figure. Explain your answers.
   a. $y = mx + 3$ **b.** $y = mx - 3$

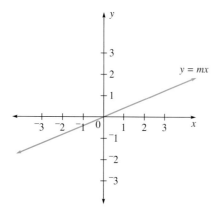

2. Sketch the graphs for each of the following equations:
   a. $y = \dfrac{-3}{4}x + 3$
   b. $y = {}^{-}3$
   c. $y = 15x - 30$
   d. $x = {}^{-}2$
   e. $y = 3x - 1$
   f. $y = \dfrac{1}{20}x$

3. Find the x-intercept and y-intercept for the equations in problem 2, if they exist.

4. In the following figure, part (a) shows a dual scale thermometer and part (b) shows the corresponding points plotted on a graph.

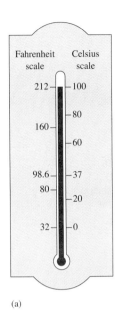

(a)

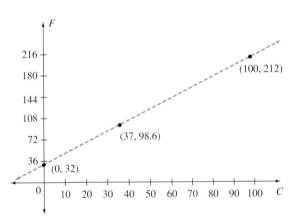

(b)

a. Use two of the points on the graph to develop a formula for conversion from degrees Celsius (C) to degrees Fahrenheit (F).
b. Use your answer in (a) to find a formula for converting from degrees F to degrees C.

5. Write each of the following equations in slope-intercept form and identify the slope and y-intercept:
   a. $3y - x = 0$
   b. $x + y = 3$
   c. $\dfrac{x}{3} + \dfrac{y}{4} = 1$
   d. $3x - 4y + 7 = 0$
   e. $x = 3y$
   f. $x - y = 4(x - y)$

6. For each of the following, write the equation of the line determined by the given pair of points in slope-intercept form or in the form $x = a$:
   a. $(^-4, 3)$ and $(1, ^-2)$
   b. $(0, 0)$ and $(2, 1)$
   c. $(0, 1)$ and $(2, 1)$
   d. $(2, 1)$ and $(2, ^-1)$
   e. $\left(0, \dfrac{-1}{2}\right)$ and $\left(\dfrac{1}{2}, 0\right)$
   f. $(^-a, 0)$ and $(a, 0)$, $a \neq 0$

7. Find the coordinates of two other points collinear (on the same line) with each of the following pairs of given points:
   a. $P(2, 2), Q(4, 2)$
   b. $P(^-1, 0), Q(^-1, 2)$
   c. $P(0, 0), Q(0, 1)$
   d. $P(0, 0), Q(1, 1)$

8. For each of the following, give as much information as possible about x and y:
   a. The ordered pairs $(^-2, 0)$, $(^-2, 1)$, and (x, y) represent collinear points.
   b. The ordered pairs $(^-2, 1)$, $(0, 1)$, and (x, y) represent collinear points.
   c. The ordered pair (x, y) is in the fourth quadrant.

9. Consider the lines through $P(2, 4)$ and perpendicular to the x- and y-axes, respectively. Find the area and the perimeter of the rectangle formed by these lines and the axes.

10. Find the equations for each of the following:
    a. The line containing $P(3, 0)$ and perpendicular to the x-axis
    b. The line containing $P(0, ^-2)$ and parallel to the x-axis
    c. The line containing $P(^-4, 5)$ and parallel to the x-axis
    d. The line containing $P(^-4, 5)$ and parallel to the y-axis

11. For each of the following, find the slope, if it exists, of the line determined by the given pair of points:
    a. $(4, 3)$ and $(^-5, 0)$
    b. $(^-4, 1)$ and $(5, 2)$
    c. $(\sqrt{5}, 2)$ and $(1, 2)$
    d. $(^-3, 81)$ and $(^-3, 198)$
    e. $(1.0001, 12)$ and $(1, 10)$
    f. (a, a) and (b, b)

12. Write the equation of each line in problem 11.

13. Wildlife experts found that the number of chirps a cricket makes in 15-sec intervals is related to the temperature T in degrees Fahrenheit as shown in the following graph:

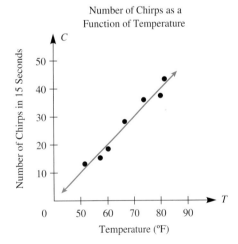

a. If C is the number of chirps in 15 sec, write a formula for C in terms of T (temperature in degrees Fahrenheit) that seems to fit the data best.
b. Use the equation in (a) to predict the number of chirps in 15 sec when the temperature is 90°.
c. If N is the number of chirps per minute, write a formula for N in terms of T.

14. a. Graph the following data and find the equation of the best-fitting line.
   b. Use your answer in (a) to predict the value of y when $x = 10$.

x	y
1	8.1
2	9.9
3	12
4	14.1
5	15.9
6	18
7	19.9

15. Solve each of the following systems, if possible. Indicate whether the system has a unique solution, infinitely many solutions, or no solution.
   a. $y = 3x - 1$
      $y = x + 3$
   b. $2x - 6y = 7$
      $3x - 9y = 10$
   c. $3x + 4y = {}^-17$
      $2x + 3y = {}^-13$
   d. $4x - 6y = 1$
      $6x - 9y = 1.5$

16. The vertices of a triangle are given by (0, 0), (10, 0), and (6, 8). Show that the segments connecting (5, 0) and (6, 8), (10, 0) and (3, 4), and (0, 0) and (8, 4) intersect at a common point.

17. The owner of a 5000-gal oil truck loads the truck with gasoline and kerosene. The profit on each gallon of gasoline is 13¢ and on each gallon of kerosene is 12¢. How many gallons of each fuel did the owner load if the profit was $640?

18. At the end of 10 mo, the balance of an account earning simple interest is $2100.
   a. If, at the end of 18 mo, the balance is $2180, how much money was originally in the account?
   b. What is the rate of interest?

19. Josephine's bank contains 27 coins. If all the coins are either dimes or quarters and the value of the coins is $5.25, how many of each kind of coin are there?

20. a. Solve each of the following systems of equations. What do you notice about the answers?
   i. $x + 2y = 3$
      $4x + 5y = 6$
   ii. $2x + 3y = 4$
       $5x + 6y = 7$
   iii. $31x + 32y = 33$
        $34x + 35y = 36$
   b. Write another system similar to those in (a). What solution did you expect? Check your guess.
   c. Write a general system similar to those in (a). What solution does this system have? Why?

Communication

21. Dahlia wanted to find what temperature has the same measure when measured in degrees Fahrenheit or degrees Celsius. To do that, she graphed the equation of the conversion formula $C = 5/9(F - 32)$. On the same coordinate system, she also graphed the equation $C = F$ and found the intersection point as $({}^-40, {}^-40)$. Explain why this procedure answers the question Dahlia asked.

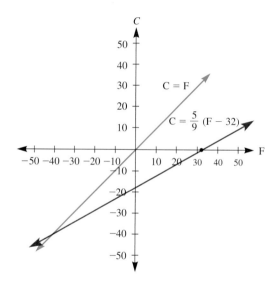

22. Explain in more than one way why two lines with the same slope are parallel.

23. Jonah tried to solve the equation ${}^-5x + y = 20$ by adding 5 to both sides. He wrote $5 - 5x + y = 5 + 20$ or $0 \cdot x + y = 25$ and finally $y = 25$. How would you help Jonah?

24. Fumio would like to know why two lines with an undefined slope are parallel. How would you respond?

Open-Ended

25. Look for data in newspapers, magazines, or books whose graphs appear to be close to linear and find the equations of the lines that you think best fit the data.

26. a. Write equations of two lines that intersect but when graphed look parallel.
   b. At what point do those two lines intersect?

Cooperative Learning

27. Play the following game between your group and another group. Each group makes up four linear equations that have a common property and presents the equations to the other group. For example, one group could present the equations $2x - y = 0$, $4x - 2y = 3$, $y - 2x = 3$, and $3y - 6x = 5$. If the second group discovers a common property that the equations share, such as the graphs of the equations are four parallel lines, they get one point. Each group takes a specified number of turns.

Review Problems

28. $ABCD$ is a trapezoid with $\overline{BC} \parallel \overline{AD}$. Find x.

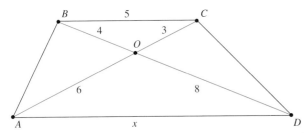

29. Draw an arbitrary triangle ABC. Let M, N, and P be the corresponding midpoints of the sides $\overline{AB}$, $\overline{BC}$, and $\overline{AC}$. Connect the points M, N, and P. Four triangles are created. Is it true that all of the smaller triangles are congruent and each is similar to $\triangle ABC$? Why or why not?

30. In triangle ABC, a square has been inscribed as shown. The lengths of the sides of $\triangle ABC$ are 3, 4, and 5 as shown. Find the length of a side of the square. (*Hint:* First show that $\triangle AED \approx \triangle ABC$.)

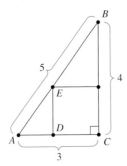

31. In the figure below, find BE.

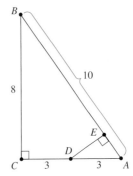

HINT FOR SOLVING THE PRELIMINARY PROBLEM

Use similar triangles to show that the ratio of the length of any segment of the wire to the length of the previous segment is constant (that is, always the same).

QUESTIONS FROM THE CLASSROOM

1. On a test, a student wrote $AB \cong CD$ instead of $\overline{AB} \cong \overline{CD}$. Is this answer correct? Why?
2. A student asks if there are any constructions that cannot be done with a compass and a straightedge. How do you answer?
3. A student asks for a mathematical definition of congruence that holds for all figures. How do you respond? Is your response the same for similarity?
4. One student claims that by trisecting $\overline{AB}$ and drawing $\overrightarrow{CD}$ and $\overrightarrow{CE}$, as shown in the following figure, she has trisected $\angle ACB$:

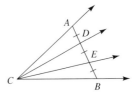

How do you convince her that her construction is wrong?

5. A student claims that polygon *ABCD* in the following drawing is a parallelogram if ∠1 ≅ ∠2. Is he correct? Why?

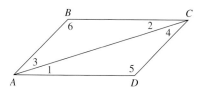

6. A student claims that when the midpoints of the sides of any polygon are connected, a polygon similar to the original results. Is this true? Why?
7. A student asks why "congruent" rather than "equal" is used to discuss triangles that have the same size and shape. What do you say?
8. A student draws the following figure and claims that because every triangle is congruent to itself, we can write △*ABC* ≅ △*BCA*. What is your response?

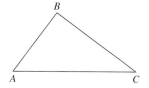

9. A student asks whether there is an AAAA similarity condition for quadrilaterals. How do you respond?
10. A student asks if, for the same *n*, all regular *n*-gons are similar. How do you respond?
11. A student says she thinks all circles are similar but would like to know why. How do you answer her?
12. A student says that a vertical line has an infinite slope. How do you respond?
13. A student claims that the lines whose equations are $y = 3x + 5$ and $y = 3x + 4$ are parallel because, when she tried to solve the equations, she got the false statement $4 = 5$. How do you respond?

CHAPTER OUTLINE

I. Congruence
   A. Two geometric figures are **congruent** if, and only if, they have the same size and shape.
   B. Two triangles are congruent if they satisfy any of the following properties:
      1. **Side, Side, Side (SSS)**
      2. **Side, Angle, Side (SAS)**
      3. **Angle, Side, Angle (ASA)**
      4. **Angle, Angle, Side (AAS)**
   C. **Triangle inequality:** The sum of the measures of any two sides of a triangle must be greater than the measure of the third side.
   D. Corresponding parts of congruent figures are congruent.
   E. A **rep-tile** is a figure that is used in the construction of a similar figure.
II. Circles
   A. An **arc** of a circle is part of the circle that can be drawn without lifting a pencil. The **center of an arc** is the center of the circle containing the arc.
   B. A **chord** is a segment whose endpoints lie on a circle.
III. Proportion
   A. If a line parallel to one side of a triangle intersects the other sides, it divides those sides into proportional segments.
   B. If parallel lines cut off congruent segments on one transversal, they cut off congruent segments on any transversal.
IV. Constructions that can be accomplished using a compass and a straightedge
   A. Copy a line segment.
   B. Copy a circle.
   C. Copy an angle.
   D. Bisect a segment.
   E. Bisect an angle.
   F. Construct a perpendicular from a point to a line.
   G. Construct a perpendicular bisector of a segment.
   H. Construct a perpendicular to a line through a point on the line.
   I. Construct a parallel to a line through a point not on the line.
   J. Divide a segment into congruent parts.
   K. Inscribe some regular polygons in a circle.
   L. Circumscribe a circle about a triangle.
   M. Inscribe a circle in a triangle.
V. Similarity and proportion
   A. Two polygons are **similar** if, and only if, their corresponding angles are congruent and their corresponding sides are proportional.

B. **AA**: If two angles of one triangle are congruent to two angles of a second triangle, respectively, the triangles are similar.
C. If a line divides two sides of a triangle into proportional segments, then the line is parallel to the third side.

VI. Lines in a Cartesian coordinate system
A. Every nonvertical line has an equation of the form $y = mx + b$, where m is the slope and b is the y-intercept.
B. The equation of any vertical line can be written in the form $x = a$ and any horizontal line in the form $y = b$ or $y = 0 \cdot x + b$.
C. Slope formula: Given two points $A(x_1, y_1)$ and $B(x_2, y_2)$, the slope m of line AB is given by the following:

$$m = \frac{y_2 - y_1}{x_2 - x_1} = \frac{\text{rise}}{\text{run}}$$

D. The equation of a line with slope m through a given point with coordinates (x_1, y_1) is $y - y_1 = m(x - x_1)$.
E. Parallel nonvertical lines have the same slope.
F. If two lines have the same slope they are parallel.
G. The slope of a horizontal line is 0 and it is impossible to define the slope of a vertical line.
H. A system of **linear equations** can be solved graphically by drawing the graphs of the equations.
I. A system of linear equations can be solved algbraically by either the **substitution method** or the **elimination method.**
J. The **best-fitting line** is an equation of a straight line that approximates data that seem to follow a straight line.

CHAPTER REVIEW

1. Each of the following figures contains at least one pair of congruent triangles. Identify them and tell why they are congruent.

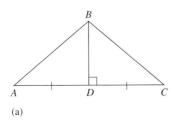
(a)

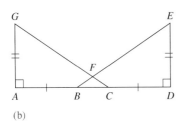
(b)

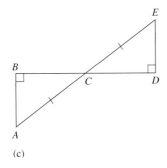
(c)

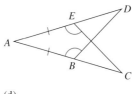
(d)

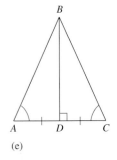
(e)

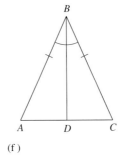
(f)

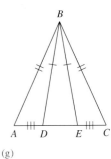
(g)

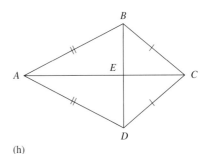
(h)

2. In the following figure, *ABCD* is a square and $\overline{DE} \cong \overline{BF}$. What kind of figure is *AECF*? Justify your answer.

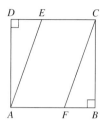

3. Construct each of the following by (1) using a compass and straightedge and (2) paper folding:

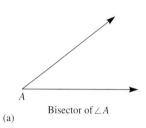
(a) Bisector of ∠*A*

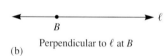
(b) Perpendicular to ℓ at *B*

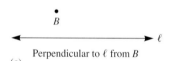
(c) Perpendicular to ℓ from *B*

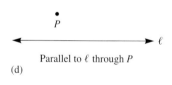
(d) Parallel to ℓ through *P*

4. For each of the following pairs of similar triangles, find the missing measures:

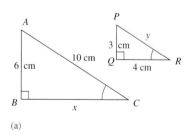
(a)

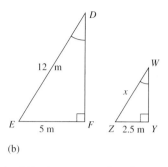
(b)

5. Divide the following segment into five congruent parts:

_____

6. If *ABCD* is a trapezoid, $\overline{EF} \parallel \overline{AD}$ and $\overline{AC}$ is a diagonal. What is the relationship between $\dfrac{a}{b}$ and $\dfrac{c}{d}$? Why?

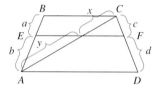

7. Given the following figure, construct a circle that contains *A* and *B* and has its center on ℓ:

8. For each of the following figures, show that appropriate triangles are similar and find *x* and *y*:

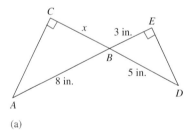
(a)

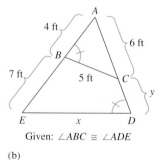
Given: ∠*ABC* ≅ ∠*ADE*
(b)

9. Determine if each of the following is true or false. If false, explain why.
   a. A radius of a circle is a chord of the circle.
   b. If a radius bisects a chord of a circle, then it is perpendicular to the chord.
10. A person 2 m tall casts a shadow 1 m long when a building has a 6-m shadow. How high is the building?
11. a. Which of the following polygons can be inscribed in a circle? Assume that all sides of each polygon are congruent and that all the angles of polygons (ii) and (iii) are congruent.
    b. Based on your answer in (a), make a conjecture about what kinds of polygons can be inscribed in a circle.

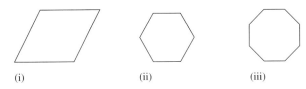

(i)   (ii)   (iii)

12. Determine the vertical height of the playground slide shown in the following figure:

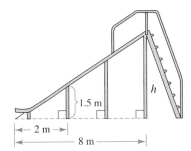

13. Find the distance d across the river sketched as follows:

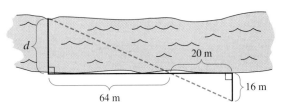

14. Is the following statement always true, always false, or true in some cases and false in others? Explain your answer.
    A quadrilateral whose diagonals are congruent and perpendicular is a square.
15. For each of the following, write the equation of the line determined by the given pair of points:
    a. $(2, ^-3)$ and $(^-1, 1)$
    b. $(^-3, 0)$ and $(3, 2)$
16. Use slope to determine if there is a single line through the points with coordinates $(4, 2)$, $(0, ^-1)$, and $(7, ^-5)$. Explain your reasoning.
17. Solve each of the following systems, if possible. If the system does not have a unique solution, explain why not.
    a. $x + 2y = 3$
       $2x - y = 9$
    b. $\dfrac{x}{2} + \dfrac{y}{3} = 1$
       $4y - 3x = 2$
    c. $x - 2y = 1$
       $4y - 2x = 0$

SELECTED BIBLIOGRAPHY

Cain, B. "The Basic Constructions." *Mathematics Teaching in the Middle School* 1 (September–October 1994): 137.

Carrol, W. "Polygon Capture: A Geometry Game." *Mathematics Teaching in the Middle School* 4 (October 1998): 90–94.

Eves, H. *An Introduction to the History of Mathematics with Cultural Connections.* Philadelphia: Saunders College Publishing (1990).

Friedlander, A., and G. Lappan. "Similarity: Investigations at the Middle Grade Level." In *Learning and Teaching Geometry, K–12.* Reston, Va.: National Council of Teachers of Mathematics (1987).

Greer, G., and L. Wantuck. "Menu of Problems." *Mathematics Teaching in the Middle School* 1 (March–April 1996): 729.

Hurd, S. "An Application of the Criteria ASASA for Quadrilaterals." *Mathematics Teacher* 81 (February 1988): 124–126.

Kennedy, J., and E. McDowell. "Geoboard Quadrilaterals." *Mathematics Teacher* 91 (April 1998): 288–290.

Lappan, G., and R. Even. "Similarity in the Middle Grades." *Arithmetic Teacher* 35 (May 1988): 32–35.

Maupin, S. "Middle-Grades Geometry Activities." *Mathematics Teaching in the Middle School* 1 (May 1996): 790–796.

Sanders, C. "Geometric Constructions: Visualizing and Understanding Geometry." *Mathematics Teacher* 91 (October 1998): 554–556.

Senk, S. L., and D. B. Hirschorn. "Multiple Approaches to Geometry: Teaching Similarity." *Mathematics Teacher* 83 (April 1990): 274–280.

Slavit, D. "Above and beyond AAA: The Similarity and Congruence of Polygons." *Mathematics Teaching in the Middle School* 3 (January 1998): 276–280.

Taylor, L. "Exploring Geometry with the Geometer's Sketchpad." *Arithmetic Teacher* 40 (November 1992): 187–191.

Vennebush, P. "Menu of Problems." *Mathematics Teaching in the Middle School* 3 (February 1998): 352–353.

Walter, M. *Boxes, Squares, and Other Things*. Reston, Va.: National Council of Teachers of Mathematics (1998).

11
Concepts of Measurement

Preliminary Problem
A farmer has a plot of land in the shape of a square that is 100 m on a side. An irrigation system can be installed with the option of one large circular sprinkler or nine small sprinklers, as shown in the accompanying figure. The farmer wants to know which plan will provide water to the greatest percentage of land in the field, regardless of the cost and the watering pattern.

▲ *Principles and Standards* discusses the attributes of objects as follows:

> *A measurable attribute is a characteristic of an object that can be quantified. Line segments have length, plane regions have area, and physical objects have mass. As students progress through the curriculum from preschool through high school, the set of attributes they can measure should expand. Recognizing that objects have attributes that are measurable is the first step in the study of measurement. Children in prekindergarten through grade 2 begin by comparing and ordering objects using language such as longer and shorter. Length should be the focus in this grade band, but weight, time, area, and volume should also be explored. In grades 3–5, students should learn about area more thoroughly, as well as perimeter, volume, temperature, and angle measure. In these grades, they learn that measurements can be computed using formulas and need not always be taken directly with a measuring tool. Middle-grade students build on these earlier measurement experiences by continuing their study of perimeter, area, and volume and by beginning to explore derived measurements, such as speed* (p. 44).

In the United States, two measurement systems are used regularly: the English system and the metric system. In *Principles and Standards*, we find the following concerning the use of the two measurement systems:

▲ *Since the customary English system of measurement is still prevalent in the United States, students should learn both customary and metric systems and should know some rough equivalences between the metric and customary systems—for example, that a two-liter bottle of soda is a little more than half a gallon. The study of these systems begins in elementary school, and students at this level should be able to carry out simple conversions within both systems. Students should develop proficiency in these conversions in the middle grades and should learn some useful benchmarks for converting between the two systems* (pp. 45–46).

Objects have attributes such as length, width, and area that can be measured. Many concepts of measurement are more confusing for students than for adults because students lack everyday measurement experiences. In this chapter, we use both systems of measurement for length, area, volume, mass, and temperature with the philosophy that students should learn to think within a system. We also develop formulas for the area of plane figures and for surface areas and volumes of solids. We use the concept of area in discussing the Pythagorean theorem.

Section 11-1 — Linear Measure

To measure a segment, we must decide on a unit of measure. Early attempts at measurement lacked a standard unit and so used hands, arms, and feet as units of measure. These early crude measurements were refined eventually and standardized by the English into a very complicated system.

The English System

Originally, in the English system, a yard was the distance from the tip of the nose to the end of an outstretched arm of an adult person and a foot was the length of a human foot. In 1893 the United States defined the yard and other units in terms of metric units. Some units of length in the English system and relationships among them are summarized in Table 11-1.

Table 11-1

Unit	Equivalent in Other Units
yard (yd)	3 ft
foot (ft)	12 in.
mile (mi)	1760 yd, or 5280 ft

Example 11-1 Convert each of the following:

a. 219 ft = _____ yd
b. 8432 yd = _____ mi
c. 0.2 mi = _____ ft
d. 64 in. = _____ yd

Solution
a. Because 1 ft = $\frac{1}{3}$ yd, 219 ft = $219 \cdot \frac{1}{3}$ yd = 73 yd.
b. Because 1 yd = $\frac{1}{1760}$ mi, 8432 yd = $8432 \cdot \frac{1}{1760}$ mi $\doteq$ 4.79 mi.
c. 1 mi = 5280 ft. Hence, 0.2 mi = $0.2 \cdot 5280$ ft = 1056 ft.
d. We first find a connection between yards and inches. We have 1 yd = 3 ft and 1 ft = 12 in. Hence, 1 yd = 3 ft = $3 \cdot 12$ in. = 36 in. Hence, 1 in. = $\frac{1}{36}$ yd; therefore 64 in. = $64 \cdot \frac{1}{36}$ yd $\doteq$ 1.78 yd.

The Metric System

At this time, the United States is the only major industrial nation in the world that continues to use the English system. However, the use of the **metric system** in the United States has been increasing, particularly in the scientific community and in industry. The *1999 World Almanac* points out:

> The Trade Act of 1988 and other legislation declare the metric system the preferred system of weights and measures of U.S. trade and commerce, call for the federal government to adopt metric specifications, and mandate the Commerce Department to oversee the program. The conversion process is currently under way; however, the metric system has not become the system of choice for most Americans' daily use.

HISTORICAL NOTE

The metric system, a decimal system, was proposed in France in 1670 by Gabriel Mouton. However, not until the French Revolution in 1790 did the French Academy of Sciences bring various groups together to develop the system. The Academy recognized the need for a standard base unit of linear measurement. The members chose $\frac{1}{10,000,000}$ of the distance from the equator to the North Pole on a meridian through Paris as the base unit of length and called it the **meter (m)**. In 1960 the meter was redefined in terms of krypton 86 wavelengths and still later as the distance traveled by light in a vacuum during $\frac{1}{299,792,458}$ sec. Since 1893 the yard in the United States has been defined as $\frac{3600}{3937}$ of a meter, whatever the definition of a meter.

Different units of length in the metric system are obtained by multiplying a power of ten times the base unit. Table 11-2 gives the prefixes for these units, the multiplication factors, and their symbols.

Table 11-2

Prefix	Symbol	Factor	
kilo	k	1000	(one thousand)
*hecto	h	100	(one hundred)
*deka	da	10	(ten)
*deci	d	0.1	(one tenth)
centi	c	0.01	(one hundredth)
milli	m	0.001	(one thousandth)

*Not commonly used

The metric prefixes, combined with the base unit meter, name the different units of length. Table 11-3 gives these units, their relationship to the meter, and the symbol for each.

Table 11-3

Unit	Symbol	Relationship to Base Unit
kilometer	km	1000 m
*hectometer	hm	100 m
*dekameter	dam	10 m
meter	**m**	**base unit**
*decimeter	dm	0.1 m
centimeter	cm	0.01 m
millimeter	mm	0.001 m

*Not commonly used

REMARK   Two other prefixes, mega (1,000,000) and micro (0.000001), are used for very large and very small units, respectively.

Estimations for a meter, a decimeter, a centimeter, and a millimeter are shown in Figure 11-1. The kilometer is commonly used for measuring longer distances. Because "kilo" stands for 1000, 1 km = 1000 m. Nine football fields, including end zones, laid end to end are approximately 1 km long.

Figure 11-1

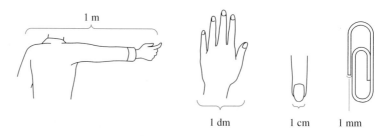

Conversions among metric lengths are accomplished by multiplying or dividing by powers of ten. As with money, we simply move the decimal point to the left or right, depending on the units. For example,

0.123 km = 1.23 hm = 12.3 dam = 123 m = 1230 dm = 12,300 cm = 123,000 mm.

It is possible to convert units by using the chart in Figure 11-2. We count the number of steps from one unit to the other and move the decimal point that many steps in the same direction.

Figure 11-2

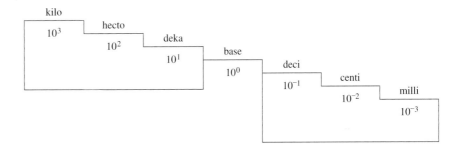

Example 11-2   Convert each of the following:

  a. 1.4 km = _____ m   **b.** 285 mm = _____ m   **c.** 0.03 km = _____ cm

  Solution   **a.** Because 1 km = 1000 m, to change kilometers to meters, we multiply by 1000. Hence, 1.4 km = 1.4 (1000 m) = 1400 m.
  b. Because 1 mm = 0.001 m, to change from millimeters to meters, we multiply by 0.001. Thus 285 mm = 285 (0.001 m) = 0.285 m.
  c. To change kilometers to centimeters, we first multiply by 1000 to convert kilometers to meters and then multiply by 100 to convert meters to centimeters. Therefore we move the decimal five places to the right to obtain 0.03 km = 3000 cm.

Linear units of length are commonly measured with rulers. Figure 11-3 shows part of a centimeter ruler. Rulers can be used to measure distance.

Figure 11-3

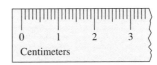

Measuring distances in the real world frequently results in errors. Because of this, many industrial plants using parts from a variety of sources rely on portable calibration units that are taken from plant to plant to test measuring instruments used in constructing the parts. This is done so that the final assembly plant can fit all the parts together to make the product. To calibrate the measuring instruments, technicians must establish the greatest possible error (GPE) allowable in order to obtain the final fit. The **greatest possible error (GPE)** of a measurement is one-half the unit used. For example, if the width of a piece of board was measured 5 cm to the nearest centimeter, the actual width must be between 4.5 cm and 5.5 cm. Therefore, the GPE for this measurement is 0.5 cm. If the width of a button was measured as 1.2 cm, then the actual width is between 1.15 cm and 1.25 cm and so the GPE is 0.05 cm or 0.5 mm.

When drawings are given, we assume that the listed measurements are accurate. When actually measuring objects in the real world, we find that such accuracy is usually impossible.

Distance Properties

A person using the expression "the shortest distance between two points is a straight line" may have good intentions, but the expression is actually false. (Why?) The shortest among all the polygonal paths connecting two points A and B is the segment $\overline{AB}$. This fact is one of the basic properties of distance listed next.

Properties

1. The distance between any two points A and B is greater than or equal to 0, written $AB \geq 0$. The length of $\overline{AB}$ is denoted by AB.
2. The distance between any two points A and B is the same as the distance between B and A, written $AB = BA$.
3. For any three points A, B, and C, the distance between A and B plus the distance between B and C is greater than or equal to the distance between A and C, written $AB + BC \geq AC$.

In the special case where A, B, and C are collinear and B is between A and C, as in Figure 11-4(a), we have $AB + BC = AC$. Otherwise, if A, B, and C are not collinear, as in Figure 11-4(b), then they form the vertices of a triangle and $AB + BC > AC$. This inequality, $AB + BC > AC$, is the **Triangle Inequality.** In general, the Triangle Inequality states that *the sum of the lengths of any two sides of a triangle is greater than the length of the third side.*

Figure 11-4

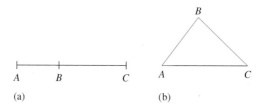

NOW TRY THIS 11-1

- If two sides of a triangle are 31 cm and 85 cm long and the measure of the third side must be a whole number of centimeters,

  a. what is the longest the third side can be?

  b. what is the shortest the third side can be?

REMARK Notice the difference between AB, $\overline{AB}$, and $\overleftrightarrow{AB}$. AB is the distance between two points A and B and therefore a nonnegative real number, $\overline{AB}$ is the segment connecting points A and B and therefore a set of points. $\overleftrightarrow{AB}$ is the line through points A and B.

Distance Around a Plane Figure

perimeter The **perimeter** of a simple closed curve is the length of the curve, that is, the distance around the figure. If a figure is a polygon, its perimeter is the sum of the lengths of the sides. A perimeter is always expressed in linear measure.

Example 11-3 Find the perimeter of each of the shapes in Figure 11-5.

Figure 11-5

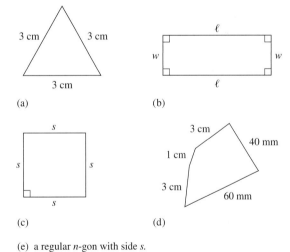

(e) a regular n-gon with side s.

Solution
a. The perimeter is $3(3) = 9$ cm.
b. The perimeter is $2w + 2l$.
c. The perimeter is $4s$.
d. Because 40 mm = 4 cm and 60 mm = 6 cm, the perimeter is $1 + 3 + 4 + 6 + 3 = 17$ cm.
e. Because all sides of a regular n-gon are congruent, the perimeter is ns.

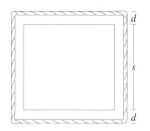

Figure 11-6

Problem Solving — Roping a Square

Given a square of any size, stretch a rope tightly around it. Now take the rope off, add 100 ft to it and put the extended rope back around the square so that the new rope makes a square around the original square. Make sure there is uniform spacing between the original square and the square formed by the rope. What is the distance between a side of the original square and the corresponding parallel side of the rope square?

- **Understanding the Problem** We are to determine the distance between a square and a new square formed by adding 100 ft of rope to a rope that was stretched around the original square. If the length of the side of the original square is s and the unknown distance is d, then Figure 11-6 shows the situation.

- **Devising a Plan** At first it seems that not enough information is given to solve the problem. However, if we use variables to represent the unknowns, we can *write an equation* to model the problem. The perimeter of the new square is $4s + 100$. Another way to represent this perimeter is $4(s + 2d)$. Therefore, we have $4s + 100 = 4(s + 2d)$. If we can solve this equation for d we have the solution.

- **Carrying Out the Plan** We solve the equation as shown below.

$$4s + 100 = 4(s + 2d)$$
$$4s + 100 = 4s + 8d$$
$$100 = 8d$$
$$12.5 = d$$

Therefore the distance between the squares is 12.5 ft.

- **Looking Back** Try some lengths of different squares to convince yourself that this solution is correct. It seems remarkable that we can solve this problem without knowing the length of the side of the original square. This problem can be extended to figures other than squares. Try the Brainteaser at the end of this section.

Circles

In Figure 11-7, the regular 24-gon resembles a circle. The more sides a regular n-gon has, the closer it resembles a circle. (This type of thinking has led to a procedure in Logo for drawing a turtle-type circle as in the following Technology Corner.) A **circle** is defined as the set of all points in a plane that are the same distance from a given point, the **center.**

circle
center

Figure 11-7

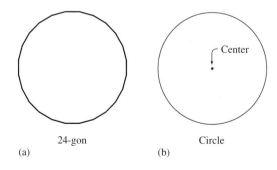

24-gon
(a)

Circle
(b)

TECHNOLOGY CORNER

To instruct the turtle to draw a figure that looks like a circle (*turtle-type circle*), we tell the turtle to move forward a little and turn a little and then repeat this sequence of motions until it comes back to its original position. For example, the turtle could go forward 1 unit and turn right 1°. This sequence of motions repeated 360 times yields a turtle-type circle, as shown in Figure 11-8.

Figure 11-8

```
TO CIRCLE1
  REPEAT 360[FD 1 RT 1]
END
```

CIRCLE1

Figure 11-8 shows a Logo procedure CIRCLE1 and its output. Explore how to draw larger and smaller turtle-type circles using similar procedures.

Circumference of a Circle

circumference

The perimeter of a circle is its **circumference**. The ancient Greeks discovered that if they divided the circumference of any circle by the length of its diameter, they always obtained approximately the same number. The number is approximately 3.14 (see the Laboratory Activity at the end of this section). Today, the ratio of circumference C to diameter d is

pi

symbolized as π (**pi**). For most practical purposes, π is approximated by $\frac{22}{7}$, $3\frac{1}{7}$, or 3.14.

These values are approximations, not exact values of π. For example, the "exact" circumference of a circle with diameter 6 cm is 6π cm.

A circumference is always expressed in linear measure. In the late eighteenth century, mathematicians proved that the ratio $\frac{C}{d}$, or π, is not a terminating or repeating decimal. Rather, it is an irrational number.

The relationship $\frac{C}{d} = \pi$ is used for finding the circumference of a circle and normally is written as $C = \pi d$ or $C = 2\pi r$ because the length of diameter d is twice the radius (r) of the circle.

HISTORICAL NOTE

$\pi = 3.14159$
26535
89793
23846
26433
83279
50288
41971
69399
37510
58209
.
.
.

Archimedes (b. 287 B.C.) found an approximation for π given by the inequality $3\frac{10}{71} < \pi < 3\frac{10}{70}$. A Chinese astronomer thought that $\pi = \frac{355}{113}$. Ludolph van Ceulen (1540–1610), a German mathematician, calculated π to 35 decimal places. The approximation was engraved on his tombstone. Leonhard Euler adopted the symbol π in 1737 and caused its wide usage. In 1761 Johann Lambert, an Alsatian mathematician, proved that π is an irrational number. In 1989 Columbia University mathematicians and Soviet émigré brothers, David and Gregory Chudnovsky, used computers to establish 480 million digits of π. If these digits were printed along a line, the line would extend 600 miles.

Arc Length

The length of an arc depends on the radius of the circle and the central angle determining the arc. If the central angle has a measure of 180°, as in Figure 11-9(a), the arc is a **semicircle**. The length of a semicircle is $\frac{1}{2} \cdot 2\pi r$, or πr. The length of an arc whose central angle is $\theta°$ can be developed as in Figure 11-9(b) by using proportional reasoning. Since a circle has 360°, an angle of $\theta°$ determines $\theta/360$ of a circle. Because the circumference of a circle is $2\pi r$, an arc of $\theta°$ has length $\frac{\theta}{360} \cdot 2\pi r$, or $\frac{\pi r \theta}{180}$.

Figure 11-9

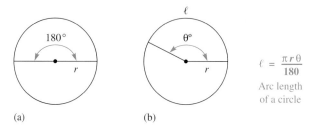

$\ell = \frac{\pi r \theta}{180}$

Arc length of a circle

Example 11-4 Find each of the following:

a. The circumference of a circle if the radius is 2 m
b. The radius of a circle if the circumference is 15π m
c. The length of a 25° arc of a circle of radius 10 cm
d. The radius of an arc whose central angle is 87° and whose length is 154 cm

Solution
a. $C = 2\pi(2) = 4\pi$; thus the circumference is 4π m.
b. $C = 2\pi r$ implies $15\pi = 2\pi r$. Hence, $r = \frac{15}{2}$ and the radius is 7.5 m.
c. The arc length is $\frac{\pi r \theta}{180} = \frac{\pi \cdot 10 \cdot 25}{180}$ cm, or $\frac{25\pi}{18}$ cm, or approximately 4.36 cm.
d. The arc length ℓ is $\frac{\pi r \theta}{180}$, so that $154 = \frac{\pi r \cdot 87}{180}$. Thus, $r \doteq 101.4$ cm.

LABORATORY ACTIVITY

To approximate the value of π, you need string, a marked ruler, and several different-sized round tin cans or jars. Pick a can and wrap the string tightly around the can. Use a pen to mark a point on the string where the beginning of the string meets the string again. Unwrap the string and measure its length. Next, determine the diameter of the can by tracing the bottom of the can on a piece of paper. Fold the circle onto itself to find a line of symmetry. The chord determined by the line is a diameter of the circle. Measure the diameter and determine the ratio of the circumference to the diameter. (Use the same units in all of your measurements.) Repeat the experiment with at least three cans and find the average of the corresponding ratios.

ONGOING ASSESSMENT 11-1

1. Use the following picture of a ruler to find each of the following lengths in centimeters:
   a. AB b. DE c. CJ d. EF
   e. IJ f. AF g. IC h. GB

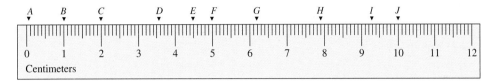

2. Convert each of the following:
   a. 100 in. = _____ yd
   b. 400 yd = _____ in.
   c. 300 ft = _____ yd
   d. 372 in. = _____ ft

3. Draw segments that you estimate to be of the following lengths. Use a metric ruler to check the estimates.
   a. 10 mm
   b. 100 mm
   c. 1 cm
   d. 10 cm
   e. 0.01 m
   f. 15 cm

4. Estimate the length of the following segment and then measure it:

   |—————————————————————————|

   Express the measurement in each of the following units:
   a. Millimeters
   b. Centimeters

5. Choose an appropriate metric unit and estimate each of the following measures:
   a. The length of a pencil
   b. The diameter of a nickel
   c. The width of the top of a desk
   d. The thickness of the top of a desk
   e. The length of this sheet of paper
   f. The height of a door

6. Redo problem 5 using English measures.

7. Complete the following table:

Item	m	cm	mm
a. Length of a piece of paper		35	
b. Height of a woman	1.63		
c. Width of a filmstrip			35
d. Length of a cigarette			100
e. Length of two meter sticks laid end to end	2		

8. For each of the following, place a decimal point in the number to make the sentence reasonable:
   a. A stack of 10 dimes is 1000 mm high.
   b. The desk is 770 m high.
   c. The distance from one side of a street to the other is 100 m.
   d. A dollar bill is 155 cm long.
   e. The basketball player is 1950 cm tall.
   f. A new piece of chalk is about 8100 cm long.
   g. The speed limit in town is 400 km/hr.

9. List the following in decreasing order:
   8 cm, 5218 mm, 245 cm, 91 mm, 6 m, 700 mm.

10. Draw each of the following as accurately as possible:
    a. A regular polygon whose perimeter is 12 cm
    b. A circle whose circumference is 4 in.
    c. A triangle whose perimeter is 4 in.
    d. A nonconvex quadrilateral whose perimeter is 8 cm

11. Guess the perimeter in centimeters of each of the following figures and then check the estimates using a ruler:

(a)

(b)

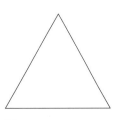

(c)

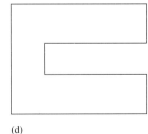

(d)

12. Complete each of the following:
    a. 10 mm = _____ cm   b. 262 m = _____ km
    c. 3 km = _____ m   d. 30 mm = _____ m
    e. 35 m = _____ cm   f. 359 mm = _____ m
    g. 647 mm = _____ cm   h. 0.1 cm = _____ mm

13. Draw a triangle *ABC*. Measure the length of each of its sides in millimeters. For each of the following, tell which is greater and by how much:
    a. $AB + BC$ or AC
    b. $BC + CA$ or AB
    c. $AB + CA$ or BC
14. Can the following be the lengths of the sides of a triangle. Why or why not?
    a. 23 cm, 50 cm, 60 cm
    b. 10 cm, 40 cm, 50 cm
    c. 410 mm, 260 mm, 14 cm
15. Do you think it is possible to draw a square whose perimeter is exactly equal to the sum of the length of its diagonals? Justify your answer.
16. Take an $8\frac{1}{2}$-×-11-in. piece of typing paper, fold it as shown in the following figure, and then cut the folded paper along the diagonal segment:

    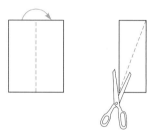

    a. Rearrange the two smaller pieces to find a triangle with the minimum perimeter.
    b. Arrange the two smaller pieces to form a triangle with the maximum perimeter.
17. The following figure made of 6 unit squares has a perimeter of 12 units. The figure is made in such a way that each square must share at least one complete side with another square.

    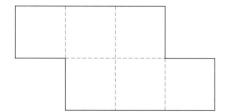

    a. Add more squares to the figure so that the perimeter of the new figure is 18.
    b. What is the minimum number of squares required to make a figure of perimeter 18?
    c. What is the maximum number of squares that can be used to make a figure of perimeter 18?

18. For each of the following circumferences, find the radius of the circle:
    a. 12π cm
    b. 6 m
    c. 0.67 m
    d. 92π cm
19. For each of the following, if a circle has the dimensions given, determine its circumference:
    a. 6 cm diameter
    b. 3 cm radius
    c. $\frac{2}{\pi}$ cm radius
    d. 6π cm diameter
20. What happens to the circumference of a circle if the length of the radius is doubled?
21. The following figure is a circle whose radius is *r* units. The diameters of the two semicircular regions inside the large circle are also *r* units long. Compute the length of the curve that separates the shaded and white regions.

    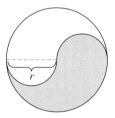

22. Astronomers use a light year to measure distance. A light year is the distance light travels in 1 yr. The speed of light is 300,000 km/sec.
    a. How long is 1 light year in kilometers?
    b. The nearest star (other than the sun) is Alpha Centauri. It is 4.34 light years from Earth. How far is that in kilometers?
    c. How long will it take a rocket traveling 60,000 km/hr to reach Alpha Centauri?
    d. How long will it take the rocket in (c) to travel to the sun if it takes approximately 8 min 19 sec for light from the sun to reach Earth?
23. Jet planes can exceed the speed of sound, so a new measurement called *Mach number* was invented to measure the speed of such planes. Mach 2 is twice the speed of sound. (Mach number is a number that indicates the ratio of the speed of an object through a medium to the speed of sound in the medium.) The speed of sound in air is approximately 344 m/sec.
    a. Express Mach 2.5 in kilometers per hour.
    b. Express Mach 3 in meters per second.
    c. Express the speed of 5000 km/hr as a Mach number.
24. Refer to the following figure and determine the perimeter of the paint lane and the semicircle determined by the free-throw line on a basketball court:

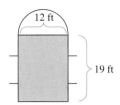

25. Find the length of the side of a square that has the same perimeter as a rectangle that is 66 cm by 32 cm.

26. Give the greatest possible error for each of the following measurements:
   a. 23 m
   b. 3.6 cm
   c. 3.12 m

27. On a circular merry-go-round, one horse is 3 m from the center and another is 6 m from the center. The merry-go-round makes 3 revolutions per minute. Are the two horses traveling at the same speed? If not, how fast is each horse traveling?

Communication

28. There has been considerable debate about whether the United States should change to the metric system. Based on your experiences with linear measure, what do you see as the advantages of changing? Which system do you think would be easier for children to learn? Why?

29. A student has a tennis ball can with a flat top and bottom containing three tennis balls. To the student's surprise, the perimeter of the top of the can is longer than the height of the can. The student wants to know if this fact can be explained without performing any measurements. Can you help?

30. In track, the second lane from the inside of the track is longer than the inside lane. Use this information to explain why, in running events that require a complete lap of the track, runners are lined up at the starting blocks as shown in the following figure:

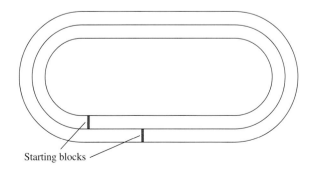

Open-Ended

31. Observe that it is possible to build a triangle with toothpicks that has sides of 3, 4, and 5 toothpicks, as shown, and answer the questions that follow.

   a. Find two other triples of toothpicks that can be used as sides of a triangle and two other triples that cannot be used to create a triangle.
   b. Describe how to tell whether a given triple of numbers a, b, c can be used to construct a triangle with sides of a, b, and c toothpicks. Explain why your rule is valid.

32. Make a scale drawing of your classroom on square centimeter graph paper.

Cooperative Learning

33. a. Help each person in the group find his or her height in centimeters.
   b. Help each person in the group find the length of his or her outstretched arms (horizontal) from fingertip to fingertip.
   c. Compute the difference between the two measurements in parts (a) and (b). Compare the results of the group members and make a conjecture about the lengths of the two measurements.
   d. Compare your group's results with other groups to determine if they have similar findings.

34. Jerry wants to design a gold chain 60-cm long made of thin gold wire circles, each of which is the same size. He wants to use the least amount of wire and wonders what the radius of each circle should be.

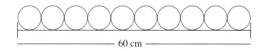

   a. Each member of the group should choose a specific number of circles and find the length of wire needed to make a 60-cm chain with the chosen number of circles.
   b. Compare your results and make a conjecture based on the results.
   c. Justify your conjecture.

BRAIN TEASER Suppose a wire is stretched tightly around Earth. (The radius of Earth is approximately 6400 km.) Then suppose the wire is cut and its length is increased by 20 m. It is then placed back around the planet so that it is the same distance from Earth at every point. Could you walk under the wire?

Section 11-2 — Areas of Polygons and Circles

Area is measured using square units and the area of a region is the number of square units that cover the region without overlapping. A square measuring 1 ft on a side has an area of one square foot, denoted by 1 ft². A square measuring 1 cm on a side has an area of one square centimeter, denoted by 1 cm².

Areas on a Geoboard

In teaching the concept of area, intuitive activities should precede the development of formulas. Many such activities can be accomplished using a *geoboard* or *dot paper*. Notice that the square unit is defined in the upper left corner of the geoboard in Figure 11-10(a). The area of the pentagon can be found by finding the sum of the areas of smaller pieces. Finding the area in this way, is the *addition method.* The region in Figure 11-10(b) has been divided into smaller pieces in Figure 11-10(c), what is the area of this shape?

Figure 11-10

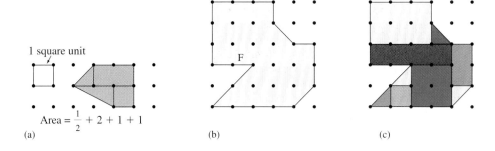

(a) Area = $\frac{1}{2}$ + 2 + 1 + 1   (b)   (c)

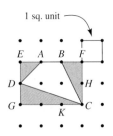

Figure 11-11

Another method of finding the area of shapes on the geoboard is the *rectangle method.* To find the area of quadrilateral $ABCD$ in Figure 11-11, we construct the rectangle $EFCG$ around the quadrilateral and then subtract the areas of the shaded triangles EAD, BFC, and DGC. The area of the rectangle $EFCG$ can be counted to be 6 square units. The area of $\triangle EAD$ is $\frac{1}{2}$ square unit, and the area of $\triangle BFC$ is half the area of rectangle $BFCK$, or $\frac{1}{2}$ of 2, or 1 square unit. Similarly, the area of $\triangle DGC$ is half the area of rectangle $DHCG$, that is, $\frac{1}{2} \cdot 3$, or $\frac{3}{2}$ square units. Consequently, the area of $ABCD$ is $6 - \left(\frac{1}{2} + 1 + \frac{3}{2}\right)$, or 3 square units.

Use the rectangle method to find the area of the shape in Figure 11-10(b).

SECTION 11-2   *Areas of Polygons and Circles*   **605**

Example 11-5   Using a geoboard, find the area of each of the shaded parts of Figure 11-12.

Figure 11-12

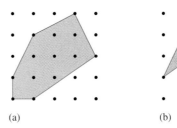

(a)   (b)

Solution   **a.** We construct a rectangle around the hexagon and then subtract the areas of regions *a*, *b*, *c*, *d*, and *e* from the area of this rectangle, as shown in Figure 11-13. Therefore the area of the hexagon is $16 - (3 + 1 + 1 + 1 + 1)$, or 9 square units. The addition method could also be used in this problem.

Figure 11-13

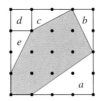

b. The area of the hexagon equals the area of the surrounding rectangle shown in Figure 11-14 minus the sum of the areas of figures *a*, *b*, *c*, *d*, *e*, *f*, and *g*. Thus the area of the hexagon is $12 - (3 + 1 + \frac{1}{2} + \frac{1}{2} + 1 + 1 + 1)$, or 4 square units.

Figure 11-14

NOW TRY THIS 11-2

- Instead of using a square for the unit of area, find how many of each of the following shapes are contained in Figure 11-15:

    a. Triangles

    b. Rhombuses

    c. Trapezoids

    Figure 11-15

Converting Units of Area

The price of new carpet for a house is quoted in terms of square yards, for example, $12.50/yd^2. The basic units of area in the English system are the square inch (in.^{2}), the square foot (ft^2), the square yard (yd^2), the square mile (mi^2), and, for land measure, the acre (A). In the metric system, the basic units are the square millimeter (mm^2), the square centimeter (cm^2), the square meter (m^2), the square kilometer (km^2), and, for land measure, the hectare (ha). It is often necessary to convert from one area measure to another within a system.

To determine how many 1-cm squares are in a square meter, look at Figure 11-16(a). There are 100 cm in 1 m, so each side of the square meter has a measure of 100 cm. Thus it takes 100 rows of 100 1-cm squares each to fill a square meter, that is, 100 · 100, or 10,000 1-cm squares. Because the area of each centimeter square is 1 cm · 1 cm, or 1 cm^2, there are 10,000 cm^2 in 1 m^2. In general, the area A of a square that is s units on a side is s^2, as shown in Figure 11-16(b).

Figure 11-16

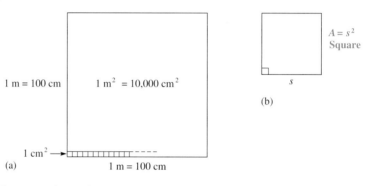

Other metric conversions of area measure can be developed similarly. For example, Figure 11-17(a) shows that 1 m^2 = 10,000 cm^2 = 1,000,000 mm^2. Likewise, Figure 11-17(b) shows that 1 m^2 = 0.000001 km^2. Similarly, 1 cm^2 = 100 mm^2 and 1 km^2 = 1,000,000 m^2.

Figure 11-17

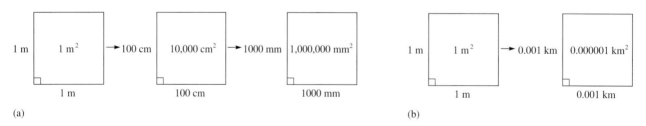

Table 11-4 shows the symbols for metric units of area and their relationship to the square meter.

Table 11-4

Unit	Symbol	Relationship to Square Meter
square kilometer	km^2	1,000,000 m^2
*square hectometer	hm^2	10,000 m^2
*square dekameter	dam^2	100 m^2
square meter	**m^2**	**1 m^2**
*square decimeter	dm^2	0.01 m^2
square centimeter	cm^2	0.0001 m^2
square millimeter	mm^2	0.000001 m^2

*Not commonly used

Example 11-6 Convert each of the following:

a. $5 \text{ cm}^2 = \underline{} \text{ mm}^2$
b. $124{,}000{,}000 \text{ m}^2 = \underline{} \text{ km}^2$

Solution a. $1 \text{ cm}^2 = 100 \text{ mm}^2$ implies $5 \text{ cm}^2 = 5 \cdot 1 \text{ cm}^2 = 5 \cdot 100 \text{ mm}^2 = 500 \text{ mm}^2$.
b. $1 \text{ m}^2 = 0.000001 \text{ km}^2$ implies $124{,}000{,}000 \text{ m}^2 = 124{,}000{,}000 \cdot 1 \text{ m}^2$
   $= 124{,}000{,}000 \cdot 0.000001 \text{ km}^2 = 124 \text{ km}^2$.

REMARK Students sometimes confuse the area of 5 cm^2 with the area of a square 5 cm on each side. The area of a square 5 cm on each side is $(5 \text{ cm})^2$, or 25 cm^2. Five squares each 1 cm by 1 cm have the area of 5 cm^2. Thus $5 \text{ cm}^2 \neq (5 \text{ cm})^2$.

Based on the relationship among units of length in the English system, it is possible to convert among English units of area. For example, because 1 yd = 3 ft, it follows that $(1 \text{ yd})^2 = 1 \text{ yd} \cdot 1 \text{ yd} = 3 \text{ ft} \cdot 3 \text{ ft} = 9 \text{ ft}^2$. Similarly, because 1 ft = 12 in., $(1 \text{ ft})^2 = 1 \text{ ft} \cdot 1 \text{ ft} = 12 \text{ in.} \cdot 12 \text{ in.} = 144 \text{ in.}^2$ Table 11-5 summarizes various relationships among units of area in the English system.

Table 11-5

Unit of Area	Equivalent of Other Units
1 ft^2	$\frac{1}{9} \text{ yd}^2$, or 144 in.^2
1 yd^2	9 ft^2
1 mi^2	$3{,}097{,}600 \text{ yd}^2$, or $27{,}878{,}400 \text{ ft}^2$

Land Measure

One application of area today is in land measure. The common unit of land measure in the English system is the **acre**. There are 4840 yd^2 in 1 acre. For very large land measures in the English system, the **square mile** (mi^2), or 640 acres, is used.

In the metric system, small land areas are measured in terms of a square unit 10 m on a side, called an **are** (pronounced "air") and denoted by **a**. Thus $1 \text{ a} = 10 \text{ m} \cdot 10 \text{ m}$, or 100 m^2. Larger land areas are measured in **hectares**. A hectare is 100 a. A hectare, denoted by **ha**, is the amount of land whose area is $10{,}000 \text{ m}^2$. It follows that 1 ha is the area of a square that is 100 m on a side. For very large land measures, the **square kilometer**, denoted by km^2, is used. One square kilometer is the area of a square with a side 1 km, or 1000 m, long. Land area measures are summarized in Table 11-6.

acre
square mile

are (a)
hectare (ha)

square kilometer

Table 11-6

Unit of Area	Equivalent in Other Units
1 a	100 m^2
1 ha	100 a, or $10{,}000 \text{ m}^2$
1 km^2	$1{,}000{,}000 \text{ m}^2$
1 acre	4840 yd^2
1 mi^2	640 acres

Example 11-7

a. A square field has a side of 400 m. Find the area of the field in hectares.
b. A square field has a side of 400 yd. Find the area of the field in acres.

Solution
a. $A = (400 \text{ m})^2 = 160{,}000 \text{ m}^2 = \dfrac{160{,}000}{10{,}000} \text{ ha} = 16 \text{ ha}$

b. $A = (400 \text{ yd})^2 = 160{,}000 \text{ yd}^2 = \dfrac{160{,}000}{4840} \text{ acre} \doteq 33.1 \text{ acre}$

Area of a Rectangle

To measure area, we may count the number of units of area contained in any given region. For example, suppose the square in Figure 11-18(a) represents 1 square unit. Then, the rectangle $ABCD$ in Figure 11-18(b) contains $3 \cdot 4$, or 12 square units.

Figure 11-18

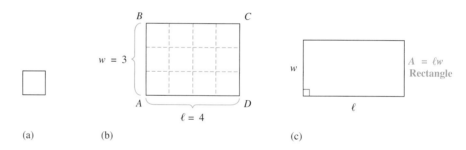

If the unit in Figure 11-18(a) is 1 cm², then the area of rectangle $ABCD$ is 12 cm². In general, the area A of any rectangle may be found by multiplying the lengths of two adjacent sides ℓ and w, or $A = \ell w$, as given in Figure 11-18(c).

Example 11-8 Find the area of each rectangle in Figure 11-19.

Figure 11-19

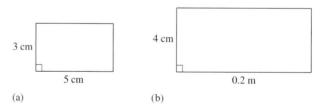

Solution
a. $A = (3 \text{ cm})(5 \text{ cm}) = 15 \text{ cm}^2$

b. First, write the lengths of the sides in the same unit of length. Because 0.2 m = 20 cm, $A = (4 \text{ cm})(20 \text{ cm}) = 80 \text{ cm}^2$. Alternatively, 4 cm = 0.04 m, so $A = (0.04 \text{ m})(0.2 \text{ m}) = 0.008 \text{ m}^2$.

NOW TRY THIS 11-3

- Estimate, without looking, the area in square centimeters of a dollar bill. Measure and calculate how close your estimate is to the actual area.

Area of a Parallelogram

The area of a parallelogram can be found by *reducing the problem to one that we already know how to solve,* in this case, finding the area of a rectangle. To develop the area formula for a parallelogram, complete Now Try This 11-4.

NOW TRY THIS 11-4

- Cut out a parallelogram *ABCD* similar to the one in Figure 11-20(a). Now cut off a shaded triangle as shown and move it to the right to obtain a rectangle.

a. How do the areas of the parallelogram and the rectangle compare? Why?

b. How does this experiment lead to the formula for finding the area of a parallelogram?

Figure 11-20

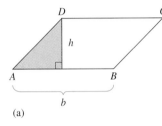
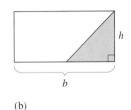

$A = bh$
Parallelogram

(a)   (b)

base · height

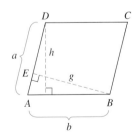

Figure 11-21

In general, any side of a parallelogram can be designated as a **base**. The **height** (h) is the distance between the bases and is always the length of a segment perpendicular to the lines containing the bases. "Now Try This 11-4" shows that the area of parallelogram *ABCD* is given by $A = bh$, that is, the length of the base times the corresponding height. Similarly, in Figure 11-21, *EB*, or g, is the height that corresponds to the bases $\overline{AD}$ and $\overline{BC}$, each of which has measure a. Consequently, the area of the parallelogram *ABCD* is ag. Similarly, its area can be expressed as bh. Therefore $A = ag = bh$.

Area of a Triangle

The formula for the area of a triangle can be derived from the formula for the area of a parallelogram. To explore this, suppose $\triangle BAC$ in Figure 11-22(a) has base b and height h. Let $\triangle BAC'$ be the image of $\triangle BAC$ when $\triangle BAC$ is rotated 180° about M, the midpoint of $\overline{AB}$, as in Figure 11-22(b). Proving that quadrilateral $BCAC'$ is a parallelogram is left as an exercise. Parallelogram $BCAC'$ has area bh and is constructed of congruent triangles *BAC* and

BAC'. So the area of $\triangle ABC$ is $\frac{1}{2}bh$. In general, the area of a triangle is equal to half the product of the length of a side and the altitude to that side.

Figure 11-22

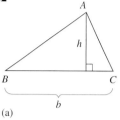
(a)

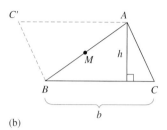
(b)

$A = \frac{1}{2}bh$
Triangle

In Figure 11-23, $\overline{BC}$ is a base of $\triangle ABC$, and the corresponding height h_1, or AE, is the distance from the opposite vertex A to the line containing $\overline{BC}$. Similarly, $\overline{AC}$ can be chosen as a base. Then h_2, or BG, the distance from the opposite vertex B to the line containing $\overline{AC}$, is the corresponding height. If $\overline{AB}$ is chosen as a base, then the corresponding height is h_3, or FC. Thus the area A of $\triangle ABC$ is

$$A = \frac{bh_1}{2} = \frac{ah_2}{2} = \frac{ch_3}{2}.$$

Figure 11-23

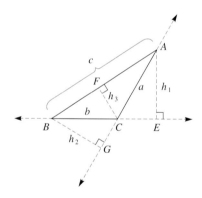

Example 11-9 Find the areas in Figure 11-24. Assume the quadrilaterals in (a) and (b) are parallelograms.

Figure 11-24

Solution
a. $A = bh = (16 \text{ cm})(4 \text{ cm}) = 64 \text{ cm}^2$
b. $A = bh = (5 \text{ cm})(8 \text{ cm}) = 40 \text{ cm}^2$
c. $A = \frac{1}{2}bh = \frac{1}{2}(10 \text{ cm})(4 \text{ cm}) = 20 \text{ cm}^2$
d. $A = \frac{1}{2}bh = \frac{1}{2}(5 \text{ cm})(4 \text{ cm}) = 10 \text{ cm}^2$
e. $A = \frac{1}{2}bh = \frac{1}{2}(2 \text{ cm})(4 \text{ cm}) = 4 \text{ cm}^2$

Area of a Trapezoid

The formula for the area of a trapezoid can also be developed informally, as shown in Now Try This 11-5.

NOW TRY THIS 11-5

- Cut out a trapezoid *ABCD* as shown in Figure 11-25. Copy and place the new trapezoid as shown. Use the figure obtained from the union of the original trapezoid and its image to derive the formula for the area of a trapezoid.

Figure 11-25

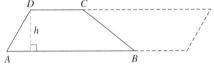

Areas of general polygons can also be found by partitioning the polygons into triangles, finding the areas of the triangles, and summing those areas. In Figure 11-26(a), trapezoid *ABCD* has bases b_1 and b_2 and height h. By connecting points *B* and *D*, as in Figure 11-26(b), we create two triangles: one with base $\overline{AB}$ and height *DE* and the other with base $\overline{CD}$ and height *BF*. Because $\overline{DE} \cong \overline{BF}$, each has length h. Thus the areas of triangles *ADB* and *DCB* are $\frac{1}{2}(b_1 h)$ and $\frac{1}{2}(b_2 h)$, respectively. Hence, the area of trapezoid *ABCD* is $\frac{1}{2}(b_1 h) + \frac{1}{2}(b_2 h), = \frac{1}{2}h(b_1 + b_2)$. That is, the area of a trapezoid is equal to half the height times the sum of the lengths of the bases.

Figure 11-26

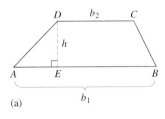
(a)

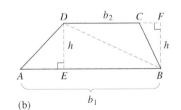
(b)

$A = \frac{1}{2}h(b_1 + b_2)$
Trapezoid

Example 11-10 Find the areas of the trapezoids in Figure 11-27.

Figure 11-27

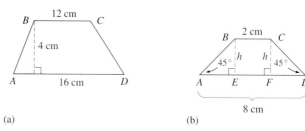

(a) (b)

Solution
a. $A = \frac{1}{2}h(b_1 + b_2) = \frac{1}{2}(4 \text{ cm})(12 \text{ cm} + 16 \text{ cm}) = 56 \text{ cm}^2$

b. To find the area of trapezoid $ABCD$, we use the strategy of determining a subgoal of finding the height, h. In Figure 11-27(b), $BE = CF = h$. Also, $\overline{BE}$ is a side of $\triangle ABE$, which has angles with measures of 45° and 90°. Consequently, the third angle in triangle ABE is $180 - (45 + 90)$, or 45°. Therefore $\triangle ABE$ is isosceles and $AE = BE = h$. Similarly, it follows that $FD = h$. Because $AD = 8 \text{ cm} = h + EF + h$, we could find h if we knew the value of EF. From Figure 11-27(b), $EF = BC = 2$ cm because $BCFE$ is a rectangle (why?) and opposite sides of a rectangle are congruent. Now $h + EF + h = h + 2 + h = 8$ cm.

Thus $h = 3$ cm and the area of the trapezoid is $A = \frac{1}{2}(3 \text{ cm})(2 \text{ cm} + 8 \text{ cm})$, or 15 cm².

Problem Solving   Equal Areas Problem

Larry purchased a plot of land surrounded by a fence. The former owner had subdivided the land into 13 equal-sized square plots, as shown in Figure 11-28. To reapportion the property into two plots of equal area, Larry wishes to build a single, straight fence beginning at the far left corner (point P on the drawing). Is such a fence possible? If so, where should the other end be?

Figure 11-28

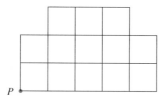

- **Understanding the Problem** We want to divide the land in Figure 11-28 into two plots of equal area by means of a straight fence starting at point P. A *subgoal* is to find the other endpoint. Because the area of the entire plot is 13 square units, the area of each part formed by the fence must be $\frac{1}{2} \cdot 13$, or $6\frac{1}{2}$ square units.

- **Devising a Plan** To find an approximate location for the fence, consider a fence connecting P with point A, as shown in Figure 11-29. The area of the land below fence PA is the sum of the areas of △APD and the rectangle DAFE. The area of △APD is 4 and the area of rectangle DAFE is 2, so the area below the fence $\overline{PA}$ is 4 + 2, or 6, square units. We want an area of $6\frac{1}{2}$ square units. Consequently, the other end of the fence should be above point A.

Figure 11-29

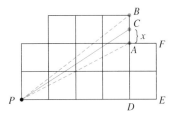

A similar argument shows that the area below $\overline{PB}$ is 8 square units and hence the end of the fence should be below B. Thus the other end of the fence should be at a point C between A and B. To find the exact location of point C, we designate CA by x. We then *write an equation* for x by finding the area below $\overline{PC}$ in terms of x, make the area equal to $6\frac{1}{2}$, and solve for x.

- **Carrying Out the Plan** The area below $\overline{PC}$ equals the area of △PCD plus the area of the rectangle DAFE. The area of △PCD is

$$\frac{PD \cdot DC}{2} = \frac{4(2 + x)}{2} = 2(2 + x) = 4 + 2x.$$

The area of rectangle DAFE is 2, so (4 + 2x) + 2 should equal half the area of the plot. Consequently, we have the following:

$$4 + 2x + 2 = \frac{13}{2}$$

$$2x = \frac{1}{2}$$

$$x = \frac{1}{4}$$

Therefore the fence should be built along the line connecting point P to the point C, which is $\frac{1}{4}$ unit directly above point A. Point C can be found by dividing $\overline{AB}$ into four congruent parts.

- **Looking Back** We check that the solution is correct by finding the area above $\overline{PC}$. The problem can be varied by changing the shape of the plot. Another variation is to ask if Larry could divide the plot into thirds, fourths, and so on. We could also approach the problem as if Larry wanted to subdivide the land but wanted to use the existing lines that mark the squares.

Area of a Regular Polygon

The area of a triangle can be used to find the area of any regular polygon, as illustrated *using a simpler case* involving a regular hexagon in Figure 11-30(a). The hexagon can be separated into 6 congruent triangles, each with a vertex at the center, with side s and height a.

The height of such a triangle of a regular polygon is the *apothem* and is denoted by a. The area of each triangle is $\frac{1}{2}as$. Because six triangles make up the hexagon, the area of the hexagon is $6(\frac{1}{2}as)$, or $\frac{1}{2}a(6s)$. However, $6s$ is the perimeter p of the hexagon, so the area of the hexagon is $\frac{1}{2}ap$. The same process can be used to develop the formula for the area of any regular polygon. That is, the area of any regular polygon is $\frac{1}{2}ap$, where a is the height of one of the triangles involved and p is the perimeter of the polygon, as shown in Figure 11-30(b).

Figure 11-30

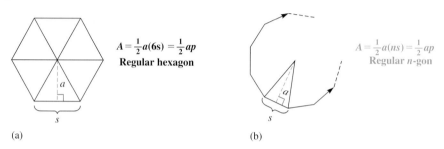

Area of a Circle

We use the strategy of *examining a related problem* to find the area of a circle. The area of a regular polygon inscribed in a circle, as in Figure 11-31, approximates the area of the circle, and we know that the area of any regular n-gon is $\frac{1}{2}ap$, where a is the height of a triangle of the n-gon and p is the perimeter. If the number of sides n is made very large, then the perimeter and the area of the n-gon are close to those of the circle.

Figure 11-31

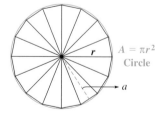

Also, the apothem a is approximately equal to the radius r of the circle, and the perimeter p approximates the circumference $2\pi r$. Because the area of the circle is approximately equal to the area of the n-gon, $\frac{1}{2}ap \doteq \frac{1}{2}r \cdot 2\pi r = \pi r^2$. In fact, the area of the circle is precisely πr^2.

Other approaches exist for leading students to discover the formula for finding the area of a circle. Consider the following student page from *Addison Wesley–Scott-Foresman Middle Grade Math* 2, 1999. In the section "Pie Are Squared" students use graph paper to estimate the area of a circle and then enter their data in a spreadsheet to calculate A/r^2 for various circles. What do you think the spreadsheet will show? In the section "Area of a Circle," notice that a circle can be cut apart and rearranged to form a new figure that looks like a parallelogram. Work through this student page to see how the formula for the area of a circle is developed from that of a parallelogram.

Area of a Circle

11-7

▶ **Lesson Link** You've used a formula to find the circumference of a circle. Now you'll develop and apply a formula for the area of a circle. ◀

You'll Learn ...
■ to find the area of a circle

... How It's Used
The radius of a radio station's signal determines the area it can reach.

Explore | Area of a Circle

Pie Are Squared

Materials: Graph paper, Compass, Spreadsheet software

You can count squares on graph paper to estimate the area of a circle.

1. Use your compass to draw several different-sized circles on graph paper. Make the radius of each circle a whole number.

2. Count squares to estimate the area of each circle.

3. Record the radius (r) and area (A) of each circle in two rows of a spreadsheet. In another row, calculate $\frac{A}{r^2}$ for each circle. What do you notice?

4. Use your conclusion from Step 3 to write an equation with $\frac{A}{r^2}$.

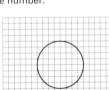

$r = 4$

Learn | Area of a Circle

A dart board is divided into sectors.

If the sectors are cut apart and arranged as shown below, the new figure looks like a parallelogram.

The height of the parallelogram is the radius of the circle, and its base is half the circumference, so its area is $A = bh = \frac{1}{2}Cr$.

Since the circumference of the circle is equal to $2\pi r$,

$A = \frac{1}{2}(2\pi r)r = \pi \cdot r \cdot r = \pi r^2$.

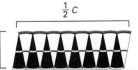

$\frac{1}{2}C$

r

Area of a Sector

sector   A **sector** of a circle is a pie-shaped region of the circle determined by an angle whose vertex is the center of the circle. This angle is a **central angle**. The area of a sector depends on the radius of the circle and the measure of the central angle determining the sector. If the angle has a measure of 90°, as in Figure 11-32(a), the area of the sector is one-fourth the area of the circle, or $\frac{90}{360}\pi r^2$. In any circle, there are 360°, so the area of a sector whose central angle has measure θ degrees is $\frac{\theta}{360}(\pi r^2)$, as shown in Figure 11-32(b).

central angle

Figure 11-32

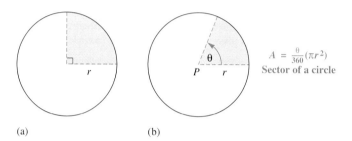

(a)   (b)

$A = \frac{\theta}{360}(\pi r^2)$
Sector of a circle

ONGOING ASSESSMENT 11-2

1. Choose the most appropriate metric units (cm^2, m^2, or km^2) and English units (in.^{2}, yd^2, mi^2) for measuring each of the following:
   a. Area of a sheet of notebook paper
   b. Area of a quarter
   c. Area of a desktop
   d. Area of a classroom floor
   e. Area of a parallel parking space
   f. Area of an airport runway

2. Estimate and then measure each of the following using cm^2, m^2, or km^2.
   a. Area of a door   b. Area of a chair seat
   c. Area of a desktop   d. Area of a chalkboard

3. Complete the following conversion table:

Item	m^2	cm^2	mm^2
a. Area of a sheet of paper		588	
b. Area of a cross-section of a crayon			192
c. Area of a desktop	1.5		
d. Area of a dollar bill		100	
e. Area of a postage stamp		5	

4. Using a calculator, complete the following conversions:
   a. 4000 ft^2 = _____ yd^2 b. 10^6 yd^2 = _____ mi^2
   c. 10 mi^2 = _____ A   d. 3 A = _____ ft^2

5. Complete each of the following:
   a. A football field is about 49 m × 100 m or _____ m^2.
   b. About _____ a are in two football fields.
   c. About _____ ha are in two football fields.

6. Find the areas of each of the following figures if the distance between two adjacent dots in a row or a column is one unit:

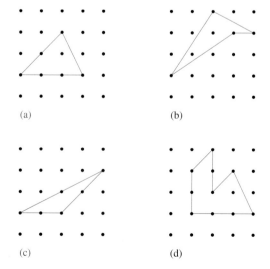

(a)   (b)

(c)   (d)

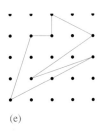
(e)

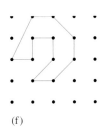
(f)

7. If all vertices of a polygon are points on square-dot paper, the polygon is called a **lattice polygon**. In 1899, G. Pick discovered a surprising theorem involving I, the number of dots *inside* the polygon, and B, the number of dots that lie *on* the polygon. The theorem states that the area of any lattice polygon is $I + \frac{1}{2}B - 1$. Check that this is true for the polygons in problem 6.

8. Find the area of $\triangle ABC$ in each of the following triangles:

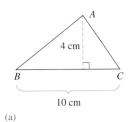
(a)

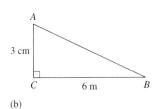
(b)

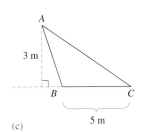
(c)

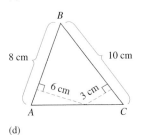
(d)

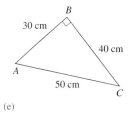
(e)

9. Find the area of each of the following quadrilaterals:

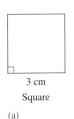
(a) Square

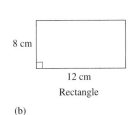
(b) Rectangle

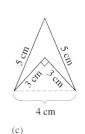
(c)

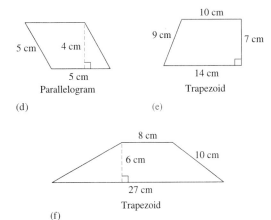
(d) Parallelogram (e) Trapezoid

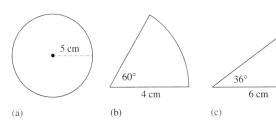
(f) Trapezoid

10. **a.** A rectangular piece of land is 1300 m × 1500 m.
    i. What is the area in square kilometers?
    ii. What is the area in hectares?
  b. A rectangular piece of land is 1300 yd × 1500 yd.
    i. What is the area in square miles?
    ii. What is the area in acres?
  c. Explain which measuring system you would rather use to solve problems like those in (a) and (b).

11. For a parallelogram whose sides are 6 cm and 10 cm, which of the following is true?
  a. The data are insufficient to enable us to determine the area.
  b. The area equals 60 cm².
  c. The area is greater than 60 cm².
  d. The area is less than 60 cm².

12. If the diagonals of a rhombus are a and b units long, find the area of the rhombus in terms of a and b.

13. Find the cost of carpeting the following rectangular rooms:
  a. Dimensions: 6.5 m × 4.5 m; cost = \$13.85/m²
  b. Dimensions: 15 ft × 11 ft; cost = \$30/yd²

14. Find the area of each of the following. Leave your answers in terms of π.

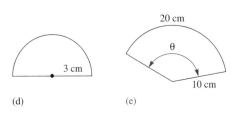
(a) (b) (c)

(d) (e)

15. Joe uses stick-on square carpet tiles to cover his 3 m × 4 m bathroom. If each tile is 10 cm on a side, how many tiles does he need?
16. A rectangular plot of land is to be seeded with grass. If the plot is 22 m × 28 m and a 1-kg bag of seed is needed for 85 m² of land, how many bags of seed are needed?
17. Find the area of each of the following regular polygons:

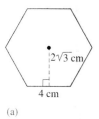
(a)

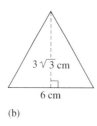
(b)

18. Suppose the largest square peg possible is placed in a circular hole as shown in the following figure and that the largest circular peg possible is placed in a square hole. In which case is there a smaller percentage of space wasted?

19. a. If a circle has a circumference of 8π cm, what is its area?
    b. If a circle of radius r and a square with a side of length s have equal areas, express r in terms of s.
20. Find the area of each of the following shaded parts. Assume all arcs are circular.

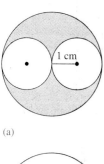
(a)

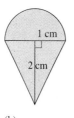
(b)

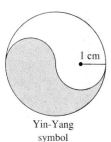
Yin-Yang symbol
(c)

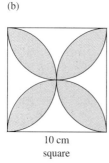
10 cm square
(d)

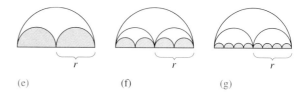
(e) (f) (g)

21. A circular flower bed is 6 m in diameter and has a circular sidewalk around it 1 m wide. Find the area of the sidewalk in square meters.
22. a. If the area of a square is 144 cm², what is its perimeter?
    b. If the perimeter of a square is 32 cm, what is its area?
23. a. What happens to the area of a square when the length of each side is doubled?
    b. If the ratio of the sides of two squares is 1 to 5, what is the ratio of their areas?
24. a. What happens to the area of a circle if its diameter is doubled?
    b. What happens to the area of a circle if its radius is increased by 10%?
    c. What happens to the area of a circle if its circumference is tripled?
25. A rectangular field is 64 m × 25 m. Shawn wants to fence a square field that has the same area as the rectangular field. How long are the sides of the square field?
26. A store has wrapping paper on sale. One package is 3 rolls of $2\frac{1}{2}$ ft × 8 ft for $6.00. Another package is 5 rolls of $2\frac{1}{2}$ ft × 6 ft for $8.00. Which is the better buy?
27. Find the shaded area enclosed by two semicircles as shown in the following figure:

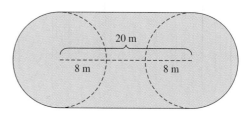

28. An aircraft company starts with a square piece of metal measuring 10 in. × 10 in. and wants to remove a strip x in. wide from all sides to form another square with an area of 64 in.² Find x.
29. The following figure consists of five congruent squares. Find a segment through point P that divides the figure into two parts of equal area.

30. a. Sketch a graph showing the relationship between the length and width of all rectangles with perimeters of 12 cm.
  b. Sketch a graph showing the relationship between the length and width of all rectangles with areas of 12 cm².

31. For a dartboard (see the following figure), Joan is trying to determine how the area of the outside shaded region compares with the area of the inside shaded region so that she can determine payoffs. How do they compare?

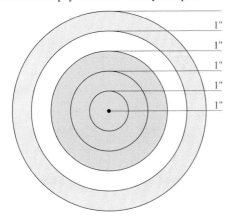

32. Complete and explain how to use geometric shapes to find an equivalent algebraic expression not involving parentheses for each of the following:
  a. $a(b + c)$

  b. $(a + b)(c + d)$

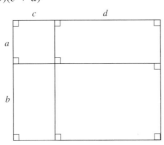

33. Draw 2 rectangles that have the same perimeter but different areas.

34. In the following figure, $\ell \parallel \overleftrightarrow{AB}$. If the area of $\triangle ABP$ is 10 cm², what are the areas of $\triangle ABQ$, $\triangle ABR$, $\triangle ABS$, $\triangle ABT$, and $\triangle ABU$? Explain your answers.

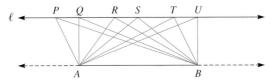

★**35.** In the following figure, quadrilateral $ABCD$ is a parallelogram and P is any point on $\overline{AC}$. Prove that the area of $\triangle BCP$ is equal to the area of $\triangle DPC$.

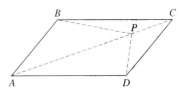

Communication

36. Describe two methods for approximating the area of a circle.

37. a. If a 10-in. (diameter) pizza costs $10, how much should a 20-in. pizza cost? Explain the assumptions you made in your answer.
  b. If the ratio between the diameters of two pizzas is $1:k$, what should the ratio be between the prices? Explain the assumptions you made in your answer.

38. a. Explain how the following drawing can be used to determine a formula for the area of $\triangle ABC$:

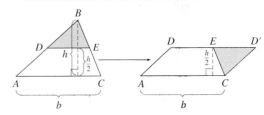

  b. Use paper cutting to reassemble $\triangle ABC$ in (a) into parallelogram $ADD'C$.

39. The area of a parallelogram can be found by using the concept of a half-turn (a turn by 180°). Consider the parallelogram $ABCD$ and let M and N be the midpoints of $\overline{AB}$ and $\overline{CD}$, respectively. Rotate the shaded triangle with vertex M about M by 180° clockwise and rotate the shaded triangle with vertex N about N by 180° counterclockwise. What kind of figure do you obtain? Now complete the argument to find the area of the parallelogram.

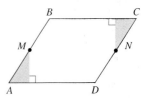

Open-Ended

40. a. Estimate the area in square centimeters that your handprint will cover.
  b. Trace the outline of your hand on square-centimeter grid paper and use the outline to obtain an estimate for the area. Explain how you arrived at your estimate.

41. a. Give dimensions of a square and a rectangle that have the same perimeter but the square has the greater area.
   b. Give dimensions of a square and a rectangle that have equal area but the rectangle has greater perimeter.

Cooperative Learning

42. Use 5 squares to build the following shape and discuss the questions that follow:

   a. What is the area of this shape?
   b. What is the perimeter of this shape?
   c. Add squares to the shape such that each square added touches a complete edge with at least one other square.
      i. What is the minimum number of squares that can be added so that the shape has a perimeter of 18?
      ii. What is the maximum number?
      iii. What is the maximum area the new shape could have and still have a perimeter of 18?
   d. Using the five squares, have members of the group start with shapes different from the one above and answer the questions in (c). Discuss your results.
   e. Explore shapes that are made up of more than 5 squares.

TECHNOLOGY CORNER

Use a geometry utility such as "Geometer's Sketchpad" to explore area concepts such as the following:

1. Complete each of the following:
   a. Construct a segment and label the endpoints A and B.
   b. Choose a point not on $\overline{AB}$ and label it C.
   c. Select $\overline{AB}$ and the point C and construct a line through C parallel to $\overline{AB}$.
   d. Choose a point on the new line and label it D.
   e. Select points A and D and construct a segment connecting points A and D.
   f. Select $\overline{AD}$ and point B and construct a line through B parallel to $\overline{AD}$.
   g. Select the two constructed lines. Construct the point at their intersection and label it E.
   h. Select the vertices (E, D, A, B) and construct the interior of the polygon.
   i. Measure the area of the polygon.
   j. Select $\angle DAB$ and measure the angle.
   k. Move point D along $\overleftrightarrow{DC}$ and find the area of all the new parallelograms formed by $D, C, B,$ and A. How are the areas related?
   l. How does this activity lead to the formula for finding the area of a parallelogram if you know how to find the area of a rectangle?

2. Complete each of the following:
   a. Repeat steps (a) through (d) from part 1.
   b. Select points $A, B,$ and D and construct segments connecting these points to form a triangle.
   c. Select the vertices $A, D,$ and B and construct the interior of the polygon.
   d. Measure the area of $\triangle ADB$.
   e. Move point D along $\overleftrightarrow{DC}$ and find the area of all the triangles that are formed by $A, D,$ and B.
   f. How do these areas compare with the area of the original triangle?
   g. How can this activity be used to motivate the formula for finding the area of a triangle?

3. Take your triangle and line in 2(b) and add a line through B parallel to $\overline{AD}$ to form a parallelogram. Find the area of the parallelogram and compare it to the area of the triangle. How can this activity be used to motivate the formula for finding the area of a triangle?

4. Devise a way to motivate the formula for finding the area of a trapezoid using the geometry utility.

LABORATORY ACTIVITY

1. On a 5 × 5 geoboard, make △DEF as shown in Figure 11-33. Keep the rubber band around D and E fixed and move the vertex F to all the possible locations so that the triangles formed will have the same area as the area of △DEF. How do the locations for the third vertex relate to D and E?

Figure 11-33

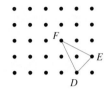

2. On a 5 × 5 geoboard, construct, if possible, squares of areas 1, 2, 3, 4, 5, 6, and 7 square units.

3. On a 5 × 5 geoboard, construct triangles that have areas $\frac{1}{2}$, 1, $1\frac{1}{2}$, 2, ..., until the maximum-sized triangle is reached.

BRAIN TEASER

The rectangle in Figure 11-34(b) was apparently formed by cutting the square in Figure 11-34(a) along the dotted lines and reassembling the pieces as pictured.

1. What is the area of the square in (a)?
2. What is the area of the rectangle in (b)?
3. How do you explain the discrepancy between the areas?

Figure 11-34

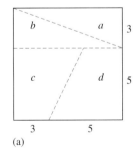
(a)

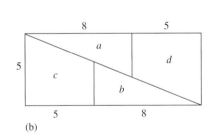
(b)

Section 11-3 — The Pythagorean Theorem

Surveyors often have to calculate distances that cannot be measured directly such as horizontal distances on mountain sides or distances across water, as illustrated in Figure 11-35.

Figure 11-35

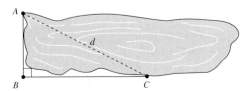

To accomplish these measurements, they use one of the most remarkable and useful theorems in geometry: the Pythagorean Theorem. This theorem was illustrated on a Greek stamp in 1955, as shown in Figure 11-36, to honor the 2500th anniversary of the founding of the Pythagorean School.

Figure 11-36

hypotenuse
legs

In the triangle on the stamp, the side opposite the right angle is the **hypotenuse**. The other two sides are **legs.** Interpreted in terms of area, the Pythagorean Theorem states that the area of a square with the hypotenuse of a right triangle as a side is equal to the sum of the areas of the squares with the legs as sides.

HISTORICAL NOTE

Pythagoras (ca. 582–507 B.C.), a Greek philosopher and mathematician, was head of a group known as the Pythagoreans. Members of the group regarded Pythagoras as a demigod and attributed all their discoveries to him. The Pythagoreans believed in the transmigration of the soul from one body to another. One of Pythagoras's most unusual discoveries was the dependence of the musical intervals on the ratio of the length of strings at the same tension, with the ratio 2:1 giving the octave, 3:2 the fifth, and 4:3 the fourth.

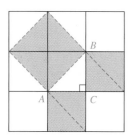

Figure 11-37

Because the Pythagoreans affirmed geometric results on the basis of special cases, mathematical historians believe it is possible they may have discovered the theorem by looking at a floor tiling like the one illustrated in Figure 11-37.

Each square can be divided by its diagonal into two congruent isosceles right triangles, so we see that the shaded square constructed with $\overline{AB}$ as a side consists of four triangles, each congruent to $\triangle ABC$. Similarly, each of the shaded squares with legs $\overline{BC}$ and $\overline{AC}$ as sides consists of two triangles congruent to $\triangle ABC$. Thus the area of the larger square is equal to the sum of the areas of the two smaller squares. The theorem is true in general and is stated as follows using Figure 11-38.

Figure 11-38

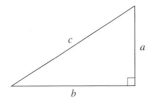

Theorem 11-1

Pythagorean Theorem   If a right triangle has legs of lengths a and b and hypotenuse of length c, then $c^2 = a^2 + b^2$.

There are hundreds of known proofs for the Pythagorean Theorem. The classic book *The Pythagorean Proposition*, by E. Loomis, contains many of these proofs. Some proofs involve the strategy of *drawing diagrams* with a square area c^2 equal to the sum of the areas a^2 and b^2 of two other squares. One such proof is given in Figure 11-39; others are discussed in Ongoing Assessment 11-3. In Figure 11-39(a), the measures of the legs of a right triangle ABC are a and b and the measure of the hypotenuse is c. We draw a square with sides of length $a + b$ and subdivide it, as shown in Figure 11-39(b). In Figure 11-39(c), another square with side of length $a + b$ is drawn and each of its sides is divided into two segments of length a and b, as shown.

Figure 11-39

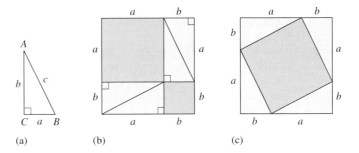

(a)   (b)   (c)

Each yellow triangle is congruent to $\triangle ABC$ (why?). Consequently, each triangle has hypotenuse c and the same area, $\frac{1}{2}ab$. Thus the length of each side of the inside blue quadrilateral in Figure 11-39(c) is c and so the figure is a rhombus. In fact, it is possible to show that the figure is a square whose area is c^2. To complete the proof, we consider the four triangles in Figure 11-39(b) and (c). Because the areas of the sets of four triangles in both Figure 11-39(b) and (c) are equal, the sum of the areas of the two shaded squares in Figure 11-39(b) equals the area of the shaded square in Figure 11-39(c), that is, $a^2 + b^2 = c^2$.

NOW TRY THIS 11-6

- Henry Perigal, a London stockbroker, discovered what has been called the "paper and scissors" proof of the Pythagorean Theorem. It is illustrated in Figure 11-40. Explain how this figure could be used to justify the theorem.

Figure 11-40

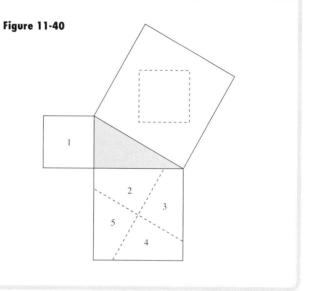

Example 11-11

a. For the drawing in Figure 11-41, find the value of x.

Figure 11-41

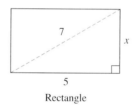

Rectangle

b. The size of a rectangular television screen is given as the length of the diagonal of the screen. If the length of the screen is 24 cm and the width is 18 cm, as shown in Figure 11-42, what is the diagonal length?

Figure 11-42

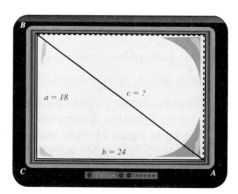

Solution **a.** In the rectangle, the diagonal partitions the rectangle into two right triangles, each with lengths 5 units and width x units. Thus we have the following:

$$5^2 + x^2 = 7^2$$
$$25 + x^2 = 49$$
$$x^2 = 24$$
$$x = \sqrt{24}, \text{ or approximately 4.9 units.}$$

b. A right triangle is formed with the diagonal as the hypotenuse and the legs of measure 24 cm and 18 cm. The Pythagorean Theorem can be used to find the length of the diagonal.

$$c^2 = 18^2 + 24^2$$
$$c^2 = 324 + 576$$
$$c^2 = 900$$
$$c = 30$$

Because all the measurements are in centimeters, the diagonal has length 30 cm.

When using the Pythagorean Theorem, we must work with a right triangle. At times, though, the segment whose length we want to find may not be a side of any known right triangle. The following examples deal with such situations.

Example 11-12 A pole $\overline{BD}$, 28 ft high, is perpendicular to the ground. Two wires $\overline{BC}$ and $\overline{BA}$, each 35 ft long, are attached to the top of the pole and to stakes A and C on the ground as shown in Figure 11-43. If points A, D, and C are collinear, how far are the stakes A and C from each other?

Figure 11-43

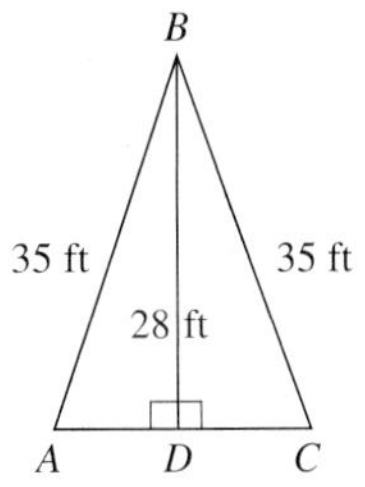

Solution $\overline{AC}$ is not a side in any known right triangle, but we want to find AC. Because a point equidistant from the endpoints of a segment must be on a perpendicular bisector of the segment, then $AD = DC$. Therefore AC is twice as long as DC. Our *subgoal* is to find DC. We may find DC by applying the Pythagorean Theorem in triangle BDC. This results in the following:

$$28^2 + (DC)^2 = 35^2$$
$$(DC)^2 = 35^2 - 28^2$$
$$DC = \sqrt{441}, \text{ or 21 ft}$$
$$AC = 2 \cdot DC = 42 \text{ ft}$$

Example 11-13 How tall is the Great Pyramid of Cheops, a right regular square pyramid, if the base has a side 775 ft and the slant height is 608 ft?

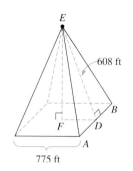

Figure 11-44

Solution In Figure 11-44, $\overline{EF}$ is a leg of a right triangle formed by $\overline{FD}$, $\overline{EF}$, and $\overline{ED}$. Because the pyramid is a right regular pyramid, $\overline{EF}$ intersects the base at its center. Thus $DF = \left(\frac{1}{2}\right)AB$, or $\left(\frac{1}{2}\right)775$, or 387.5 ft. Now ED, the slant height, has length 608 ft, and we can apply the Pythagorean Theorem as follows:

$$(EF)^2 + (DF)^2 = (ED)^2$$
$$(EF)^2 + (387.5)^2 = (608)^2$$
$$(EF)^2 = 219{,}507.75$$
$$EF \doteq 468.5 \text{ ft}$$

Thus the Great Pyramid is approximately 468.5 ft tall.

Special Right Triangles

An isosceles right triangle has two legs of equal length and two 45° angles. Any such triangle is a **45°-45°-90° right triangle**. Drawing a diagonal of a square forms two of these triangles, as shown in Figure 11-45.

45°-45°-90° right triangle

Figure 11-45

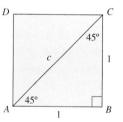

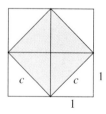

Figure 11-46

In Figure 11-46, we see several 45°-45°-90° triangles. Each side of the shaded square is a hypotenuse of a 45°-45°-90° triangle. The area of the shaded square is 2 square units (why?). Therefore $c^2 = 2$ and $c = \sqrt{2}$. Another way to see that $c = \sqrt{2}$ is to apply the Pythagorean Theorem to one of the nonshaded triangles. Because $c^2 = 1^2 + 1^2 = 2$, then $c = \sqrt{2}$.

In the isosceles right triangle pictured in Figure 11-46, each leg is 1 unit long and the hypotenuse is $\sqrt{2}$ units long. This property is generalized when the isosceles right triangle has a leg of length a, as follows.

Property

Property of 45°-45°-90° triangle: In an isosceles right triangle, if the length of each leg is a, then the hypotenuse has length $a\sqrt{2}$.

Similarly, Figure 11-47(a), shows that a 30°-60°-90° triangle is half of an equilateral triangle. When the equilateral triangle has side 2 units long, then in the 30°-60°-90° triangle, the leg opposite the 30° angle is 1 unit long and the leg opposite the 60° angle has a length of $\sqrt{3}$.

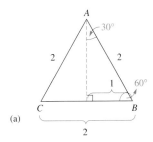

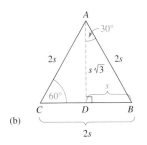

Figure 11-47

This example may also be generalized using the triangle in Figure 11-47(b). When the side of the equilateral triangle ABC is $2s$, then in triangle ABD, the side opposite the 30° angle, $\overline{BD}$, is s units long, and AD may be found using the Pythagorean Theorem to have a length of $s\sqrt{3}$ units. This discussion is summarized in the following property.

Property

Property of 30°-60°-90° triangle In a 30°-60°-90° triangle, the length of the hypotenuse is two times as long as the leg opposite the 30° angle and the leg opposite the 60° angle is $\sqrt{3}$ times the shorter leg.

Converse of the Pythagorean Theorem

The converse of the Pythagorean Theorem is also true. It provided a useful way for early surveyors, sometimes called Egyptian rope stretchers, to determine right angles. Figure 11-48(a) shows a knotted rope with 12 equally spaced knots. Figure 11-48(b) shows how the rope might be held to form a triangle with sides of lengths 3, 4, and 5. The triangle formed is a right triangle and contains a 90° angle.

Figure 11-48

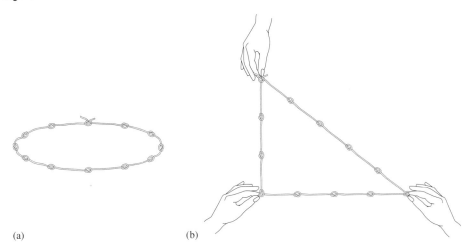

(a)   (b)

Given a triangle with sides of lengths a, b, and c such that $a^2 + b^2 = c^2$, must the triangle be a right triangle? The answer is yes, and we state the following theorem without proof.

Theorem 11-2

Converse of the Pythagorean Theorem If $\triangle ABC$ is a triangle with sides of lengths a, b, and c such that $a^2 + b^2 = c^2$, then $\triangle ABC$ is a right triangle with the right angle opposite the side of length c.

Example 11-14 Determine if the following can be the lengths of the sides of a right triangle:

a. 51, 68, 85
b. 2, 3, $\sqrt{13}$
c. 3, 4, 7

Solution
a. $51^2 + 68^2 = 7225 = 85^2$, so 51, 68, and 85 can be the lengths of the sides of a right triangle.
b. $2^2 + 3^2 = 4 + 9 = 13 = (\sqrt{13})^2$, so 2, 3, and $\sqrt{13}$ can be the lengths of the sides of a right triangle.
c. $3^2 + 4^2 \neq 7^2$, so the measures cannot be the lengths of the sides of a right triangle. In fact, since $3 + 4 = 7$, these segments do not form a triangle.

NOW TRY THIS 11-7

a. Draw three segments that could be used to form the sides of a right triangle and discuss how you would show that these three lengths determine a right triangle.

b. Multiply the lengths of the three segments in (a) by a fixed number and determine if the resulting three lengths could be sides of a right triangle.

c. Using three new numbers, repeat the experiment in (a) and (b). Form a conjecture based on your experiments.

The Distance Formula: An Application of the Pythagorean Theorem

Given the coordinates of two points A and B, we can find the distance AB. We first consider the special case in which the two points are on one of the axes. For example, in Figure 11-49(a), $A(2, 0)$ and $B(5, 0)$ are on the x-axis. The distance between these two points is 3 units:

$$AB = OB - OA = 5 - 2 = 3$$

Figure 11-49

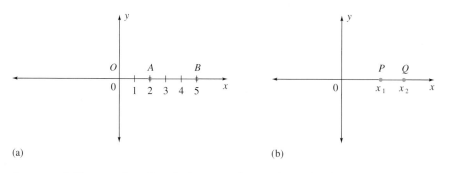

(a) (b)

In general, if two points P and Q are on the x-axis, as in Figure 11-49(b), with x-coordinates x_1 and x_2, respectively, and $x_2 > x_1$, then $PQ = x_2 - x_1$. In fact, *the distance between two points on the x-axis is always the absolute value of the difference between the x-coordinates of the points* (why?). A similar result holds for any two points on the y-axis.

Figure 11-50 shows two points in the plane: $C(2, 5)$ and $D(6, 8)$. The distance between C and D can be found by using the strategy of *looking at a related problem*. We know how to find the length of a segment if the segment is a side or the hypotenuse of a right triangle. We obtain a right triangle by drawing perpendiculars from the points to the x-axis and to the y-axis, respectively, thus defining triangle CDE. The lengths of the legs of triangle CDE are found by using horizontal and vertical distances and properties of rectangles.

$$CE = |6 - 2| = 4$$
$$DE = |8 - 5| = 3$$

Figure 11-50

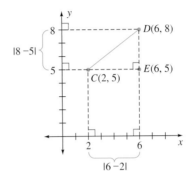

The distance between C and D can be found by applying the Pythagorean Theorem to the triangle.

$$CD^2 = DE^2 + CE^2$$
$$= 3^2 + 4^2$$
$$= 25$$
$$CD = \sqrt{25}, \text{ or } 5$$

The method can be generalized to find a formula for the distance between any two points $A(x_1, y_1)$ and $B(x_2, y_2)$. Construct a right triangle with $\overline{AB}$ as one of its sides by drawing a segment through A parallel to the x-axis and a segment through B parallel to the y-axis, as shown in Figure 11-51. The lines containing the segments intersect at point C, forming right triangle ABC. Now, apply the Pythagorean Theorem.

Figure 11-51

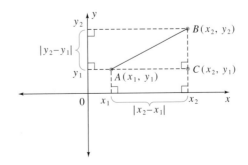

In Figure 11-51, we see that $AC = |x_2 - x_1|$ and $BC = |y_2 - y_1|$. By the Pythagorean Theorem, $(AB)^2 = |x_2 - x_1|^2 + |y_2 - y_1|^2$, and consequently $AB = \sqrt{|x_2 - x_1|^2 + |y_2 - y_1|^2}$. Because $|x_2 - x_1|^2 = (x_2 - x_1)^2$ and $|y_2 - y_1|^2 = (y_2 - y_1)^2$, $AB = \sqrt{(x_2 - x_1)^2 + (y_2 - y_1)^2}$. This result is known as the **distance formula**.

distance formula

Distance Formula

The distance between the points $A(x_1, y_1)$ and $B(x_2, y_2)$ is given by

$$AB = \sqrt{(x_2 - x_1)^2 + (y_2 - y_1)^2}.$$

NOW TRY THIS 11-8

- Investigate whether it makes any difference in the distance formula if $(x_1 - x_2)$ and $(y_1 - y_2)$ are used instead of $(x_2 - x_1)$ and $(y_2 - y_1)$, respectively.

Example 11-15 For each of the following, determine the distance between P and Q:

a. $P(2, 7), Q(3, 5)$
b. $P(0, 0), Q(3, {}^-4)$

Solution
a. $PQ = \sqrt{(3 - 2)^2 + (5 - 7)^2} = \sqrt{1 + 4} = \sqrt{5}$
b. $PQ = \sqrt{(0 - 3)^2 + [0 - ({}^-4)]^2} = \sqrt{9 + 16} = \sqrt{25} = 5$

Example 11-16
a. Show that $A(7, 4), B({}^-2, 1)$, and $C(10, {}^-5)$ are the vertices of an isosceles triangle.
b. Show that $\triangle ABC$ in (a) is a right triangle.

Solution a. Using the distance formula, we find the lengths of the sides.

$$AB = \sqrt{({}^-2 - 7)^2 + (1 - 4)^2} = \sqrt{({}^-9)^2 + ({}^-3)^2} = \sqrt{90}$$
$$BC = \sqrt{[10 - ({}^-2)]^2 + ({}^-5 - 1)^2} = \sqrt{12^2 + ({}^-6)^2} = \sqrt{180}$$
$$AC = \sqrt{(10 - 7)^2 + ({}^-5 - 4)^2} = \sqrt{3^2 + ({}^-9)^2} = \sqrt{90}$$

Thus $AB = AC$, and so the triangle is isosceles.

b. Because $(\sqrt{90})^2 + (\sqrt{90})^2 = (\sqrt{180})^2$, $\triangle ABC$ is a right triangle with $\overline{BC}$ as hypotenuse and $\overline{AB}$ and $\overline{AC}$ as legs.

ONGOING ASSESSMENT 11-3

1. Use the Pythagorean Theorem to find x in each of the following:

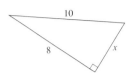
(a)

(b)

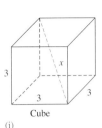
Cube
(i)

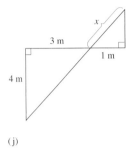
(j)

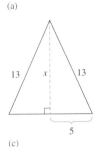
(c)

Equilateral triangle
(d)

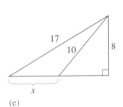
(e)

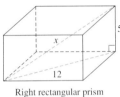
Right rectangular prism
(f)

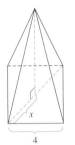
Right square pyramid
(g)

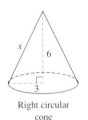
Right circular cone
(h)

2. If the hypotenuse of a right triangle is 30 cm long and one leg is twice as long as the other, how long are the legs of the triangle?

3. For each of the following, determine whether the given numbers represent lengths of sides of a right triangle:
   a. 10, 24, 16
   b. 16, 34, 30
   c. $\sqrt{2}, \sqrt{2}, 2$
   d. $\frac{3}{2}, \frac{4}{2}, \frac{5}{2}$

4. What is the longest line segment that can be drawn in a right rectangular prism that is 12 cm wide, 15 cm long, and 9 cm high?

5. Two airplanes depart from the same place at 2:00 P.M. One plane flies south at a speed of 376 km/hr, and the other flies west at a speed of 648 km/hr. How far apart are the airplanes at 5:30 P.M.?

6. Starting from point A, a boat sails due south for 6 mi, then due east for 5 mi, and then due south for 4 mi. How far is the boat from A?

7. a. In the cartoon below, is Hobbes's square really a square? Why?
   b. If Hobbes's figure in the second panel is a rectangle that is 6 units by 3 units, how long is y?
   c. If Hobbes draws a rectangle with each dimension doubled, how does the length of the diagonal change?

8. A 15-ft ladder is leaning against a wall. The base of the ladder is 3 ft from the wall. How high above the ground is the top of the ladder?

9. In the following figure, two poles are 25 m and 15 m high. A cable 14 m long joins the tops of the poles. Find the distance between the poles.

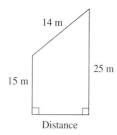

10. Find the area of each of the following:

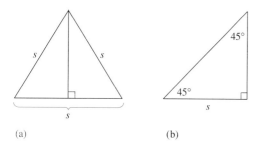

(a)   (b)

11. For each of the following, solve for the unknowns:

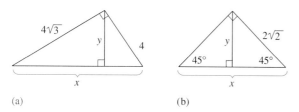

(a)   (b)

12. A builder needs to calculate the dimensions of a regular hexagonal window. Assuming the height CD of the window is 1.3 m, find the width AB (O is the midpoint of $\overline{AB}$) in the following figure:

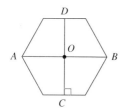

13. The length of the diagonal $\overline{AC}$ of a rhombus $ABCD$ is 20 cm. The distance between $\overline{AB}$ and $\overline{DC}$ is 12 cm. Find the length of the sides of the rhombus and the length of the other diagonal.

14. If $\overline{AB}$, a diameter of circle O, has length 10 cm, point C is on circle O, and AC is 6 cm, how long is $\overline{BC}$?

15. Georgette wants to put a diagonal brace on a gate that is 3 ft wide and 5 ft high. If she uses a board that is 6 in. wide and 8 ft long, how much will she have left?

16. If a third baseman on the base throws to first base, how far is the ball thrown? (*Hint:* The distance from home plate to first base is 90 ft.)

17. What is the longest piece of straight spaghetti that will fit in a cylindrical can that has a radius of 2 in. and height of 10 in.?

18. If possible, draw a square with the given number of square units on a geoboard grid. (You will have to draw your own geoboard grid.)
    a. 5
    b. 7
    c. 8
    d. 14
    e. 15

19. Use the following drawing to prove the Pythagorean Theorem by using corresponding parts of similar triangles $\triangle ACD$, $\triangle CBD$, and $\triangle ABC$. Lengths of sides are indicated by a, b, c, x, and y. (*Hint:* Show that $b^2 = cx$ and $a^2 = cy$.)

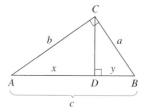

20. An access ramp enters a building 1 m above ground level and starts 3 m from the building. How long is the ramp?

21. To make a homeplate for a neighborhood baseball park, we can cut the plate from a square, as shown in the following figure. If A, B, and C are midpoints of the sides of the square, what are the dimensions of the square to the nearest tenth of an inch?

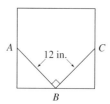

22. A company wants to lay cable across a lake. To find the length of the lake, they made the following measurements. What is the length of the lake?

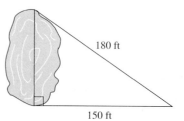

23. A CB radio station C is located 3 mi from the interstate highway h. The station has a range of 6.1 mi in all directions from the station. If the interstate is along a straight line, how many miles of highway are in the range of this station?

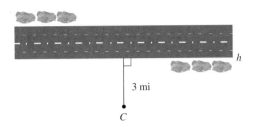

24. Before James Garfield was elected President of the United States, he discovered a proof of the Pythagorean Theorem. He formed a trapezoid like the one that follows and found the area of the trapezoid in two ways. Can you discover his proof?

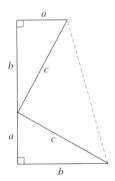

25. Use the following figure to prove the Pythagorean Theorem by first proving that the quadrilateral with side c is a square. Then, compute the area of the square with side $a + b$ in two ways: (a) as $(a + b)^2$ and (b) as the sum of the areas of the 4 triangles and the square with side c.

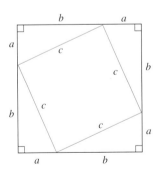

26. Construct semicircles on right triangle ABC with $\overline{AB}$, $\overline{BC}$, and $\overline{AC}$ as diameters. Is the area of the semicircle on the hypotenuse equal to the sum of the areas of the semicircles on the legs? Why?

27. On each side of a right triangle, construct an equilateral triangle. Is the area of the triangle constructed on the hypotenuse always equal to the sum of the areas of the triangles constructed on the legs? Why?

28. For each of the following, find the length of $\overline{AB}$:
    a. $A(0, 3), B(0, 7)$
    b. $A(0, 3), B(4, 0)$
    c. $A(^-1, 2), B(3, ^-4)$
    d. $A(4, ^-5), B\left(\dfrac{1}{2}, \dfrac{^-7}{4}\right)$

29. Find the perimeter of the triangle with vertices at $A(0, 0)$, $B(^-4, ^-3)$, and $C(^-5, 0)$.

30. Show that $(0, 6)$, $(^-3, 0)$, and $(9, ^-6)$ are the vertices of a right triangle.

31. Show that the triangle whose vertices are $A(^-2, ^-5)$, $B(1, ^-1)$, and $C(5, 2)$ is isosceles.

32. Find x if the distance between $P(1, 3)$ and $Q(x, 9)$ is 10 units.

33. If the hypotenuse in a 30°-60°-90° triangle is $c/2$ units, what is the length of the side opposite the 60° angle? Explain your answer.

Communication

34. Given the following square, describe how to use a compass and a straightedge to construct a square whose area is as follows:
    a. Twice the area of the given square
    b. Half the area of the given square

35. If the hypotenuse and a leg of one right triangle are congruent to the hypotenuse and a leg of another right triangle, respectively, must the triangles be congruent? Explain.

36. Gail tried the Egyptian method of using a knotted rope to determine a right angle so that she could build a shed. She placed her knots so that each was 1 ft from the next. She stretched out her rope in the form of a triangle whose sides were of lengths 5, 12, and 13 ft. Did she have a right angle? Explain why or why not.

Open-Ended

37. Draw several kinds of triangles including a right triangle. Draw a square on each of the sides of the triangles. Compute the areas of the squares and use this information to investigate whether the Pythagorean Theorem works for only right triangles. Use a geometry utility if available.

38. Find an application from real life in which knowing the Pythagorean Theorem would be useful. Write a problem about the application to share with the class.

39. **Pythagorean triples** are three natural numbers a, b, and c that satisfy the relationship $a^2 + b^2 = c^2$. The least three

numbers that are Pythagorean triples are 3-4-5. Another triple is 5-12-13 because $5^2 + 12^2 = 13^2$.
a. Find two other Pythagorean triples.
b. Does doubling each number in a Pythagorean triple result in a new Pythagorean triple? Why or why not?
c. Does adding a fixed number to each number in a Pythagorean triple result in a new Pythagorean triple? Why or why not?
d. Suppose $a = 2uv$, $b = u^2 - v^2$, and $c = u^2 + v^2$, where u and v are whole numbers. Determine whether a-b-c is a Pythagorean triple.

Cooperative Learning

40. There are more than 300 proofs of the Pythagorean Theorem. Have each person in the group find a proof not given in the text and present the proof to the group. Decide on your favorite proof and be prepared to present it to the class.
41. Have each person in the group use a 1-m string to make a different right triangle. Measure each side to the nearest centimeter. Use these measurements to see if the Pythagorean Theorem holds for your measurements. If not, explain why the results may not be exact.

Review Problems

42. Arrange the following in decreasing order: 3.2 m, 322 cm, 0.032 km, 3.020 mm.
43. Find the area of each of the following figures:

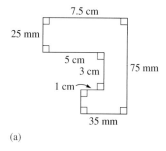

(a)

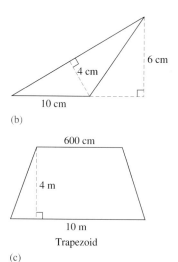

44. Complete the following table, which concerns circles:

	Radius	Diameter	Circumference	Area
a.	5 cm			
b.		24 cm		
c.				$17\pi m^2$
d.			20π cm	

45. A 10-m wire is wrapped around a circular region. If the wire fits exactly, what is the area of the region?

BRAIN TEASER A spider is sitting at A, the midpoint of the edge of the ceiling in the room shown in Figure 11-52. It spies a fly on the floor at C, the midpoint of the edge of the floor. If the spider must walk along the wall, ceiling, or floor, what is the length of the shortest path the spider can travel to reach the fly?

Figure 11-52

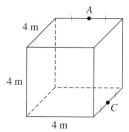

TECHNOLOGY CORNER

Use a geometry utility to determine the relationship between the length of the hypotenuse of a 45°-45°-90° triangle and the length of a leg.

a. Construct a 45°-45°-90° triangle, label the vertices as in Figure 11-53, and measure the lengths of the sides. Record the data for triangle 1 in the following table and compute the ratio.

Figure 11-53

	AC	CB	AB	AB/CB
Triangle 1				
Triangle 2				
Triangle 3				
Triangle 4				

b. Repeat (a) for three other triangles.
c. Make a conjecture about the relationship between the length of the hypotenuse and the length of a leg for these triangles.
d. Given a 30°-60°-90° triangle, determine the relationship between the lengths of the hypotenuse and the shorter leg and the relationships between the lengths of the longer and shorter legs.

Section 11-4 — Surface Areas

Painting houses, buying roofing, seal-coating driveways, and buying carpet are among the common applications that involve computing areas. In many real-world problems, we must find the surface areas of such three-dimensional figures as prisms, cylinders, pyramids, cones, and spheres. Formulas for finding these areas are usually based on finding the area of two-dimensional pieces of the three-dimensional figures. In this section, we use the notion of a **net**, a two-dimensional pattern that can be used to construct three-dimensional figures, to aid in determining surface areas of the figures.

Surface Area of Right Prisms

Consider the cereal box shown in Figure 11-54(a). If we ignore the flaps for gluing the box together, to find the amount of cardboard necessary to make the box, we cut the box along the edges and make it lie flat as shown in Figure 11-54(b). When we do this we obtain a *net* for the box. The box is composed of a series of rectangles. We find the area of each rectangle and sum those areas to find the surface area of the box.

Figure 11-54

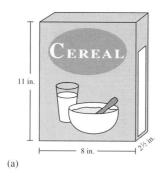
(a)

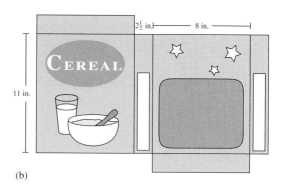
(b)

NOW TRY THIS 11-9

- **a.** Find the surface area of the box in Figure 11-54.
- **b.** Could the box be made from a rectangular piece of cardboard 21 in. by 15 in.? If not, what size rectangle could you use and how would you do it?

A similar process can be used for many three-dimensional figures. For example, the surface area of the cube in Figure 11-55(a) is the sum of the areas of the faces of the cube. Because each of the six faces is a square of area 16 cm², the surface area is 6 · (16 cm²), or 96 cm², or in general, for a cube whose edge is e units as in Figure 11-55(b), the surface area is $6e^2$.

Figure 11-55

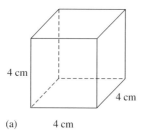
(a)

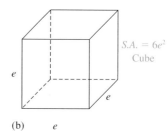
(b)

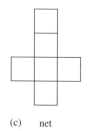
(c) net

To find the surface area of a right prism, we find the sum of the areas of the rectangles that make up the lateral faces and the areas of the top and bottom. The sum of the areas of the lateral faces is the **lateral surface area**. The **surface area** (S.A.) is the sum of the lateral surface area and the area of the bases.

NOW TRY THIS 11-10

● Figure 11-56 shows a right pentagonal prism with a net for the prism. If B stands for the area of each of the prism's bases, show that the surface area of the prism could be computed as $S.A. = ph + 2B$, where p is the perimeter of the base of the prism and h is the height. Does this formula hold for all right prisms? Why or why not? ●

Figure 11-56

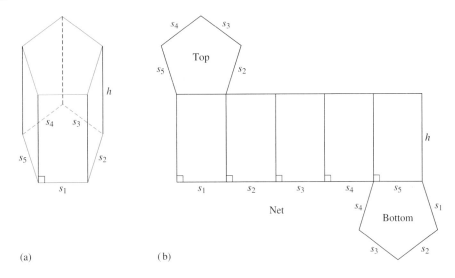

(a)   (b)

Example 11-17   Find the surface area of each of the right prisms in Figure 11-57.

Figure 11-57

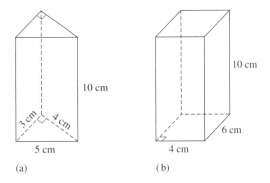

(a)   (b)

Solution   **a.** Each base is a right triangle. The area of the bases is $2(1/2 \cdot 3 \cdot 4)$, or 12 cm^2. The area of the three lateral faces is $4 \cdot 10 + 3 \cdot 10 + 5 \cdot 10$, or 120 cm^2. Thus the surface area is $12 \text{ cm}^2 + 120 \text{ cm}^2$, or 132 cm^2.

b. The area of the bases is 2(4 · 6), or 48 cm². The lateral surface area is 2 · (10 · 6) + 2 · (4 · 10), or 200 cm². Thus the surface area is 248 cm².

Surface Area of a Cylinder

To find the surface area of the right circular cylinder shown in Figure 11-58, we cut off the bases and slice the lateral surface open by cutting along any line perpendicular to the bases. Such a slice is shown as a dotted segment in Figure 11-58(a). Then we unroll the cylinder to form a rectangle, as shown in Figure 11-58(b). To find the total surface area, we find the area of the rectangle and the areas of the top and bottom circles. The length of the rectangle is the circumference of the circular base $2\pi r$, and its width is the height of the cylinder h. Hence, the area of the rectangle is $2\pi r h$. The area of each base is πr^2. Because the surface area is the sum of the areas of the two circular bases and the lateral surface area, we have

Figure 11-58

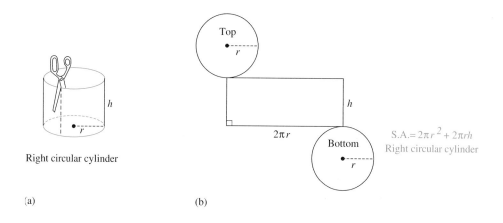

Surface Area of a Pyramid

The surface area of a pyramid is the sum of the lateral surface area of the pyramid and the area of the base. A right regular pyramid is a pyramid such that the segments connecting the apex to each vertex of the base are congruent and the base is a regular polygon. The lateral faces of the right regular pyramid pictured in Figure 11-59 are congruent triangles. Each triangle has an altitude of length ℓ, called the *slant height*. Because the pyramid is right regular, each side of the base has the same length b. To find the lateral surface area of a right regular pyramid, we need to find the area of one face $\frac{1}{2}b\ell$ and multiply it by n, the number of faces. Adding the lateral surface area $n\left(\frac{1}{2}b\ell\right)$ to the area of the base B gives the surface area.

Figure 11-59

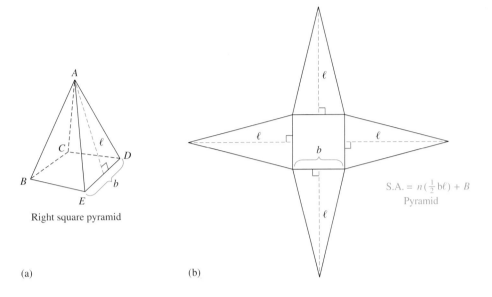

Right square pyramid

S.A. = $n(\frac{1}{2}b\ell) + B$
Pyramid

(a)   (b)

Example 11-18 Find the surface area of the right regular pyramid in Figure 11-60.

Solution The surface area consists of the area of the square base plus the area of the four triangular faces. Hence, the surface area is

$$4 \text{ cm} \cdot 4 \text{ cm} + 4 \cdot \left(\frac{1}{2} \cdot 4 \text{ cm} \cdot 5 \text{ cm}\right) = 16 \text{ cm}^2 + 40 \text{ cm}^2$$
$$= 56 \text{ cm}^2.$$

Figure 11-60

Example 11-19 The Great Pyramid of Egypt is a right square pyramid with a height of 148 m and a square base with perimeter of 930 m. The altitude of each triangular face is 188 m. The basic shape of the Transamerica Building in San Francisco is a right square pyramid that has a height of 260 m and a square base with a perimeter of 140 m. The altitude of each triangular face is 261 m. How do the lateral surface areas of the two structures compare?

Solution The length of one side of the square base of the Great Pyramid is $\frac{930}{4}$, or 232.5, m. Likewise the length of one side of the square base of the Transamerica Building is 35 m. The lateral surface area (*L.S.A.*) of the two are computed below.

(Great Pyramid) $L.S.A. = 4 \cdot (1/2 \cdot 232.5 \cdot 188) = 87{,}420 \text{ m}^2$
(Transamerica) $L.S.A. = 4 \cdot (1/2 \cdot 35 \cdot 261) = 18{,}270 \text{ m}^2$

Therefore the lateral surface area of the Great Pyramid is approximately 4.8 times greater than that of the Transamerica Building.

Surface Area of a Cone

It is possible to find a formula for the surface area of a cone by approximating the cone with a pyramid. As shown in Figure 11-61, we inscribe in the circular base of the cone a regular polygon with many sides. The polygon can be used as the base of a regular right pyramid. The lateral surface area of the pyramid is close to the lateral surface area of the cone. The greater the number of faces of the pyramid, the closer the surface area of the pyramid is to that of the cone. The lateral surface of the pyramid is $\frac{1}{2}p \cdot h$, where p is the perimeter of the base and h is the height of each triangle. With many sides in the pyramid, the perimeter of its base is close to the perimeter of the circle, $2\pi r$. The height of each triangle of the pyramid is close to the slant height ℓ, a segment that connects the vertex of the cone with a point on the circular base, as shown in Figure 11-61(b). Consequently, it is reasonable that the lateral surface of the cone becomes $\frac{1}{2} \cdot 2\pi r \cdot \ell$, or $\pi r \ell$. To find the total surface area of the cone, we add πr^2, the area of the base. Thus $S.A. = \pi r^2 + \pi r \ell$.

Figure 11-61

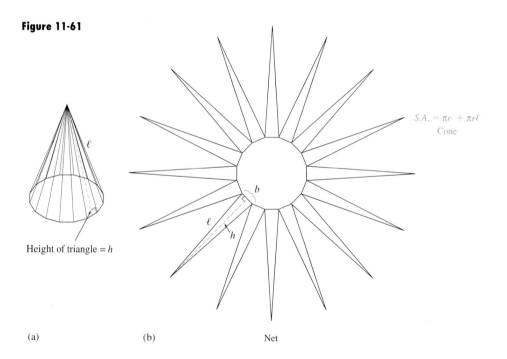

(a)   (b)   Net

Example 11-20 Given the cone in Figure 11-62, find the surface area of that cone.

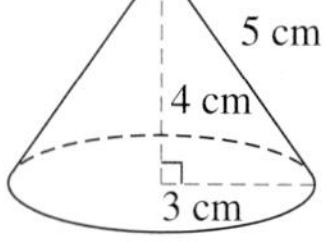

Figure 11-62
Right circular cone

Solution The base of the cone is a circle with radius 3 cm and area $\pi(3 \text{ cm})^2$, or $9\pi \text{ cm}^2$. The lateral surface has area $\pi(3 \text{ cm})(5 \text{ cm})$, or $15\pi \text{ cm}^2$. Thus we have the following surface area:

$$S.A. = \pi(3 \text{ cm})^2 + \pi(3 \text{ cm})(5 \text{ cm})$$
$$= 9\pi \text{ cm}^2 + 15\pi \text{ cm}^2$$
$$= 24\pi \text{ cm}^2$$

Surface Area of a Sphere

Finding a formula for the surface area of a sphere is a simple task using calculus, but it is not easy in elementary mathematics. The surface area of a sphere is four times the area of a great circle of the sphere. Therefore the formula is $S.A. = 4\pi r^2$, as pictured in Figure 11-63.

Figure 11-63

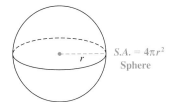

$S.A. = 4\pi r^2$
Sphere

LABORATORY ACTIVITY

Create different cones from sectors of a circle. Use a compass to draw a sector of a circle whose diameter is almost as large as the width of a page of paper. Draw two such sectors that have the same radii but different central angles. In one sector, make the central angle measure smaller than 180°, and in the other, make it measure greater than 180°. Then make a cone from each sector by gluing the edges of each sector together. Can you predict which cone will be taller? Without performing the experiment, can you explain why that cone will be taller?

ONGOING ASSESSMENT 11-4

1. Find the surface area of each of the following:

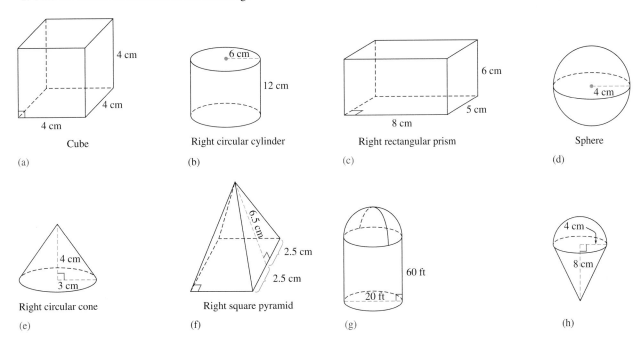

(a) Cube
(b) Right circular cylinder
(c) Right rectangular prism
(d) Sphere
(e) Right circular cone
(f) Right square pyramid
(g)
(h)

2. How many liters of paint are needed to paint the walls of a room that is 6 m × 4 m × 2.5 m if 1 L of paint covers 20 m²? (Assume there are no doors or windows.)

3. The napkin ring pictured in the following figure is to be resilvered. How many square millimeters must be covered?

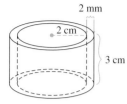

4. Assume the radius of Earth is 6370 km and Earth is a sphere. What is its surface area?

5. Two cubes have sides of length 4 cm and 6 cm, respectively. What is the ratio of their surface areas?

6. Suppose one cylinder has radius 2 m and height 6 m and another has radius 6 m and height 2 m.
   a. Which cylinder has the greater lateral surface area?
   b. Which cylinder has the greater total surface area?

7. The base of a right pyramid is a regular hexagon with sides of length 12 m. The altitude of the pyramid is 9 m. Find the total surface area of the pyramid.

8. A soup can has a $2\frac{5}{8}$-in. diameter and is 4 in. tall. What is the area of the paper that will be used to make the label for the can if the paper covers the entire lateral surface area?

9. A square piece of paper 10 cm on a side is rolled to form the lateral surface area of a cylinder and then a top and bottom are added. What is the surface area of the cylinder?

10. Approximately how much material is needed to make the tent illustrated in the following figure (both ends and the bottom should be included)?

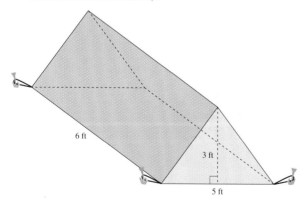

11. The top of a rectangular box has an area of 88 cm². The sides have area 32 cm² and 44 cm². What are the dimensions of the box?

12. How does the surface area of a box (including top and bottom) change if
    a. each dimension is doubled?
    b. each dimension is tripled?
    c. each dimension is multiplied by a factor of k?

13. How does the lateral surface area of a cone change if
    a. the slant height is tripled but the radius of the base remains the same?
    b. the radius of the base is tripled but the slant height remains the same?
    c. the slant height and the radius of the base are tripled?

14. What happens to the surface area of a sphere if the radius is
    a. doubled?
    b. tripled?

15. Find the surface area of a square pyramid if the area of the base is 100 cm² and the height of the pyramid is 20 cm.

16. Suppose a structure is composed of unit cubes with at least one face of each cube connected to the face of another cube, as shown in the following figure:

   a. If one cube is added, what is the maximum surface area the structure can have?
   b. If one cube is added, what is the minimum surface area the structure can have?
   c. Is it possible to design a structure so that one can add a cube and yet add nothing to the surface area of the structure? (*Hint:* Cubes might have to be glued together.) Explain your answer.

17. The sector shown in the following figure is rolled into a cone so that the dotted edges just touch. Find the following:
    a. The lateral surface area of the cone
    b. The total surface area of the cone

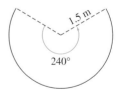

18. Each region in the following figure revolves about the indicated axis. For each case, sketch the three-dimensional figure obtained and find its surface area.

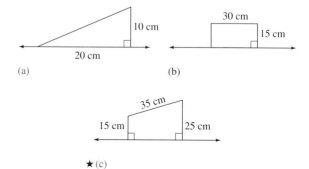

★ (c)

19. The total surface area of a cube is 10,648 cm². What is the length of each of the following?
   a. One of the sides
   b. A diagonal that is not a diagonal of a face
★20. Find the total surface area of the following stand, which was cut from a right circular cone:

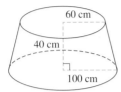

★21. In the following figure, a cylinder is inscribed in a cone. Find the lateral surface area of the cylinder if the height of the cone is 40 cm, the height of the cylinder 30 cm, and the radius of the base of the cone is 25 cm.

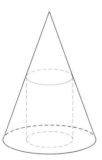

Communication

22. Which do you think would melt faster in an ice chest: a block of ice made from a gallon of water or ice cubes made from a gallon of water? Explain why.
23. A student wonders if she doubles each measurement of a cereal box, will she need twice as much cardboard to make the new box. How would you help her decide?
24. Tennis balls are packed tightly three to a can that is shaped like a cylinder.
   a. Estimate how the surface area of the balls compares to the lateral surface area of the can. Explain how you arrived at your estimate.
   b. See how close your estimate in (a) was by actually computing the surface area of the balls and the lateral surface area of the can.

Open-Ended

25. One method of estimating body surface area in burn victims uses the fact that 100 handprints will approximately cover the whole body.
   a. What percentage of the body surface area is the surface area of two handprints?
   b. Estimate the percentage of the body surface area of one arm. Explain how you arrived at your estimate.
   c. Estimate your body surface area in square centimeters. Explain how you arrived at your estimate.
   d. Find the area of the flat part of your desk. How does the area of the desk compare with the surface area of your body?
26. Design a net for a polyhedron in such a way that the surface area of the polyhedron is 10 cm². Explain what polyhedron the net will form and why its surface area is 10 cm².

Cooperative Learning

27. a. Shawn used small cubes to build a bigger cube that was solid and was three cubes long on each side. He then painted all the sides of the new, large cube red. He dropped the newly painted cube and all the little cubes came apart. He noticed that some cubes had only one side painted, some had two sides painted, and so on. Describe the number of cubes with 0, 1, 2, 3, 4, 5, or 6 sides painted. Have each member of the group choose a different number of sides and then combine your data to see if it makes sense. Look for any patterns that occur.
   b. What would the answers be if the large cube was four small cubes long on a side?
   c. Make a conjecture about how to count the cubes if the large cube were n small cubes long on a side.

Review Problems

28. Complete each of the following:
   a. $10 \text{ m}^2 = \underline{\hspace{1cm}} \text{ cm}^2$
   b. $13{,}680 \text{ cm}^2 = \underline{\hspace{1cm}} \text{ m}^2$
   c. $5 \text{ cm}^2 = \underline{\hspace{1cm}} \text{ mm}^2$
   d. $2 \text{ km}^2 = \underline{\hspace{1cm}} \text{ m}^2$
   e. $10^6 \text{ m}^2 = \underline{\hspace{1cm}} \text{ km}^2$
   f. $10^{12} \text{ mm}^2 = \underline{\hspace{1cm}} \text{ m}^2$
29. The sides of a rectangle are 10 cm and 20 cm long. Find the length of a diagonal of the rectangle.
30. The length of the side of a rhombus is 30 cm. If the length of one diagonal is 40 cm, find the length of the other diagonal.
31. Find the perimeters and the areas of the following figures:

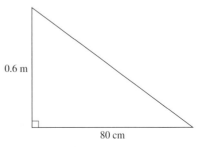

(a)

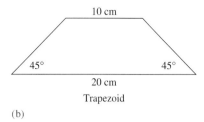

(b)

32. In the following figure, the length of the longer diagonal $\overline{AC}$ of rhombus $ABCD$ is 40 cm; $AE = 24$ cm. Find the length of a side of the rhombus and the length of the other diagonal.

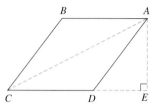

TECHNOLOGY CORNER

Find various-shaped cardboard containers and cut them apart to form nets. Make sure the nets are smaller in size than your computer screen.

a. Estimate the surface area of the nets.
b. Trace the perimeter of each net on a transparency sheet and hang the sheet on the monitor of your computer. Use a geometry utility to trace the outline from the transparency and compute the area.
c. Compare your estimate in (a) with the answer in (b).

BRAIN TEASER

A manufacturer of paper cups wants to produce paper cups in the form of truncated cones 16 cm high, with one circular base of radius 11 cm and the other of radius 7 cm, as shown in Figure 11-64. When the base of such a cup is removed and the cup is slit and flattened, the flattened region looks like a part of a circular ring. To design a pattern to make the cup, the manufacturer needs the data required to construct the flattened region. Find these data.

Figure 11-64

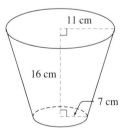

Section 11-5 — Volume, Mass, and Temperature

In Section 11-4, we investigated surface areas of various-shaped containers. In this section, we explore how much the containers will hold. This distinction is sometimes confused by elementary-school students. Whereas the surface area is the number of square units covering a three-dimensional figure, volume describes how much space a three-dimensional figure will hold. The unit of measure for volume must be a shape that tessellates the space. Cubes tessellate space; that is, they can be stacked so that they leave no gaps and fill space. Standard units of volume are based on cubes and are *cubic units*. A cubic unit is the amount of space enclosed within a cube that measures 1 unit on a side. The distinction between surface area and volume is demonstrated in Figure 11-65.

Figure 11-65

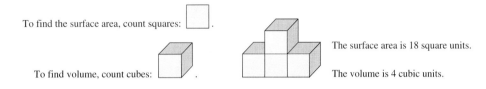

To find the surface area, count squares: ☐.

To find volume, count cubes: ☐.

The surface area is 18 square units.

The volume is 4 cubic units.

NOW TRY THIS 11-11

- In Figure 11-66, the purple block is moved from one position to another.

Figure 11-66

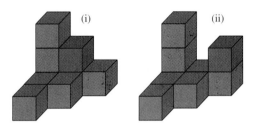

(i) (ii)

a. How does the volume in figure (i) compare with the volume in figure (ii)?
b. Do figures (i) and (ii) have the same surface area? If not, estimate which has the greatest surface area.
c. Find the surface area of each figure.

Volume of Right Rectangular Prisms

The volume of a right rectangular prism can be measured by determining how many cubes are needed to build it. To find the volume, count how many cubes cover the base and then how many layers of these cubes are used to fill the prism. As shown in Figure 11-67(a), there are $8 \cdot 4$, or 32, cubes required to cover the base and there are five such layers. The volume of the rectangular prism is $(8 \cdot 4) \cdot 5$, or 160 cubic units. For any right rectangular prism with dimensions ℓ, w, and h measured in the same linear units, the volume of the prism is given by the area of the base, ℓw, times the height, h, or $V = \ell w h$, as shown in Figure 11-67(b).

Figure 11-67

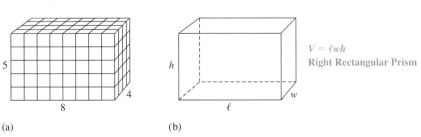

(a) (b)

$V = \ell w h$
Right Rectangular Prism

Converting Metric Measures of Volume

cubic centimeter · cubic meter

The most commonly used metric units of volume are the **cubic centimeter** and the **cubic meter**. A cubic centimeter is the volume of a cube whose length, width, and height are each 1 cm. One cubic centimeter is denoted by 1 cm^3. Similarly, a cubic meter is the volume of a cube whose length, width, and height are each 1 m. One cubic meter is denoted by 1 m^3. Other metric units of volume are symbolized similarly.

Figure 11-68 shows that since 1 dm = 10 cm, 1 dm^3 = (10 cm) · (10 cm) · (10 cm) = 1000 cm^3.

Figure 11-68

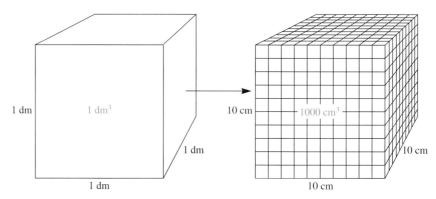

Figure 11-69 shows that 1 m^3 = 1,000,000 cm^3 and that 1 dm^3 = 0.001 m^3.

Figure 11-69

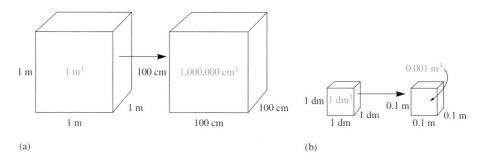

(a)   (b)

Each metric unit of length is 10 times as great as the next smaller unit. Each metric unit of area is 100 times as great as the next smaller unit. Each metric unit of volume is 1000 times as great as the next smaller unit. For example:

$$1 \text{ cm} = 10 \text{ mm}$$
$$1 \text{ cm}^2 = 100 \text{ mm}^2$$
$$1 \text{ cm}^3 = 1000 \text{ mm}^3$$

Because 1 cm = 0.01 m, then 1 cm^3 = (0.01 · 0.01 · 0.01) m^3, or 0.000001 m^3. To convert from cubic centimeters to cubic meters, we move the decimal point six places to the left.

Example 11-21 Convert each of the following:
a. $5 \text{ m}^3 = \underline{} \text{ cm}^3$ b. $12{,}300 \text{ mm}^3 = \underline{} \text{ cm}^3$

Solution a. 1 m = 100 cm, so 1 m³ = (100 cm)(100 cm)(100 cm), or 1,000,000 cm³. Thus $5 \text{ m}^3 = (5)(1{,}000{,}000 \text{ cm}^3) = 5{,}000{,}000 \text{ cm}^3$.
b. 1 mm = 0.1 cm, so 1 mm³ = (0.1 cm)(0.1 cm)(0.1 cm), or 0.001 cm³. Thus $12{,}300 \text{ mm}^3 = 12{,}300(0.001 \text{ cm}^3) = 12.3 \text{ cm}^3$.

liter   In the metric system, cubic units may be used for either dry or liquid measure, although units such as liters and milliliters are usually used for liquid measures. By definition, a **liter**, symbolized by L, equals, or is the capacity of, a cubic decimeter; that is, 1 L = 1 dm³. (In the United States, L is the symbol for liter, but this is not universally accepted.)

Because 1 L = 1 dm³ and 1 dm³ = 1000 cm³, it follows that 1 L = 1000 cm³ and 1 cm³ = 0.001 L. Also, 0.001 L = 1 milliliter = 1 mL. Hence, 1 cm³ = 1 mL. These relationships are summarized in Figure 11-70 and Table 11-7.

Figure 11-70

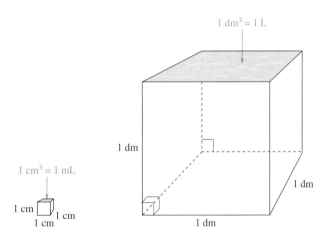

Table 11-7

Unit	Symbol	Relation to Liter
kiloliter	kL	1000 L
*hectoliter	hL	100 L
*dekaliter	daL	10 L
liter	**L**	**1 L**
*deciliter	dL	0.1 L
centiliter	cL	0.01 L
milliliter	mL	0.001 L

*Not commonly used

Example 11-22 Convert each of the following as indicated:

a. 27 L = _____ mL
b. 362 mL = _____ L
c. 3 mL = _____ cm^3
d. 3 m^3 = _____ L

Solution
a. 1 L = 1000 mL, so 27 L = 27 · 1000 mL = 27,000 mL.
b. 1 mL = 0.001 L, so 362 mL = 362(0.001 L) = 0.362 L.
c. 1 mL = 1 cm^3, so 3 mL = 3 cm^3.
d. 1 m^3 = 1000 dm^3 and 1 dm^3 = 1 L, so 1 m^3 = 1000 L and 3 m^3 = 3000 L.

Converting English Measures of Volume

Basic units of volume in the English system are the cubic foot (1 ft^3), the cubic yard (1 yd^3), and the cubic inch (1 in.^{3}). In the United States, 1 gal = 231 in.^{3}, which is about 3.8 L, and 1 qt = $\frac{1}{4}$ gal, or about 58 in.^{3}

Relationships among the one-dimensional units enable us to convert from one unit of volume to another, as shown in the following example.

Example 11-23 Convert each of the following, as indicated:

a. 45 yd^3 = _____ ft^3
b. 4320 in.^{3} = _____ yd^3
c. 10 gal = _____ ft^3
d. 3 ft^3 = _____ yd^3

Solution
a. Because 1 yd^3 = (3 ft)^3 = 27 ft^3, 45 yd^3 = 45 · 27 ft^3, or 1215 ft^3.

b. Because 1 in. = $\frac{1}{36}$ yd, 1 in.^{3} = $\left(\frac{1}{36}\right)^3$ yd^3. Consequently, 4320 in.^{3}

$= 4320 \cdot \left(\frac{1}{36}\right)^3$ yd$^3 \doteq 0.0926$ yd^3, or approximately 0.1 yd^3.

c. Because 1 gal = 231 in.^{3} and 1 in.^{3} = $\left(\frac{1}{12}\right)^3$ ft^3, 10 gal = 2310 in.^{3}

$= 2310\left(\frac{1}{12}\right)^3$ ft$^3 \doteq 1.337$ ft^3, or approximately 1.3 ft^3.

d. As seen in (a), 1 ft^3 = $\frac{1}{27}$ yd^3. Hence 3 ft^3 = 3 · $\frac{1}{27}$ yd^3 = $\frac{1}{9}$ yd^3.

Volumes of Prisms and Cylinders

We have shown that the volume of a right rectangular prism, as shown in Figure 11-71, involves multiplying the area of the base times the height. If we denote the area of the base by B and the height by h, then $V = Bh$.

Figure 11-71

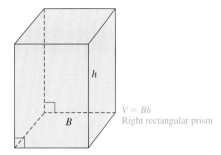

Formulas for the volumes of many three-dimensional figures can be derived using the volume of a right prism. In Figure 11-72(a), a rectangular solid box has been sliced into thin layers. If the layers are shifted to form the solids in Figure 11-72(b) and (c), the volume of each of the three solids is the same as the volume of the original solid. This idea is the basis for **Cavalieri's principle.**

Cavalieri's Principle

Cavalieri's Principle

Two solids each with a base in the same plane have equal volumes if every plane parallel to the bases intersects the solids in cross-sections of equal area.

Figure 11-72

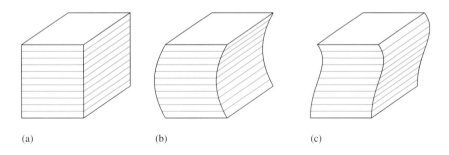

HISTORICAL NOTE

Bonaventura Cavalieri (1598–1647), an Italian mathematician and disciple of Galileo, contributed to the development of geometry, trigonometry, and algebra in the Renaissance. He became a Jesuit at an early age and later, after reading Euclid's *Elements,* was inspired to study mathematics. In 1629, Cavalieri became a professor at Bologna and held that post until his death. Cavalieri is best known for his principle concerning the volumes of solids.

650 CHAPTER 11 *Concepts of Measurement*

NOW TRY THIS 11-12

- **a.** The two right prisms in Figure 11-73 have the same height. How do their volumes compare? Explain why.

Figure 11-73

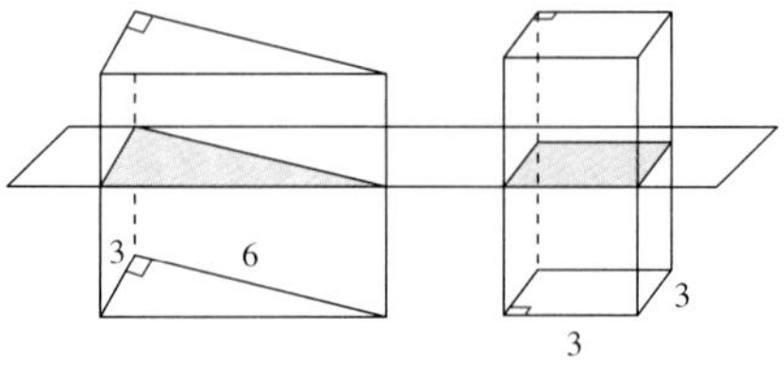

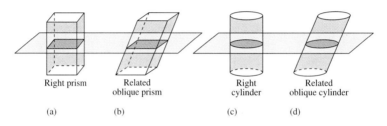

b. Consider the right prism and right cylinder in Figure 11-74(a) and (c) as stacks of papers. If the papers are shifted as shown in Figure 11-74(b) and (d), an oblique prism and an oblique cylinder, respectively, are formed.

  (i) Explain how the volume of the oblique prism is related to the volume of the right prism.
  (ii) Explain how the volume of the oblique cylinder is related to the volume of the right cylinder.

Figure 11-74

Right prism (a)   Related oblique prism (b)   Right cylinder (c)   Related oblique cylinder (d)

The volume of a cylinder can be approximated using prisms with increasing numbers of sides in their bases. The volume of each prism is the product of the area of the base and the height. Similarly, the volume V of a cylinder is the product of the area of the base B and the height h; that is, $V = Bh = \pi r^2 h$.

Example 11-24 Find the volume of each figure in Figure 11-75.

Figure 11-75

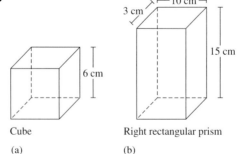

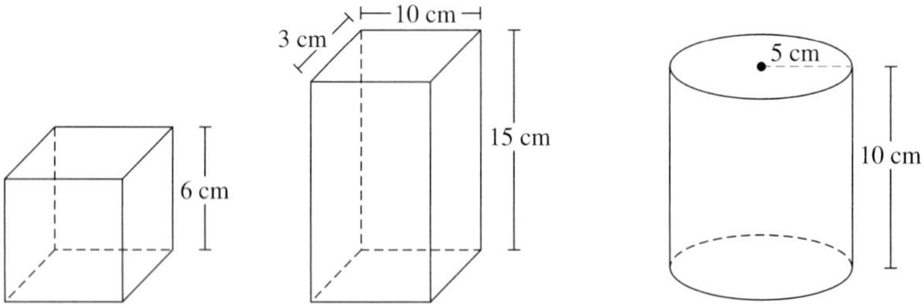

Cube (a)   Right rectangular prism (b)   Right circular cylinder (c)

Solution
  a. $V = Bh = (6 \text{ cm} \cdot 6 \text{ cm}) \cdot 6 \text{ cm} = 216 \text{ cm}^3$
  b. $V = Bh = (10 \text{ cm} \cdot 3 \text{ cm}) \cdot 15 \text{ cm} = 450 \text{ cm}^3$
  c. $V = \pi r^2 h = \pi (5 \text{ cm})^2 \cdot 10 \text{ cm} = 250\pi \text{ cm}^3$

Volumes of Pyramids and Cones

Figure 11-76(a) and (b) show a right prism and a right pyramid with congruent bases and equal heights.

Figure 11-76

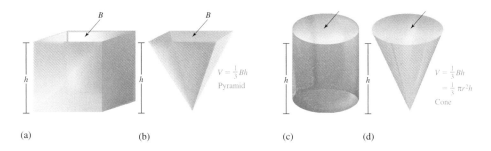

How are the volumes of these containers related? Students may explore the relationship by filling the pyramid with water, sand, or rice and pouring the contents into the prism. They should find that it takes three full pyramids to fill the prism. Therefore the volume of the pyramid is equal to one-third the volume of the prism. This relationship between prisms and pyramids with congruent bases and heights, respectively, is true in general; that is, for a pyramid $V = (1/3)Bh$, where B is the area of the base and h is the height. The same relationship holds between the volume of a cone and the volume of a cylinder, where they share congruent bases and equal heights, as shown in Figure 11-76(c) and (d). Therefore the volume of a cone is given by $V = (1/3)Bh$, or $V = (1/3)\pi r^2 h$.

Another way to determine the area of a pyramid in terms of a prism is to start with a cube and three diagonals from one vertex drawn to other vertices, as shown in Figure 11-77(a). We can see that there are three pyramids formed inside the cube as shown in Figure 11-77(b), (c), and (d).

Figure 11-77

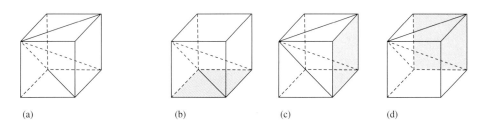

The three pyramids are identical in size and shape, do not overlap, and their union is the whole cube. Therefore, each pyramid has volume one-third that of the cube. This result is true in general, and once again we see that for a pyramid $V = (1/3)Bh$, where B is the area of the base and h is the height. This can be demonstrated by building three paper models of the pyramids and fitting them together into a prism. A net that can be enlarged and used for the construction is given in Figure 11-78.

Figure 11-78

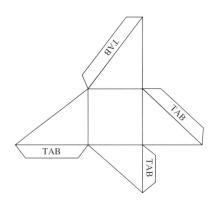

Example 11-25 Find the volume of each figure in Figure 11-79.

Figure 11-79

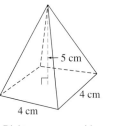

Right square pyramid
(a)

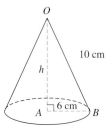

Right circular cone
(b)

Solution
a. The figure is a pyramid with a square base whose area is 4 cm · 4 cm and whose height is 5 cm. Hence, $V = \frac{1}{3}Bh = \frac{1}{3}(4 \text{ cm} \cdot 4 \text{ cm})(5 \text{ cm}) = \frac{80}{3} \text{ cm}^3$.

b. The base of the cone is a circle of radius 6 cm. Because the volume of the cone is given by $V = \frac{1}{3}\pi r^2 h$, we need to know the height. In the right triangle OAB, $OA = h$ and by the Pythagorean Theorem, $h^2 + 6^2 = 10^2$. Hence, $h^2 = 100 - 36$, or 64, and $h = 8$ cm. Thus $V = \frac{1}{3}\pi r^2 h = \frac{1}{3}\pi(6 \text{ cm})^2(8 \text{ cm}) = 96\pi \text{ cm}^3$.

Example 11-26 Figure 11-80 is a net for a pyramid. If each triangle is equilateral, find the volume of the pyramid.

Figure 11-80

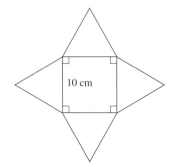

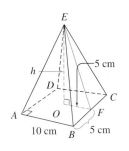

Figure 11-81

Solution The pyramid obtained from the folded model is shown in Figure 11-81. The volume of the pyramid is $V = \frac{1}{3}Bh = \frac{1}{3} \cdot 10^2 h$. We must find h. Notice that h is a leg in the right triangle EOF, where F is the midpoint of $\overline{CB}$. We know that $OF = 5$ cm. If we knew EF, we could find h by applying the Pythagorean Theorem to $\triangle EOF$. To find the length of $\overline{EF}$, notice that $\overline{EF}$ is a leg in the right triangle EBF. ($\overline{EF}$ is the perpendicular bisector of $\overline{BC}$ in the equilateral triangle BEC.) In the right triangle EBF, we have $(EB)^2 = (BF)^2 + (EF)^2$. Because $EB = 10$ cm and $BF = 5$ cm, it follows that $10^2 = 5^2 + (EF)^2$, or $EF = \sqrt{75}$ cm $\doteq 8.66$ cm. In $\triangle EOF$, we have $h^2 + 5^2 = (EF)^2$, or $h^2 + 25 = 75$. Thus, $h = \sqrt{50}$ cm $\doteq 7.07$ cm, and $V \doteq \frac{1}{3} \cdot 10^2 \cdot 7.07 \doteq 235.7$ cm^3.

Volume of a Sphere

To find the volume of a sphere, imagine that a sphere is composed of a great number of congruent pyramids with apexes at the center of the sphere and that the vertices of the base touch the sphere, as shown in Figure 11-82. If the pyramids have very small bases, then the height of each pyramid is nearly the radius r. Hence, the volume of each pyramid is $\frac{1}{3}Bh$ or $\frac{1}{3}Br$, where B is the area of the base. If there are n pyramids each with base area B, then the total volume of the pyramids is $V = \frac{1}{3}nBr$. Because nB is the total surface area of all the bases of the pyramids and because the sum of the areas of all the bases of the pyramids is very close to the surface area of the sphere, $4\pi r^2$, the volume of the sphere is given by $V = \frac{1}{3}(4\pi r^2)r = \frac{4}{3}\pi r^3$.

Figure 11-82

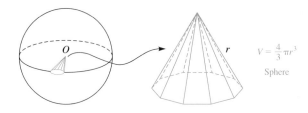

Example 11-27 Find the volume of a sphere whose radius is 6 cm.

Solution $V = \frac{4}{3}\pi(6 \text{ cm})^3 = \frac{4}{3}\pi(216 \text{ cm}^3) = 288\pi \text{ cm}^3$

Problem Solving   **Volume Comparisons: Cylinders and Boxes**

A metal can manufacturer has a large quantity of rectangular metal sheets 20 cm × 30 cm. Without cutting the sheets, the manufacturer wants to make cylindrical pipes with circular cross-sections from some of the sheets and box-shaped pipes with square cross-sections from the other sheets. The volume of the box-shaped pipes is to be greater than the volume of the cylindrical pipes. Is this possible? If so, how would the pipes be made and what are their volumes?

- **Understanding the Problem** We are to use 20 cm × 30 cm rectangular sheets of metal to make some cylindrical pipes as well as some box-shaped pipes with square cross-sections that have a greater volume than do the cylindrical pipes. Is this possible, and if so, how should the pipes be designed and what are their volumes?

  Figure 11-83 shows a sheet of metal and two sections of pipe made from it, one cylindrical and the other box-shaped. A model for such pipes can be designed from a piece of paper by bending it into a cylinder or by folding it into a right rectangular prism, as shown in the figure.

Figure 11-83

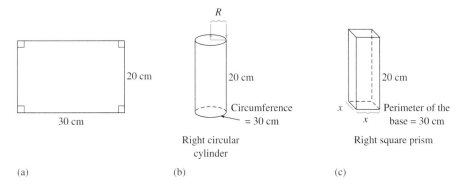

(a) (b) (c)

- **Devising a Plan** If we compute the volume of the cylinder in Figure 11-83(b) and the volume of the prism in Figure 11-83(c), we can determine which is greater. If the prism has a greater volume, the solution of the problem will be complete. Otherwise, we look for other ways to design the pipes before concluding that a solution is impossible.

  To compute the volume of the cylinder, we find the area of the base. The area of the circular base is πr^2. To find r, we note that the circumference of the circle $2\pi r$ is 30 cm. Thus $r = \dfrac{30}{2\pi} \doteq 4.77$ cm, and the area of the circle is $\pi r^2 \doteq \pi(4.77)^2 \doteq 71.48$ cm^2.

  With the given information, we can also find the area of the base of the rectangular box. Because the perimeter of the base of the prism is $4x$, we have $4x = 30$, or $x = 7.5$ cm. Thus the area of the square base is $x^2 = (7.5)^2$, or 56.25 cm^2.

- **Carrying Out the Plan** Denoting the volume of the cylindrical pipe by V_1 and the volume of the box-shaped pipe by V_2, we have $V_1 \doteq 71.48 \cdot 20$, or approximately 1429.6 cm^3. For the volume of the box-shaped pipe, we have $V_2 \doteq 56.25 \cdot 20$, or 1125 cm^3. We see that in the first design for the pipes, the volume of the cylindrical pipe is greater than the volume of the box-shaped pipe. This is not the required outcome.

  Rather than bending the rectangular sheet of metal along the 30-cm side, we could bend it along the 20-cm side to obtain either pipe, as shown in Figure 11-84. Denoting the radius of the cylindrical pipe by r, the side of the box-shaped pipe by y, and their volumes by V_3 and V_4, respectively, we have $V_3 = \pi r^2 \cdot 30 = \pi(20/2\pi)^2 \cdot 30 = (10^2 \cdot 30)/\pi$, or approximately 954.9 cm^3. Also, $V_4 = y^2 \cdot 30 = \left(\dfrac{20}{4}\right)^2 \cdot 30 = 25 \cdot 30$, or 750 cm^3. Because $V_2 = 1125$ cm^3 and $V_3 = 945.9$ cm^3, we see that the volume of the box-shaped pipe with an altitude of 20 cm is greater than the volume of the cylindrical pipe with an altitude of 30 cm.

Figure 11-84

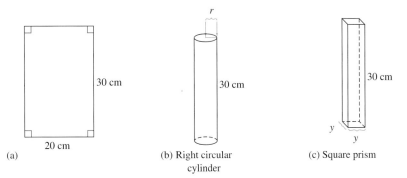

(a) (b) Right circular cylinder (c) Square prism

• **Looking Back** We could ask for the volumes of other three-dimensional objects that can be obtained by bending the rectangular sheets of metal. Also, because the lateral surface areas of the four types of pipes were the same but their volumes were different, we might want to investigate whether there are other cylinders and prisms that have the same lateral surface area and the same volume. Is it possible to find a circular cylinder with lateral surface area of 600 cm^2 and smallest possible volume? Similarly, is there a circular cylinder with the given surface area and greatest possible volume?

Mass

Three centuries ago, Isaac Newton pointed out that in everyday life the word *weight* is used for what is really mass. *Mass* is a quantity of matter as opposed to *weight,* which is a force exerted by gravitational pull. When astronauts are in orbit above Earth, their weights have changed even though their masses remain the same. In common parlance on Earth, *weight* and *mass* are still used interchangeably. In the English system, weight is measured in avoirdupois units such as tons, pounds, and ounces. One pound (lb) equals 16 ounces (oz) and 2000 lb equals 1 English ton.

gram   In the metric system, the unit for mass that prefixes are used with is the **gram,** denoted by g. An ordinary paper clip or a thumbtack each has a mass of about 1 g. As with other base metric units, prefixes are added to gram to obtain other units. For example, a kilogram (kg) is 1000 g. Two standard loaves of bread have a mass of about 1 kg. A person's mass also is measured in kilograms. A newborn baby has a mass of about 4 kg. Another unit of mass is the metric ton (t), which is equal to 1000 kg. The metric ton is used to record the masses of objects such as cars and trucks. A small foreign car has a mass of about 1 t. Mega (1,000,000) and micro (0.000001) are other prefixes used with the base unit.

Table 11-8 lists metric units of mass. Conversions that involve metric units of mass are handled in the same way as conversions that involve metric units of length.

Table 11-8

Unit	Symbol	Relationship to Gram
ton (metric)	t	1,000,000 g
kilogram	kg	1000 g
*hectogram	hg	100 g
*dekagram	dag	0 g
gram	**g**	**1 g**
*decigram	dg	0.1 g
*centigram	cg	0.01 g
milligram	mg	0.001 g

*Not commonly used

Example 11-28 Complete each of the following:

a. 34 g = _____ kg
b. 6836 kg = _____ t

Solution
a. 34 g = 34(0.001 kg) = 0.034 kg
b. 6836 kg = 6836(0.001 t) = 6.836 t

The relationship among the units of volume, capacity, and mass in the metric system is illustrated in Figure 11-85.

Figure 11-85

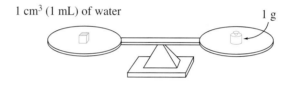

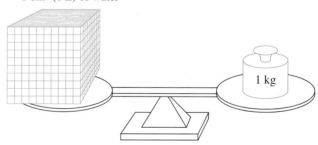

NOW TRY THIS 11-13

- Find the following relationships:

  a. 1 cm³ of water has a mass of 1 _____.
  b. 1 dm³ of water has a mass of 1 _____.
  c. 1 L of water has a volume of 1 _____.
  d. 1 cm³ of water has a capacity of 1 _____.
  e. 1 mL of water has a mass of 1 _____.
  f. 1 m³ of water has a capacity of 1 _____.
  g. 1 m³ of water has a mass of 1 _____.

Example 11-29 A waterbed measures 180 cm × 210 cm × 20 cm.

a. Approximately how many liters of water can it hold?
b. What is its mass in kilograms when it is full of water?

Solution **a.** The volume of the waterbed is approximated by multiplying the length ℓ times the width w times the height h.

$$V = \ell w h$$
$$= 180 \text{ cm} \cdot 210 \text{ cm} \cdot 20 \text{ cm}$$
$$= 756{,}000 \text{ cm}^3, \text{ or } 756{,}000 \text{ mL}$$

Because 1 mL = 0.001 L, the volume is 756 L.

b. Because 1 L of water has a mass of 1 kg, 756 L of water has a mass of 756 kg, which is 0.756 t.

REMARK To see one advantage of the metric system, suppose the bed in Example 11-29 is 6 ft × 7 ft × 9 in. Try to approximate the volume in gallons and the weight of the water in pounds.

Temperature

degree Kelvin The base unit of temperature for the metric system, the **degree Kelvin,** is used only for scientific measurements and is an absolute temperature. The freezing point of water is 273° on this scale. For normal temperature measurements in the metric system, the base unit is the **degree Celsius**

degree Celsius, named for Anders Celsius, the Swedish scientist who invented the system. The Celsius scale has 100 equal divisions between 0 degrees Celsius (0°C), the freezing point of water, and 100 degrees Celsius (100°C), the boiling point of water, as seen in Figure 11-86. In the English system, the Fahrenheit scale has 180 equal divisions between 32°F, the freezing point of water, and 212°F, the boiling point of water.

Figure 11-86

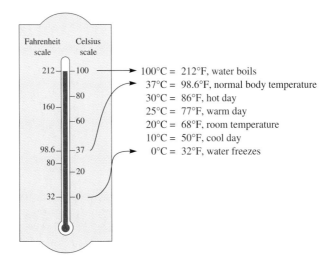

Figure 11-86 gives other temperature comparisons of the two scales and further illustrates the relationship between them. Because the Celsius scale has 100 divisions between the freezing point and the boiling point of water, whereas the Fahrenheit scale has 180 divisions, the relationship between the two scales is 100 to 180, or 5 to 9. For every 5 degrees on the Celsius scale, there are 9 degrees on the Fahrenheit scale, and for each degree on the Fahrenheit scale, there is $\frac{5}{9}$ degree on the Celsius scale. Because the ratio between the number of degrees above freezing on the Celsius scale and the number of degrees above freezing on the Fahrenheit scale remains the same and equals $\frac{5}{9}$, we may convert temperature from one system to the other.

For example, suppose we want to convert 50° on the Fahrenheit scale to the corresponding number on the Celsius scale. On the Fahrenheit scale, 50° is 50 − 32, or 18°, above freezing, but on the Celsius scale, it is $\frac{5}{9} \cdot 18$, or 10°, above freezing. Because the freezing temperature on the Celsius scale is 0°, 10° above freezing is 10° Celsius. Thus 50°F = 10°C. In general, F degrees is $F - 32$ above freezing on the Fahrenheit scale, but only $\frac{5}{9}(F - 32)$ above freezing on the Celsius scale. Thus we have the relation $C = \frac{5}{9}(F - 32)$. If we solve the equation for F, we obtain $F = \frac{9}{5}C + 32$.

NOW TRY THIS 11-14

- Does it ever happen that the temperature measured in Celsius degrees is the same if it is measured in Fahrenheit degrees? If so, when?

ONGOING ASSESSMENT 11-5

1. Complete each of the following:
   a. $8 \text{ m}^3 = $ _____ dm^3
   b. $500 \text{ cm}^3 = $ _____ m^3
   c. $675{,}000 \text{ m}^3 = $ _____ km^3
   d. $3 \text{ m}^3 = $ _____ cm^3
   e. $7000 \text{ mm}^3 = $ _____ cm^3
   f. $0.002 \text{ m}^3 = $ _____ cm^3
   g. $400 \text{ in.}^3 = $ _____ yd^3
   h. $25 \text{ yd}^3 = $ _____ ft^3
   i. $0.2 \text{ ft}^3 = $ _____ in.^3
   j. $1200 \text{ in.}^3 = $ _____ ft^3

2. If a faucet is dripping at the rate of 15 drops/min and there are 20 drops/mL, how many liters of water are wasted in a 30-day month?

3. Find the volume of each of the following:

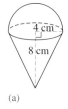

(a)

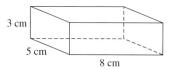

(b) Right rectangular prism

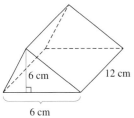
(c) Right triangular prism

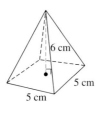
(d) Square pyramid

(e) Right circular cone

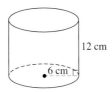
(f) Right circular cylinder

(g) Sphere

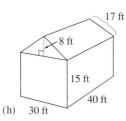
(h)

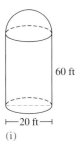
(i)

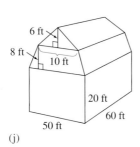
(j)

4. Complete the following chart:

	a.	b.	c.	d.	e.	f.
cm³		500			750	4800
dm³	2					
L				1.5		
mL				5000		

5. Place a decimal point in each of the following to make it an accurate sentence:
   a. A paper cup holds about 2000 mL.
   b. A regular soft drink bottle holds about 320 L.
   c. A quart milk container holds about 10 L.
   d. A teaspoonful of cough syrup is about 500 mL.

6. Determine the volume of silver needed to make the napkin ring in the following figure out of solid silver. Give your answer in cubic millimeters.

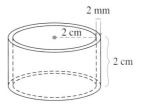

7. Two cubes have sides of lengths 4 cm and 6 cm, respectively. What is the ratio of their volumes?

8. What happens to the volume of a sphere if the radius is doubled?

9. Complete the following chart for right rectangular prisms with the given dimensions:

	a.	b.	c.	d.
Length	20 cm	10 cm	2 dm	15 cm
Width	10 cm	2 dm	1 dm	2 dm
Height	10 cm	3 dm		
Volume (cm³)				
Volume (dm³)				7.5 dm³
Volume (L)			4 L	

10. Determine how many liters a right cylindrical tank holds if it is 6 m long and 13 m in diameter.

11. Earth's diameter is approximately four times the moon's and both bodies are spheres. What is the ratio of their volumes?

12. An Olympic-sized pool in the shape of a right rectangular prism is 50 m × 25 m. If it is 2 m deep throughout, how many liters of water does it hold?

13. A standard straw is 25 cm long and 4 mm in diameter. How much liquid can be held in the straw at one time?

14. a. What happens to the volume of an aquarium that is in the shape of a rectangular prism if the length, width, and height are all doubled?
    b. From your answer to (a), conjecture what happens to the volume of the aquarium if all the measurements are tripled.
    c. When you multiply each linear dimension of an aquarium by a positive value n, what happens to the volume?

15. The Great Pyramid of Egypt is a right square pyramid with height of 148 m and a square base with a perimeter of 930 m. The Transamerica Building in San Francisco has the basic shape of a right square pyramid that has a square base with a perimeter of 140 m and a height of 260 m. Which one has the greater volume and by how many times greater?
16. The Great Pyramid of Egypt has a square base of 756 ft on a side and a height of 481 ft. How many apartments 35 ft × 20 ft × 8 ft would be needed to have a volume equivalent to that of the Great Pyramid's?
17. A rectangular-shaped swimming pool with dimensions 10 m × 25 m is being built. The pool has a shallow end that is uniform in depth and a deep end that drops off as shown in the following figure. What is the volume of this pool in cubic meters?

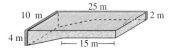

18. If 50 steel marbles that are 1 cm in diameter are melted down, will enough steel result to build a marble that is 4 cm in diameter? Explain.
19. A cone-shaped paper water cup has a height of 8 cm and a radius of 4 cm. If the cup is filled with water to half its height, what portion of the volume of the cup is filled with water?
20. If each edge of a cube is increased by 30%, by what percent does the volume increase?
21. One freezer measures 1.5 ft × 1.5 ft × 5 ft and sells for $350. Another freezer measures 2 ft × 2 ft × 4 ft and sells for $400. Which freezer is the better buy in terms of dollars per cubic foot?
22. A tennis ball can in the shape of a cylinder holds three tennis balls snugly. If the radius of a tennis ball is 3.5 cm, what percentage of the tennis ball can is occupied by air?
23. A box is packed with six soda cans, as in the following figure. What percentage of the volume of the interior of the box is not occupied by the cans?

24. A right rectangular prism with base *ABCD* as the bottom is shown in the following figure.

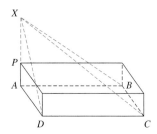

Suppose *X* is drawn so that $AX = 3 \cdot AP$, where *AP* is the height of the prism and *X* is connected to *A*, *B*, *C*, and *D* to form a pyramid. How do the volumes of the pyramid and the prism compare?

25. A right cylindrical can is to hold exactly 1 L of water. What should be the height of the can if the radius is 12 cm?
26. A theater decides to change the shape of its popcorn container from a regular box to a right regular pyramid and charge only half as much as shown in the following figure.

If the containers are the same height and the tops are the same size, is this a bargain for the customer? Explain.

27. Which is the better buy: a grapefruit 5 cm in radius that costs 22¢ or a grapefruit 6 cm in radius that costs 31¢? Explain.
28. Two spherical cantaloupes of the same kind are sold at a fruit and vegetable stand. The circumference of one is 60 cm and that of the other is 50 cm. The larger melon is $1\frac{1}{2}$ times as expensive as the smaller. Which melon is the better buy and why?
29. An engineer is to design a square-based pyramid whose volume is to be 100 m³.
    a. Find the dimensions (the length of a side of the square and the altitude) of one such pyramid.
    b. How many (noncongruent) such pyramids are possible? Why?
30. A square sheet of cardboard measuring *y* cm on a side is to be used to produce an open-top box when the maker cuts off a small square *x* cm × *x* cm from each corner and bends up the sides.
    a. Find the volume of the box if *y* = 200 cm and *x* = 20 cm.
    ★b. Assume *y* is known. Find the expression for the volume *V* as a function of *x*.
31. Suppose a scoop of ice cream that is a sphere with radius 5 cm fits exactly along one of its great circles in a sugar cone, as shown in the following figure.

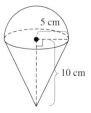

Suppose the ice cream melts and the cone does not absorb any of it. If the cone is 10 cm tall, will it hold the melted ice cream? If not, how tall would the cone have to be to hold the melted ice cream?

★32. Half of the air is let out of a spherical balloon. If the balloon remains in the shape of a sphere, how does the radius of the smaller balloon compare to the original radius?

33. For each of the following, select the appropriate metric unit of measure (gram, kilogram, or metric ton):
   a. Car
   b. Adult
   c. Can of frozen orange juice
   d. Elephant
   e. Jar of mustard
   f. Bag of peanuts
   g. Army tank
   h. Cat

34. For each of the following, choose the correct unit (milligram, gram, or kilogram) to make each sentence reasonable:
   a. A staple has a mass of about 340 ___.
   b. A professional football player has a mass of about 110 ___.
   c. A vitamin tablet has a mass of about 1100 ___.
   d. A dime has a mass of 2 ___.
   e. The recipe said to add 4 ___ of salt.
   f. One strand of hair has a mass of 2 ___.

35. Complete each of the following:
   a. 15,000 g = ___ kg
   b. 8000 kg = ___ t
   c. 0.036 kg = ___ g
   d. 72 g = ___ kg
   e. 4320 mg = ___ g
   f. 5 kg 750 g = ___ g
   g. 0.03 t = ___ kg
   h. 2.6 lb = ___ oz
   i. 25 oz = ___ lb
   j. 3.8 lb = ___ oz

36. A paper dollar has a mass of approximately 1 g. Is it possible to lift $1,000,000 in the following denominations:
   a. $1 bills
   b. $10 bills
   c. $100 bills
   d. $1000 bills
   e. $10,000 bills

37. A fish tank, which is a right rectangular prism, is 40 cm × 20 cm × 20 cm. If it is filled with water, what is the mass of the water?

38. Convert each of the following from degrees Fahrenheit to the nearest integer degree Celsius:
   a. 10°F
   b. 0°F
   c. 30°F
   d. 100°F
   e. 212°F
   f. ⁻40°F

39. Answer each of the following:
   a. The thermometer reads 20°C. Can you go snow skiing?
   b. The thermometer reads 26°C. Will the outdoor ice rink be open?
   c. Your body temperature is 39°C. Are you ill?
   d. It is 40°C. Will you need a sweater at the outdoor concert?
   e. The temperature reads 35°C. Should you go water skiing?
   f. Your bath water is 16°C. Will you have a hot, warm, or chilly bath?
   g. It's 30°C in the room. Are you comfortable, hot, or cold?

40. a. Rainfall is usually measured in linear measure. Suppose St. Louis received 2 cm of rain on a given day. If a certain lot in St. Louis has measure 1 ha, how many liters of rainfall fell on the lot?
    b. What is the mass of the water that fell on the lot?

Communication

41. a. Which will increase the volume of a circular cylinder more: doubling its height or doubling its radius? Explain.
    b. Is your answer the same for a circular cone? Why?

42. Write a one-page paper to a sixth-grade student explaining the difference between surface area and volume.

43. Explain how you would find the volume of an irregular shape.

44. Read the following problems (i) and (ii):
    i. A tank in the shape of a cube 5 ft 3 in. on a side is filled with water. Find the volume in cubic feet, the capacity in gallons, and the weight of the water in pounds.
    ii. A tank in the shape of a cube 2 m on a side is filled with water. Find the volume in cubic meters, the capacity in liters, and the mass of the water in kilograms.
    Discuss which problem is easier to work and why the metric system has an advantage over the English system in this case.

45. If only the metric system were taught in elementary school, what implications would this have for the topics that are taught in mathematics?

Open-Ended

46. A right circular cylinder has a 4-in. diameter, is 6 in. high, and is completely full of water. Design a right rectangular prism that will hold the water as exactly as possible.

47. Circular-shaped cookies are to be packaged 48 to a box. Each cookie is approximately 1 cm thick and has a diameter of 6 cm. Design a box that will hold this volume of cookies and has the least amount of surface area.

48. Design a cylinder that will hold 1 L of juice. Give the dimensions of your cylinder and tell why you designed the shape as you did.

Cooperative Learning

49. a. Find many different types of cans that are in the shape of a cylinder. Measure the height and diameter for each can.
    b. Find the surface area of each can.
    c. Find the volume of each can.
    d. Compute the ratio of surface area to volume for each can.
    e. Compare your results with those of other groups.
    f. Based on the information collected, write recommendations to the manufacturers of the cans about an ideal surface-area-to-volume ratio.

Review Problems

50. Find the perimeter and the area of the following figures:

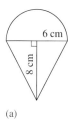
(a)

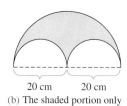
(b) The shaded portion only

51. Complete the following:
    a. 350 mm = _____ cm
    b. 1600 cm² = _____ m²
    c. 0.4 m² = _____ mm²
    d. 5.2 m³ = _____ cm³
    e. 5.2 m³ = _____ L
    f. 3500 cm³ = _____ m³

52. Determine whether each of the following is a right triangle:

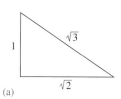
(a)

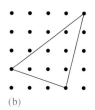
(b)

53. Find the surface area of each of the following:

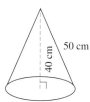

(a) Right circular cone

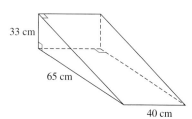

(b) Right prism in which the faces are either rectangles or right triangles

---

LABORATORY ACTIVITY

Record the mass of each U.S. coin. Which coin has the greatest mass? Which of the following have the same mass?

a. A half-dollar vs. two quarters
b. A quarter vs. two dimes and a nickel
c. A dime vs. two nickels
d. A dime vs. ten pennies
e. A nickel vs. five pennies

HINT FOR SOLVING THE PRELIMINARY PROBLEM

The area of the square is 100 m · 100 m, or 10,000 m². We can compute the area of the large circular area and calculate the percentage of the square covered by the larger sprinkler. Next, we can compute the area of one of the small circles, multiply this by nine, and then compute the percentage of the square covered by the nine smaller sprinklers. We can then compare the percents and make a decision on the two systems.

QUESTIONS FROM THE CLASSROOM

1. A student asks if the units of measure must be the same for each term in order to use the formulas for volumes. How do you respond?
2. As part of the discussion of the Pythagorean Theorem, squares were constructed on each side of a right triangle. A student asks, "If different similar figures are constructed on each side of the triangle, does the same type of relationship still hold?" How do you reply?
3. A student asks, "Can I find the area of an angle?" How do you respond?
4. A student argues that a square has no area because its interior can be thought of as the union of infinitely many points, each of which has no area. How do you react?
5. A student asks whether the volume of a prism can ever be the same number as its surface area. How do you answer?
6. A student asks, "Why should the United States switch to the metric system?" How do you reply?
7. A student claims that in a triangle with 20° and 40° angles, the side opposite the 40° angle is twice as long as the side opposite the 20° angle. How do you reply?
8. A student interpreted 5 cm^3 as shown in Figure 11-87. What is wrong with this interpretation?
9. A student claims that because *are* and *hectare* are measures of area, we should say "square are" and "square hectare." How do you respond?
10. A student claims that the area of his hand does not exist because it cannot be found by any formula. How do you respond?
11. A student claims that since a circular cylinder has a curved surface, its lateral surface area should not be expressed in square units. How do you respond?
12. A student claims that it does not make any difference if his temperature is 2 degrees above normal Fahrenheit or 2 degrees above normal Celsius because in either case he is only 2 degrees above normal. How do you respond?
13. Larry and Gary are discussing whose garden has the most area to plant flowers. Larry claims that all they have to do is walk around the two gardens to get the perimeter and the one with the greatest perimeter has the greatest area. How would you help these students?
14. Andrea claims that if she doubles the length and width of the base of a rectangular prism and triples the height, she has increased the volume by a factor $2 \cdot 2 \cdot 3 = 12$. What would you tell her?

Figure 11-87

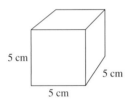

CHAPTER OUTLINE

I. The English system of measure
   A. Linear measure
      1 ft = 12 in.
      1 yd = 3 ft
      1 mi = 5280 ft = 1760 yd
   B. Area measure
      1. Units commonly used are the **square inch** (in.^{2}), **square yard** (yd^2), and **square foot** (ft^2).
      2. Land can be measured in **acres**.
   C. Volume measure
      Units commonly used are the **cubic inch** (in.^{3}), **cubic foot** (ft^3), **cubic yard** (yd^3), and **gallon.**
   D. Mass
      Units of mass commonly used are **pound** (lb), **ounce** (oz) (1 oz = $\frac{1}{16}$ lb), and **ton** (1 ton = 2000 lb).

II. The metric system of measure
   A. A summary of relationships among prefixes and the base unit of linear measure follows:

Prefix	Unit	Relationship to Base Unit	Symbol
kilo	kilometer	1000 m	km
*hecto	hectometer	100 m	hm
*deka	dekameter	10 m	dam
	meter	**1 m**	**m**
*deci	decimeter	0.1 m	dm
centi	centimeter	0.01 m	cm
milli	millimeter	0.001 m	mm

*Not commonly used

B. Area measure
   1. Units commonly used are the **square kilometer** (km^2), **square meter** (m^2), **square centimeter** (cm^2), and **square millimeter** (mm^2).
   2. Land can be measured using the **are** ($100\ m^2$) and the **hectare** ($10{,}000\ m^2$).
C. Volume measure
   1. Units commonly used are the **cubic meter** (m^3), **cubic decimeter** (dm^3), and **cubic centimeter** (cm^3).
   2. $1\ dm^3 = 1\ L$ and $cm^3 = 1\ mL$.
D. Mass
   1. Units of mass commonly used are the **milligram** (mg), **gram** (g), **kilogram** (kg), and **metric ton** (t).
   2. 1 L and 1 mL of water have masses of approximately 1 kg and 1 g, respectively.
E. Temperature
   1. In the metric system the unit commonly used is the **degree Celsius.** In the English system, the unit of temperature is the **degree Fahrenheit.**
   2. Basic temperature reference points are the following:
      100°C—boiling point of water
      37°C—normal body temperature
      20°C—comfortable room temperature
      0°C—freezing point of water
   3. $C = \frac{5}{9}(F - 32)$ and $F = \frac{9}{5}C + 32$

III. Distance
   1. **Distance properties.** Given points A, B, and C,
      a. $AB \geq 0$
      b. $AB = BA$
      c. $AB + BC \geq AC$
   2. The distance around a two-dimensional figure is the **perimeter.** The distance C around a circle is the **circumference.** $C = 2\pi r = \pi d$, where r is the radius of the circle and d is the **diameter.**
   3. **Distance Formula.** The distance between the points $A\ (x_1, y_1)$ and $B\ (x_2, y_2)$ is given by
      $AB = \sqrt{(x_2 - x_1)^2 + (y_2 - y_1)^2}$

IV. Areas
   A. Formulas for areas
      1. **Square:** $A = s^2$, where s is the length of a side.
      2. **Rectangle:** $A = \ell w$, where ℓ is the length and w is the width.
      3. **Parallelogram:** $A = bh$, where b is the length of the base and h is the height.
      4. **Triangle:** $A = \frac{1}{2}bh$, where b is the length of the base and h is the altitude to that base.
      5. **Trapezoid:** $A = \frac{1}{2}h(b_1 + b_2)$, where b_1 and b_2 are the lengths of the bases and h is the height.
      6. **Regular polygon:** $A = \frac{1}{2}ap$, where a is the apothem and p is the perimeter.
      7. **Circle:** $A = \pi r^2$, where r is the radius.
      8. **Sector:** $A = \theta \pi r^2/360$, where θ is the measure of the central angle forming the sector and r is the radius of the circle containing the sector.
   B. **The Pythagorean Theorem:** In any right triangle, the square of the length of the hypotenuse is equal to the sum of the squares of the lengths of the legs.
   C. Triangle relations
      1. **Property of 30°-60°-90° triangle:** The length of the hypotenuse in a 30°-60°-90° triangle is 2 times the length of the leg opposite the 30° angle, and the length of the leg opposite the 60° angle is $\sqrt{3}$ times the length of the short leg.
      2. **Property of 45°-45°-90° triangle:** The length of the hypotenuse of a 45°-45°-90° triangle is $\sqrt{2}$ times the length of a leg.
   D. **Converse of the Pythagorean Theorem:** In any triangle ABC with sides of lengths a, b, and c such that $a^2 + b^2 = c^2$, $\triangle ABC$ is a right triangle with the right angle opposite the side of length c.

V. Surface areas and volumes
   A. Formulas for areas
      1. **Right prism:** $S.A. = 2B + ph$, where B is the area of a base, p is the perimeter of the base, and h is the height of the prism.
      2. **Right circular cylinder:** $S.A. = 2\pi r^2 + 2\pi rh$, where r is the radius of the circular base and h is the height of the cylinder.
      3. **Right circular cone:** $S.A. = \pi r^2 + \pi r\ell$, where r is the radius of the circular base and ℓ is the slant height.
      4. **Right regular pyramid:** $S.A. = B + \frac{1}{2}p\ell$, where B is the area of the base, p is the perimeter of the base, and ℓ is the slant height.
      5. **Sphere:** $S.A. = 4\pi r^2$, where r is the radius of the sphere.
   B. Formulas for volumes
      1. **Right prism:** $V = Bh$, where B is the area of the base and h is the height.
         a. **Right rectangular prism:** $V = \ell wh$, where ℓ is the length, w is the width, and h is the height.
         b. **Cube:** $V = e^3$, where e is an edge.
      2. **Right circular cylinder:** $V = \pi r^2 h$, where r is the radius of the base and h is the height of the cylinder.
      3. **Pyramid:** $V = \frac{1}{3}Bh$, where B is the area of the base and h is the height of the pyramid.
      4. **Circular cone:** $V = \frac{1}{3}\pi r^2 h$, where r is the radius of the circular base and h is the height.
      5. **Sphere:** $V = \frac{4}{3}\pi r^3$, where r is the radius of the sphere.

CHAPTER REVIEW

1. Complete the following.
   a. 50 ft = ____ yd
   b. 947 yd = ____ mi
   c. 0.75 mi = ____ ft
   d. 349 in. = ____ yd
   e. 5 km = ____ m
   f. 165 cm = ____ m
   g. 52 cm = ____ mm
   h. 125 m = ____ km

2. Given three segments of length p, q, and r, where $p > q$, determine if it is possible to construct a triangle with sides of length p, q, and r in each of the following cases. Justify your answers.
   a. $p - q > r$
   b. $p - q = r$

3. Determine the area of the shaded region on each of the following geoboards if the unit of measure is 1 cm²:

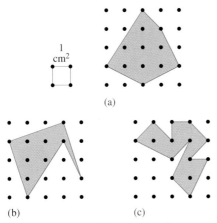

(a)

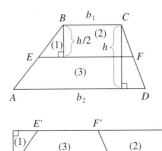

(b) (c)

4. Explain how the formula for the area of a trapezoid can be found by using the following figures:

5. Use the following figure to find each of the following areas:
   a. The area of the hexagon
   b. The area of the circle

6. Find the area of each shaded region in the following figures:

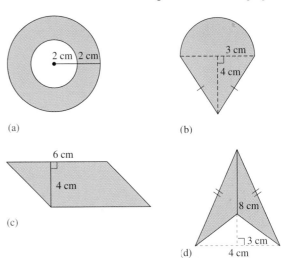

(a) (b)

(c)

(d)

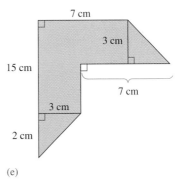

(e)

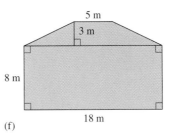

(f)

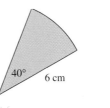

(g)

7. For each of the following, determine whether the measures represent sides of a right triangle. Explain your answers.
   a. 5 cm, 12 cm, 13 cm
   b. 40 cm, 60 cm, 104 cm

8. Find the surface area and volume of each of the following figures:

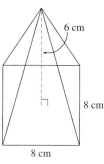

Right square pyramid
(a)

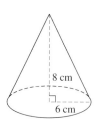

Right circular cone
(b)

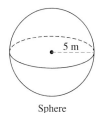

Sphere
(c)

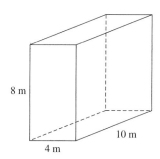

Right circular cone
(d)

Right rectangular prism
(e)

9. Find the lateral surface area of the following right circular cone:

10. Doug's Dog Food Company wants to impress the public with the magnitude of the company's growth. Sales of Doug's Dog Food doubled from 1999 to 2000, so the company is displaying the following graph, which shows the radius of the base and the height of the 2000 can to be double those of the 1999 dog food can. What does the graph really show with respect to the company's growth? Explain your answer.

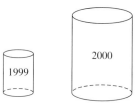

11. Find the area of the kite shown in the following figure:

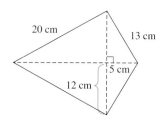

12. The diagonal of a rectangle has measure 1.3 m, and a side of the rectangle has measure 120 cm. Find the following:
   a. Perimeter of the rectangle
   b. Area of the rectangle
13. Find the area of a triangle that has sides of 3 m, 3 m, and 2 m.
14. A poster is to contain 0.25 m² of printed matter, with margins of 12 cm at top and bottom and 6 cm at each side. Find the width of the poster if its height is 74 cm.
15. Complete each of the following:
   a. A very heavy object has mass that is measured in _____.
   b. A cube whose length, width, and height are each 1 cm has a volume of _____.
   c. If the cube in (b) is filled with water, the mass of the water is _____.
   d. Which has a larger volume: 1 L or 1 dm³?
   e. If a car uses 1 L of gas to go 12 km, the amount of gas needed to go 300 km is _____ L.
   f. 20 ha = _____ a
   g. 51.8 L = _____ cm³
   h. 10 km² = _____ m²
   i. 50 L = _____ mL
   j. 5830 mL = _____ L
   k. 25 m³ = _____ dm³
   l. 75 dm³ = _____ mL
   m. 52,813 g = _____ kg
   n. 4800 kg = _____ t
16. Two cones are defined to be similar if the ratio between their heights equals the ratio between their radii. If two similar cones have heights h_1 and h_2, find the ratio between their volumes in terms of h_1 and h_2.
17. a. A tank that is a right rectangular prism is 1 m × 2 m × 3 m. If the tank is filled with water, what is the mass of the water?

b. Suppose the tank is exactly half full of water and then a heavy metal sphere of radius 30 cm is put into the tank. How high is the water now if the height of the tank is 3 m?

18. For each of the following, fill in the correct unit to make the sentence reasonable:
    a. Anna filled the gas tank with 80 ____ .
    b. A man has a mass of about 82 ____ .
    c. The textbook has a mass of 978 ____ .
    d. A nickel has a mass of 5 ____ .
    e. A typical adult cat has a mass of about 4 ____ .
    f. A compact car has a mass of about 1.5 ____ .
    g. The amount of coffee in the cup is 180 ____ .

19. For each of the following, decide if the situation is likely or unlikely:
    a. Carrie's bath water has a temperature of 15°C.
    b. Anne found 26°C too warm and so lowered the thermostat to 21°C.
    c. Jim is drinking water that has a temperature of $^-5$°C.
    d. The water in the teakettle has a temperature of 120°C.
    e. The outside temperature dropped to 5°C, and ice appeared on the lake.

20. Complete each of the following:
    a. 2 dm^3 of water has a mass of ____ g.
    b. 1 L of water has a mass of ____ g.
    c. 3 cm^3 of water has a mass of ____ g.
    d. 4.2 mL of water has a mass of ____ kg.
    e. 0.2 L of water has a volume of ____ m^3.

SELECTED BIBLIOGRAPHY

Battista, M. "How Many Blocks?" *Mathematics Teaching in the Middle School* 3 (March-April 1998): 404–411.

Battista, M., and D. Clements. "Finding the Number of Cubes in Rectangular Cube Buildings." *Teaching Children Mathematics* 4 (January 1998): 258–264.

Binswanger, R. "Discovering Perimeter and Area with Logo." *Arithmetic Teacher* 36 (September 1988): 18–24.

Clopton, E. "Sharing Teaching Ideas: Area and Perimeter Are Independent." *Mathematics Teacher* 84 (January 1991): 33–35.

Cohen, D. "Estimating the Volumes of Solid Figures with Curved Surfaces." *Mathematics Teacher* 84 (May 1991): 392–395.

Collier, P., and T. Pateracki. "Geometry in the Middle School: An Exchange of Ideas and Experiences. *Mathematics Teaching in the Middle School* 3 (March-April 1998): 412–415.

Hayes, N. "Cardboard Tubes Bring Geometry from Home." *Mathematics Teaching in the Middle School* 4 (October 1998): 120–122.

Lamphere, P. "Geoboard Patterns and Figures." *Teaching Children Mathematics* 1 (January 1995): 282–287.

LaSaracina, B., and S. White. "The Restless Rectangle and the Transforming Trapezoid." *Teaching Children Mathematics* 5 (February 1999): 336–337, 366.

Manouchehri, A., M. Enderson, and L. Pagnucco. "Exploring Geometry with Technology." *Mathematics Teaching in the Middle School* 3 (March-April 1998): 436–442.

Miller, W., and L. Wagner. "Pythagorean Dissection Puzzles." *Mathematics Teacher* 86 (April 1993): 302–308, 313–314.

Moore, D. "Some Like It Hot: Promoting Measurement and Graphical Thinking by Using Temperature." *Teaching Children Mathematics* 5 (May 1999): 538–543.

Naraine, B. "If Pythagoras Had a Geoboard." *Mathematics Teacher* 86 (February 1993): 137–140, 145–148.

Nitabach, E., and R. Lehrer. "Developing Spatial Sense through Area Measurement." *Teaching Children Mathematics* 2 (April 1996): 473–476.

Nowlin, D. "Practical Geometry Problems: The Case of the Ritzville Pyramids." *Mathematics Teacher* 86 (March 1993): 198–200.

Parker, J., and C. Widmer. "Patterns in Measurement." *Arithmetic Teacher* 40 (January 1993): 292–295.

Pudelka, P. "Sharing Teaching Ideas: Formulas and Sugar Cubes." *Mathematics Teacher* 83 (February 1990): 119–120.

Robertson, S. "Getting Students Actively Involved in Geometry." *Teaching Children Mathematics* 5 (May 1999): 526–529.

Shultz, J. "Area Models—Spanning the Mathematics of Grades 1–9." *Arithmetic Teacher* 39 (October 1991): 42–46.

Smith, L. "Areas and Perimeters of Geoboard Polygons." *Mathematics Teacher* 83 (May 1990): 392–398.

Spangler, D. "A Case of the Smash Hit." *Mathematics Teaching in the Middle School* (March-April 1998): 424–427.

Stone, M. "Teaching Relationships between Area and Perimeter with the Geometer's Sketchpad." *Mathematics Teacher* 87 (November 1994): 590–594.

Stover, D. "Sharing Teaching Ideas: Area of a Triangle," *Mathematics Teacher* 83 (February 1990): 120.

Taylor, L. "Exploring Geometry with the Geometer's Sketchpad." *Arithmetic Teacher* 40 (November 1992): 187–191.

Usnick, V., P. Lamphere, and G. Bright. "A Generalized Area Formula." *Mathematics Teacher* 85 (December 1992): 752–754.

12

Motion Geometry and Tessellations

Preliminary Problem
Suppose you are looking in a mirror hung flat on a wall and your image goes from the top of the mirror to the bottom. How does the length of the part of your body that you see compare with the length of the mirror?

E uclid envisioned moving one geometric figure in a plane and placing it on top of another to determine if the two figures were congruent. Intuitively, we know this can be done by making a tracing of one figure, then sliding, turning, or flipping the tracing, and finally placing it back down atop the other figure. Elementary school students seem to be able to identify congruences by this type of predeductive activity.

Symmetries are fundamental in the study of geometry, nature, and shapes. Many symmetries are the results of sliding, flipping, and turning shapes. A study of these "motions" and symmetries leads to tessellations of the plane or space. A **tessellation** is the filling of a plane or space with repetitions of a figure (or figures) in such a way that none overlap and there are no gaps. This chapter contains sections on motions, symmetries, and tessellations.

tessellation

In the *Principles and Standards* for grades PreK–2, we find the following:

▲ *Students can naturally use their own physical experiences with shapes to learn about transformations such as slides (translations), turns (rotations), and flips (reflections)* (p. 99).

At grades 3–5, we find:

▲ *An understanding of congruence and similarity will develop as students explore shapes that in some way look alike. They should come to understand congruent shapes as those that exactly match and similar shapes as those that are related by "magnifying" or "shrinking"* (p. 166).

Finally at grades 6–8, we see:

▲ *Transformational geometry offers another lens through which to investigate and interpret geometric figures. To help them form images of shapes through different transformations, students can use physical objects, figures traced on tissue paper, mirrors or other reflective surfaces, figures drawn on graph paper, and dynamic geometry software* (p. 235).

HISTORICAL NOTE

In 1872, at age 23, Felix Klein (1849–1925) was appointed to a chair at the University of Erlangen, Germany. His inaugural address, referred to as the *Erlanger Programm*, described geometry as the study of properties of figures that do not change under a particular set of transformations. Specifically, Euclidean geometry was described as the study of such properties of figures as area and lengths, which remain unchanged under a set of transformations called *isometries*.

One of the first attempts to introduce geometric transformations into the elementary school in the United States was by the University of Illinois Committee on School Mathematics (UICSM). UICSM published *Motion Geometry* in four volumes in 1969.

Section 12-1 — Translations and Rotations

Translations

Figure 12-1 shows a two-dimensional representation of a child moving down a slide without twisting or turning. This type of motion is a **translation,** or **slide.**

translation/slide

Figure 12-1

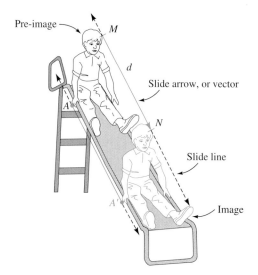

pre-image
slide line · image
slide arrow · vector

In Figure 12-1, the child (**pre-image**) at the top of the slide moves a certain distance in a certain direction along a **slide line** to obtain the **image** at the bottom of the slide.

Figure 12-1 shows a translation that takes the pre-image to its image. The translation is determined by the **slide arrow**, or **vector**, from M to N. The vector determines the image of any point in a plane in the following way: The image of a point A in the plane is the point A' obtained by sliding A along a line parallel to $\overrightarrow{MN}$ in the direction from M to N by the distance MN. (MN is also denoted by d in Figure 12-1.) Notice that $AA' = MN = d$ and $\overleftrightarrow{MN}$ is parallel to $\overleftrightarrow{AA'}$. It appears that under the translation, figures change neither their shapes nor sizes. In fact, a translation preserves both length and angle size, and thus congruence of figures.

isometry
rigid motion

Any motion that preserves distance is an **isometry** (derived from Greek and meaning "equal measure"), or **rigid motion.** For example, a translation is an isometry.

> **Definition of a Translation**
>
> A **translation** is a motion of a plane that moves every point of the plane a specified distance in a specified direction along a straight line.

Constructions of Translations

The image of a figure under a translation can be constructed easily with tracing paper or by using only a compass and straightedge. Using tracing paper may be the more natural way to construct a "motion." However, since traditionally most constructions in geometry are accomplished with a compass and a straightedge, we consider first how that might be done and then leave the actual construction as an activity in "Now Try This 12-1."

To construct the image of an object under a translation, we first need to know how to construct the image A' of a single point A. In Figure 12-1, we know that $MN = AA'$ and $\overleftrightarrow{MN}$

is parallel to $\overleftrightarrow{AA'}$. Thus, $MAA'N$ is a parallelogram. (Why?) Thus, to construct the image A' of a point A with a compass and straightedge, we need only to construct a parallelogram $MAA'N$ so that $\overrightarrow{AA'}$ is in the same direction as $\overrightarrow{MN}$.

NOW TRY THIS 12-1

a. Use a compass and straightedge to find the image of A under a translation that takes M to N as in Figure 12-2.

b. There are two parallelograms with vertices A, M, and N. How do you know which vertex to use for the construction?

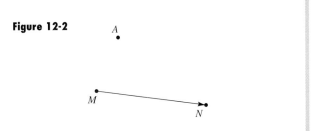

Figure 12-2

To find the image of a triangle under a translation, we find the images of the three vertices by using a process similar to that used in Figure 12-1 or Figure 12-2 and connect these images with segments to form the triangle's image.

It also is often possible to use a geoboard or a grid to find an image of a segment, as the following example shows.

Example 12-1   Find the image of $\overline{AB}$ under the translation from X to X' pictured on the dot paper in Figure 12-3.

Figure 12-3

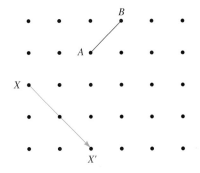

Solution X' is the image of X under the translation. So it could be obtained from X by shifting X two units vertically down and then two units horizontally to the right, as shown in Figure 12-4. This shifting determines the slide arrow from X to X'. The image of each point on the dot paper can be obtained by first shifting it 2 units down and then 2 units to the right. The image of $\overline{AB}$ is found in this way in Figure 12-4.

Figure 12-4

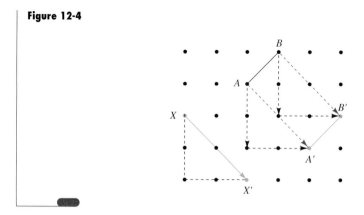

Coordinate Representation of Translations

In many applications of mathematics such as computer graphics, it is necessary to use translations in a coordinate system. In Figure 12-5, $\triangle A'B'C'$ is the image of $\triangle ABC$ under the translation defined by the slide arrow from O to O', where O is the origin and O' has coordinates $(5, {}^-2)$. Point O' is the image of point O under the given translation. The point $O'(5, {}^-2)$ can be obtained by moving O horizontally to the right 5 units and then 2 units down. As each point in the triangle is translated in the direction from O to O' by the distance O', we can obtain the image of any point by moving horizontally to the right 5 units and then vertically 2 units down. This is shown in Figure 12-5 for points A, B, C and their corresponding images A', B', and C'.

Figure 12-5

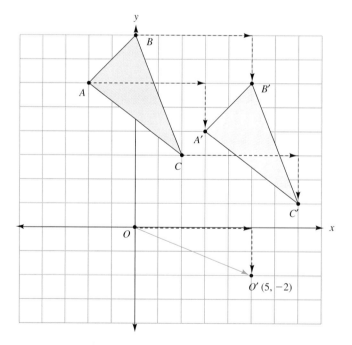

Table 12-1 shows how the coordinates of the image vertices A', B', and C' in Figure 12-5 are obtained from the coordinates of A, B, and C.

Table 12-1

Point (x, y)	Image Point (x + 5, y − 2)
A (−2, 6)	A′ (3, 4)
B (0, 8)	B′ (5, 6)
C (2, 3)	C′ (7, 1)

This discussion suggests that we could describe a translation by showing how the coordinates of any point (x, y) are changed. The translation described in Table 12-1 can be written symbolically as $(x, y) \to (x + 5, y - 2)$ where "$\to$" denotes "moves to."

Definition of a Translation in a Coordinate System

A translation is a function from the plane to the plane such that to every point (x, y) corresponds the point $(x + a, y + b)$ for real numbers a and b.

In general, the translation is symbolized as $(x, y) \to (x + a, y + b)$ because the point $(x + a, y + b)$ is the image of the point (x, y).

Example 12-2 Find the coordinates of the image of quadrilateral $ABCD$ in Figure 12-6 under each of the translations in parts (a) through (c). Draw the image in each case.

Figure 12-6

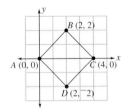

a. $(x, y) \to (x - 2, y + 4)$
b. A translation determined by the slide arrow from $A(0, 0)$ to $A'(-2, 4)$
c. A translation determined by the slide arrow from $S(4, -3)$ to $S'(2, 1)$

Solution a. Because $(x, y) \to (x - 2, y + 4)$, the images A', B', C', and D' of the corresponding points A, B, C, and D can be found as follows:

$$A(0, 0) \to A'(0 - 2, 0 + 4), \text{ or } A'(-2, 4)$$
$$B(2, 2) \to B'(2 - 2, 2 + 4), \text{ or } B'(0, 6)$$
$$C(4, 0) \to C'(4 - 2, 0 + 4), \text{ or } C'(2, 4)$$
$$D(2, -2) \to D'(2 - 2, -2 + 4), \text{ or } D'(0, 2)$$

The square ABCD and its image $A'B'C'D'$ are shown in Figure 12-7.

Figure 12-7

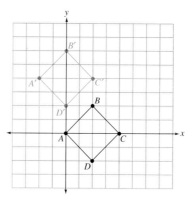

b. To obtain the image of any point in the plane under the translation from A to A', we also move 2 units to the left and then 4 units up. Consequently, the image of any point (x, y) is $(x + (^-2), y + 4)$, or $(x - 2, y + 4)$. This is the same translation as in (a) and hence the graph is the same as that in (a).

c. We could move from $S(4, ^-3)$ to $S'(2, 1)$ by moving a certain number of units horizontally and then a certain number of units vertically. Because $4 + (^-2) = 2$ and $^-3 + 4 = 1$, moving 2 units to the left and then 4 units up will take us from S to S'. The image of any point (x, y) under the translation from S to S' can be obtained in the same way; that is, the image of any point (x, y) is $(x - 2, y + 4)$. Thus this graph also is the same as that obtained in (a).

Rotations

rotation/turn

A **rotation,** or **turn,** is another kind of isometry. Figure 12-8 illustrates congruent figures that resulted from a rotation about point O. The image of the letter **F** is shown in green.

Figure 12-8

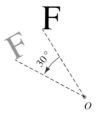

turn center
turn angle

A rotation can be constructed by using tracing paper, as in Figure 12-9. In Figure 12-9(a), $\triangle ABC$ and point O are traced on tracing paper. Holding point O fixed, we turn the tracing paper to obtain the image, $\triangle A'B'C'$, as shown in Figure 12-9(b). Point O is the **turn center,** and $\angle COC'$ is the **turn angle.**

Figure 12-9 Construction of a rotation using tracing paper

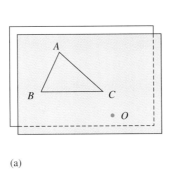
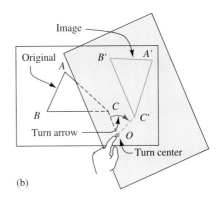

(a)   (b)

To determine a rotation, we must know three pieces of data: the turn center; the direction of the turn, either clockwise or counterclockwise; and the amount of the turn. The amount and the direction of the turn can be illustrated by a **turn arrow,** or they can be specified in numbers of degrees.

turn arrow

This discussion leads to the following definition.

Definition of Rotation

A **rotation** is a motion of the plane determined by holding one point—the center—fixed and rotating the plane about this point by a certain amount in a certain direction.

To construct an image of a figure under a rotation, observe that any point on the tracing-paper construction moves along a circle. Also the angle formed by any point, the center of the rotation O, and the image of the point is the angle of the turn. (Why?) With this in mind, we construct the image of a point under a rotation in the following "Now Try This" activity.

NOW TRY THIS 12-2

- Use a compass and a straightedge to construct the image of point P under a rotation with center O through the angle and in the direction given in Figure 12-10.

Figure 12-10

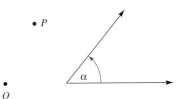

Because a rotation is an isometry, the image of a figure under a rotation is congruent to the original figure. It can be shown that *under any isometry, the image of a line is a line, the image of a circle is a circle, and the images of parallel lines are parallel lines.*

For certain angles like 90°, rotations may be constructed on a geoboard or dot paper, as demonstrated in Example 12-3.

Example 12-3   Find the image of △ABC under the rotation with center O, as shown in Figure 12-11.

Figure 12-11

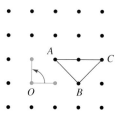

Solution   △A'B'C', the image of △ABC, is shown in Figure 12-12. The image of A is A' because ∠A'OA is a right angle (why?) and OA = OA'. Similarly B' is the image of B. To find the location of C', we use the fact that △A'B'C' ≅ △ABC and hence ∠B ≅ ∠B'. The location of point C' shown makes ∠B ≅ ∠B', C'B' = CB (why?), and the direction of the rotation is counterclockwise as specified.

Figure 12-12

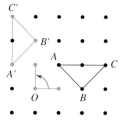

half-turn

A rotation of 360° about a point will move any figure onto itself. A rotation of 180° about a point is also of particular interest. Such a rotation is a **half-turn.** Because a half-turn is a rotation, it has all the properties of rotations. Figure 12-13 shows some shapes and their images under a half-turn about point O.

Figure 12-13

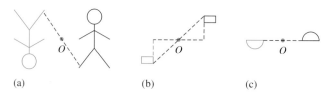

(a)   (b)   (c)

Rotations are useful in determining turn symmetries, which we investigate later. A different application of rotations appears when we consider the slopes of perpendicular lines as seen in the following.

Slopes of Perpendicular Lines

Transformations can be used to investigate various mathematical relationships. For example, consider the relationship between the slopes of two perpendicular lines, neither of which is vertical.

We first consider a special case in which the lines go through the origin. Suppose the slopes of the lines ℓ_1 and ℓ_2, shown in Figure 12-14, are m_1 and m_2, respectively. Because the slope of a line is equal to rise over run, the slope of ℓ_1 can also be determined from $\triangle OBA$, in which we choose $OA = 1$. We have $m_1 = \dfrac{\text{rise}}{\text{run}} = \dfrac{BA}{1} = BA$.

Figure 12-14

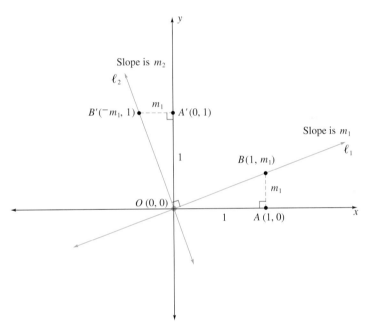

We then rotate the plane 90° counterclockwise about center O. The image of ℓ_1 is ℓ_2. To determine the image of $\triangle OBA$, we need only determine the images of each of the vertices of the triangle. The image of O is O itself. The image of A is A' on the y-axis (why?), and the image of B is B' on ℓ_2 (why?). Because rotation preserves congruence, $\triangle OB'A' \cong \triangle OBA$; consequently, $\angle B'A'O'$ is a right angle, $A'B' = m_1$, and $OA' = 1$. Thus as shown in Figure 12-14, point B' is at $(^{-}m_1, 1)$. We can use the slope formula to find the slope of ℓ_2 as follows:

$$m_2 = \dfrac{1 - 0}{^{-}m_1 - 0} = \dfrac{1}{^{-}m_1} = \dfrac{^{-}1}{m_1}.$$

Thus $m_2 = {}^{-}1/m_1$, or $m_1 m_2 = {}^{-}1$.

The relationship between the slopes m_1 and m_2 of two perpendicular lines (neither of which is vertical) that do not intersect at the origin can always be found using two lines parallel to the original lines but that pass through the origin. Because parallel lines have equal slopes the relationship between the slopes of the perpendicular lines is the same as the relationship between the slopes of the perpendicular lines through the origin, that is,

$m_1 m_2 = {}^-1$.

It is also possible to prove the converse statement, that is, if the slopes of two lines satisfy the condition $m_1 m_2 = {}^-1$, then the lines are perpendicular. We summarize these results in the following property.

Property of Slopes of Perpendicular Lines

Two lines, neither of which is vertical, are perpendicular if, and only if, their slopes m_1 and m_2 satisfy the condition $m_1 m_2 = {}^-1$. Any vertical line is perpendicular to a line with slope 0.

Example 12-4 Find the equation of line ℓ through point $({}^-1, 2)$ and perpendicular to the line $y = 3x + 5$.

Solution If m is the slope of ℓ as in Figure 12-15, then ℓ can be written as $y = mx + b$.

Figure 12-15

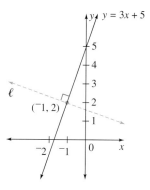

Because the line $y = 3x + 5$ has slope 3 and is perpendicular to ℓ, we have $m \cdot 3 = {}^-1$; therefore $m = -\frac{1}{3}$. Consequently, the equation of ℓ is

$$y = -\frac{1}{3}x + b.$$

Because the point $({}^-1, 2)$ is on ℓ, we can substitute $x = {}^-1$, $y = 2$ in $y = -\frac{1}{3}x + b$ and solve for b as follows:

$$2 = -\frac{1}{3} \cdot ({}^-1) + b$$

$$\frac{5}{3} = b$$

Consequently, the equation of ℓ is

$$y = -\frac{1}{3}x + \frac{5}{3}.$$

ONGOING ASSESSMENT 12-1

1. For each of the following, find the image of the given quadrilateral under a translation from A to B:

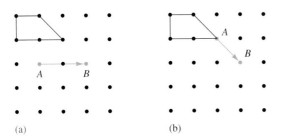

(a)   (b)

2. Find the figure whose image is given in each of the following under a translation from X to X':

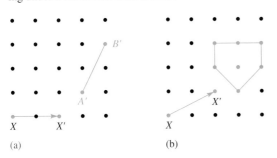

(a)   (b)

3. Construct the image of $\overline{BC}$ under the translation pictured in the following figure by using the following:
   a. Tracing paper
   b. Compass and straightedge

4. Find the coordinates of the image for each of the following points under the translation defined by $(x, y) \rightarrow (x + 3, y - 4)$:
   a. $(0, 0)$
   b. $(^-3, 4)$
   c. $(^-6, ^-9)$
   d. $(7, 14)$
   e. (h, k)

5. Find the coordinates of the points whose images under the translation $(x, y) \rightarrow (x - 3, y + 4)$ are the following:
   a. $(0, 0)$
   b. $(^-3, 4)$
   c. $(^-6, ^-9)$
   d. $(7, 14)$
   e. (h, k)

6. Consider the translation $(x, y) \rightarrow (x + 3, y - 4)$. In each of the following, draw the image of the figure under the translation and find the coordinates of the points of the image.

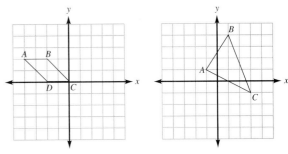

(a)   (b)

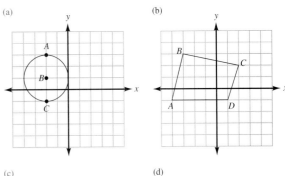

(c)   (d)

7. Consider the translation $(x, y) \rightarrow (x + 3, y - 4)$. In each of the following, draw the figure whose image is shown:

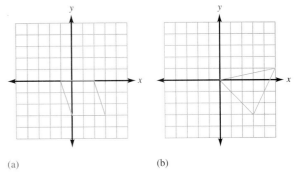

(a)   (b)

8. Find the image of the following quadrilateral in a 90° counterclockwise rotation about O:

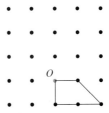

9. If P' is the image of point P under a half-turn about its center O, what can be said about points P', P, and O? Why?
10. Use a compass and a straightedge to find the image of line ℓ under a half-turn about point O as shown.

11. Using printed capital letters of the English alphabet, answer the following:
    a. Are there letters that can be their own images under a rotation? If so, sketch the letter and identify the center.
    b. Are there letters in part (a) that are their own images under a half-turn? If so, identify them.
12. The images of $\overline{AB}$ under various rotations are given in the following figures. Find $\overline{AB}$ in each case.

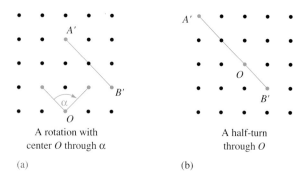

A rotation with center O through α
(a)

A half-turn through O
(b)

13. a. The image of NOON is still NOON after a special half-turn. List some other words that have the same property. What letters can such words contain?
    b. 1961 is the image of 1961 after a special half-turn. What other natural numbers less than 10,000 have this property?
14. a. Refer to the following figure and use paper folding or any other method to show that if P' is the image of P under rotation about point O by a given angle, then O is on the perpendicular bisector of $\overline{PP'}$.

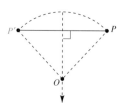

   b. $\triangle A'B'C'$ shown in the following figure was obtained by rotating $\triangle ABC$ about a certain point O. Explain how to find the point O and the angle of rotation.

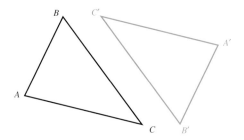

15. The images of any point under a rotation by certain angles can be found with only a compass and straightedge (without the use of a protractor). Construct P', the image of P when it is rotated about O, as shown in the following figure, for angles with the following measures and direction:
    a. 90° counterclockwise
    b. 90° clockwise
    c. 45° counterclockwise
    d. 60° clockwise
    e. 30° counterclockwise

• P

• O

16. For each of the following points, find the coordinates of the image point under a half-turn about the origin:
    a. (4, 0)
    b. (0, 3)
    c. (2, 4)
    d. ($^-$2, 5)
    e. ($^-$2, $^-$4)
    f. (a, b)
    g. (^-a, ^-b)
17. In each of the following figures, find the image of the figure under a half-turn about O:

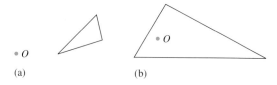

(a)   (b)

18. Draw any line and label it ℓ. Use tracing paper to find ℓ', the image of ℓ under each of the following rotations. In each case, describe in words how ℓ' is related to ℓ.
    a. Half-turn about point O on ℓ
    b. Half-turn about a point O, not on ℓ
    c. A 90° turn counterclockwise about point O, not on ℓ
    d. A 60° turn counterclockwise about point O, not on ℓ

19. a. For each of the following, find the coordinates of A', the image of A under rotation about the origin counterclockwise by a right angle.

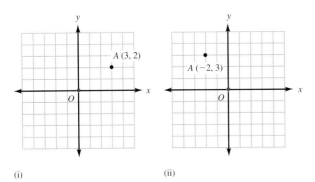

(i)   (ii)

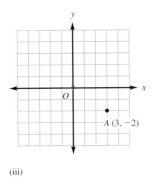

(iii)

b. Based on your answers in (a), conjecture the coordinates of A', the image of A under the rotation about the origin counterclockwise by a right angle if the coordinates of A are (a, b).

c. Justify your conjecture in (b).

20. a. Find the final image of $\triangle ABC$ by performing two rotations in succession each with center O, one by angle α and the other by angle β in directions as shown in the figure.

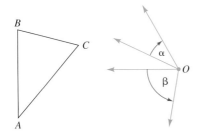

b. Is the order of the rotations important?
c. Could the result have been accomplished in one rotation?

21. When $\triangle ABC$ in the following figure is rotated about a point O by $360°$, each of the vertices traces a path.

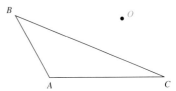

a. What geometric figure does each vertex trace?
b. Identify all points O for which two vertices trace an identical path. Justify your answer.
c. Given any $\triangle ABC$, is there a point O such that the three vertices trace an identical path? If so, describe how to find such a point. Justify your answer.

22. a. Translations may be explored in Logo by using a figure called an EE. Edit the following programs to fit the particular Logo you have and type it into your computer. Then run the following with the turtle starting at home with heading 0:
 (i) SLIDE 40 45
 (ii) SLIDE 200 57
 (iii) SLIDE (−50) (−75)

```
TO SLIDE :DIRECTION :DISTANCE
  EE
  PENUP
  SETHEADING :DIRECTION
  FORWARD :DISTANCE
  PENDOWN
  SETHEADING 0
  EE
END
TO EE
  FORWARD 50 RIGHT 90
  FORWARD 25 BACK 25
  LEFT 90 BACK 25
  RIGHT 90 FORWARD 10
  BACK 10 LEFT 90
  BACK 25 RIGHT 90
  FORWARD 25 BACK 25
  LEFT 90
END
```

b. Edit the SLIDE procedure in (a) so that it will slide an equilateral triangle.

23. Write a Logo procedure called ROTATE that will draw a square and produce the image of the square when the square is rotated by an arbitrary angle :A about one of its vertices.

24. Write a Logo procedure called TURN.CIRCLE that will draw a circle passing through the home of the turtle and produce the image of the circle under the following transformations:
a. A half-turn about the turtle's home
b. A 90° counterclockwise rotation about the turtle's home

Communication

25. If we are given two congruent nonparallel segments, is it always possible to find a rotation so that the image of one segment will be the other segment? Explain why or why not.
26. a. If you rotate an object 180° clockwise or counterclockwise, using the same center, is the image the same in both cases?
    b. Answer part (a) if you rotate the object 360°.
27. For each of the following figures, trace the figure on tracing paper, rotate the tracing by 180° about the given point O, sketch the image, and then make a conjecture about the kind of figure that is formed by the union of the original figure and its image. In each case, explain why you think your conjecture is true.

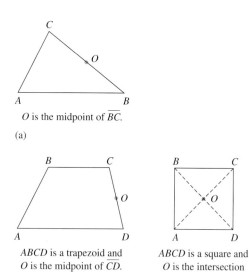

28. If $\overline{AB}$ is rotated about point O by 90° clockwise, explain whether its image is perpendicular to $\overline{AB}$, regardless of the location of point O.

Open-Ended

29. A drawing of a cube, shown in the following figure, can be created by drawing a square $ABCD$, finding its image under translation defined by the slide arrow from A to A' so that $AA' = AB$, and connecting the points A, B, C, and D with their corresponding images.

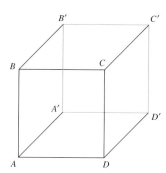

Draw several other perspective geometric figures using translations. In each case, name the figures and indicate the slide arrow that defines the translation.

30. Wall stenciling has been used to obtain an effect similar to that of wallpapering. The stencil pattern shown in the following figure can be used to create a border on a wall.

Measure the length of a wall of a room and design your own stencil pattern to create a border. Cut the pattern from a sheet of plastic or cardboard. Define the translation that will accomplish creating an appropriate border for the wall.

31. The following pattern can be created by rotating figure A about O by the indicated angle, then rotating the image B about O by the same angle, and then rotating the image C about O by the same angle, and so on until one of the images coincides with the original figure A.

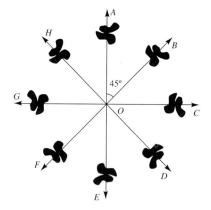

Make several designs with different numbers of congruent figures around a circle in which the image of each figure under the same rotation is the next figure and so that one of the images coincides with the original figure.

Cooperative Learning

32. Mark a point A on a sheet of paper and set a straightedge through A as shown in the following figure. Find a circular shape (a jar lid is a good choice) and mark point P on the edge of the shape. Place the shape on the straightedge so that P coincides with A. Consider the path traced by P as the circle rolls so that its edge stays in contact with the straightedge all the time and until P comes in contact with the straightedge again at point B.

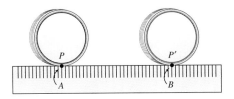

a. Have one member of your group roll the circular shape and another draw the path traced by P as accurately as possible.
b. Have each member of the group identify the transformations that are involved in the experiment. Compare your answers.
c. Discuss how to check if the path traced by point P is an arc of a circle.
d. Find the length of $\overline{AB}$.

TECHNOLOGY CORNER

Use a geometry utility to draw an equilateral triangle and two altitudes of the triangle. Let O be the point at which the altitudes intersect. Rotate the triangle by 120° about O in any direction. Make a conjecture based on this experiment. Do you think your conjecture may be true for some triangles that are not equilateral? Why?

BRAIN TEASER

In Figure 12-16, a coin is shown above and touching another coin. Suppose the top coin is rotated around the circumference of the bottom coin until it rests directly below the bottom coin. Will the head be straight up or upside down?

Figure 12-16

684   CHAPTER 12   *Motion Geometry and Tessellations*

Section 12-2 — **Reflections and Glide Reflections**

Reflections

reflection · flip

Another isometry is a **reflection,** or **flip.** One example of a reflection often encountered in our daily lives is a mirror image. Figure 12-17 shows a figure with its mirror image.

Figure 12-17

Another reflection is shown in the following "B.C." cartoon.

reflecting line
mirror image

We can simulate reflections in a plane in various ways. Consider the half tree shown in Figure 12-18(a). Folding the paper along the **reflecting line** and drawing the image gives the **mirror image,** or *image,* of the half tree. In Figure 12-18(b), the paper is shown unfolded. Another way to simulate a reflection in a line involves using a Mira, as illustrated in Figure 12-18(c).

Figure 12-18

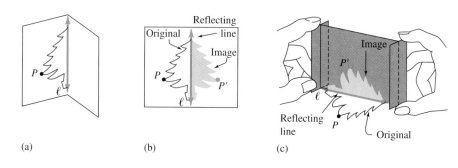

(a)   (b)   (c)

In Figure 12-19(a), the image of P under a reflection in line ℓ is P'. $\overline{PP'}$ is both perpendicular to and bisected by ℓ, or equivalently, ℓ is the perpendicular bisector of $\overline{PP'}$.

Figure 12-19

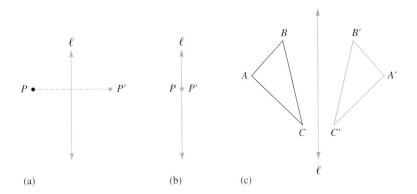

(a)　　(b)　　(c)

In Figure 12-19(b), P is its own image under the reflection in line ℓ. If ℓ were a mirror, then P' would be the mirror image of P. This leads us to the following definition of a reflection.

Definition of Reflection

A **reflection** in a line ℓ is a motion of a plane that pairs each point P of the plane with a point P' in such a way that ℓ is the perpendicular bisector of $\overline{PP'}$, as long as P is not on ℓ. If P is on ℓ, then $P = P'$.

In Figure 12-19(c), we see another property of a reflection. In the original triangle ABC, if we walk clockwise around the vertices, starting at vertex A, we see the vertices in the order A-B-C. However, in the reflection image of triangle ABC, if we start at A' (the image of A) and walk clockwise, we see the vertices in the following order: A'-C'-B'. Thus a reflection does something that neither a translation nor a rotation does; it reverses the

orientation　**orientation** of the original figure.

There are many methods of constructing a reflection image. We already illustrated such constructions with paper folding and a Mira. Next, we illustrate the construction of the image of a figure under a reflection in a line with tracing paper.

Constructing a Reflection by Using Tracing Paper

Figure 12-20(a) shows the use of tracing paper. We trace the original figure, the *reflecting line,* and a point on the reflecting line, which we use as a *reference point.* When we flip the tracing paper over to perform the reflection, we align the reflecting line and the reference point, as in Figure 12-20(b). Aligning the reference point ensures that no translating occurs along the reflecting line when the reflection is performed. If we wish the image to be on the paper with the original, we may indent the tracing paper or acetate sheet to mark the images of the original vertices.

Figure 12-20

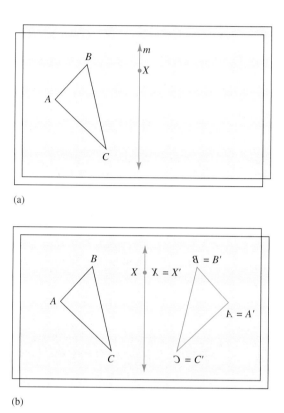

(a)

(b)

NOW TRY THIS 12-3

- Use the definition of a reflection in a line and properties of a rhombus to construct the image P' of point P in Figure 12-21 under reflection in line m using only a compass and straightedge.

Figure 12-21

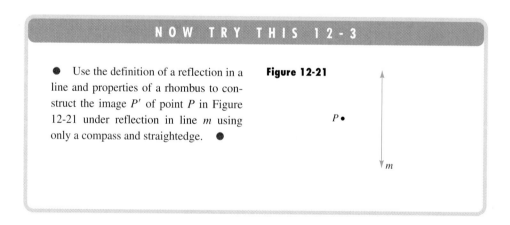

Graph paper and mirrors are often used in elementary school to motivate and study reflections in a line and an application of a reflection, line symmetry. An example is seen on the following student page from *Scott Foresman-Addison Wesley Middle School Math Course 2*, 1999.

Reflections and Line Symmetry

11-11

▶ **Lesson Link** You've transformed figures by sliding them. Now you'll investigate transformations made by flipping figures. ◀

You'll Learn ...
- to identify lines of symmetry
- to reflect figures on a coordinate plane

Explore Reflections

Mirror, Mirror, on the Graph

Materials: Graph paper, Markers

1. Set up x- and y-axes on your graph paper. Use a marking pen to sketch a simple cartoon character or irregular design in the second quadrant of your coordinate system.

2. Fold your paper along the y-axis so the original figure is on the outside. Turn your paper over and trace your figure onto the other half of the paper.

3. Unfold your paper. Compare your original figure to the tracing. Are the figures identical? If not, what differences do you see?

4. Choose a point on your original figure. How far is the point from the y-axis? How far is the matching point on your tracing from the y-axis?

... How It's Used

Judges at dog shows look for symmetry when choosing champion dogs.

Vocabulary
symmetry
line symmetry
line of symmetry
reflection

Learn Reflections and Line Symmetry

A balance, or **symmetry**, is often found in nature and in art.

When one half of an object is a mirror image of the other, the object has **line symmetry**, and the imaginary "mirror" is the **line of symmetry**.

Since kaleidoscopes use several mirrors, the patterns they produce have many lines of symmetry.

Example 12-5 Describe how to construct the image of $\overleftrightarrow{AB}$ under a reflection in line m in Figure 12-22.

Figure 12-22

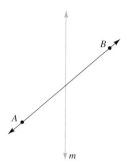

Solution Under a reflection, the image of a line is a line. Thus, to find the image of $\overleftrightarrow{AB}$, it is sufficient to choose any two points on the line and find their images. The images determine the line that is the image of $\overleftrightarrow{AB}$. We choose two points whose images are easy to find. Point X, the intersection of $\overleftrightarrow{AB}$ and m, is its own image. If we choose point A and use a compass and straightedge, we produce the construction shown in Figure 12-23.

Figure 12-23

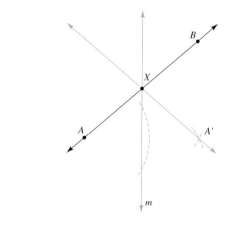

Constructing a Reflection on Dot Paper or a Geoboard

On dot paper or a geoboard, the images of figures under a reflection can sometimes be found by inspection, as seen in Example 12-6.

Example 12-6 Find the image of $\triangle ABC$ under a reflection in line m, as in Figure 12-24.

Figure 12-24

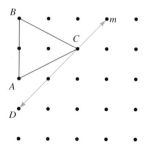

Solution The image $A'B'C'$ is given in Figure 12-25. Note that C is the image of itself and the images of the vertices A and B are A' and B' such that m is the perpendicular bisector of $\overline{AA'}$ and $\overline{BB'}$.

Figure 12-25

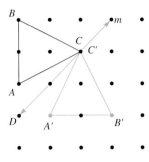

It is possible to find the reflecting line, if we are given an original figure and its reflection image. This is left as an exercise.

> **NOW TRY THIS 12-4**
>
> • Describe how to find the line of reflection using paper folding. •

Reflections in a Coordinate System

For some reflection lines like the x-axis and y-axis and the line $y = x$, it is quite easy to find the coordinates of the image, given the coordinates of the point. In Figure 12-26, the line $y = x$ bisects the angle between the x-axis and y-axis. The image of $A(1, 4)$ is the point $A'(4, 1)$. Also the image of $B(^-3, 0)$ is $B'(0, ^-3)$. It is left as an exercise to show that in general the image of $P(a, b)$ is the point $P'(b, a)$. Consequently, the reflection in the line $y = x$ exchanges the coordinates of the point. Is this still true if the point is not above line $y = x$?

Figure 12-26

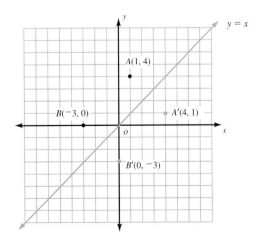

Glide Reflections

glide reflection

Another basic isometry is a **glide reflection.** An example of a glide reflection is shown in the footprints of Figure 12-27. We consider the footprint labeled F_1 to have been translated to footprint F_2 and then reflected over line m (parallel to $\overline{F_1F_2}$) to yield F_3, the image of F_2. F_3 is the final image of F_1.

Figure 12-27

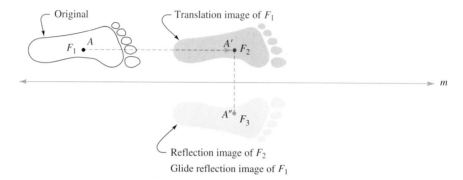

Note that point A is translated to find point A' and then point A' is reflected over line m (parallel to $\overline{F_1F_2}$) to obtain point A''. Thus, A'' is the image of A in the glide reflection.

The illustration in Figure 12-27 leads us to the following definition.

Definition of Glide Reflection

A **glide reflection** is a motion consisting of a translation followed by a reflection in a line parallel to the slide arrow.

Because constructing a glide reflection involves constructing a translation and a reflection, the task of constructing a glide reflection is not a new problem. Exercises involving the construction of images of figures under glide reflections are given in the Ongoing Assessment 12-2.

We have seen that under an isometry, the image of a figure is a congruent figure. Also, given two congruent figures, it is possible to show that one can be transformed to the other by isometries. The following example shows one illustration of such a transformation.

Example 12-7

$ABCD$ in Figure 12-28 is a rectangle. Describe a sequence of isometries to show

a. $\triangle ADC \cong \triangle CBA$;
b. $\triangle ADC \cong \triangle BCD$; and
c. $\triangle ADC \cong \triangle DAB$.

Figure 12-28

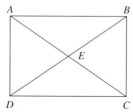

Solution a. A half-turn of △ADC with center E is one such transformation.
b. A reflection in a line passing through E and parallel to $\overline{AD}$ is one such transformation.
c. A reflection of △ADC in a line passing through E and parallel to $\overline{DC}$ is one such transformation.

Light Reflecting from a Surface

angle of incidence
angle of reflection

When a ray of light bounces off a mirror or when a billiard ball bounces off the rail of a billiards table, the **angle of incidence,** the angle formed by the incoming ray in Figure 12-29 and a line perpendicular to the mirror, is congruent to the **angle of reflection,** the angle between the reflected ray and the line perpendicular to the mirror.

Figure 12-29

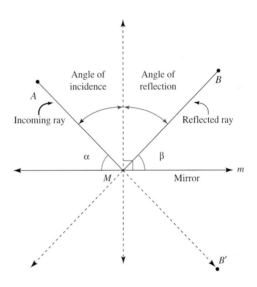

ONGOING ASSESSMENT 12-2

1. For each of the following figures, find the image of the given quadrilateral under a reflection in ℓ.

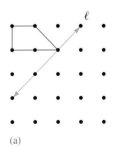
(a)

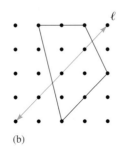
(b)

2. In the following figure, find the image of △ABC under a reflection in line ℓ.

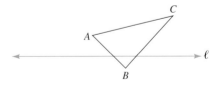

3. Draw a line and then draw a circle whose center is not on the line. Find the image of the circle under a reflection in the line.
4. Determine which of the following figures have a reflecting line such that the image of the figure under the reflecting line is the figure itself. In each case, find as many such reflecting lines as possible, sketching appropriate drawings.
   a. Circle
   b. Segment
   c. Ray

d. Square
  e. Rectangle
  f. Scalene triangle
  g. Isosceles triangle
  h. Equilateral triangle
  i. Trapezoid whose base angles are not congruent
  j. Isosceles trapezoid
  k. Arc
  l. Kite
  m. Rhombus
  n. Regular hexagon
  o. Regular *n*-gon

5. Determine the final result when △ABC is reflected in line ℓ and then its image is reflected again in ℓ.

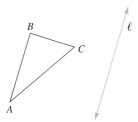

6. a. Refer to the following figure and suppose lines ℓ and *m* are parallel and △ABC is reflected in ℓ to obtain △A'B'C' and then △A'B'C' is reflected in *m* to obtain △A"B"C". Determine whether the same final image is obtained if △ABC is reflected first in *m* and then its image is reflected in ℓ.

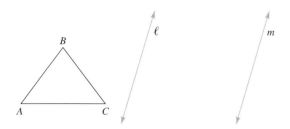

  b. Conjecture what single transformation will take △ABC directly to △A"B"C". Check your conjecture using tracing paper.

7. a. For the following figure, use any construction method to find the image of △ABC if △ABC is reflected in ℓ to obtain △A'B'C' and then △A'B'C' is reflected in *m* to obtain △A"B"C" (ℓ and *m* intersect at O).

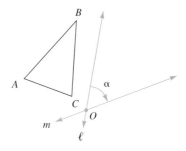

  b. Conjecture what single transformation will take △ABC directly to △A"B"C". Check your conjecture using tracing paper.
  c. Answer the question in (a) for the case in which ℓ and *m* are perpendicular.

8. Use a Mira if available to investigate problems 6 and 7.

9. Given △ABC and its reflection image △A'B'C' find the line of reflection.

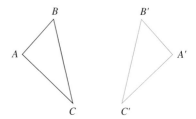

10. a. The word TOT is its own image when it is reflected through a vertical line through O, as shown in the following figure. List some other words that are their own images when reflected similarly.

  b. The image of BOOK is still BOOK when it is reflected through a horizontal line. List some other words that have the same property. Which uppercase letters can you use?

  c. The image of 1881 is 1881 after reflection in either a horizontal or vertical line, as shown in the following figure. What other natural numbers less than 2000 have this property?

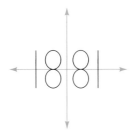

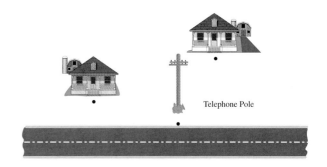

Telephone Pole

11. A glide reflection was defined as a translation followed by a reflection in appropriate lines.
    a. Determine whether the same final image is obtained if the reflection is followed by the translation.
    b. Use your answer in (a) to determine whether the reflection and translation involved in the glide reflection are commutative.
12. For the following figure numbered 1, decide whether a reflection, a translation, a rotation, or a glide reflection will transform the figure into each of the other numbered figures. (There may be more than one possible answer.)

Communication

16. If you look into a hinged mirror (as depicted below) and see an image with six sides, what is the minimum number of sides that the original could have? Explain your reasoning.

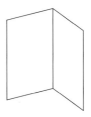

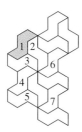

17. When a billiard ball bounces off a side of a pool table, the angle of incidence is usually congruent to the angle of reflection. In the following figure showing a scale drawing of a pool table, a cue ball is at point A. Show how a player should aim to hit two sides of the table and then the ball at B. Justify your solution.

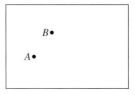

13. Given points $A(3, 4)$, $B(2, {}^-6)$, and $C({}^-2, 5)$, find the coordinates of the images of these points under each of the following transformations.
    a. Reflection in the x-axis
    b. Reflection in the y-axis
    c. Reflection in the line $y = x$
    d. Reflection in the line $y = {}^-x$
14. a. Conjecture what the image of a point with coordinates (x, y) will be under each of the transformations in problem 13.
    b. Suppose a point P with coordinates (x, y) is reflected in the x-axis and then its image P' is reflected in the y-axis to obtain P''. What are the coordinates of P'' in terms of x and y? Justify your answer.
★15. Two farm houses are located away from a road, as shown. A telephone company wants to construct a telephone pole at the edge of the road so that the telephone cable connecting the houses to the pole is as short as possible. Where should the pole be located?

18. a. Draw an isosceles triangle ABC and then construct a line such that the image of $\triangle ABC$ when reflected in the line is $\triangle ABC$ though every point is not necessarily its own image. Explain why the line you constructed has the required property.
    b. For what kind of triangles is it possible to find more than one line with the property in (a)? Justify your answer.
    c. Given a scalene triangle ABC, is it possible to find a line ℓ such that when $\triangle ABC$ is reflected in ℓ, its image is itself? Explain your answer.
    d. Draw a circle with center O and a line with the property that the image of the circle, when reflected in the line, is the original circle. Identify all such lines. Justify your answer.

19. Use the following drawing to explain how a periscope works.

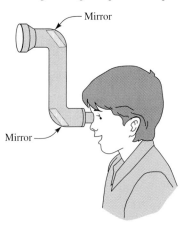

Open-Ended

20. In the following figure representing a miniature golf course, explain and justify the procedure showing how to aim the ball so that it gets in the hole if it is to bounce off
   a. one wall only.
   b. two walls.

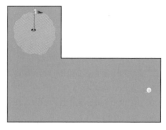

21. Design several wall stencil patterns using a reflection (see problem 30 in Ongoing Assessment 12-1). In each case, explain how you would use the stencil in practice.
22. Design wall stencil patterns using a glide reflection.
23. If a right triangle △ABC is reflected in one of its legs as shown in the following triangle, the triangle and its image form an isosceles triangle △ABA'.

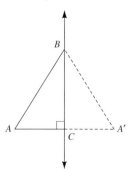

From the properties of reflection, we can deduce that the base angles in an isosceles triangle are congruent and that the altitude to the base bisects ∠ABA' as well as $\overline{AA'}$.

Apply the concept of reflection to deduce properties of other geometric figures by reflecting (**a**) a scalene triangle in one of its sides, (**b**) a right angle trapezoid in one of its sides, and (**c**) other figures. In each case, define the reflection, list the geometric properties of the figure obtained from the union of your original figure and its image, and justify the properties.

Cooperative Learning

24. In the following figure representing a pool table, ball *B* is sent on a path that makes a 45° angle with the table wall, as shown. It bounces off the wall five times and returns to its original position.

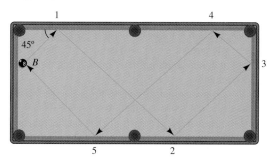

   a. Have each member of your group use graph paper to construct rectangular models of different-sized pool tables. Simulate the experiment using any tools (such as a straightedge, compass, and protractor) by choosing different positions for ball *B*.
   b. Share the results of your experiments with the rest of the group and together conjecture for which dimensions of the pool table and for what positions of *B* the experiment described in the problem will work.

Review Exercises

25. Which single digits are their own images under a rotation?
26. What is the image of a point (a, b) under a half-turn in the origin?
27. MOW is an example of a word that could be transformed into itself by which isometry?
28. a. Find all possible rotations that transform a circle into itself.
    b. By what other kinds of transformations can a circle be transformed onto itself?
29. Explain how a translation can be used to construct a rectangle whose area is equal to that of the parallelogram *ABCD* in the following figure.

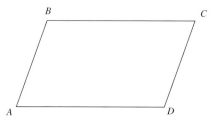

TECHNOLOGY CORNER

Edit the following to fit your particular Logo software. Enter the procedures into your computer:

```
TO SQ
  REPEAT 4 [FD 40 RT 90]
END
TO FSQ
  REPEAT 4 [FD 40 LT 90]
END
```

Describe the transformations illustrated in each of the following:

a.
```
TO MOVE1
  SQ
  RT 150
  SQ
END
```
b.
```
TO MOVE2
  SQ
  FSQ
END
```
c.
```
TO MOVE3
  SQ
  PU RT 45 FD 60 PD
  SQ
END
```

BRAIN TEASER

Two cities are on opposite sides of a river, as shown in Figure 12-30.

Figure 12-30

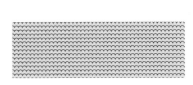

The cities' engineers want to build a bridge across the river that is perpendicular to the banks of the river and access roads to the bridge so that the distance between the cities is as short as possible. Where should the bridge and the roads be built?

Section 12-3 — Size Transformations

We have investigated transformations that preserved distance. Consequently, the image of a figure under one of these transformations was a figure congruent to the original. A different type of transformation happens when a slide is projected on a screen. All objects on the slide are enlarged on the screen by the same factor. Figure 12-31 is another example of such a transformation.

The point O is the *center* of the *size transformation* and 2 is the *scale factor*. Points O, A, and A' are collinear and $OA' = 2 \cdot OA$; also, O, C, and C' are collinear and $OC' = 2 \cdot OC$. Similarly, O, B, and B' are collinear and $OB' = 2 \cdot OB$. It can be shown that each side of $\triangle A'B'C'$ is twice as long as the corresponding side of $\triangle ABC$.

Figure 12-31

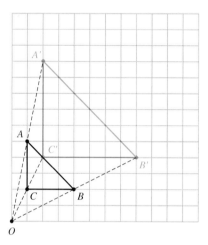

In general, we have the following definition.

Definition of Size Transformation

A size transformation from the plane to the plane with center O and scale factor $r (r > 0)$ is a transformation that assigns to each point A in the plane a point A', such that O, A, and A' are collinear and $OA' = r \cdot OA$ and so that O is not between A and A'.

Example 12-8

a. In Figure 12-32(a), find the image of point P under a size transformation with center O and scale factor $\frac{2}{3}$.

Figure 12-32

(a) (b)

b. Find the image of the quadrilateral $ABCD$ in Figure 12-32(b) under the size transformation with center O and scale factor $\frac{2}{3}$.

Solution **a.** In Figure 12-33(a), we connect O with P and divide $\overline{OP}$ into three congruent parts. The point P' is the image of P because $OP' = \frac{2}{3}OP$.

b. We find the image of each of the vertices and connect the images to obtain the quadrilateral $A'B'C'D'$ shown in Figure 12-33(b).

Figure 12-33

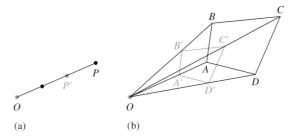

In Figure 12-33(b), the sides of the quadrilateral $A'B'C'D'$ are all parallel to the corresponding sides of the original quadrilateral and the angles of the quadrilateral $A'B'C'D'$ are congruent to the corresponding angles of quadrilateral $ABCD$. Also, each side in the quadrilateral $A'B'C'D'$ is $\frac{2}{3}$ as long as the corresponding side of quadrilateral $ABCD$. These properties are true for any size transformation and are summarized in the following theorem.

Theorem 12-1

A size transformation with center O and scale factor $r (r > 0)$ has the following properties:

1. The image of a line segment is a line segment parallel to the original segment and r times as long.
2. The image of an angle is an angle congruent to the original angle.

From Theorem 12-1, it follows that the image of a polygon under a size transformation is a similar polygon (why?). However, for any two similar polygons it is not always possible to find a size transformation so that the image of one polygon under the transformation is the other polygon. But, given two similar polygons, we can "move" one polygon to a place so that it will be the image of the other under a size transformation. The following examples show such instances.

Example 12-9 Show that △ABC in Figure 12-34 is the image of △ADE under a size transformation. Identify the center of the size transformation and the scale factor.

Figure 12-34

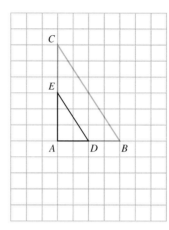

Solution Because $\frac{AB}{AD} = \frac{AC}{AE} = 2$, we choose A as the center of the size transformation and 2 as the scale factor. Notice that under this transformation, the image of A is A itself. The image of D is B, and the image of E is C.

Example 12-10 Show that △ABC in Figure 12-35 is the image of △APQ under a succession of isometries with a size transformation.

Figure 12-35

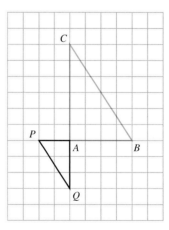

Solution We use the strategy of *looking at a related problem*. In Example 12-9, the common vertex served as the center of the size transformation. This was possible because the corresponding sides of the triangles were parallel. To achieve a similar situation, we first transform △APQ by a half-turn in A and obtain △AP'Q', as shown in Figure 12-36.

Figure 12-36

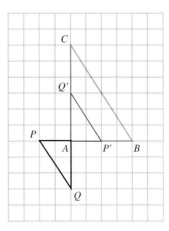

Now C is the image of Q' under a size transformation with center at A and scale factor 2. B is the image of P', and A is the image of itself under this transformation. Thus $\triangle ABC$ can be obtained from $\triangle APQ$ by first finding the image of $\triangle APQ$ under a half-turn in A and then applying a size transformation with center A and a scale factor 2 to that image.

Examples 12-9 and 12-10 are a basis for an alternative definition of similar figures.

Definition of Similar Figures

Two figures are similar if it is possible to transform one onto the other by a sequence of isometries followed by a size transformation.

Applications of Size Transformations

perspective drawing One way to make an object appear three-dimensional is to use a **perspective drawing.** For example, to make a letter appear three-dimensional we can use a size transformation with an appropriate center O and a scale factor, as shown in Figure 12-37, for the letter L.

Figure 12-37

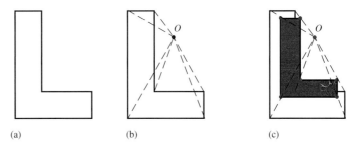

(a)   (b)   (c)

When a picture of an object is taken, the object appears upside down on the negative. The picture of the object on the negative can be interpreted as an image under composition of a half-turn and a size transformation. Figure 12-38(a) illustrates the image of an arrow from A to B under a composition of a half-turn followed by a size transformation with scale factor $\frac{1}{2}$.

Figure 12-38

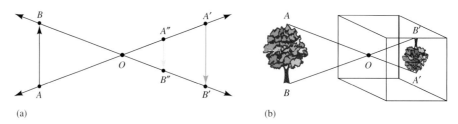

The image A' of A under the half-turn with center O is found on the ray opposite $\overrightarrow{OA}$ so that $OA' = OA$. The point B', the image of B under the half-turn, is found similarly on the ray opposite $\overrightarrow{OB}$. The images of A' and B' under the size transformation are A'' and B'', respectively. Consequently, the image of the arrow from A to B under the composition of the half-turn followed by the size transformation is the arrow from A'' to B''. Figure 12-38(b) illustrates another composition of a half-turn and a size transformation in a simple box camera.

ONGOING ASSESSMENT 12-3

1. In the following figures, describe a sequence of isometries followed by a size transformation so that the larger triangle is the final image of the smaller one.

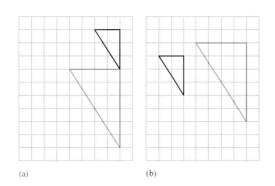

(a) (b)

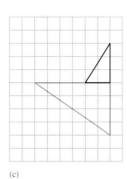

(c)

2. In the following drawing, find the image of $\triangle ABC$ under the size transformation with center O and scale factor $\frac{1}{2}$:

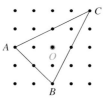

3. In each of the following drawings, find transformations that will take $\triangle ABC$ to its image, $\triangle A'B'C'$, which is similar:

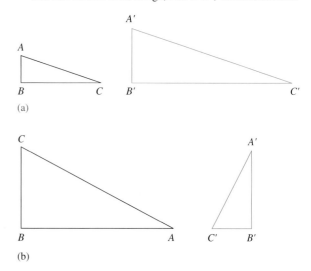

(a)

(b)

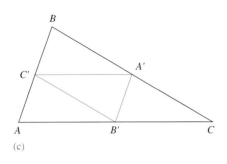
(c)

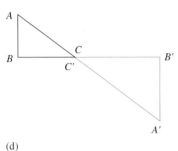
(d)

6. Each of the following figures describes a size transformation with center O and images in blue. Find the scale factor and the lengths designated by x and y.

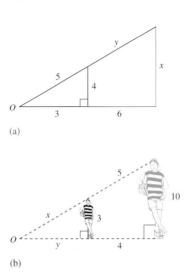
(a)

(b)

4. In each of the following figures, the smaller one is the image of the larger under a size transformation. In each case, find the scale factor and the length of x and y as pictured.

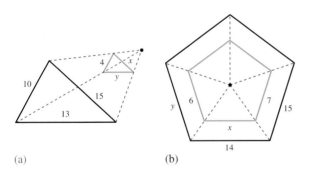
(a)   (b)

5. $\overline{A'B'}$ is the image of a candle $\overline{BA}$ produced by a box camera. Given the measurement of the figure, find the height of the candle.

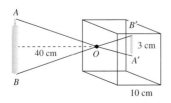

7. a. Find the coordinates of the images of $A\,(2, 3)$, $B\,(3, 4)$ and $C\,(-2, 3)$ under a size transformation with the center at the origin and a scale factor of 3.
   b. From your answer to (a), conjecture the coordinates of the image of the point (x, y) under a size transformation with the center at the origin and a scale factor r.

8. If a size transformation with center O and scale factor r takes a quadrilateral $ABCD$ to $A'B'C'D'$, what size transformation will take $A'B'C'D'$ back to $ABCD$?

Communication

9. Which of the following properties do not change under a size transformation? Explain how you can be sure of your answers.
   a. Distance between points
   b. Angle measure
   c. Parallelism; that is, if two lines are parallel to each other, then their images are parallel to each other.

10. Given two similar figures, explain how to tell if there is a size transformation that transforms one of the figures onto the other.

11. a. Consider two consecutive size transformations, each with center O and corresponding scale factors $\frac{1}{2}$ and $\frac{1}{3}$, respectively. Suppose the image of figure F under the first transformation is F' and the image of F' under the second transformation is F''. What single transformation will map F directly onto F''? Explain why.
    b. What would be the answer to (a) if the scale factors were r_1 and r_2?

12. Copy the following figure onto grid paper and determine the center and the scale factor of the size transformation. Explain why there is only one possibility for the center.

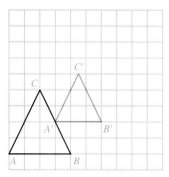

13. Is the image of a circle with center O under a size transformation with center O always a circle? Explain why or why not.

Open-Ended

14. Describe several real-life situations other than the ones discussed in this section in which size transformations occur.
15. Use a sheet of graph paper with a coordinate system. Locate the origin as the center of a transformation with a scale factor of $^{-}1$ and find the image of some triangle located in the first quadrant. What other transformation that you've studied has the same image as this transformation?

Cooperative Learning

16. A *frieze pattern* is a pattern that extends indefinitely in both directions and the image of the pattern under a translation is the original pattern. Pictured below are seven frieze patterns, each of which is the image of itself under a translation and also by the other isometries indicated. Every frieze pattern can be classified into one of these seven categories.

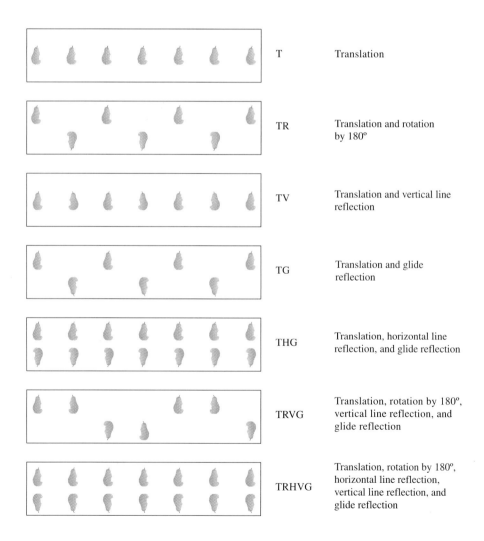

a. Have each member of your group draw one or two of the patterns (depending on the size of the group) on tracing paper or a transparency with a corresponding label using the letters T, R, H, V, or G, as in the figure. Exchange the drawings so that each member of your group checks how each of the patterns can be transformed onto itself by each of the isometries. Compare your answers with those of your group members.
b. Create other more elaborate patterns for each of the seven categories and present each pattern to a partner for identification. (Label the patterns with numbers 1 through 7, keeping to yourself the corresponding T, R, H, V, G labeling.) Compare your partner's answers with yours.

17. Have members of your group draw several figures and find their images under a size transformation with a scale factor of 3.
    a. How does the perimeter of each image compare to the perimeter of the original figure? Compare your answers.
    b. How does the area of each image compare to the area of the original figure? Compare your answers.
    c. Make a conjecture concerning the relationship between the perimeter of each image and the perimeter of the original figure under a size transformation with a scale factor r.
    d. Repeat (c) for the area of each figure.
    e. Discuss your findings and come up with a group conjecture.

Review Problems

18. Describe a transformation that would "undo" each of the following:
    a. A translation determined by slide arrow from M to N
    b. A rotation of 75° with center O in a clockwise direction
    c. A rotation of 45° with center A in a counterclockwise direction
    d. A glide reflection that is the composition of a reflection in line m and a translation that takes A to B
    e. A reflection in line n.

19. In the following coordinate plane, find the images of each of the given points in the transformation that is the composition of a reflection in line m followed by a reflection in line n.
    a. (4, 3)
    b. (0, 1)
    c. ($^-$1, 0)
    d. (0, 0)

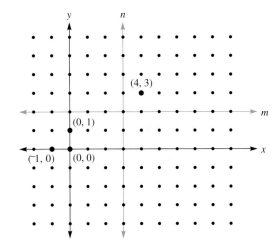

LABORATORY ACTIVITY

Consider △ABC and its image after it is reflected in lines m, n, and p in order in Figure 12-39.

Figure 12-39

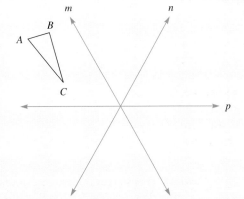

Find a single line q that could be used to reflect the original △ABC onto the final image. Explain whether you think it is always possible to find such a line.

Section 12-4 — Symmetries

Line Symmetries

The concept of a reflection can be used to identify line symmetries of a figure. All the drawings in Figure 12-40 have symmetries about the dashed lines.

Figure 12-40

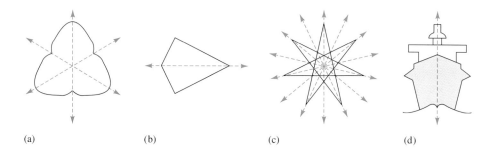

(a) (b) (c) (d)

line of symmetry Mathematically, a geometric figure has a **line of symmetry** ℓ if it is its own image under a reflection in ℓ. A method of creating a symmetrical figure is seen in Example 12-11.

Example 12-11 In Figure 12-41, we are given a figure and a line m. Do the minimum amount of drawing to create a figure from the given figure so that the result has line m as its line of symmetry.

Figure 12-41

Solution For the resulting figure both to be symmetric about line m and to incorporate the existing figure, we need to reflect the existing figure about line m. The desired result of doing that is the combination of the original figure and the image. The resulting figure is shown in Figure 12-42.

Figure 12-42

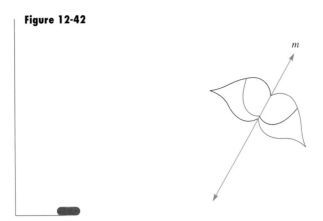

Example 12-12 How many lines of symmetry does each drawing in Figure 12-43 have?

Figure 12-43

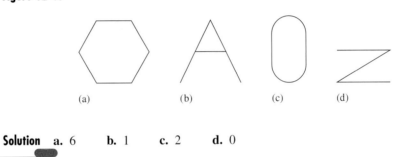

(a)　　(b)　　(c)　　(d)

Solution　a. 6　　b. 1　　c. 2　　d. 0

Problem Solving　**Archaeological Find**

At the site of an ancient settlement, archaeologists found a fragment of a saucer as shown in Figure 12-44. To restore the saucer, the archaeologists need to determine the radius of the original saucer. How can they do this?

Figure 12-44

- **Understanding the Problem**　The border of the shard shown in Figure 12-44 was part of a circle. To reconstruct the saucer, we are to determine the radius of the circle of which the shard is a part.

- **Devising a Plan** We can use a *model* to determine the radius. We trace an outline of the circular edge of the three-dimensional shard on a piece of paper. The result is an arc of a circle, as shown in Figure 12-45. To determine the radius, we find the center O. A circle has infinitely many lines of symmetry and each line passes through the center of the circle, where all the lines of symmetry intersect.

Figure 12-45

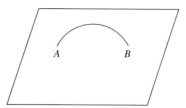

- **Carrying Out the Plan** To find a line of symmetry, fold the paper containing $\widehat{AB}$ so that a portion of the arc is folded onto itself. Then unfold the paper and draw the line of symmetry on the fold mark, as shown in Figure 12-46(a). By refolding the paper in Figure 12-46(a) so that a different portion of the arc $\widehat{AB}$ is folded onto itself, we can determine a second line of symmetry, as shown in Figure 12-46(b). The two dotted lines of symmetry intersect at O, the center of the circle of which $\widehat{AB}$ is an arc. To complete the problem, measure the length of either $\overline{OB}$ or $\overline{OA}$. (They should be the same.)

Figure 12-46

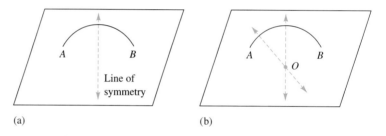

(a)   (b)

- **Looking Back** In the first fold, endpoint B of the arc was folded onto another point of the arc. Label this other point X. The result is shown in Figure 12-47. Because the center of the circle lies on the perpendicular bisector of a chord (why?) and the fold line ℓ is a line of symmetry of the circle containing $\widehat{AB}$, the fold line must be the perpendicular bisector of $\overline{XB}$ and it must contain the center of the circle. We could have used this property to determine the center of the circle by choosing two chords on the arc and finding the point on which the perpendicular bisectors of the chords intersect. Alternatively, we could have used a compass and a straightedge.

Figure 12-47

A related problem is, What would happen if the piece of pottery had been part of a sphere? Would the same ideas still work?

Rotational (Turn) Symmetries

rotational symmetry · turn symmetry

A figure has **rotational symmetry,** or **turn symmetry,** when the traced figure can be rotated less than 360° about some point so that it matches the original figure. Note that the condition "less than 360°" is necessary because any figure will coincide with itself if it is rotated 360° about any point. In Figure 12-48, the equilateral triangle coincides with itself after a rotation of 120° about point O. Hence, we say that the triangle has 120° rotational symmetry. Also in Figure 12-48, if we were to rotate the triangle another 120°, we would find again that it matches the original. So we can say that the triangle also has 240° rotational symmetry.

Figure 12-48

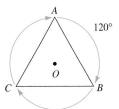

In general, if a figure has $\alpha°$ rotational symmetry, it also will coincide with itself when rotated by $n\alpha°$ for any non-zero integer n. For this reason, in rotational symmetry the smallest possible positive angle measure that turns the figure onto itself is reported. Notice that a circle has a rotational symmetry by any turn around its center.

Other examples of figures that have rotational symmetry are shown in Figure 12-49. Figures 12-49(a), (b), (c), and (d) have 72°, 90°, 180°, and 180° rotational symmetries, respectively [(a) and (b) also have other rotational symmetries].

Figure 12-49

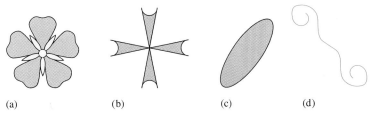

In general, we can determine whether a figure has rotational symmetry by tracing it and turning the tracing about a point (the center of the figure) to see if it aligns on the figure before the tracing has turned in a complete circle, or 360°. The amount of the rotation can be determined by measuring the angle $\angle POP'$ through which a point P is rotated around a point O to match another point P' when the figures align. Such an angle, $\angle POP'$, is labeled with points P, O, and P' in Figure 12-50 and has measure 120°. Point O, the point held fixed when the tracing is turned, is the *turn center.*

Figure 12-50

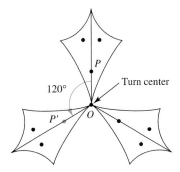

Example 12-13 Determine the amount of the turn for the rotational symmetries of each part of Figure 12-51.

Figure 12-51

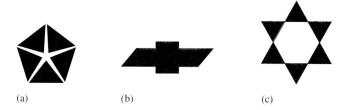

(a) (b) (c)

Solution
a. The amounts of the turns are $\frac{360°}{5}$ or 72°, 144°, 216°, and 288°.
b. The amount of the turn is 180°.
c. The amounts of the turns are 60°, 120°, 180°, 240°, and 300°.

The rotations in Figure 12-51(b) and (c) exemplify yet another type of symmetry, namely, point symmetry.

Point Symmetry

point symmetry Any figure that has 180° rotational symmetry is said to have **point symmetry** about the turn center. Some figures with point symmetry are shown in Figure 12-52.

Figure 12-52

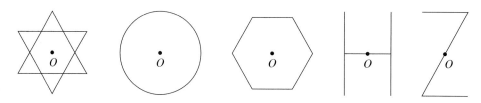

As illustrated in Figure 12-52, any figure with point symmetry is its own image under a half-turn. This makes the center of the half-turn the midpoint of a segment connecting a point and its image.

Other Notions of Symmetry

plane of symmetry A three-dimensional figure has a **plane of symmetry** when every point of the figure on one side of the plane has a mirror image on the other side of the plane. Examples of figures with plane symmetry are shown in Figure 12-53. Solids can also have point symmetry, line symmetry, and turn symmetry. These symmetries are analogous to the two-dimensional symmetries and are investigated in Ongoing Assessment 12-4.

Figure 12-53

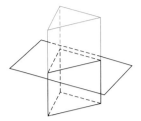

Geometric figures in a plane can be classified according to the number of symmetries they have. Consider a triangle described as having exactly one line of symmetry and no turn symmetries. What could the triangle look like? The only possibility is a triangle in which two sides are congruent, that is, an isosceles triangle. The line of symmetry passes through a vertex, as shown in Figure 12-54.

Figure 12-54

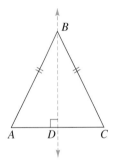

Just as we used the number of lines of symmetry to describe an isosceles triangle, we can describe equilateral and scalene triangles in terms of the number of lines of symmetry they have. This is left as an exercise.

A square, as in Figure 12-55, can be described as a four-sided polygon with four lines of symmetry—d_1, d_2, h, and v—and three turn symmetries about point O. In fact, we can use lines of symmetry and turn symmetries to define various types of quadrilaterals normally used in geometry. It is left as an exercise to see how these definitions differ from those in Chapter 9.

Figure 12-55

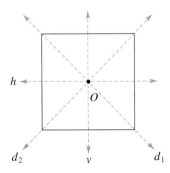

ONGOING ASSESSMENT 12-4

1. Why is "SOS" the international distress symbol?
2. Various international signs have symmetries. Determine which of the following have (i) line symmetry, (ii) rotational symmetry, and/or (iii) point symmetry.

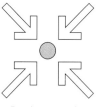

Rendezvous point
(a)

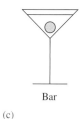

Light switch
(b)

Switzerland
(i)

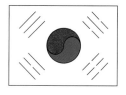

South Korea
(ii)

Bar
(c)

Observation deck
(d)

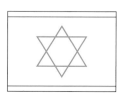

Israel
(iii)

Barbados
(iv)

3. Design symbols that have each of the following symmetries, if possible:
   a. Line symmetry but not rotational symmetry
   b. Rotational symmetry but not point symmetry
   c. Rotational symmetry but not line symmetry
4. In each of the following figures, complete the sketches so that they have line symmetry about ℓ.

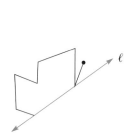

(a) Line symmetry about ℓ

(b) Line symmetry about ℓ

5. a. Determine the number of lines of symmetry in each of the following flags.
   b. Sketch the lines of symmetry for each.

6. a. Determine how many lines of symmetry the following figure has.

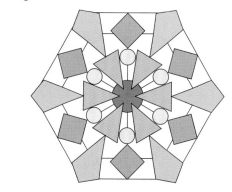

   b. Does the figure have rotational symmetry?
7. Find the lines of symmetry, if any, for each of the following trademarks.

Volkswagen of America
(a)

The Yellow Pages
(b)

Chevrolet
(c)

Chrysler Corporation
(d)

8. In each of the following figures, complete the sketches so that they have the indicated symmetry.

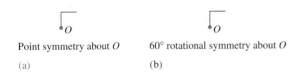

Point symmetry about O
(a)

60° rotational symmetry about O
(b)

9. Determine how many planes of symmetry, if any, each of the following three-dimensional vehicle controls has.

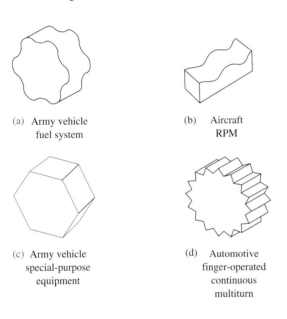

(a) Army vehicle fuel system

(b) Aircraft RPM

(c) Army vehicle special-purpose equipment

(d) Automotive finger-operated continuous multiturn

10. Write a Logo procedure that draws a square and produces a figure with rotational symmetry of (a) 60°; (b) 120°; (c) 180°; (d) 240°; (e) 300°.
11. Write a Logo procedure that draws an equilateral triangle and produces a figure with rotational symmetry of (a) 60°; (b) 120°; (c) 240°; (d) 300°.

Communication

12. Answer each of the following. If your answer is no, provide a counterexample.
    a. If a figure has point symmetry, must it have rotational symmetry? Why?
    b. If a figure has rotational symmetry, must it have point symmetry? Why?
    c. Can a figure have point, line, and rotational symmetry? If so, sketch a figure that has these properties.
    d. If a figure has point symmetry, must it have line symmetry? Is the converse true? Why?
    e. If a figure has both point and line symmetry, must it have rotational symmetry? Why?

Open-Ended

13. If possible, sketch a triangle that satisfies each of the following:
    a. It has no lines of symmetry.
    b. It has exactly one line of symmetry.
    c. It has exactly two lines of symmetry.
    d. It has exactly three lines of symmetry.
14. Sketch a figure that has point symmetry but no line symmetry.
15. a. In the following figure, ABCD is a rectangular sheet of paper. Fold the paper so that the opposite edges $\overline{AB}$ and $\overline{DC}$ coincide (the crease $\overline{EF}$ is created). Then fold the resulting rectangle ABEF so that $\overline{BE}$ and $\overline{AF}$ coincide (the crease $\overline{HG}$ is created). The rectangle AFGH is obtained. Now cut a curved piece of paper out of the corner G as shown and unfold the paper. Describe all the symmetries that the unfolded figure has.

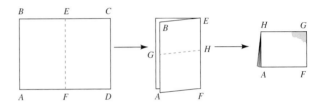

   b. Repeat the experiment in (a) by successively folding a new sheet of paper three times and cutting out a curved piece containing a corner that resulted from the three folds. Predict all the symmetries of the unfolded figure. Check your answer by unfolding the paper.
16. For each of the following, use paper folding or any other method to create, if possible, two nonsimilar figures that have the specified symmetries:
    a. Point symmetry but no other symmetries
    b. Exactly one line of symmetry
    c. Exactly two lines of symmetry but no rotational symmetry
    d. 45° rotational symmetry

Cooperative Learning

17. With a partner, design a three-mirror kaleidoscope by fastening three mirrors together, each perpendicular to a flat surface so that they form an equilateral triangular prism. Place colored paper with a pattern (or design your own pattern) in the base of the kaleidoscope. Peer over the edge of the kaleidoscope to view the generated figure. Repeat the experiment for different patterns.
18. With another person or group, play the following game several times. First, draw a polygon that has different kinds of symmetries. Without revealing your polygon to your partner, tell your partner all you know about the symmetries of the figure. Next, from this information, your opponent

attempts to draw the type of figure you drew. If your opponent produces a figure of the type you drew, he or she earns two points. If your opponent produces a different figure having the symmetries that you reported or if you failed to reveal some of the symmetries of your polygon, your opponent gets three points. Alternate roles several times.

Review Problems

19. For each of the following cases, find the image of the given figure using paper folding:

(a) Reflection about ℓ

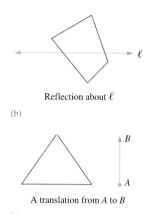

(b) Reflection about ℓ

(c) A translation from A to B

20. Construct each image in problem 19 using a compass and a straightedge.

TECHNOLOGY CORNER

Edit the following INSPI procedure for your Logo software. The procedure produces drawings that have various symmetries, depending on the inputs for :ANGLE:

```
TO INSPI :SIDE :ANGLE
  FD :SIDE RT :ANGLE
  INSPI :SIDE :ANGLE + 5
END
```

Some of these drawings are shown in Figure 12-56.

Figure 12-56

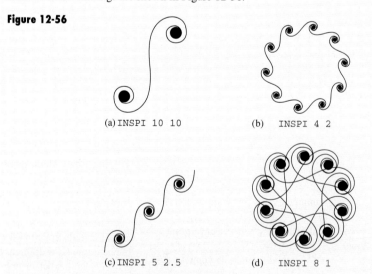

(a) INSPI 10 10
(b) INSPI 4 2
(c) INSPI 5 2.5
(d) INSPI 8 1

Determine what kinds of symmetries each figure has. Try some of your own inputs and see what drawings they produce and what kinds of symmetries the resulting figures have.

*Section 12-5 — Tessellations of the Plane

tessellation

In this section we use concepts from motion geometry to study *tessellations* of the plane. A **tessellation** of a plane is the filling of the plane with repetitions of figures in such a way that no figures overlap and there are no gaps. (Similarly, one can tessellate space.) The tiling of a floor and various mosaics are examples of tessellations. Maurits C. Escher, born in the Netherlands in 1902, was a master of tessellations. Many of his drawings have fascinated mathematicians for decades. An example of his work, *Study of Regular Division of the Plane with Reptiles* (pen, ink, and watercolor), 1939, contains an exhibit of a tessellation of the plane by a lizardlike shape, as shown in Figure 12-57.

Figure 12-57

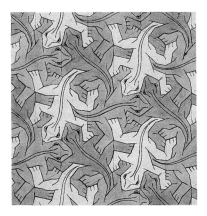

At the heart of the tessellation in the figure, we see a regular hexagon. But perhaps the simplest tessellation of the plane can be achieved with squares. Figure 12-58 shows two different tessellations of the plane with squares.

Figure 12-58

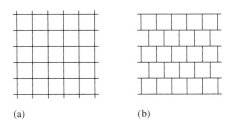

(a)   (b)

Regular Tessellations

Tessellations with regular polygons are appealing and interesting because of their simplicity. Figure 12-59 shows portions of tessellations with equilateral triangles (a) and with regular hexagons (b).

Figure 12-59

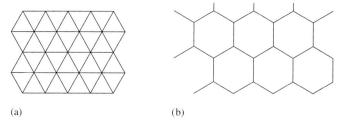

(a)   (b)

To determine other regular polygons that tessellate the plane, we investigate the possible size of the interior angle of a tessellating polygon. If n is the number of sides of a regular polygon, then because the sum of the measures of the exterior angles of the regular polygon is $360°$, the measure of a single exterior angle of the polygon is $360°/n$. Hence, the measure of an interior angle is $180° - 360°/n$. Table 12-2 gives some values of n, the type of regular polygon related to each, and the angle measure of an interior angle found by using the expression $180° - 360°/n$.

Table 12-2

Number of Sides (n)	Regular Polygon	Measure of Interior Angle
3	Triangle	60°
4	Square	90°
5	Pentagon	108°
6	Hexagon	120°
7	Heptagon	900/7°
8	Octagon	135°
9	Nonagon	140°
10	Decagon	144°

If a regular polygon tessellates the plane, the sum of the congruent angles of the polygons around every vertex must be $360°$. Thus 360 divided by the angle measure gives the number of angles around a vertex and hence must be an integer. If we divide $360°$ by each of the angle measures in the table, we find that of these measures only $60°$, $90°$, and $120°$ divide $360°$; hence of the listed polygons, only an equilateral triangle, a square, and a regular hexagon can tessellate the plane.

Can other regular polygons tessellate the plane? Notice that $\frac{360}{120} = 3$. Hence, 360 divided by a number greater than 120 also is smaller than 3. However, the number of sides of a polygon cannot be less than 3. Because a polygon with more than six sides has an interior angle greater than $120°$, it actually is not necessary to consider polygons with more than six sides.

Tessellating with Other Shapes

Next, we consider tessellating the plane with arbitrary convex quadrilaterals. Before reading on, you may wish to investigate the problem yourself, with the help of cardboard quadrilaterals. Figure 12-60 shows an arbitrary convex quadrilateral and a way to tessellate the plane with the quadrilateral. Successive $180°$ turns of the quadrilateral about the midpoints *P*, *Q*, *R*, and *S* of its sides will produce four congruent quadrilaterals around a common vertex. Notice that the sum of the measures of the angles around vertex *A* is $a + b + c + d$. This is the sum of the measures of the interior angles of the quadrilateral, or 360°.

Figure 12-60

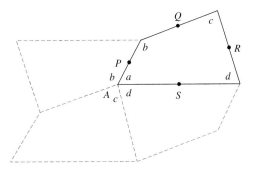

As we saw earlier in this section, a regular pentagon does not tessellate the plane. However, some nonregular pentagons do. One is shown in Figure 12-61, along with a tessellation of the plane by the pentagon.

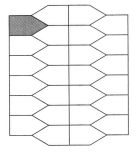

Figure 12-61

HISTORICAL NOTE

Determining which irregular pentagons tessellate is a surprisingly rich problem. Mathematicians thought they had solved it when they had classified eight types of pentagons that would tessellate. They believed they had all of them. But then in 1975, Marjorie Rice, a woman with no formal training in mathematics, discovered a ninth type of tessellating pentagon. She went on to discover four more by 1977. Her interest was piqued by reading an article in *Scientific American* by Martin Gardner. Two of the pentagons she found are shown in Figure 12-62. The problem of how many types of pentagons tessellate remains unsolved.

Figure 12-62

Type 9 discovered in February 1976

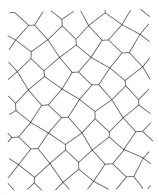

Type 13 discovered in December 1977

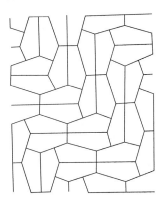

What other types of designs can be made that tessellate a plane? The plane geometry and motions studied earlier give some clues on how to design shapes that work. Consider one of the methods used in Chapter 10 to determine the area of a parallelogram. In Figure 12-63(a), triangle *ABE* was removed from the left of parallelogram *ABCD* and slid to the right forming the rectangle $BB'E'E$ of Figure 12-63(b). This same notion can be used to create a tessellating shape.

Figure 12-63

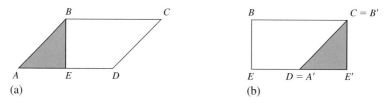

Consider any polygon known to tessellate a plane, such as rectangle *ABCD* in Figure 12-64(a). On the left side of the figure draw any shape in the interior of the rectangle as in Figure 12-64(b). Cut this shape from the rectangle and slide it to the right by the slide that takes *A* to *B* as shown in Figure 12-64(c). The resulting shape will tessellate the plane. (Why?)

Figure 12-64

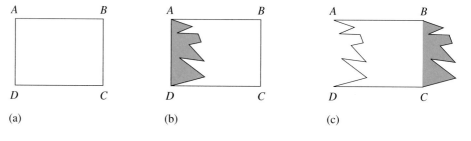

A second method of forming a tessellation involves a series of rotations of parts of a figure. In Figure 12-65(a), we start with an equilateral triangle *ABC*, choose the midpoint *O* of one side of the triangle and cut out a shape as in Figure 12-65(b) being careful not to cut away more than half of angle *B* and then rotate the shape 180° clockwise around point *O*.

Figure 12-65

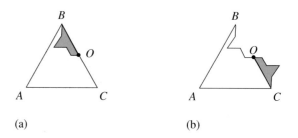

If we continue this process on the other two sides, then we obtain a shape that can be rotated around point *A* to tessellate the plane. Complete the tessellating shape and tessellate the plane with it in the following "Now Try This" activity.

> **NOW TRY THIS 12-5**
>
> • Continue the drawing of the tessellating shape in Figure 12-65. Cut out the shape and use it to draw a tessellation of the plane. •

ONGOING ASSESSMENT 12-5

1. On dot paper, draw a tessellation of the plane using the following figures.

   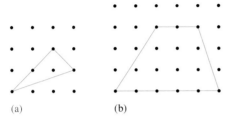

   (a) (b)

2. **a.** Tessellate the plane with the following quadrilateral.

   b. Is it possible to tessellate the plane with any quadrilateral? Why or why not?

3. On square-dot paper, use each of the following four pentominoes, one at a time, to make a tessellation of the plane, if possible. (A pentomino is a polygon composed of five congruent, nonoverlapping squares.) Which of the pentominoes tessellate the plane?

   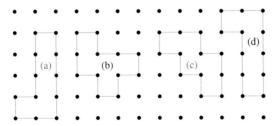

4. We have seen that equilateral triangles, squares, and regular hexagons are the only regular polygons that will tessellate the plane by themselves. However, there are many ways to tessellate the plane by using combinations of these and other regular polygons, as shown in the following figure.

   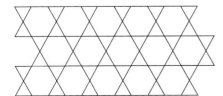

   Try to produce other such tessellations by using the following.
   a. Only equilateral triangles, squares, and regular hexagons
   b. Regular octagons (8-gons) and squares

5. To determine if a shape created using a glide reflection will tessellate the plane, complete the following:
   a. Start with a rectangle. Determine some shape that you might use with a slide to form a tessellating shape. Slide it as shown below. Determine the horizontal line of symmetry of the rectangle, and reflect as shown below.

   b. Explain why the described series of motions is a glide reflection.
   c. Determine whether the final shape will tessellate the plane.

6. The **dual of a tessellation** is the tessellation obtained by connecting the centers of the polygons in the original tessellation that share a common side. The dual of the tessellation of equilateral triangles is the tessellation of regular hexagons, shown in color in the following figure.

   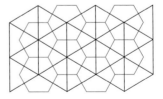

   Describe and show the dual of each of the following:
   a. The regular tessellation of squares shown in Figure 12-58(a)
   b. The tessellation of squares in Figure 12-58(b)
   c. A tessellation of regular hexagons

7. Write Logo procedures to draw tessellations with the following figures. Have each tessellation appear on the screen in the form of two vertical strips.
   a. Squares
   b. Equilateral triangles
   c. Regular hexagons

8. A sidewalk is made of tiles of the type shown in the following figure:

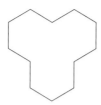

Each tile is made of three regular hexagons from which three sides have been removed. Write a Logo procedure to draw a tessellation composed of four such figures.

Communication

9. The following figure is a partial tessellation of the plane with the trapezoid *ABCD*:

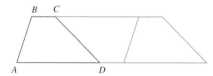

   a. Explain how the tessellation can be used to find a formula for the area of the trapezoid.
   b. Tessellate the plane with a triangle and show how the tessellation can be used to find the relationship between the length of the segment connecting the midpoints of the two sides of a triangle and the length of the third side.

10. Explain in your own words why only three types of regular polygons tessellate the plane.

Open-Ended

11. There are endless numbers of figures that tessellate a plane. In the following drawing, the shaded figure is shown to tessellate the plane.

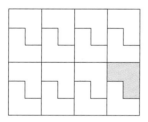

   Design several different polygons and show how each can tessellate the plane. What transformations are used in each of your designs? Explain how they are used to tessellate the plane.

12. Examine different quilt patterns or floor coverings and make a sketch of those you found that tessellate a plane.

13. A cube will tessellate space but a sphere will not. List several other solids that will tessellate space and several that will not.

Cooperative Learning

14. Each member of a small group is to find a drawing by M. C. Escher that does not appear in this text and in which the concept of tessellation is used. Each then shows the other members of the group, in detail, how he or she thinks Escher created the tessellation in his or her drawing.

15. a. Convince the members of your group that the following figure containing six equilateral triangles tessellates the plane:

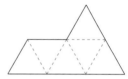

   b. As a group, find different figures that contain six equilateral triangles. How many such figures can you find? Discuss the meaning of "different."
   c. Find some of the figures in (b) that are *rep-tiles*. (A rep-tile is a figure whose copies can be used to form a larger figure similar to itself.) Convince other members of your group that your figures are rep-tiles and that they tessellate the plane.

LABORATORY ACTIVITY

Use pattern blocks to construct tessellations using each of the following types of pieces:

1. Squares
2. Equilateral triangles
3. Octagons and squares
4. Rhombuses

HINT FOR SOLVING THE PRELIMINARY PROBLEM

Draw the segments determined as follows in Figure 12-66: Connect your eye (E) to the point marking the highest part of your body seen in the mirror (A); connect point (A) to the corresponding part of your body (H). Next, connect point E to the point marking the lowest part of your body seen in the mirror (D), and then connect point D to the corresponding point of your body (B). These segments, along with the vertical line of your body $\overleftrightarrow{HB}$, determine several triangles. Use these triangles and their altitudes from the points marked on the mirror to determine the relationship.

Figure 12-66

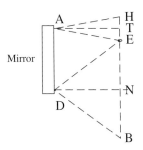

QUESTIONS FROM THE CLASSROOM

1. A student asks, "If I have a point and its image, is that enough to determine whether the image was found using a translation, reflection, rotation, or glide reflection?" How do you respond?
2. Another student asks a question similar to question 1 but is concerned about a segment and its image. How do you respond to this student?
3. A student claims that a kite has no lines of symmetry. How do you respond?
4. A student says that every three-dimensional figure that has plane symmetry automatically has line symmetry. Do you agree? Why or why not?
5. A student says that in a size transformation with the scale factor 0, we do not have a transformation. Is that true?
6. A student asks if every translation on a grid can be accomplished by a translation along a vertical direction followed by a translation along a horizontal direction. How do you respond?
7. A student asks why the images of two perpendicular lines will also be perpendicular under a size transformation. How do you respond?
8. When asked if symmetries are used in occupations other than mathematics, how do you respond?
9. A student asks why some wheel covers have symmetry. How do you respond?

CHAPTER 12 — Motion Geometry and Tessellations

CHAPTER OUTLINE

I. Motions of the plane
   A. **Isometries** are transformations that preserve distance.
      1. A **translation** is a motion of the plane that moves every point a specified distance in a specified direction along a straight line.
      2. A **rotation** is a motion of the plane determined by holding one point (the center) fixed and rotating the plane about this point by a certain amount in a certain direction.
      3. A **half-turn** is a rotation of 180°.
      4. Properties of a rotation can be used to show that two lines, neither of which is vertical, are perpendicular if, and only if, their slopes m_1 and m_2 satisfy the condition $m_1 m_2 = {}^-1$.
      5. A **reflection** in a line m is a motion among points of the plane that pairs each point P of the plane with a point P' in such a way that m is the perpendicular bisector of $\overline{PP'}$, as long as P is not on m. If P is on m, then $P = P'$.
      6. A **glide reflection** is the composition of a translation and a reflection in a line parallel to the slide arrow of the translation.
   B. A **size transformation** S from the plane to the plane is defined as follows: Some point O, the center of the size transformation, is its own image. For any other point Q of the plane, its image Q' is such that $OQ'/OQ = r$, where r is a positive real number and O, Q, and Q' are collinear.
   C. Two figures are **similar** if it is possible to transform one onto the other by a sequence of isometries followed by a size transformation.

II. Symmetries
   A. A figure has a **line symmetry** if it is its own image under a reflection.
   B. A figure has **rotational symmetry** if it is its own image under a rotation of less than 360° about a turn center.
   C. A figure has **point symmetry** if it has 180° rotational symmetry.
   D. A three-dimensional figure has a **plane of symmetry** when every point of the figure on one side of the plane has a mirror image on the other side of the plane.

*III. Tessellations
   A **tessellation** of a plane is the filling of the plane with repetitions of figures in such a way that no figures overlap and there are no gaps.

CHAPTER REVIEW

1. Complete each of the following motions.

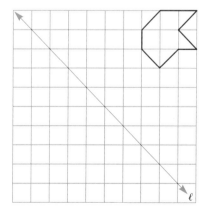

A reflection in ℓ

(a)

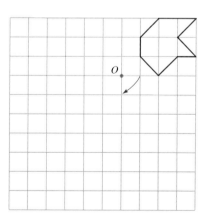

A rotation in O through the given arc

(b)

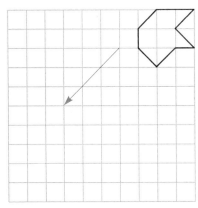

(c) A translation, as pictured

2. For each of the following figures, construct the image of △ABC.

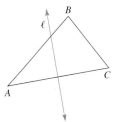

(a) Through a reflection in ℓ

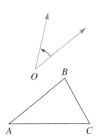

(b) Through the given rotation in O

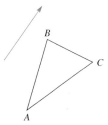

(c) Through the translation arrow pictured

3. Determine how many lines of symmetry, if any, each of the following figures has.

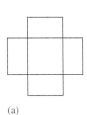   

(a)   (b)

   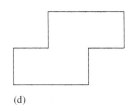

(c)   (d)

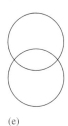   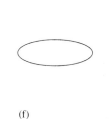

(e)   (f)

4. For each of the following figures, identify the types of symmetry (line, rotational, or point) it possesses.

(a)   (b)

(c)

5. Determine how many planes of symmetry each of the following has.
   a. A ball
   b. A right cylindrical water pipe
   c. A box that is a right rectangular prism but not a cube
   d. A cube

6. What type of symmetry (line, rotational, or point) does each of the lowercase letters of the printed English alphabet have?
7. In the following figure, △A'B'C' is the image of △ABC under a size transformation.

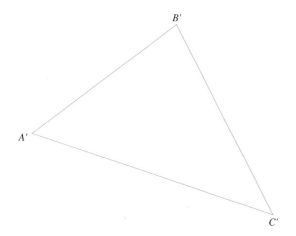

Locate points A, B, and C such that A' is the center of the size transformation and $BC = \frac{1}{2}B'C'$.

8. Given that STAR in the following figure is a parallelogram, describe a sequence of isometries to show the following:
   a. △STA ≅ △ARS
   b. △TSR ≅ △RAT

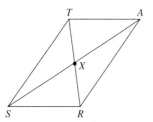

9. Given that BEAUTY in the following figure is a regular hexagon, describe a sequence of isometries that will transform the following:
   a. BEAU into AUTY
   b. BEAU into YTUA

10. Given that △SNO ≅ △SWO in the following figure, describe one or more isometries that will transform △SNO into △SWO.

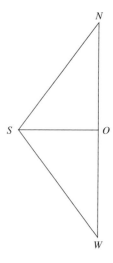

11. Show that △SER in the following figure is the image of △HOR under a succession of isometries with a size transformation.

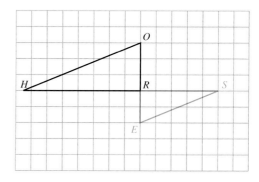

12. The triangle A'B'C' with A'(0, 7.91), B'(−5, −4.93), C'(4.83, 0) is the image of triangle ABC under the translation $(x, y) \rightarrow (x + 3, y - 5)$. Find the coordinates of A, B, and C.
13. Show that △TAB in the following figure is the image of △PIG under a succession of isometries with a size transformation:

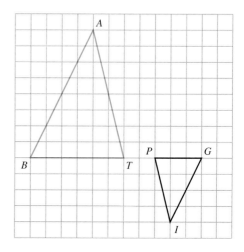

14. If a translation is determined by $(x, y) \to (x + 3, y - 2)$ is followed by another translation determined by $(x, y) \to (x - 3, y + 2)$, describe in detail a motion that would achieve the same thing.
*15. Explain why a regular octagon cannot tessellate the plane.
*16. Write a Logo procedure called RHOMSTRIP that will draw a vertical strip of rhombuses.
*17. A square will tessellate a plane; can we always find a square that will tessellate a rectangle?
*18. Can any tessellating shape be used as a unit for measuring area? If so, explain why.

SELECTED BIBLIOGRAPHY

Bannon, T. "Fractals and Transformations." *Mathematics Teacher* 84 (February 1991): 178–185.

Bidwell, J. "Using Reflections to Find Symmetric and Asymmetric Patterns." *Arithmetic Teacher* 34 (March 1987): 10–15.

Billstein, R., S. Libeskind, and J. Lott. *Logo: MIT Logo for the Apple.* Menlo Park, Calif.: Benjamin/Cummings, 1985.

Clason, R., D. Ericksen, and C. Ericksen. "Cross-and-Turn Tile Patterns." *Mathematics Teaching in the Middle Schools* 2 (May 1997): 430–437.

Dayoub, I., and J. Lott. *Geometry: Constructions and Transformations.* White Plains, NY: Dale Seymour Publishing Company, 1977.

DeTemple, D. "Reflection Borders for Patchwork Quilts." *Mathematics Teacher* 80 (February 1986): 138–143.

Eddins, S., et al. "Geometric Transformations: Part 1." *Mathematics Teacher* 87 (March 1994): 177–180.

Eddins, S., et al. "Geometric Transformations: Part 2." *Mathematics Teacher* 87 (April 1994): 258–260.

Fossnaugh, L., and M. Harrell. "Covering the Plane with Rep-Tiles." *Mathematics Teaching in the Middle Schools* 1 (January-February 1996): 666–670.

Gardner, M. "On Tessellating the Plane with Convex Polygonal Tiles." *Scientific American* (July 1975): 112–117.

Harrell, M., and L. Fossnaugh. "Allium to Zircon: Mathematics & Nature." *Mathematics Teaching in the Middle Schools* 2 (May 1997): 380–389.

Harris, J. "Using Literature to Investigate Transformations." *Teaching Children Mathematics* 4 (May 1998): 510–513.

Lappan, G., and R. Even. "Research into Practice: Similarity in the Middle Grades." *Arithmetic Teacher* 35 (May 1988): 32–35.

May, B. "Reflections on Miniature Golf." *Mathematics Teacher* 78 (May 1985): 351–353.

Phillips, J. M. and R. E. Zwoyer. *Motion Geometry Book 2: Congruence.* New York: Harper & Row, Publishers, 1969.

——— *Motion Geometry Book 1: Slides, Flips, and Turns.* New York: Harper & Row, Publishers, 1969.

——— *Motion Geometry Book 3: Symmetry.* New York: Harper & Row, Publishers, 1969.

——— *Motion Geometry Book 4: Area, Similarity, and Constructions.* New York: Harper & Row, Publishers, 1969.

Ranucci, E., and J. Teeters. *Creating Escher-type Drawings.* Palo Alto, Calif.: Creative Publications (1977).

Reesink, C. "Crystals: Through the Looking Glass with Planes, Points, and Rotational Symmetry." *Mathematics Teacher* 80 (May 1987): 377–388.

Seidel, J. "Symmetry in Season." *Teaching Children Mathematics* 4 (January 1998): 244–249.

Sellke, D. "Geometric Flips via the Arts." *Teaching Children Mathematics* 4 (February 1999): 379–383.

Sicklick, F., B. Turkel, and F. R. Curcio. "The Transformation Game." *Arithmetic Teacher* 36 (October 1988): 37–41.

Speer, W., and J. Dixon. "Reflections of Mathematics." *Teaching Children Mathematics* 2 (May 1996): 537–543.

Walter, M. *The Mirror Puzzle Book.* New York: Parkwest Publications (1985).

Willcutt, B. "Triangular Tiles for Your Patio." *Arithmetic Teacher* 34 (May 1987): 43–45.

Woods, J. "Let the Computer Draw the Tessellations That You Design." *Mathematics Teacher* 81 (February 1988): 138–141.

Zaslavsky, C. "Symmetry in American Folk Art." *Arithmetic Teacher* 37 (January 1990): 6–12.

Appendix I: Logo Turtle Graphics

Introducing the Turtle

turtle Turtle graphics, implemented using the computer language Logo, are especially suited for studying geometry. Students can draw geometric figures by giving instructions to a **turtle,** a triangular figure on the display screen. Different versions of Logo exist and commands may vary depending on the version. You may need to consult your user's manual if commands do not function exactly the same way as they are presented here. If an abbreviation can be used in place of a command, then it is given in parentheses immediately after the command when the command is introduced.

HISTORICAL NOTE

The computer language Logo was developed in 1967 at Bolt, Beranek, and Newman, Inc., of Cambridge, Massachusetts, and the Massachusetts Institute of Technology (MIT) by Daniel Bobrow, Wallace Feurzeig, and Seymour Papert. The name "Logo" is derived from the Greek word for "thought." The developers of Logo were influenced by the field of artificial intelligence, the computer language LISP, and the theories of Jean Piaget. The tradition of calling the display creature a turtle can be traced to early experiments involving robotlike creatures referred to as "tortoises." When computer graphics were implemented, the screen creature inherited the turtle terminology. In some versions, the figure on the screen actually appears as a turtle.

draw mode/DRAW To execute turtle graphics commands, enter **draw mode** by typing **DRAW** and pressing RETURN or ENTER . In the draw mode, the turtle appears in the center of the screen, as shown in Figure AI-1. The turtle's position in the center of the screen is called "home."

Figure AI-1

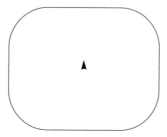

Moving the Turtle

FORWARD (FD)
BACK (BK)

To make the turtle change position, we use the primitives **FORWARD (FD)** and **BACK (BK)**, *followed by a space* and a numerical input. The numerical input tells the turtle how far to move. For example, in the draw mode typing FORWARD 100 or FD 100 and pressing RETURN causes the turtle to move 100 "turtle units" in the direction it is pointing. Figure AI-2 shows a series of directions and drawings at each stage when the turtle starts at home pointing upwards. Similarly, the BACK command may be used with a numerical input. For example, BACK 75 or BK 75 causes the turtle to move backwards 75 units. Giving the turtle too great an input causes the turtle to "wrap around" the screen. To explore how the turtle wraps, try FD 250 and observe what happens.

Figure AI-2

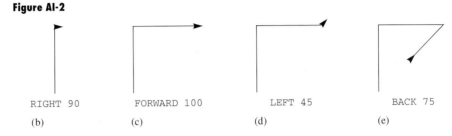

DRAW FORWARD 100	RIGHT 90	FORWARD 100	LEFT 45	BACK 75
(a)	(b)	(c)	(d)	(e)

Turning the Turtle

RIGHT (RT) · LEFT (LT)

To make the turtle change direction, we used the commands **RIGHT (RT)** and **LEFT (LT)**. The RIGHT and LEFT commands, along with numerical inputs, cause the turtle to turn in place the specified number of degrees. For example, typing RIGHT 90 or RT 90 and pressing RETURN causes the turtle to turn 90° to the right of the direction it previously pointed. A sequence of moves illustrating these commands is given in Figure AI-2.

Logo accepts a sequence of commands written on one line. For example, Figure AI-2(e) could be drawn by typing the following and pressing RETURN :

```
DRAW FD 100 RT 90 FD 100 LT 45 BK 75
```

PENUP (PU)
PENDOWN (PD)
HIDETURTLE (HT)
SHOWTURTLE (ST)
HOME

To move the turtle without leaving a trail, we use the command **PENUP (PU)**. To make the turtle leave a trail again, type **PENDOWN (PD)**. It is possible to hide the turtle by typing **HIDETURTLE (HT)**. To make the turtle reappear, type **SHOWTURTLE (ST)**.

To return the turtle to the center of the screen with heading 0, type the command **HOME**. Unless the command PENUP is used before HOME, a trail to the center of the screen will be drawn from the position the turtle occupied before HOME was typed.

To start a new drawing with a clear screen, we type DRAW. This returns the turtle to its initial position and direction in the center of the screen and clears the screen. Any time the turtle points straight north (up), we say it has heading 0. A heading of 90 is directly east, 180 is directly south, and 270 is directly west. The screen could be marked as shown in Figure AI-3.

Figure AI-3

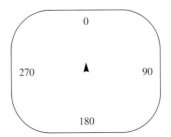

HEADING

PRINT (PR)

SETHEADING (SETH)

primitives

To learn the turtle's heading, we type **HEADING** with no inputs. Typing HEADING in the draw mode and pressing RETURN causes the computer to output the turtle's heading. To have the computer print only the value of the heading, we use **PRINT (PR)** along with HEADING, as in PRINT HEADING. For example, if the turtle is at home with heading 0 and we type RT 45 PR HEADING, then 45 will be displayed. If we execute RT 45 PR HEADING again, then 90 will be displayed.

The command **SETHEADING (SETH)** requires one input and can be used to turn the turtle in a direction from the 0 heading. For example, SETH 100 turns the turtle so that it has a heading of 100. This command can be used no matter where the turtle is located or what its heading is at the time. Figure AI-4 gives an example of the use of the SETH and HOME commands. A summary of commands, or **primitives,** introduced thus far is shown in Table AI-1.

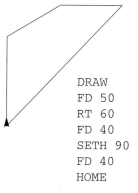

DRAW
FD 50
RT 60
FD 40
SETH 90
FD 40
HOME

Figure AI-4

Table AI-1

Command (Primitives)	Abbreviation	Logo Instruction
BACK	BK	BK 60
DRAW		
FORWARD	FD	FD 50
HEADING		PR HEADING
HIDETURTLE	HT	
HOME		
LEFT	LT	LT 45
PENDOWN	PD	
PENUP	PU	
PRINT	PR	PR "LOGO
RIGHT	RT	RT 90
SETHEADING	SETH	SETH 270
SHOWTURTLE	ST	

Creating Figures

People studying Logo are encouraged to "play turtle" and act out their commands. For example, to act out drawing a square, we may walk around the square by moving forward 50 units, turning right 90°, moving forward 50 units, turning right 90°, moving forward 50 units, turning right 90°, and finally moving forward 50 units. The sequence of commands for these moves is summarized in Figure AI-5(a), with the resulting square and final position of the turtle shown.

Figure AI-5

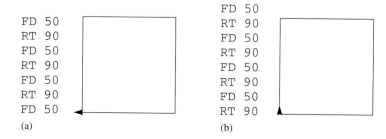

(a)   (b)

In Figure AI-5(b) an additional command of RT 90 is included and the turtle's final position is the same as its initial position and the turtle's final heading is the same as its initial heading. When drawing a figure, you will often find it convenient to have the turtle's final state be the same as its initial state where "state" refers to position and direction. When the final and initial states are the same, we say that the set of commands is **state transparent.**

state transparent

The sequence of commands in Figure AI-5(b) contains the instructions FD 50 and RT 90 repeated four times. Logo allows us to use the REPEAT command to repeat a list of instructions. For example, to draw the square in Figure AI-5(b), we would type the following:

```
REPEAT 4 [FD 50 RT 90]
```

REPEAT In general, **REPEAT** takes two inputs: a number and a list of commands. The commands in the brackets are repeated the designated number of times.

Example AI-1 Predict the results of each of the following, indicating the initial and final turtle states. Then, check your answers with a computer. In each case, assume the turtle starts at home with heading 0.

a. FD 100
   RT 135
   FD 100
   RT 45
   FD 100
   RT 135
   FD 100
   RT 45
b. REPEAT 2 [FD 100 RT 135 FD 100 RT 45]
c. REPEAT 8 [FD 50 RT 45]
d. BK 100 SETH 270 FD 100 HOME
e. SETH 90 REPEAT 5 [FD 10 PU FD 5 PD]

Solution The results are depicted in Figure AI-6.

Figure AI-6

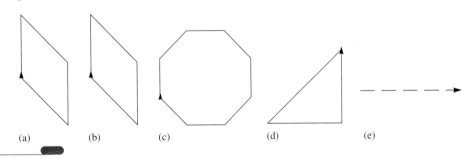

(a)　(b)　(c)　(d)　(e)

Defining Procedures

The sequence of commands in Figure AI-5(b) instruct the turtle to draw a square. If the screen is cleared, the figure is lost. To redraw the square, the entire sequence of commands must be retyped. Fortunately, with Logo, it is possible to store instructions in the computer's memory by creating a procedure. A **procedure** is a group of one or more instructions to the computer that the computer can store to be used later.

To create a procedure, we type **TO,** followed by the name we wish to call the procedure, and then press RETURN . When RETURN is pressed, the computer enters **edit mode,** or the teaching mode. In this mode, the lines that follow are not executed but may be stored in memory under the given name. The name must be a sequence of symbols with no spaces, and it may not be the name of a primitive. For example, to create a procedure called SQUARE1 to draw a square, the following is entered.

```
TO SQUARE1
  REPEAT 4 [FD 50 RT 90]
END
```

To signify the end of a procedure, we type **END** as the last line of the procedure. Typing END at the end of a procedure is necessary if you intend to define another procedure without leaving edit mode. (It is not necessary otherwise.)

In Logo, one procedure can call another, as shown in Example AI-2. *Note:* If the SQUARE1 procedure has not been defined on your computer, be sure to define it before working the example. In the rest of this section, we assume that the SQUARE1 procedure and all subsequent procedures are stored in the computer's memory and can be reused.

Example AI-2

Predict the figures that will be drawn by defining and executing each of the following procedures. Assume the turtle starts at home with heading 0.

a.
```
TO SQUARE2
  RT 90
  SQUARE1
END
```

b.
```
TO SQUARESTACK
  SQUARE1
  RT 90
  SQUARE1
END
```

c.
```
TO STAIR
  SQUARE1
  RT 180
  SQUARE1
END
```

d.
```
TO TURNSQUARE
  SQUARE1
  RT 45
  SQUARE1
END
```

Solution The results are depicted in Figure AI-7.

Figure AI-7

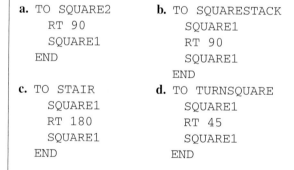

(a)   (b)   (c)   (d)

Problem Solving Drawing an Equilateral Triangle

Write a procedure for drawing a triangle whose sides are each 50 turtle steps long and whose angles each measure 60°. This type of triangle is an *equilateral triangle*.

- **Understanding the Problem** We are to write a procedure to draw a triangle with all sides of length 50 turtle units and all angles of measure 60°. We can start at any position with any heading.

- **Devising a Plan** It is helpful to sketch the triangle to determine the angle the turtle needs to turn at each vertex. Suppose the turtle starts at point *A* with heading 0 and moves 50 turtle steps to point *B*, as shown in Figure AI-8. This can be done by telling the turtle to move FD 50. At point *B*, the turtle still has heading 0. To walk on $\overrightarrow{BC}$, the turtle must turn 120° to the right. Thus the next command should be RT 120. The triangle has three sides of equal length, so three turns are necessary to achieve the turtle's initial heading. Repeating the sequence FD 50 RT 120 three times should cause the turtle to walk around the triangle and finish in its original position with its original heading.

Figure AI-8

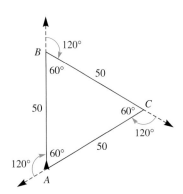

- **Carrying Out the Plan** A procedure called TRIANGLE1 based on the preceding discussion follows:

```
TO TRIANGLE1
  REPEAT 3 [FD 50 RT 120]
END
```

- **Looking Back** Executing the TRIANGLE1 procedure yields the desired figure. Additional investigations include writing a procedure to draw the same type of triangle by turning left instead of right or writing a procedure to draw a triangle with one horizontal side. Procedures for drawing other polygons could also be explored.

One of the great advantages of Logo is its ability to use procedures to define new procedures. Consider the following problem.

Problem Solving Drawing a House

Write a procedure to draw the "house" shown in Figure AI-9.

- **Understanding the Problem** The top of the house appears to be a triangle similar to the one in Figure AI-8, and the bottom appears to be a square. We must write a super

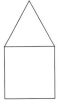

Figure AI-9

procedure in which procedures for drawing a square and a triangle will be incorporated to draw the house.

- **Devising a Plan**   One way to solve the problem is to break down the problem of drawing a house into *simpler problems,* that of drawing the bottom of the house (the square) and that of drawing the roof (the triangle). The type of programming that starts with a general idea and breaks down the problem into smaller parts is **top-down programming.** We have a procedure SQUARE1 for drawing a square of length 50 units and a procedure TRIANGLE1 for drawing a triangle of length 50 units. If we use these two procedures, then we should be able to draw the house.

- **Carrying Out the Plan**   If the turtle has heading 0, it may seem that typing SQUARE1 followed by TRIANGLE1 would draw the desired house. The result of this effort is shown in Figure AI-10(a). Why did it not produce the desired figure?

Figure AI-10

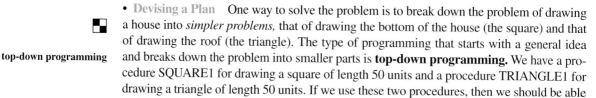

To draw the roof in proper position, we need the turtle to be at the upper left vertex of the square. This can be achieved by typing SQUARE1 FD 50. But, if we now type TRIANGLE1, we obtain the shape in Figure AI-10(b), which is still not the desired house.

After we type SQUARE1 and FD 50, the turtle is at point A with heading 0. For it to form the roof shown in Figure AI-11(a), and to walk on $\overline{AB}$, the turtle needs to turn right by $90° - 60°$, or $30°$. With the turtle's having this heading, typing TRIANGLE1 should cause the turtle to draw the desired roof. The complete procedure, called HOUSE, is shown in Figure AI-11(b).

Figure AI-11

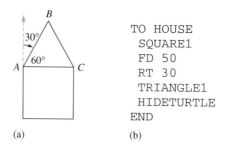

```
TO HOUSE
  SQUARE1
  FD 50
  RT 30
  TRIANGLE1
  HIDETURTLE
END
```

(a)   (b)

- **Looking Back**   If the HOUSE procedure is executed, the desired figure is obtained. Alternative techniques for drawing the figure could also be explored. Houses of other sizes could be drawn, and windows and doors could be added.

As the Drawing a House Problem shows, trial and error helps the user to get acquainted with the problem and eventually to find the correct solution. This process of rewriting a program that does not do what we want it to do is called *debugging*. Being able to experiment with Logo is part of its power. Students need to understand that sometimes bugs or experimentation can lead to unexpected results or pictures that are sometimes better than what was intended.

To help you understand how Logo works when a procedure calls another procedure, Figure AI-12 gives a telescoping model of the HOUSE procedure in Figure AI-11(b).

Figure AI-12

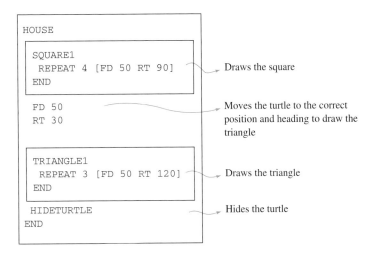

When HOUSE is run, it encounters the call for SQUARE1. At this point, all the lines of SQUARE1 are inserted. After SQUARE1 has been completed, control is returned to the procedure that called it, namely, HOUSE. Now, HOUSE continues where it left off and executes FD 50 RT 30. Then it calls the TRIANGLE1 procedure. After TRIANGLE1 has been executed, control returns to HOUSE, which hides the turtle and encounters its own END statement. (Remember to clear the screen before trying HOUSE again.)

In working through the HOUSE procedure, we went through several steps. These are summarized next. They might be useful in solving a variety of problems presented in this text.

1. Sketch your drawing on paper (preferably graph paper) to get an idea of the scale to be used and of how the final picture should look.
2. Divide the drawing into parts that are repeated, that you already know how to draw, or that are smaller parts of the whole. Separate procedures for drawing each part are easier to debug than a single procedure for the whole drawing.
3. Decide how your procedures are going to fit together to form the complete picture. Some procedures might be necessary just to move the turtle to the right position for drawing the individual parts.
4. Write your procedures. One approach is to write individual procedures for separate pieces, make sure they work, and then try to put them all together to form the complete picture. Another approach is to fit the procedures together as they are completed. Either approach is an acceptable problem-solving strategy, and each has advantages in different situations.

Problem Solving — Drawing a Glass

Write a procedure to draw the figure sketched on the graph paper in Figure AI-13.

Figure AI-13

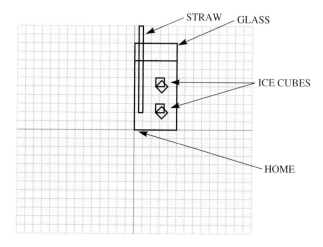

- **Understanding the Problem** We are to write a procedure to draw a figure similar to the one shown in Figure AI-13. Each length of a side of a small square on the grid paper represents 10 turtle steps.

- **Devising a Plan** The figure can be broken into three parts: the glass, the straw, and the ice cubes. Using top-down programming, we can write a procedure called DRINK that draws the figure.

```
TO DRINK
  GLASS
  ICE.CUBES
  ICE.CUBES
  STRAW
  HT
END
```

To complete the problem, we must write procedures for each portion of the DRINK procedure.

- **Carrying Out the Plan** First, we design a procedure called GLASS for drawing the glass. If the turtle starts at home with heading 0, one possible procedure and its output are given in Figure AI-14.

Figure AI-14

```
TO GLASS
  REPEAT 2 [FD 100 RT 90 FD 50 RT 90]
  FD 80 RT 90
  FD 50 BK 50
  LT 90 BK 80
END
```

Likewise, we can design procedures called STRAW and ICE.CUBES to draw the other two parts, as shown in Figure AI-15(a) and (b).

Figure AI-15

```
TO STRAW
  REPEAT 2 [FD 100 RT 90 FD 5 RT 90]
END

TO ICE.CUBES
  SQUARE3
  RT 45
  SQUARE3
  LT 45
END

TO SQUARE3
  REPEAT 4 [FD 10 RT 90]
END
```

(a)

(b)

If we now execute DRAW and attempt to execute the DRINK procedure as defined, the result is as shown in Figure AI-16.

Figure AI-16

To correct the DRINK procedure so that it will draw the desired figure, we must keep track of the turtle's position and heading. Sometimes, it is convenient to move the turtle to the required positions and headings by using a set of procedures. The following procedures—SETUP.CUBES1, SETUP.CUBES2, and SETUP.STRAW—move the turtle to the correct position and heading to draw each part. Notice the use of PU and PD to keep the transitions invisible.

```
TO SETUP.CUBES1
  PU FD 50 RT 90 FD 25 LT 90 PD
END

TO SETUP.CUBES2
  PU BK 30 PD
END

TO SETUP.STRAW
  PU LT 90 FD 20 RT 90 PD
END
```

If we edit DRINK and add these new procedures, we obtain the procedures and figure shown in Figure AI-17.

Figure AI-17

```
TO DRINK
  GLASS
  SETUP.CUBES1
  ICE.CUBES
  SETUP.CUBES2
  ICE.CUBES
  SETUP.STRAW
  STRAW
  HT
END
```

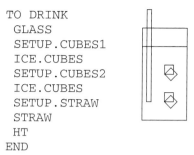

- **Looking Back** When the DRINK procedure is executed, it yields the desired figure. The procedure could have been written in many different ways. Although various strategies could be used to develop the procedure, we see that the top-down strategy can be very useful. One advantage of this strategy is that it is easier to debug smaller portions of the figure rather than to try to do the complete figure all at one time.

Writing Procedures with Variables

The SQUARE1 procedure allows us to draw only squares of side length 50. If we want to draw smaller or larger squares, we must write a new procedure. It would be more convenient if we could write one procedure that would work for a square of any size. This can be accomplished in Logo by using a variable as input, rather than a fixed number such as 50 in FD 50. To use a variable input in Logo, we need to warn the computer that the "thing" we are going to type is a variable. We do this by using a colon before the variable name. For example, a variable input to the SQUARE1 procedure might be called :SIDE, where :SIDE stands for the length of a side of the square. Note that there is no space between the colon and the word SIDE. We define a new SQUARE procedure with variable input :SIDE and place the name of the variable in the title line.

```
TO SQUARE :SIDE
  REPEAT 4 [FD :SIDE RT 90]
END
```

If we want the turtle to draw a square of size 40, we type SQUARE 40. Notice that we do not type SQUARE :40 because 40 is not a variable. Investigate what happens if SQUARE is typed with no inputs.

REMARK We call the new variable square procedure SQUARE instead of SQUARE1. If we attempt to enter the edit mode to define a new SQUARE1 procedure and the old procedure has not been erased, the computer will display the old SQUARE1 procedure on the screen for us to edit. Consult your Logo manual for directions on how to edit procedures.

A procedure may have more than one input. In the following procedure, two variables are used so that two inputs can be accepted.

```
TO RECTANGLE :HEIGHT :WIDTH
  REPEAT 2 [FD :HEIGHT RT 90 FD :WIDTH RT 90]
END
```

Figure AI-18 shows rectangles drawn by the RECTANGLE procedure with different inputs for the sides.

Figure AI-18

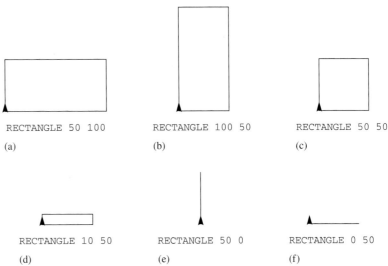

Logo Recursion

recursion **Recursion** (as used in Logo) is the process of a procedure's calling a copy of itself. As a first example of recursion, we write a procedure called CIRC for drawing a "turtle-type" circle. This could be done by having the turtle move forward "a little," then turn right "a little," and then continuing this process until a closed figure is obtained. Thus we could start the procedure with FD 1 RT 1 and then have the turtle start the procedure anew each time the instruction is executed. Such a procedure follows:

```
TO CIRC
  FD 1 RT 1
  CIRC
END
```

To stop the procedure, press CTRL G (or ⌘ G). (This key sequence varies depending on the version of Logo and on the brand of computer.) To understand how CIRC works and, in general, what happens when a procedure calls itself, we use the telescoping model in Figure AI-19.

Figure AI-19

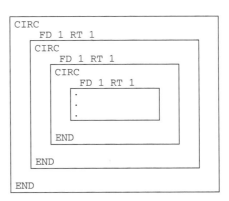

When CIRC is executed, FD 1 RT 1 causes the turtle to move forward one unit and then turn right 1°. CIRC then calls a copy of the CIRC procedure, which again executes FD 1 RT 1 and in turn calls another copy of CIRC, and so on. The process continues because we have made CIRC one of the instructions in the CIRC procedure. The END statement is never reached, and the instruction FD 1 RT 1 is executed indefinitely.

The repetitive process shown in the CIRC procedure occurs in the type of recursion called **tail-end recursion.** In tail-end recursion, only one recursive call is made in the body of the procedure, and it is the final step before the END statement.

tail-end recursion

NOW TRY THIS AI-1

- Write a procedure called CIRCLE1 that uses a REPEAT command rather than recursion to draw a Logo-type circle similar to the one drawn by the CIRC procedure.

Recursion is particularly valuable when we do not know how many times to repeat a set of instructions to accomplish some goal. For example, consider the shapes that can be drawn by repeating the instruction "Go forward some fixed distance and turn right some fixed angle." A recursive procedure called POLY that does this is as follows:

```
TO POLY :SIDE :ANGLE
  FD :SIDE RT :ANGLE
  POLY :SIDE :ANGLE
END
```

To execute the POLY procedure, we need two numerical inputs, one for :SIDE and the other for :ANGLE. Figure AI-20 shows shapes drawn by POLY with different inputs. The drawings were stopped using CTRL-G.

Figure AI-20

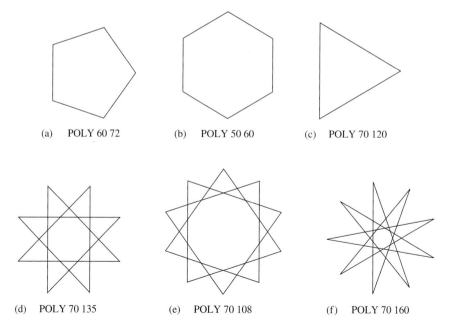

(a) POLY 60 72   (b) POLY 50 60   (c) POLY 70 120

(d) POLY 70 135   (e) POLY 70 108   (f) POLY 70 160

The POLY procedure draws regular polygons (polygons that have congruent sides and congruent angles), as in Figure AI-20(a), (b), and (c), and also star shapes as in Figure AI-20(d), (e), and (f). Try other executions of POLY, such as POLY 50 180, POLY 50 181, POLY 60 288, POLY 6000 300, and POLY 7000 135. Try to predict which inputs produce regular polygons and which produce star shapes.

All the figures drawn by the POLY procedure in Figure AI-20 are closed; that is, they can be drawn by starting and stopping at the same point. Will all figures drawn by POLY be closed? We can also ask the following questions:

1. Given the value of :ANGLE in the POLY procedure, is it possible to predict (before the figure is drawn) how many vertices the figure will have?
2. If we wish the POLY procedure to draw a figure with a given number of vertices, can we determine the correct angle input?

With the help of recursion, we can accomplish tasks that cannot be done easily with just the REPEAT command, especially if we do not know how many times to repeat a sequence of instructions. Consider drawing a square-type spiral as shown in Figure AI-21.

Suppose each side of the figure is five units longer than the preceding side. If the turtle starts at home, the figure can be drawn by telling the turtle to move forward a certain length :LEN, turn right 90°, move forward five units more than the previous value of :LEN, and so on. A recursive procedure called SQSPI shows how this can be done.

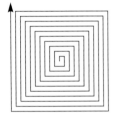

Figure AI-21

```
TO SQSPI :LEN
  FD :LEN RT 90
  SQSPI :LEN + 5
END
```

Each time SQSPI calls itself, the length of :LEN is increased by five units. When SQSPI is run, the sides grow too large to fit on the screen. Rather than stop SQSPI with CTRL-G, we can write a "stop" instruction in the procedure. This can be done with the IF and STOP primitives. **IF** is a primitive that tests one of three conditions: equal (=), less than (<), or greater than (>). The IF primitive is used in the following form:

IF

IF *(Condition)*   *(Action to be taken if condition is true)*

The parentheses should not be typed.

For example, if we do not want the turtle to draw any segment longer than 100 units, we insert the following instruction:

```
IF :LEN > 100 STOP
```

When this line is inserted into a procedure, the IF statement causes the computer to check whether the value of :LEN is greater than 100. If it is, the procedure stops; if not, the next line is executed. The **STOP** primitive causes the current procedure to stop and returns control to the calling procedure, if there is one. An edited form of the SQSPI procedure is as follows:

STOP

```
TO SQSPI :LEN
  IF :LEN > 100 STOP
  FD :LEN RT 90
  SQSPI :LEN + 5
END
```

The SQSPI procedure can be generalized to draw other spiral-type figures. Investigate the following POLYSPI procedure for various inputs:

```
TO POLYSPI :SIDE :ANGLE
  IF :SIDE > 100 STOP
  FD :SIDE RT :ANGLE
  POLYSPI :SIDE + 5 :ANGLE
END
```

What inputs should be given to POLYSPI in order to achieve the same effect that SQSPI does? Also, investigate what happens when :ANGLE, rather than :SIDE, is incremented each time the recursive call is made.

Table AI-2 gives a summary of commands in this section.

Table AI-2
Summary of Commands

Command	Description
BACK (BK)	Takes one input. Positive input moves the turtle backwards the number of turtle units that are input. For example, BK 40 moves the turtle backwards 40 units.
DRAW	Needs no input. It sends the turtle home and clears the graphics screen.
END	Used at the end of a procedure. Tells the computer that there are no more instructions to be given in the procedure.
FORWARD (FD)	Takes one input. Positive input moves the turtle forward (in the direction the turtle is facing) the number of turtle units that are input. For example, FD 20 moves the turtle forward 20 units.
HEADING	Needs no input. In draw mode, it outputs the turtle's heading.
HIDETURTLE (HT)	Needs no input. It causes the turtle to disappear.
HOME	Needs no input. It returns the turtle to the center of the screen and sets its heading to 0. If the pen is down, it leaves a track from the turtle's present location to the home position.
IF	Takes two inputs. The first must be either true or false. The second contains instructions that are carried out if, and only if, the first is true.
LEFT (LT)	Takes one input. Positive input turns the turtle left from its present heading the number of degrees that are input. For example, LT 90 turns the turtle left 90°.
PENDOWN (PD)	Needs no input. In graphics mode, it causes the turtle to leave a track.
PENUP (PU)	Needs no input. In graphics mode, it enables the turtle to move without leaving a track.
PRINT (PR)	Takes one input. It causes the input to be printed on the screen and moves the cursor to the next line.
REPEAT	Takes a number and a list as input. It executes the instructions in the list the designated number of times.
RIGHT (RT)	Takes one input. Positive input turns the turtle right from its present heading the number of degrees that are input. For example, RT 90 turns the turtle right 90°.
SETHEADING (SETH)	Takes one input. It turns the turtle to the heading indicated by the input.
SHOWTURTLE (ST)	Needs no input. It causes the turtle to reappear.
STOP	Takes no inputs. It causes the current production to stop and then returns control to the calling procedure.
TO	Takes the name of a procedure as input and causes Logo to enter edit mode.

ONGOING ASSESSMENT A-I

1. Sketch figures drawn by the turtle using each of the following sets of instructions. Check your sketches by executing the instructions on a computer. Type DRAW after each lettered part.
   a. FD 50
      RT 90
      FD 50
      RT 45
      FD 50
      RT 135
      FD 50
   b. FD 50
      RT 90
      BK 50
      RT 60
      FD 50
   c. FD −50 FD 50
   d. LT −90 BK −50 RT 40 PR HEADING
   e. RT 360 PR HEADING
   f. SETH 30 REPEAT 3 [FD 50 RT 120]
   g. FD 100/2 RT 5*6 BK 100 + 20
2. Experiment with the turtle to find the dimensions of the screen.
3. Predict what the turtle will draw with the following sets of instructions. Check your answers by executing the instructions on the computer. (SQUARE1 and TRIANGLE1 are defined in the text.)
   a. REPEAT 8 [SQUARE1 RT 45]
   b. REPEAT 6 [TRIANGLE1 RT 60]
   c. REPEAT 36 [SQUARE1 RT 10]
4. Write procedures to draw figures similar to each of the following:

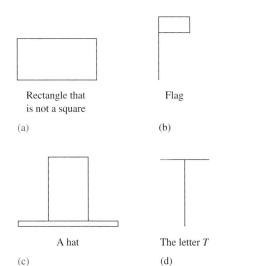

Rectangle that is not a square
(a)

Flag
(b)

A hat
(c)

The letter *T*
(d)

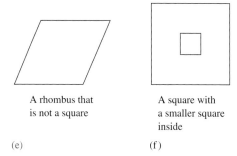

A rhombus that is not a square
(e)

A square with a smaller square inside
(f)

5. Write a procedure to draw the following:

6. Use any procedures in this section to write new procedures that will draw each of the following figures:

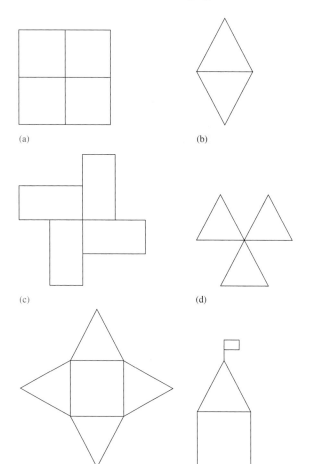

(a) (b)

(c) (d)

(e) (f)

7. Write procedures to draw figures similar to each of the following:

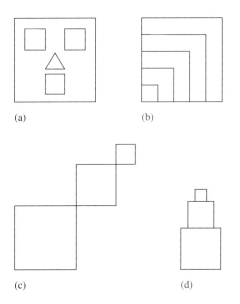

(a) (b)

(c) (d)

8. Use top-down programming to write a procedure called DOG to draw a figure similar to the following:

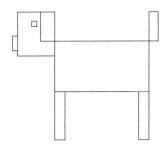

9. Use top-down programming to write a procedure called KITE to draw a figure similar to the following:

10. Write procedures to draw figures similar to those in problem 4, but of variable size.

11. Write a procedure called BLADES to draw a figure similar to the following, but of variable size:

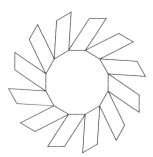

12. Write a procedure called RECTANGLES to draw a figure similar to the following, but of variable size:

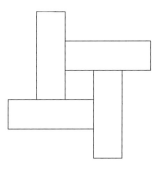

13. Predict the shapes that will be drawn by the following procedures and then check your predictions on the computer. The SQUARE and TRIANGLE procedures are defined as follows:

```
TO TRIANGLE :SIDE
  REPEAT 3 [FD :SIDE RT 120]
END
TO SQUARE :SIDE
  REPEAT 4 [FD :SIDE RT 90]
END
```

a.
```
TO FIGURE :SIDE
   TRIANGLE :SIDE
   RT 10
   FIGURE :SIDE
END
```

b.
```
TO FIGURE1 :SIDE
   IF :SIDE < 5 STOP
   TRIANGLE :SIDE
   RT 10
   FIGURE1 :SIDE  −5
END
```

c. TO TOWER :SIDE
   SQUARE :SIDE
   FD :SIDE
   TOWER :SIDE * 0.5
   END

d. TO TOWER1 :SIDE
   IF :SIDE < 2 STOP
   SQUARE :SIDE
   FD :SIDE
   TOWER1 :SIDE * 0.5
   END

e. TO SQ :SIDE
   IF :SIDE < 2 STOP
   SQUARE :SIDE
   SQ :SIDE −5
   END

f. TO SPIRAL :SIDE
   IF :SIDE > 50 STOP
   FD :SIDE
   RT 30
   SPIRAL :SIDE +3
   END

14. Given the following NEWPOLY, POLYSPIRAL, and INSPI procedures, predict the shapes that will be drawn by each and then check your predictions on the computer:

    TO NEWPOLY :SIDE :ANGLE
      FD :SIDE RT :ANGLE
      FD :SIDE RT :ANGLE * 2
      NEWPOLY :SIDE :ANGLE
    END
    TO POLYSPIRAL :SIDE :ANGLE :INC
      FD :SIDE RT :ANGLE
      POLYSPIRAL (:SIDE + :INC) :ANGLE :INC
    END
    TO INSPI :SIDE :ANGLE :INC
      FD :SIDE RT :ANGLE
      INSPI :SIDE (:ANGLE + :INC) :INC
    END

    a. NEWPOLY 50 30
    b. NEWPOLY 50 144
    c. NEWPOLY 50 125
    d. POLYSPIRAL 2 85 3
    e. POLYSPIRAL 1 119 2
    f. POLYSPIRAL 1 100 5
    g. INSPI 10 2 20
    h. INSPI 2 0 10
    i. INSPI 10 5 10

15. Write recursive procedures to draw figures similar to the following six figures. Use the STOP command in your procedures.

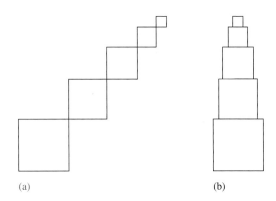
(a)   (b)

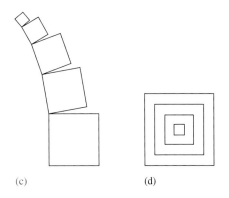
(c)   (d)

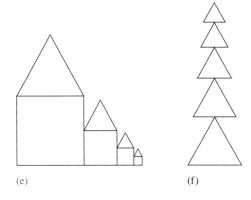
(e)   (f)

16. Write a recursive procedure with a STOP command to draw a figure similar to the following:

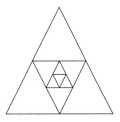

17. Write a procedure called SPIN.SQ that uses recursion and a STOP command to spin a variable-sized square while "shrinking" its size, as shown in the following figure.

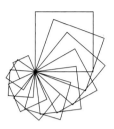

18. Write a recursive procedure with a STOP statement that draws the following variable-sized figure made of squares:

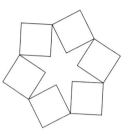

Appendix II: Graphing Calculators

Introduction

In this appendix, we introduce many features of the graphing calculator by working through problems in various ways to show how you might approach a problem by using a graphing calculator. One advantage of the graphing calculator is the display, not only of graphs but also text. With this feature, computations and answers can be viewed, various methods can be compared, and changes can be made easily.

Graphing calculators are now being marketed for the middle-school level. In this appendix, we assume your calculator has at least the capabilities of the Texas Instruments TI-73 middle-school graphing calculator. If you use other calculators, you will have to determine whether those features are available on them.

Problem Solving   **Molly's CD Problem**

Molly belongs to a club that sells CDs by mail. The CDs sell for $12.95 each, and she can order as many as she wants each month. Each order has a shipping charge of $5.00 no matter how many CDs are ordered. She would like to develop a table that tells her how much money she owes when she orders from 1 to 10 CDs. Develop such a table.

This problem is worked using a variety of techniques to show some of the capabilities of the graphing calculator.

Using the Replay Feature

Suppose we want to solve Molly's CD Problem using paper and pencil to build a table and using the calculator to perform the computations. First, we must realize that the cost for any number of CDs is given by multiplying the number of CDs by $12.95 and adding $5.00. To find the cost of one CD using this process, enter the following:

$$\boxed{1}\ \boxed{\times}\ \boxed{1}\ \boxed{2}\ \boxed{.}\ \boxed{9}\ \boxed{5}\ \boxed{+}\ \boxed{5}$$

and press $\boxed{\text{ENTER}}$. The answer, 17.95, is displayed. This number can then be recorded in a table. It is important for elementary school students to see the whole problem as well as the answer. This allows students to check visually whether they have entered their numbers and operations correctly.

replay feature   To find the cost for other quantities of CDs, we use the **replay feature** of the calculator. To activate this feature, press $\boxed{\text{2nd}}\ \boxed{\text{ENTRY}}$; the previous entry is displayed. We then use the left arrow to move the cursor over the first 1 and replace it with a 2. When $\boxed{\text{ENTER}}$ is pressed, the next answer, 30.90, is displayed. We could shorten the left arrow strokes even further by pressing 2nd followed by the left arrow; the cursor will go to the beginning of the line. If we continue in this manner, as shown in Figure AII-1, we could find the cost of any number of CDs and enter these costs into a table.

Figure AII-1

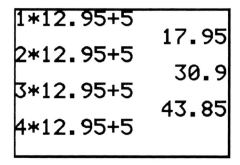

Using Data Lists

Many graphing calculators can work with lists. To clear existing lists, press STAT and choose the **CLRLIST** (under OPS on the TI-73) feature with the name of the list as input. For example, to clear list 1 (2nd L1), we enter CLRLIST L1 and press ENTER . A message is displayed telling us that the list is cleared. Other lists are cleared in the same way.

CLRLIST

We use L1 to represent the number of CDs and then use L2 to represent the total cost for the CDs. To enter the values for L1, we press STAT and choose the **EDIT** feature. We then press ENTER , and the lists are shown in columns as in Figure AII-2(a). (On a TI-73 press LIST.) On the bottom edit line where L1(1) = is displayed, type the numeral 1 followed by ENTER . A 1 will appear as the first entry in L1, as shown in Figure AII-2(a).

EDIT

Figure AII-2

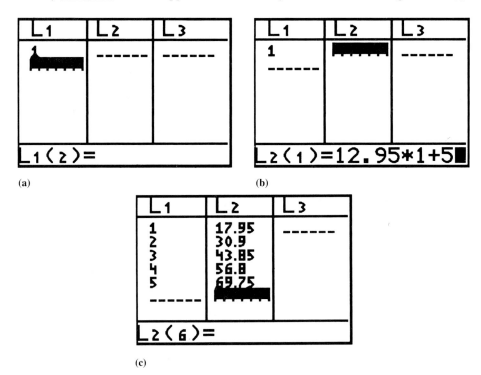

By pressing the right arrow, we move to list L2. The first entry in L2 can be entered directly or it can be computed by entering the expression L2(1) = 12.95 × 1 + 5 in the edit line at the bottom of the screen, as shown in Figure AII-2(b). When ENTER is pressed, 17.95 appears as the first entry in L2. We next use the left and down arrows to return to the second entry in L1 and enter 2. The remainder of the table as shown in Figure AII-2(c) can be completed.

This method using lists is a hard way to complete the table. An easier method for finding the various costs is to define a general pattern and have the calculator do all the computations at once. To do this, enter the numbers 1 through 10 in L1. Move the cursor to the top of L3 so that L3 is highlighted. When L3 is highlighted, anything that is done in the edit line on the bottom will happen to every element in L3. In the edit line, enter L3 = 12.95L1 + 5, as shown in Figure AII-3(a), and press ENTER . The entire list in L3 is then computed based on the entries in L1 and the rule given in the edit line (see Figure AII-3b). Lists L2 and L3 should be the same even though they were obtained in two different ways.

Figure AII-3

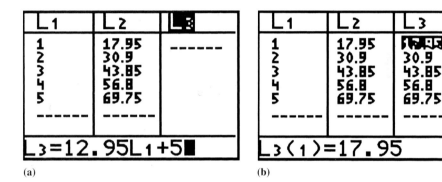

(a)   (b)

Using Lists on the Home Screen

Computations involving lists can be done on the home screen using the braces { and } located above the parentheses keys (or in the TEXT editor on a TI-73). A list can be entered by using a left brace, followed by the list, with each element separated by a comma. Figure AII-4(a) shows a list representing the number of possible CDs ordered: {1, 2, 3, 4, 5, 6, 7, 8, 9, 10}. We then multiply each element in this list by 12.95 and add 5 to the result; the answers are given on the screen (see Figure AII-4b). To see the entire list of answers, scroll to the right using the right arrow key.

Figure AII-4

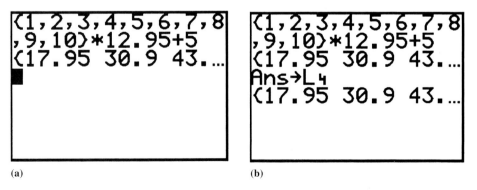

(a)   (b)

You can store the list of answers in one of the list memories by using STO and the list number you choose. If we press STO L4 , the answers are stored in L4. The list of answers still appears on the home screen, as shown in Figure AII-4(b), even though the answers are stored in L4. If we press STAT, select EDIT, and use the right arrow to move to the L4 column, we can see that the list of answers has been stored in L4. (On a TI-73, press LIST.) Notice that this list matches the ones in L2 and L3.

Using STAT Plot Graphing Capabilities

Next we investigate how Molly's CD Problem could be solved using a graphical representation. Before we do any graphing, we need to set an appropriate graphing window. We must decide how large a window is required in this particular problem. If we plot the number of CDs ordered on the x-axis and the total costs on the y-axis, we need only look at our lists to determine appropriate values. The number of CDs ranges from 1 to 10, and the cost ranges from $12.95 to $134.50. We have several choices. If we press WINDOW, then we can enter the choices, as shown in Figure AII-5(a). After we select WINDOW, we can enter the minimum and maximum values for the x- and y-axes and the scale that gives the distance between the marks on each axis. (On certain calculators you can have the calculator choose the window for you.)

Figure AII-5

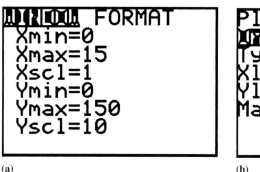
(a)

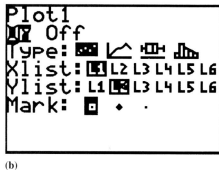
(b)

We can plot these points by using 2nd to choose STAT PLOT (or PLOT) and then select Plot1, as shown in Figure AII-5(b). We see in Figure AII-5(b) that Plot1 is on, the first choice of a scatterplot is chosen, the x-values come from L1, and the y-values come from L2. We can choose 5 plots and each plot can be a scatterplot, a line graph, a box-and-whisker plot, or a histogram.

By pressing GRAPH, we obtain the graph shown in Figure AII-6(a). If we then press TRACE and use the right and left arrow keys, we can see the trace cursor move along the dots, displaying the x- and y-values at each point, as shown in Figure AII-6(b).

Figure AII-6

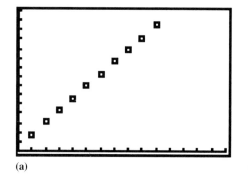
(a)

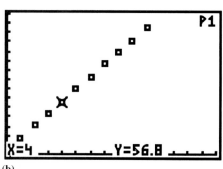
(b)

NOW TRY THIS AII-1

- a. How are the values that are displayed using the TRACE feature related to L1 and L2?
- b. Why are just the points shown rather than a line containing the points?
- c. What pattern is shown in the graphical representation that is not shown when working with only numbers or lists?
- d. Use the graph to approximate the cost of 12 CDs.

Connecting Graphing and Algebra

Earlier in this appendix, we generated a mathematical rule for computing the total cost for any number of CDs. The rule was to multiply the number of CDs by 12.95 and add 5. We can enter this rule as a function by pressing $\boxed{Y=}$. We can enter the rule for Y_1 using x to represent the number of calculators. The variable x has a separate key, either $\boxed{X}$, $\boxed{X, T}$, or $\boxed{X, T, \theta}$. Figure AII-7(a) shows the function, and Figure AII-7(b) shows the graph obtained when $\boxed{GRAPH}$ is pressed.

Figure AII-7

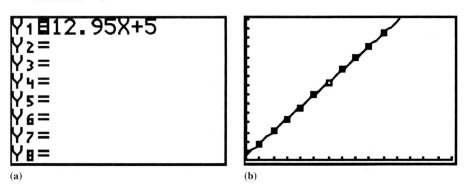

(a)   (b)

Press $\boxed{TRACE}$ and a P1 (Plot1) appears in an upper corner. The P1 indicates that the STAT PLOT (P1) is being traced. The values that appear are the same as those we generated earlier when we built the cost tables. If we press the up arrow, a 1 appears in the corner. This indicates that the line Y_1 from the $\boxed{Y=}$ menu is being traced. You can toggle back and forth between the two graphs by pressing the up arrow.

When the line Y_1 is being traced, the x-values can be nonintegers. The graph of the line Y_1 is not really an appropriate graph for this problem because the data in this case are not continuous, that is, the y-values exist only for nonnegative integer values of x. There is no such thing as a cost for 1.5 CDs as implied by the graph. However, the graph can be used to find values of y (costs) when x is a nonnegative integer.

Using Tables

We can build a table based on the algebraic rule that we developed for Molly's CD Problem. To do this, we press $\boxed{2nd}$ to choose $\boxed{TblSet}$ and set the table menu, as shown in Figure AII-8(a). The TblMin set at 1 starts the table at 1 and the $\triangle$Tbl setting of 1 sets the increment between x-values at 1. If we next press $\boxed{2nd}$ to select $\boxed{TABLE}$, we see the table of values generated by the rule $y = 12.95x + 5$. Notice that the x-values are incremented by 1 in each case because this is what we selected in the TblSet menu. Also notice that if we use the up and down arrow keys, we can continue to obtain y-values for any integer x-value, as shown in Figure AII-8(b). From the result in Figure AII-8(b), determine the cost of 15 CDs.

Figure AII-8

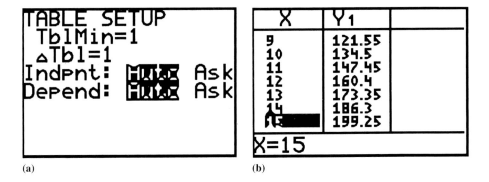

(a)   (b)

Problem Solving   **Choosing a CD Company**

Another company is advertising CDs for $7.95 per CD with no shipping charge. However, they charge a processing fee of $20 for each order. Molly is considering switching companies. Compare the prices of the two companies and discuss who has the better price.

The rule for computing the cost under the second plan is $Y_2 = 7.95x + 20$. If we enter this rule as Y_2 in the $\boxed{Y =}$ menu, as shown in Figure AII-9(a), we can compare the two plans using tables. If we keep the same table setup as in the earlier problem, then Figure AII-9(b) shows a comparison of the two rules. It follows that the first plan is better if we order either one or two CDs. If we order three CDs, it makes no difference which plan we use. If we order more than three CDs, then the second plan is better.

Figure AII-9

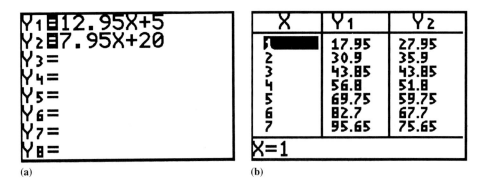

(a)   (b)

We can compare the two plans graphically by pressing $\boxed{\text{GRAPH}}$. The two graphs are shown in Figure AII-10(a). The same comments that were made about continuous data are also true in this case. If $\boxed{\text{TRACE}}$ and the arrow keys are used, then we see in Figure AII-10(b) that the two lines appear to intersect at point (3, 43.85), which is the point on which the two plans match. After $x = 3$, we see that the second line illustrates the less expensive cost.

Figure AII-10

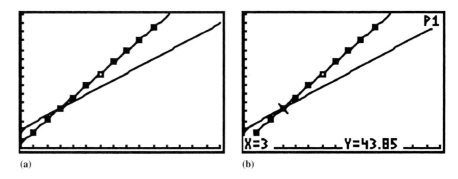

(a)   (b)

Using ZOOM

We can use various zoom features to adjust the viewing window to see more of the features of the graph. If we press ZOOM, we see the menu shown in Figure AII-11(a). The first item on the menu, ZBox, lets us use the cursor to select opposite corners of a box to define a new viewing window. To use this feature, we move the zoom cursor to any point on the screen on which we want to locate a corner of the box and then press ENTER. As we move the cursor away from the selected point, we see a small square dot indicating the selected corner. We then move the cursor to the diagonal corner of the box we want to define. As we use the arrow keys to move the cursor, we see the box change on the screen. When we get the box where we want it, as shown in Figure AII-11(b), we press ENTER to replot the graph as shown in Figure AII-11(c). The TRACE feature can now be used to find values of various points. We can cancel ZBox at any time by pressing CLEAR before pressing ENTER.

Figure AII-11

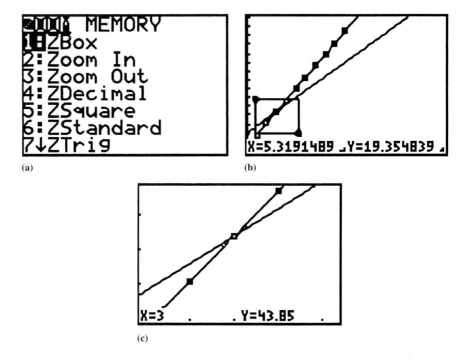

(a)   (b)

(c)

There are other zoom features in the ZOOM menu. For example, we can select Zoom In from the menu and then press ENTER. We then can move the cursor to the point on which we want the center of the new viewing window to be and press ENTER. The graph will be replotted. We can also select Zoom Out in a similar manner.

These are only a few of the features available on a graphing calculator. Many other important features are contained in menus under keys such as MATH. Consult your user's manual to see how you can use these other tools to solve mathematical problems.

ONGOING ASSESSMENT A-II

1. Use trial and error to determine when $1000 invested at 6% annual compounded interest will double in value. (*Hint:* The value after 10 yr is given by 1000 × 1.06^10.)
2. Given the sequence 3, 6, 9, 12, 15, 18, ..., use the TABLE feature to find the 32nd term by entering the appropriate equation using the Y= menu.
3. Enter $y = 3x^2 + 5x + 7$ in the Y= menu and use the TABLE feature to evaluate the function at 10 and $^-10$.
4. Set the WINDOW to graph on each axis from $^-10$ to 10 with increments of 1 unit, then graph each of the following, and answer the questions that follow:
   i. $y = x + 3$
   ii. $y = 2x + 3$
   iii. $y = 3x + 3$
   iv. $y = 4x + 3$
   a. What do the graphs have in common?
   b. How do they differ?
   c. How does changing the slope, m, change the shape of the graph in the equation $y = mx + b$?
5. Leave the WINDOW the same as it was in problem 4, graph each of the following, and answer the questions that follow:
   i. $y = x + 3$
   ii. $y = x + 4$
   iii. $y = x + 5$
   iv. $y = x + 6$
   a. What do the graphs have in common?
   b. How do they differ?
   c. How does changing the value, b, in the equation $y = mx + b$ change the shape of the graph?
6. Graph the following system. Make sure to choose an appropriate window. Use ZOOM and TRACE features to find the point of intersection to two decimal places.

$$y = 2x - 3$$
$$y = {}^-7x + 8$$

7. The distance from Missoula to Billings is 350 mi. To investigate how long it takes to drive this distance, Joan entered the following equations for Y=.

$$Y_1 = 50x, \quad Y_2 = 60x, \quad Y_3 = 70x, \quad Y_4 = 350$$

She set the window for Xmin = 0, Xmax = 10, Xscl = 1, Ymin = $^-100$, Ymax = 600, and Yscl = 20 and then graphed the functions.
   a. How could she tell the time it takes to make the trip traveling at 50, 60, and 70 mph?
   b. How much time is saved traveling at 70 mph rather than 50 mph?
8. a. Evaluate the function $y = x^2 - 5x + 4$ at each integer value between $^-10$ and 10. How many sign changes are there in this range, and where do they occur?
   b. Set the window from $^-10$ to 10 on the x-axis and $^-10$ to 10 on the y-axis with increments of 1 and graph the equation.
   c. Use the TRACE feature to examine where the function crosses the x-axis.
9. Linda has 33 coins in dimes and quarters. The value of the coins is $5.55. Determine how many of each coin she has.

Appendix III: Using a Geometry Drawing Utility

Introduction

In the time of Plato, the use of only a compass and straightedge became the norm for classical geometry. Today's world demands that we consider other tools. In this appendix, we present a series of problems appropriate for use with a computer geometry utility. Most geometry utilities allow both drawing and construction. Some educators have noted the difference between the two methods. In a drawing, an "eyeballing" approach is used to place a figure to look as it should, but the figure may not be constrained by elements of its geometric properties. For example, a segment may be drawn in a circle to look like a diameter, as in Figure AIII-1(a). However, if it is not constructed to pass through the center of the circle and one moves the circle as in Figure AIII-1(b), the segment may no longer move with the circle and may not appear to be a diameter.

The constraint of having the diameter pass through the center was not used. If the segment is constructed as a chord to contain the center as a part of the construction, then when the circle is moved, the segment moves accordingly (as a diameter should). For most purposes in this appendix, we are using constructions with all the geometric constraints applied.

In the Ongoing Assessment in this appendix, the problems are roughly arranged in the order they might be used with the geometry chapters of the book. But they may be used independently if the teacher provides the language and definitions as needed.

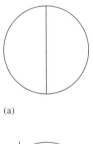
(a)

(b)

Figure AIII-1

About a Geometry Utility

Figure AIII-2 is an example of a drawing window with the Toolbox and Menu Bar shown from *The Geometer's Sketchpad (GSP)*. Other utilities have different windows, but we use *GSP* for illustrations in this appendix.

Figure AIII-2

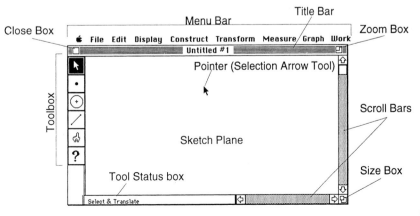

751

The Menu bar of Figure AIII-2 shows the following menu options:

- **File,** for opening, closing, saving, and printing documents
- **Edit,** for selecting objects and editing drawings and scripts
- **Display,** for changing the appearance of drawings and setting preferences
- **Construct,** for constructing figures
- **Transform,** for translating, rotating, dilating, and reflecting figures
- **Measure,** for displaying measurements and making calculations
- **Graph,** for options with coordinate axes and graphs
- **Work,** for options with open sketches and scripts

In the Toolbox, located along the left side (see Figure AIII-2), are the following tools:

▶ *Selection tool for choosing objects and translating, rotating, and dilating them.* This tool is used for clicking on an object by using the mouse and moving it about the screen by dragging the object and then clicking the mouse again when the desired location is reached.

• *Point tool, for creating points.* We click on this tool and move the mouse pointer to a desired location of a point on the screen and then click the mouse again to place a point on the screen at that location.

⊙ *Circle tool (also called the* Compass *tool) for creating circles.* This tool does exactly what the title suggests. By clicking this tool and moving the mouse pointer about the screen, we can draw circles.

/ *Segment tool, for creating segments, rays, and lines.* When you click on this tool, a small box appears at the left of the tool to allow you to choose a segment, a ray, or a line. After the choice is made, you may move the mouse pointer on the screen to decide on the placement of the desired object.

✋ *Text tool, for creating labels* and captions for drawings.

❓ *Information tool, for inspecting and altering characteristics of a selection.* This tool is used to show information about an object or group of objects in a drawing.

In this appendix, we suggest a series of investigations or problem constructions that are to be used to try out features of *GSP*. Both types of activities are labeled as investigations. Both include suggested steps for the drawing or construction using *GSP*.

Investigation AIII-1: Regions of a Circle

A problem in Chapter 1 deals with the maximum number of regions into which a circle can be separated by chords that are drawn by connecting given points on the circle. A geometry utility is a convenient tool for examining this problem because segments or chords can be moved easily without recreating the picture. Study this problem by drawing sketches with a geometry utility.

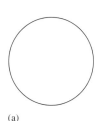
(a)

(b)

Figure AIII-3

Steps to Proceed

1. Use the *Circle* tool to draw a large circle, as in Figure AIII-3(a).
2. We know that one point does not separate the circular region into other regions. We begin recording the information as in Table AIII-1. (A table could be created on *GSP*, but for our purposes record the information in a table on paper.)

Table AIII-1

Number of Points	Number of Regions
1	1
2	2

3. Place two points on the circle as in Figure AIII-3(b) using the *Point* tool and then draw the segment connecting the points. Record the information in the table.
4. Continue the process by adding more points on the circle, determining the number of regions, and recording the maximum number of regions of the circle determined by connecting the points.
5. Predict how many regions are determined with *n* points on the circle.
6. Does your record help to verify the formula

$$\frac{n(n-1)(n-2)(n-3)}{24} + \frac{n(n-1)}{2} + 1$$

for the maximum number of regions? If not, reexamine your drawings.

Extension: Duplicate Investigation AIII-1 for any simple closed figure. Are your results the same?

Investigation AIII-2: Inscribing a Circle in a Triangle

This investigation is built in stages, with the construction of a bisector of an angle of a triangle first.

Steps to Proceed

1. Construct any triangle by following these steps:
   a. Choose the *Point* tool.
   b. Move the mouse pointer to the screen and click anywhere. This chooses the first vertex.
   c. The *Point* tool is still active, so move to another point on the screen and click to choose the second vertex.
   d. Choose the *Segment* tool. Move the mouse until the crosshair pointer is centered on one of the points. Press the mouse button down and hold it down. Drag the mouse pointer to the second point until the crosshair pointer is centered over the second point. Release the mouse pointer, and the segment is drawn.
   e. Repeat the process to choose the third vertex and to complete the triangle.
2. Construct a bisector of one angle of the triangle. First, click on the **Construct** menu as shown in Figure AIII-4. Next pull down the menu and choose *Angle Bisector*.

Figure AIII-4

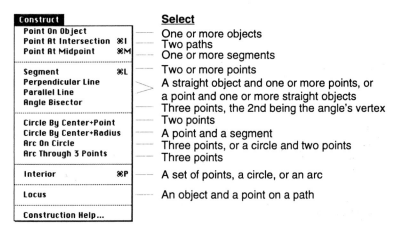

To use this construction tool, we must first highlight three points, which determine the angle with the vertex as the second point chosen. To highlight the three points, click on the first point and hold down ⎡Shift⎤ as you click on each of the other points in order. (This process is used when naming a series of objects needed for a construction.) With the three points highlighted, you can then use the *Angle Bisector* construction.

3. Choose any point on the angle bisector with the **Construct** menu item *Point on Object*. Construct perpendicular segments from the point to the sides of the angle using the *Perpendicular Line* selection on the **Construct** menu.
4. To identify the perpendicular segments constructed, we use the **Construct** menu item *Point at Intersection* to find the points on which the perpendiculars intersect the sides of the angle.
5. Measure the lengths of the perpendicular segments using the **Measure** menu shown in Figure AIII-5.

Figure AIII-5

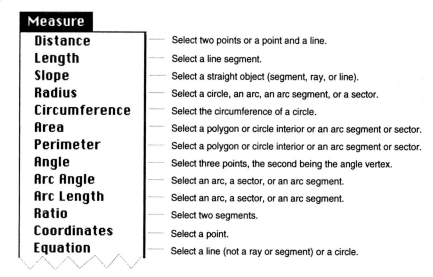

6. Move the point along the angle bisector by clicking the mouse pointer on it and dragging it along the bisector. As the point is dragged, consider the measures shown.
7. Make a conjecture about any observed results.

incircle of the triangle

8. Construct the **incircle of the triangle.** The incircle is the circle that is inside the triangular region and is tangent to each of the sides of the triangle. The point at which the three angle bisectors intersect is the center of the incircle. The radius is the length of the perpendicular segment from the center to one of the sides.

Investigation AIII-3: Circumscribing a Circle about a Triangle

This investigation is built in stages, beginning with the construction of a perpendicular bisector of a side of a triangle.

Steps to Proceed

1. Construct any triangle using the process in Investigation AIII-2.
2. Construct the midpoint of the sides of the triangles and then construct the perpendicular bisector of one side of the triangle using the **Construct** menu item *Perpendicular Line*.

3. Choose any point on the perpendicular bisector using the **Construct** menu item *Point of Object* and draw segments connecting the point to the endpoints of the original side of the triangle.
4. Measure the lengths of the segments drawn in (3) using the **Measure** menu.
5. Move the point along the perpendicular bisector (in the same manner we moved a point along the angle bisector in Investigation AIII-2), observing the measurements of the segments as in (4).
6. Make a conjecture about any observed results.
7. Construct the **circumscribed circle** (a circle containing each of the vertices of the triangle) about the triangle by constructing all the perpendicular bisectors of the sides. Then find the point on which they intersect—the **circumcenter of the circle**—and use the distance from the circumcenter to one of the vertices as the radius of the circle.

Extension: Suppose the triangle being circumscribed is a right triangle. Where is the center of the circumscribed circle located?

Investigation AIII-4: The Pythagorean Theorem

The Pythagorean Theorem is a classic theorem of Euclidean geometry. It is probably the most useful of all geometry theorems.

Steps to Proceed

1. Construct any right triangle *ABC* with the right angle at vertex *C*. To do this, we may construct perpendicular segments using the **Construct** menu.
2. Construct a square on each side of the triangle, again using items from the **Construct** menu. (We may use a combination of constructions using the items *Perpendicular Line* and *Parallel Line*.)
3. Find the areas of the squares using the **Measure** menu. To do this, we need to identify a polygon by clicking on the vertices in order while holding down Shift . Then use the **Construct** menu item *Polygon Interior*. With the polygon interior defined, the area can be found using the **Measure** menu.
4. Find the sum of the areas of the smaller squares and compare this sum to the area of the largest square. To compute the sums, we may want to use the *Calculate* feature under the **Measure** menu. To use this feature, highlight the measurement items we want to use in the calculations. Then when the *Calculate* feature is chosen, we may use any highlighted items and the calculator shown on the screen.
5. Change the sizes of the right triangle by moving around any vertex.
6. Make a conjecture about the results. (Note that the use of a geometry utility allows us to make conjectures. Verifying a conjecture with a drawing or a series of drawings does not constitute a proof of the conjecture.)

Extension: Construct any similar figures, for example, equilateral triangles or semicircles, on the sides of the original triangle. Use an entire side as a side of each similar figure. Perform steps similar to (3), (4), (5), and (6). Again make a conjecture based on the results.

Investigation AIII-5: Angles in Circles

Some angles with vertices on a circle may be related to angles with vertices at the circle's center. This investigation explores such a relationship.

Steps to Proceed

1. Construct any circle, marking its center by using the **Construct** menu item *Circle by Center + Point.*
2. Construct a central angle (one whose vertex is at the center of the circle).
3. Measure the angle and its intercepted arc by using the **Measure** menu item *Arc Length.*
4. Draw any inscribed angle whose vertex is a point of the circle and whose sides intersect the circle in the same points as the central angle.
5. Measure the inscribed angle by using the **Measure** menu item *Arc Angle.*
6. Move the vertex of the inscribed angle and make a conjecture about the relationship between the measure of an inscribed angle and the central angle that intercepts the same arc.

Extension: If the points of intersection of the sides of the inscribed angle intersect the circle at the ends of a diameter, what is the measure of the inscribed angle?

Investigation AIII-6: Finding the Area of a Figure

In Chapter 11, areas are estimated using the grid method. Another technique for doing this involves both similar figures and the use of a geometry utility.

Steps to Proceed

1. Draw any two-dimensional shape on the screen for which the area is desired.
2. Click on a point on the boundary of the shape. Then hold down $\boxed{\text{Shift}}$ to identify points in a sequence and continue to trace around the shape, outlining it as closely as possible with a polygon by clicking on successive points and then returning to the starting point. Use the **Construct** menu to construct the *Polygon Interior.*
3. Find the area of the polygon using the **Measure** menu item *Area.* This process allows the approximation of the area of the shape.

Extension: Draw on transparent paper any shape smaller than your computer screen. Tape the drawing to the screen. Draw a polygon on the screen that approximates the shape by clicking on successive points around the shape while holding down $\boxed{\text{Shift}}$ and finally returning to the point on which you started. Then measure as before.

Consider how to use a copy machine, a transparency, the process described in the extension and properties of similar figures to determine the area of any size shape.

Investigation AIII-7: Ratios and Slopes

This investigation explores the relationship between slopes and the tangent ratio.

Steps to Proceed

1. Use the **Graph** menu items *Create Axes* and *Show Grid* as seen in Figure AIII-6 to construct a coordinate system.
2. On the coordinate grid, construct any nonvertical line through the origin O and passing through the first and third quadrants.
3. Choose any point P not on the axes and construct a perpendicular segment $\overline{PX}$ to point X on the *x*-axis. (Follow the construction procedure outlined in Investigation AIII-3.)
4. Measure $\overline{OX}, \overline{PX},$ and $\overline{OP}.$
5. Record the following ratios in a chart: $\dfrac{OX}{OP}, \dfrac{PX}{OP}, \dfrac{OP}{OX}.$

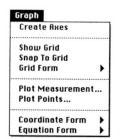

Figure AIII-6

6. Move point *P* up and down the line. What happens to the ratios?
7. Describe the ratios in terms of slope.

Extension: Drag the original line about the origin while continuing to find the ratios. What are the maximum and minimum values, if any, for the ratios?

Investigation AIII-8: Symmetry

The *GSP* can be used to design a logo for a company or for fun.

Steps to Proceed

1. Use the *Segment* tool to draw a line on the computer screen.
2. Use any *GSP* construction tools to design half of a logo on one side of the line.
3. Use the **Transform** menu item *Reflect,* shown in Figure AIII-7, to mark a mirror (reflecting line).
4. Select the object (the half of the design created) to be reflected.
5. Select the *Reflect* item from the **Transform** menu again, and *GSP* will construct the reflected image to complete the design.

Extension: Try other features of the **Transform** menu to design a logo with three-turn or rotational symmetries. Construct one-third of the design on the drawing utility and use the *Rotate* transformation to create the rest of the design.

Figure AIII-7

ONGOING ASSESSMENT A-III

The following items are suggested for further investigation using a geometry drawing utility. Note that some parts of the assessment items may be directions for completing the investigations and not separate problems.

1. Study the relationship between the areas of a rectangle and a parallelogram, both with the same base and height.
2. Study the relationship between the areas of a triangle and a parallelogram, both with the same base and height.
3. Draw any two similar figures. Find the ratio of the corresponding sides and the ratios of the corresponding areas of the figures. Make a conjecture. Check the conjectures for several similar figures.
4. a. Construct any regular polygon inscribed in a circle.
   b. Connect vertices to the center.
   c. Find the height of one of the triangles formed by a side and the center of the circle.
   d. Find the area of each triangle formed.
   e. Find the area of the regular polygon.
   f. Find the perimeter of the regular polygon.
   g. What is the ratio of the area in (e) to the height in (c)? (To find this, you may want to investigate the *Calculate* item on the **Measurement** menu.)
5. a. Draw any circle.
   b. Inscribe a regular polygon in the circle.
   c. Find the measure of the longest diagonal of the polygon and the perimeter of the polygon. Find the ratio of the perimeter to the length of the diagonal.
   d. Repeat the process using a polygon with twice as many sides.
   e. Find the ratio again.
   f. Repeat the process in (d) two more times. What do you expect the ratio to be? Why?
6. a. Inscribe a circle in a square. Find the percentage of the area in the square not covered by the circular region.
   b. Inscribe a square in a circle. Find the percentage of the area of the circle not covered by the square region.
   c. Use the answers from (a) and (b) to determine whether a square peg might fit better into a round hole or a round peg might fit better in a square hole.

7. a. Draw a rectangle with two circles inside, pictured as follows:

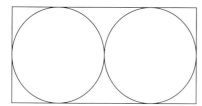

Find the percentage of the area of the rectangle not covered by the circular regions.
   b. Repeat the process in (a) using four circles.
   c. Make a conjecture about the percentage of the area of any rectangle not covered by circular regions when constructed in this manner.
8. a. Draw any shape with the utility.
   b. Draw any line and move the line through the shape.
   c. Stop moving the line when you estimate the shape is divided into two equal areas.
   d. Find the areas of the shapes to check the estimate.
   e. Repeat (a) through (d) to practice your area estimation skills.
9. a. Draw any rectangle and a segment, shown as follows:

   b. Move the segment across the rectangle and estimate when 60% of the area is to the right of the segment.
   c. Measure the areas to check.
   d. Repeat steps (a) through (c) to practice area estimation skills with other percents.

Appendix IV: Using a Spreadsheet

Introduction

spreadsheet

An electronic **spreadsheet** is a table of rows and columns in which each cell may contain a value that can be operated on and changed at any time. A spreadsheet has the capacity to allow a change in one cell to be reflected in any other cell that relies on the data in that original cell.

Table AIV-1 depicts certain values for the function $f(x) = x^2$. In Table AIV-1, each item in the $f(x)$ column depends on the value in the x column. A spreadsheet, as it might appear on a computer, is shown in Table AIV-2, depicting the information in Table AIV-1. (The labels x and $f(x) = x\text{\textasciicircum}2$ are for headings only.)

Table AIV-1

x	$f(x)$
1	1
2	4
3	9
4	16
5	25

Table AIV-2

	A	B	C	D	E	F
1	x	f(x) = x^2				
2	1	1				
3	2	4				
4	3	9				
5	4	16				
6	5	25				

In Table AIV-2, an index column down the left-hand side numbers the rows of the table. Across the top, an index row of letters identifies the columns. In the body of the table, the value 3 is in column A and row 4. As a result, the value 3 could be identified by label A4, where A is the column heading and 4 is the row number. The labels can be used to write formulas for the values of column B in terms of column A. For example, the value in B2 could be written as A2 times A2, or in computer language as A2^2, depicting $(A_2)^2$. Similarly, the value in B5 could be written as A5^2. In Table AIV-2, the items in row 1 are used as headings to indicate what the table is depicting.

The user bars on an opening of the *Microsoft Excel* spreadsheet are shown in Figure AIV-1.

Figure AIV-1

In the figure, three bars are depicted: a **Menu** bar, a **Standard** toolbar, and a **Formula** bar. The **Menu** bar offers such options as **File, Edit, Format,** and so on. For example, as a beginning user you might choose to click on **File** and choose *New* to obtain a new worksheet on the screen. The standard Toolbar shows different options available to the user, including file folder, disk, and printing options on the left along with other types of options, such as **B** for bold type and the *I* for italics. The **Formula** bar is for entry of items in a worksheet and is discussed later in this appendix.

Developing a Spreadsheet

To create the spreadsheet in Table AIV-2, we first open a new worksheet and then type the entries in cells A1 and B1 by typing x and $f(x) = x^2$, respectively, in the entry line (also known as the **Formula** bar) and pressing RETURN after each. These are simply headings for our reference.

1. To create the column of values listed under x in the A column, we enter the first item as 1 in A2 and press RETURN. Then we use a formula to create the value for A3 by assigning the values for A3 using a formula. We tell the spreadsheet that a formula is being used by highlighting the cell to be filled and typing = in the entry line followed by the formula we wish to use. In this case, we want A3 to have the value A2 + 1. We signal this by highlighting cell A3, typing = A2 + 1 and pressing RETURN. That done, the value 2 appears in A3, as shown in Table AIV-3. (Near the **Formula** bar are two boxes, one containing an X and one containing a tic mark. The X is used to cancel any changes you have made in a formula. The tic is used in the same way as the RETURN key.)

Table AIV-3

	A	B	C	D	E	F
1	x	f(x) = x^2				
2	1					
3	2					

2. To complete the column under A, we use the *Fill Down* command by first highlighting A3 and all the cells to be filled and then using the *Fill Down* command in the **Menu** bar normally found under the **Edit** menu. The *Fill Down* command fills successive cells in column A by adapting the created formula to accommodate the cell number. For example, the entry in A4 is automatically created as the value of A3 + 1, and so on. The *Fill Down* command will fill all highlighted cells. If the cells A3 through A6 are highlighted, the result is seen in Table AIV-4. (If there is no data in a cell to create other cells, the program assumes that the value in the cell is 0.)

Table AIV-4

	A	B	C	D	E	F
1	x	f(x) = x^2				
2	1					
3	2					
4	3					
5	4					
6	5					

3. To complete the column under the heading B and $f(x) = x^{\wedge}2$, we highlight cell B2, type = A2^2, and press $\boxed{\text{RETURN}}$. This should cause a 1 to be placed in cell B2. We then highlight cell B2 and the rest of the column through row 6 and use the *Fill Down* command to complete column B.

REMARK Most spreadsheets also have a *Fill Right* command to fill in cells in rows as well as columns.

To clear a table or a set of values in *Excel,* highlight the desired values to be deleted and pull down the **Edit** menu. If we choose *Clear,* then we are asked about clearing *All, Formats, Formulas,* or *Notes.* To clear all highlighted values, choose *All* and press $\boxed{\text{RETURN}}$.

Most spreadsheets allow the use of various functions, including $+, -, \div, \cdot, \wedge$, trigonometric functions, square roots, and absolute values. To determine what your spreadsheet can do, consult the software manual. (Note that many graphing calculators have the capability of acting like a spreadsheet.)

Graphing with a Spreadsheet

In addition to allowing the use of arithmetic operations, most spreadsheets can create graphs of the data presented in a table. For example, to create a graph of the data in Table AIV-2, we follow these steps:

1. Highlight the information to be graphed. In the graph in Figure AIV-2, the data from rows A2 through A6 and columns B2 through B6 were highlighted. After the data is highlighted, then we choose the *Graph* icon from the **Toolbar.** (On some spreadsheets, when you choose the *Graph* icon, you can size the graph as you want; other spreadsheets will do this automatically.) In *Excel,* clicking on the icon once causes the highlighted information to be placed in a flashing dashed rectangle. If we move the mouse pointer on the worksheet, we see a + symbol. Click the mouse, and drag it while holding down the mouse button to size the graph. Once the graph is sized, most spreadsheets ask for a variety of information, such as what you want for the *x*-values, what you want for the *y*-values, what legends you want, and what maximum and minimum values you want. Because each spreadsheet is different, we suggest that you consult your user's manual to see what features are available. For example, in *Excel* the first question is about the range of data. With the data we are graphing, the range is written as \$A\$2:\$B\$6 automatically. We can change or accept this range. The \$A\$2 indicates that the first values from our table to be used in the graph start at A2 and this will not change in the graph; \$B\$6 tells the computer that cell B6 contains the last value of the graph.

REMARK The $ is used before and after a variable to indicate that the variable is fixed and does not change in this application. This feature is useful particularly when we want to use a specific value of a variable over and over in an application. An example is seen in Ongoing Assessment problem 4, where we use the calculated mean of a set of data over and over to find the standard deviation of the set of data.

2. Because the range from A2 to B6 contains the values we intend to graph, we move to the next screen, where we must select a chart type, usually from several options. *Excel* presents these options in icon form. Figure AIV-2 was created as a scatterplot. (See Chapter 8.) Once we choose the scatterplot style, by double-clicking on an icon, the next screen asks about the type of scatterplot desired. Again, in *Excel,* the choices are presented as icons. We chose the connected scatterplot (option 2), after which a graph is drawn. But there are other options to consider. To use the data as ordered pairs with the *x* data as the first column, we are using the data series in columns. Finally, we are asked if we want to use the first row as a legend or as data. Because we started with A2, we want this as the first value of *x* data. Next, we are asked if we want to add a legend to the graph. Figure AIV-2 shows the data from Table AIV-2 as a connected scatterplot. Using a title of Chart 1, we see a finished product with the $f(x)$ values and *x*-values labeled along the axes.

Figure AIV-2

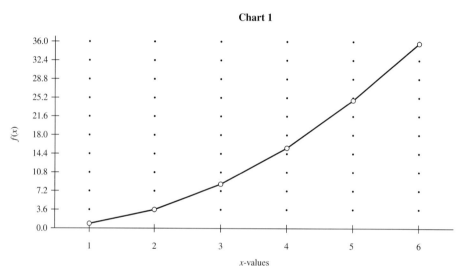

The same information could be depicted as a bar graph, as shown in Figure AIV-3, by using other graph icons.

Figure AIV-3

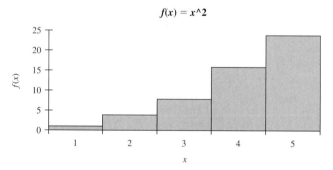

Many spreadsheets allow the options of using scatterplots, line plots, bar graphs, stem and leaf plots, and three-dimensional plots. Again, check the user's manual to see what is available for your spreadsheet.

Explicit and Recursive Formulas

When using a spreadsheet as in the previous paragraphs, most users discover they need to know the difference between explicit and recursive formulas. In Table AIV-4, a recursive formula was used—a value in a cell was determined by the value in the cell immediately above it. For example, the value in cell A5 is 1 plus the value in cell A4. Whenever a value in any cell is expressed in terms of a value in a previous cell (or cells), we say that the expression is a *recursive expression* or *formula*. Recursive formulas are typically used by most young students to describe patterns. For example, most students would describe the following pattern as adding 3, meaning that to find the next item, a person would add 3 to the previous term:

$$3, 6, 9, 12, 15, 18, 21, 24, \ldots$$

On a spreadsheet, the pattern could be depicted using two columns. Row A may serve to count the numbers of the terms, as seen in Table AIV-5. (Note that column A serves the same purpose as the index column, but we can use the numbers in the A cells to create the terms in the B column.)

Table AIV-5

	A	B
1	1	3
2	2	6
3	3	9
4	4	12
5	5	
6	6	

To fill in additional rows, we can highlight cell B5 and type = B4 + 3 in the entry line and then use the *Fill Down* command by highlighting the number of entries needed. Clearly, the value of a cell is determined by the cell immediately above it.

If we decided to write a formula to describe a cell value using number of the term (as in Chapter 1), then we would need to look at the number of the term and the values in column B to see how they are related. In this example, the values in column B are multiples of 3. If n is the term number, then the corresponding value is $3n$. The expression $3n$ is an *explicit formula* for finding the value of an expression as a function of the number of the term n.

To use the explicit formula $3n$ to create the entries in column B, we use the formula = 3*A:A to fill in the entry in B1 and then use the *Fill Down* command to fill all highlighted cells of column B where there is a corresponding A entry. The use of :A at the end of the formula tells the spreadsheet to use column A to fill column B entries.

Other Features of a Spreadsheet

Spreadsheets can act as simple calculators. For example, determine the number of handshakes at a party that 25 people attended. Each person shook hands with each other person, and no one shook hands with himself or herself. This problem can be solved using a spreadsheet. We consider a *simpler case* whereby there is first only one person in the room, in which case there are no handshakes. Then another person enters for a total of two persons, and so one handshake occurs. When a third person enters the room, two more handshakes take place, and so on. This could be recorded as in Table AIV-6, in which column A records the number of people in the room and is developed by using the *Fill Down* command and the formula = A2 + 1. Column B depicts the number of additional handshakes when another person enters the room. Column B is 1 less than the number in column A and can be set up using the formula = A2 − 1.

Table AIV-6

	A	B
1	No. people	No. of additional handshakes
2	1	0
3	2	1
4	3	2
5	4	3
6		
...	...	...
26	25	24
27		300

To find the total number of handshakes, we need to find the total number of handshakes in column B. The spreadsheet does this when we highlight the cell in which we want the total, in this case B27, and then click on the Σ button. (On some spreadsheets, you may need to use the *SUM* feature.) When we click on Σ, the spreadsheet surrounds the numbers immediately above the highlighted cell with a flashing rectangle to show what will be added. (In some cases, we may have to click and drag a rectangle around the numbers that are to be added. This is true especially if there is an empty cell among the other cells to be added.) When the RETURN or ENTER key is pressed, the total appears in the highlighted cell. As Table AIV-6 shows, the total is 300 handshakes for the 25 people.

Many other features are available on most spreadsheets but are not discussed here. We suggest that you consult the user's manual that comes with your spreadsheet program and try the following problems.

ONGOING ASSESSMENT A-IV

1. **a.** Write any arithmetic sequence.
   b. In column A of your spreadsheet, list the number of the term of the arithmetic sequence.
   c. What formula did you use to fill down the column?
   d. List the first 25 terms of the arithmetic sequence in column B.
   e. What formula could be used to fill in the terms in column B?

f. Find the sum of the first 25 terms in the arithmetic sequence. Describe how that was done with a spreadsheet.
g. Plot the number of the terms of the arithmetic sequence versus the actual terms using the graphing or chart option of your spreadsheet. Describe the graph of the arithmetic sequence.

2. a. Write any geometric sequence.
b. In column A of your spreadsheet, list the number of the term of the geometric sequence.
c. List the first 25 terms of the geometric sequence in column B.
d. What formula could be used to fill in the terms of column B?
e. Find the sum of the first 25 terms of the geometric sequence. Describe how that was done by using a spreadsheet.
f. Plot the number of the terms of the geometric sequence versus the actual terms using the graphing or chart option of your spreadsheet. Describe the graph of the geometric sequence.

3. Given the following set of data, use a spreadsheet to find the arithmetic mean:

23, 45, 67, 78, 98, 54, 36, 76, 75, 24, 43, 54, 100, 99

4. Use the data in problem 3 to develop a spreadsheet to find the standard deviation of the data. Use the columns of the spreadsheet to represent the number of the term, the term, and the difference of the mean and the term. Then find the square of each of the differences, the sum of those squares, and the quotient of the sum and n, where n is the number of terms. Finally, find the square root of the quotient.

5. Businesspeople use spreadsheets for the calculation of interest on loans or outstanding bills. Consider a debt of $1000 with payments of $40 per month, which includes 1.5% interest per month on the unpaid balance. Develop a spreadsheet that shows the number of the month the payment was made, the amount of payment in each month, and the outstanding balance. If no other debts accrue, how many months will it take to pay off the debt?

6. Use a spreadsheet to show the first 100 multiples of 13. Explain all steps in developing this spreadsheet.

7. Develop a spreadsheet for finding your college grade-point average. Explain all steps used in developing the spreadsheet.

8. The sequence of Fibonacci numbers is 1, 1, 2, 3, 5, 8, ... , where each successive term after the first two is the sum of the two preceding terms.
a. Develop a spreadsheet to find the first 25 Fibonacci numbers.
b. Extend your spreadsheet to find the square of each term of the Fibonacci sequence and the sum of those squares. Make a conjecture about the sum of the squares of the first n terms of the sequence.
c. To examine the updating feature of your spreadsheet, change the first two terms of the Fibonacci sequence and observe how each cell that was written based on these terms is changed. Does the conjecture in (b) still hold?

9. Develop a spreadsheet for finding $n!$. Explain the steps used to develop the spreadsheet.

Answers to Selected Problems

CHAPTER 1

Ongoing Assessment 1-1

1. (a) [rectangle figure]
   (c) [triangle figure]
   (e) [cube stack figure]

2. (a) 11, 13, 15 arithmetic   (c) 96, 192, 384 geometric
   (e) 33, 37, 41 arithmetic
4. 2, 7, 12
5. (a) Answers vary, for example, one possibility is to notice that the sum of the first n odd numbers is given by n^2; that is, $1 + 3 + 5 + 7 = 4^2 = 16$.
7. (b) 10,100
8. (a) 51
9. (a) 41
10. (a) 10,000
11. (a) 42
13. 15 liters
14. 23rd year
16. (a) 299, 447, 644   (b) 56, 72, 90
17. (a) 101   (d) 87
18. (a) 3, 6, 11, 18, 27   (c) 9, 99, 999, 9999, 99999
19. (a) 1, 1, 2, 3, 5, 8, 13, 21, 34, 55, 89, 144   (c) 143
21. The sequence in (b) becomes greater than the sequence in (a) on the 12th term.
23. Two solutions are possible: 64, 128, 256, or ⁻64, 128, ⁻256.

Communication

25. (a) You must produce at least one rectangle that does not have diagonals that are perpendicular. Any nonsquare rectangle will do.   (b) You must produce two even numbers whose sum is not divisible by 4. For example $2 + 4 = 6$, and 6 is not divisible by 4.

28. (a) Yes. The difference between terms in the new sequence is the same as in the old sequence because a fixed number was added to each number in the sequence.

Open-Ended

29. Answers vary, for example, two more patterns follow.

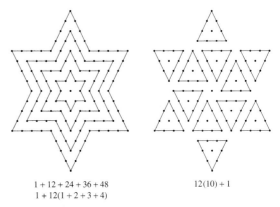

$1 + 12 + 24 + 36 + 48$
$1 + 12(1 + 2 + 3 + 4)$

$12(10) + 1$

33. (a) 81   (b) 40

Ongoing Assessment 1-2

1. (a) 4950   (c) 251,001
2. (b) 5
3. 12
4. 12
6. Dandy, Cory, Alababa, Bubba
8. 18
10. 275,000,000
11. (a) (i) 541×72 (ii) divide 754 by 12
12. 17 terms
14. (a) $2^6 - 1$
16. $2.45
18. (a) 10,500 squares   (b) $n^2 + 5n$ squares
20. width = 230 ft; length = 310 ft
22. (b) 20,503
23. (a) 204 squares
24. (a) Answers vary, for example, we could weigh 4 marbles against 4 marbles and then pick the heavier side. We could then weigh 2 marbles against 2 marbles and pick the heavier side. Finally we could weigh 1 marble against 1 marble and the heavier one is the one we are looking for.

26. (a) If both numbers were less than or equal to 9, then their product would be less than or equal to $9 \cdot 9 = 81$, which is not greater than 82.

Communication

27. Answers vary, but the discussion should point out that the two methods give the same answer and that the advantage of this method is that it works with an even or odd number of terms.

Cooperative Learning

32. (a) Play second and make sure that the sum showing when you hand the calculator to your opponent is a multiple of 3. **(c)** Play first and press 3. After that, make sure that each time you hand the calculator to your opponent it displays 3 more than a multiple of 10. **(e)** Play second. Make sure the calculator displays a multiple of 4 each time you hand it to your opponent.

Review Problems

33. (a) 21, 24, 27   **(b)** 243, 2, 729
34. $22 + (n - 1)10$, or $10n + 12$
35. 21 terms
36. 903
37. (a) The digits in the product always sum to 9. The tens digit in the answer is always one less than the number that is multiplied by 9. This pattern works with the other exercises presented.   **(b)** The pattern can be used to check if you remembered the product of a digit and 9 correctly.

Ongoing Assessment 1-3

1. (b) $20 + 25h$ **(c)** $175d$ **(d)** $3x + 3$
2. (a) $P = 0.04E$ **(c)** 5.4 lb
3. $g = 0.607165d$ or approximately $0.61d$
4. (a) 220
6. (b) $C = K - 273.15$
7. (c) $P = 20 + 10t$ **(e)** $C = 30 + 0.35m$
8. (a) $x = 18$
10. Rick has $100 and David has $300.
12. Eldest, $30,000; middle, $24,000; youngest, $10,000

Communication

13. Both are correct. For the first student, x is the first of the three consecutive integers. The second chose x to be the second of the three consecutive integers.
15. (a) $5x$ **(b)** The sum of any five consecutive natural numbers is 5 times the middle number.

Review Problems

18. (a) 35, 42, 49   **(b)** 1, 16, 1
19. $20n - 8$
20. 21 terms
21. 9 ways

Ongoing Assessment 1-4

1. (a) False statement   **(e)** Not a statement   **(h)** Not a statement
2. (a) There exists a natural number x such that $x + 8 = 11$.
(b) For all x, $x + 0 = x$.
4. (a) This book does not have 500 pages.   **(d)** No people have blond hair.   **(g)** Not all squares are rectangles, or some squares are not rectangles.
5. (a)

p	$\sim p$	$\sim(\sim)p$
T	F	T
F	T	F

(c) Yes
6. (a) $q \wedge r$ **(c)** $\sim(q \wedge r)$
7. (a) False   **(c)** True   **(e)** False   **(g)** False
(i) False
9. (a) No   **(c)** No
11. (a) Either today is not Wednesday, or the month is not June.   **(c)** It is not true that it is both raining and the month is July.
12. (a) $p \to q$ **(c)** $p \to \sim q$ **(e)** $\sim q \to \sim p$
13. (a) Converse: If $2x = 10$, then $x = 5$. Inverse: If $x \neq 5$, then $2x \neq 10$. Contrapositive: If $2x \neq 10$, then $x \neq 5$.
(c) Converse: If you have cavities, then you do not use Ultra Brush toothpaste. Inverse: If you use Ultra Brush toothpaste, then you do not have cavities. Contrapositive: If you do not have cavities, then you use Ultra Brush toothpaste.
15. (a) No   **(c)** No
17. (a) Valid   **(c)** Valid
18. (a) Helen is poor.   **(c)** If I study for the final, then I will look for a teaching job.
19. (a) If a figure is a square, then it is a rectangle.   **(c)** If a figure has exactly three sides, then it may be a triangle.
20. $\sim(p \vee q)$ is equivalent to $\sim p \wedge \sim q$, and $\sim(p \wedge q)$ is equivalent to $\sim p \vee \sim q$.

Communication

23. (a) Therefore we go shopping. (Going shopping is equivalent to getting a bonus.)   **(c)** Let p be the statement "It is sunny," q the statement "We go hiking," and r the statement "It is freezing." Symbolically, $p \to q$ and $r \to \sim q$. The second statement gives $q \to \sim r$. So the chain rule tells us that $p \to \sim r$; that is, if it is sunny then it is not freezing.

Chapter Review

1. (a) 15, 21, 28   **(c)** 400, 200, 100   **(e)** 17, 20, 23
(g) 16, 20, 24
3. (a) $3n + 2$ **(c)** 3^n
4. (a) 5, 8, 11, 14, 17   **(c)** 3, 7, 11, 15, 19
5. (a) 10,100
6. (a) 123456, 1234567, 12345678, …
8. 26

10. 21 posts
12. $19{,}305 = 3 \cdot 5 \cdot 9 \cdot 11 \cdot 13$.
14. 20 students
16. 45 triangles
18. Width is 10 ft, length is 24 ft.
20. 4 questions
22. **(a)** 4 **(c)** Impossible
23. **(a)** Yes **(c)** No
24. **(a)** No women smoke. **(c)** Some heavy-metal rock is not loud, or not all heavy-metal rock is loud.
26. **(a)** Joe Czernyu loves Mom and apple pie. **(c)** Albertina passed Math 100.
27. Let the following letters represent the given sentences.
   p: You are fair-skinned.
   q: You will sunburn.
   r: You do not go to the dance.
   s: Your parents want to know why you didn't go to the dance.
Symbolically, $p \to q, q \to r, r \to s$. The argument given is valid: $\sim s \to \sim r, \sim r \to \sim q, \sim q \to \sim p$. By the chain rule, $\sim s \to \sim p$; that is, if your parents do not want to know why you didn't go to the dance, then you are not fair-skinned.

CHAPTER 2

Ongoing Assessment 2-1

1. **(a)** $\{m, a, t, h, e, i, c, s\}$ **(c)** $\{x \mid x \text{ is a natural number, and } x > 20\}$, or $\{21, 21, 23, \ldots\}$
3. **(d)** No
4. **(b)** 720
5. **(a)** 24
7. **(e)** 3
9. **(a)** 7
11. **(c)** $\notin$ **(e)** $\notin$
13. **(a)** Yes. **(d)** No. Consider $A = \{1\}$ and $B = \{1, 2\}$.
14. **(a)** Let $A = \{a, b\}$ and $B = \{a, b, c, d\}$. Then $n(A) = 2$ and $n(B) = 4$. Since $A \subset B$, $n(A)$ is less than $n(B)$, i.e., 2 is less than 4.

Communication

16. **(a)** This set is not well defined, since we do not know what is meant by "wealthy." **(b)** This set is not well defined, since we do not know what is meant by "great." **(c)** This set is well defined. Given a number, we can easily tell if it is a natural number greater than 100. **(d)** This set is well defined. We could just list them all. **(e)** This set is well defined; it is the empty set.
19. **(b)** $\overline{B} \subset \overline{A}$.

Open-Ended

21. **(a)** Let A be the set of all natural numbers not equal to 1. Then $\overline{A} = \{1\}$ is finite. **(b)** Let A be the set of even natural numbers. Its complement, $\overline{A}$, is the set of odd natural numbers and so is infinite. (Here U is the set of natural numbers.)

Cooperative Learning

23. **(a)** There are $2^{64} \doteq 1.84 \times 10^{19}$ subsets of $\{1, 2, 3, \ldots, 64\}$. If a computer can list one every millionth of a second, then it would take $1.84 \times 10^{19} \times 0.000001 \text{ sec} \times \dfrac{1 \text{ yr}}{31{,}536{,}000 \text{ sec}} \doteq 580{,}000 \text{ yr}$ to list all the subsets.

Ongoing Assessment 2-2

1. **(a)** Yes **(d)** Yes **(e)** Yes
2. **(b)** False. Let $A = \{a, b, c\}$ and $B = \{a, b\}$. Then $A - B = \{c\}$, but $B - A = \emptyset$. **(e)** True. **(f)** False. Let $A = \{1, 2, 3\}, B = \{3, 4, 5\}$, and $U = \{1, 2, 3, \ldots, 10\}$. Then $\{A \cup B\} - A = \{1, 2, 3, 4, 5\} - \{1, 2, 3\} = \{4, 5\} \neq B$.
3. **(a)** $A \cap B = B$.
4. **(a)** 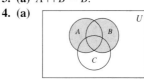 **(b)**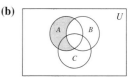

   (h)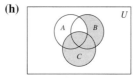

5. **(a)** $S \cup \overline{S} = U$.
6. **(a)** $A - B = A$. **(b)** $A - B = \emptyset$.
8. **(c)** $(A \cap B) - C$ **(d)** $A \cap C$
9. **(a)** 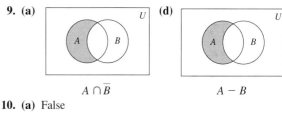 **(d)**

   $A \cap \overline{B}$ $A - B$

10. **(a)** False

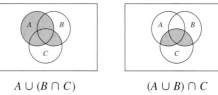

$A \cup (B \cap C)$ $(A \cup B) \cap C$

11. (a) $A \cap B \cap C \subseteq A \cap B$ (c) $(A \cup B) \cap C \subseteq A \cup B$
12. (a) (i) 5; (ii) 2; (iii) 2; (iv) 3
13. (a) Greatest is 15, least is 6.
16. (a) The set of all Paxson 8th graders who are members of the band but not the choir (b) The set of all Paxson 8th graders who are members of both the band and the choir (c) The set of all Paxson 8th graders who are members of the choir but not the band (d) The set of all Paxson 8th graders who are members of neither the band nor the choir
18. 4
20. (c) 10
21. The following Venn diagram indicates that only 490 cardholders are accounted for:

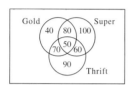

So either there is some other type of credit card the remaining 10 people could have, or else the editor was right.
22. 3. Using the following Venn diagram and the fact that the set of people who are O-negative is $100 - n(A \cup B \cup C)$, we see that the answer is 3.

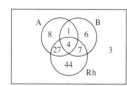

23. (b) False. Let $A = B$ (c) False. Let $A = \{1, 2, 3\}$ and $B = \{1, 2, 3, 4\}$
25. (a) $A \times B = \{(x, a), (x, b), (x, c), (y, a), (y, b), (y, c)\}$.
(d) $(A \cup B) \times C = \{(x, 0), (y, 0), (a, 0), (b, 0), (c, 0)\}$.
26. (a) $C = \{a\}, D = \{b, c, d, e\}$.

Communication

30. (a) Yes. $A \cap B \subseteq A \subseteq A \cup B$

Review Problems

37. (a) 6 (b) 6
38. (a) These are all the subsets of $\{2, 3, 4\}$. There are $2^3 = 8$ such subsets. (b) There are 8 subsets that contain the number 1. Every subset either contains 1 or it does not. So exactly half of the $2^4 = 16$ subsets contain 1. (c) Twelve subsets contain 1 or 2 (or both). There are 4 subsets of $\{3, 4\}$. We can form subsets that contain 1 or 2 or both by adding 1 to each, 2 to each, or 1 and 2 to each. By the fundamental counting principle then, there are $3 \cdot 4 = 12$ possibilities. (It's also not hard to list them.) (d) Four subsets contain neither 1 nor 2, since 12 subsets do contain 1 or 2. (e) B has $2^5 = 32$ subsets. Half contain 5 and half do not. (f) Every subset of A is a subset of B. The others can be listed by adding the number 5 to each subset of A. So there are twice as many subsets of B as subsets of A ($2^5 = 32$).
39. (a) A and B are equal. (b) C is a proper subset of A and a proper subset of B, since $C = \{4, 8, 12, 16, \ldots\}$.
40. Answers vary.

Ongoing Assessment 2-3

2. (a) Yes (c) Yes (e) Yes
4. (a) $x + 7 = 21$.
5. (a) Commutative property of addition (c) Commutative property of addition
7. (a) (i) For any whole numbers a and b, $a < b$ if and only if there exists a natural number k such that $b - k = a$, or equivalently if and only if $b - a$ is a natural number.
8. (a) 33, 38, 43
9. (a) 9 (c) 3 (e) 5
10. (a) 1 (c) 8 or 9
12. (a)

8	1	6
3	5	7
4	9	2

15. (a) 28
19. (a) Kent is the shortest and Vera is the tallest.

Communication

20. An arrow starting at 0 and ending at 3 represents the same number as an arrow starting at 4 and ending at 7. One way to explain this to students is to make physical models of each and show that the lengths are the same by matching.

Review Problems

31. (a) Yes, because there are 1000/2 elements in the first one and 1500/3 elements in the second one, so each has 500 elements. (b) Yes, because each has 1000 elements. (c) Yes, to each element n in the first set corresponds the element $n + 1$ in the second set. (d) Yes, to each element x in the first set corresponds the element $2x$ in the second set.
32. (a) True, can be shown with Venn diagrams (b) True. $A \subseteq A \cup B = B$. (c) True. $\overline{A \cup B} = \overline{A} \cap \overline{B} = \overline{\varnothing} = U$.
(d) False, let $A = \{1, 2, 3\}$ and $B = \{a, b, c\}$ and $C = \{1, 2, 3, 4\}$ then A is not a subset of B, B is not a subset of C, but A is a subset of C.
33. (a) 6 (b) 9
34. (a) False, let $A = \{a, b, c, d\}$, and $B = \{c\}$, and $C = \{b, d\}$, then $A \cup B = A \cup C$ but B does not equal C. (b) False, let $A = \{a, b, c, d\}, B = \{x, a, y\}$, and $C = \{z, a\}$, then $A \cap B = A \cap C$ but B does not equal C. (c) False, let $A = \{a, b, d\}, B = \{x, y, z\}$, and $C = \{c, d, e\}$, then $A \cap B = \varnothing, B \cap C = \varnothing$, but $A \cap C = \{d\}$.
35. 15

36. (a) 2200 people, because if all those people saw *The Lion King* as well, then they saw both, but 3100 saw *The Lion King*, and not all of them saw *Apollo 13*. **(b)** The smallest number of people that could have seen both movies is 500. Let x be the number of people who saw both movies. Then $3100 + 2200 - x$, or $5300 - x$, is the number of people who saw one or the other movie (or both). This number cannot exceed 4800, the total number of people in town. Thus the smallest value of x is $5300 - 4800$, or 500. (Any value smaller than 500 will make $5300 - x$ greater than 4800.)

37. (a) $\{a, b, c, d\}$ **(b)** $\{a, d\}$ **(c)** $\emptyset$ **(d)** $\emptyset$ **(e)** $\emptyset$

Ongoing Assessment 2-4

1. **(b)** 4
2. Each possible pairing of two of the sets is disjoint.
3. **(a)** Yes **(b)** Yes **(c)** Yes **(e)** Yes
4. **(a)** No, $2 + 3 = 5$ **(b)** Yes
5. **(a)** $ac + ad + bc + bd$ **(d)** $x^2 + 2xy + xz + y^2 + yz$
6. **(a)** $(4 + 3) \cdot 2 = 14$ **(c)** $(5 + 4 + 9) \div 3 = 6$
8. **(a)** 6 **(c)** 4
9. **(a)** $40 = 8 \cdot 5$ **(c)** $48 = x \cdot 16$
11. **(a)** $(8 \div 4) \div 2 \neq 8 \div (4 \div 2)$
12. **(a)** $(a + b)^2 = (a + b)(a + b) = (a + b)a + (a + b)b = a^2 + ba + ab + b^2 = a^2 + 2ab + b^2$
14. **(b)** Let $a \div c = x$ and $b \div c = y$. Then $a = xc$ and $b = yc$. Consequently
$$a + b = xc + yc$$
$$= (x + y) \cdot c$$
Now by definition of division, $x + y = (a + b) \div c$. When you substitute for x and y, the property follows.
15. After 5 months Samantha will have $10 more than Millie.
16. The first string makes 0 intersections, the second makes one, the third makes two, and so on, so there are $0 + 1 + 2 + \ldots + 8 + 9 = 45$ intersections.
17. 2; 3 left
19. A possible answer is given, resulting in $4 \cdot 3$, or 12, color schemes.

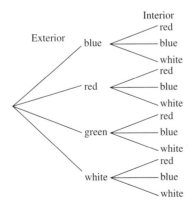

21. **(a)** 3 **(c)** 2 **(e)** 4
22. The answers depend upon the keys available on your calculator.
(b) For example $4 \cdot 4 - 4 + 4 \div 4$

23. **(a)** Subtract 18. **(c)** Add 11 and 48.
25. This is the case when x is either 0 or 1.
27. **(c)** $60h$
29. **(a)** Yes **(c)** Yes, a

Communication

31. The distributive property works with two operations. Only one operation is used in this example. Multiplication is not distributive over multiplication.

Review Problems

39. For example, $\{0, 1\}$
40. No. For example, $5 - 2 \neq 2 - 5$.
41.

42. Because $a < b$, we have $a + k = b$ for some nonzero whole number k. Multiply both sides of the equation by $c (c \neq 0)$ to obtain:
$$(a + k)c = bc,$$
$$ac + kc = bc.$$
Because $k \neq 0$ and $c \neq 0$, $kc \neq 0$. Consequently $ac > bc$.

Ongoing Assessment 2-5

1. **(a)** Double the input number. **(d)** Square the input number and add 1.
2. **(a)** This is not a function, since the input 1 is paired with 2 outputs (a and d). **(d)** This is not a function, since the input 1 is paired with several outputs.

4. **(c)**

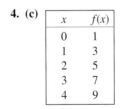

x	$f(x)$
0	1
1	3
2	5
3	7
4	9

5. **(b)** This is a function. **(e)** This is a function.
7.

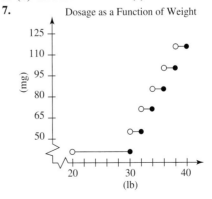

9. (a) 8 dollars
10. (c) $L(n) = n(n + 1)$.
11. (a) 7 (b) 55
12. (c) $2n$
13. (c) 65
14. (c) 2, 12, 2550
15. (a) $2 \cdot 1 + 2 \cdot 7 = 16; 2 \cdot 2 + 2 \cdot 6 = 16; 2 \cdot 6 + 2 \cdot 2 = 16;$
$2 \cdot 5 + 2 \cdot 5 = 20$
16. (b) The same each month
17. (b) Between 6:00 A.M. and 6:30 A.M. (c) 0
19. (a) $H(2) = 192$. $H(6) = 192$. $H(3) = 240$. $H(5) = 240$.
Some of the heights correspond to the ball going up, some to the ball coming down.

(b)
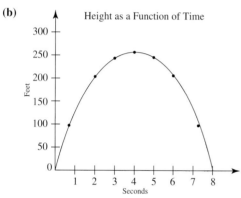

At $t = 4$ seconds, the ball's height is $H(4) = 256$ feet above the ground.
(c) 8 seconds (d) $0 \leq t \leq 8$. (e) $0 \leq H(t) \leq 256$.
20. (a) $A(x) = x \cdot \frac{1}{2}(900 - x)$.
21. (a) (i) 40
(ii) 38
(iii) 22
(b) (i) $S(n) = 2n(n + 1)$
(ii) $S(1) = 4$, $S(n) = 2n(n + 1) - 2$ when $n \geq 2$
(iii) $S(n) = 6n - 2$
23. (b) $\{(A, B), (A, C), (A, D), (C, A), (C, B), (C, D), (D, A),$
$(D, B), (D, C), (F, G), (G, F), (I, J)\}$
24. (a) Function (b) Relation, but not a function
26. (a) None (f) Reflexive and symmetric (g) Transitive

Communication

28. Yes, since each element of A is paired with exactly one element of B.
30. (a) This is not a function, since a faculty member may teach more than one class. (b) This is a function (assuming only one teacher per class).

Review Problems

39. (a) $109 = 98 + 11$ (b) $x = z + y$ (c) $60 = 4 \cdot 15$
(d) $x = y \cdot 3$ (e) $x = 10 \cdot 5 + 3$

40. No, because if $A \cap B \neq \emptyset$ then we would be counting those elements twice.
41.
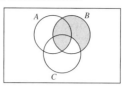

Chapter Review

2. (a) $A \cup B = A$. (d) $A \cap \overline{D} = \{r, v\}$.
3.
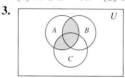

(a) $A \cap (B \cup C)$
7.
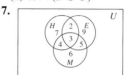

(a) 36 (b) 6
9. (a) False. Consider the sets $\{a\}$ and $\{2\}$. (d) False. This is in one-to-one correspondence with the set of natural numbers. (g) True
10. (a) 17 (d) 17
13. (a) Distributive property of multiplication over addition
(c) Identity property of multiplication for whole numbers
(e) Commutative property of multiplication
14. (a) Let $A = \{1, 2, 3, \ldots 13\}$ and $B = \{1, 2, 3\}$ then B is a proper subset of A. Therefore B has fewer elements than A so $n(B)$ is less then $n(A)$. Thus $3 < 13$.
15. (a) 15, 14, 13, 12, 11, or 10 (c) Any whole number
16. (a) $15a$ (c) $xa + xb + xy$
20. $6000 for 80 people is cheaper
22. $400
23. (a) Function (b) Not a function (c) Function
24. (a) Range = $\{3, 4, 5, 6\}$. (d) Range = $\{5, 9, 15\}$.
25. (a) This is not a function, since one student can have two majors. (e) This is a function. The range is N.
26. (c) After the ninth month, the cost exceeds $600.

CHAPTER 3

Ongoing Assessment 3-1

1. (a) $\overline{\overline{MCDXXIV}}$; The double bar over M represents $1000 \cdot 1000 \cdot 1000$. (b) 46,032; The 4 in 46,032 represents 40,000 while the 4 in 4632 represents only 4000. (c) ⋖ ▼▼; The space in the latter number indicates ⋖ is multiplied by

10 · 60 rather than by 10.   **(d)** The 𒁹 represents 1000 while 𒐝 represents only 100.   **(e)** 𒀯 represents three groups of 20 plus zero 1s while ☰ represents three 5s and three 1s.

2. (a) MCML; MCMXLVIII   **(c)** M; CMXCVIII
(e) 𒁹991; 𒁹9∩∩∩∩∩∩∩∩∩∩IIIIIIIII
3. 1922
4. (a) CXXI   **(c)** LXXXIX,
5. (a) ∩∩∩∩II   **(c)** ⌒III
6. (a) ▼ ◁▼▼; ∩∩∩∩∩∩∩II; LXXII; ⋮
    (c) 1223; ≪ ≪▼▼▼; MCCXXIII; ⋮
7. (a) Hundreds   **(c)** Thousands
8. (a) 3,004,005   **(c)** 3,560
9. (a) 86
10. 811 or 910
11. (a) (1, 10, 11, 100, 101, 110, 111, 1000, 1001, 1010, 1011, 1100, 1101, 1110, 1111)$_{two}$ **(c)** (1, 2, 3, 10, 11, 12, 13, 20, 21, 22, 23, 30, 31, 32, 33)$_{four}$
12. 20
14. (a) 111_{two} **(c)** 999_{ten}
15. (a) ETE_{twelve}; EEl_{twelve} **(c)** 554_{six}; 1000_{six} **(e)** 444_{five}; 1001_{five}
16. (a) There is no numeral 4 in base four.   **(c)** There is no numeral T in base three.
17. (a) 3212_{five} **(c)** 12110_{four}
19. (a) 117   **(c)** 1331   **(e)** 157
21. 1 prize of $625, 2 prizes of $125, and 1 of $25
23. (a) 8 weeks, 2 days   **(c)** 1 day, 5 hours
25. (a) 6   **(c)** nine
27. Above the bar are depicted 5s, 50s, 500s, and 5000s. Below the bar are 1s, 10s, 100s, and 1000s. Thus there are 1 · 5000, 1 · 500, 3 · 100, 1 · 50, 1 · 5, and 2 · 1 depicted for a total of 5857. The number 4869 could be depicted as follows:

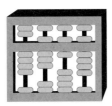

28. Assume an eight-digit display without scientific notation.
(a) 98,765,432   **(c)** 99,999,999

Communication

30. Answers will vary. Ben is incorrect. Zero is a place holder in the Hindu-Arabic system. It is used to differentiate between numbers like 54 and 504. If zero were nothing, then we could eliminate it without changing our number system. Zero is the cardinal number of the empty set.

32. This is primarily for readability. It has been proposed with the metric system to drop the commas and simply use spaces instead.

Open-Ended

33. 4; 1, 2, 4, 8; 1, 2, 4, 8, 16

Ongoing Assessment 3-2

1. (a)   981
          + 421
         ─────
          1402

2. (a)   383
          − 159
         ─────
           224

3. (a) One possibility: 863
                            + 752
                           ─────
                            1615

4. Only if positive numbers are used: **(a)**   876
                                                   − 235
                                                  ─────
                                                    641

5. No, not all at dinner. He can have either the steak or the salad.
7.   3428
       + 5631
      ─────
        9059

8. (a) About 121 weeks   **(c)** Answers vary.
9. (a) (i) 1,236   (ii) 1,032
10. Answers may vary, for example,   **(a)** The tens digit was not carried.   **(c)** The units minuend is subtracted from the subtrahend.
11. 1 hour 34 minutes 15 seconds
12. (a) 121_{five} **(c)** 1010_{five} **(e)** 1001_{two}
14. (b) 1 hour 39 minutes 40 seconds
15. (a) 2 quarts, 1 pint, 0 cups, or 1 half-gallon, 0 quarts, 1 pint, 0 cups   **(c)** 2 quarts, 1 pint, 1 cup
16. $2\frac{1}{2}$, so buy 3 gallons
17. (a) The method produces a palindrome in each case: (i) 363   (ii) 9339   (iii) 5005.
19. It is doubling the second number in the operation.
20. (a) 34; 34; 34   **(c)** 34   **(e)** Yes
21. (b) $^3\cancel{7}_1 2$
             1 $\cancel{3}_0$
             2 2
             $\cancel{4}_3 \cancel{3}_0$
             $\cancel{2}_0$ 3
             1 $\cancel{2}_0$
            ─────
             3 1 0$_{five}$
22. (a) 3 gross 10 dozen 9 ones
23. (a) 22 students on Tuesday

24. (a) 70

26. (a)
$$230_{\text{five}}$$
$$-\ 22_{\text{five}}$$
$$203_{\text{five}}$$

Communication

29. For example, the words *regroup* and *trade* more accurately reflect the actions that are taken when performing a subtraction problem. The words *borrow* and *carry* seem to reflect mechanics and not the mathematical ideas used in the algorithm.

Cooperative Learning

34. The US monetary system is not a true base-ten system because of the coins and bills involved. A true base-ten system would have only pennies, dimes, dollars, ten dollars, 100 dollars, etc. The system would not have nickels, quarters, half-dollars, five-dollar bills, etc. Money in other countries varies.

Review Problems

35. This will be studied in detail in later chapters. However with the meter as a basic unit of length, we have 10 meters = 1 decameter, 10 decameters = 1 hectometer, 10 hectometers = 1 kilometer.

36. $5280 = 5 \cdot 10^3 + 2 \cdot 10^2 + 8 \cdot 10^1 + 0 \cdot 1$

37. 1410

38. (a) 1594   (b) 11   (c) 23

Ongoing Assessment 3-3

1. (a)
$$\begin{array}{r} 426 \\ \times\ 783 \\ \hline 1278 \\ 3408 \\ 2982 \\ \hline 333558 \end{array}$$

2. (a)

5. (a) 5^{19} (c) 10^{313}

6. (a) 2^{100} because $2^{80} + 2^{80} = 2^{80}(1+1) = 2^{80} \cdot 2 = 2^{81}$

8. (c)
$$\begin{array}{r} 363 \\ \times\ 84 \\ \hline 2904\ \ (8 \times 363) \\ 1452\ \ (4 \times 363) \\ \hline 30492 \end{array}$$

11. (a) 1332 calories   (c) Maurice, 96 more calories

13. $60

15. (a) $3\overline{)876}$

17. 3

18. 8 cars (remember the match)

19.

2	11
4	15
0	7
6	19
12	31

21. (b) $(10a + b) \cdot (10c + d) = (10b + a) \cdot (10d + c)$ implies $100ac + 10bc + 10ad + bd = 100bd + 10ad + 10bc + ac$ or $99ac = 99bd$, which implies that $ac = bd$.

22. 3 hrs

24. 58 buses needed, not all full

27. (a) 233_{five} (c) 2144_{five} (e) 110_{two}

28. (a) Nine   (c) Six

30. (a)
$$\begin{array}{r} 763 \\ \times\ 8 \\ \hline 6104 \end{array}$$

31. (a)
$$\begin{array}{r} 762 \\ \times\ 83 \\ \hline 63{,}246 \end{array}$$

32. (a)
$$\begin{array}{r} 37 \\ \times\ 43 \\ \hline 111 \\ 1480 \\ \hline 1591 \end{array}$$

33. (a) 1; 121; 12,321; 1,234,321. For n 1s in each of the two factors, the product "counts up" to n, then back down to 1. The next product should be 123,454,321.

Review Problems

41. 999999∩∩∩∩∩∩∩III

42. 300,260

43. For example, $3 + 0 = 3 = 0 + 3$.

44. (a) $x(a + b + 2)$ (b) $(3 + x)(a + b)$

45. 6979 miles

Ongoing Assessment 3-4

2. Associative property of addition; commutative property of addition; associative property of addition; addition algorithm; addition algorithm

3. (a) 605

4. (a) 36

7. (a) 5300   (c) 120,000

9. Answers vary, for example, (a) $2 + 3 + 5 = 10$, so 10,000 is an initial estimate. $10,000 + 2000$ (adjustment) = 12,000 for the final estimate.

10. (a) The first set of numbers is not clustered. The second set is clustered about 500 so an estimate is 2500.

11. (a) The range is 600 (20 · 30) to 1200 (30 · 40).
14. Answers vary, for example, 3300 − 100 − 300 − 400 − 500 = 2000. The estimate is high because the amounts that were taken away were rounded up and resulted in more being taken away than the $13 that was added to the $3287.
16. (a) Different answers since the estimates of 800 and 220 are way off.
18. The clustering strategy gives 6 · 70,000 or 420,000.
19. One possibility is that to find $x5^2$ we could write $x(x + 1)$ and append the digits 25. For example, in 65^2, we take 6 · 7 = 42 and append 25 to obtain 4225.

Communication

21. Mental mathematics is the process of producing an exact answer to a computation without using external aids. Computational estimation is the process of forming an approximate answer to a numerical problem.

Open-Ended

25. Answers vary, for example, when you are determining the amount of a tip for a waiter in a restaurant.

Chapter Review

1. **(a)** 400,044 **(c)** 1704 **(e)** 1448
2. **(a)** CMXCIX **(c)** ⋮ **(e)** 11011_{two}
3. **(a)** 3^{17} **(c)** 3^5
4. 1119
5. 60,074
6. **(a)** 5 remainder 243 **(c)** 120_{five} remainder 2_{five}
8. **(a)** tens **(c)** hundreds
9. $395
11. 2600 cases
13. 40 cans
14. 26
16. $6000
18. $214
20. Selling pencils by the units, dozens, and gross is an example of the use of base 12.
21. **(a)** The Egyptian system had seven symbols. It was a *tally* system, a *grouping* system, and it used the *additive property*. It did not have a symbol for zero but this was not very important because they did not use place value.
22. **(a)** 1003_{five} **(c)** $T8_{twelve}$
23. **(a)** 212_{five} **(c)** 1442_{five}
24. **(a)** For example, (26 + 24) + (37 − 7) = 50 + 30 = 80.
26. Answers vary. For example, **(a)** 2300 + 300 (adjustment) = 2600. **(b)** 2600
27. 2400 · 4 = 9600.

CHAPTER 4

Ongoing Assessment 4-1

1. **(b)** 5 **(f)** ⁻a + ⁻b or ⁻(a + b)
3. **(a)** 5 **(d)** ⁻5
4. **(a)**

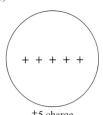
+5 charge

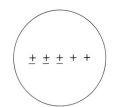
Add 3 negative charges. Net result: 2 positive charges

(d)

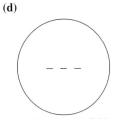
⁻3 charge on the field

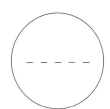
Add 2 negative charges. Net result: 5 negative charges

5. **(c)** ⁻1
7. **(a)** ⁻7 **(d)** ⁻$150
10. **(b)** 3 − 1 = 2; 2 − 1 = 1; 1 − 1 = 0; 0 − 1 = ⁻1; ⁻1 − 1 = ⁻2; ⁻2 − 1 = ⁻3
11. **(b)** 3 **(e)** ⁻13
13. **(c)** 192°F
14. **(a)** 10W−40 or 10W−30 **(c)** 10W−40, 5W−30 or 10W−30 **(e)** 10W−30 or 10W−40
15. **(b)** $2x + y$
17. **(b)** All positive integers **(e)** There are none. **(g)** There are none.
18. **(a)** I **(c)** $I - \{0\}$ **(e)** ∅ **(g)** $\{0\}$ **(i)** I
20. **(a)** 0 **(c)** 1
21. **(c)** 0 or 2
23. **(b)** 19 **(c)** 19
25. **(b)** $d = {}^-4$, next terms: ⁻9, ⁻13 **(d)** $d = 2x$, next terms: $1 + 3x, 1 + 5x$
26. **(b)** 3775
28. **(b)** True **(c)** True **(e)** False; let $x = {}^-1$
30. **(b)** ⁻106 **(d)** 22 **(f)** 2 **(h)** 23
31. **(b)** 516 **(c)** 10,894

Communication

34. **(a)** $(a + b) + ({}^-a + {}^-b) = a + b + {}^-a + {}^-b$
$= a + {}^-a + b + {}^-b = (a + {}^-a) + (b + {}^-b) = 0 + 0 = 0$

37. Two numbers are opposite of each other if, and only if, their sum is 0. We have:
$$(b - a) + (a - b) = b + (^-a) + a + (^-b)$$
$$= b + 0 + {}^-b$$
$$= b + {}^-b$$
$$= 0$$

Ongoing Assessment 4-2

4. (a) $^-20 \cdot 4$ **(c)** ^-20n
5. (b) $^-13$ **(d)** 0 **(f)** Impossible; division by 0 is not defined.
6. (d) a; If $b \neq 0$ **(f)** Not defined **(g)** 0
7. (b) $0°C + 25 \cdot (4°C)$ **(d)** $25°C - 20 \cdot (3°C)$
(e) $^-5 + (m - 1)d$
8. (b) $^-66$ divided by 11 = $^-6$; He lost 6 yards.
10. (a) $^-1(^-5 + {}^-2) = {}^-1(^-7) = 7$;
$(^-1)(^-5) + (^-1)(^-2) = 5 + 2 = 7$
11. (b) 16 **(d)** 81 **(f)** $^-1$ **(h)** $^-1$
12. (a) 12 **(c)** $^-5$ **(f)** $^-9$ **(g)** $^-13$ **(h)** $^-8$
14. (b) = (c); (d) = (e); (g) = (h)
15. (a) Commutative property of multiplication
(d) Distributive property of multiplication over addition
16. (b) $2xy$ **(f)** b **(h)** y
17. (b) 2 **(d)** $^-6$ **(f)** 6 **(h)** All integers except 0
(j) 3 or $^-3$ **(l)** All integers except 0 **(n)** All integers
18. (d) $^-x^2 + xy$ **(f)** $^-x^2 + xy + 3x$
(h) $x^2 - y^2 - 1 - 2y$
19. (a) $(50 + 2)(50 - 2) = 50^2 - 2^2 = 2500 - 4 = 2496$
(d) $4 - 9x^2$
20. (a) $8x$ **(e)** $x(x + y)$ **(h)** $x(3x + y - 1)$
(j) $(a + b)c$ **(m)** $(2x + 5y)(2x - 5y)$
22. (b) True **(c)** True
23. (a) False. Let $x = 3, y = {}^-5$, then $|x + y| = 2$ but $|x| + |y| = 8$. **(c)** True. Because $x^2 \geq 0, |x^2| = x^2$.
24. (a) The sums are 9 times the middle number.
25. (b) $^-8, ^-11, d = {}^-3$, nth term is $^-3n + 13$
(d) $^-128, 256, r = {}^-2$, nth term is $(^-2)^n$
27. (a) $^-9, ^-6, ^-1, 6, 15$ **(e)** $^-1, 4, ^-9, 16, ^-25$
30. (c) Sometimes true. True if, and only if, $x = 0$. If $x > 0$, $^-x^2$ is negative and x^2 is positive and hence the expressions cannot be equal. **(d)** Always true. For all $x, |x|$ is nonnegative and hence greater than $^-1$.

Communication

34. $^-(a + b) = (^-1)(a + b)$ Nancy's assumption
$= (^-1)a + (^-1)b$ by distributive property
$= {}^-a + {}^-b$ Nancy's assumption
35. (a) $(^-1) \cdot a + a = (^-1)a + 1 \cdot a$
$= (^-1 + 1)a$
$= 0 \cdot a$
$= 0$

Review Problems

44.

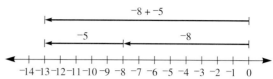

45. (a) 5 **(b)** $^-7$ **(c)** 0
46. (a) 14 **(b)** 21 **(c)** $^-4$ **(d)** 22
47. 400 lb

Ongoing Assessment 4-3

1. (b) True **(d)** True **(f)** False, 6 is a factor of 30 not a multiple of 30.
2. (a) Yes **(c)** Yes **(e)** Yes
3. (b) 2, 3, 6, 9 **(d)** 2, 4 **(f)** none of them
4. (a) No, 17 | 34000 and 17∤15, so 17∤34,015. **(c)** No, 19 |19000 and 19∤31, so 19∤19,031. **(e)** No, $5 | 2 \cdot 3 \cdot 5 \cdot 7$ and 5∤1, so $5 \nmid (2 \cdot 3 \cdot 5 \cdot 7) + 1$.
5. (a) True by Theorem 4-1 **(c)** None **(e)** True by Theorem 4-1
6. (a) True **(c)** False **(e)** True **(g)** True
8. (a) A number is divisible by 16 if, and only if, the last four digits form a number divisible by 16. **(b)** A number is divisible by 25 if and only if the number formed by the last two digits is divisible by 25.
10. (a) 7
11. (c) 1, 3, 5, 7, 9 **(e)** 7
13. Each bar costs 19¢.
14. (a) Yes
15. (a) 1, 2, 4, 5, 8, 11
17. (a) $12{,}343 + 4546 + 56 = 16{,}945$; $4 + 1 + 2 = 7$ has a remainder of 7 when divided by 9, as does $1 + 6 + 9 + 4 + 5$.
(c) $10{,}034 + 3004 + 400 + 20 = 13{,}458$;
$8 + 7 + 4 + 2 = 21$ has a remainder 3 when divided by 9, as does $1 + 3 + 4 + 5 + 8$. **(e)** $1003 - 46 = 957$; $4 - 1 = 3$ has a remainder of 3 when divided by 9, as does $9 + 5 + 7 = 21$.
18. (a) False; 2 | 4, but 2∤1 and 2∤3 **(c)** False; 12 | 72 but 12∤8 and 12∤9
21. (a) The result is always 9.

Communication

24. (a) Yes, $d | n$ because a display of 32 implies that $n = 32 \cdot d$ and this implies that $d | n$.
28. (a) Yes, 4 | 52,832, so 4 divides any integer times 52,832. Therefore 4 divides $52{,}832 \cdot 324{,}518$, which is the area.
29. (a) 1, 3, and 7 divide n. 1 divides every number. Also, because $n = 21 \cdot d, d \in I$, then $n = (3 \cdot 7)d = 3(7d) = 7(3d)$, which implies that n is divisible by 3 and by 7.

31. (a) No. If $5 \nmid d$ for any integer d, then there is no integer m such that $5m = d$. If we assume $10 \mid d$, this means there exists n such that $10n = d$ or $5(2n) = d$. This contradicts the original assumption that d is not divisible by 5.
33. (a) All of the numbers are divisible by 11.

Review Problems

38. See Example 4-10 in Section 4-2
39. (a) $^{-}2$ **(b)** No such integers exist **(c)** No such integers exist **(d)** All integers **(e)** No such integers exist **(f)** All positive integers
40. (a) $5x - 1$ **(b)** x^2 **(c)** $x - y$ **(d)** 1

Ongoing Assessment 4-4

1. 30
3. (a) Prime **(c)** Not prime **(e)** Prime
5. (a)

```
        504
       /   \
      2    252
          /   \
         2    126
             /   \
            2    63
                /  \
               3   21
                  /  \
                 3    7
```
$504 = 2^3 \cdot 3^2 \cdot 7$

(c)
```
      11250
      /   \
     2   5625
         /   \
        3   1875
            /   \
           3    625
               /   \
              5    125
                  /   \
                 5    25
                     /  \
                    5    5
```
$11250 = 2 \cdot 3^2 \cdot 5^4$

6. (a) $1 \times 48, 2 \times 24, 3 \times 16, 4 \times 12$ **(b)** Only one, 1×47.
8. (a) 3, 5, 15, 29 people
9. (a) 1, 2, 3, 4, 6, 9, 12, 18, or 36 **(c)** 1 or 17
12. (a) The fundamental theorem of arithmetic says that n can be written as a product of primes in one and only one way. Since $2 \mid n$ and $3 \mid n$ and 2 and 3 are both prime, they must be included in the unique factorization.
That is, $2 \cdot 3 \cdot p_1 \cdot p_2 \cdot \ldots \cdot p_m = n$
Therefore, $(2 \cdot 3)(p_1 \cdot p_2 \cdot \ldots \cdot p_m) = n$
Thus, $6 \mid n$.

14. 101, 103, 107, 109, 113, 127, 131, 137, 139, 149, 151, 157, 163, 167, 173, 179, 181, 191, 193, 197, 199
15. 3, 5; 5, 7; 11, 13; 17, 19; 29, 31; 41, 43; 59, 61; 71, 73; 101, 103; 107, 109; 137, 139; 149, 151; 179, 181; 191, 193; 197, 199
19. (a) 4, 6, 8, 0, or 5 **(c)** 23, 29, 31, 37, 53, 59, 71, 73, 79
20. There are infinitely many composites of the form 1, 11, 111, 1111, 11111, 111111 … since every third member of this sequence will be divisible by 3.
24. (a) $3 \cdot 5 \cdot 7 \cdot 11 \cdot 13$ is composite because it is divisible by 3, 5, 7, 11, and 13. **(c)** $(3 \cdot 5 \cdot 7 \cdot 11 \cdot 13) + 5 = 5((3 \cdot 7 \cdot 11 \cdot 13) + 1)$ and so it is composite. **(e)** $10! + k$ can be factored as in part (d) for the given values of k and so it is composite.

Communication

28. No, they are not both correct. Using 3 and 4 is correct because they have no common divisors. Using 2 and 6 there is a common divisor of 2. Using this test will ensure only that the number is divisible by 6.
31. 91 eggs. Let n be the number of eggs in the basket, then $n - 1$ is a multiple of 3 and of 5. Hence $n - 1 = 3 \cdot 5k$ for some integer k. Thus, $n = 15k + 1$. We know that $n \leq 100$ and that $7 \mid n$. We substitute $k = 1, 2, 3 \ldots$ to obtain the following values for n that are less than or equal to 100.

$$16, 31, 46, 61, 76, 91$$

Among these values, only 91 is divisible by 7. Hence, $n = 91$.
33. It can be conjectured and shown that only perfect squares have an odd number of divisors.
   The perfect squares less than 1000 are $1^2, 2^2, 3^2, \ldots 31^2$. Therefore there are 31 perfect squares between 1 and 1000, and hence 31 numbers with an odd number of divisors. Consequently $1000 - 31$, or 969, numbers between 1 and 1000 have an even number of divisors. Clearly the student is wrong.

Review Problems

40. (a) False **(b)** True **(c)** True **(d)** True
41. (a) 2, 3, 6 **(b)** 2, 3, 5, 6, 9, 10
42. If $12 \mid n$, there exists an integer a such that $12a = n$.
   $(3 \cdot 4) a = n$
   $3(4a) = n$
   Thus, $3 \mid n$.
43. Yes, among 8 people. Each would get $422.

Ongoing Assessment 4-5

1. (a) $D_{18} = \{1, 2, 3, 6, 9, 18\}$
   $D_{10} = \{1, 2, 5, 10\}$
   $GCD(18, 10) = 2$
   $M_{18} = \{18, 36, 54, 72, 90, \ldots\}$
   $M_{10} = \{10, 20, 30, 40, 50, 60, 70, 80, 90, \ldots\}$
   $LCM(18, 10) = 90$

(b) $D_{24} = \{1, 2, 3, 4, 6, 8, 12, 24\}$
$D_{36} = \{1, 2, 3, 4, 6, 9, 12, 18, 36\}$
GCD(24, 36) = 12
$M_{24} = \{24, 48, 72, 96, 120, 144, 168, \ldots\}$
$M_{36} = \{36, 72, 108, 144, 180, \ldots\}$
LCM(24, 36) = 72

2. (a) $132 = 2^2 \cdot 3 \cdot 11$
$504 = 2^3 \cdot 3^2 \cdot 7$
GCD(132, 504) = $2^2 \cdot 3 = 12$
LCM(132, 504) = $2^3 \cdot 3^2 \cdot 7 \cdot 11 = 5544$
(b) $65 = 5 \cdot 13$
$1690 = 2 \cdot 5 \cdot 13^2$
GCD(65, 1690) = $5 \cdot 13 = 65$
LCM(65, 1690) = $2 \cdot 5 \cdot 13^2 = 1690$

3. (a) GCD(2924, 220) = GCD(220, 64) = GCD(64, 28) = GCD(28, 8) = GCD(8, 4) = GCD(4, 0) = 4
(c) GCD(123, 152, 122, 368) = GCD(122, 368, 784) = GCD(784, 64) = GCD(64, 16) = GCD(16, 0) = 16

4. (a) 72 **(c)** 630

5. (a) $220 \cdot 2924/4$, or 160,820 **(c)** $123,152 \cdot 122,368/16$, or 941,866,496

7. (a) LCM(15, 40, 60) = 120 minutes = 2 hours. So the clocks alarm again together at 8:00 A.M.

8. (a) \$60 **(c)** 30

12. After 7 1/2 hours, or 2:30 A.M.

14. (a) ab **(c)** GCD(a^2, a) = a; LCM(a^2, a) = a^2
(e) GCD(a, b) = 1; LCM(a, b) = ab **(g)** $b|a$

15. (a) True. If a and b are even, then GCD(a, b) ≥ 2.
(c) False. The GCD could be a multiple of 2; for example, GCD(8, 12) = 4. **(e)** True, by Theorem 4–8 **(g)** True. If LCM(a, b) < a, then the LCM could not be a multiple of a.

17. (a) $4 = 2^2$. Since 97,219,988,751 is odd, it has no prime factors of two. Consequently, 1 is their only common divisor and they are relatively prime.

20. 120 days

22. 12 revolutions

24. (a)

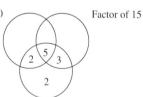

Factor of 10, Factor of 15, Factor of 60

25. $1, 2, 2^2, 2^3, \ldots, 2^{20}$

Communication

29. No; for example, consider GCD(2, 4, 10) = 2, LCM(2, 4, 10) = 20, and the GCD · LCM = (2 · 20) = 40 while $abc = 2 \cdot 4 \cdot 10 = 80$.

31. No. Let $a = 2 \cdot 3, b = 3 \cdot 5, c = 5 \cdot 7$. Then GCD($a, b, c$) = 1 but GCD($a, b$) = 3 and GCD($b, c$) = 5.

Review Problems

35. $x = 15{,}625; y = 64$
36. (a) 83,151; 83,451; 83,751 **(b)** 86,691 **(c)** 10,396
37. $17 \cdot 183 = 3111$; thus 3111 is not prime.
38. Answers may vary. $30{,}030 = 2 \cdot 3 \cdot 5 \cdot 7 \cdot 11 \cdot 13$
39. 27,720
40. 43

Ongoing Assessment 4-6

1. 2:00 P.M.
2. (a) 3 **(c)** 6 **(e)** 3 **(g)** Does not exist
3. (a) 2 **(c)** 2 **(g)** 2
4. (a)

⊕	1	2	3	4	5	6	7
1	2	3	4	5	6	7	1
2	3	4	5	6	7	1	2
3	4	5	6	7	1	2	3
4	5	6	7	1	2	3	4
5	6	7	1	2	3	4	5
6	7	1	2	3	4	5	6
7	1	2	3	4	5	6	7

5. (a)

⊗	1	2	3	4	5	6	7
1	1	2	3	4	5	6	7
2	2	4	6	1	3	5	7
3	3	6	2	5	1	4	7
4	4	1	5	2	6	3	7
5	5	3	1	6	4	2	7
6	6	5	4	3	2	1	7
7	7	7	7	7	7	7	7

6. (a) 10 **(c)** 7 **(e)** 1
7. (a) 2, 9, 16, 30 **(c)** $366 \equiv 2 \pmod{7}$; Wednesday
8. (a) 4 **(c)** 0
9. (a) $x = 2k$, k is an integer **(c)** $x - 3 = 5k$ implies $x = 3 + 5k$ where k is an integer.

Chapter Review

1. (a) ⁻3 **(e)** $x - y$ **(g)** 32
2. (a) ⁻7 **(d)** 0 **(e)** 8
3. (a) 3 **(c)** Any integer except 0 **(e)** ⁻41
5. (a) $10 - 5 = 5$ **(b)** $1 - (⁻2) = 3$
6. (a) $(x - y)(x + y) = (x - y)x + (x - y)y$
$= x^2 - yx + xy - y^2$
$= x^2 - xy + xy - y^2$
$= x^2 - y^2$
7. (a) ⁻x **(c)** $3x - 1$ **(f)** ⁻$x^2 - 6x - 9$
8. (a) ⁻$2x$ **(b)** $x(x + 1)$ **(e)** $5(1 + x)$ **(f)** $(x - y)x$
9. (a) False, it is not positive for $x = 0$. **(c)** False, if $b < 0$.
(e) False, it is equal to ab.
10. (b) $3 - (4 - 5) \neq (3 - 4) - 5$ **(d)** $8/(4 - 2) \neq 8/4 - 8/2$

11. (a) False (c) True (e) False; 9, for example
12. (a) False; 7|7 and 7∤3 yet 7|3·7 (c) True (e) True
13. (a) Divisible by 2, 3, 4, 5, 6, 8, 9, 11
15. (a) 87<u>2</u>4; 86<u>5</u>4; 87<u>8</u>4 (c) 87,<u>1</u>74; 87,<u>4</u>64; 87,<u>7</u>54
17. (a) Composite
19. (a) 4
20. (a) $2^4 \cdot 5^3 \cdot 7^4 \cdot 13 \cdot 29$
23. (a) $2^2 \cdot 43$ (c) $2^2 \cdot 5 \cdot 13$
26. 9:30 A.M.
28. 5 packages
30. We first show that among any three consecutive odd integers, there is always one that is divisible by 3. For that purpose, suppose that the first integer in the triplet is not divisible by 3. Then by the division algorithm that integer can be written in the form $3n + 1$ or $3n + 2$ for some integer n. Then the three consecutive odd integers are $3n + 1$, $3n + 3$, $3n + 5$ or $3n + 2$, $3n + 4$, $3n + 6$. In the first triplet $3n + 3$ is divisible by 3, and in the second $3n + 6$ is divisible by 3. This implies that if the first odd integer is greater than 3 and not divisible by 3, then the second or the third must be divisible by 3, and hence cannot be prime.
33. mod 360. It would cover all the area encircling the lighthouse.

CHAPTER 5

Ongoing Assessment 5-1

1. (a) The solution to $8x = 7$ is 7/8. (c) The ratio of boys to girls is seven to eight.
2. (a) 1/6 (c) 2/6 or 1/3 (e) 5/16
3. (a) 2/3 (c) 6/9 or 2/3
4. (a)

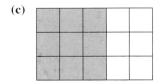

(c)

(e)

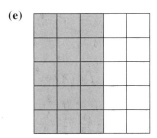

5. (a) 9/24 or 3/8 (c) 4/24 or 1/6
6. (a) 4/18, 6/27, 8/36 (c) 0/1, 0/2, 0/4
7. (a) 52/31 (c) ⁻5/7 (e) 144/169
9. (a) undefined (c) 0 (e) cannot be simplified (g) 5/3
10. (a) 1 (c) $a/1$
11. (a) equal (c) equal
12. (a) not equal (c) equal
13. Yes, 1/32 in.
15. 36/48
17.

18. (a) $2\frac{7}{8}$ in. (c) $1\frac{3}{8}$ in.
19. (a) 32/3 (c) x is any rational number except 0
20. (a) $a = b, c \neq 0$
21. $N \subseteq W \subseteq I \subseteq Q$
23. (a) > (c) < (e) =
24. (a) $\frac{11}{13}, \frac{11}{16}, \frac{11}{22}$
25. $\frac{a}{b} < 1$ and $\frac{c}{b} > 0$ imply $\frac{a}{b} \cdot \frac{c}{d} < 1 \cdot \frac{c}{d}$ or $\frac{a}{b} \cdot \frac{c}{d} < \frac{c}{d}$.
27. Answers may vary. The following are possible answers.
(a) $\frac{10}{21}, \frac{11}{21}$ (c) $\frac{997}{1200}, \frac{998}{1200}$
28. (a) 1 (c) The ratios are the same. To show this, let x equal the top circled number. Then the sum of the circled numbers is $x + (x + 12) + (x + 19) + (x + 31) = 4x + 62$. The sum of the four interior numbers is $(x + 10) + (x + 11) + (x + 20) + (x + 21) = 4x + 62$. Here the ratio is always 1.
29. 456 mi

Communication

32. Each class does not have to have the same number of females. One class could have 8 females in a class of 24 while another could have 12 females in a class of 36.
33. The new fraction is equal to $\frac{1}{2}$. The principle can be generalized as follows:
$$\frac{a}{b} = \frac{ar_1}{br_1} = \frac{ar_2}{br_2} = \frac{ar_3}{br_3} = \ldots = \frac{ar_n}{br_n} =$$
$$\frac{a(r_1 + r_2 + r_3 + \ldots + r_n)}{b(r_1 + r_2 + r_3 + \ldots + r_n)} = \frac{a}{b}.$$
35. (a) Suppose the rational numbers are $\frac{2}{16}$ and $\frac{1}{4}$. $\frac{1}{4} > \frac{2}{16}$.
Iris is incorrect.
37. Because 36 in. = 1 yd, then to convert x in. to yards, we could consider how to write equivalent fractions 36/1 and $x/1$ unknown yards. By seeing how many 36's are in x, we could determine the unknown number of yards. Because 1 in. = 1/36 yd, the process to convert yards to inches could be accomplished using similar methods.

Open-Ended

38. Frequently in recipes, measurements are found in fractional parts of cups, teaspoons, tablespoons. For example, a recipe might call for 1/2 teaspoon of salt and 2/3 cup of flour.

Cooperative Learning

40. In this assessment, students must work together to determine the heights and to order the people according to height. Once that is done, the students may decide how many are in the class, for example 24, and then number the people in the class from 1/24 to 24/24, or 1. Many other rational numbers may be used.

Ongoing Assessment 5-2

1. (a)

$\frac{1}{2} + \frac{2}{3} = \frac{14}{12}$, or $\frac{7}{6}$

$\frac{1}{2} = \frac{6}{12}$ $\frac{2}{3} = \frac{8}{12}$

0 ———————— 1 $\frac{14}{12}$

(c) $\frac{5y - 3x}{xy}$ **(e)** $\frac{71}{24}$ or $2\frac{23}{24}$

2. (a) $18\frac{2}{3}$ **(c)** $-2\frac{93}{100}$

3. (a) $\frac{27}{4}$ **(c)** $\frac{-29}{8}$

5. (a) 1/3, high   **(c)** 3/4, low

6. (a) Beavers   **(c)** Bears   **(e)** Lions

8. (a) 1/2, high   **(c)** 3/4, high   **(e)** 1, low   **(g)** 3/4, low

9. (a) 2   **(c)** 0

10. (a) 1/4   **(c)** 0

11. (a) A   **(c)** T

13. (a) $\frac{1}{30}$ **(c)** $\frac{1}{60}$

14. $6\frac{7}{12}$ yards

16. $2\frac{5}{6}$ yards

18. (a) Team 4, $76\frac{11}{16}$ pounds

19. (a) $\frac{1}{2} + \frac{3}{4} \in Q$

(c) $\left(\frac{1}{2} + \frac{1}{3}\right) + \frac{1}{4} = \frac{1}{2} + \left(\frac{1}{3} + \frac{1}{4}\right)$

20. (a) $\frac{6}{4}, \frac{7}{4}, 2$; arithmetic, $\frac{1}{2} - \frac{1}{4} = \frac{3}{4} - \frac{1}{2} = 1 - \frac{3}{4} = \frac{5}{4} - 1$

(c) $\frac{17}{3}, \frac{20}{3}, \frac{23}{3}$; arithmetic; $\frac{5}{3} - \frac{2}{3} = \frac{8}{3} - \frac{5}{3} = \frac{11}{3} - \frac{8}{3} = \frac{14}{3} - \frac{11}{3}$

22. (a) 622/985

23. $1, \frac{7}{6}, \frac{8}{6}, \frac{9}{6}, \frac{10}{6}, \frac{11}{6}, 2$

24. (a) (i) $\frac{3}{4}$ (ii) $\frac{25}{12}$ or $2\frac{1}{12}$ (iii) 0

25. (a) $f(0) = {}^-2$ **(c)** $f({}^-5) = \frac{1}{2}$

26. (b) $\frac{1}{n} = \frac{1}{n+1} + \frac{1}{n(n+1)}$

Communication

27. It might be easier but she would be incorrect. Think of the numerator as the number of pieces of pie cut into the denominator's value of slices. Then to add the pieces of pie, we would add the numerators and have that number of pieces.

28. (a) Yes. If a, b, c, and d are integers, then $\frac{a}{b} - \frac{c}{d} = \frac{ad - bc}{bd}$ is a rational number.   **(c)** No. For example, $\frac{1}{2} - \left(\frac{1}{4} - \frac{1}{8}\right) \neq \left(\frac{1}{2} - \frac{1}{4}\right) - \frac{1}{8}$.   **(e)** No. Since there is no identity, an inverse cannot be defined.

29. (a) Like digits are being canceled.   **(c)** Numerators and denominators of fractional portions are both being subtracted.

Open-Ended

30. (a) It is not feasible to add the numbers in the table for Montana and Russia to determine the population density of the combined country and state because the population density is a ratio of the number of people per square mile in each case. The numbers of square miles are not the same in both cases.   **(c)** Answers will vary. For example, one question might be "If the population of Montana is removed from the United States' population, is the population density of the remaining portion of the United States 62?"

Cooperative Learning

31. Depending on the people interviewed, students may hear an answer like the following from a teacher: I use fractions in determining total grades for my classes. For example, if a paper is 1/2 of the grade and a test is another 1/3 of the grade, I need to know what fractional part of the grade is yet to be determined.

Review Problems

32. (a) The triangles created will vary.   **(b)** The ratio of the directed segments is always 1/1.   **(c)** The conjecture should be that all triangles created in this manner will have sides in the ratio of 1/1. This ratio of change in y-coordinates to change in x-coordinates is the slope of the line created.

33. (a) $\frac{2}{3}$ (b) $\frac{13}{17}$ (c) $\frac{25}{49}$ (d) $\frac{a}{1}$ or a (e) reduced

34. (a) equal (b) not equal (c) equal (d) not equal

35. (a) February (b) The answer depends upon whether the year is a leap year. If it is a leap year, the answer is 185/366; if not, the answer is 184/365. (c) Most people consider there to be 365 1/4 days in a year. As an improper fraction, this number is 1461/4.

36. We are considering $\frac{a}{b}$ and $\frac{a+x}{b+x}$ when $a < b$. $\frac{a}{b} < \frac{a+x}{b+x}$ because $ab + ax < ab + bx$.

Ongoing Assessment 5-3

1. (a) $\frac{1}{4} \cdot \frac{1}{3} = \frac{1}{12}$

2. (a) 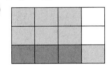 (c)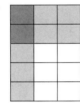

3. (a) $\frac{1}{5}$ (c) $\frac{za}{x^2y}$ (e) $\frac{44}{3}$ or $14\frac{2}{3}$

4. (a) $10\frac{1}{2}$ (c) $24{,}871\frac{1}{20}$

5. (a) $^-3$ (c) $\frac{y}{x}$

6. (a) 26 (c) 92 (e) 6 (g) 9

7. (a) 20 (c) 2

8. (a) 18 (c) 7

9. (a) Less than 1 (c) Greater than 2 (e) Greater than 4

10. (c)

11. $\frac{29}{36}$

13. 400

15. (a) 39 uniforms

16. $240

17. (a) $121,000 (c) $300,000

19. $225

20. (a) Peter, 30 min.; Paul, 25 min.; Mary, 20 min.

21. (a) $89\frac{3}{5}°$F

22. $2253\frac{1}{8}$

24. The arithmetic is not true. There are 3600 seconds in an hour. $3600/2264 \neq 1\ 1/2$

26. 120 1/4 lb

27. (a) $1/3^{13}$ (c) 5^{11} (e) $1/(^-5)^2$ or $1/5^2$ (g) a^2

28. (a) $(1/2)^{10}$ (c) $(2/3)^9$ (e) $(3/5)^{11}$

29. (a) False. $2^3 \cdot 2^4 \neq (2 \cdot 2)^{3+4}$ (c) False. $2^3 \cdot 2^3 \neq (2 \cdot 2)^{2 \cdot 3}$ (e) False. $(2+3)^2 \neq 2^2 + 3^2$ (g) False. $2^{2 \cdot 3} \neq 2^2 \cdot 2^3$

30. (a) 5 (c) $^-2$ (e) 0

31. (a) $2 \cdot 10^{11}$

32. (a) $x \leq 4$ (c) $x \geq 2$

33. (a) $\left(\frac{1}{2}\right)^3$ (c) $\left(\frac{4}{3}\right)^{10}$ (e) $\left(\frac{4}{3}\right)^{10}$

34. (a) 10^{10} (b) $10^{10} \cdot \left(\frac{6}{5}\right)^2 = 1.44 \cdot 10^{10} = 14.4$ billion

35. (a) $2S = 2\left(\frac{1}{2} + \frac{1}{2^2} + \ldots + \frac{1}{2^{64}}\right) = 1 + \frac{1}{2} + \frac{1}{2^2} + \ldots + \frac{1}{2^{63}}$ (c) $1 - \frac{1}{2^n}$

36. (a) $1\frac{49}{99}$

37. (i) $\frac{1}{32}, \frac{1}{64}$; geometric ratio $= \frac{1}{2}$ (ii) $\frac{-1}{32}, \frac{1}{64}$; geometric ratio $= \frac{-1}{2}$ (iii) $\frac{81}{256}, \frac{243}{1024}$; geometric ratio $= \frac{3}{4}$ (iv) $\frac{5}{3^5}, \frac{6}{3^6}$; not geometric, $\frac{2}{3^2} \div \frac{1}{3} \neq \frac{3}{3^3} \div \frac{2}{3^2}$

38. (a) $\frac{3}{4}$ (c) $\frac{3}{128}$

39. (a) $\frac{3}{2}, \frac{3}{4}, \frac{3}{8}, \frac{3}{16}, \frac{3}{32}$ (c) $\frac{3}{1024}$

40. (a) 32^{50}, since $32^{50} = (2^5)^{50} = 2^{250}$ and $4^{100} = (2^2)^{100} = 2^{200}$

41. (a) $n(n+1) + \left(\frac{1}{2}\right)^2$

42. (a) (i) $\frac{-4}{5}$ (ii) $\frac{-26}{17}$ (iii) $\frac{-14}{33}$ (c) $\frac{5}{4}$

43. (a) First 3, second 4, third 5. Guess 6. The guess is correct since $\left(1 + \frac{1}{1}\right)\left(1 + \frac{1}{2}\right)\left(1 + \frac{1}{3}\right)\left(1 + \frac{1}{4}\right)\left(1 + \frac{1}{5}\right) = 5\left(1 + \frac{1}{5}\right) = 6$ (c) $n + 2$

Communication

45. Never less than n. $0 < \frac{a}{b} < 1$ implies $0 < 1 < \frac{b}{a}$. The last inequality implies $0 < n < n \cdot \left(\frac{b}{a}\right)$. Also $n \div \left(\frac{a}{b}\right) = n \cdot \frac{b}{a} > n$.

47. One estimate might be 5 1/2 because 1/7 of 35 is 5 and 1/7 of 42 is 6; and 39 is approximately halfway between 35 and 42. Another reasonable estimate might be found by finding 1/13 of 39 to be 3 and then use two of these as 6. 2/13 is reasonably close to 2/14 or 1/7.

49. (a) $2 \div 1 \neq 1 \div 2$ (c) There is no rational number a such that $2 \div a = a \div 2 = 2$.

Open-Ended

50. Answers will vary about class use. 7 ounces

Cooperative Learning

51. The answers depend on the size of the bricks and the size of the joints. In all likelihood, the measurements will be made in fractions of inches for the size of the joints. The size of the bricks may be done in inches. An alternative is to measure in centimeters. Thus, all measurement is approximate and some rounding or estimation may occur.

Review Problems

52. (a) 17 minutes after the experiment started (b) $^-108°C$
53. (a) $\frac{25}{16}$ or $1\frac{9}{16}$ (b) $\frac{25}{18}$ or $1\frac{7}{18}$ (c) $\frac{5}{216}$ (d) $\frac{259}{30}$ or $8\frac{19}{30}$ (e) $\frac{37}{24}$ or $1\frac{13}{24}$ (f) $\frac{^-39}{4}$ or $^-9\frac{3}{4}$
54. 120 students

Ongoing Assessment 5-4

1. (a) 5:21
2. (a) 30 (c) $23\frac{1}{3}$
3. 36 lbs
5. $1.19
7. 64
9. (a) 42, 56
11. $14,909.09; $29,818.18; $37,272.73
13. 135
14. (a) 5/7
15. 120 ft
17. (a) 27
19. 312 lbs
20. (a) 2:5. Because the ratio is 2:3, there are $2x$ boys and $3x$ girls; hence, the ratio of boys to all students is $2x/(2x + 3x) = 2/5$.
21. (a) 2/3 tsp mustard seeds, 1 c scallions, $2\frac{1}{6}$ c beans.
(c) 7/13 tsp mustard seeds, $1\frac{8}{13}$ c tomato sauce, $\frac{21}{26}$ c scallions.
23. 35 ft
25. (b) $WL = 10$ (d) $L = 10/W$
26. (b) Let $\frac{a}{b} = \frac{c}{d} = \frac{e}{f} = r.$
Then $a = br$
$c = dr$
$e = fr.$
So, $a + c + e = br + dr + fr$
$a + c + e = r(b + d + f)$
$\frac{a + c + e}{b + d + f} = r.$

27. (a) $\frac{a}{b} = \frac{c}{d}$ implies $\frac{a}{b} + 1 = \frac{c}{d} + 1$, which implies $\frac{a + b}{b} = \frac{c + d}{d}$. (c) $\frac{a}{b} = \frac{c}{d}$ implies $\frac{a}{b} - 1 = \frac{c}{d} - 1$, which implies $\frac{a - b}{b} = \frac{c - d}{d}$. From part (a) and this last result, we have $\frac{a + b}{b} + \frac{a - b}{b} = \frac{c + d}{d} + \frac{c - d}{d}$, which implies $\frac{a + b}{a - b} = \frac{c + d}{c - d}.$

Communication

28. (a) 40/700 or 4/70 or 2/35 (c) For the first set, $\frac{\text{footprint length}}{\text{thighbone length}} = \frac{40}{100} = \frac{20}{50}$; i.e., a 50-cm thighbone would correspond to a 20-cm footprint. Thus it is not likely that the 50-cm thighbone is from the animal which left the 30 cm footprint. (Notice that $\frac{20}{50} \neq \frac{30}{50}$.)

29. No, the ratio of the prices is proportional to the ratio of the areas and not to the ratio of the diameters.

31. Yes. From $\frac{H}{M} = \frac{M}{S}$ and $M^2 = HS = (2.9 \cdot 10^{32})(1.7 \cdot 10^{-29}) = 4.93 \cdot 10^3$ kg. But then M is approximately 70 kg.

Open-Ended

33. (a) 57.6 lb per sq in.

Cooperative Learning

35. Just as financial analysts disagree upon how to rate stocks, students will as well.

Review Problems

36. (a) 0 (b) 5 (c) 588/13 or 45 3/13 (d) $x \leq 588/13$
37. All rational numbers except 0 have a multiplicative inverse.
38. The ratio of the areas of the squares is $r^2/1$.
39. 2, 2 1/4, 2 1/2, 2 3/4, 3
40. 256/9 or 28 4/9
41. 1/10
42. (a) $\frac{3 + 3}{3} \neq \frac{3}{3} + 3$ (b) $\frac{4}{2 + 2} \neq \frac{4}{2} + \frac{4}{2}$ (c) $\frac{ab + c}{a} \neq \frac{ab + c}{a}$ (d) $\frac{a \cdot a - b \cdot b}{a - b} \neq \frac{a \cdot a - b \cdot b}{a - b}$
(e) $\frac{a + c}{b + c} \neq \frac{a + e}{b + e}$

Chapter Review

1. (a)

   (c)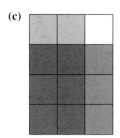

3. (a) $\dfrac{6}{7}$ (c) $\dfrac{0}{1}$ (e) $\dfrac{b}{1}$
4. (a) = (c) >
5. (a) $^-3, \dfrac{1}{3}$ (c) $\dfrac{^-5}{6}, \dfrac{6}{5}$
7. 17 pieces, $\dfrac{11}{6}$ yards left
8. (a) 15 (c) 4
9. $\dfrac{a}{b} \div \dfrac{c}{d} = x$ if and only if $\dfrac{a}{b} = \dfrac{c}{d} \cdot x$. $x = \dfrac{d}{c} \cdot \dfrac{a}{b}$ is the solution of the equation because $\dfrac{c}{d} \cdot \left(\dfrac{d}{c} \cdot \dfrac{a}{b}\right) = \dfrac{a}{b}$.
11. 76/100, 78/100, but answers may vary.
13. $333\dfrac{1}{3}$ calories
15. $240/1000 = 6/25$
17. The numerators of the rational numbers are integers and follow the properties of integers; the same is true of the denominators. Thus both the numerator and denominator of the answer are integers, and we can apply another property of integers to determine the sign of the answer.
19. 4/15
20. $^-12/10$ is greater than $^-11/9$ because $^-12/10 - (^-11/9)$ is a positive number.
21. (a) 3
22. $1\dfrac{7}{9}$ oz
23. (a) 6 mi
24. 560 fish

CHAPTER 6

Ongoing Assessment 6-1

1. (a) $0 \cdot 10^0 + 0 \cdot 10^{-1} + 2 \cdot 10^{-2} + 3 \cdot 10^{-3}$
   (c) $3 \cdot 10^2 + 1 \cdot 10^1 + 2 \cdot 10^0 + 0 \cdot 10^{-1} + 1 \cdot 10^{-2} + 0 \cdot 10^{-3} + 3 \cdot 10^{-4}$
2. (a) 4356.78 (c) 40,000.03
3. (b) 3.008 (d) 5,000,000.2
4. (a) $\dfrac{436}{1000} = \dfrac{109}{250}$ (c) $\dfrac{^-316,027}{1000}$ (e) $\dfrac{^-43}{10}$
5. (a) yes (c) yes (e) yes (g) no (i) no
7. (a) 13.492, 13.49199, 13.4919, 13.49183
8. (a) 0.0000000032 (c) 0.42
9. (b) $5.797 \cdot 10^6$
10. (b) 19,900
11. (a) $4.8 \cdot 10^{28}$ (c) $2 \cdot 10^2$
13. $100,000^3 = 1000^5$ and these are the greatest. To see this, write the numbers in scientific notation. Then it is easy to compare them:

$$100,000^3 = (1 \cdot 10^5)^3 = 1 \cdot 10^{15}$$
$$1000^5 = (1 \cdot 10^3)^5 = 1 \cdot 10^{15}$$
$$100,000^2 = (1 \cdot 10^5)^2 = 1 \cdot 10^{10}$$

15. Rhonda, Martha, Kathy, Molly, Emily

Communication

16. Answers vary; for example, you would use 2 blocks to represent the units, 3 flats to represent the tenths, 4 longs to represent the hundredths, and 5 cubes to represent the thousandths. In this way each base-ten block represents 1/10 of the base-ten block to its left and there are enough base-ten blocks for all the numbers used.
17. Answers vary; for example, a fraction can be written as a terminating decimal if it can be written as a fraction with a denominator that is a power of 10. The denominator can be written as a power of 10 if it contains only factors of 2 and 5. Other factors may appear in the denominator if the fraction is not in simplest form. For example, in 28/35 the denominator of 35 has a factor of 7 but in its simplest form 4/5, there is no factor of 7.

Ongoing Assessment 6-2

1. $231.24
2. 62.298 lb
4. $8.00/lb
6. (a) Approx. 6391 cm^3 (rounded to nearest cm^3)
8. (a) 5.4, 6.3, 7.2 (c) 0.0625, 0.03125, 0.015625
10. (a) 200 (c) 204 (e) 203.65
11. 19
14. Profit of $2098 (rounded to nearest dollar)
15. (b) $464
17. (a) 49,736.5281 (b) 41,235.6789
18. Answers vary; for example, $40 \cdot 6 = $240 plus 40\left(\dfrac{1}{4}\right) = $10 so her salary is $250.
21. (a) 181.56 (b) 148.551
22. (a) Yes (b) No
24. (a) $1.5^2, 2.5^2, 3 \cdot 4 + 0.25 = 3.5^2; 4 \cdot 5 + 0.25 = 4.5^2$

Communication

25. Answers vary; for example, the multiplication of decimals is exactly like the multiplication of whole numbers if the decimal points are ignored. The difference occurs when the decimal point is placed to obtain the final answer.

27. Answers vary; for example, many of the estimation techniques that work for whole-number division also work for decimal division. The long division algorithm is more efficient when good estimate are used. Also, estimates are important to determine whether an answer obtained by long division is reasonable. Estimation techniques can also be used to place the decimal point in the quotient when decimals are divided.

Review Problems

35. $14.0479 = 1 \cdot 10^1 + 4 \cdot 10^0 + 0 \cdot 10^{-1} + 4 \cdot 10^{-2} + 7 \cdot 10^{-3} + 9 \cdot 10^{-4}$

36. $\frac{35}{56}$ is a terminating decimal

37. (a) $3.32 \cdot 10^6$ **(b)** $2.367 \cdot 10^{-4}$

38. 0.625

Ongoing Assessment 6-3

1. (a) $0.\overline{4}$ **(c)** $0.\overline{27}$ **(e)** $0.02\overline{6}$ **(g)** $0.8\overline{3}$

2. (a) $\frac{4}{9}$ **(c)** $\frac{7}{5}$ **(e)** $\frac{-211}{90}$

4. (a) $1.\overline{6}, 2, 2.\overline{3}$

5. (a) 6 digits

6. (a) $0.\overline{076923}$ **(c)** $0.\overline{157894736842105263}$

7. (a) Answers vary; for example, 3.21, 3.213, 3.214

8. (a) 0.45

9. (a) Answers vary; for example, 0.751, 0.752, 0.753

10. (a) 8

11. (b) $\frac{1}{9999}$

12. (a) $\frac{2}{9}$ **(c)** $\frac{5}{9}$ **(e)** 10

13. (a) $\frac{5}{99}$ **(c)** $\frac{322}{99}$

Communication

14. Yes. For example, if there is a decimal portion replace the rightmost nonzero decimal digit, d, with $d - 1$ and tack on the repetend $\overline{9}$. It all decimal digits are zero, subtract 1 from the number and tack on $\overline{9}$. For example, $0.265 = 0.264\overline{9}$ and $34 = 33.\overline{9}$.

Open-Ended

16. (a) They each have six-digit repetends using each of the digits 1, 2, 4, 5, 7, 8 exactly once. **(c)** For example, $\frac{n}{14}$

Review Problems

18. $22,761.95

19. $2.35 \cdot 10^{13}$ miles

20. 0.077. The rule says that the placement should be four places or 0.0770. Because 0.077 and 0.0770 are equivalent, the rule still works.

Ongoing Assessment 6-4

1. (b) 3.2% **(d)** 20% **(f)** 15% **(h)** 37.5% **(j)** $16\frac{2}{3}\%$
(l) 2.5%

2. (b) 0.045 **(d)** $0.00\overline{285714}$ **(f)** 1.25 **(h)** 0.0025

3. (a) 4 **(c)** 25 **(e)** 12.5

5. (a) 2.04 **(c)** 60 **(e)** 300%

6. 63 boxes

8. $25,500

10. 20%

12. 18.4%

14. 100%

16. $5.10

18. $336

20. $3200

22. $\frac{325}{500}$; $\frac{325}{500} = \frac{650}{1000} = \frac{65}{100} = 65\%$, while $\frac{600}{1000} = \frac{60}{100} = 60\%$.

24. It is less than 30. 30 is the number plus 50% of the number.

26. $16.\overline{6}\%$

28. 11.1% (approximately)

30. $33\frac{1}{3}\%$

32. (a) $3.30 **(c)** $1.90

36. (a) 4% **(b) (i)** 44 **(ii)** 8.8%

Communication

38. Yes. 40% of $30 = \left(\frac{40}{100}\right)\left(\frac{30}{1}\right) = \frac{40 \cdot 30}{100} = \frac{30 \cdot 40}{100}$
$= \left(\frac{30}{100}\right)\left(\frac{40}{1}\right) = 30\%$ of 40.

40. Answers vary; for example, a price can go up 150%. If an item is bought for $100 it can be sold for $250. The price can't go down 150% because 100% is all there is and it can't go lower. A price cannot be less than 0.

42. Let x be the amount invested. The first stock option will yield $(1.15x) \cdot 0.85$ after two years. The second stock will yield $(0.85x) \cdot 1.15$. Because each yield equals $(1.15 \cdot 0.85)x$, the investments are equally good.

Review Problems

46. (a) 2.5 (b) 2.5
47. $\dfrac{6544}{900} = \dfrac{1636}{225}$
48. Answers vary, for example, $0.2\overline{1}$.
49. $\dfrac{5}{3}, 2.0\overline{5}, 2.\overline{15}, \dfrac{7}{3}, 2.5$
50. Answers vary; for example, $\dfrac{2}{3} = 0.\overline{6}$ and $\dfrac{1}{3} = 0.\overline{3}$ so $0.\overline{9} = 0.\overline{6} + 0.\overline{3} = \dfrac{2}{3} + \dfrac{1}{3} = 1$; $n = 0.\overline{9}$, so $10n = 9.\overline{9}$. $10n - n = 9.\overline{9} - 0.\overline{9}$, so $9n = 9$ and $n = 1$; $0.\overline{4} = \dfrac{4}{9}, 0.\overline{5} = \dfrac{5}{9}$, $0.\overline{6} = \dfrac{6}{9}, 0.\overline{7} = \dfrac{7}{9}, 0.\overline{8} = \dfrac{8}{9}$ so $0.\overline{9} = \dfrac{9}{9}$ or 1.

Ongoing Assessment 6-5

2. $5,460.00
3. $24.45
6. $64,800
8. $1944
10. Approximately $2.53
12. Approximately $3592.89
14. Approximately $7.026762 \cdot 10^8$
16. $81,628.83
18. $10,935

Communication

19. Let a be the original value of the house. Because it depreciates 10% each year for the first 3 years, using compound depreciation the price after 3 years will be $a(1 - 0.10)^3$ or $a \cdot 0.9^3$. Because of compound appreciation, after another 3 years the value of the house will be $a(0.9^3) \cdot 1.1^3$ or $a(0.9^3 \cdot 1.1^3)$, which equals approximately $a \cdot 0.9703$. Because $a \cdot 0.9703 < a$, the value of the house decreased after 6 years. The value of the house decreases by approximately 3%.

21. No. The percentages cannot be added because each time the percent is of a different quantity. The 15% is a savings of the original price of the fuel. The second savings of 35% is of the new price after the first savings. We could find the percent of savings as follows. For each $100 of the cost of fuel, the new cost after a 15% savings will be $85. When the second device is installed, an additional 35% is saved and the new cost is $(65/100) \cdot 85$. After the third device is installed an additional 50% is saved and the new cost is $(50/100) \cdot (65/100) \cdot 85$. The savings on a $100 initial cost is $100 - (50/100) \cdot (65/100) \cdot 85$, or $72.375. Because the savings were based on a $100 initial cost, the percent of total savings with all three devices is 72.375%.

Open-Ended

23. (b) Let a be an initial cost of an item that depreciates at a rate of $p\%$ each period of time. If n is the number of periods and $C(n)$ is the cost (as a function of n) after n periods, then $C(n) = a(1 - P/100)^n$.

Ongoing Assessment 6-6

2. (a) $\sqrt{6}$ (c) 3
3. $0.\overline{9}, 0.9\overline{98}, 0.\overline{98}, 0.9\overline{88}, 0.9, 0.\overline{898}$
4. (a) Yes (c) No (e) Yes
5. (a) 15 (c) 13 (e) Impossible
7. (b) False; $^-\sqrt{2} + \sqrt{2}$ (d) False; $\sqrt{2} - \sqrt{2}$
9. Answers vary. For example, assume the following pattern continues: 0.54544544454444 . . .
10. (a) R (c) Q (e) R
11. (a) Q, R (c) S, R (e) Q, R
13. (a) 64 (c) $^-64$ (e) All real numbers greater than zero.
14. 6.4 ft
15. (b) $\sqrt{363} = 11\sqrt{3}$.
16. (b) $2\sqrt[5]{3}$ (d) $^-3$
17. (b) $2, 2\sqrt[4]{1/2}, 2\sqrt[4]{1/4}, 2\sqrt[4]{1/8}, 1$
18. (a) 2^{10} (c) 2^{12}
19. (a) $\sqrt{3}$
20. (a) 4 (c) $^-4/7$
21. (a) Rational (c) Irrational

Communication

24. Answers vary; for example, to be rational $\sqrt{2}$ must be able to be written in the form $\dfrac{a}{b}$ where a and b are integers. When $\sqrt{2}$ is written as $\dfrac{\sqrt{2}}{1}$ it still fails to have a numerator that is an integer.
26. False: $\sqrt{64 + 36} \neq \sqrt{64} + \sqrt{36}$.
30. (a) Sometimes (if $a \geq 0$) (c) Always (e) Sometimes (if $a \geq 0$)

Review Problems

36. (a) 21.6 lb (b) 48 lb
37. (a) 418/25 (b) 3/1000 (c) $^-507/100$ (d) 123/1000
38. (a) $4.\overline{9}$ (b) $5.0\overline{9}$ (c) $0.4\overline{9}$
39. 3/12,500
40. 8/33
41. (a) 208,000 (b) 0.00038

Chapter Review

2. (a) $3 \cdot 10^1 + 2 \cdot 10^0 + 0 \cdot 10^{-1} + 1 \cdot 10^{-2} + 2 \cdot 10^{-3}$
4. 8
5. (a) $0.\overline{571428}$ (c) $0.\overline{6}$
6. (a) 7/25 (c) 1/3
7. (b) 307.6 (d) 300
8. (a) $4.26 \cdot 10^5$ (c) $2.37 \cdot 10^{-6}$
9. $1.451\overline{9}, 1.45\overline{19}, 1.4\overline{519}, 1.4519, {}^-0.134, {}^-0.13401, {}^-0.13\overline{401}$
10. (b) $1/\sqrt[4]{4}, 1/\sqrt[4]{16}, 1/\sqrt[4]{64}$
11. (a) $1.78341156 \cdot 10^6$ (c) $4.93 \cdot 10^9$ (e) $4.7 \cdot 10^{35}$
13. (b) 192 (d) 20%
14. (a) 12.5% (c) 627% (e) 150%
15. (a) $0.\overline{60}$ (b) $0.00\overline{6}$ (c) 1
17. $3.\overline{3}\%$
19. $5750
21. $80
22. $15,000
23. $15,110.69
24. (a) Irrational (c) Rational (e) Irrational
25. (a) $11\sqrt{2}$ (c) $6\sqrt{10}$
26. (a) No; $\sqrt{2} + \left({}^-\sqrt{2}\right)$ is rational. (c) No; $\sqrt{2} \cdot \sqrt{2}$ is rational.

CHAPTER 7

Ongoing Assessment 7-1

2. (b), (c)
3. (a) {0, 1, 2, 3, 4, 5, 6, 7, 8, 9} (c) {1, 3, 5, 7, 9}
4. (a) 5/26
5. (a) 3/8 (c) 4/8 or 1/2 (e) 0 (g) 1/8
6. (a) 26/52 or 1/2 (c) 28/52 or 7/13 (e) 48/52 or 12/13 (g) 3/52
7. (a) 4/12 or 1/3 (c) 0
8. (a) 1
10. (a) 8/36 or 2/9 (c) 24/36 or 2/3 (e) 0 (g) 10 times
12. (a) 18/38 or 9/19 (c) 26/38 or 13/19
14. (a) No (c) Yes (e) No (g) No
15. (a) 2/4 or 1/2 (c) 3/4
18. Assuming that the dealt cards are not put back into the deck, the answers are: (a) 20/50 or 2/5 (c) 0
19. The answers may vary depending on how the 6, 7, and 9 are formed. The following answers are based on a Casio digital watch. (b) 8/10 (d) 10/10 or 1
20. (a) The probability of students taking Algebra or Chemistry (b) The probability of students taking Algebra and Chemistry
22. (a) 45/80 or 9/16 (d) 30/80 or 3/8
23. 0.4
24. (a) 22

Cooperative Learning

33. (a) 1 and 2 and then 0 and 3

Ongoing Assessment 7-2

1. (a)

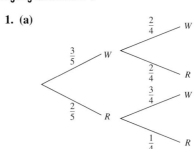

The probability that two balls are of different colors is
$$\frac{3}{5} \cdot \frac{2}{4} + \frac{2}{5} \cdot \frac{3}{4} = \frac{3}{5}$$

2. (a) {(1, 1), (1, 2), (1, 3), (2, 1), (2, 2), (2, 3)}
(c) {(1, 2), (2, 1), (2, 2), (2, 3)}
3. (a) 1/216
4. (a) 1/24 (c) 1/84
6. (a) Box 1, with probability 1/3 (Box 2 has probability 1/5)
7. (a) 64/75
9. (a) 1/5 (c) 11/15
11. 1/16
12. (a) 1/4 (b) 5/8 (c) 1/8
13. (a) 8/20 or 2/5
14. 2/16 or 1/8
15. (a) 1/320 (c) 0
16. (a) The first spinner. If you choose the first spinner, you win if and only if the spinning combinations are as follows:

Outcome on spinner A	Outcome on spinner B
4	3
6	3 or 5
8	3 or 5

The probability of this happening is
$$\frac{1}{3} \cdot \frac{1}{3} + \frac{1}{3} \cdot \frac{2}{3} + \frac{1}{3} \cdot \frac{2}{3} = \frac{5}{9}$$
If you choose spinner B, the probability of winning is only $\frac{4}{9}$.

17. (a) $\frac{5}{8}$
18. 1/32
22. (a) 1/25 (b) 8/25
23. (a) 100 square units (c) 1/625
24. 0.7
27. 2/5
28. 25/30 or 5/6
30. 69/3000 or 23/1000
31. (b) $\frac{1}{3} \cdot \frac{3}{11} + \frac{1}{3} \cdot \frac{3}{11} + \frac{1}{3} \cdot \frac{3}{11}$ or $\frac{3}{11}$
32. Each room has the same probability of being chosen. Hence it does not matter where the car is placed.

33. Billie-Bobby-Billie because the probability of winning two in a row is greater this way. Note, it does not say win two out of three.

Communication

37. The red ball in one of the boxes and the three white balls in the other box. In this case the probability of getting the red ball is $\frac{1}{2} \cdot 1 + \frac{1}{2} \cdot 0$ or $\frac{1}{2}$. In the three other arrangements, the probability of getting the red ball is smaller.

Cooperative Learning

41. If the first player chooses *TT* and the second *HT*, then the first player wins the game if and only if "Tails" appears on the first and second flips. The probability of this happening is $\frac{1}{4}$. The probabilities of winning the game for other choices are summarized below:

		\multicolumn{4}{c}{First Player's Choice}			
		HH	HT	TH	TT
Second	HH	—	0.50	0.75	0.50
Player's	HT	0.50	—	0.50	0.25
Choice	TH	0.25	0.50	—	0.50
	TT	0.50	0.75	0.50	—

Of the 12 possible games, only eight result in choices with equally likely probabilities. Therefore, the game is not fair.
42. (a) The game is not fair. You should choose spinner *A*.

Review Problems

43. (a) v   **(b)** iii   **(c)** ii   **(d)** i   **(e)** iv
44. (a) 1/30   **(b)** 0   **(c)** 19/30

Ongoing Assessment 7-3

1. Answers vary, for example, a black card might represent the birth of a boy and a red card might represent the birth of a girl. Choose a card to represent a birth.
2. (a) Answers vary, for example, you could use a random digit table. The numbers 1 through 9 could represent rain, and 0 could represent no rain.   **(c)** Approximately 0.52.
3. (a) Let 1, 2, 3, 4, 5, and 6 represent the numbers on the die and ignore the numbers 0, 7, 8, 9.   **(c)** Represent Red by the numbers 0, 1, 2, 3, 4; Green by the numbers 5, 6, 7; Yellow by the number 8; and White by the number 9.
5. To simulate Monday, let the digits 1 through 8 represent rain and 0 and 9 represent no rain. If rain occurred on Monday, repeat the same process for Tuesday. If it did not rain on Monday, let the digits 1 through 7 represent rain and 0, 8, and 9 represent dry. Repeat a similar process for the rest of the week.

6. Answers vary, for example, mark off blocks of two digits and let the digits 00, 01, 02, ..., 13, 14 represent contracting the disease and 15 to 99 represent no disease. Mark off blocks of six digits to represent the three children. If at least one of the numbers is in the range 00 to 14, then this represents a child in the three-child family having strep.
8. 1200 fish
9. (a) 7

Communication

12. One way is to turn left when an event number comes up, turn right when the number 1 comes up, and move straight when 3 or 5 come up.

Open-Ended

13. Answers vary, one could use a random digit table with blocks of two digits, or a spinner could be designed with 12 sections representing the different months. The spinner could be spun five times to represent the birthdays of five people. We could then keep track of how many times at least two people have the same birthday.
14. Answers vary, for example, use a random digit table. Let the digits 1–8 represent a win and the digits 0 and 9 represent losses. Mark off blocks of three. If only the digits 1–8 appear then this represents three wins in a row.
15. Let the 10 ducks be represented by the digits 0, 1, 2, 3, ..., 8, 9. Then pick a starting point in the table and mark off 10 digits to simulate which ducks the hunters shoot at. Count how many of the digits 0 through 9 are not in the 10 digits and this represents the ducks that escaped. Do this experiment many times and take the average to determine an answer. See how close your simulation comes to 3.49 ducks.

Cooperative Learning

16. (d) The probability of two boys and two girls is 6/16 or 3/8.
17. (d) $(1/2)^{10} = 1/1024$

Review Problems

20. No, you will win about 6/16 or 3/8 of the games.
21. (a) 1/4   **(b)** 1/52   **(c)** 48/52 or 12/13   **(d)** 3/4   **(e)** 1/2   **(f)** 1/52   **(g)** 16/52 or 4/13   **(h)** 1
22. (a) 15/19   **(b)** 56/361   **(c)** 28/171

Ongoing Assessment 7-4

1. (a) 12 to 40 or 3 to 10
3. 15 to 1
4. (a) 1/2   **(c)** 1023 to 1
5. 5/8
6. 1 to 1
7. 4 to 6 or 2 to 3
9. 1/27

10. $E = 1/6 (10) + 5/6 (^-2) = 10/6 + (^-10/6) = 0$.
Therefore, you should come out about even if you play a long time.
11. (c) $^-2/38$ or $^-1/19$ dollars
13. $10,000
15. (a) $7 (b) No

Communication

16. Odds are determined from probabilities. The *odds in favor* of an event, E, are determined by $P(E)/P(\overline{E})$. The *odds against* an event are determined by $P(E)/P(\overline{E})$.
17. Don't believe it. If the odds of getting AIDS are 68,000 to 1, then the probability of getting AIDS is 68,000/68,001. Therefore the probability of getting AIDS is almost certain. The article should have talked about the odds against getting AIDS.

Review Problems

22. (a) $\{1, 2, 3, 4\}$ (b) $\{\text{Red}, \text{Blue}\}$ (c) $\{(1, \text{Red}), (1, \text{Blue}), (2, \text{Red}), (2, \text{Blue}), (3, \text{Red}), (3, \text{Blue}), (4, \text{Red}), (4, \text{Blue})\}$
(d) $\{(\text{Blue}, 1), (\text{Blue}, 2), (\text{Blue}, 3) (\text{Blue}, 4), (\text{Blue}, 5), (\text{Blue}, 6), (\text{Red}, 1), (\text{Red}, 2), (\text{Red}, 3), (\text{Red}, 4), (\text{Red}, 5), (\text{Red}, 6)\}$
(e) $\{(1, 1), (1, 2), (1, 3), (1, 4), (2, 1), (2, 2), (2, 3), (2, 4), (3, 1), (3, 2), (3, 3), (3, 4), (4, 1), (4, 2), (4, 3), (4, 4)\}$
(f) $\{(\text{Red}, \text{Red}), (\text{Red}, \text{Blue}), (\text{Blue}, \text{Red}), (\text{Blue}, \text{Blue})\}$
23. The blue section must have 300°; the red has 60°.
24. 25/676

Ongoing Assessment 7-5

1. 224
2. 32
6. (a) True (c) False (e) True (g) True
8. 15
9. (a) 12 (c) 3360 (e) 3780
11. (a) 24,360
12. 792
13. 1/20
15. 1260
16. (a) 6
18. (a) 1/13
19. 2,598,960 different five-card hands (Order within the hand is not important.)
21. 1/25,827,165
23. 13440/59049 or approx. 0.228
24. (b) $\left(\frac{1}{6}\right)^2 \cdot \left(\frac{5}{6}\right)^6 \cdot {}_8C_2 \doteq 0.260476$
25. (a) $\left(\frac{6}{36}\right)^5$ or $\frac{1}{6^5}$
26. (a) $\frac{9}{133}$ (b) $1 - \frac{{}_{10}C_4}{{}_{22}C_4}$ or $\frac{203}{209}$
27. (a) $({}_{20}C_2 \cdot {}_{21}C_4 \cdot {}_4C_2) \div {}_{45}C_8$ (b) $\frac{{}_{25}C_8}{{}_{45}C_8}$

Communication

28. Answers vary. For example, the Fundamental Counting Principle (FCP) says that to find the number of ways of making several decisions in a row, multiply the number of choices that can be made for each decision. The FCP can be used to find the number of permutations. A permutation is an arrangement of things in a definite order. A combination is a selection of things in which the order is not important. We could find the number of combinations by using the FCP and then dividing by the number of ways in which the things can be arranged.
30. (a) There are ten choices
32. (a) $8! \cdot 3!$. If the family is considered a unit and each of the remaining people also a unit, we have 8 units. There are 8! ways to arrange the 8 units. For each of the 8! ways the family unit can be arranged in 3! ways and hence the number of seating arrangements is $8! \cdot 3!$.
33. 3840. Consider each couple as a unit. The five units can be arranged in 5! ways. For each of the 5! arrangements, each of the five couples can be arranged in 2 ways. Consequently there are $5! \cdot 2^5$ arrangements.

Open-Ended

34. (a) 10^6 or 1,000,000 (c) Answers vary, for example, you would first find the population of California and then experiment with using letters in the license plates. This would help because the choice is for 26 letters in a slot rather than 10 numbers.

Cooperative Learning

35. (c) The sums of the numbers in the rows are 1, 2, 4, 8, 16, 32, 64. The sum in the 10th row is $2^{10} = 1024$. (d) Yes, a similar relationship holds in all the rows for the entries in Pascal's triangle.

Review Problems

36. (a) 396/2652 or 33/221 (b) 1352/2652 or 26/51
37. 3/36 or 1/12
38. $E = \$0$ so the game is fair.

Chapter Review

2. (a) {Monday, Tuesday, Wednesday, Thursday, Friday, Saturday, Sunday} (c) 2/7
3. There are 800 blue ones, 125 red ones, and 75 that are neither blue nor red.
4. (a) Approximately 0.501 (c) 34,226,731 to 34,108,157
5. (a) 5/12 (c) 5/12 (e) 0
6. (a) 13/52 or 1/4 (c) 22/52 or 11/26
7. (a) 64/729
8. 6/25
9. 14/80 or 7/40

10. 7/45
11. 4 to 48 or 1 to 12
13. 3/8
14. $0.30
16. The expected value is $1.5. The expected average earnings are ⁻$0.5.
17. 900
19. 5040
21. (a) $5 \cdot 4 \cdot 3$ or 60   (c) 1/60
23. 2/5
27. (a) 1/8   (c) 1/16

12.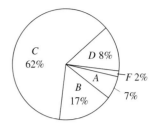

CHAPTER 8

Ongoing Assessment 8-1

2. (a) 225 million   (c) 550 million
3.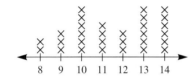

4. (a) 72, 74, 81, 81, 82, 85, 87, 88, 92, 94, 97, 98, 103, 123, 125   (c) 125 lbs
5.

7. (b) 50 cm
8. (a)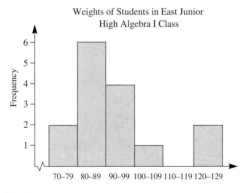

   3 | 4 represents 34 years old

(c) 20
9. (a) Approximately 3800 km

13. (b)

Classes	Tally	Frequency
$15–19	I	1
$20–24	II	2
$25–29		0
$30–34	II	2
$35–39	IIII	5
$40–44	IIII	4
$45–49	III	3
$50–54	IIII	4
$55–59	I	1
$60–64	III	3
		25

(c)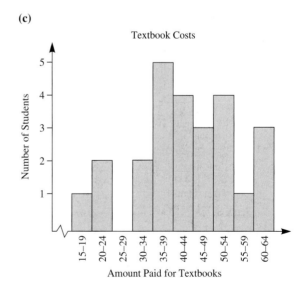

14. (a) Chicken   (c) Cheetah

15. (a) Women (c) Approximately 7 years
16. (a) Approximately $8400 (c) Approximately $7000
17. (a) Asia (c) It is about 2/3 as large. (e) 5:16
18. (a)

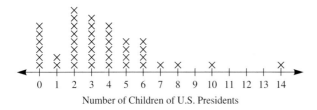

(c) 2
19. (a)

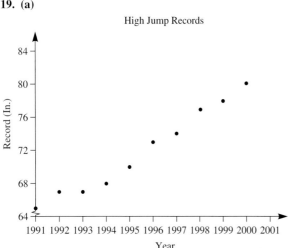

(b) Positive
20. (a) Negative (c) 22 years old

Communication

28. Answers vary, for example, the line graph is more useful because we can approximate the point midway between 8:00 A.M. and 12:00 noon and then draw a vertical line upward until it hits the line graph. An approximation for the 10:00 A.M. temperature can be obtained from the vertical line.

Cooperative Learning

33. Answers vary, for example, students may not be able to make definitive decisions other than for their class. You may want to check in department stores to see if they can determine the desired size.

Ongoing Assessment 8-2

1. (a) Mean = 6.625, median = 7.5, mode = 8
(c) Mean ≐ 19.9, median = 18, modes = 18 and 22
(e) Mean = 5.83, median = 5, mode = 5
2. (a) The mean, median, and mode are all 80.
3. 1500
5. $78.\overline{3}$
6. (a) $\bar{x} = 18.4$ years (c) 28.4 years
8. Approximately 2.59
10. $1880
11. (a) $41,275 (c) $38,000
12. (a) Balance beam—Olga (9.575); Uneven Bars—Lisa (9.85); Floor—Lisa (9.925)
14. 58 years old
15. (a) A (c) C
17. 96, 90, and 90
19. (a) (i) $\bar{x} = 5$, median = 5 (ii) $\bar{x} = 100$, median = 100 (iii) $\bar{x} = 307$, median = 307
20.

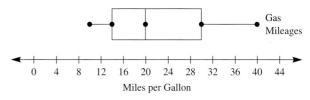

21. (a) A—$25, B—$50 (c) $80 at B
23. (a)

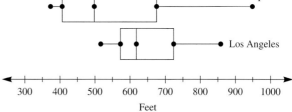

(b) There are no outliers.
26. (a) $s = 0$
27. (a) Approximately 76.8 (c) 71 (f) Approximately 12.5
28. (a) (i) Increase by $1000 (ii) Increase by $1000 (iii) Increase by $1000 (iv) Increase by $1000 (v) Stays the same (vi) Stays the same

Communication

30. One might use the mode. If you collect the data of all states and consider the most common age at which one could get a license, that would be the mode. Both the mean and median might be decimals, and no state would worry about a decimal age for a driver's license.
32. The government probably uses the mean of data collected over a period of time.
33. (a) Mean = 90; Median = 90; Mode = 90 (c) Mean

35. No. To find the average speed we divide the distance traveled by the time it takes to drive it. The first part of the trip took $\frac{5}{30}$, or $\frac{1}{6}$, of an hour. The second part of the trip took $\frac{5}{50}$, or $\frac{1}{10}$, of an hour. Therefore, to find the average speed we compute $\frac{10}{\frac{1}{6} + \frac{1}{10}}$ to obtain 37.5 mph.

Cooperative Learning

38. Answers will vary. To do this problem, the class will have to be divided in such a way that the choice of newspapers does not overlap. A teacher may want to discuss how one might do a random sample of newspapers.

Review Problems

39. Answers vary; for example, the number of men in the work force has increased by a factor of about 4 over the last hundred years while the number of women in the work force has increased by more than a factor of 15. As late as 1960 there were more than 200% more men in the work force than there were women. In 1970 this was no longer true and now in 1990 there are only about 20% more men.
40. Approximately, Midsize—176°, Large—33°, Luxury—54°, Small—97°.
41. (a) Everest, approximately 8500 m **(b)** Aconcagua, Everest, McKinley
42. (a)

History Test Scores

5	5
6	48
7	2334679
8	0255567889
9	00346

7 | 2 represents a score of 72

(b)

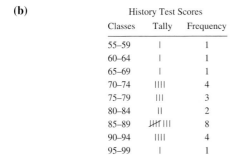

History Test Scores

Classes	Tally	Frequency
55–59	I	1
60–64	I	1
65–69	I	1
70–74	IIII	4
75–79	III	3
80–84	II	2
85–89	ИН III	8
90–94	IIII	4
95–99	I	1

(c)

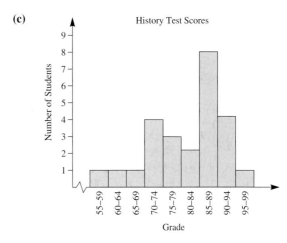

(d) Approximately 115°

Ongoing Assessment 8-3

1. Answers may vary in all parts of this question.
(a) A question to ask is whether the car is running. If it is not running, then there is no sound and the car would be quieter.
(c) 11% more fruit solids than 10% is very little. **(e)** "Up to" is very indefinite, as are the conditions under which the 30 was obtained. 6 mpg is "up to" 30 mpg. **(g)** How many dentists responded? **(i)** A question to ask is whether another airline flies to the city.
3. She could have taken a different number of quizzes during the first part of the quarter than the second part.
4. When the radius of a circle is doubled, the area is quadrupled, which is misleading since the population has only doubled.
5. The horizontal axis does not have uniformly sized intervals and both the horizontal axis and the graph are not labeled.
7. It could very well be that most of the pickups sold in the last 10 years were actually sold during the last two years. In such a case, most of the pickups have been on the road for only two years, and therefore the given information might imply but would not substantiate that the average life of a pickup is around ten years.
9. The three-dimensional drawing distorts the graph. The result of doubling the radius and the height of the can is to increase the volume by a factor of 8.
10. No labels, so we cannot compare actual sales. Also, there is no scale on the vertical axis.
12. (a) False, prices vary only by $30. **(c)** True
13. (a) This bar graph could have perhaps 20 accidents at the point where the scale starts. Then 38 in 1996 would appear to be almost double the 24 of 1988, when in fact it is only 58% higher.
14. Answers may vary, but one such would be 5, 5, 5, 5, 5, 5, 100, 100. The mean would be 28.75 and the median 5.

15. You could not automatically conclude correctly that the population of the West Coast has increased since 1790. However, based on the westward movement of the mean center of population, there would be strong suspicion that was the case.
18. A student would need to know the highest possible score that a person could make. Also the scores of other students would be important.
20. You could report the mode of a selected number of spots if enough spots were chosen at random. It is also possible that the mode would not exist. A median might be misleading, depending on the number of data points given. Also the mean would not be sufficient. A report of the mean, median, and standard deviation would be the most helpful of all "averages" studied.
22. Answers may vary. A sample size is needed. Then, one way to pick a random sample of adults in the town is to use the telephone book or a voter registration list. These methods will not list all adults in the town, but these are probably the most accessible sets of data. To pick the sample, one might roll a die and consider the number n that appears on it. Then starting at some point in the adult list, choose every nth person after the start on the list.

Review Problems

24. (a) 74.17 **(b)** 75 **(c)** 65 **(d)** 237.97 **(e)** 15.43
25. 27.74
26. 76.6
27.

Men's Olympic
100-m Run Times
1896–1964

```
 9 | 789
10 | 00011233334568888
11 | 00                    10 | 0 represents
12 | 0                          10.0 seconds
```

28.

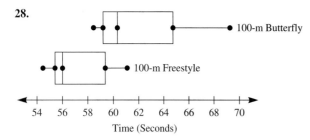

From examining the box-and-whisker plot, we can see that the times on the 100-m butterfly are much greater (relatively speaking) than the times on the 100-m freestyle.

Chapter Review

1. If the average is 2.41 children, then the mean is being used. If the average is 2.5, then the mean or the median might have been used.

3. (a) Mean = 30, median = 30, mode = 10
5. (a)

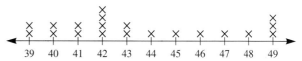

(b)

Miss Rider's Class
Masses in Kilograms

```
3 | 99
4 | 00112222334567899          4 | 0 represents
                                      40 kg
```

(c)

Miss Rider's Class
Masses in Kilograms

Mass	Tally	Frequency
39	II	2
40	II	2
41	II	2
42	IIII	4
43	II	2
44	I	1
45	I	1
46	I	1
47	I	1
48	I	1
49	III	3
		20

(d)

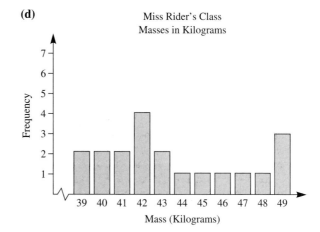

6. (a)

Test Grades

Classes	Tally	Frequency
61–70	⊥╫ I	6
71–80	⊥╫ ⊥╫ I	11
81–90	⊥╫ II	7
91–100	⊥╫ I	6
		30

(b)

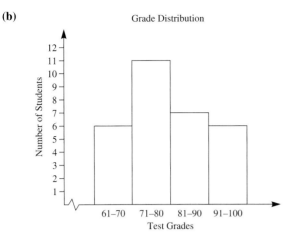

8. The widths of the bars are not uniform.
9. $2840
11. **(a)**

Life Expectancies
of Males and Females

Females		Males
	67	1446
	68	28
	69	156
	70	0049
	71	02235578
	72	01145
	73	16
7	74	
9310	75	
86	76	
88532	77	
9999854332211	78	
210	79	

7 | 74 | represents | 67 | 1 represents
74.7 years old   67.1 years old

(b)

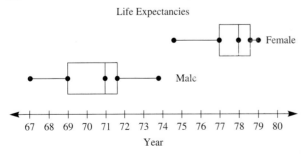

12. Larry was correct because his average was $3.2\overline{6}$ while Marc's was $2.7\overline{3}$.

13. **(a)** 360   **(c)** 350
14. **(a)** 67

(c)

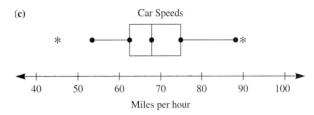

(e) 30%
15. **(a)** Positive   **(c)** 67 in.   **(e)** 50 lb
18. The advertisement is not reasonable. One would expect the snow to be okay, but with reports like those given, there is little information about the snow in the middle of the mountain. If it is cold enough to have 23 in. at the bottom of a hill, one would expect there to be snow all the way to the top. However, if the bottom and the top are shaded, then there could be much variation if a part of the hill were sunny.

CHAPTER 9

Ongoing Assessment 9-1

1. **(a)** For example: $\overleftrightarrow{BC}$ and $\overleftrightarrow{DH}$ or $\overleftrightarrow{AE}$ and $\overleftrightarrow{BD}$
   (c) No   **(e)** Point H **(g)** For example: $\overleftrightarrow{BF}$ and $\overleftrightarrow{DH}$
2. **(a)** $\angle EFB$
3. **(a)** $\varnothing$ **(c)** $\{A\}$ **(e)** $\overleftrightarrow{AC}$ and $\overrightarrow{DE}$ or $\overrightarrow{AD}$ and $\overleftrightarrow{CE}$
4. 20 pairs
6. **(a)** 110°   **(c)** 20°
7. **(a)** Approximately 36°
8. **(a) (i)** 41°31'10"   **(b) (i)** 54'
9. **(b)** 52°30'
10. 2:27
11. **(a)** 4   **(c)** 8
12. **(b)** $n(n-1)/2$
14. Answers vary depending on your Logo. For example:
   (a) TO ANGLE :SIZE
       FD 100 BK 100
       RT :SIZE FD 100
       BK 100 LT :SIZE
       END
   (c) TO PERPENDICULAR :LENGTH1 :LENGTH1/2
       FD :LENGTH1 BK :LENGTH2
       RT 90 FD :LENGTH2
       BK :LENGTH2 LT 90
       BK :LENGTH1/2
       END
16. **(a)** No   **(c)** Yes.
17. **(a)** 1   **(c)** $_nC_3$, or $\dfrac{n(n-1)(n-2)}{3 \cdot 2 \cdot 1}$, or $\dfrac{n(n-1)(n-2)}{6}$

Communication

18. (b) The fire is located near the intersection of the two bearing lines.
19. (a) Yes. Suppose the line is in the plane.

Cooperative Learning

21. (a) An angle of 20° can be drawn by tracing a 50° angle and a 30° angle, as shown in the following figure. Another 20° angle adjacent to the first 20° angle can be drawn in a similar way, thus creating a 40° angle.

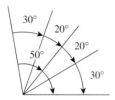

(b) Yes; answers vary. **(c)** All multiples of 10° from 10° to 180°.

Ongoing Assessment 9-2

1. (a) 1, 4, 6, 7, 8 **(c)** 6, 7
3. A concave polygon
5. (d) and **(e)** are impossible because each angle of an equilateral triangle has measure 60°.
6. (a) 35 **(c)** 4850
8. (a) and **(b)** represent rhombuses and rectangles.
9. (a) T, Q, R, H, G, I, F, J **(c)** W, D, A, Z, U, E **(e)** Y
10. (b) Answers vary; for example:
```
TO RECTANGLE :WIDTH :LENGTH
  REPEAT 2 [FD :WIDTH RT 90 FD
  :LENGTH RT 90]
END
```

Cooperative Learning

14. (a) The trapezoid definition is different.
15. (a) Reuleaux triangles aren't truly triangles. They are curves of constant width.
16. (b) A square cannot exist.

Review Problems

17. (a) 45 **(b)** $n(n-1)/2$
18. ∅, 1 point, 2 points, ray
19. (a) False. A ray has only one endpoint. **(b)** True
(c) False. Skew lines cannot be contained in the single plane.
(d) False. $\overrightarrow{MN}$ has endpoint M and extends in the direction of point N; $\overrightarrow{NM}$ has endpoint N and extends in the direction of point M. **(e)** True **(f)** False. Their intersection is a line.

Ongoing Assessment 9-3

2. (a) 60° **(c)** 60°
3. (a) Yes. A pair of corresponding angles are 50° each.
5. (a) 20
6. (a) 70° **(c)** 65°
7. (a) $x = 40°$ and $y = 50°$ **(c)** $x = 50°$ and $y = 60°$
8. (a) 60°
9. (a) 360° **(c)** 360°
10. 60, 84, 108, 132, 156
12. 111°
14. 135°
15. (a) 100°
16. (a) $m(\angle ACB) = 40°, m(\angle ABD) = 40°$

Communication

17. Place a long ruler or board so that it touches all the stairs and the ground. The angle α that the board makes with the ground measures the steepness of the stairs.

19. No, it is not possible. The sum of all the measures of the angles in $\triangle ABC$ must equal 180°, but $m(\angle ABC) + m(\angle ACB) + m(\angle BAC) = 90° + 90° + m(\angle BAC) > 180°$. Hence, the situation shown in the diagram is not possible.
20. (a) Five triangles will be constructed in which the sum of the angles of each triangle is 180°. The sum of the measures of the angles of all the triangles equals 5(180°), from which we subtract 360° (the sum of all the measures of the angles of the triangles with vertex P). Thus $5(180°) - 360° = 540°$.
22. (a) Choose a vertex and fold the triangle at that vertex so that the other two vertices fall on top of each other. Two angles of the triangle should fall on top of each other. Repeat at a second vertex.
23. (a) Divide the quadrilateral into two triangles: $\triangle ABC$ and $\triangle ACD$. The sum of the measures of the interior angles of each triangle is 180°, so the sum of the measures of the interior angles of the quadrilateral is 2(180°), or 360°.
24. No. Regular hexagons fit because the measure of each vertex angle is 120°, three hexagons fit to form 360°, and the plane can be filled. For a regular pentagon, the measure of each vertex angle is 108°, so pentagons cannot be placed

together to form 360° (360 is not divisible by 108) and the plane cannot be filled.

25. Answers vary. For example:
(a) `TO PARALLELOGRAM :L :W :A`
     `REPEAT 2 [FD :L RT 180- :A FD :W RT :A]`
   `END`
(c) `TO RHOMBUS :L :A`
     `PARALLELOGRAM :L :L :A`
   `END`
(e) Execute `RHOMBUS 50 90`

Open-Ended

27. Answers vary. The following are some properties of figures in a plane and corresponding properties on a sphere:

Property in a Plane	Property on a Sphere
Two lines in a plane may intersect in one point or may be parallel.	Two great circles (lines on a sphere) always intersect in two points.
Two points determine a unique line.	Two points do not always determine a unique great circle.
A line has no length, and any point on the line separates it into two parts.	A great circle has a finite length ($2\pi r$ where r is the radius of the sphere) and is not separated into two parts by a point on the circle.

Cooperative Learning

28. (a) If $m(\angle A) = \alpha$ and $m(\angle B) = \beta$, then $m(\angle D) = 180 - \frac{\alpha}{2} + \frac{\beta}{2}$. **(c)** Answers vary. The following is a possible solution. From (a):

$$m(\angle D) = 180 - \left(\frac{\alpha + \beta}{2}\right) = \frac{360 - (\alpha + \beta)}{2}.$$ Because $\alpha + \beta = 180 - m(\angle C)$, we get the following:

$$m(\angle D) = \frac{360 - (180 - m(\angle C))}{2}$$
$$= \frac{360 - 180 + m(\angle C)}{2}$$
$$= 90 + \frac{1}{2} m(\angle C).$$

Because the answer for $m(\angle D)$ depends only on $m(\angle C)$, the conjecture is justified.

29. (a) Because congruent supplementary angles are formed **(c)** When A and C are folded, congruent supplementary angles are formed and hence, each is a right angle. When B is folded along $\overline{BB'}$, the crease $\overline{DE}$ formed is perpendicular to $\overline{BB'}$ (again because congruent supplementary angles are formed), consequently, $\overline{DE} \parallel \overline{GF}$. Thus D and E are also right angles.

Review Problems

30. (a) All angles must be right angles, and all diagonals are the same length. **(b)** All sides are the same length, and all angles are right angles. **(c)** This is impossible because all squares are parallelograms.
31. Yes. Thinking like this will help in the development of circumference and area formulas.
32. (a) Two sets of parallel sides—A, B, C, D, E, F, G, I; one set of parallel sides—I and J (exactly one);
(b) Four right angles—D, F, G; exactly two right angles—I
(c) 4 congruent sides—A, B, C, F, G;
Two pairs of congruent sides only—E, H;
One pair of congruent sides only—J
No sides congruent—I
(d) D, F, G, J
33. Answers will vary but should include the following:
- Congruent opposite sides describe a parallelogram.
- All sides congruent describe a rhombus (and square).
- Four right angles describe a rectangle (and square).
- Two pairs of consecutive sides congruent describe a kite.

Ongoing Assessment 9-4

1. (a) Quadrilateral pyramid **(c)** Pentagonal pyramid
2. (a) A, D, R, W **(c)** $\triangle ARD, \triangle AWD, \triangle AWR, \triangle WDR$
4. (a) 5 **(c)** 4
5. (a) True **(c)** True **(e)** False **(g)** False
8. Answers may vary.
(a)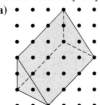

9. (a) Hexagonal pyramid **(c)** Cube **(e)** Hexagonal prism
10. (a) iv
11. (a) i, ii, and iii
13. (a)

14. (a) (2)
16. (a) Parallelogram, rectangle, square, triangle
17. (a) $10 + 7 - 15 = 2$

19.

	Pyramid	Prism
(a)	$n+1$	$n+2$
(c)	$2n$	$3n$

Communication

20. 3. Each pair of parallel faces could be considered bases.
22. Both could be drawings of a quadrilateral pyramid. In **(a)**, we are directly above the pyramid, and in **(b)**, we are directly below the pyramid.

Cooperative Learning

26. (a) Parallelogram, rectangle, square, scalene triangle, isosceles triangle, equilateral triangle

Review Problems

27. Yes because the sum of the angles of a triangle is 180°.
28. $m(\angle BCD) = 60°$.
29. 140°
30. (a) True **(b)** True **(c)** False (e.g., equilateral triangle)
31. Lines in the same plane and perpendicular to the same line are parallel. This is because corresponding angles are right angles and hence are congruent.

Ongoing Assessment 9-5

1. All except (d), (f), and (i) are traversable; (a), (b), and (j) are Euler circuits (starting and stopping points are the same)

(a)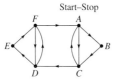

Path: *ABCACDEFDFA*; any point can be a starting point.

(c)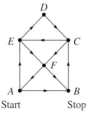

Path: *ABCFAEDCEFB*; only points *A* and *B* can be starting points.

(e)

Start
Path: *ABCBDCAD*; only points *A* and *D* can be starting points.

(g)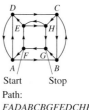

Start   Stop
Path: *FADABCBGFEDCHEHG*; only points *F* and *G* can be starting points.

(i) Not traversable; has more than two odd vertices.

3.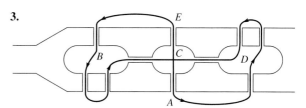

Path: *CEBABCADEDC*; any point can be a starting point.

4. (b) Network (i) is not traversable because it has four odd vertices. Network (ii) has two odd vertices, so it is traversable, as shown in the following figure:

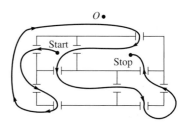

6. It is not possible.

8.

All vertices are even, so the trip is possible. It makes no difference where she starts.

Communication

9. Answers vary. Possibilities are as follows:

(a)   (b)

Open-Ended

11. (a) A postal worker would have to travel each route twice (both sides). A traveling salesperson would only be interested in vertices (cities or stores). A highway inspector would travel each arc only once.

12. Answers vary; for example:

Chapter Review

1. (a) $\overleftrightarrow{AB}$, $\overleftrightarrow{BC}$, and $\overleftrightarrow{AC}$ **(c)** $\overline{AB}$
2. (a) Answers may vary. **(c)** $\overrightarrow{AQ}$
4. (a) No. The sum of the measures of two obtuse angles is greater than 180°, which is the sum of the measures of the angles of any triangle.
6. (a) Given any convex n-gon, pick any vertex and draw all possible diagonals from this vertex. This will determine $n - 2$ triangles. Because the sum of the measures of the angles in each triangle is 180°, the sum of the measures of the angles in the n-gon is $(n - 2)180°$.
9. $m(\angle 3) = m(\angle 4) = 45°$.
10. 35°8′35″
11. (a) 60° **(c)** 120°
12. (a) Alternate interior angles are congruent.
(c) $m(\angle B) + m(\angle C) + m(\angle BAC) = m(\angle BAD) + m(\angle DAE) + m(\angle BAC) = 180°$.
13. Sketches may vary, but the possibilities are the empty set, a single point, a segment, a quadrilateral, a triangle, a pentagon, and a hexagon. Various types of quadrilaterals are possible.
14. 8

16. Answers may vary.

(a)   (c)   (e)

17. $_nC_3$, or $\dfrac{n(n-1)(n-2)}{3 \cdot 2 \cdot 1} = \dfrac{n(n-1)(n-2)}{6}$.

19. No; see definition.

21.

(b) (i)

Path: ABCDEFACEA; any point can be used as a starting point.

(ii)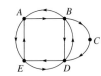

Path: ABCDAEDBE; points A and E are possible starting points.

(iii)

Path: BDEABAEDBCD; points B and D are possible starting points.

CHAPTER 10

Ongoing Assessment 10-1

1. (b) The side of greater length is opposite the angle of greater measure.
2. (c) Scalene right **(d)** No triangle is possible.
4. 22 triangles
5. (a) Yes; SAS
7. The lengths of the wires must be the same because they are congruent parts of the congruent triangles formed.
11. (b) Because $\triangle ABC \cong \triangle BCA$, $\angle A \cong \angle B$ and $\angle B \cong \angle C$. Hence $\angle A \cong \angle B \cong \angle C$.
12. (a) One method of placing the six points on the circle is to use a compass with opening the length of a radius of the circle. **(c)** The triangles are congruent by SSS. The sides are the same length as a radius of the circle.
13. (a) F is the midpoint of both diagonals. We can show $\triangle ABD \cong \triangle CBD$ by SSS and then $\angle BDC \cong \angle BDA$ by CPCTC. $\triangle AFD \cong \triangle CFD$ by SAS, so $\overline{AF} \cong \overline{FC}$; thus F is the midpoint of $\overline{AC}$. A similar argument will show $\overline{BF} \cong \overline{FD}$.

14. (a) A parallelogram. Let $ABCD$ be the quadrilateral, with E the intersection point of its diagonals. Show that $\triangle AED \cong \triangle CEB$ and that $\triangle BEA \cong \triangle DEC$. Use congruent alternate interior angles to show $\overline{BC} \| \overline{AD}$ and $\overline{AB} \| \overline{CD}$.
(c) Rhombus
15. (a) The angles formed by the diagonals of a rhombus are right angles.
16. A parallelogram. In quadrilateral $ABCD$ let $\overline{AB} \cong \overline{CD}$ and $\overline{BC} \cong \overline{AD}$. Prove that $\triangle ABC \cong \triangle CDA$ and conclude that the $\overline{BC} \| \overline{AD}$. Similarly show $\triangle ABD \cong \triangle DCB$ and conclude that $\overline{AB} \| \overline{DC}$.
20. (a) 6   **(c)** $n(n-1)(n-2) \ldots 3 \cdot 2 \cdot 1$ or $n!$

Communication

27. (a) $\triangle ABC \cong \triangle ADC$ by SSS. Hence $\angle BAC \cong \angle DAC$ and $\angle BCM \cong \angle DCM$ by CPCTC. Therefore $\overleftrightarrow{AC}$ bisects $\angle A$ and $\angle C$.

Open-Ended

31. (i) Answers vary. For example, each of the following constitutes sufficient conditions to define a parallelogram:
  a1. A quadrilateral in which each pair of opposite sides is parallel.
  a2. A quadrilateral in which each pair of opposite sides is congruent.
  a3. A quadrilateral in which each pair of opposite angles is congruent.
  a4. A quadrilateral whose diagonals bisect each other.
  a5. A quadrilateral in which a pair of opposite sides is parallel and congruent (?).

Ongoing Assessment 10-2

1. (d) Infinitely many triangles are possible.
3. (a) Yes; ASA   **(c)** No; SSA does not assure congruence.
5. (b) None   **(f)** Square
8. (a) If one leg and an acute angle of one right triangle are congruent, respectively, to a leg and an acute angle of another right triangle, the triangles are congruent.
 Also if the hypotenuse and an acute angle of a right triangle are congruent, respectively, to the hypotenuse and an acute angle of another right triangle, the triangles are congruent.
9. (b) True   **(e)** True   **(f)** False; A counterexample can be seen in a trapezoid in which two consecutive angles are right angles but the other two are not.
10. (c) No; any parallelogram with a pair of right angles must have right angles as its other pair of angles and hence be a rectangle.
12. The quadrilateral formed must be a rhombus because all the sides are congruent.
15. (a) The sides which are not bases are congruent.
 (b) The diagonals are congruent.

16. $\dfrac{a-b}{2}$

18. (b) Parallelogram. **(c)** Suppose $ADCB$ in part (a) is a parallelogram. Use SAS to show that $\triangle EDH \cong \triangle GBF$ and conclude that $\overline{EH} \cong \overline{GF}$. Similarly, show that $\triangle ECF \cong \triangle GAH$ and hence that $\overline{EF} \cong \overline{GH}$. Next use SSS to prove that $\triangle EFG \cong \triangle GHE$. Now conclude that $\angle GEH \cong \angle EGF$ and consequently that $\overline{FG} \| \overline{EH}$. Similarly, show that $\overline{EF} \| \overline{HG}$.
19. (b) The lengths of the sides of two perpendicular sides of the rectangles must be equal.
21. (a) Use the definition of a parallelogram and ASA to prove that $\triangle ADB \cong \triangle CDB$ and $\triangle ADC \cong \triangle CBA$.   **(c)** Hint: Prove that $\triangle ABF \cong \triangle CDF$.
22. (a) Answers may vary.
```
    TO RHOMBUS :SIDE :ANGLE
      REPEAT 2 [FD :SIDE RT (180 -
      :ANGLE) FD
      :SIDE RT :ANGLE]
    END
```
(c)
```
    TO SQ.RHOM :SIDE
      RHOMBUS :SIDE 90
    END
```

Review Problems

27. The triangles that are congruent to triangle ABC are triangles AED, CDE, and BCD. They are all congruent by SAS. Students may need to cut these triangles out and compare shapes.
28.–29. Constructions
30. (a) Yes: SAS   **(b)** Yes: SSS   **(c)** No

Ongoing Assessment 10-3

2. Constructions. The advantages and disadvantages of each may be discussed. The Mira is easy to use when the paper on which the constructions are to be performed may not be altered. The compass and straightedge is the classical way to do constructions. Paperfolding adds a tactile approach to the problem. The geometric drawing utility demands that exact measurements must be used on the screen unless you want similar figures.
3. (b) The altitude of the triangle is along the cable.
4. (a) The perpendicular bisectors of the sides of an acute triangle meet inside the triangle.   **(c)** The perpendicular bisectors of the sides of an obtuse triangle meet outside the triangle.
8. (a) The perpendicular bisector of a chord of a circle contains the center of the circle.   **(c)** Hint: Construct two non-parallel chords and find their perpendicular bisectors. The intersection of the perpendicular bisectors is the center of the circle.
12. (a) $\overrightarrow{PQ}$ is the perpendicular bisector of $\overline{AB}$.   **(c)** $\overrightarrow{PQ}$ is the angle bisector of $\angle APB$; $\overrightarrow{QC}$ is the angle bisector of $\angle AQB$.

14. (b) Construct two perpendicular segments bisecting each other and congruent to the given diagonal. **(d)** Without the angle between the sides being given, there is no unique parallelogram. **(f)** This is impossible because the sum of the measures of the angles would be greater than 180°.
15. (a) Construct an equilateral triangle and bisect one of its angles.
17. (a) Since the triangles are congruent, the acute angles formed by the hypotenuse and the line are congruent. Since the corresponding angles are congruent, the hypotenuses are parallel (the line formed by the top of the ruler is the transversal).
20. Hint: Construct two perpendicular diameters. The ends of the diameters are the vertices of the square.

Communication

24. Answers may vary. Most students will probably lobby to have other tools included as construction tools.

Open-Ended

25. Most students will probably say that there are more perpendiculars from a point to a line using the North Pole as a point and the equator as a line. All lines of longitude intersect at the North Pole and all are perpendicular to the equator.

Review Problems

29. $\triangle ABC \cong \triangle DEC$ by ASA. ($\overline{BC} \cong \overline{CE}$, $\angle ACB \cong \angle ECD$ as vertical angles, and $\angle B \cong \angle E$ as alternate interior angles formed by the parallels $\overline{AB}$ and $\overline{ED}$ and the transversal $\overleftrightarrow{EB}$.) $\overline{AB} \cong \overline{DE}$ by CPCTC.
30. Construction
31. (a) Through C draw a line parallel to $\overleftrightarrow{AB}$. Let E be the point where the line intersects $\overline{AD}$. Because $ABCE$ is a parallelogram ($\overleftrightarrow{BC} \| \overleftrightarrow{AE}$ and $\overleftrightarrow{AB} \| \overleftrightarrow{EC}$) we have $AB = CE$. (Recall that in a parallelogram opposite sides are congruent, per properties of a parallelogram.) This, along with the hypothesis $AB = CD$, implies that $CE = CD$. Thus $\triangle ECD$ is an isosceles triangle. Hence $\angle E \cong \angle D$. Because $\angle E$ and $\angle A$ are corresponding angles formed by the parallel lines $\overleftrightarrow{AB}$ and $\overleftrightarrow{CE}$ and the transversal $\overleftrightarrow{AD}$, it follows that $\angle A \cong \angle E$. Consequently $\angle A \cong \angle D$. **(b)** $\triangle ABD \cong \triangle DCA$ by SAS since $AB = DC, AD = DA$ and $\angle BAD \cong \angle CDA$. Consequently $\overline{BD} \cong \overline{CA}$ (CPCT).
32. If $\angle A$ is not the right angle the triangles are congruent. If $\angle A$ is the right angle, the triangles are not necessarily congruent.

Ongoing Assessment 10-4

1. (a) Yes; AAA   **(e)** Yes; radii are proportional.   **(f)** No
3. (c) The triangles are similar if the corresponding sides are proportional.
5. Answers may vary.   **(a)** Two rectangles, one of which is a square and the other is not.

7. (b) (i) 2/3   **(ii)** 1/2   **(iii)** 3/4   **(iv)** 3/4
8. (b) 24/7   **(c)** 3
10. (a) (1) $\triangle ABC \sim \triangle ACD$ by AA since $\angle ADC$ and $\angle ACB$ are right angles, and $\angle A$ is common to both. (2) $\triangle ABC \sim \triangle CBD$ by AA since $\angle CDB$ and $\angle ACB$ are right angles and $\angle B$ is common to both. (3) Using (1) and (2), $\triangle ACD \sim \triangle CBD$ by the transitive property.
12. 15 m
13. 9 m
16. Place the projector so that the slide is 23 ft 9 in. from the screen.
17. (a) (i) Connect B with D and apply Theorem 10-8 to $\triangle ABD$ and then $\triangle BCD$. **(ii)** Theorems 10-7 and 10-8 imply that $MP = \frac{1}{2}a$ and $PN = \frac{1}{2}b$. Thus $MN = MP + PN = \frac{1}{2}a + \frac{1}{2}b = \frac{1}{2}(a+b)$.
18. (c) rhombus

20. The perimeters have ratio $1/k$, because all sides are in this proportion. The sum must be in the same proportion.
23. 10 cm.
24. Answers may vary.
(a)
```
TO RECTANGLE :LEN :WID
    REPEAT 2 [FD :LEN RT 90 FD :WID RT
    90]
END
TO SIM.RECT :LEN :WID
    RECTANGLE :LEN*2 :WID*2
END
```

Communication

25. Any two cubes are similar because they have the same shape.

Cooperative Learning

29. (b) The following are two different size triangles with the given data (the given angle measures are approximate). The triangles are approximately similar but not congruent. (The ratio of the corresponding sides is $\frac{80}{100}$ or $\frac{4}{5}$.) Hence, the surveyor and the architect could both have been correct in their conclusions.

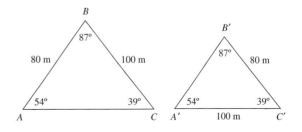

Review Problems

30. No; the image is two-dimensional while the original person is three-dimensional.

31. Start with the given base and construct the perpendicular bisector of the base. The vertex of the required triangle must be on that perpendicular bisector. Starting at the point where the perpendicular bisector intersects the base, mark on the perpendicular bisector a segment congruent to the given altitude. The endpoint of the segment not on the base is the vertex of the required isosceles triangle.

32. Construct an equilateral triangle with the given side, then construct the perpendicular from any vertex to the opposite side.

33. Answers may vary. Students may suggest that angles of measure 45° be constructed with the endpoints of the hypotenuse as vertices of the 45° angles and the hypotenuse as one of the sides of the angles. Both angles need to be constructed on the same side of the hypotenuse.

34. Yes. Use the vertical angles to justify this result.

Ongoing Assessment 10-5

2. (b)   (e)

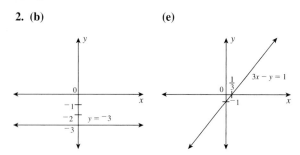

4. (b) $C = 5/9(F - 32)$
5. (c) $y = 4x/3 + 4$ (e) $y = (1/3)x$
6. (c) $y = 1$ (e) $y = x - 1/2$
8. (b) x is any real number; $y = 1$.
9. Perimeter = 12 units, Area = 8 sq units
10. (c) $y = 5$
11. (a) 1/3   (c) 0   (e) 20,000
13. Answers vary. For example: (a) Using (60, 18) and (50, 8), the slope of the line is $(18 - 8)/(60 - 50) = 1$. So $C = T + b$. Substitute in the point (50, 8) to see $b = {}^-42$. Thus $C = T - 42$.
15. (a) (2, 5), unique solution   (d) Infinitely many solutions; x can be any real number and $y = \dfrac{4x - 1}{6}$.
18. (a) $2000
19. 17 quarters, 10 dimes

Communication

23. Tell Jonah that $5 - 5x$ is not equal to 0 (his order of operations is incorrect). We have $5 - 5x = 5 \cdot 1 - 5x = 5(1 - x)$ which is not 0.

Cooperative Learning

28. $x = 10$

29. Yes. Because M, N, and P are midpoints of the corresponding sides the *Midsegment Theorem* (Theorem 10-7) implies that $\overline{MN} \parallel \overline{AC}$, $\overline{NP} \parallel \overline{AB}$ and $\overline{MP} \parallel \overline{BC}$. Thus $AMNP$, $MNCP$ and $MPNB$ are parallelograms. Because a diagonal of a parallelogram divides it into two congruent triangles, we have $\triangle AMP \cong \triangle NPM$, $\triangle NPM \cong \triangle PNC$ and $\triangle NPM \cong \triangle MBN$. Thus the four triangles are congruent. Because $\overline{MP} \parallel \overline{BC}$, corresponding angles created by the parallels and the transversals $\overleftrightarrow{AB}$ and $\overleftrightarrow{BC}$ are congruent. Hence by AA, $\triangle AMP \sim \triangle ABC$.

30. $\dfrac{12}{7}$

31. $BE = 8.2$

Chapter Review

1. (b) $\triangle GAC \cong \triangle EDB$ by SAS   (g) $\triangle ABD \cong \triangle CBE$ by SSS

2. A parallelogram. $\overline{EC} \parallel \overline{AF}$ because these segments are on parallel sides of the square. We will show that $\overline{AE} \parallel \overline{CF}$. Notice that $\triangle ADE \cong \triangle CBF$ by SAS. Hence, $\angle DEA \cong \angle BFC$. Since $\angle DEA \cong \angle EAF$ (alternate interior angles between the parallels $\overleftrightarrow{DC}$ and $\overleftrightarrow{AB}$ and the transversal $\overleftrightarrow{AE}$), it follows that $\angle EAF \cong \angle CFB$. Consequently, $\overline{AE} \cong \parallel \overline{CF}$. Thus $AECF$ is a parallelogram.

4. (a) $x = 8$ cm; $y = 5$ cm.
8. (a) $\triangle ACB \sim \triangle DEB$ by AA. $x = 24/5$ in.
11. (a) (ii) and (iii)
12. 6 m
13. 256/5 m
15. (b) $y = \dfrac{1}{3}(x + 3)$
16. There is no single line through the points with the given coordinates because using two of the points the slope is 3/4 while using one of those points with the third, the slope is $^-4/7$.
17. (a) (4.2, $^-$0.6)   (b) $\left(\dfrac{10}{9}, \dfrac{4}{3}\right)$ (c) no solution, parallel lines

CHAPTER 11

Ongoing Assessment 11-1

1. (a) 0.9 cm   (c) 8 cm   (e) 0.7 cm   (g) 7.3 cm
2. (b) 14,400   (d) 31
7. (a) 0.35, 350   (c) 0.035, 3.5   (e) 200, 2000
8. (a) 10.00   (c) 10.0   (e) 195.0   (g) 40.0
9. 6 m, 5218 mm, 245 cm, 700 mm, 91 mm, 8 cm
12. (a) 1 cm   (c) 3000 m   (e) 3500 cm   (g) 64.7 cm
14. (a) Can be   (c) Cannot be. In (b) and (c) the numbers cannot be the lengths of the sides of a triangle because in

(b) $10 + 40 = 50$ and in (c) $260 + 14 < 410$, each of which contradicts the triangle inequality.
17. (a) Answers vary; for example, add 4 squares to one row to form a 7 by 2 rectangle.   (b) 8 squares
18. (b) $3/\pi$ m   (d) 46 cm
19. (b) 6π cm   (d) $6\pi^2$ cm
21. πr
22. (b) About $4.1 \cdot 10^{13}$ km   (d) 2495 hr, or about 104 days
23. (b) 1032 m/sec
26. (a) 0.5 m   (c) 0.005 m or 5 mm

Communication

29. The height is $3d$, where d is the diameter of a tennis ball. The perimeter is πd or about $3.14d$, which is greater than $3d$.

Open-Ended

31. (a) Answers vary. For example: 1-1-1 and 2-2-1 work, 2-3-6 and 1-2-3 do not.

Ongoing Assessment 11-2

1. (a) cm^2, in^2.   (c) m^2 or cm^2, yd^2 or in^2.   (e) m^2, yd^2
3. (a) 0.0588 m^2, $58,800$ mm^2 (c) $15,000$ cm^2, $1,500,000$ mm^2 (e) 0.0005 m^2, 500 mm^2
4. (a) $444.\overline{4}$ yd^2 (c) 6400 acres
5. (a) 4900 m^2 (c) 0.98 ha
6. (a) 3 sq. units   (c) 2 sq. units   (e) 6 sq. units
8. (a) 20 cm^2 (c) 7.5 m^2 (e) 600 cm^2
9. (a) 9 cm^2 (c) $(2\sqrt{21} - 2\sqrt{5})$ cm^2, or approx. 4.69 cm^2 (e) 84 cm^2
10. (a) (i) 1.95 km^2 (ii) 195 ha
11. (a) True   (c) Don't know. It could be 60 cm^2.
12. $(1/2)ab$
13. (a) $405.11
14. (a) 25π cm^2 (c) $(18/5)\pi$ cm^2 (e) 100 cm^2
15. 1200 tiles
17. (a) $24\sqrt{3}$ cm^2
19. (a) 16π cm^2
20. (a) 2π cm^2 (c) 2π cm^2 (e) $(1/4)\pi r^2$ (g) $(1/16)\pi r^2$
22. (a) 48 cm
23. (a) The area is quadrupled.
24. (a) The area is quadrupled.   (c) The area is increased by a factor of 9.
26. The first is a better buy at 10 ft^2 per dollar versus 9.375 ft^2 per dollar.
27. $(320 + 64\pi)$ m^2
28. 1 in.

30. (a)

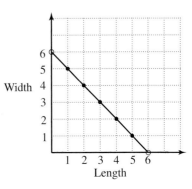

31. The area of the outside shaded portion is the same as the area of the inside shaded region, 9π $in.^2$.
32. (a)

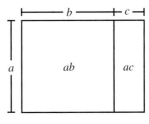

$a(b + c) = ab + ac$

34. The area of each triangle is 10 cm^2 because the base of each triangle is $\overline{AB}$ and the height of each triangle is the perpendicular distance between the two lines. Because each triangle has the same base and height, the areas of the triangles are the same.

Communication

37. (a) The area of the 10-in. pizza is 25π $in.^2$ The area of the 20-in. pizza is 100π $in.^2$. Because the area of the 20-in. pizza is four times as great, this pizza might cost four times as much, or $40. However, this is not the case because other factors are considered rather than just the area of the pizza.
39. After the rotations, a rectangle is formed. The area of the rectangle is length times width which in the case of the parallelogram is the same as the base times the height.

Cooperative Learning

42. (a) 5 square units   (b) 12 units

Ongoing Assessment 11-3

1. (a) 6   (c) 12   (e) 9   (g) $2\sqrt{2}$ (i) $3\sqrt{3}$
3. (a) No   (c) Yes
5. About 2622 km
7. (a) No, in order for it to be a square the sides must be the same length and the angles must be right angles. The sides are

of different lengths. (c) The length of the diagonal is doubled.
9. 9.8 m approx.
10. (a) $(s^2\sqrt{3})/4$
11. (a) $x = 8, y = 2\sqrt{3}$
13. 12.5 cm; 15 cm
14. 8 cm
16. $90\sqrt{2}$, or about 127.28 ft
19. $\triangle ACD \sim \triangle ABC$; $AC/AB = AD/AC$ implies $b/c = x/b$ which implies $b^2 = cx$; $\triangle CBD \sim \triangle ABC$; $AB/CB = CB/DB$ implies $c/a = a/y$ which implies $a^2 = cy$; $a^2 + b^2 = cx + cy = c(x + y) = cc = c^2$
20. $\sqrt{10}$, or approximately 3.16 m
22. Approximately 99.5 ft.
23. Approximately 10.6 mi.
24. The area of the trapezoid is equal to the sum of the areas of the three triangles. Thus,
$1/2(a + b)(a + b) = (1/2)ab + (1/2)ab + (1/2)c^2$
$1/2(a^2 + 2ab + b^2) = ab + (1/2)c^2$
$a^2/2 + ab + b^2/2 = ab + c^2/2$
Subtracting ab from both sides and multiplying both sides by 2, we have $a^2 + b^2 = c^2$. The reader should also verify that the angle formed by the two sides of the length c has measure 90°.
26. Yes
28. (a) 4 (c) $2\sqrt{13}$
29. $10 + \sqrt{10}$
31. The side lengths are 5, $7\sqrt{2}$, and 5 and so the triangle is isosceles.
32. $x = 9$ or $^-7$

Communication

34. (a) Let the length of the side of the square be s. Draw the diagonal of the square and make the new square have side lengths equal to the diagonal. Then the area is $(\sqrt{2}s)^2 = 2s^2$.
36. Yes, she had a right triangle because the converse of the Pythagorean Theorem implies that if $13^2 = 12^2 + 5^2$, then the triangle is a right triangle. Because this is true, she has a right triangle and therefore a right angle.

Open-Ended

39. (b) Yes, we know that if a-b-c is a Pythagorean triple, then $a^2 + b^2 = c^2$. This implies that $4(a^2 + b^2) = 4c^2$, and that $(2a)^2 + (2b)^2 = (2c)^2$.
(d) $a^2 + b^2 = (2uv)^2 + (u^2 - v^2)^2$
$= 4u^2v^2 + u^4 - 2u^2v^2 + v^4$
$= u^4 + 2u^2v^2 + v^4$
$= (u^2 + v^2)^2$
$= c^2$

Review Problems

42. 0.032 km, 322 cm, 3.2 m, 3.020 mm.
43. (a) 33.25 cm^2 (b) 30 cm^2 (c) 32 m^2
44. (a) 10 cm, 10π cm, 25π cm^2 (b) 12 cm, 24π cm, 144π cm^2 (c) $\sqrt{17}$ m, $2\sqrt{17}$ m, $2\pi\sqrt{17}$ m (d) 10 cm, 20 cm, 100π cm^2
45. $25/\pi$ m^2

Ongoing Assessment 11-4

1. (a) 96 cm^2 (c) 236 cm^2 (e) 24π cm^2 (g) 1500π ft^2
3. 2688π mm^2
5. 4:9
6. (a) They have equal lateral surface areas.
8. Approx. 32.99 in.^{2}
10. Approx. 91.86 ft^2
11. $\ell = 11$ cm, $w = 8$ cm, $h = 4$ cm
12. (a) The surface area is multiplied by 4. (c) The surface area is multiplied by k^2.
13. (a) The lateral area is multiplied by 3. (c) The lateral area is multiplied by 9.
14. (a) The surface area is multiplied by 4.
16. (a) 44 (c) Yes, for example, place five cubes in the shape of a C. Then adding a cube to the center of the C would add no surface area.
17. (a) 1.5π m^2
18. (a) $100\pi(1 + \sqrt{5})$ cm^2 (c) 2250π cm^2
19. (a) Approx. 42 cm
21. 375π cm^2

Communication

22. The ice cubes would melt faster because they have a greater surface area that is exposed to the air.
23. She would need four times as much cardboard. If each face is doubled, then the area of each face is increased by a factor of 4, that is, $A_1 = \ell w$ and $A_2 = (2\ell)(2w) = 4\ell w = 4A_1$. Because this is true for all faces, the surface area is multiplied by 4.

Open-Ended

25. (a) 2%

Review Problems

28. (a) 100,000 (b) 1.3680 (c) 500 (d) 2,000,000 (e) 1 (f) 1,000,000
29. $10\sqrt{5}$ cm
30. $20\sqrt{5}$ cm
31. (a) 240 cm; 2400 cm^2 (b) $(10\sqrt{2} + 30)$ cm, 75 cm^2
32. Length of side is 25 cm. Diagonal BD is 30 cm.

Ongoing Assessment 11-5

1. (a) 8000 (c) 0.000675 (e) 7 (g) 0.00857 approx. (i) 345.6
2. 32.4 L

3. (a) $\left(\dfrac{256}{3}\right)\pi$ cm^3 (c) 216 cm^3 (e) 21π cm^3
(g) (4000/3)π cm^3 (i) (20,000/3)π ft^3
4. (a) 2000, 2, 2000 (c) 1500, 1.5, 1500 (e) 0.75, 0.75, 750
5. (a) 200.0 (c) 1.0
6. 1680π mm^3
8. It is multiplied by 8.
9. (a) 2000, 2, 2 (c) 2 dm, 4000, 4
10. 253,500π L
12. 2,500,000 L
13. π mL
14. (a) It is multiplied by 8. (b) It is multiplied by 27.
15. The Great Pyramid has the greater volume. It is approx. 25.12 times greater.
19. 1/8 of the cone is filled.
21. The 2 × 2 × 4 ft freezer is a better buy at \$25/ft^3.
23. About 21.5%
24. They are equal.
26. No, it is only 1/3 of the volume for 1/2 the price.
28. The larger is the better buy. The volume of the larger melon is 1.728 times the volume of the smaller but is only 1.5 times as expensive.
30. (a) 512,000 cm^3
31. It won't hold the cream at 10 cm tall; it would have to be 20 cm tall.
33. (a) Kilograms or metric tons (c) Grams (e) Grams
(g) Metric tons
34. (a) Milligrams (c) Milligrams (e) Grams
35. (a) 15 (c) 36 (e) 4.320 (g) 30 (i) 1.5625
36. (a) No (c) Yes (e) Yes
37. 16 kg
38. (a) $^-$12°C (c) $^-$1°C (e) 100°C
39. (a) No (c) Yes (e) Yes (g) Hot
40. (a) 200,000 L

Communication

41. (a) Doubling the height will only double the volume. Doubling the radius will multiply the volume by 4. This happens because the value of the radius is squared after it is doubled.

Review Problems

50. (a) (20 + 6π) cm; (48 + 18π) cm^2 (b) 40π cm; 100π cm^2
51. (a) 35 (b) 0.16 (c) 400,000 (d) 5,200,000
(e) 5200 (f) 0.0035
52. (a) Yes (b) No
53. (a) 2400π cm^2 (b) (6065 + 40$\sqrt{5314}$) cm^2

Chapter Review

1. (a) $16\dfrac{2}{3}$ yd (c) 3960 ft (e) 5000 m (g) 520 mm

2. (a) Not possible, $p > q + r$
3. (a) 8 1/2 cm^2 (c) 7 cm^2
4. The pieces of the trapezoid are rearranged to form a rectangle with width $h/2$ and length $(b_2 + b_1)$. The area is $A = h/2(b_2 + b_1)$ which is the area of the initial trapezoid.
5. (a) 54$\sqrt{3}$ cm^2
6. (a) 12π cm^2 (c) 24 cm^2 (e) 64.5 cm^2 (g) 4π cm^2
7. (a) Yes
8. (a) S. A. = 32(2 + $\sqrt{13}$) cm^2; V = 128 cm^3
(c) S. A. = 100π m^2; V = (500 π)/3 m^3 (e) S. A. = 304 m^2; V = 320 m^3
9. 65π m^3
10. The graph on the right has 8 times the volume of the figure on the left, rather than double as it should be.
11. 252 cm^2
12. (a) 340 cm (b) 6000 cm^2
13. 2$\sqrt{2}$ m^2
14. 62 cm
15. (a) Metric tons (c) 1 g (e) 25 (g) 51,800
(i) 50,000 (k) 25,000 (m) 52.813
16. $h_1^3/h_2^3 = V_1/V_2$
17. (a) 6000 kg
18. (a) L (c) g (e) kg (g) mL
19. (a) Unlikely (c) Unlikely (e) Unlikely
20. (a) 2000 (c) 3 (e) 0.0002

CHAPTER 12

Ongoing Assessment 12-1

1. (a)

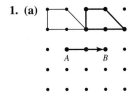

2. Reverse the translation so that the image completes a slide from X' to X (to its pre-image). Then check by carrying out the given motion in the "forward" direction; that is, see if $\overline{AB}$ goes to $\overline{A'B'}$.

(a)

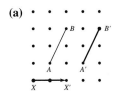

4. (a) (3, $^-$4) (c) ($^-$3, $^-$13) (e) (h + 3, k − 4)
5. (a) (3, $^-$4) (c) ($^-$3, $^-$13) (e) (h + 3, k − 4)

6. (a)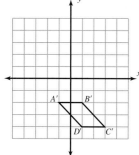

(c)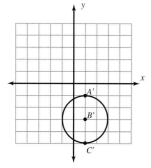

7. (a)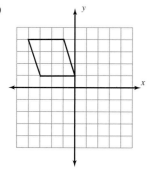

9. P', P, and O are collinear because the measure of angle POP' must be 180°.

11. (a) Yes; all with 360° rotation with any center; If the O is perfectly circular, then the center of the circle is the center of rotation; any number of degrees may be used. Letters such as H, I, S, N, X, and Z may have 180° of rotation depending on the way letters are drawn.

12. Reverse the rotation (to the counterclockwise direction) to locate $\overline{AB}$, that is, the pre-image.

(a)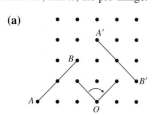

13. (a) Answers may vary, but H, I, N, O, S, X, or Z could appear in such rotational words. Examples include *SOS*.

Variations could use M and W in rotational images; for example, *MOW*.

16. (a) $(^-4, 0)$ **(c)** $(^-2, ^-4)$ **(e)** $(2, 4)$ **(g)** (a, b)
19. (a) (i) $A'(^-2, 3)$ (ii) $A'(^-3, ^-2)$ (iii) $A'(2, 3)$
21. (a) A circle
23.
```
TO ROTATE :A :SIDE
   SQUARE :SIDE
   RIGHT :A
   SQUARE :SIDE
END

TO SQUARE :SIDE
   REPEAT 4 [FORWARD :SIDE RIGHT 90]
END
```
24. (a)
```
TO TURN.CIRCLE :A
   CIRCLE
   LEFT :A
   CIRCLE
END

TO CIRCLE
   REPEAT 360 [FORWARD 1 RT 1]
END
```
To produce the desired transformation, execute `TURN.CIRCLE 180`.

Communication

26. (a) The image is the same. The result is a half-turn.
28. The original and the image lie along perpendicular lines.

Cooperative Learning

32. (c) The path is not an arc of a circle. The perpendicular bisectors of all the chords (segments connecting two points on the arc) do not intersect on a single point.

Ongoing Assessment 12-2

1. Locate the image of vertices directly across (perpendicular to) ℓ on the geoboard.

(a)

2.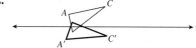

4. Reflecting lines are described for each. **(a)** All diameters (infinitely many) **(c)** The line containing the ray **(e)** Perpendicular bisectors of pairs of parallel sides

(g) Perpendicular bisector of the side that is not congruent to the other two **(i)** None **(k)** Perpendicular bisector of the chord connecting the endpoints of the arc **(m)** The lines containing the diagonals **(o)** There will be n reflecting lines in all. If n is even, the lines are determined as in part (n). If n is odd, the lines are the perpendicular bisectors of the sides.

5. The original figure

6. (b) A translation determined by a slide arrow from P to R is determined as follows: Let P be any point on ℓ and Q on m such that $\overline{PQ} \perp \ell$. Point R is on $\overrightarrow{PQ}$ such that $PQ = QR$.

7. (b) A rotation about O by 2α, from ℓ to m as shown

10. (a) Examples include MOM, WOW, TOOT, and HAH.

11. (a) The images are the same.

12. None of the images has a reverse orientation, so there are no reflections or glide reflections involved. Thus

1 to 2 is a counterclockwise rotation.
1 to 3 is a clockwise rotation.
1 to 4 is a translation down.
1 to 5 is a rotation (with an exterior point as the center of rotation).
1 to 6 is a translation (sides are parallel to 1).
1 to 7 is a translation (sides are parallel).

13. (a) A' (3, $^-$4), B' (2, 6), C' ($^-$2, $^-$5) **(c)** A' (4, 3), B' ($^-$6, 2), C' (5, $^-$2)

14. (a) (i) $(x, {}^-y)$ **(ii)** $({}^-x, y)$ **(iii)** (y, x) **(iv)** $({}^-y, {}^-x)$

15. If H represents the house, P the pole, and T the other house, then the shortest path connecting H and T' (the reflection of T in r) is a segment. The intersection of $\overline{HT'}$, and r determines the point on the road at which the pole should be placed, as in (a).

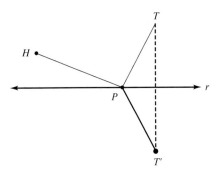
(a)

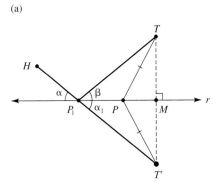
(b)

To prove that the path from H to P_1 to T is the shortest possible, we need to prove that $HP_1 + P_1T < HP + PT$, where P is any point on r different from P_1, as in (b). Because $HP_1 + P_1T = HP_1 + P_1T' = HT'$ (why?) and $HP + PT = HP + PT'$ (why?), the inequality that we need to prove is equivalent to $HT' < HP + PT'$. This last inequality follows from the Triangle Inequality.

Communication

16. The minimum number of sides is two as shown.

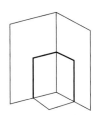

18. (a) If $\overline{AB} \cong \overline{BC}$, then the perpendicular bisector of $\overline{AC}$ is the required line. The image of B is B, the image of A is C, and the image of C is A_1. Hence, the image of $\triangle ABC$ is $\triangle CBA$.
(c) No. Since no sides (or angles) are congruent, bisecting any side or angle will leave noncongruent portions of the triangle on opposite sides of the bisector.

19. The angle of incidence is the same as the angle of reflection. With the mirrors tilted 45°, the object's image reflects to 90° down the tube and then 90° to the eyepiece. The two reflections "counteract" each other, leaving the image upright.

Open-Ended

20. (a) Reflect the location P of the ball in $\overline{AB}$. The intersection of $\overline{P'Q}$ with $\overline{AB}$ is the point at which the ball should be aimed. The justification is similar to the one given in problem 16.

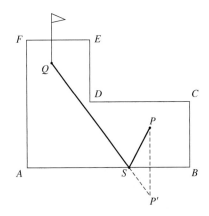

Review Problems

25. Depending on how they are drawn: 0, 1, and 8.
26. $(^-a, ^-b)$
27. A half-turn about the center of the letter O
28. (a) A rotation by any angle about the center of the circle will result in the same circle. **(b)** Reflections about lines containing diameters
29. Construct $\overline{BE}$ perpendicular to $\overline{AD}$ as shown. Next translate $\triangle ABE$ by the slide arrow from B to C. The image of $\triangle ABE$ is $\triangle DCE'$. The rectangle $BCE'E$ is the required rectangle.

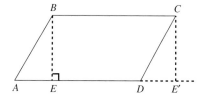

Ongoing Assessment 12-3

1. (a) Slide the smaller triangle down two units (translation). Then complete a size transformation with scale factor 2 using the top-right vertex as the center.   **(c)** Rotate 90° counterclockwise with the lower-right vertex of the smaller triangle as the center of rotation. Then complete a size transformation with scale factor 2 using the same point as center.
3. (a) Translation taking B to B' followed by a size transformation with center B' (and scale factor 2)
(c) Half-turn with the midpoint of $\overline{AA'}$ as center, followed by a size transformation with scale factor 1/2 and center A'.
4. (a) $x = 6, y = 5.2$. Scale factor 2/5.
6. (a) Scale factor 3, $x = 12, y = 10$
7. (a) $A'(6, 9), B'(9, 12), C'(^-6, 9)$
8. The size transformation with center O and scale factor $1/r$

Communication

9. (a) It does change. For example, consider the segment whose endpoints are (0, 0) and (1, 1) and has length $\sqrt{2}$. Under the size transformation with center at (0, 0) and scale factor 2, the image of the segment is a segment whose endpoints are (0, 0) and (2, 2). That segment has length $2\sqrt{2}$.
(c) It does not change. Given two parallel lines, draw a transversal that intersects each line. Because the lines are parallel, the corresponding angles are congruent. From (b), the images of the angles will also be congruent and hence the image lines will be parallel.
11. (a) A single size transformation with center O and scale factor $\frac{1}{2} \cdot \frac{1}{3}$ or $\frac{1}{6}$. Let P be any point and P' its image under the first size transformation and P'' the image of P' under the second size transformation. Then $\frac{OP'}{OP} = \frac{1}{2}$ and $\frac{OP''}{OP'} = \frac{1}{3}$. Consequently, $\left(\frac{OP'}{OP}\right) \cdot \left(\frac{OP''}{OP'}\right) = \frac{1}{2} \cdot \frac{1}{3}$, or $\frac{OP''}{OP} = \frac{1}{2} \cdot \frac{1}{3}$. Thus P' can be obtained from P by a size transformation with center O and scale factor $\frac{1}{2} \cdot \frac{1}{3}$.
13. Yes. Suppose the size transformation with center O has a scale factor r. The image of any point P on the circle with radius d is P' such that $\frac{OP'}{OP} = r$. Thus $OP' = r(OP)$ or $OP' = rd$. This means that the image of every point on the circle is at the same distance rd from O and hence on a circle with radius rd.

Open-Ended

15. The result is equivalent to a half-turn through the origin.

Cooperative Learning

17. (a) If p is the perimeter of the figure and p' the perimeter of the image, then $p' = 3p$.   **(c)** $p' = rp$ **(e)** Answers vary.

Review Problems

18. (a) The translation given by slide arrow from N to M
(b) A counterclockwise rotation of 75° about O **(c)** A clockwise rotation of 45° about A **(d)** A reflection in m and translation from B to A **(e)** A second reflection in n
19. (a) (4, 3) reflects about m to (4, 1); (4, 1) reflects about n to (2, 1).   **(b)** $(0, 1) \to (0, 3) \to (6, 3)$ **(c)** $(^-1, 0) \to (^-1, 4) \to (7, 4)$ **(d)** $(0, 0) \to (0, 4) \to (6, 4)$

Ongoing Assessment 12-4

2. (a) (i) Yes. A line may be drawn through the center of the circle, either horizontally or vertically. A line may also be drawn through any of the sets of arrows. (ii) Yes. The figure will match the original figure after rotations of 90°, 180°, or 270° about the center of the circle. (iii) Yes. Any figure having 180° rotational symmetry has point symmetry about the turn center.   **(c)** (i) Yes. A vertical line through the stem is a line of symmetry. (ii) No (iii) No
4. Reflect the given portions about ℓ.

(a)

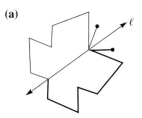

5. (a) (i) Four lines of symmetry; the diagonals and horizontal or vertical lines through the center (ii) No lines of symmetry (iii) Two lines of symmetry; horizontally and vertically through the center (iv) One line of symmetry; vertically through the center
6. (a) 6
7. (a) One; vertically through the center **(c)** None

8. (a)

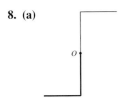

9. (a) Seven; three through the "peaks," three through the "valleys," and one perpendicular to the others through the width of the figure **(c)** Seven; three through the vertices, three through the faces, and one perpendicular to the others through the width of the figure

10.
```
TO TURN.SYM :S :N :A
   REPEAT :N [SQUARE :S RIGHT :A]
END

TO SQUARE :S
   REPEAT 4 [FORWARD :S RIGHT 90]
END
```
(a) Execute TURN.SYM 50 6 60 **(c)** Execute TURN.SYM 50 2 180 **(e)** Execute TURN.SYM 50 6 300

11.
```
TO TURN.SY :S :N :A
   REPEAT :N [EQTRI :S RIGHT :A]
END

TO EQUITRI :S
   REPEAT 3 [FORWARD :S RIGHT 120]
END
```
(a) Execute TURN.SY 50 6 60 **(c)** Execute TURN.SY 50 3 240

Communication

12. (a) Yes. The definition of point symmetry is that it is rotational symmetry of 180°. **(c)** Yes. A circle is an example, as is the figure in problem 1. **(e)** Yes. Point symmetry implies 180° rotational symmetry.

Open-Ended

13. (a) A scalene triangle **(c)** Not possible

Cooperative Learning

18. Answers vary; for example, consider a rectangle. You would report that your figure has two lines of symmetry and a rotational symmetry of 180°.

Review Problems

19. One method is to trace over the figure. Then fold at ℓ and trace along the figure as seen through the paper.
20. Find the images of the vertices.

Ongoing Assessment 12-5

1. (a)

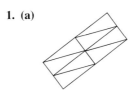

2. (b) Yes. If a polygon tessellates the plane, the sum of the angles around every vertex must be 360°. Successive 180° turns of a quadrilateral about the midpoints of its sides will produce four congruent quadrilaterals around a common vertex, with each of the quadrilateral's angles being represented at each vertex. These angles must add up to 360°, as angles of any quadrilateral do.

3. Experimentation by cutting shapes out and moving them about is one way to learn about these types of problems.

(a)

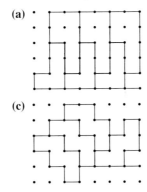

(c)

5. (b) It combines a translation and a reflection.
6. (a) The dual is another tessellation of squares (congruent to those given). **(c)** The tessellation of equilateral hexagons is illustrated in the statement of the problem
7. (a)
```
TO TESSELSQUARE
   PENUP BACK 70 PENDOWN
   REPEAT 9 [SQUARE 20 FORWARD 20]
   PENUP BACK 180 RIGHT 90
   FORWARD 20 LEFT 90 PENDOWN
   REPEAT 9 [SQUARE 20 FORWARD 20]
END
TO SQUARE :SIDE
   REPEAT 4 [FORWARD :SIDE RIGHT 90]
END
```
(c)
```
TO TESSELHEX
  PENUP BACK 70 LEFT 90 PENDOWN
   REPEAT 4 [HEXAGON 20 RIGHT 120
   FORWARD 20 LEFT 60 HEXAGON 20
```

```
      FORWARD 20 LEFT 60]
END

TO HEXAGON :SIDE
  REPEAT 6 [FORWARD :SIDE RIGHT 60]
END
```

Communication

9. (a) The image *ABCD* under a half-turn in *M* (the midpoint of *CD*) is the trapezoid *FEDC*. Because the trapezoids are congruent, *ABFE* is a parallelogram. The area of the parallelogram is $AE \cdot h$ or $(a + b) \cdot h$. The parallelogram is the union of two nonoverlapping congruent trapezoids. The area of each trapezoid is $(a + b)(h/2)$.

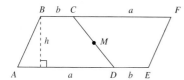

Open-Ended

11. Answers vary. In the given design, the tessellation can be obtained by half-turns followed by reflections.
13. Answers vary. For example, a right rectangular prism, as well as a tetrahedron, will tile the space, but a cylinder or a square pyramid will not.

Cooperative Learning

15. (a) Two such figures can be put together to form a parallelogram. Because parallelograms tessellate the plane, the original figure tessellates the plane. **(c)** Answers vary. For example the following figure is a rep-tile.

Chapter Review

1. (a)

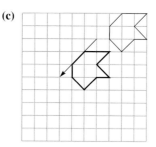

(c)

2. In each part find the images of the vertices.
3. (a) 4 **(c)** 1 **(e)** 2
4. (a) Line and rotational **(c)** Line
5. (a) Infinitely many **(c)** 3
7. $A = A'$, *B* is the midpoint of $\overline{A'B'}$, and *C* is the midpoint of $\overline{A'C'}$.
9. (a) Rotation by 120° about the center of the hexagon
10. Reflection in $\overleftrightarrow{SO}$
11. Let $\triangle H'O'R'$ be the image of $\triangle HOR$ under a half-turn about *R*. Then $\triangle SER$ is the image of $\triangle H'O'R'$ under a size transformation with center *R* and scale factor $\frac{2}{3}$. Thus $\triangle SER$ is the image of $\triangle HOR$ under the half-turn about *R* followed by the size transformation described above.
13. Rotate $\triangle PIG$ 180° (half-turn) about the midpoint of $\overline{PT}$, then perform a size transformation with scale factor 2 and center $P'(= T)$.
14. $(x, y) \rightarrow (x, y)$. It is an "identity" transformation with every point its own image.
15. The measure of each exterior angle of a regular octagon is $\frac{360°}{8}$, or 45°. Hence the measure of each interior angle is $180° - 45°$ or 135°. Because 135 ∤ 360, a regular octagon does not tessellate the plane.
*17. No.

APPENDIX I

Note: Different versions of Logo may affect answers.

Ongoing Assessment A-I-1

1. (a)

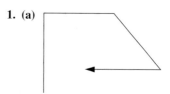

(c)

(e)

0

(g)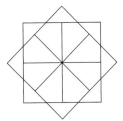

2. The answer may vary depending upon the type of computer being used.

3. (a)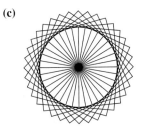

(c)

4. Answers may vary.
(a) TO RECT
     REPEAT 2 [FORWARD 30 RIGHT 90
       FORWARD 60 RIGHT 90]
   END
(c) TO HAT
     REPEAT 2 [FORWARD 60 RIGHT 90
       FORWARD 30 RIGHT 90]
     PENUP LEFT 90 FORWARD 30 PENDOWN
     REPEAT 2 [LEFT 90 FORWARD 6 LEFT 90
       FORWARD 90]
   END
(e) TO RHOMBUS
     RIGHT 20
     REPEAT 2 [FD 40 RT 70 FD 40 RT 110]
   END

6. Answers may vary.
   TO SQUARE1
     REPEAT 4 [FORWARD 50 RIGHT 90]
   END
   TO TRIANGLE1
     REPEAT 3 [FORWARD 50 RIGHT 120]
   END

(a) TO SQUARE.PILE
     REPEAT 4 [SQUARE1 RIGHT 90]
   END
(c) TO RECT1
     REPEAT 2 [FORWARD 60 RIGHT 90
       FORWARD 30 RIGHT 90]
   END
   TO RECT.SWIRL
     RIGHT 30
     REPEAT 4 [RECT1 LEFT 90]
   END
(e) TO STAR
     RIGHT 30
     REPEAT 4 [TRIANGLE1 RIGHT 60
       FORWARD 50 RIGHT 30]
   END

7. Answers may vary.
(b) TO BUILD.SQR :S
     SQUARE :S
     SQUARE :S + 10
     SQUARE :S + 20
     SQUARE :S + 30
     SQUARE :S + 40
   END
(d) TO TOWER :S
     SQUARE :S
     FORWARD :S RIGHT 90 FORWARD :S/4
       LEFT 90
     SQUARE :S/2
     FORWARD :S/2 RIGHT 90 FORWARD :S/8
       LEFT 90 SQUARE :S/4
   END

9. TO KITE
     LEFT 45
     REPEAT 4 [FORWARD 40 RIGHT 90]
     RIGHT 45
     REPEAT 3 [BACK 20 K.TAIL RIGHT 60]
     BACK 20
   END
   TO K.TAIL
     RIGHT 60
     REPEAT 3 [FORWARD 10 RIGHT 120]
     LEFT 120
     REPEAT 3 [FORWARD 10 LEFT 120]
   END

12. TO RECTANGLES :S
     LEFT 90
     REPEAT 4 [RECTANGLE :S/3 :S RIGHT 90
       FORWARD :S/3]
   END
   TO RECTANGLE :S1 :S2
     REPEAT 2 [FORWARD :S1 RIGHT 90
       FORWARD :S2 RIGHT 90]
   END

15. Answers may vary.
(a) TO STRETCH :S
    IF :S < 5 STOP
    SQUARE :S
    FORWARD :S RIGHT 90
    FORWARD :S LEFT 90
    STRETCH :S-10
END
TO SQUARE :S
    REPEAT 4 [FD :S RT 90]
END
(c) TO PISA :S :A
    IF :S < 5 STOP
    SQUARE :S
    FORWARD :S LEFT :A
    PISA :S*0.75 :A
END
TO SQUARE :S
    REPEAT 4 [FD :S RT 90]
END
(e) TO ROW.HOUSE :S
    IF :S < 5 STOP
    HOUSE :S
    SETUP :S
    ROW.HOUSE :S/2
END
TO HOUSE :S
    SQUARE :S
    FORWARD :S
    RIGHT 30 TRIANGLE :S
    LEFT 30
END
TO SETUP :S
    BACK :S RIGHT 90
    FORWARD :S LEFT 90
END
TO SQUARE :S
    REPEAT 4 [FD :S RT 90]
END
16. TO NEST.TRI :S
    IF :S < 10 STOP
    RIGHT 30 TRIANGLE :S
    FD :S/2 RIGHT 30
    NEST.TRI :S/2
END
TO TRIANGLE :S
    REPEAT 3 [FORWARD :S RIGHT 120]
END
17. Answers may vary.
TO SPIN.SQ :S
    IF:S < 5 STOP
    SQUARE :S
    RIGHT 20
    SPIN.SQ :S-5
END
TO SQUARE :S
    REPEAT 4 [FORWARD :S RIGHT 90]
END

APPENDIX II

Ongoing Assessment A-II

1. Approximately 11.896 or 11.9 years.
3. ($^{-}10$, 257) and (10, 357)
4. (a) The graphs all pass through the point (0, 3) and all are linear. **(c)** As the slope becomes greater, the graph becomes steeper. The lines have slopes of 1, 2, 3, and 4, respectively.
5. (b) They cross the y-axis at different points.
6. Approximately (1.22, $^{-}0.55$).
7. (a) She needs to find the intersection points of Y_1, Y_2, and Y_3 with Y_4.
8. (a) Sign changes occur at x = 1 and x = 4. **(c)** The graph crossed at x = 1 and x = 4.

APPENDIX III

Ongoing Assessment A-III

1. The areas of a rectangle and a parallelogram with the same base and height are the same.
3. The ratio of the areas of two similar figures is the square of the ratio of two corresponding sides of the similar figures.
5. (f) π
6. (a) The percent of area in the square not covered by the circle is approximately 21%. **(b)** The percent of area in the circle not covered by the square is approximately 36%.
(c) The answers in parts (a) and (b) suggest that the circular peg will fit better in the square hole than vice versa.
7. (a) Approximately 21% **(b)** Approximately 21%
(c) The answer doesn't change.
9. This exercise can be used to practice estimation of areas and the estimation of percentages.

APPENDIX IV

Ongoing Assessment A-IV

1. Answers will vary. An example follows:
(a) 5, 9, 13, 17, 21, … **(b)** Column A should contain the numbers 1, 2, 3, 4, 5, 6, … **(c)** The formula =A1+1 should be used to find A2 and then to fill down the rest of the column.
(d) Column B for this example will start with 5, and each successive term is 4 more than the previous one. **(e)** The formula =B1+4 could be used here. **(f)** The Σ key should be used when the cell where the sum is wanted is highlighted.
3. If all the data is listed in the first 14 cells of column A, the Σ key can be used to find the sum of all the data, and the sum can be placed in cell A15. The formula =A15/14 can be used to place the mean in cell A16.
6. The first step is to create a column A that is the number of the term by placing 1 in cell A1 and using the formula =A1+1

to fill down for a total of 100 cells in column A. Next we could use the formula =13*A:A and fill down for a total of 100 cells in column B.

7. Answers may vary depending upon the method used for computing grade points at your college. In general, you may use 4 quality points for each hour of A, 3 quality points for each hour of B, 2 for each hour of C, 1 for each hour of D, and 0 for each hour of F. After multiplying each hour by the respective number of quality points, find the total of all quality points, then divide by the total number of hours.

9. The A column can be created by entering 1 in cell A1 and using the formula =A1+1 and Fill Down to fill as many cells as wanted. The B column can be created by entering =A1 in cell B1 and =B1*A2 with the Fill Down feature to complete the wanted cells.

Index

A

Abscissa, 567
Absolute value, 174–175
Acres, 607, 608
Acute angles, 470
Acute triangles, 480
Addition
 clock, 235, 236
 distributive property of multiplication over, 95–96, 147
 estimation strategies for, 160–163
 of integers, 172–173, 175–176
 of rational numbers, 256–262
 of whole numbers, 83–88
Addition algorithms
 in bases other than ten, 141–142
 explanation of, 135–138
 scratch, 138–139
Additive inverse property
 of integers, 175
 of rational numbers, 261
 of real numbers, 341
 of uniqueness, 175–176
Additive property
 of Egyptian system, 126
 of equality, 39, 262
Adjacent angles, 468
Adleman, Leonard, 217
Algebra, 37
Algebraic thinking
 application problems and, 41–43
 developing skills in, 36–38
 overview of, 35–36
 properties of equations and, 38–39
 solving equations and, 39–40
Algorithms
 for addition, 135–139, 141–142
 for addition and subtraction of decimals, 303
 for division, 98–99, 148–153
 for division of decimals, 305
 for division of rational numbers, 275–276
 Euclidean, 228–229
 explanation of, 135
 for multiplication, 146–148, 150–153
 for subtraction, 139–142
Altitude, 502
Ancient Egyptians, 124, 126
Angle, angle, side (AAS), 534–535

Angle, angle (AA), 552–555
Angle, side, angle (ASA), 533–534
Angle of incidence, 691
Angle of reflection, 691
Angles
 acute, 470
 adjacent, 468
 bisecting, 542–543, 546–547
 congruent, 479, 520, 526–528
 explanation of, 467–468
 historical background of, 468
 included, 526
 measurement of, 468–469
 in triangles, 487–489
 types of, 469–470, 484–486
Apex, 497
Application problems, 41
Archaeological Find Problem, 705–706
Archimedes, 599
Arcs
 explanation of, 507, 520
 length of, 600
Are, 607
Area. See also Surface area
 of circle, 614–615
 converting units of, 606–607
 of land, 607–608
 of parallelogram, 609
 of rectangle, 608
 of regular polygon, 613–614
 of sector, 616
 of trapezoid, 611–613
 of triangle, 609–611
 using geoboard, 604–605
Area model, 372
Arithmetic
 clock, 234–237
 modular, 237–239
Arithmetica (Diophantus), 37, 195
Arithmetic mean, 429
Arithmetic sequences, 6–11
Array model, 92–93
Associative property
 of addition of whole numbers, 86
 of integer addition, 175
 of multiplication of integers, 186
 of multiplication of whole numbers, 94, 95
 of real numbers, 341
Average
 choosing most appropriate, 432–433
 explanation of, 428, 429

B

Babylonian numeration system, 127, 247
Babylonians, 103, 125, 468
Back-to-back stem-and-leaf plots, 415
Balance point, 429–430
Bar graphs, 417–418
Base-five system
 addition table for, 141
 explanation of, 130, 131
 multiplication table in, 150, 152
Base-ten blocks, 135–137
Base-ten system, 130, 131
Base-twelve system, 132–133
Base-two system. See Binary system
Bernoulli, Johann, 103
Best-fitting line, 580–582
Bhramagupta, 171
Biconditionals, 49
Bimodal, 431
Binary system, 131, 132
Bobrow, Daniel, 725
Box-and-whisker plots
 explanation of, 434–435
 use of, 437–439

C

Calculators. See also Graphing calculators
 to add mixed numbers, 261
 base-ten numbers using, 131
 to change improper fractions to mixed numbers, 260
 change-of-sign key on, 179
 checking divisibility using, 197
 Euclidean algorithm using, 225
 as function machine, 106
 greatest common divisor using, 222–223
 modular arithmetic using, 239
 radicals and rational exponents using, 342
Cancellation properties of equality, 39
Cantor, Georg, 59
Cardano, Gerolamo, 171
Cardinal numbers, 64–65
Cartesian coordinate system
 elimination method and, 579–580
 equations of lines and, 568–573
 equations of vertical and horizontal lines and, 567–568

814 *Index*

explanation of, 567
fitting line to data and, 580–582
substitution method and, 578–579
systems of linear equations and, 576–578
using similar triangles to determine slope and, 573–576
Cartesian products
  explanation of, 78–79, 93–94
  multiplication of whole numbers and, 95
Cavalieri, Bonaventura, 649
Cavalieri's Principle, 649
Celsius scale, 657–658
Central angles, 616
Central tendency
  mean and, 429–430
  measures of, 428
  median and, 430–431
  mode and, 431–432
Certain event, 353
Ceulen, Ludolph van, 599
Chain rule, 51
Charged-field model
  of integer addition, 172
  of integer multiplication, 184
  of integer subtraction, 176
Checkerboard Problem, 32
Checker Games, 33
Chip model
  of integer addition, 172
  of integer multiplication, 184
  of integer subtraction, 176
Chudnovsky, David, 599
Chudnovsky, Gregory, 599
Circle graphs
  explanation of, 419–421
  misleading, 451
Circles
  area of, 614–615
  circumference of, 337, 599
  circumscribed about triangle, 529
  construction of, 520–521
  explanation of, 598
  sector of, 616
Circular cylinders, 502
Circumference, 337, 599
Clarkson, Roland, 215
Classes, 416
Clock arithmetic, 234–237
Closed curves, 477
Closure property
  of addition of whole numbers, 85
  of integer addition, 175
  of multiplication of integers, 186
  of multiplication of whole numbers, 94
  of real numbers, 341
Clustering, 161
Coding systems, 217
Coin-tossing Game, 390–391
Collinear points, 464
Colored rods model
  greatest common divisor and, 221

least common multiple and, 226, 227
Combinations
  explanation of, 396–398
  problem solving using, 399–401
  rule to count, 397
Commutative property
  of addition of whole numbers, 85–86
  of integer addition, 175
  of multiplication of integers, 186
  of multiplication of whole numbers, 94–96
  of real numbers, 341
  of set intersection, 74
  of set union, 74
Comparison model, 89
Compasses, 520
Compatible numbers, 163
Complement
  event as, 354–355
  of *A* relative to *B*, 73–74
  of sets, 65–66, 73
Complementary angles, 485
Composite numbers, 207–208
Composition of two functions, 112–113
Compound interest, 333–335
Compound statements, 46–47
Concave curves, 477
Conclusion, 47
Concurrent lines, 464
Conditionals, 47–48
Cones
  explanation of, 502
  surface area of, 640
  volume of, 651–653
Congruence
  angle, 526–528
  angle, side, angle, 533–536
  angle bisectors, 542–543, 546–547
  circle circumscribed about triangle, 529
  explanation of, 519
  geometric constructions and, 520–521
  modular, 237
  parallel lines, 540–542
  perpendicular bisector of segment, 528–529
  perpendicular lines, 543–545
  quadrilateral, 536, 537
  segment, 521
  side, side, side property, 523–524
  triangle, 521–522, 524–525
  two sides and angle of triangle, 529–530
Congruent angles, 479, 486, 520
Congruent parts, 478
Congruent segments, 478–479, 520
Conjecture, 4
Conjunctions, 46
Constructions
  congruency through, 519–530 (*See also* Congruence)
  geometric, 520–521
  of translations, 670–672
Contrapositive, 49

Convex curves, 477
Convex polyhedron, 498
Cooperative learning, 2
Coordinate system
  reflections in, 689
  translations in, 672–674
Coplanar, 464
Counterexample, 4
Counting on technique, 86
Cox, Gertrude Mary, 409
Cubic units, 645
Curves, 476, 477
Cylinders
  explanation of, 501–502
  surface area of, 638
  volume of, 650
Cylindrical Boxes Problem, 653–655

D

Danielson, Charlotte, 23
Decimal points, 295
Decimals
  conversion of rational numbers to, 297–298
  division with, 305–307
  estimating with, 310–311
  explanation of, 295–299
  historical background of, 294
  mental computation with, 308–309
  multiplication with, 303–305
  nonterminating, 314–319
  operations on, 302–303
  ordering terminating, 299–300
  overview of, 294
  repeating, 314–319
  rounding, 309–311
  scientific notation and, 300–301
  terminating, 298–299
  writing percent as, 322
Deductive reasoning
  Euclid and, 462
  explanation of, 4–6, 246, 487
Degree Celsius, 657
Degree Kelvin, 657
Degrees, 468
Denominators, 246
Denseness property, 341
Descartes, René, 103, 567
Diagonal, of polygon, 478
Diagrams, 27–28
Difference, 6
Difference-of-squares formula, 188
Diophantus, 37
Direct reasoning, 33, 50
Disjunctions, 47
Dispersion, 433–434
Disraeli, Benjamin, 447
Distance
  around place figure, 597–598
  properties of, 596–597
  Pythagorean theorem and, 628–630

Distance formula, 630
Distributive property
    of multiplication over addition, 95–96, 147, 341
    of multiplication over addition for integers, 186
    of multiplication over addition for rational numbers, 270
    of multiplication over subtraction for integers, 187, 188
    of set intersection over union, 75
Dividend, 98
Divides, 195
Divine Comedy (Dante), 349
Divisibility
    explanation of, 194–197
    problem solving and, 202–203
    rules of, 197–198
Divisibility tests
    for 2, 5, and 10, 198
    for 3 and 9, 200–201
    for 4 and 8, 198–199
    for 11 and 6, 201–202
Division
    clock, 235
    of decimals, 305–307
    estimation strategies for, 163
    of integers, 189
    of rational numbers, 274–277
    of whole numbers, 97–99
Division algorithms
    in different bases, 152–153
    explanation of, 98–99, 148–149
    short division and, 149
    with two-digit divisor, 149–150
Division-by-primes method, 229
Divisors
    explanation of, 98, 206–207
    number of, 211–212
Dog Cartoon Problem, 284–285
Domain, 105, 111
Dot paper, 604–605, 688–689
Double-bar graphs, 417, 418
Doubles technique, 87
Duodecimal system, 132–133

E

Efron, Bradley, 378
Egyptian numeration system, 126, 247
Einstein, Albert, 349
Elements, 59
The Elements (Euclid), 462
Elimination method, 579–580
Empty sets, 65, 67
English system
    area in, 606–607
    explanation of, 592–593
    volume in, 648
Enigma cipher machine, 217
Equal Areas Problem, 612–613
Equal fractions, 248–249

Equality
    addition property of, 39, 262
    cancellation properties of, 39
    multiplication property of, 39
Equations
    problem solving using, 27
    properties of, 38–39
    solving, 39–40
Equivalent fractions, 248–249
Equivalent sets
    cardinal numbers and, 64–65
    explanation of, 64
    subsets and, 66–67
Eratosthenes, 214
Escher, Maurits C., 713
Estimation
    with addition, 160–163
    front-end, 160–161, 163
    with multiplication and division, 163
    with percent, 327–328
    with rational numbers, 263–264
Euclidean algorithm method
    greatest common divisor and, 224–226
    least common multiple and, 228–229
Euclid of Alexandria, 462
Euler, Leonhard, 103, 501, 507, 599
Euler circuit, 507
Euler diagram, 49
Even numbers, 194
Events
    certain, 353
    complementary, 354–355
    impossible, 352
    mutually exclusive, 354
    nonmutually exclusive, 355–358
    probability of, 351, 353–354
Expanded form, 124
Expected value, 389–391
Experimental Designs (Cox), 409
Experiments, 349
Explicit formulas, 765
Exponents
    division of rational numbers related to, 276–277
    extending notion of, 271–274
    properties of, 342
    rational, 342
Exterior angle, 478

F

Faces, 496, 497
Face value, 124
Factor, 94, 124
Factored form, 188
Factorization, 208
Factor tree, 208
Fahrenheit scale, 657–658
Fermat, Pierre de, 171, 195, 349
Fermat's last theorem, 195
Feurzeig, Wallace, 725
Fibonacci (Leonardo of Pisa), 37, 269

Figurate numbers, 13–15
Finite sets, 65
First quartile, 434
Fisher, Ronald, 409
Flip. *See* Reflections
Fourier, Jean Joseph, 103
Fractions
    algorithm for division of, 275–276
    converted to decimals, 298
    equal, 248–249
    equality of, 250–252
    explanation of, 246
    fundamental law of, 246, 249
    historical background of, 247
    improper, 248, 260
    proper, 248
    simplifying, 249–250
Frequency tables, 416
Front-end estimation
    for addition, 160–161
    for multiplication and division, 163
Function machine
    calculator as, 106
    example of, 104
    explanation of, 103
    use of, 105
Functions
    applications of, 110–112
    composition of two, 112–113
    definition of, 105–106
    graphs and, 108–112
    historical background of, 103
    operations on, 112–114
    overview of, 102–103
    representation of, 107–108
Fundamental counting principle, 63, 393, 397
Fundamental law of fractions, 246, 249
Fundamental theorem of arithmetic, 209, 339

G

Galton, Sir Francis, 409
Gardner, Martin, 715
Gauss, Carl, 20, 171
Gauss's Problem, 20–21
Geoboards, 604–605, 688–689
The Geometer's Sketchpad, 765–766
Geometric probability, 372–373
Geometric sequences, 12–13
Geometry. *See also* specific geometric figures
    angle measurement and, 468–469
    basic notions in, 462–463
    constructions and, 520–521
    cylinders and cones and, 501–502
    historical background of, 462
    linear notions in, 463–464
    networks and, 506–510
    parallel lines and, 486–487
    perpendicular lines and, 470–472

planar notions in, 464–468
polygons and, 476–482
regular polyhedra and, 498–501
simple closed surfaces and, 496–497
sum of measures of angles of triangle and, 487–489
sum of measures of interior angles of convex polygon with *n* sides and, 490–492
symmetry and, 669
triangles and quadrilaterals and, 479–481
types of angles and, 469–470, 484–486
Geometry drawing utilities, 753–759
Germain, Sophie, 216
Glide reflections, 690–691
Grams, 655
Graphing calculators. *See also* Calculators
connecting graphing and algebra with, 749
data lists and, 746–747
graphing functions using, 113–114
lists on home screen of, 747–748
problem solving using, 30
replay feature on, 745
STAT plot graphing capabilities and, 748
using tables on, 750–751
using ZOOM on, 751–752
Graph of set, 567–568
Graphs. *See also* Statistical graphs
explanation of, 410
misleading, 448–451
as representations of functions, 108–112
Graunt, John, 409
Greater than, 68
Greatest common divisor (GCD)
calculator method and, 222–223
colored rods model and, 221
Euclidean algorithm method and, 224–226
explanation of, 220
intersection-of-sets method and, 221
prime factorization method and, 221–222
Greatest possible error (GPE), 596
Grouped frequency tables, 416
Grouping to nice numbers, 161
Group work, 2
Guess-and-check strategy, 29–30
Guess my rule game, 102–103

H

Hectares, 607, 608
Herigone, Pierre, 468
Hilbert, David, 203
Hilton, Peter, 217
Hindu-Arabic system, 124
Hindus, 172

Hippasus, 338
Histograms, 416–417
How to Lie with Statistics (Huff), 452
Huff, Darrell, 452
Hypothesis, 47

I

Identity element, 175
Identity property
of addition of whole numbers, 87
of multiplication of whole numbers, 94
of real numbers, 341
Implication, 47–48
Impossible event, 352
Improper fractions, 248, 260
Included angle, 526–527
Index, 338
Indirect reasoning, 32, 51, 339
Inductive reasoning, 4–6, 487
Inequalities, 68–69
Infinite sets, 65
Integers
absolute value of, 174–175
addition of, 172–173, 175–176
divisibility of, 194–203
division of, 189
explanation of, 170, 246
multiplication of, 183–188
negative, 170, 171
order of operations on, 180, 189–191
representations of, 171–172
subtraction of, 176–180
Interest
compound, 333–335
computation of, 331–333
explanation of, 331
simple, 331
Interest rate, 331
Interior angles
of convex polygon with *n* sides, 490–492
explanation of, 478
Interquartile range (IQR), 434, 439
Intersecting lines, 464
Intersection, 72, 74, 75
Intersection-of-sets method, 221, 227
Irrational numbers
explanation of, 337
historical background of, 338
square roots as, 339–340
Isometry
explanation of, 670, 674
rotation and, 675
Isosceles trapezoids, 480
Isosceles triangles, 480

K

Kelvin scale, 657
Kerchner, Charles, III, 216
Kerrich, John, 350

al-Khowarizmi, Mohammed, 37, 135, 172
Kites, 480
Klein, Felix, 669
Kolmogorov, Andrei Nikolaevich, 409
Köningsberg Bridge Problem, 506–510

L

Lagrange, Joseph Louis, 103
Lambert, Johann, 599
Land area, 607–608
Land measure, 607–608
Laplace, Simon de, 349
Lateral faces, 496, 497
Lateral surface area, 636
La Thiende (Stevin), 294
Lattice multiplication algorithms, 148, 152
Law of detachment, 50–51
Least common multiple (LCM)
colored rods model and, 226, 227
division-by-primes method and, 229
Euclidean algorithm method and, 228–229
explanation of, 226
intersection-of-sets method and, 227
prime factorization method and, 227
Leibnitz, Gottfried Wilhelm von, 92, 103
Leonardo of Pisa (Fibonacci), 37, 269
Less than, 68, 190–191
Linear equations
elimination method to solve, 579–580
explanation of, 572–573
solutions to systems of, 580
substitution method to solve, 578–579
systems of, 576–578
Linear measurement
circles and, 598–600
distance around plane figure and, 597–598
distance properties and, 596–597
English system and, 592–593
metric system and, 593–596
Linear notions, 463–464
Line graphs, 418
Line plots, 411–412
Lines
best-fitting, 580–582
equations of, 568–572
equations of vertical and horizontal, 567–568
perpendicular, 470
perpendicular to plane, 471–472
properties of, 464
Line segments
congruent, 478–479, 520
construction of, 521
explanation of, 464
perpendicular bisector of, 528–529
Lines Through Points Problem, 466

Line symmetry
  example of problem finding, 705
  explanation of, 704–705
  reflections and, 686, 687
Logic
  compound statements and, 46–47
  conditionals and biconditionals and, 47–49
  explanation of, 45–46
  valid reasoning and, 49
Logically equivalent statements, 47
Logo, 725. *See also* Turtle
Loomis, E., 623
Lower quartile, 434
Luo people, 130

M

Magic Square Problem, 24–25
Major arc, 520
Making 10 technique, 87
Mass, 655–657
Matching Letters to Envelopes Problem, 400–401
Matchstick patterns, 26–27
Mathematical symbols, 92, 172
Matijasevic, Yuri, 203
Mayan numeration system, 127–128
Mayans, 124
Mean
  as balance point, 429–430
  computing, 429
  definition of, 429
  misleading use of, 452
  use of, 432
Measurement
  linear, 592–600
  mass, 655–657
  polygon and circle, 604–616
  Pythagorean theorem and, 621–630
  surface area, 635–641
  temperature, 657–658
  volume, 645–655
Measures of central tendency, 428
Median
  computing, 430–431
  misleading use of, 452
  use of, 432
Members, 59
Mendel, Gregor, 409
Mental mathematics
  with addition, 157–158
  with decimals, 308–311
  distributive property of multiplication over addition for, 96
  with division, 160
  with multiplication, 159–160
  with percent, 326–327
  with subtraction, 158–159
Méré, Chevalier de, 349
Mersenne prime, 215
Meters, 593, 594

Metric system
  area in, 606–607
  explanation of, 593–596
  mass in, 655
  volume in, 646–648
Microsoft Excel, 761–762
Midsegment theorem, 557
Minor arc, 520, 521
Minutes, 468
Mira, 544, 684
Mirror image, 684
Missing-addend model, 88
Missing Grades Problem, 433
Mission-factor model, 97–98
Mixed numbers, 259–261
Mode
  computing, 431–432
  explanation of, 431
  misleading use of, 452
  use of, 432–433
Modeling games, 367–368
Modular arithmetic, 237–239
Modus ponens, 50–51
Modus tollens, 51
Motion Geometry (UICSM), 669
Mouton, Gabriel, 593
Multiplication
  clock, 235, 236
  of decimals, 303–305
  estimation strategies for, 163
  of integers, 183–188
  of rational numbers, 268–274
  of whole numbers, 92–97
Multiplication algorithms
  by 10^n, 146–147
  in different bases, 150–153
  explanation of, 146
  lattice, 148, 152
  with two-digit factors, 147–148
Multiplicative identity, 186, 269
Multiplicative inverse
  of rational numbers, 269, 270
  of real numbers, 341
Multiplicative property
  of equality, 39, 270
  use of bars and, 129
  of zero for rational numbers, 270
Mutually exclusive events, 354
Mysterious Triangles Problem, 554–555

N

Natural numbers
  explanation of, 3
  set of, 60
  sums of even, 21–22
Negation, 45, 46
Negative integers, 170, 171
Networks, 507–510
Newspaper Delivery Problem, 42–43
Newton, Sir Isaac, 655
n factorial, 393

Nightingale, Florence, 409
Noether, Emmy, 186
Noncoplanar points, 464
Nonmutually exclusive events, 355–358
Nonterminating decimals, 314–319
nth roots, 338–339
Null sets, 65
Number-line model
  of addition, 84–85
  explanation of, 89
  of integer addition, 173–174
  of integer multiplication, 184–185
  of integer subtraction, 177
Number theory, 170–171
Numerals, 124, 125
Numeration systems
  Babylonian, 127, 247
  Egyptian, 126, 247
  explanation of, 124, 125
  Hindu-Arabic, 124–125, 247
  Mayan, 127–128
  miscellaneous, 130–133
  Roman, 129
Numerators, 246

O

Oblique circular cone, 502
Oblique cylinders, 502
Oblique prisms, 496
Obtuse angles, 470
Obtuse triangles, 480
Odd numbers, 194
Odds
  against, 386
  computing, 385–389
  definition of, 386
  in favor, 385
One-to-one correspondence, 61–63
Ordered pairs
  explanation of, 78
  functions and, 107, 108, 111
Ordered stem-and-leaf plots, 413
Order of operations
  difficulties with, 97
  on integers, 180, 189–191
Ordinate, 567
Origin, 567
Oughtred, William, 92
Outcome, 349
Outliers, 435–438
Overdue Books Problem, 41–42

P

Paper-measuring Problem, 27–28
Papert, Seymour, 725
Parallel lines
  construction of, 486–487, 540–544
  explanation of, 464
Parallelograms, 480, 609
Pascal, Blaise, 349

Passing a Senate Measure Problem, 68–69
Patterns
   arithmetic sequences as, 6–11
   figure numbers as, 13–15
   geometric sequences as, 12–13
   inductive and deductive reasoning as, 4–6
   overview of, 3, 4
   problem solving using, 20–21
   recursive, 7
Patterns model
   of integer addition, 172–173
   of integer multiplication, 183–184
   of integer subtraction, 176–177
Pearson, Karl, 350
Percent
   applications involving, 323–326
   converting to, 322
   estimates with, 327–328
   explanation of, 321
   mental math with, 326–327
Perigal, Henry, 624
Permutations
   combinations and, 396, 397
   of like objects, 395–396
   of unlike objects, 392–395
Perpendicular bisector of segment, 528–529
Perpendicular lines
   construction of, 543–545
   explanation of, 470
   slopes of, 677–678
Perspective drawings, 699
Piaget, Jean, 725
Pictographs, 411
Pie charts
   explanation of, 419–421
   misleading, 451
π(pi), 337, 599, 600
Place value, 124
Planar notions, 464, 467
Plane of symmetry, 708–709
Planes
   line perpendicular to, 471–472
   properties of, 464
Plato, 498, 520
Platonic solids, 498
Points, 464
Point symmetry, 708
Polya, George, 2, 18, 41
Polygonal regions, 477
Polygons
   area of regular, 613–614
   classification of, 478
   explanation of, 477
   hierarchy among, 481
   regular, 479
   sum of measures of interior angles of convex, 490–492
   tessellations with, 712, 713, 716
Polyhedra
   explanation of, 496
   regular, 498–500

Postcards and Letters Problem, 29–30
Pre-image, 670
Prime factorization method
   example of, 210
   explanation of, 208–209
   greatest common divisor and, 221–222
   least common multiple and, 227
Prime numbers
   coding systems and, 217
   determining if number is, 212–214
   explanation of, 207
   Mersenne, 215
   Sieve of Eratosthenes process and, 214–215
   Sophie Germain, 215–216
Principal, 331
Principal square root, 337
Prisms
   explanation of, 496
   volume of, 645, 648–650
Probability
   combinations and, 396–401
   of complementary events, 354–355
   computing odds with, 385–388
   determining, 350–353
   empirically determined, 351
   of event, 351, 353–354
   expected value and, 389–391
   explanation of, 349–350
   geometric, 372–373
   historical background of, 349
   modeling games with, 367–371
   multiplication rule for, 364
   multistage experiments with, 362–373
   of mutually exclusive events, 354
   of nonmutually exclusive events, 355–358
   permutations of like objects and, 395–396
   permutations of unlike objects and, 392–395
   properties of, 356
   theoretical, 351
   uses for, 349
   using simulations in, 378–383
Problem solving
   avoiding mind sets for, 19
   four-step process of, 19, 41
   overview of, 2–3
   using algebraic thinking for, 35–43
   using logic for, 45–52
Problem-solving strategies
   drawing diagrams as, 27–28
   examining related problems as, 26–27
   examining simpler cases as, 23
   explorations with patterns as, 3–15
   guess and check as, 29–31
   identifying subgoals as, 24–25
   looking for patterns as, 20–21
   making tables as, 21–22
   using direct reasoning as, 33, 50

using indirect reasoning as, 32, 51
working backwards as, 31
writing equations as, 27
Product, 94
Proper fractions, 248
Proper subset, 66
Proportional reasoning, 281–286
Protractors, 468
Ptolemy, 247
Pyramids
   explanation of, 497
   surface area of, 638–639
   volume of, 651–653
Pythagoras, 338, 622
Pythagoreans, 338, 622
Pythagorean theorem
   converse of, 627–628
   distance formula and, 628–630
   explanation of, 339, 622–623
   proofs for, 623–625
   right triangles and, 626–627
   use of, 625–626
The Pythagorean Proposition (Loomis), 623

Q

Quadrilaterals
   classification of, 479–481
   midsegments of, 557–558
   properties of, 536–537
   tessellations with, 714–715
Quantifiers, 45–46
Quetelet, Adolph, 409
Quinary system, 130–131
Quiz-Show Game, 372–373
Quotient, 98

R

Radicals, 342
Random-digit tables, 378–379
Range
   estimation using, 163
   explanation of, 434
   of function, 105, 106
Rational exponents, 342
Rational numbers
   addition of, 256–262
   additive inverse property of, 261
   denseness of, 253
   division of, 274–277
   estimation with, 263–264
   explanation of, 246
   mixed numbers as, 259–261
   multiplication of, 268–274
   ordering, 252–253
   proportional reasoning and, 281–286
   representations of, 248–249
   square roots as, 339
   subtraction of, 262–264
   uses of, 247–248

Ratios
  explanation of, 281
  probabilities as, 349
  in scale drawings, 285–286
  use of, 282–285
Ray, 464
Real numbers
  explanation of, 337
  properties of exponents and, 342–343
  square roots and other roots and, 337–341
  system of, 341
Reasoning
  deductive, 4–6, 246
  direct, 33, 50
  indirect, 32, 51, 339
  inductive, 4–6
  proportional, 281–286
  valid, 49–50
Reciprocal, 269
Rectangles, 480, 608
Recursive formulas, 7, 765
Recursive pattern, 7
Reflections
  construction using dot paper or geoboard of, 688–689
  construction using tracing paper of, 685–688
  in coordinate system, 689
  explanation of, 685
  glide, 690–691
Regrouping
  with base-ten blocks, 136–137
  to perform subtractions using standard algorithm, 140
Regular polygons
  area of, 613–614
  explanation of, 479
  tessellations with, 712, 713
Regular polyhedrons, 498
Relative complement, 73–74
Relative frequency, 350
Relatively prime numbers, 222
Remainder, 98
Repeated-addition model, 92
Repeated-subtraction model, 98, 152
Repeating decimals
  explanation of, 314–315
  ordering, 319
  writing, 316–318
Repetend, 314
Rep-tile, 519
Rhombus
  constructing parallel lines and, 540–544
  explanation of, 480
Rice, Marjorie, 715
Right angles, 469
Right circular cone, 502
Right cylinders, 502
Right prisms
  explanation of, 496
  surface area of, 635–638
  volume of, 645, 650

Right rectangular prisms, 645
Right triangles
  explanation of, 479
  Pythagorean theorem and, 625–627
Rigit motion, 670
Rings, 186
Rivest, Ronald, 217
Robinson, Julia, 203
Roman numeration system, 129
Roping a Square Problem, 598
Rotation
  explanation of, 674–676
  perpendicular lines and, 677–678
Rotational symmetry, 707–708
Rounding
  of decimals, 309–311
  explanation of, 161–162
RSA system, 217
Rudolff, Christoff, 338
Rulers, 595

S

Sample space, 350
Scale drawings, 285–286
Scale factor, 552
Scalene triangles, 480
Scatterplots, 418–419
Scientific calculators, 342. *See also* Calculators; Graphing calculators
Scientific notation, 300–301
Scratch addition algorithms, 138–139
Seconds, 468
Sectors, of circle, 616
Segments. *See* Line segments
Semicircle, 520, 600
Sequences
  arithmetic, 6–11
  explanation of, 6
  figurate numbers as, 13–15
  geometric, 12–13
Set-builder notation, 60
Set model
  addition of whole numbers and, 83–84
  division of whole numbers and, 97
Set operations, 74–75
Sets
  Cartesian products and, 78–79
  complement of, 65–66
  empty or null, 65, 67
  equal, 61
  equivalent, 64–67
  explanation of, 59–60
  finite, 65
  inequalities and, 68–69
  infinite, 65
  intersection of, 72, 74, 75
  number in subsets of, 69–70
  one-to-one correspondence and, 61–63
  union of, 73–75
  universal, 65

Venn diagrams and, 76–79
  well-defined, 60
Set theory, 59
Shamir, Adi, 217
Side, angle, side property (SAS), 526–528
Side, side, side property (SSS), 523–524
Similar, 519
Similar triangles
  angle, angle property and, 552–555
  to determine slope, 573–576
  explanation of, 552
  indirect measurements and, 558–559
  midsegments of triangles and quadrilaterals and, 557–558
  properties of proportion and, 555–557
Simple closed surfaces, 496–497
Simple curves, 477
Simple interest, 331
Simplifying fractions, 249–250
Simulations, 378–383
Size transformations
  applications of, 699–700
  explanation of, 696–697
  properties of, 697–699
Skew lines, 464
Slide. *See* Translations
Slide arrow, 670
Slide line, 670
Slope
  explanation of, 570
  of perpendicular lines, 677–678
  using similar triangles to determine, 573–576
Somerville, Mary Fairfax, 40
Sonja's Deck Problem, 271
Sophie Germain prime, 215–216
Spheres
  explanation of, 496
  surface area of, 641
  volume of, 653–655
Spreadsheets
  development of, 762–763
  explanation of, 761–762
  explicit and recursive formulas and, 765
  features of, 765–766
  graphing with, 763–764
  recursive patterns in, 7
Square kilometer, 607
Square miles, 607
Square numbers, 14–15
Square roots
  estimating, 340–341
  explanation of, 337–338
  irrationality of, 339–340
Squares, 480
St. Andrew's cross, 92
Standard deviation, 441–442
Statements
  compound, 46–47
  conditional and biconditional, 47–49
  in logic, 45
  logically equivalent, 47

Statistical graphs
    bar graphs, 417–418
    choosing data display for, 421–422
    circle graphs, 419–421
    explanation of, 410
    frequency tables, 416
    histograms, 416–417
    line graphs, 418
    line plots, 411–412
    misleading, 448–451
    pictographs, 411
    scatterplots, 418–419
    stem-and-leaf plots, 412–416
Statistics
    explanation of, 409
    historical background of, 409
    measures of central tendency and variation and, 427–442
    misuse of, 447–452
Stem-and-leaf plots, 412–416, 438
Stevin, Simon, 294
Straight angles, 470
String-Tying Game, 369–370
Subgoals, 24–26
Subsets
    explanation of, 66–67
    number in set of, 69–70
    proper, 66
Substitution method, 578–579
Substitution property, 39
Subtraction
    clock, 235, 236
    definition of, 178
    of integers, 176–180
    as inverse of addition, 178–179
    of rational numbers, 262–264
    of whole numbers, 88–89
Subtraction algorithms
    in bases other than ten, 141–142
    explanation of, 139–140
Supplementary angles, 485
Surface area
    of cones, 640
    of cylinders, 638
    lateral, 636
    of pyramids, 638–639
    of right prisms, 635–638
    of spheres, 641
Symmetries
    explanation of, 669
    line, 686, 687, 704–705
    point, 708
    rotational, 707–708
    three-dimensional, 708–709

T

Tables, 21–22
Take-away model, 88
Tangrams, 481
Temperature, 657–658
Terminating decimals
    explanation of, 298–299
    ordering, 299–300
Tessellations
    explanation of, 669, 713
    with other shapes, 714–716
    regular, 713–714
Thales of Miletus, 558
Theoretical probability, 351
Tracing paper, 685–688
Translations
    constructions of, 670–672
    coordinate representation of, 672–674
    explanation of, 669–670
Transversal, 485
Trapezoids
    area of, 611–613
    explanation of, 480, 481
Traversable network, 507–508
Tree diagrams
    example of, 93
    probability and, 362, 363–365, 367, 373
Trend line, 418–419
Triangle inequality, 596–597
Triangles
    acute, 480
    area of, 609–611
    circle circumscribed about, 529
    classification of, 479–481
    congruent, 521–525, 529–530
    midsegments of, 557–558
    similar, 551–559
    sum of measures of angles of, 487–489
Triangular numbers, 13
True-False Test Problem, 399–400
Truth tables, 46
Tukey, John, 409
Turing, Alan, 217
Turn angle, 674
Turn arrow, 675
Turn center, 674
Turn symmetry, 707–708
Turtle
    commands with, 739
    creating figures with, 727–728
    defining procedures with, 728–735
    moving, 726
    overview of, 725
    recursing and, 736–739
    turning, 726–727
    writing procedures with variables using, 735–736

U

Union, of sets, 73–75
Universal sets, 65
Universe, 65
University of Illinois Committee on School Mathematics (UICSM), 669
Upper quartile, 434

V

Valid reasoning, 49–50
Variance, 441–442
Vector, 670
Venn, John, 65
Venn diagrams
    explanation of, 65
    as problem-solving tool, 76–79
Vertex
    of cone, 502
    even, 508
    odd, 508
    of polyhedron, 496
Vertical angles, 484
Vertical bar, 434
Viéte, Francois, 37
Volume
    confusion regarding, 644–645
    converting English units of, 648
    converting metric units of, 646–648
    of prisms and cylinders, 648–650
    of pyramids and cones, 651–653
    of right rectangular prisms, 645
    of spheres, 653–655

W

Weight, 655
Whole-number computation
    algorithms for addition and subtraction and, 135–142
    algorithms for multiplication and division and, 146–153
    estimation and, 160–163
    mental mathematics and, 157–160
    numeration systems and, 124–133
Whole numbers
    addition of, 83–88
    division of, 97–99
    explanation of, 83
    multiplication of, 92–97
    subtraction of, 88–89
Wiles, Andrew, 195
Word problems, 41

X

x-coordinate, 567

Y

y-coordinate, 567

Z

Zero, 99
Zero multiplication property, 94, 186

The Cover Solution

Back Cover

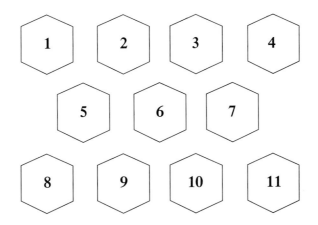

Front Cover

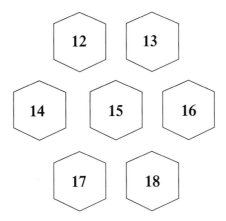

The three kaleidoscope designs that do not have complete line symmetry are number 6, number 8 (the white ring around the large green sphere at 10 o'clock is too thin) and number 16 (the white bar at 9 o'clock goes under the green bar instead of over).

Cover Images: From Kaleidoscope Symmetry Poster by Dale Seymour. © 1993 by Dale Seymour Publications. Used by permission of Pearson Learning. All quotations from PROFESSIONAL STANDARDS FOR TEACHING MATHEMATICS © 2000 National Council of Teachers of Mathematics; All student pages from SCOTT FORESMAN–ADDISON WESLEY MIDDLE SCHOOL MATH, COURSE 2, © 1999 Addison Wesley Longman, Inc. Used by permission of Prentice Hall; All student pages from SCOTT FORESMAN–ADDISON WESLEY MATH © 1999 Addison Wesley Longman, Inc.; All screen captures from The *Geometer's Sketchpad*® are available from Key Curriculum Press, P.O. Box 2304, Berkeley, CA 94702, 1-800-995-MATH. The *Geometer's Sketchpad* is a registered trademark of Key Curriculum Press; Page 2, SHOE cartoon © Tribune Media Services, Inc. All rights reserved. Reprinted with permission; Page 31, page from SCOTT FORESMAN–ADDISON WESLEY PROBLEM SOLVING HANDBOOK, © 1999 Addison Wesley Longman, Inc. Used by permission of Prentice Hall; Pages 59, 82, 280, 381, 399, PEANUTS reprinted by permission of United Feature Syndicate, Inc.; Pages 76, 251, 313, from MATH THEMATICS BOOK 1 by Rick Billstein and Jim Williamson. Copyright © 1999 by McDougal Littell Inc. All rights reserved. Reprinted by permission of McDougal Littell Inc.; Pages 109, 422, from MATH THEMATICS BOOK 3 by Rick Billstein and Jim Williamson. Copyright © 1999 by McDougal Littell Inc. All rights reserved. Reprinted by permission of McDougal Littell Inc.; Page 125, WIZARD OF ID cartoon is reprinted by permission of Johnny Hart and Creator's Syndicate, Inc.; Page 132, FOX TROT © Bill Amend. Reprinted with permission of UNIVERSAL PRESS SYNDICATE. All rights reserved; Pages 157, 631, CALVIN AND HOBBES © Watterson. Reprinted with permission of UNIVERSAL PRESS SYNDICATE. All rights reserved; Page 212, cartoon © 2000 by Sidney Harris; Page 266, "Who's Soiling the Nest?" diagram data: United Nations HUMAN DEVELOPMENT REPORT, 1994; Page 285, Mother Goose and Grimm cartoon © Tribune Media Services, Inc. All rights reserved. Reprinted with permission; Page 297, BLONDIE cartoon reprinted with special permission of King Features Syndicate; Page 308, HI & LOIS cartoon reprinted with special permission of King Features Syndicate; Page 318, ARLO & JANIS reprinted by permission of Newspaper Enterprise Association, Inc.; Page 359, KIT 'N' CARLYLE reprinted by permission of Newspaper Enterprise Association, Inc.; Page 388, cartoon © Carole Gable; Page 410, Herman® is reprinted with permission from LaughingStock Licensing Inc., Ottawa, Canada. All rights reserved; Page 427, RHYMES WITH ORANGE cartoon reprinted with special permission of King Features Syndicate; Pages 428, 448, THE FAR SIDE © FARWORKS, INC. Used by permission. All rights reserved; Pages 462, 684, B.C. cartoons are reprinted by permission of Johnny Hart and Creator's Syndicate, Inc.; Page 519, M.C. Escher's "Symmetry Drawing E22" © 2000 Cordon Art B.V.–Baarn–Holland. All rights reserved; Page 523, photo © 1996 Peter M. Bagdigian; Page 562, HAGAR THE HORRIBLE cartoon reprinted with special permission of King Features Syndicate; Page 565, fractal from Peitgen & Richter, BEAUTY OF FRACTALS, © 1986. Reprinted with permission of Springer-Verlag; Page 710, Chevrolet logo courtesy of Chevrolet Motor Division, Chrysler logo courtesy of Chrysler Corporation, Bell Symbol courtesy of Pacific Telesis Group, Volkswagen of America logo courtesy of Volkswagen of America; Page 713, M.C. Escher's "Study of Regular Division of the Plane with Reptiles" © 2000 Cordon Art B.V.–Baarn–Holland. All rights reserved; Page 761, Toolbar from Microsoft Excel manual reprinted with permission of Microsoft Corporation.